Past Glacial Environments

Sediments, forms and techniques

To the memory of my parents

The companion volume to this book is also available
from Butterworth-Heinemann

Modern Glacial Environments: Sediments, forms and techniques
(Glacial Environments: Volume 1)
edited by John Menzies
ISBN: 0 7506 2351 9

Other books of interest:

Physics of Glaciers, third edition
by W. S. B. Paterson
ISBN: 0 080 37944 3 (flexicover)
ISBN: 0 080 37945 1 (hardback)

Continental Deformation
edited by Paul L. Hancock
ISBN: 0 080 37930 3 (flexicover)
ISBN: 0 080 37931 1 (hardback)

To order please contact your nearest bookseller, or in the case
of difficulty, write to Butterworth-Heinemann Sales Department,
Linacre House, Jordan Hill, Oxford OX2 8DP, UK.

Past Glacial Environments

Sediments, forms and techniques

Editor
John Menzies

Glacial Environments: **Volume 2**

Butterworth–Heinemann Ltd
Linacre House, Jordan Hill, Oxford OX2 8DP

 A member of the Reed Elsevier plc group

OXFORD   LONDON   BOSTON
NEW DELHI   SINGAPORE   SYDNEY
TOKYO   TORONTO   WELLINGTON

First published 1996

© Butterworth-Heinemann Ltd 1996

British Library Cataloguing in Publication Data

Past Glacial Environments: Sediments,
Forms and Techniques. – (Glacial
Environments Series; Vol. 2)
  I. Menzies, John II. Series
  551.31

ISBN 0 7506 2352 7

Library of Congress Cataloguing in Publication Data

Past glacial environments: sediments, forms, and
techniques/editor, John Menzies
  p.   cm. – (Glacial environments; v. 2)
  Includes bibliographical references and index.
  ISBN 0 7506 2352 7 (pbk.)
  1. Glacial epoch. I. Menzies, John. II. Series.
  QE697.P38   1995   95-34315
  551.7'92–dc20   CIP

Composition by Genesis Typesetting, Rochester, Kent

Printed and bound in Great Britain by
Martins the Printers Ltd, Berwick upon Tweed

CONTENTS

Preface   xi
Acknowledgements   xiii
List of Contributors   xvii
List of Symbols   xix

Chapter 1   Past glacial environments   **1**
J. Menzies

1.1.   Introduction   1
1.2.   Impact of past glacial conditions on global habitats and earth systems   1
1.3.   Research in past glacial environments   3
1.4.   Research issues in past glacial environments   9
1.5.   Synopsis of chapters   12

Chapter 2   Subglacial environments   **15**
J. Menzies and W. W. Shilts

2.1.   Introduction   15
2.2.   Definition   15
2.3.   The subglacial interface – bed types and conditions   18
2.4.   Spacial variations in subglacial polythermal – rheological bed conditions   24
2.5.   Subglacial sediments and landforms   29
2.6.   Sediment structures and related sedimentological and geotechnical characteristics
       within subglacial sediments   46
2.7.   Subglacial – proglacial transition environments   55
2.8.   Subglacial landforms and bedforms   62
2.9.   Subglacial erosional forms under active ice   84
2.10. Sediments and landforms of passive ice flow   110
2.11   Repetitive event histories in subglacial environments   134
2.12. Summary   135

Chapter 3   Pleistocene supraglacial and ice-marginal deposits and landforms   **137**
W.H. Johnson and John Menzies

3.1.	Introduction	137
3.2.	Sediment and sediment associations	139
3.3.	Landforms	147
3.4.	Summary	160

Chapter 4   Ice scour as an indicator of glaciolacustrine environments   **161**
C.M.T. Woodworth-Lynas

4.1.	Introduction	161
4.2.	Scouring in the geological record	164
4.3	Discussion	177

Chapter 5   Glaciomarine environments 'ancient glaciomarine sediments'   **179**
A. Elverhøi and R. Henrich

5.1.	Introduction	179
5.2.	Glaciomarine sediments: classification and identification	181
5.3.	Examples of Pre-Cenozoic glaciomarine sequences	191
5.4.	Glaciomarine sediments and the sedimentary environment of a passive margin and its adjacent ocean – exemplified by the Norwegian–Greenland Sea and the Norwegian/Barents Sea margin	195
5.5	Conclusions	211

Chapter 6   Glacioaeolian processes, sediments and landforms   **213**
E. Derbyshire and L. A. Owen

6.1.	Introduction	213
6.2	Sediment production and sources	214
6.3	Wind action around glaciers	215
6.4	Glacioaeolian sediments and landforms	221
6.5.	Facies	229
6.6.	Conclusion	238

Chapter 7   Glacial environments of Pre-Pleistocene age   **239**
G.M. Young

7.1.	Introduction	239
7.2.	Distribution of Pre-Pleistocene glaciations in time and space	241
7.3.	Lithofacies and sedimentology of Pre-Pleistocene rocks – a brief review	242
7.4.	Geochemical aspects	249
7.5.	Tectonic setting of glacial deposits	249
7.6.	Summary	251

Chapter 8   Glacial stratigraphy   253
J. Rose and J. Menzies

8.1.	Introduction	253
8.2.	Rationale	254
8.3	Stratigraphy within glacial environments	257
8.4.	Stratigraphic nomenclature	261
8.5.	Glacial stratigraphic procedures and methods	264
8.6	Conclusion	284

Chapter 9   Lithofacies associations for terrestrial glacigenic successions   285
T.J. Kemmis

9.1.	Introduction	285
9.2.	Examples of sedimentary assemblages	287
9.3.	Conclusions	300

Chapter 10   Paleosols   301
J. Boardman

10.1.	Soils and glacial deposits	301
10.2.	Paleosols	302
10.3.	Parent materials	306
10.4.	Disturbance and survival	307
10.5.	The value of paleosols	309
10.6.	Conclusions	314

Chapter 11   Glacio-isostasy, glacio-eustasy and relative sea-level change   315
J.M. Gray

11.1.	Introduction	315
11.2.	Glacio-eustasy	316
11.3.	Glacio-isostasy	318
11.4.	Interplay of glacio-eustasy and glacio-isostasy	321
11.5.	Illustration of the theoretical principles	324
11.6.	Implications of the theoretical principles	332
11.7.	Towards an integrated model of global relative sea change	333

Chapter 12   Micromorphology   335
J.J.M. van der Meer

12.1.	Introduction	335
12.2.	Sampling and sample preparation	335

12.3.   Studying thin sections, nomenclature   336
12.4.   Micromorphological features of glacial sediments   337
12.5.   Discussion   354
12.6.   Future work   355

Chapter 13   Scanning electron microscopy   357
W. B. Whalley

13.1.   Introduction   357
13.2.   SEM and sediments   360
13.3.   Examination of bulk samples of soils and tills   374
13.4.   Other investigations   374
13.5.   Future developments   375
13.6.   Conclusions   375

Chapter 14   Geochronology of glacial deposits   377
J. Brigham-Grette

14.1.   Introduction   377
14.2.   Resolution ans stratigraphic context   379
14.3.   Geochronological terminology   380
14.4.   Methods and approaches   380
14.5.   Summary   410

Chapter 15   Drift exploration   411
W.W. Shilts

15.1.   Introduction   411
15.2.   Some constraining principles and concepts   411
15.3.   Geochemical patterns on glaciated landscapes   412
15.4.   Dispersal patterns   423
15.5.   Scales of dispersal   431
15.6.   Some important exceptions to classical dispersal patterns   434
15.7.   Conclusion   438

Chapter 16   Geology of placer deposits in glaciated environments 441
V. M. Levson and S. R. Morison

16.1.   Introduction   441
16.2.   Sedimentary environments   442
16.3.   Regional distribution of placer deposits in glaciated areas 451
16.4.   Buried placer deposits   470
16.5   Conclusions   476

Chapter 17   Problems and perspectives   **479**
J. Menzies

17.1.   Introduction   479
17.2.   Paradigm shifts   480
17.3.   Questions illustrative of problems and issues   480
17.4.   Epilogue   484

References   487

Index   569

PREFACE

This book sets out to provide a contemporary and extensive survey of the sediments and landforms of past glacial environments. The emphasis in the text is three-fold, first, to examine these environments from the perspective of sediments and forms through an understanding of the processes and dynamics associated with sediment deposition and landform development; secondly, to establish a fundamental understanding of the many separate environments within the glacial system by considering their erosional, transportation and depositional histories as manifested in present-day landscapes and glacial sediments in those areas of the world glaciated in the past; and finally to consider some of the important techniques and methods used in the study of past glacial environments.

Evidence of past glacial environments of the Pleistocene abound in those areas of the northern hemisphere from mid-latitudes poleward and likewise in the southern hemisphere in the South Island, New Zealand and in the southern regions of South America. Similarly in most mountainous regions of the world evidence of past more extensive glaciation is readily apparent. When considering pre-Pleistocene glaciations the evidence is typically muted in present-day landscapes but bedrock from as diverse locations, for example, as the Scottish Highlands, or the desert plains of Mauritania and the tropical forests of Amazonia are all of glacial origin and thereby effect to lesser or greater degrees human activities in those regions. The impact, therefore, of past glacial environments is virtually global.

It is reasonable to state that almost no human activity that utilises bedrock materials and soils, and thus farming, mining and construction in most areas of the world does not involve glacial materials in some manner. The importance, consequently, of past glacial environments is all too apparent.

This book forms the second part of a two volume work on glacial environments. The first, companion volume deals with *Modern Glacial Environments – Processes, Dynamics and Sediments*. This present text covers the sediments and landforms of past glacial environments both terrestrial and marine, general topics on stratigraphy, lithofacies, paleosols and glacio-isostasy and -eustacy, as well as discussing related methods and techniques in the study of past glacial environments, namely, micromorphology, SEM, dating methods, drift exploration and placer mining.

The serious study of past glacial environments has been in progress for the past 150 years yet our overall understanding of these environments and associated processes is still comparatively limited. Just as the study of modern glaciers and ice sheets has rapidly advanced at astonishing speed in the last two decades, our grasp of past glacial environments has virtually kept apace with modern ideas of glacial processes and dynamics. As comprehension of glacial environments has increased there has been an attendant awareness that these systems are much more complex than perhaps previously contemplated. With this increasing complexity an almost dizzying array of hypotheses have sprung up that require even greater understanding. Research into past glacial environments consequently continues at an ever-increasing pace.

The purpose of this book is to provide university students with a comprehensive examination of modern research and ideas on past glacial environments. This book was conceived from a series of university courses that attempted to provide students with a greater understanding of glaciated landscapes through an appreciation of the processes and sediments that form the glacial system.

The underlying tenet of this text is that only by understanding the complex geomorphical (geological) processes of erosion, transport and deposition within glacial systems that are, in themselves, an intricate balance of glaciodynamics and other aspect of glaciology, can a fully integrated grasp of glacial landscape evolution be achieved. Whereas in the past there was perhaps an over-emphasis on landforms (and spatial association) within glaciated landscapes and a decided lack of interest shown in glacial sediments this imbalance has, to some small degree, tipped in the other direction giving too much credence to sedimentology as compared with landscape appraisal and landform morphology. It is likely in the next decade that a broader understanding of glaciated landscapes will develop where both landform morphology and sedimentology in association with a sound appraisal of glaciology and glaciodynamics will achieve an approach that may provide the paradigms needed to better understand these complex environments.

The production of this book has only been possible thanks to the extraordinary dedication of the individual authors of each chapter. It is difficult to express sufficiently my individual thanks to each of the authors for all the time and effort they have given in producing their chapters. To John Boardman, Julie Brigham-Grette, Ed Derbyshire, Anders Elverhøi, Murray Gray, Rudiger Henrich, Hilt Johnson, Tim Kemmis, Vic Levson, Jaap van der Meer, Steve Morrison, Lewis Owen, Jim Rose, Bill Shilts, Brian Whalley, Chris Woodworth-Lynas and Grant Young, I extend my sincerest thanks. The authors have revised chapters, answered many queries and have persevered, and, like me, have waited, what, at times, must have seemed an interminable period for the arrival of this book. Numerous people throughout the world have kindly given me their time to review chapters, find photographs, answer questions and generally have encouraged the creation of this book—without their generosity and goodwill this book would have remained only an idea. I would like to thank Don Coates who was instrumental in initiating this project. I am indebted to numerous members of the academic and library faculty and staff at Brock University for answering queries, reviewing chapters, finding references and generally assisting me. I am especially grateful to Kamil Zaniewski for his superb skills in checking references. I owe thanks to Loris Gasparotto for his expertise in cartography, to Divino Mucciante for his expertise in photography and to Colleen Catling for her secretarial work. Eilleen Kraatz is to be thanked and praised for editing and refining the chapters in this book. I also have a debt of gratitude to Peter Henn, my editor at Pergamon Press, who has kept faith with this project throughout and has never ceased to support and encourage my endeavours. Also my thanks to Duncan Enright of Butterworth-Heineman who so graciously assumed editorship of both books as Elsevier Science Publishers internally restructured their publishing enterprises.

It is difficult to know, on a personal level, where my abiding interest in past glacial environments began. At an early age my late parents never ceased to inspire me with an appreciation of landscape, later a young teacher, Bill Chapman, furthered that growing interest and then while as a student Ken Walton, David Sugden, Chalmers Clapperton and Bill Ritchie at the University of Aberdeen and later Brian Sissons at the University of Edinburgh convinced me that nothing else was worth studying; to all of them I acknowledge an abiding gratitude.

As authors, we are only too aware of the enormous debt we owe to so many fellow earth scientists over the past years who have all, in some form or other, contributed to our understanding and appreciation of glaciated landscapes and sediments and whose work is embodied in this text.

Finally, my wife, Teresa, and daughters, Erica, Fiona and Rebecca deserve my deepest appreciation for their love, encouragement, patience and immense support while writing this text. My wife has given unceasing assistance, read innumerable chapters and provided space and time when I needed it most.

ACKNOWLEDGEMENTS

The editor, authors and publisher wish to thank the following for their kind permission to reproduce copyright material. Full citation can be found by consulting the captions on figures, plates and references:

The authors wish to thank the following individuals and publishing companies for their kind permission to use figures and plates:

FIGURES

S.M. Colman and K.L. Pierce for Fig. 14.20: R. LeB Hooke for Fig. 2.17; C. Kaszycki for Figs. 3.8, 9.5; C. Hällestrand for Fig. 2.25; R.R. Parizek for Fig. 3.12; J.W. Mulholland for Fig. 3.17; K.L. Pierce et al., for Fig. 14.16; S. Porter for Fig. 14.1; J.C. Ridge et al. for Fig. 14.26; A.M. Sarna-Wojcicki and J.O. Davis for Fig. 14.22; Sawkins et al. for Fig. 14.1; D.R. Sharpe and J. Shaw for Fig. 2.51a,b, 2.52a,b; J. Shaw for Figs. 2.49, 2.50, 9.7; G.M. Young for Fig. 1.1b.

PLATES

R. Aario for Plates 2.9, 2.10b, 2.33a,b, 2.34a; B.C. Ferries for Plate 2.14a; P. Birkeland for Plate 3.5; Blackwell Publishers for Plate 4.4a; University of Cape Town, South Africa for Plate 4.2; L. Clayton for Plate 3.8; Geological Survey of Canda for Plates 2.26b, 2.28, 2.29a,b, 2.31; Geological Survey of Finland for 2.22a; P.L. Gibbard for Plates 2.3b,c, 2.7;

Greenland Geoscience (Meddelesler om Groeland) for Plate 4.4b; W.B. Hall for Plate 3.6; G. Hamelin for Plates 2.17a,b, 2.19a,b, 2.21d, 2.23c; J. Harvey for Plate 16.10; S. Hicock for Plate 2.8; R. Lagerbäck for Plates 2.33c, 2.34b; P. Moller for Plate 3.2; D.W. Moore for Plate 3.1; National Air Photo Library of Canada for Plates 2.13, 2.26a, 2.30; J. Paterson for Plate 2.7b; R. Powell for Plate 4.10; M. Rappol et al. for Plate 12.5c; J.C. Ridge for Plate 14.1; A.C. Rocha-Campos for Plate 4.9; J. Shaw for Plates 2.21a,b; Society of Economic Paleontologists and Mineralogists for Plate 4.8; D. Stetz for Plate 2.10a; USDA for Plate 3.9; W.H. Wolfestan for Plates 2.14b,c; C. Zadarowicz for Plates 2.27a,b.

PUBLISHERS

A.A. Balkema Publishers for Figs. 2.14, 2.15, 2.23, 2.64, 3.10, 9.3, and 9.6; Academic Press for Fig. 11.18: Allen and Unwin for Fig. 10.4; American Association for the Advancement of Science for Fig. 14.10: American Association of Petroleum Geologists for Fig. 8.6; American Geophysical Union for Figs. 2.20 and 14.24a; Arctic and Alpine Research and the Regents of the University of Colorado for Figs. 14.18 and 14.21; British Columbia Ministry of Energy, Mines and Petroleum Resources for Figs. 16.2: British Museum Press for Fig. 8.2; Cambridge University Press for Figs. 5.8 and 8.16; Edward Arnold Publishers for Fig. 14.1; Elsevier Science Publishers Ltd, Oxford (formerly Pergamon Press,

Oxford) for Figs. 3.14, 5.7, 8.14, 9.1, 14.15 and 14.19; Elsevier Science Publishers BV, Amsterdam for Table 6.2, and Figs. 2.2a, 2.3, 2.7, 2.9, 2.10, 2.24, 2.26, 2.27, 2.29, 2.37, 2.45, 5.3, 5.9, 5.13, 5.14, 5.18, 6.2, 6.3, 6.6, 6.7, 6.9, 6.10, 6.11, 6.12, 7.2 and 10.6; Fennia (Geographical Society of Finland) for Fig. 2.31; Geological Magazine for Fig. 2.18; Geological Society of America for Figs. 3.10, 4.1e, 5.10, 8.3, 14.16 and 14.20; Geological Society of Denmark for Figs. 8.8, 8.9 and 8.10; Geological Society of France for Fig. 6.4; Geological Society of London for Figs. 2.19, 10.7, 11.12, 11.23 and 11.24; Geological Survey of Canada for Figs. 2.58, 2.35, 2.59, 2.60, 2.61 and 2.62; Geological Survey of Finland for Fig. 2.47; Geological Survey of Norway for Fig. 2.41; Illinois State Geological Survey for Figs. 10.3 and 10.8; Institute of British Geographers for Figs. 11.13, 11.14, 11.15, 11.17, 11.20, 11.21 and 11.22; International Glaciological Society for Figs. 2.4, 2.6, 2.32, 2.33, 2.34, 2.39, 2.48 and 2.54; International Association of Sedimentologists for Figs. 3.1, 6.5, 14.11 and 14.12; Kluwer Academic Publishers for Figs. 5.15, 5.16, 5.17, 5.21, 5.22 and 5.23; Methuen Publishers for Fig. 2.38; National Academy of Sciences for Fig. 3.21; National Research Council of Canada for Figs. 2.49, 2.50, 2.51, 9.4 and 9.5; Nature for Figs. 2.46, 8.1, 8.13 and 14.6; Norsk Polarinstitutt for Fig. 5.11; University of Oulu for Figs. 2.21 and 2.63; Precambrian Research for Fig. 5.8; Progress in Physical Geography for Fig. 14.17; Royal Scottish Geographical Society for Fig. 2.53c; Royal Society of Canada for Fig. 10.2; Royal Society of Edinburgh for Fig. 8.11; Scandinavian University Press for Figs. 2.11, 2.43, 3.6, 3.11, 4.1b and 10.1; Science Progress for Fig. 8.12; Scottish Academic Press for Fig. 8.5; Societas Upsaliensis pro Geologica Quaternia for Fig. 2.36; Society of Economic Paleontologists and Mineralogists for Figs. 4.1d and 9.2; Springer-Verlag for Fig. 2.40; State University of New York (SUNY) Binghampton for Figs. 3.11 and 3.20; University of Washington for Figs. 14.1, 14.2 and 14.23; John Wiley and Sons Ltd. for Figs. 2.8, 2.22, 8.15 and 14.9.; and the Yukon Exploration and Geological Services for Figs. 16.4.

The authors wish to acknowledge the following individuals and organisations:

Chapter 2: The authors wish to thank in particular Jan Aylsworth, Shirley McCaig, Mark Roberts, Martin Rappol and several others who gave advice in the development of this chapter.

Chapter 3: The author wishes to thank numerous colleagues over the years for discussions of glacial sedimentology and geomorphology. I appreciate being able to use personal photographs of Peter Birkeland and Lee Clayton, and copyright material from a number of authors and publishers.

Chapter 4: Funding for the analysis presented in this study is provided under a Natural Sciences and Engineering Research Council (NSERC) (Canada) Strategic Grant (No. 0032716) entitled 'An integrated investigation into the processes of ice keel/soil interaction'. I thank Dr R. N. Hiscott (Department of Earth Sciences, Memorial University of Newfoundland), Dr W. Jamison (Centre for Earth Resources Research, Memorial University of Newfoundland), and an anonymous reviewer for their very helpful critiques of the manuscript. I also thank Dr A.C. Rocha-Campos and Dr Paulo Santos (Instituto ge Geociencias, Universidade de Sao Paulo) and Dr R. Powell (Department of Geology, Northern Illinois University) for their kind permission to allow the reproduction of photographs, and Desirée King for drafting the figures.

Chapter 6: The authors wish to acknowledge Justin Jacyno for drafting the diagrams.

Chapter 7: The author would like to acknowledge the great educational benefit derived from participation in numerous IGCP projects related to glacial topics and to thank individuals such as Georges Choubert, Anne Faure-Muret, Kalervo Rankama, W.B. Harland, Janine Sarfati, Max Deynoux and Julia Miller who initiated and administered these valuable projects.

Chapter 9: The author wishes to thank the individual reviewers who so kindly gave critical advice and encouragement in reading an early draft of this chapter.

Chapter 10: The author thanks Dr R.A. Kemp and Dr R.F. Smith for valuable comments on a draft of this chapter.

Chapter 11: The author wishes to thanks Ed Oliver for drawing figures for this chapter.

Chapter 12: The material presented here would have been impossible without the work of Cees Zeegers, who prepared the thin sections. The many discussions with Dr H.J. Mücher are gratefully acknowledged. Several colleagues have allowed me to use unpublished observations on thin sections collected for combined research, I greatly appreciate their permission. Mr Ch. Snabilié and Mr H. van Maaren assisted with the preparation of the illustrations, Mrs M. Shani-Veels with the preparation of the manuscript.

Chapter 13: The author would like to thank the staff of the Electron Microscopy Unit of Queen's University of Belfast for their assistance in producing high quality photographs for the figures in this chapter.

Chapter 14: The author wishes to thank Peter Clark, Steve Colman, Darrell Kaufman, Steve Roof, and Al Werner for their careful and timely reviews of the manuscript. Their suggestions helped to clarify wording and balance in the chapter. I want to especially thank Steve Roof for assisting with several of the computer-generated figures.

Chapter 15: The author is grateful to my colleagues, both in and outside the Geological Survey of Canada for their ideas and discussions that have contributed to the necessarily brief distillation of ideas presented here. In particular, I thank my colleagues in the former S and MT section, J. Aylsworth, S. Courtney, R. DiLabio, C. Kaszycki, I. Kettles, R. Klassen, M. Lamothe, M. Rappol, and H. Thorleifson, P. Wyatt; W. Coker and E. Hornbrook of geochemistry subdivision; and R. Ridler, presently of Ridler and Associates, Inc. The following individuals and companies either provided samples described here or permitted me to collect them on their properties: M. Bouchard, J. Burzynski, W. Coker, A. Dyke, I. Kettles, A. Lecheminant, A. Miller, and C. Schlüchter; Dejour Mines, Falconbridge Mines (Formerly Kidd Creek Mines), and Golden Hope Mines.

Chapter 16: The authors would like to acknowledge the British Columbia Geological Survey, the Geological Survey of Canada and the Yukon Exploration and Geological Services Division, Northern Affairs program, for permission to use a number of published sources in preparation of this chapter, including papers by Morison (1983a, 1983b, 1985a and 1989), Levson *et al.* (1990), Levson and Giles (1991, 1993), and Levson (1991a, 1992a, 1992b). The authors also acknowledge contributions made by anonymous external reviewers and editorial revisions by Brian Grant of the British Columbia Geological Survey.

Chapter 17: The author would like to thank Jim Rose for his sound and sage advice and helpful comments in reviewing this chapter.

CONTRIBUTORS

John Boardman
School of Geography and Environmental Change Unit, University of Oxford, Oxford, UK

Julie Brigham-Grette
Department of Geography and Geology, University of Massachusetts, Amherst, Massachusetts, USA

Ed Derbyshire
Department of Geography, Royal Holloway, University of London, Egham, Surrey, UK

Anders Elverhøi
Department of Geology, University of Oslo, formerly of Norsk Polarinstuut, Oslo, Norway

J.M. Gray
Department of Geography, Queen Mary and Westfield College, University of London, London, UK

Ruediger Henrich
Department of Geology, University of Bremen, formerly of GeoMar – Research Center for Marine Geosciences, Kiel, Germany

W.H. Johnson
Department of Geology, University of Illinois, Urbana-Champaign, Illinois, USA

Tim Kemmis
Rust Environment and Infrastructure, Sheboygan Wisconsin; formerly of Iowa Geological Survey Bureau, Iowa, USA

Vic Levson
Department of Energy, Mines and Petroluem Resources, Victoria, BC, Canada

Jaap van der Meer
Fysisch Geografisch en Bodemkundig Lab., Univeristy of Amsterdam, Amsterdam, The Netherlands

John Menzies
Departments of Geography and Earth Sciences, Brock University, St Catharines, Ontario, Canada

Steve Morison
Department of Energy, Mines and Petroluem Resources, Victoria, BC, Canada

Lewis Owen
Department of Geography, Royal Holloway, University of London, Egham, Surrey, UK

Jim Rose
Department of Geography, Royal Holloway, University of London, Egham, Surrey, UK

W.W. Shilts
Illinois State Geological Survey, Champaign, USA, formerly of Terrain Sciences Division, Geological Survey of Canada, Ottawa, Ontario, Canada

Brian Whalley
School of Geosciences and Department of Geography, The Queen's University, Belfast, Northern Ireland

Chris Woodworth-Lynas
C-Core, Memorial University of Newfoundland, St. John's, Newfoundland, Canada

G.M. Young
Department of Earth Sciences, University of Western Ontario, London, Ontario, Canada.

SYMBOLS

a	Ablation	$\dot{c}$	Accumulation rate
$\dot{a}$	Ablation rate	c_b	Debris concentration in ice
a_g	Geometric factor in clast 'ploughing'	c_m	Debris concentration in moraine septa
a_p	Sinusoidal bed amplitude	c_r	Constant related to regelation process
a_s	Summer ablation	c_s	Summer accumulation
a_{sp}	Bed asperity roughness	c_t	Total accumulation
a_{sr}	Measure of surface roughness	c_w	Winter accumulation
a_t	Total ablation	C	Cohesion
a_w	Winter ablation	C_r	Areal concentration of clasts in contact with the bed
A	Ice hardness coefficient		
b	Mass balance	C_3	Empirical factor related to the mode of water flow in a conduit, the conduit geometry and ice flow conditions
b_L	Balance rate at the ice terminus		
b_n	Net balance		
b_o	Stream channel width	d	Obstacle size
b_s	Summer balance	d_c	Controlling obstacle size
b_w	Winter balance	d_o	Mean particle size diameter
B	Ice sliding coefficient	d_w	Depth of water film at the ice/bed interface
B_e	Energy balance on ice mass surface	D	Constant related to ice and water density
B_f	Buoyancy flux	D_c	Conduit diameter
B_p	Back pressure within an ice shelf	D_e	Longitudinal strain or extension
c	Accumulation	E_i	Young's modulus of ice
		E^*	Activation energy of the creep of ice

xix

f Area of ice/bed interface occupied by meltwater

f_{sh} Shape factor affecting frictional drag in valley glaciers and ice streams

f_t Fraction of the bed that is thawed

F_r Froude number

F_c Effective contact force between an indenting clast and the bed

F_w Longitudinal normal force within a sediment wedge

F_{wd} Channel width/depth ratio

g Acceleration due to gravity

G Hydraulic gradient

G_{eo} Geometric factor relating mean basal normal stress to basal shear stress and that fraction of the ice/bed interface covered in meltwater

h Ice elevation above grounding line of floating ice margin

h_b Depth of deforming layer

h_{bc} Height of cavity backwall

h_d Thickness of debris-rich ice layer

h_i Ice thickness

h_m Thickness of debris septa taken as equal to ice thickness

h_R Present-day elevation of bedrock above or below the margin of the ice sheet

h_s Sediment thickness in a wedge

h_{sp} Thickness of supraglacial debris cover

h_{sw} Height of submerged ice-cliff wall

h_w Water depth

H Maximum surface elevation of an ice mass

H_o Channel depth at stream mouth

i Hydraulic gradient

i_c Critical hydraulic gradient

I^*_p Ploughing index on a clast in a deforming sediment

J_s Approximation of sediment flux along a channel

k Permeability of sediment

k_m Reciprocal of the Manning roughness parameter

k_o Critical wave number

k_r Hydraulic roughness parameter for conduits

k_{tc} Thermal conductivity

k_w Wave number = $2\pi/\lambda$

K Hydraulic conductivity

K_i Fracture toughness of ice

K^* Material constant

K_b Constant related to deforming sediment

K_f Constant related to ice shelf ice fabric, temperature and ratio of components of strain rate

K_p Apparent conductivity of a particle array in relation to ice

l Glacier length

L Distance from ice divide to ice margin

L_c Length of main channel

L_h Latent heat of fusion of ice

L_{ic} Length of ice cliff face

L_m Latent heat of melting

L_v Length of valley axis

m Constant of ice sliding related to ice creep

$\dot{m}$ Rate of melting of conduit walls

$\dot{m}_b$ Rate of basal ice melting

m^* Constant related to spectral density of the bed in terms of mean quadratic slope in the direction of the ice bed slip

$\dot{m}_c$ Rate of conduit closure by volume

M Manning roughness coefficient

M Annual mass balance

M_1 Melting rate of ice cliffs

M_2 Mean melting rate of ice cliffs

M_f Momentum flux of a meltwater jet

n Visco-plastic component for ice creep deformation

n_K Number of cavities/passageways across the width of a glacier

n_p Porosity of sediment

P Solid precipitation

P_c Cavity roof perimeter

P_{cs} Channel sinuosity

$P_{cs(tot)}$ Total channel sinuosity

P_o Normal stress distribution across a glacier bed

q_s Proportion of sand-size particles (0.063–0.5 mm diameter) to total sediment weight

q_z Proportion of quartz particles to all other mineral particles in the 0.063–0.125 mm size range

Q Volumetric flux (discharge) of meltwater

Q_{max} Peak discharge from a glacier during a jökulhlaup

Q_o Stream discharge

Q_s Suspended load carried by stream

Q_x Water influx along a section of channel x metres in length

$\dot{r}$ Rate of closure of a conduit

r Radius of a conduit

r_c Radius of curvature of a clast

r_p Particle radius

r_s Grain shape

r_{sr} Isostatic ratio variable

R Universal gas constant

R_e Reynolds number

R_I The Cailleaux–Tricart roundness index

R_n Richardson number

R_u Bed roughness parameter

s Areal fraction of the bed cavity/passageway uncoupled

s_r Bed roughness slope

S Cross-sectional area of channel

S_g Stream gradient

S_j Area of clast imposed upon by ploughing pressure

t_h Time in years for an ice mass to equilibriate

T_n Response time of an ice mass to changes in equilibrium status

T_s Surface water temperature

T_∞ Far-field ambient water temperature

TCL Total length of all braid channels

u Spreading ice velocity

$\bar{u}$ Average ice velocity in ice column

u_+ Sliding velocity of a surging ice mass

u_b Basal sliding velocity of ice

u_{by} Vertical ice flow vector

u_{bz} Transverse ice flow vector

u_c Velocity of a clast

u_{cs} Creep velocity of sediment

$\dot{u}_\varepsilon$ Effective rate of cavity closure

u_i Overall ice mass velocity where $u_i = u_s + u_b$

u_L Terminus ice velocity

u_m Sliding velocity of mobile sediment at its upper interface

u_n Component of ice velocity normal to the bed

u_o Component of u_n due to melting of ice caused by geothermal heat and sliding friction

u_s Surface velocity of ice

u_t Component of ice velocity parallel to the bed interface

u_w Dynamic viscosity of water

U_m Mean centreline velocity of stream at conduit mouth

U_{mo} Initial near-centreline stream velocity

U_w Water velocity

v_p Radius of a particle

$\bar{v}_w$ Average water velocity within a meltwater film at the ice/bed interface

V_m Mean flow velocity of sediment in a channel

V_{max} Total maximum volume of meltwater drained from an ice mass due to a jökulhlaup

V_w Specific discharge along a channel where inflow and outflow are in equilibrium

w Width of ice column

W Width of deforming sediment layer, transverse to maximum ice flow

W_b Width of channel bed

w_m Width of debris septa

W_s Stream power

w_{sp} Width of supraglacial debris cover

z Some vertical thickness of ice

z_{cs} Depth within a deforming sediment

z_s Thickness of deforming sediment

$z_{s(crit)}$ Depth within a deforming sediment at which point basal shear stress = r_s

Z_o Constant related to sediment flux along a channel

α Ice surface slope

α_b Slope at the basal ice/sediment (bed) interface

α_c Slope of a conduit

α_d Constant dependent upon sediment layer thickness and water pressure

α_{hc} Constant dependent on hardness of clasts, the bed and the geometry of the striator point

α_i Incremental ice surface slope

β Slope of glacier bed

Γ Constant related to ice regulation processes

Γ_{pg} Pressure gradient of meltwater in a channel in the direction of the mean slope

γ Shear strain

γ_f Finite longitudinal strain

γ_g Density of sediment

γ_s Shear strain rate of sediment in plane of shearing

γ_w Specific weight of water

Δ Dilation within a sediment

Δ_h Incremental ice thickness steps

Δ_x Incremental length steps from ice divide to margin

ΔP Difference in pressure in ice adjacent to a conduit and the water pressure within a conduit

ΔT Temperature difference between ice and water

$\Delta\sigma_p$ Stress fluctuation

δ Parameter in Fowler's surging theory

δ_f Loss of water pressure in a conduit due to friction

δ_h Incremental steps of ice thickness

δ_p Change in local water pressure in a conduit

δ_s Length of straight conduit

δ_u Incremental lateral spreading of ice due to internal deformation

δ_z Incremental vertical motion of ice due to internal deformation

γ_i	Incremental longitudinal strain	$\bar{\rho}_i$	Average density of ice
$\dot{\varepsilon}_d$	Strain rate for dilatant deforming sediment	ρ_o	Initial discharge density
$\dot{\varepsilon}_{sh}$	Strain rate within an ice shelf	ρ_R	Rock density
η	Ice viscosity	ρ_s	Density of sediment
η_s	Newtonian viscosity	ρ_{sw}	Density of sea water
η_{sed}	Sediment viscosity	ρ_w	Porewater pressure
η_w	Viscosity of water	ρ_{w}	Density of water
Φ	Angle of internal friction of sediment	σ	Standard deviation of the Gaussian distribution for stream velocity across plume width
Φ_s	Slope of conduit		
κ	Thermal diffusivity of ice	σ_b	Ice overburden normal pressure at the base of an ice mass
κ_b	Bulk density of deforming sediment		
κ_s	Fluid density	$\bar{\sigma}_b$	Mean stress level at ice bed
λ	Bed wavelength	σ_{cj}	Local effective stress upon a clast of jth size class
λ_f	'Frictional parameter'		
λ_{rs}	Removal rate constant for suspended particles	σ_ε	Effective stress
λ_t	Transition wavelength at ice/bed interface where regelation and ice deformation are equally efficient	σ_i	Normal stress at a point within an ice mass
		σ_{zs}	Effective stress within deforming sediment
		τ	Shear stress
λ_{wr}	Characteristic bed wavelength – 'white roughness'	τ_0	Shear stress of ice along flow line
λ_1	Principal direction of finite extension of sediment under shear	τ_b	Basal shear stress of ice
		τ_c	Yield strength of sediment
λ_3	Shortest direction of finite extension of sediment under shear	τ_D	Shear stress for a dry bed
		τ_g	Gravitational shear stress in sediment wedge
λ_R	Bed roughness term	τ_i	Yield strength of ice
μ	Coefficient of rock friction	τ_j	Basal shear stress applied to a clast of jth size class
μ_{cs}	Coefficient of friction of sediment		
Ξ	Melting-stability parameter	τ_s	Shear strength of deforming sediment
ξ	Balance velocity	τ_w	Shear stress for a wet bed
ξ_j	Controlling obstacle factor	τ_{xy}	Internal ice shear stress
ρ_a	Average compressive stress	τ_{xz}	Applied shear stress in z-plane
ρ_{ad}	Ambient fluid density	Ψ	Drag coefficient applied to a clast at the ice/bed interface
ρ_i	Density of ice		

Ψ_c Rate of channel closure due to ice creep

Ψ_e Rate of erosion in evacuating a till channel

Ψ_{el} Rate of erosion by laminar flow evacuation in conduit

Ψ_{et} Rate of erosion by turbulent flow evacuation in conduit

Ω Conduit shape factor

Ω_w Constant between 0 and 1 that depends on λ and the transition wavelength (λ_t)

$\bar{\omega}$ Angle made by equipotential lines with the ice surface

Chapter 1

PAST GLACIAL ENVIRONMENTS

J. Menzies

1.1. INTRODUCTION

Throughout Earth's history, the interaction of oceans, continents and the atmosphere in response to solar forcing has resulted in repeated global glaciations. Testimony to these glaciations can be found in the geological stratigraphic record stretching from the Precambrian to the Quaternary (Deynoux *et al.*, 1994) (Fig. 1.1). Among other evidence, for example, are roche moutonées due to subglacial streamlining of bedrock that can be recognised from the Ordovician in Mauritania in Saharan Africa; glacial striations and floating ice mass scourings in the Sturtian of the Precambrian in South Australia; or dropstones and fissure patterns that can be observed in the diamictites of the Precambrian Gowganda in Ontario. Likewise, sediments of the Pleistocene glaciations of the Quaternary are spread across most of the northern hemisphere (approximately north of 40°) and the southern hemisphere in the southern extremity of South America. In mountainous areas of the World, today, can be found extensive glacial sediments and associated landforms, proof of past colder climatic conditions during the Pleistocene. These sediments and landforms stretch from the mountain ranges to the foothills and beyond as for example, along the flanks of Kiliminjaro, Kenya, or in the deep mountain ravines of the Andes or in the plains skirting the Southern Alps in New Zealand. More recent evidence of the cold climate periods in medieval times, known as the Little Ice Age, are well documented. For example, in Europe Swiss parish priests, Icelandic saga writers and Flemish painters all recorded the very cold winters and resultant famines of the period. Geological evidence, of this Little Ice Age, exists in the form of small moraines in Scottish glens, in extensive outwash fans in Norwegian fjord heads (Grove, 1988); and in North America, as trim lines and boulder fields in high valleys in the Rocky Mountains. Likewise in South America glacial deposits on the high alps of the Peruvian Andes record this same cold phase (Clapperton, 1993).

1.2. IMPACT OF PAST GLACIAL CONDITIONS ON GLOBAL HABITATS AND EARTH SYSTEMS

Rather than view these global glaciations as aberrations, it is now apparent that the Earth is essentially a Glacial Planet punctuated by periods of ameliorative conditions similar to or occasionally warmer than the period we live in today. Almost all aspects of life on Earth are influenced to a greater or lesser extent by the impact and persisting effects of glaciation. The distribution of plants, animals, early humans, soil types, their fertility and coastal morphology are a few examples of the direct influence of global glaciation. Even in the tropics, where climatic conditions have remained almost unchanged for at least the past 15 million years, the northern and southern boundaries

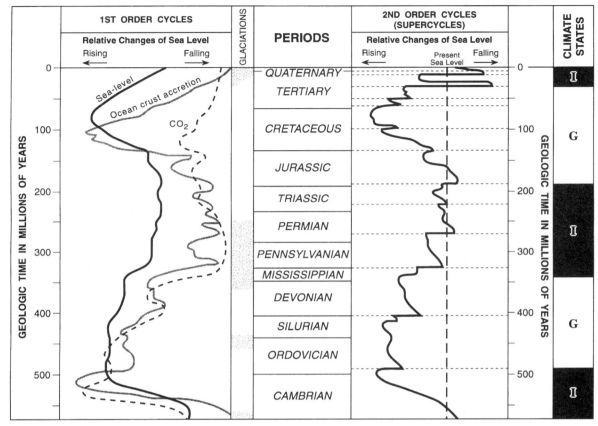

FIG. 1.1. (a) The Earth's glacial record from the Cambrian to the Quaternary including sea level variations (after Vail *et al.*, 1977), atmospheric CO2 (after Berner, 1990), rate of ocean crust accretion (after Gaffin, 1987), and climate 'states' (G – Greenhouse; I – Icehouse) (after Fischer, 1984). (Diagram modified from Deynoux *et al.*, 1994) (reprinted with permission from *Cambridge University Press*)

on land and the repeated changes in ocean sea level have resulted in climatic, and biogeographic changes all as a response, however imperceptible and subtle, to global glaciation.

Considering the enormity of this influence, the relevance today of past glaciations and their associated sediments and forms cannot be undervalued. It is more than likely that the Earth will experience further global glaciations. It is also possible that human activities, especially over the past 200 years, have exacerbated and possibly accelerated some of the complex oceanic/atmospheric and solar forcing interrelationships but to an extent that remains unknown. To be able to predict and be prepared for future global change, a profound knowledge of past

glacial environments must be gleaned from the vast record that past glaciations have left behind.

In the shorter time scale, human activity in those lands of the northern and southern hemispheres is constantly influenced by past glacial environments whether in building roads or housing foundations, planting crops or disposing of waste materials. As a case example, in 1990 an enormous fire of automobile tires occurred on the edge of a small Ontario town. The blaze was of such magnitude and intensity that it took days to extinguish. The first, and most immediate, impact of the fire was the effects of toxic fumes drifting over the town and nearby farmland. However, a second, and equally dangerous but much less controllable, consequence was the seepage of toxic

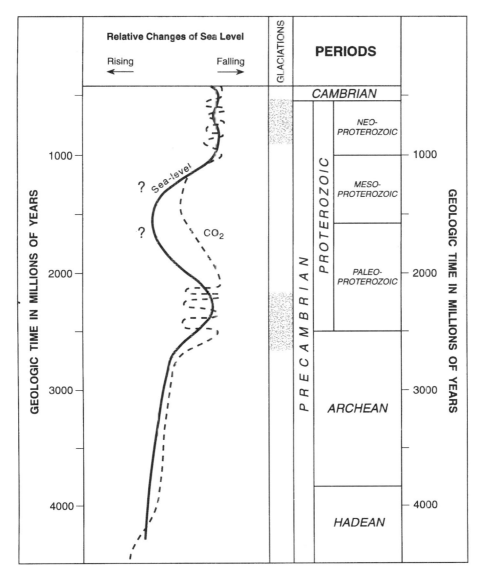

FIG. 1.1. (b) Record of Precambrian glacial record (diagram and interpretation by Young, 1994)

chemicals into the ground from the melting tires. The extent of seepage, the volume of chemical involved and the likely effects on ground water supply are largely unknown. The sediments through which these chemical percolated are glacial. Such an example illustrates, all too pertinently, the necessity to fully understand the sedimentology of glacial sediments.

1.3. RESEARCH IN PAST GLACIAL ENVIRONMENTS

Research into past glacial environments has reached a new intensity over the past two decades. The reasons for this acceleration in research interest are a growing awareness of the relevance of past glacial events and

TABLE 1.1. Developments in Glacial Studies up to 1986

10th Century AD		Recognition in Icelandic sagas of the power and impact of glaciers upon the landscape
1740	D. Tilas	Finnish prospector recognised the concept of drift prospecting tracing ore bodies from dispersal trains
1795	James Hutton	Identifies erratics in the Jura Mountains as being transported by glaciers
1802	John Playfair	Supports Hutton's concepts and suggests glaciation had occurred in Scotland
1815	J.P. Perraudin	Suggests from observations in Val de Bagnes, Switzerland greater extension of glaciers
1821	I. Venetz	A Swiss highway engineer agrees with Perraudin and elaborates these ideas scientifically to the Society of Natural History, Luzern
1824	J.Esmark	Recognition of former extent of glaciation in Norway
1829	I. Venetz	Argues that most of Europe was glaciated
1830	C. Lyell	Puts forward his 'drift theory'
1832	Berhardi	Recognises former continental glaciation in Germany
1837	K. Schimper	A German botanist introduces the term Ice Age (*Eiszeit*)
	L. Agassiz	Convinced by Venetz and J. de Charpentier of the validity of the 'glacial theory', announces his theory of the 'Great Ice Age' to the Swiss Society of Natural Sciences in Neuchâtel
1838	G. Martins	A French geologist suggests concept of vast continental ice sheets, based on work from Spitsbergen
	W. Buckland	Renounces his belief in a biblical flood to explain drift sediments
1839	C. Lyell	Introduces the term *Pleistocene*
	Conrad	Accepts glacial theory in North America
1840	L. Agassiz	Publishes *Etudes sur les glaciers*, Neuchâtel. He travels in Scotland with Buckland and Lyell convincing them that the surficial sediments are of glacial origin
1841	C. McLaren	Argues from evidence in Scotland for eustatic changes in sea-level during the Ice Age.
	J. de Charpentier	Publishes *Essai sur le glaciers et sur le terrain erratique du bassin du Rhône*, in Switzerland
1842–43	J. Adhemar and U. Leverrier	In France the concept of an astronomical theory to explain the origin of glaciation is put forward
1843	J.D. Forbes	Following expeditions to the Swiss and French Alps and to Norway, publishes the first 'glaciological' style text on glacier movement, mass balance and erosional processes
1845	W. Hopkins	Recognition of mechanism of glacier sole sliding
1847	L. Agassiz	Recognition that North European and Alpine glaciations separate
1854	-	The term *Quaternary*, generally accepted in Europe and North America
1859	O. Torell	Postulation of a Fenno-Scandian Ice Sheet
1863	A. Geikie	Publishes first scientific paper on glacial deposits mapped in Scotland
1864	J. Croll	Publishes his theory on astronomical causes of glaciation
1865	T. Jamieson	Argues that isostatic depression of the land surface occurred due to overlying weight of Pleistocene Ice Sheets, evidence from raised beaches
1870	G.K. Gilbert	Shows from mapping in Utah, USA the past existence of large proglacial and ice-dammed lakes (Lake Bonnevile)
	F.von Richtofen	Concludes that loess is of glacio-aeolian origin
1871	A. Worthen	Shows that more than one glaciation occurred in Illinois
1874	J. Geikie	Publishes the *Great Ice Age*
1875	-	HMS Challenger returns from circum-global oceanographic expedition with extensive deep-sea deposit data
	J.G. Goodchild	Recognition of melt-out tills.

TABLE 1.1. Continued

1883	T.C. Chamberlin	Develops a 'till' classification
	W.J. McGee	Recognises the concept of U-shaped valleys and links to glacial erosion
1892	W. Upham	Introduces and describes 'lodgement till'
1894	J. Geikie	Publishes new edition of the 'Great Ice Age' with glacial maps of North America, Europe and Asia
1897	G. De Geer	Recognition of use of varves in glacial chronology
1906	B. Brunhes	Discovery of palaeomagnetic evidence of polar wanderings and reversals
	G.K. Gilbert	Major scientific paper on glacial erosional forms and processes
1909	A. Penck and E. Brückner	Mapping of Alpine foreland terraces and reconstructing a Pleistocene age succession (Günz, Mindel, Riss, Würm)
1914	R.S. Tarr and L. Martin	Recognition of surging behaviour in glaciers
1920	M. Milankovitch	Publishes astronomical theory of ice ages based upon solar forcing
1926	H. Mothes	Uses seismograph and dynamite charges to estimate ice thickness on the Hintereisferner, Switzerland
1929	M. Matuyama	Discovery of polar reversals
1933	E. Sorge	Estimation of Greenland Ice Sheet thickness
1935	W. Schott	German meteor expedition uncovers evidence of Pleistocene period from equatorial Atlantic ocean cores
1941	C.D. Holmes	Extensive discussion on use of till fabrics
1944	S. Thorarinsson	Introduction of use of tephrochronology
1945	C.M. Mannerfelt	Recognition of value of glacial meltwater channels in interpreting deglaciation
1947	H. Urey	Publishes work on oxygen isotope ratio dating method
	H. Carol	Suggested glacial erosion by plucking may be due to regelation processes
	R.F. Flint	Publication of *Glacial Geology and the Pleistocene Epoch*
1948	H.W. Ahlmann	Develops thermal classification of ice masses
1950	R. Beschel	Introduction of use of lichenometry
1951	W. Libby	Develops radiocarbon (^{14}C) dating method
1952	J. Nye	Development of theories on 'glacier flow'
1953	G. Hoppe and V. Schytt	Describe 'fluttings' on subglacial sediment surfaces
1954	J. Glen and M.F. Perrutz	Experimental work on ice rheology
	G de Q Robin	Estimation of Antartic Ice Sheet thickness
1955	J. Glen	Development of law of ice deformation 'Glens' Law
	R.F. Sitler and C.A. Chapman	Early use of micromorphology in glacigenic sediments
	C. Emilliani	Use of oxygen isotopes in determination of sea paleotemperatures
1957–58	–	International Geophysical Year – major initiatives in glaciology
1957	J. Weertman	Discussion of ice basal motion by slippage on a film of water
	J.K. Charlesworth	Publication of the *Quarternary Era*
	L. Liboutry	Discussion of ice basal motion by basal cavity development
	J.B. Sissons	Re-interpretation of glacial meltwater channels systems and their significance in deglaciation
	J.H. Hartshorn	Introduces term 'flowtill'
1961	D.B. Ericson *et al.*	Early compilation of deep sea sediment cores showing evidence of warm and cold phases in synchroneity with on-land warming and cooling phases
	J.A. Elson	Recognises 'deformation' till

TABLE 1.1. Developments in Glacial Studies up to 1951

1962	J.B. Sissons	Re-interpretation of raised shorelines and the significance of glacial isostasy in Britain
1963	A. Gow	Reports details of penetration of Antarctic Ice Sheet at Byrd Station
1964	A.T. Wilson	Suggestion that Anarctic Ice Sheet may surge leading to a trigger for global glaciation
1965	J. Gjessing	Suggest role of wet till in forming P-forms
1966	I.J. Smalley	Introduces idea of dilatant subglacial landforms
1967	J.B. Bird	Regional synthesis of glacial landforms
1969	IGCP	Iniation of the International Geological Correlation programme Project 38 on pre-Pleistocene tillites
	J. Lunqvist	Recognises regional patterns of subglacial landforms
	J.A.T. Young	Recognition of extreme local variability of till fabrics
1970	W.F. Budd et al.	Early attempts at mathematical ice sheet modelling (Antarctica)
1971	A. Dreimanis and U. Vagners	Develop idea of 'terminal' grade in tills
1972	H. Röthlisberger R.L. Shreve	Major papers on subglacial hydrology
1973	D. Krinsley and J. Doornkamp	Publication of Atlas of Quartz grains using SEM
1974	G.S. Boulton	Develops an abrasion rate equation and links glacial erosion to lodgement process
1976	G. de Q. Robin	Advocates a 'heat pump' effect beneath ice masses – polythermal bed condition
1977	R. Aario	Recognizes possibility of subglacial bedform continuum – rogen moraine→drumlin→fluted moraine
1978	D.E. Sugden	Recognition of continental and regional patterns of glacial erosion related to bed thermal states
1979	G.S. Boulton and A.S. Jones	Theoretical discussion on model advocating deformable subglacial bed conditions
	N. Eyles	Recognition of the use of lithofacies in glacigenic sediments
	B. Hallet	Develops an abrasion rate model
	J. Menzies	Recognises role of porewater in subglacial sediment deposition and related drumlin formation
1981	G.H. Denton and T.J. Hughes	Compilation of Late Wisconsinan Ice Sheets Chronology and extents
1983	J. Shaw	From recognition of stratified drumlin core suggests 'flood' hypothesis for formation
1984	S. Manabe and A.J. Broccoli	Recognition of the role of continental plate positions and motion in timing of glacial periods
1986	V. Šibrava et al.	Major correlation of Quarternary glaciations of the northern hemisphere
	R.B. Alley et al.	Report of soft deformable bed beneath Ice Stream B, West Antarctica

*[Table modified following Imbrie and Imbrie (1979) and Bowen, (1984)]

environments to modern habitats and life in those areas of the Earth glaciated in the past.

1.3.1. Recognition of 'Ice Ages'

From medieval times there have been innumerable explanations of features that we now recognize as glacial such as erratic boulders viewed as the putting stones of giants, potholes as devils' punch bowls and other demonic interpretations for glacial phenomena. Where and when precisely a glacial explanation of many of these features was first enunciated is difficult to tell but by the mid-18th century in Scandinavia, Germany, Iceland and Switzerland, several individuals had begun to suggest that glaciers had been more extensive in the past (cf. Flint, 1947, 1957, 1971; Charlesworth, 1957; Embleton and King, 1968; Imbrie and Imbrie, 1979; Nilsson, 1983). In the 9th century, Charpentier, Agassiz, Buckland and Esmark, to name a few, had begun to realize that large areas of Europe had been glaciated by vast ice sheets and thus the concept of the 'Ice Age' was proposed. This suggestion did not become established, however, in many parts of Europe and North America until the late 1800s and even up to the 1920s there were individuals who still questioned the very idea of an Ice Age (Table 1.1) (Bowen, 1978). As early as 1863, Archibald Geikie interpreted the unlithified sediments and landforms of Scotland as evidence of glaciation and, from that beginning in Britain, a rapid period of geological mapping and stratigraphic interpretation spread throughout most of the northern hemisphere. By the mid-1930s knowledge of the extent and details of multiple glaciations to have affected Europe and North America was well established (Flint, 1947; Charlesworth, 1957).

1.3.2. Development of Glacial Chronologies

Penck and Brückner (1909), in the European Alps, had established a four-fold sequence of major glaciations based upon interpretations of the extent and distribution of outwash fans and terraces in the northward trending Bavarian, alpine-foreland, river valleys of the Günz, Mindel, Riss and Würm (oldest to youngest glaciation; see Chapter 8). The younger three glacials of the four-fold sequence of Alpine Pleistocene glaciations were adopted in northern Europe as the Elster, Saale and Weischsel Glaciations, and in Britain as the Lowestoft, Gipping and Devensian Glaciations. In North America, four glacial phases, similar to the European Alpine Model, were primarily used – the Nebraskan, Kansan, Illinoian and Wisconsin an. Today, throughout the world, this simple four-fold sequence has been shown to be rudimentary and a larger number of glacial and interglacial events have been shown to exist (see Menzies, 1995a, Chapter 2, fig. 2.18). At present, over 17 major continental glaciations appear to have occurred during the Pleistocene and this figure may increase as new data are uncovered (cf. Andrews, 1987a) (see Chapter 8).

1.3.3. Multidisciplinary Nature of Glacial Studies

The study of past glacial environments has generated a multidisciplinary strategy of scientific inquiry (Fig. 1.2). The underlying thesis of all these separate, yet connected, studies is to understand past glacial events and processes, global climatic and oceanic circulation patterns, and botanical and zoological adaptions and adjustments. With that knowledge, predictions can be made of possible future global events, patterns and responses.

Central to the study of past glacial environments are glacial sediments. Their properties, characteristic structures, fossil content, age, stratigraphic position, landform association, morphology and location are characteristically the sole evidence from which reconstruction of past glacial environments can be made. To aid in reconstruction, surrogate and long-distance evidence must be gathered to augment what may, at times, be scanty data. In recent years, for example, radiometric dating techniques, and oxygen isotope records from deep ocean sediments and ice sheets have supplied objective and precise information. Similarly, models that can be run repeatedly with ever changing parameters such as those for weather patterns or oceanic circulation provide additional clues as to conditions during incipient, full, and waning global glacial and interglacial phases. Such models reveal theoretical possibilities and feasibilities, providing scientific support for explanations of past environmental conditions and constraints.

1.3.4. Glacial Sediments and Geo(morpho)logy

This textbook takes glacial sediments as central to any explanation of past glacial environments. The study of these sediments comes under the heading of glacial geomorphology or geology (geo(morpho)logy). Glacial geo(morpho)logical studies, in the past, have been of two types those interested in sediments and landforms and those interested in the chronological sequence of glacial events. In many cases these two approaches were combined thus developing a series of chronostratigraphic models of glacial events for a particular location (Hollingworth, 1931; Sissons, 1958b, 1963, 1982; Price, 1960; Muller, 1963; Karrow, 1967; Clapperton, 1971; Dreimanis and

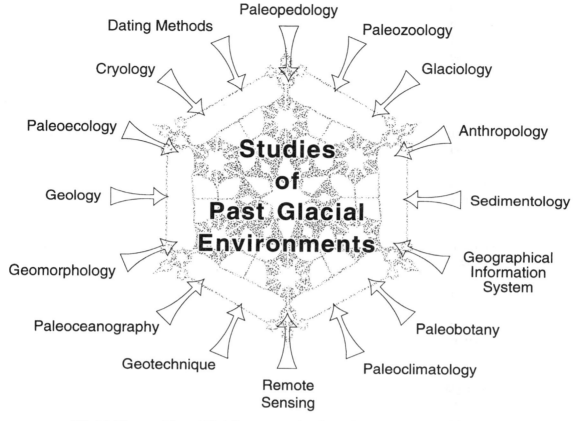

FIG. 1.2. Diagram of the multidisciplinary nature of studies pertinent to past glacial environments.

Karrow, 1972). Prior to the Second World War, research in glacial studies was strongly geological with an emphasis on processes and sedimentology but often within spatially limited areas. After 1945, a morphological approach was adopted in which the geographical distribution of landforms was used foremost in developing explanations of the glacial events for a specific site. Emphasis on sediment types and stratigraphic relationships was only used where stratigraphy was of interest and, too often, little attention was given to glacial sedimentology. The literature on drumlins, for example, details this change in research emphasis before and after the Second World War. Prior to 1939 considerable relevance was placed on the internal sedimentology of drumlins while after 1945, their spatial distribution and interrelationships received greater consideration

until the 1970s when internal sedimentology was again viewed as important (Menzies, 1984).

By the 1970s, a search for explanations led research to again rely more heavily on glacial sedimentology (with a dependence on glaciological conditions) thereby de-emphasing the once strongly geographical paradigm (Boulton, 1987b). This glacio-sedimentological approach (Menzies, 1989b) seeks to find answers by considering all glacial sediments and landforms within the framework of known sedimentological and glaciological conditions before, during and after specific events within a glacial system. The success of this approach therefore hinges upon knowledge of those 'events' and 'conditions'. Glacial geo(morpho)logy has grown as a science in terms of the detailed understanding of glacial dynamics and processes, and in breadth of methodology. An increasing sophistication of techniques can be brought to bear on problems within the field and laboratory that have only become available in the last twenty years. Likewise, knowledge of global events in climate, ocean levels, and ice core stratigraphy can now be directly evaluated and integrated within he context of local site findings.

1.4. RESEARCH ISSUES IN PAST GLACIAL ENVIRONMENTS

As Figure 1.2 illustrates, the breadth of scientific interest in past glacial environments encompasses many diverse fields of inquiry and each one has pertinent research issues. At the general level, however, there are several issues that transcend discipline boundaries, for example:

- The problems of recognising glacial from non-glacial sediments.
- Discriminating between different facies, sub-facies and facies associations within one or adjacent glacial environments.
- Recognising the contribution of water in the many sub-environments of glacial systems.
- Identifying and elucidating boundary interface processes and related bed- and land-form initiation in several sub-environments of the glacial system.
- Identifying those sediment characteristics that are indicative of diagenesis.
- Distinguishing and verifying the impact of freezing conditions on sediments.
- Characterizing habitat changes close to ice masses in relation to climate change, plant colonisation, and faunal and human migration; resolving, at the more local scale, the interplay between land and sea levels before, during and after global glaciations in relation to rapid changes in ice sheet volumes, ice marginal positions and oceanic circulation; developing objective multiple taxonomic criteria in glacial stratigraphy.
- Acquiring dating techniques and refining dating resolution to permit even more precise determination of events and process rates; determining modes of transport in glacial systems from better definition of transport signatures on individual

grain surfaces and understanding of erosion, transport and deposition sequences as manifest in sediment placer bodies.

A persistent problem in all glacial studies is the recognition of glacial sediments (Harland *et al.*, 1966; Dreimanis and Lundqvist, 1984). Although many characteristics have been suggested, there still remains concerns when interpreting lacustrine, marine and distal proglacial sediments (cf. Gravenor *et al.*, 1984; Rappol, 1983; Dreimanis and Schlüchter, 1985; Dreimanis, 1988).

Once determination has been made that a particular sediment or form is glacial, controversy often surrounds the origin of a particular facies or sub-facies (Evenson *et al.*, 1977; Dreimanis, 1979, 1982; Gibbard, 1980; Haldorsen and Shaw, 1982; Shaw 1987). Within glacial systems there are several environments which produce sediments, forms and internal structures that are virtually identical and indistinguishable (see Chapter 9). Under these circumstances, facies associations may often be able to aid in the resolution of a particular problem, for example, in the interpretation of the glacial sequence at Scarborough Bluffs near Toronto, Ontario (cf. Eyles *et al.*, 1983; Dreimanis, 1984a,b; Karrow, 1984a,b; Sharpe and Barnett, 1985), or as to the nature of Precambrian Port Askaig sediments of Scotland (see Chapter 5, Section 5.3.2). However, since glacial sediments may progress through repeated cycles of erosion and deposition, an equifinality is commonly encountered in specific facies units. Only by using related diagnostic attributes possibly linked to stratigraphic position, location or other facies associations can a facies unit be designated in certain cases. Discrimination is especially problematic between sediments in adjacent facies environments such as proglacial proximal and subaqueous (cf. Cheel and Rust, 1982; Domack and Lawson, 1985; Ashley *et al.*, 1991). For example, the distinction between subglacial and subaqueous diamictons remains unresolved (cf. Gibbard, 1980; Frakes, 1985; Lawson, 1988; Weddle, 1992).

It has become apparent that water, as meltwater and porewater, has played a much greater role in most glacial sub-environments than hitherto assumed. Processes of subglacial erosion, for example, are much more widespread than recognised in the past both at the micro- and macro-scale levels (Sharpe and Shaw, 1989; Shaw, 1988c; Kor *et al.*, 1991). More controversial has been the concept of ice sheet stability and subglacial bed conditions controlled, to some extent, by massive subglacial floods (Shoemaker, 1992a,b; Rains *et al.*, 1993). The central role that meltwater and porewater plays in varying effective stress levels within glacial sediments has profound effects on sediment strength and mobility (cf. Menzies, 1979a, 1989a; Boulton, 1987a; Clarke, 1987b; Alley, 1991). However, precise details of sediment geotechnical changes as controlled by porewater content remain rudimentary.

The relative importance of porewater within glacial debris has been largely ignored. However, in many sub-environments where stress-sensitive sediments occur such as flow tills, porewater content is the controlling variable in determining rates of deposition and transport, and effective stress levels (cf. Menzies, 1979a; Paul, 1981). The porewater content of subglacial deformable beds control the rate of debris mobilization and possible bedform development and/ or survival (Boulton and Hidmarsh, 1987; Menzies, 1989a; Alley, 1991). From modern glacial environments it is known, for example, that massive jökulhlaups occur causing devastating effects in the proglacial zone moving vast quantities of debris from the subglacial, terminal and proximal areas of ice masses (Menzies, 1995a, Chapters 6 and 12), yet stratigraphic recognition remains imprecise and problematic.

Much of geomorphology is concerned with the interaction of earth surface processes across the boundary interface between the atmosphere and the earth's surface or the base of an ice mass and its sole, or the bed of a lake or sea or river and flowing water. It is at these interfaces that landforms and bedforms develop. Of intrinsic interest therefore in glacial environments is the reaction between ice, meltwater and the earth's surface and the formation of glacial landforms and bedforms. Studies in glacial geo(morpho)logy have, in the past, attempted to explain the origin of glacial landforms as unique entities (Embleton and King, 1968; Sugden and John, 1976). With insufficient knowledge of many glacial processes, these explanations, although often remarkably accurate, were imprecise concerning processes and rates of

landform development. As understanding of glacial processes increased details of landform development have likewise become more complex (cf. Menzies, 1979b; Bouchard, 1989; Zilliacus, 1989). At the same time it has become apparent that many glacial landforms are not unique but are closely related in origin. Many landforms may be part of a continuum of bedforms that alter morphology in a spatially graduated manner as conditions change. This concept of bedforms has been applied to subglacial stream-lined forms (Aario, 1977a,b; Menzies, 1987; Rose, 1987a) bringing a new perspective on subglacial landforms and their possible origins. Finally, it is evident that many glacial landforms can be constructed by a combination of differing processes producing the same end result.

In examining and analyzing past glacial sediments, it must be remembered that these sediments have been deposited or emplaced for many thousands of years resulting in subsequent changes in their properties. These changes may be the result of exposure at the earth's surface and the influence of subaerial processes, possibly exhumation due to surface erosion, uplift from a subaqueous to subaerial position due to isostatic or eustatic changes, the impact of vegetation colonisation and soil development and the impact of climatic change. Any of these effects may act alone or in concert over various periods of time and to varying depths within the sediments. All of these influences are generally referred to as the processes of diagenesis but the degree and extent of impact is often poorly understood or even recognised at present. Diagenesis may manifest as geotechnical alterations, the result of consolidation, removal of fine sediments, particle to particle interrelationships of fabric or structure, and/or new fracture geometries. Geochemical changes may occur that cause authigenic mineralisation, mineral weathering, pore cementation, and/or mineral precipitation. These changes to the original sediments may be visible and significant while in other sediments the changes are imperceptible and minor. The perception, however, that what is not visible must be of minor effect is dangerously erroneous and new research can only aid in exploring and understanding the effects of diagenesis.

Within the glacial environment transient freezing conditions are commonplace. Sediments are frozen on a seasonal basis or occasionally for longer time periods. Freezing may occur due to changes in stress levels, glacial ice thickness variations, meltwater surcharges or other temporary thermal fluctuations. As sediments transit the glacial system it is likely that they may undergo several freeze/thaw cycles. Indirect indicators are used to try to establish that past frozen conditions have occurred (see Chapter 10 and 12). These indicators are, for example, the preservation of delicate bedding structures, brecciation within subfacies units, the presence of fossil fauna, or features indicative of periglacial conditions such as silt droplets and/or sand wedges. However, at present, recognition of the existence of past freezing conditions is difficult to substantiate. The affirmation of frozen conditions would aid in the recognition of particular environments and associated processes within glacial environments (see Chapter 2).

As ice masses creep forward, or slowly retreat, or surge into a proglacial lake, or float as a tidewater glacier, or join an ice shelf, the margins of these ice bodies are in constant change both sedimentologically and in terms of stress level fluctuations. Marginal conditions, therefore, are of immense importance in the glacial system since many sediments and landforms are associated with these marginal environments and fluctuations. Recognition of marginal environments is often indirectly obtained from sediment facies types and associations, and related landform sequences. Although the sedimentology of land-based ice marginal fluctuations is fairly well established (see Chapter 3), there remains a poor understanding of land/water glacier margins (see Chapters 4 and 5). This limitation is first, due to a restricted knowledge of processes occurring immediately at the ice land/water grounding line positions – largely the result of past inaccessibility, and second due to the inability to clearly discriminate in the glacial sediment record evidence of lithofacies and facies associations indicative of the land/water margins. For example, it has been hypothesised that drumlin development, in some instances, may be linked to near marginal subglacial/subaqueous ice conditions. Evidence from Donegal, Ireland (Dardis, 1987; Hanvey, 1987) and Ontario, Canada (Menzies, 1986) both suggest that drumlin formation could be

linked to nearby grounding line positions in the sea or in a large proglacial lake.

Although present ice marginal environments are usually associated with tundra-like environments, the ice margins of the mid-latitude ice sheets during the Pleistocene were often less extreme than is witnessed today. Likewise the margins of the vast Fennoscandian and Laurentide Ice Sheets varied enormously in terms of climate, flora and fauna along their southern and northern edges. Much remains to be learned of the margins of the past ice sheets to permit a greater understanding of plant colonisation, animal movement and migration, as well as plant and animal evolution and extinction.

A persistent problem in studying past glacial environments is in establishing the relative position of land and sea in local areas (see Chapter 11). Often isostatic and eustatic changes have lead to repeated inundations and re-emergence of land surfaces, generating complex stratigraphies and landform assemblages. Although the general framework of land/sea changes are known much remains to be elucidated at the local scale due to regional variations in mantle viscosity, ice mass volumes, ice mass marginal movements and local topography.

As knowledge of glacial processes and depositional mechanics is refined, the ability to improve the stratigraphic resolution at different sites should be enhanced (see Chapter 8). Allied to improvements in stratigraphic definition is the need to increase the resolution of dating techniques and the number of different dating methods that can be used on glacial materials (see Chapter 14). Both stratigraphic definition and dating resolution demand a better understanding of glacial processes, process rates and the recognition of hiatus in erosional and depositional events before a greater comprehension of the chronology and sequence of glacial events can be resolved.

As a final example of the many research issues in glacial environments, the question of tracing and delineating transport pathways remains a persistent problem in drift prospecting (see Chapters 15 and 16). The ability to unravel the many pathways a particle may have taken en route through repeated glacial events is probably unresolvable. However, under certain unique instances of specific geology or topography, delineation of pathways may be possible from a clearer knowledge of particle surface morphology and imprints peculiar to certain environments and stress conditions (see Chapter 13).

1.5 SYNOPSIS OF CHAPTERS

The record of past glacial environments is immense yet our understanding of the processes and sediments of these environments, stretching from the Pleistocene back in geological time to the Precambrian, remains incomplete. For the past 150 years research has been largely directed at the Quaternary but now considerable data has been acquired from older, and often more extensive, global glaciations. Past glacial environments are a vast repository of information not only of a geological and sedimentological nature but of data pertinent to global climatic, biological and astronomical problems. Predictions for global warming and climate change can be refined from an understanding of past global environmental change. The chapters in this book attempt to summarize and synthesize the most recent research into past glacial environments, indicate where problems remain and where new research might be directed.

In Chapter 2, Menzies and Shilts summarize the vast literature on subglacial environments from a sedimentological perspective. This environment produces the most pervasive of all land-based glacial sediments. The chapter focuses upon the sedimentology of subglacial sediments, diagnostic properties of the different subfacies, and the relationship between these sediments and the many landform types generated within the subglacial environment.

The marginal areas of major land-based ice sheets produce a distinctive suite of sediments and landforms. Johnson, in Chapter 3, considers this environment using the mid-west of the United States for many examples. This particular environment is often misinterpreted and facies types unrecognised within the broad spectrum of ice retreat conditions that overprint and/or rework sediments from subglacial and proglacial environments.

Glaciolacustrine environments are typically recognised by the presence of several characteristic structures and bedding types found in fine-grained laminated sediments. However, another distinctive criterion is the presence of ice scours due to ice mass scouring over the bed of glacial lakes. Similar phenomena occur in shallow seas and on continental shelves. In Chapter 4, Woodworth-Lynas, discusses evidence of ice scours from the Pleistocene to the Precambrian.

Whereas in the past emphasis was largely upon the glacial sediments of land-based ice masses, it has become apparent in the last decade that the total volume and areal spread of glaciomarine sediments in the World far exceeds terrigenous glacial sediments. Elverhøi and Henrich, in Chapter 5, bring a wealth of new data from the North Atlantic and Pacific Oceans to evaluate the current status of our understanding of glaciomarine environments.

In considering glacial environments all too often the contribution of glacioaeolian sediments around the margins of large ice sheets is forgotten or given minor recognition in comparison to those areas actually glaciated. Yet vast areas of Asia, Europe and North America have been blanketed in glacioaeolian sediments. Derbyshire and Owen, in Chapter 6, redress this imbalance and discuss the sedimentology of these sediments, and the mechanics of their generation and deposition in relation to mountain massifs, atmospheric circulation and ice sheet limits.

It is remarkable that in the past most textbooks on glacial environments have concentrated solely on Pleistocene glaciations, yet a vast store of data on glacial environments exist from the Pre-Pleistocene geological record. Young, in Chapter 7, details the many global glaciations before the Pleistocene, the characteristic sediments and possible explanations for this repetitive occurrence of glaciations throughout geologic time.

Stratigraphy of glacial sediments is a central theme in any discussion of glacial environments in order to establish a chronological sequence of geo(morpho-)logical events. The development of a sound stratigraphic framework in glacial environments is fraught with many specific problems not the least of which is the interplay between erosional and depositional processes, the rapidity of these processes and their potentially immense spatial and temporal variations. Rose and Menzies, in Chapter 8, discuss the impact of glacial processes and events on the development of glacial stratigraphic frameworks.

In developing an objective method of sediment characterization, the use of the facies concept has become an invaluable tool. Kemmis, in Chapter 9, illustrates the use of this concept in glacial environments, showing the value of facies association especially when individual similar glacial lithofacies types can be found in several diverse sub-environments.

Boardman, in Chapter 10, summarizes the existing knowledge on paleosols, emphasizing their importance in understanding past glacial environments. Throughout geologic time glaciations have been interrupted by both short and long time periods when ice margins retreated and ice may have withdrawn to levels of glaciation similar to today. These interstadial and interglacial periods lasted from a few to tens of thousands of years during which climatic amelioration of peripheral areas and once glaciated terrains occurred. In either period, subaerial pedological processes of varying intensities were active on the ground surface glacial sediments. The effect of pedological processes has been to develop soil horizons of differing degrees of maturity that on ice readvance or subsequent glaciation were buried. These paleosols are of immense value in determining glacial stratigraphy and environmental conditions at a particular site.

Perhaps the greatest global impact of glaciation has been the effect of ice sheets in changing ocean and land surface levels. The mechanics of glacio-eustasy and glacio-isostasy, respectively, are discussed by Gray in Chapter 11. The complex interrelationships between ice sheet extents, global climates, and oceanic currents as functions of glacial isostasy and eustacy are especially pertinent to any understanding of glacial environments. Sea and land level changes affected the migration of fauna and flora, isolated zoogeographic and botanical regions, and influenced the migration of early humans. The interaction of land and sea affected marginal activity of the vast ice sheets in terms of balance gradients, basal ice motion, rates of advance and retreat, grounding line positions and glacial lithofacies types.

At the macro or visible scale many sediments appear monolithic and devoid of structure. However, in microscopic thin section these sediments may reveal an immense variability in morphology and internal structures. The study of these microscale features termed micromorphology is discussed by van der Meer (see Chapter 12). Micromorphhology provides a wealth of new and objective data from which new hypotheses on the mechanisms of sediment deposition within a glacial environment can be tested.

At even greater magnification using Scanning Electron Microscopy (SEM), Whalley discusses, in Chapter 13, features and surface characteristics of glacial sediments at 300 to over 1200 times magnification. These artefacts viewed using SEM are of immense benefit to understanding processes of glacial erosion and transportation.

One of the great advances in glacial studies was the development and use of radio carbon14 dating in 1951. Since then, as summarized by Brigham-Grette in Chapter 14, there have been further major advances in the use of a variety of radiometric and other dating techniques. In the last decade the development of thermoluminescence and related techniques have seen an additional tool in the search for greater resolution and consistency in dating. The establishment of a rigorous stratigraphic framework can only be tested for reliability and validated with the use of multiple dating techniques.

Studies of past glacial environments have a strongly applied aspect in terms of their influence, for example, on urban areas, agriculture, ground water, waste disposal. However, the importance of these influences have only been recently realised from an applied viewpoint. In contrast, as Shilts points out, in Chapter 15, the use of glacial sediments as a means of tracing mineral sources has been known since the eighteenth century. Drift prospecting is an important and, depending upon World economics, at times, lucrative business. Shilts notes the use these techniques can be in verifying ice sheet flow directions, changing and cross ice lobe pathways and other glaciodynamics of ice masses as well as supplying indirect knowledge of erosional and depositional phases within subglacial environments.

One of the surprising effects of glaciation is the emplacement of placer deposits of gold and other precious minerals. In Chapter 16, Levson and Morison describe the mechanisms of placer formation within the context of glacial environments and review the location and possible emplacement processes for glacially-influenced placer deposits throughout the World.

As knowledge of past glacial environments has developed so the need to apply that knowledge to human and environmental issues has become evermore pertinent. Glacial geo(morpho)logy is entering a new phase of applied studies that must rely on a basic fundamental understanding of glacial processes, sediments and landforms. Much research remains to be done. The vast storehouse of data that still remains untapped within past glacial sediments of the Pleistocene in middle and higher latitudes as well as the extensive sediments of Pre-Pleistocene age should supply additional new information and an even better understanding of the glacial system over the next several decades.

Chapter 2

SUBGLACIAL ENVIRONMENTS

J. Menzies and W.W. Shilts

2.1. INTRODUCTION

No other glacial environment has left its imprint so indelibly on glaciated areas as the subglacial environment. Other than processes active in the glaciomarine environment, subglacial processes and sediments dominate previously glaciated continental land surfaces and epicontinental margins. Subglacial erosional, depositional and bedforming processes are among the most complex, yet least understood set of glacial processes. Our poor knowledge stems from the inaccessible nature of what occurs beneath an ice mass and the limited extent of modern analogs accessible when attempting to comprehend Pleistocene and pre-Pleistocene subglacial environments.

The products of the subglacial environment and its impact upon the land surface are of paramount importance to society. The spatial and volumetric dominance of subglacial sediments in those countries once glaciated cannot be understated. Most building foundations, roads, railroads, aircraft runways, sewage disposal sites, toxic waste sites, agricultural land, groundwater aquifers and construction materials in those areas last glaciated in the Pleistocene, utilise and are in direct contact with subglacial sediments. As our increasing awareness of environmental hazards and rationally sound land use develops into the twentyfirst century, the demand for expert and applied knowledge of glacial sediments and environments will dramatically rise with a prime place given to knowledge of the subglacial environment (cf. De Mulder and Hageman, 1989; Coates, 1991). Perhaps no other glacial environment impacts on the daily lives of millions of people to such a high degree as does the subglacial.

2.2. DEFINITION

Any definition is fraught with problems in terms of defining and delimiting where the subglacial environment ceases and the proglacial, submarginal and/or subaqueous glacial environments commence (Boulton and Eyles, 1979; Eyles, 1983a; Goldthwait and Matsch, 1988; Brodzikowski and Van Loon, 1987, 1991). The subglacial environment is that glacial subsystem lying directly beneath an ice mass and in close contact with the overlying ice, including those cavities and channels beneath the ice that are not influenced by subaerial processes. On land, the subglacial environment may continue for some distance beyond the apparent surface ice margin since buried active ice may underlie debris that has been deposited either from supraglacial sources or from various meltwater streams exiting the ice (Fig. 2.1a). Likewise, on entering a body of water an ice mass may be sufficiently thick or the water shallow enough to permit ice to remain in contact with the bed, the basal environment thereby remaining subglacial.

However, where ice begins to rise off its bed, the subglacial environment can be said to cease down-ice

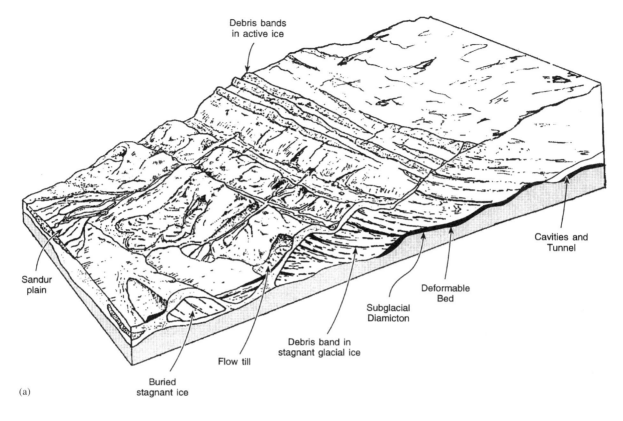

(a)

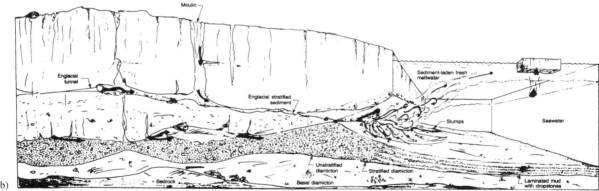

(b)

FIG. 2.1. (a) Model of a ice mass with a land-based margin (adapted from Boulton, 1972; Drewry, 1986) (b) Model of an ice mass with an aquatic margin (adapted from Edwards, 1986)

of this dynamic hinge point, known as a grounding-line (Fig. 2.1b) (Menzies, 1995a, Chapter 14). The nature of the frontal zone of an ice mass, whether tidal or floating, whether in the form of an ice cliff or a thinning tapered snout, impacts upon subglacial conditions and environments. Down-ice from a grounding line an extensive proglacial subaqueous proximal environment may exist. Such a proximal subaquatic zone may become transiently subglacial for short periods of time and at different parts of the

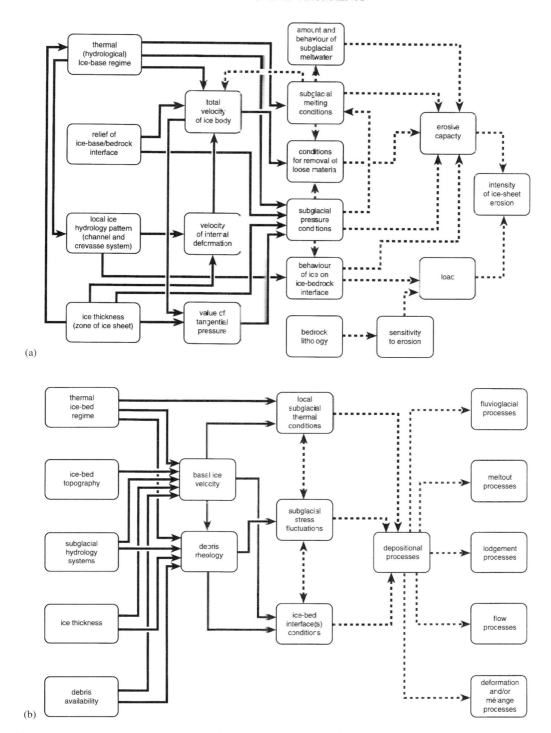

(a)

(b)

FIG. 2.2. (a) Model of interrelationships between main parameters influencing subglacial processes and erosion. (Based upon Sugden and John, 1976; Embleton and King, 1977 and Embleton and Thornes, 1977, modified from Brodzikowski and van Loon, 1991 (b) Model of interrelationships between main parameters influencing subglacial processes and deposition bed (Brodzikowski and Van Loon, 1991; Powell, 1991).

Finally, the subglacial environment is a boundary zone, the interface between the overlying ice and its bed. A complex set of processes interact across this boundary altering the morphological, thermal and rheological state of the interface, evolving into several and, at times, a single interface junction. The nature of this interface(s) controls the erosional, transportational and depositional processes that act across and along this complex and fluctuating boundary. The key to understanding subglacial bed types and associated environmental conditions lies in comprehending the mechanics of this interface (Menzies, 1987). These complex boundary zones migrate across the landscape with every advance and retreat of the ice masses. Therefore all terrains covered by glaciers have been affected and altered by the passage of this boundary zone. Subglacial depositional and erosional phenomena are often destroyed, radically altered or obscured by later processes associated with proglacial environments following gradual ice retreat or by more complex subglacial overriding and subsequent ice retreat, or by subaerial diagenesis (see Chapter 10).

2.3. THE SUBGLACIAL INTERFACE – BED TYPES AND CONDITIONS

The subglacial environment at any one point in time and space is the product of a large group of interrelated and diverse factors (Fig. 2.2). Many factors which directly influence the subglacial environment are indirectly affected by external factors, for example, climate, rates of snow transformation and tectonic setting (cf. Hindmarsh et al., 1989). The end product of these diverse influences reduces to a series of subglacial interface bed types that can be distinguished by thermal and rheological variations which are then subject to an enormous dynamic range of temporal and spatial variations. Precise knowledge of the factors that contribute to the nature of any given subglacial interface environment remain the subject of continuing research.

Any subglacial interface and related bed type is a function of the prevailing basal ice and bed conditions. These conditions are the result of complex relationships between basal ice dynamics, subglacial

sediments and bedrock, subglacial hydraulics and the ambient thermal state for any given area of glacier bed (Boulton and Hindmarsh, 1987; Clarke, 1987a; Lingle and Brown, 1987; Menzies, 1989a; Alley, 1991; Kamb, 1991). Changes in basal ice and/or bed conditions may be widespread or local and may develop rapidly or slowly. These changing conditions may be of enormous magnitude or may simply be minor variations within the interface boundary zone, the former being sometimes detectable whilst the latter may leave little or no imprint in the sedimentary record.

Subglacial thermal conditions have been described, in the past, in simple terms, for instance cold- and warm-based conditions (Paterson, 1981). However, work by Sugden (1977) and Hughes (1981) suggested a more complex subglacial thermal regime. It is possible that frozen and melted bed conditions occur at the subglacial interface as part of a spectrum of thermal states (Fig. 2.3 see also colour plates) (Menzies, 1995a, Chapters 3, 4 and 5). Not only do thermal conditions vary areally across a bed interface but they change over time, creating a transient subglacial environment in which a large and complex array of differing sedimentological conditions associated with distinct thermal regimens can pass repeatedly over a given point on the glacier bed. Directly connected with these thermal phase changes

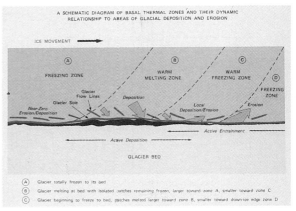

FIG. 2.3. Models of subglacial thermal regimes, their spatial relationships and processes of subglacial erosion, transportation and deposition.

are rheological changes associated with unconsolidated bed sediments (Boulton and Jones, 1979; Alley, 1989a,b; MacAyeal, 1989a; Kamb, 1991). As Hutter (1983) has pointed out, our understanding of glacier bed conditions must be viewed as thermo-rheological in nature, and as such is a complex problem to interpret and model (Menzies, 1995a, Chapter 5).

Recent research from West Antarctica suggests that parts of that ice sheet, in particular Ice Stream B, is underlain by a layer of deforming sediment (Alley et al., 1986, 1987a; Blankenship et al., 1986, 1987; MacAyeal, 1989a). The deforming debris layer extends downstream for about 200 km and averages 6 m thick with a variation of between ≤1–12 m. The upper surface of the layer appears comparatively smooth but flutes (≈10 m high by 300–1000 m wide) were detected at the lower interface. Typically, the debris has a high porosity (n≈0.4), high porewater pressure (ρ_w = 50 kPa), debris cohesion <15 kPa, low effective pressure and is probably a water-saturated unconsolidated diamicton.

Additional proof of deformable bed conditions under present-day ice masses has been reported from Iceland (Boulton and Dent, 1974; Boulton et al., 1974; Boulton, 1987a), Alaska (Engelhardt et al., 1978; Kamb et al., 1985), China (Echelmeyer and Wang, 1987), Sweden (Brand et al., 1987); Ellesmere Island, Canada (Koerner and Fisher, 1979) and Washington State, USA (Brugman, unpublished 1985). The data gathered from these diverse locations show actual or calculated values for strain rate, deforming layer depth and basal ice velocities (Table 2.1). Evidence of deformable debris layers beneath present-day ice masses confirms what has been

advocated by several workers; that deforming bed conditions possibly existed beneath the Pleistocene ice sheets (for example, the southern margins of the Laurentide Ice Sheet along ice stream trajectories such as Lake Michigan and parts of Lake Ontario; or along the eastern margin of the Devensian Ice Sheet in England in the area of Dimlington, Yorkshire). If so a new and markedly different set of conditions and criteria for analysing subglacial sediments, sediment structures, facies associations and stratigraphy must be considered (cf. Boulton and Jones, 1979; Fisher et al., 1985) (Chapter 8). Under marginal and lowland areas of Pleistocene ice sheets it is thought probable that soft sediment or deformable bed conditions were widespread (Boulton et al., 1977; Fisher et al., 1985; Beget, 1986, 1987; Fulton and Andrews, 1987; Brown et al., 1987; Hart et al., 1990; Menzies, 1989a,b; Ridky and Blankenship, 1990; Hart, 1990; Hart and Boulton, 1991; Hicock, 1992; Hicock and Dreimanis, 1992a). This reassessment of subglacial sedimentological conditions raises the fundamental question as to whether these mobile sediment layers are indicative of a common, pervasive environmental condition, or are local, isolated aberrations. Only new research can answer this question.

Several subglacial bed types would seem to best typify environmental conditions at the subglacial interface (Menzies, 1995a, Chapter 4). The bed types discussed below should be viewed in terms of our present state of knowledge and are likely to be modified and constrained as new data and understanding emerge. There is no agreed upon subglacial bed classification and several slightly different

TABLE 2.1.

Location	u_s (m yr-1)	h_s (m)	e_s (yr^{-1})
Urumqui Glacier No. 1 China (a)	~3 00	~0.35	≈8.57
			≈8.39
Blue Glacier, Washington, USA (b)	~4 00	~0.10	≈40.00
Breiðamerkujökull, Iceland (c)	~16 00	~0.50	≈32.00
Upstream B. Camp, Ice Stream B, West Antarctica (d)	~450.00	~6.00	≈75.00
Pleistocene Puget Lobe, Washington, USA (e)	~500.00	~15 undilated	≈33.33
		~22.5 dilated	≈22.22

(a) Echelemeyer and Wang, 1987; (b) Engelhardt et al., 1978; (c) Boulton, 1979; (d) Alley et al., 1987a; (e) Brown et al., 1987

taxonomies exist (cf. Shoemaker, 1986b; Röthlisberger and Lang, 1987; Boulton and Hindmarsh, 1987; Menzies, 1989a).

Four basic bed types appear to exist: (1) Polar Active, (2) Polar Passive, (3) Temperate Active, and (4) Temperate Passive. Most research efforts to date have centred around Temperate Active bed types since these are probably the most widespread and pervasive. However, some consideration of the other forms is necessary.

It has never been defined specifically when an ice mass is characterized by widespread polar, as opposed to temperate, bed conditions. The standard definition of either condition is based on the thermal state of the specific ice mass but no definitive spatial, volumetric and/or temporal parameters have been employed (Menzies, 1995a, Chapter 3). As a general guide where an ice mass exhibits a specific thermal condition areally over several tens of square kilometres and from close to the surface to its bed, over at least a decade of observations, perhaps a specific thermal state can be said to prevail. However, polythermal bed conditions, as a general encompassing state, best typify most ice masses, other than the extreme cases of either polar or temperate states.

2.3.1. Polar Bed Types

Under active Polar glacier ice conditions it has been assumed in the past that no movement occurs at the ice/bed interface that frozen conditions prevail and thus no free meltwater is present and that the glacier ice is below the pressure melting point (cf. Sugden and John, 1976; Drewry, 1986) (Fig. 2.3). However, Shreve (1984) demonstrated theoretically that movement at the frozen interface can occur. Echelmeyer and Wang (1987) observed movement along the ice/bed interface in a subpolar glacier (Urumqui Glacier No.1, China) confirming Shreve's theory. Boulton and Spring (1986) have also suggested that where frozen bed conditions prevail subglacial interface tectonics may occur, resulting in both ice and basal sediment being tectonically deformed upward into the moving ice mass (Fig. 2.4) (Menzies, 1995a, Chapter 10). The rate of ice movement under polar conditions in comparison to that under temperate ice masses is magnitudes less. In terms of the potential volume of tectonised sediment, however, a considerable amounts of sediment and/or bedrock may be tectonically stacked and 're-transported' under polar thermal conditions (e.g., Christiansen and Whitaker, 1976; Berthelsen, 1979; Moran *et al.*, 1980; Aber *et al.*, 1988; Croot, 1987, 1988; van Wateren, 1987).

2.3.2. Temperate Bed Types

Temperate or warm glacier ice conditions are the result of glacier ice being at or above the pressure melting point. Under these conditions a spectrum of environments reflecting both thermal and rheological fluctuations can be expected to exist. Many thermal fluctuations are of either short-term spatial and/or temporal variations. Localised patches of a frozen bed state may occur temporarily as may passive stagnant conditions. However, the major distinction between Polar and Temperate subglacial bed states is that meltwater in some form is present in the latter. Three dominant bed states best typify our present understanding of temperate bed conditions (1) Hard or Rigid or 'H' beds, (2) Soft or Mobile or 'M' beds, and (3) Quasi-Rigid/Quasi-Soft or 'Q' beds. These three bed states include a series of variants that are spatially correlative to transient changes in subglacial interface thermal, rheological, geological and morphological constraints (see Section 2.4). In considering these interfaces, it is possible that several interfaces may be found below the uppermost interface that exists immediately below the mobile ice mass. As a consequence, a complex series of diverging and interconnecting interface boundaries are thought to evolve beneath these ice masses (cf. Menzies, 1987; Menzies, 1995a, Chapter 4).

2.3.2.1. 'H' Beds

'H-beds' are characterised by a hard or rigid bed state in which almost all basal motion is attributable to overlying ice stresses. Movement is not facilitated by the development of lowered effective stress levels in sediment or bedrock below the upper ice/bed interface due to meltwater infiltration. These bed types may develop where an ice mass directly overlies bedrock of low permeability ('Ha') (Kamb and La Chapelle, 1964; Vivian and Bocquet, 1973; Souchez and

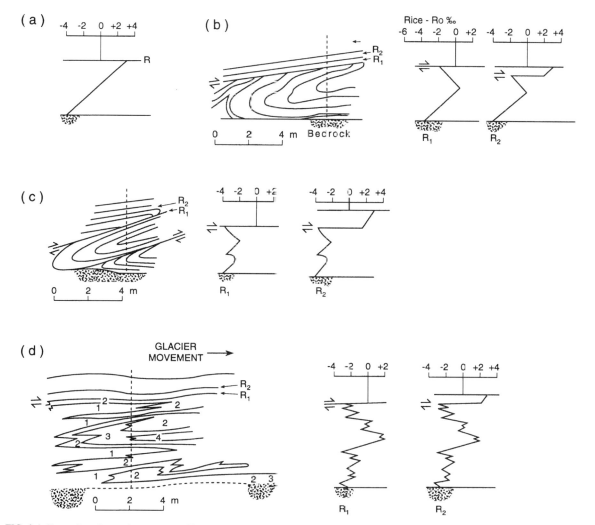

FIG. 2.4. Examples of complex patterns of basal ice flow on isotopic profiles. (a) Isotopic pattern in basal regelation ice in which simple laminar flow has occurred; (b) pattern of basal ice folding beneath a sharp décollement in Sørbreen. Spitsbergen; (c) basal ice folding and faulting in Aavatsmarkbreen, Spitsbergen; (d) multiple folding in the basal ice of Lower Wright Glacier, Antarctica
Numbers identify different folded strata. The direction of displacement of the isotopic profile at the top of regelation ice (shaded) depends upon whether the top of the regelation ice coincides with the décollement (position R1), or whether it lies above the décollement (position R2). Vertical dashed lines show the positions of hypothetical isotopic profiles in the folded ice. (After Boulton and Spring, 1986; reproduced by courtesy of the International Glaciological Society.)

Lorrain, 1991), frozen debris ('Hb') of low or zero permeability (Freeze and Cherry, 1979; Chinn and Dillon, 1987; Gordon et al., 1988), or unfrozen debris ('Hc') of low hydraulic conductivity (for example, $<10^{-6}$ ms^{-1}). In all instances, most free meltwater moves at the ice/bed interface and is prevented from penetrating below this boundary. In some cases,

permeable debris ('Hd') subjacent to the ice/bed interface overlying an aquitard is quickly saturated preventing any further meltwater movement except at the ice/bed interface (Shoemaker, 1986). Finally, debris ('He') with a high hydraulic conductivity may overlie an aquifer of similar or higher conductivity permitting high rates of porewater movement through the debris layer (Menzies, 1981, 1987). In this instance, provided meltwater production at the upper ice/bed interface does not produce discharges greater than the through-flux of porewater within the debris layer, a process of ice/bed interface meltwater evacuation can be established (see Shoemaker, 1986, Type II) (Engelhardt *et al.*, 1987; Hodge, 1979; Boulton, 1987). The persistence and integrity of an 'He' bed state is largely a function of the basal debris layer's permeability and the maintenance of a hydraulic gradient below the critical hydraulic gradient (Terzaghi and Peck, 1969; Freeze and Cherry, 1979).

In all of the above conditions (except 'He') the ice/bed interface is effectively sealed off from the underlying bedrock or debris thus meltwater flows along the ice/bed interface leading to the possible development of decoupling instability (Menzies, 1995a, Chapter 6). Both Shoemaker (1986) and Boulton and Hindmarsh (1987) have suggested that a large percentage of this ice/bed interface meltwater flow is in channels and not sheets (cf. Walder, 1986; Weertman, 1986).

A rigid sediment structure, in which there is zero intergranular movement, is assumed to exist under 'Hd' and 'He' bed conditions. Under these conditions porewater, derived from the ice/bed interface, passes through the sediment in a manner described by Darcy's equation for flow through a porous medium.

$$Q = iK \qquad (2.1)$$

where Q is the specific discharge, assuming that the inflow rate and outflow rate of porewater is equal through a unit area of sediment, i is the hydraulic gradient defined as the excess hydrostatic pressure in the meltwater as it flows through a specific length of sediment and K the hydraulic conductivity. Shoemaker (1986) and Lingle and Brown (1987) have estimated flow rates in thousands of years except under shallow 'He' bed conditions. Also, and more critical, are rapid increases in porewater pressures that could result in a critical hydraulic gradient developing. With a critical hydraulic gradient ($i_c = 1.2$), the effective stresses become zero or negative causing porewaters to move upward to the ice/bed interface inducing liquefaction of sediment. If liquefaction

becomes widespread beneath an ice mass rapid ice/bed decoupling could take place. Although this condition may occur beneath ice masses and may be an important process in subglacial deformation, it can be neither widespread nor typical. It seems unlikely that subglacial debris, even under 'He' conditions, can, exclusively, evacuate subglacially produced meltwaters. Therefore, under all 'H' bed states it is probable that substantial meltwater discharge must take place at the ice/bed interface (Boulton and Hindmarsh, 1987; Lingle and Brown, 1987;).

The style and nature of subglacial channel flow at the upper ice/bed interface (Menzies, 1995a, Chapter 6) remains the subject of considerable controversy (Shoemaker, 1986; Walder, 1986; Weertman, 1986; Clarke, 1987b). Both 'N' and 'R' type channels are likely to develop or a combination of both ('C'-channels). Under 'H' bed conditions, tunnel valleys (entrenched 'N' channels) may develop across the ice/bed interface (Wright, 1973; Attig *et al.*, 1989; Booth and Hallet, 1993; Piotrowski, 1994; see Section 2.9.2b).

2.3.2.2. 'M' Beds

In what is a radical new phase in understanding subglacial environments, the presence and possible widespread occurrence of mobile, soft deforming or deformable beds required a major reappraisal of temperate subglacial conditions and the likely effects such deformable beds may have upon bedforming and sedimentation processes (Menzies, 1995a, Chapter 5). Typically, these beds are composed of saturated debris of a slurry-like consistency in which free meltwater discharge is less dominant and porewater movement critical (Boulton and Dent, 1974; Boulton *et al.*, 1974; Boulton and Jones, 1979; Alley *et al.*, 1986). Under 'H' bed conditions, removal of subglacial meltwater solely by pore space flux within the subglacial sediment layer is virtually impossible due to constraints of permeability and hydraulic gradients within the debris. However, if bulk deformation of the upper parts of the subglacial debris layer occurs, via intergranular deformation in the form of a saturated debris flow (en masse), then meltwater discharge may be possible (Alley *et al.*, 1986, 1987; Lingle and Brown, 1987; Menzies, 1989). Debris slurry flow direction would, generally, be parallel to the principal subglacial pressure head direction (down the ice flow line toward the ice front). Where major obstructions at the ice/bed interface are encountered locally deviations from the main flow direction occur and are evident in the deformation structures and fabric extant within the deformed sediment (Hart, 1992; van der Meer, 1992; van Wateren, 1992; Menzies and Woodward, 1993).

Subglacial debris flows in a manner similar to a Bingham visco-plastic material; thus, mobilization, once begun, becomes a function of the viscosity of the slurry and its strain rate (Johnson, 1970; Cheng and Richmond, 1978; Iverson, 1985; Clarke, 1987b; Kamb, 1991) (Menzies, 1995a, Chapter 5). In argillaceous sediments such as glacial diamictons, it is thought that the rate of deformation is more a function of viscosity (controlled largely by debris porewater content) instead of strain rate (Maltman, 1988).

Deformation may occur as a pervasive or non-pervasive process. In the former, the sediment as a whole (each individual grain) will deform en masse therefore existing structures and fabrics within the sediment will be grossly affected, if not erased and homogenised, by the deformation process. In the latter process, in contrast, discrete shear planes or zones develop within the subglacial debris layer that 'carry' the principal components of applied stress resulting in shear planes or zones separated by areas of sediment that remain largely unaffected by the deformation process, itself. This process of deformation thus permits pre-deformation structures and fabric between the sheared areas, if any, to remain intact. Non-pervasive deformation has been observed both experimentally and in Quaternary sediments (Maltman, 1987, 1988; Talbot and Von Brunn, 1987; Menzies, 1990a; Hicock, 1992; Hicock and Dreimanis, 1992; Menzies and Maltman, 1992; van der Meer, 1994). The fundamental difference between these two styles of deformation has important implications for the analyses of sediment structures and subglacial stratigraphy (Menzies and Maltman, 1992; van der Meer, 1994).

At present understanding of subglacial debris rheology is rudimentary, specifically in terms of those factors that influence and control viscosity and debris porosity. The thickness of the debris layer is con-

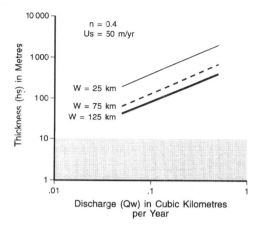

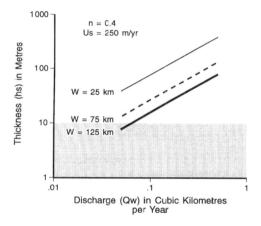

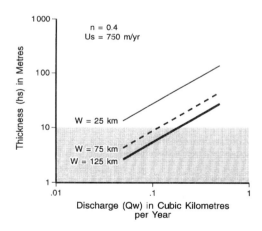

FIG. 2.5.Hypothetical examples of changing debris thickness of a subglacial mobile deforming layer in response to varying meltwater fluxes, and bed widths with constant sediment porosity and upper interface sediment velocity

trolled by the applied stresses and the rheology of the debris. As applied stresses, from the overlying ice, rise across the upper ice/bed interface, more subjacent debris will increasingly be mobilized and vice versa. For example, in order that debris thicknesses (h_s >2 m) be mobilized, Lingle and Brown (1987) have shown that basal ice velocities of $\approx 700\,myr^{-1}$ are necessary. This computation is based upon Lingle and Brown's equation:

$$h_s = \frac{2Q_w}{n\ W\ u_s} \qquad (2.2)$$

where n is debris porosity of 0.6, glacier bed width (W) is 125 km, u_s is the surface velocity of the debris, equal, for this calculation, to the basal ice velocity, and subglacial meltwater flux rate (Q_w) is 0.05 km^3 yr^{-1}. A range of subglacial debris thicknesses to debris surface velocities are illustrated in Figure 2.5. It is apparent that for a substantial thickness of subglacial debris (>1 m) to be mobilized, a high velocity is necessary at the upper ice/bed interface. If it is assumed that the surface debris velocity can be equated with the basal ice velocity, then velocities are those typically found, today, beneath ice streams or fast outlet lobes of large ice sheets (Fisher et al., 1985; Brown et al., 1987; Alley, 1991). The broad implications of this style of ice/bed interface environment affect not only the rate of movement of ice masses and their surface profiles (Mathews, 1974; Boulton and Jones, 1979; Beget, 1986) but also the sedimentological conditions and bedforming processes at the subglacial interface.

2.3.2.3. 'Q' Beds

It is probable that 'Q' bed subglacial states in which hard and soft bed conditions alternate both temporally and spatially at the ice/bed interface, are the most common beneath polythermal ice masses. The importance of 'Q' bed states is apparent in the variability of subglacial environmental conditions where changes in subglacial debris rheology (Boulton, 1987; Boulton and Hindmarsh, 1987; Clarke, 1987; Alley, 1989; Kamb, 1991), and in basal ice thermo-mechanical behaviour (Budd et al., 1971; Boulton, 1983; Hutter, 1983; Boulton and Spring, 1986; Zotikov, 1986) can

occur with varying degrees of frequency, duration and spatial distribution. It has been demonstrated that beneath differing areas of a major ice sheet certain conditions prevail depending upon mass balance state (Hindmarsh et al., 1989). These fluctuating subglacial conditions have major sedimentological implications for Quaternary research where terrain is examined that has been overprinted by several glaciations (Boulton and Eyles, 1989; Eyles, 1983; Ehlers, 1983a,b; Šibrava et al., 1986; Fulton and Andrews, 1987; Boulton and Clark, 1990; Clark, 1991).

2.4. SPATIAL VARIATIONS IN SUBGLACIAL POLYTHERMAL-RHEOLOGICAL BED CONDITIONS

It is apparent that most ice masses are underlain by polythermal bed states. These polyphase beds might be compared to a chequered quilt where each square represents a thermal and/or rheological state that is in continual state of transition (cf. Sugden, 1977; Hughes, 1981; Eyles and Menzies, 1983; Kleman, 1994). Any one patch may, over time, alter from hard to soft bed conditions and vice versa. From a sedimentological viewpoint, these subglacial variations can provide glaciological explanations of the spatially complex geomorphological processes that take place. Polyphase subglacial bed conditions account for distinctive beforming processes, spatial bedform associations, subglacial sediments and associated structures (Eyles and Miall, 1984; Hicock, 1990). A 'problem' in the past has been the inability to explain localised glacial deposition. Previously, it had been thought that erosion took place in some distant up-ice location while deposition occurred in marginal areas of ice sheets. Even as late as 1971, debate still persisted concerning short and long distance subglacial transport (cf. Goldthwait, 1971, editorial). One reason for this perception is that metamorphic and igneous rocks tend to produce limited debris and mostly boulders whereas sedimentary rock produces much more debris (but few boulders). It is now generally accepted that most subglacial deposition (the bulk of the till matrix) is the result of relatively short distance transport in the order of 10–15 km or less. To explain this apparent juxtaposition of erosion, transport and deposition, a model analogous to the geomorphic conditions found on opposite banks of a river meander bend are prerequisite. Polythermal bed conditions essentially provide this state in which local erosion, transport and deposition can occur simultaneously across a glacier bed varying both in time and space as a function of the varying subglacial bed conditions discussed in the above section.

Sugden (1977, 1978) and Hughes (1981) suggested that traversing the bed of an ice sheet from its centre to its margins would reveal transient polyphase bed conditions (Fig. 2.6). Recent work by Aylsworth and Shilts (1989) demonstrated the style and nature of the sedimentological changes encountered across a section of the Laurentide Ice Sheet in Canada (Fig. 2.7) by relating landform zones to the geology of the bed and to changing ice dynamics.

At the centre of an ice sheet, where basal ice velocities are at a minimal and polar bed conditions may prevail, limited sediment erosion and, thus, transportation is expected to occur (Boulton, 1982; Shilts et al., 1987; Kleman, 1994). Therefore, central areas of ice sheets can be expected to have limited glacial erosion or deposition at the stage of maximum or near maximum ice extent. The geological results of zero velocity and/or polar bed conditions are, how-

ever, rarely observed for most ice sheets. This is because ice flow centres continually migrate thus terrain where flow centres existed for some time are affected by previous or later ice movements (Boulton and Clark, 1990).

Under ideal conditions a series of polyphase bed states should be found sequentially outward from an ice flow centre. Each individual bed state will exhibit, as the distance from the ice centre extends, increasing basal ice velocity and frictional heat. From the central portions of an ice sheet, changes adjacent and interfingering with one another can be expected to pass into a succession of bed states viz:

polar → temperate 'Q' beds with a dominance of temperate 'H' beds → temperate 'Q' beds with a dominance of 'M' beds → temperate 'M' beds → temperate 'Q' beds with a dominance of 'M' beds → temperate 'Q' beds with a dominance of temperate 'H' beds → polar beds (Fig. 2.8).

Such a succession may not always occur nor does this sequence necessarily always follow but, in many instances, such an outward developing series of subglacial bed conditions can be anticipated. The series may alter according to local variations in the geology of the substrate including lithology and

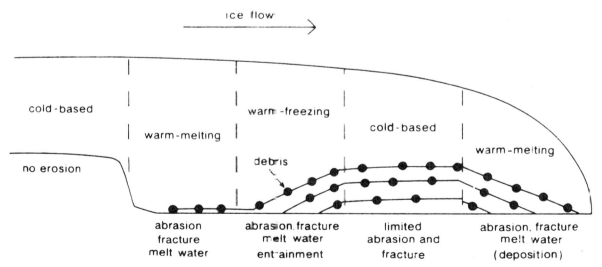

FIG. 2.6. Model of relationship between processes of subglacial erosion and basal thermal regime beneath an ice sheet. (After Sugden, 1978; reproduced by courtesy of the *International Glaciological Society*.)

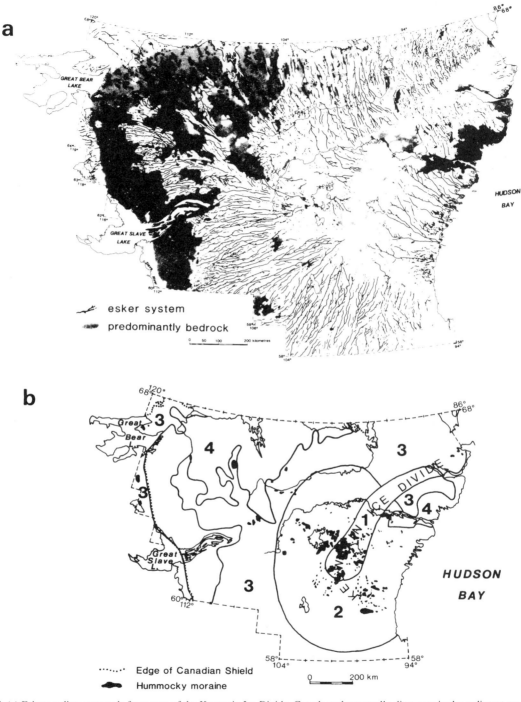

FIG. 2.7. (a) Eskers radiate outwards from area of the Keewatin Ice Divide, Canada and eventually disappear in the sediment-poor region along the western edge of the Canadian Shield. (b) Landform/sediment zones around the Keewatin Ice Divide. Dark areas represent hummocky moraine, the characteristic landform of the Ice Divide. (After Aylsworth and Shilts, 1989; reproduced with permission from *Elsevier Science*.)

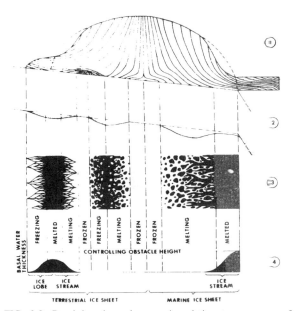

FIG. 2.8. Basal ice thermal zones in relation to processes of subglacial erosion and deposition beneath a steady-state ice sheet having terrestrial and marine portions. 1, internal ice flow trajectories; 2, basal topography created by glacial erosion; 3, areal distribution of subglacial meltwater. Subglacial meltwater shown in black in 3 and 4. (After Hughes, 1981; reprinted from Denton and Hughes, 1981a by permission John Wiley and Sons, Ltd.

If subglacial meltwater production increases to the point where the ice is decoupled from its bed, frictional resistance will be reduced locally to near zero. Once this state is reached, frictional heat that produces meltwater will drop and meltwater production will decrease, allowing the ice to contact its bed once again. With renewed contact, debris production increases, frictional heat increases and the cyclic production of meltwater begins again. This cycle will vary with changing ice flow velocities, in a general way outward from the ice flow centre and locally due to topographic irregularities such as depressions where constricted ice flow leads to increased velocities. Such local controls are evident in the high Canadian Arctic where 'normal' tills were produced immediately adjacent to marine channels (Sugden, 1978).

Likewise, in areas or zones of faster moving ice such as ice streams, sufficient debris may possibly exist at the bed to enable deformable bed conditions to be generated. In this situation 'Q' beds with a dominance of 'M' bed conditions may begin to prevail locally. In basal environments of rapid ice flow, subglacial bedform development under

structure of outcrops, local topography and the physical nature of pre-existing unconsolidated sediments. It may also vary according to basal sediment grain size and therefore rheological parameters, variations in ice basal stress fields, basal ice thermal patterns, interface morphology and topology, basal ice velocity, and subglacial meltwater flux (Fig. 2.9). From the central areas of ice sheets outwards, as basal ice velocities increase, production of subglacial meltwater should also increase, perhaps to the point where sufficient portions of the ice/bed interface begin to slip over the bed (Lliboutry, 1968; Weertman, 1972). Close to ice flow centres where limited debris has been produced and fast zones of ice flow are unlikely to be found, deformable bed conditions are unlikely to develop (Fig. 2.9). Instead, bed conditions similar to those defined as 'H' bed states may exist. In these inner reaches of an ice sheet some sediment transport by meltwater sheets and channels can be expected but significant fluvioglacial deposition is unlikely (Fig. 2.7).

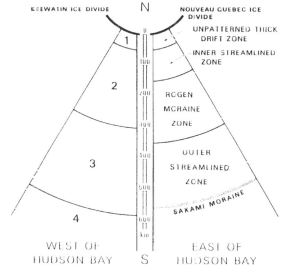

FIG. 2.9. Landform/bedform zonation around the Keewatin Ice Divide, west of Hudson Bay (from Shilts et al., 1987), and around Nouveau-Québec Ice Divide, east of Hudson Bay Canada. (After Bouchard, 1989; reproduced with permission from *Elsevier Science Publishers*)

28 SUBGLACIAL ENVIRONMENTS

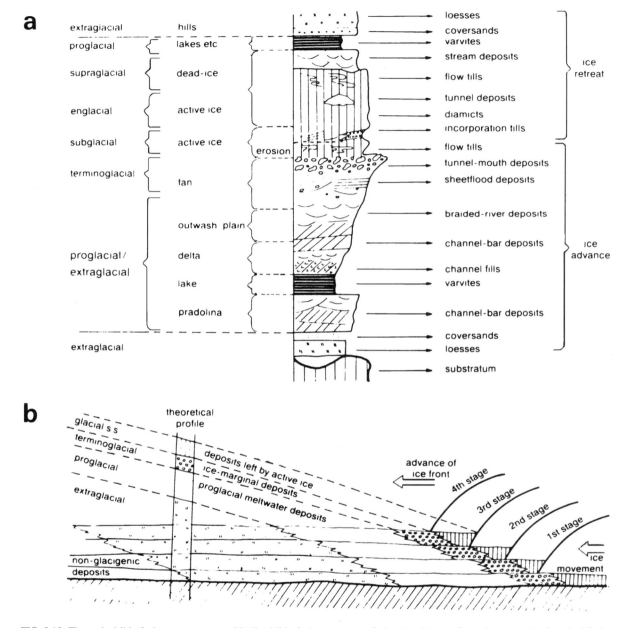

FIG. 2.10. Theoretical lithofacies sequences: (a) Idealized lithofacies sequence of glacial sediments formed as a result of gradual facies environment changes following periods of ice advance and retreat. (b) A schematic model of spatial changes in lithofacies types as a result of ice advance and retreat. (Reproduced from Brodzikowski and van Loon, 1991 with permission from Elsevier Science Publishers)

deforming sediment conditions may transmutate through a sequence or continuum of forms (Aario, 1977a; Menzies, 1987; Rose, 1989b). Where subglacial environments are unsuitable for 'M' bed conditions, 'H' bed states may exist, with large volumes of subglacial sediment being transported within meltwater channels. Conversely, meltwater streams may carve erosion marks or remnants of previously deposited sediment into fluvial bedforms (Shaw et al., 1989).

It is under 'Q' beds that subglacial sedimentation of both unsorted and sorted sediments occurs. The timing of deposition remains enigmatic since sedimentation is likely to occur both during advance and retreat of the ice margin. Where a clear stratigraphic record can be observed it is possible that the sequence of sedimentation following ice advance and retreat can be distinguished (Fig. 2.10). Where overriding by later ice movement has disrupted strata or where syndepositional allokinetic or autokinetic deformation occurred it may be difficult to separate the sequence of depositional phases (see Chapter 8). Where stresses develop locally within deposited sediment, either due to ice overburden loading and/or internal sediment deformation, sediment structures indicative of over-shearing ice are likely to evolve (cf. McClintock and Dreimanis, 1964; Hicock, 1992; Dreimanis, 1993; Menzies, 1995a, Chapter 10).

Near ice sheet margins, where similar polyphase sequences of bed conditions occur, increasing debris and meltwater volumes develop progressively, creating widespread examples of bedforms and thicker sediment piles. Also near the ice front, where ice is thinner, topographic controls begin to influence ice flow and meltwater directions, often allowing radial deposition of various subglacial bedforms (for example, drumlins in Wisconsin, USA, Alden, 1911). Not all radial flow patterns, of course, are indicative of ice marginal positions, for example, eskers and drumlins in the areas of northern Manitoba, Ungava, Quebec, and the Northwest Territories, Canada (Shilts et al., 1987; Bouchard and Marcotte, 1986; Aylsworth and Shilts, 1989) may be representative rather of basal ice motion away from ice flow centres and divides and, in places, major topographic control.

Finally, where ice margins float, basal ice dynamic conditions alter down-ice of the grounding-line (Menzies, 1995a, Chapter 14). Changes in subglacial stresses are transmitted back up-ice from the grounding-line for some distance, thereby affecting bed conditions and influencing subglacial processes. 'Draw-down' effects may channel ice, sediment and meltwater flow toward a floating or grounded margin in preference to more stable subaerial marginal sites, thus modifying the spatial distribution of sub-mar-

ginal subglacial sediments and bedforms (Krüger, 1994).

It is apparent from the above discussion that similar spatial down-ice distributions cannot be expected within valley glaciers (cf. Boulton and Eyles 1979; Gurnell and Clark, 1987). In general, due to the short distances of transport, as compared to ice sheets, basal ice thermal zonation patterns are unlikely to develop. Within the middle and lower reaches of valley glaciers, polyphase conditions may prevail with 'H' bed states dominant. Occasionally, where large volumes of debris have been derived or overrun, deformable 'M' beds may occur (Brand et al., 1987; Echelmeyer and Wang, 1987). Sequences of subglacial sediment and bedform continua are unlikely to develop as depicted beneath ice sheets. Also, unlike large ice sheets, the impact of bedrock topographic and geological control is probably far greater under valley glacier subglacial conditions.

2.5. SUBGLACIAL SEDIMENTS AND LANDFORMS

Sediments and landforms of subglacial environments can be subdivided into those developed under 1) active ice flow (advance and active retreat phases); and 2) passive or 'dying' ice flow (retreat phase). In each group the forms related to both erosion and deposition will be discussed as will the sediments and their main distinguishing characteristics.

It must be pointed out that rarely a single or a few characteristics can be used that are, in themselves, diagnostic, especially when studying glacial sediments in the field. Rather, it is a combination of properties, stratigraphic relationships and both lateral and vertical lithofacies associations that aids in the identification of a particular sediment type and formative process(es) (cf. Albino and Dreimanis, 1988; Hicock, 1992).

2.5.1. Sediments and Landforms of Active Ice Flow

The processes and mechanisms that lead to sediment deposition beneath an active land-based ice sheet can be subdivided into those processes producing: a) *in situ* tills (diamictons), b) glaciofluvial and glaciola-

custrine sediments, and c) glacial diamicton mélange
units indicative of syndepositional deformation (Flint,
1971; Goldthwait, 1971; Legget, 1976; Stankowski,
1976; Schlüchter, 1979; Evenson *et al.*, 1983; Eyles,
1983b; Eyles and Miall, 1984; Königsson, 1984; van
der Meer, 1987; Goldthwait and Matsch, 1988; Hart,
1990, 1992; Hicock, 1990, 1992; Hicock and Dreimanis, 1992b; Menzies, 1990c; Brodzikowski and Van
Loon, 1991).

2.5.1.1. Tills – diamictons

Till (often termed Diamict or Diamicton, today, to
avoid the genetic connotation that the term 'Till'
carries) has been defined by Dreimanis and Lundqvist
(1984, p.9) as '... a sediment that has been transported and is subsequently deposited by or from
glacier ice, with little or no sorting by water'. Tills
exhibit a wide range of grain size and may be very

poorly to well-sorted (Plate 2.1) (Menzies, 1995a,
Chapters 8 and 9). Typically, tills contain a variable
percentage of exotic clasts and mineral grains that
have been transported considerable distances from
up-ice sources (Chapter 15). In general, however, the
majority of the clasts and finer particles are of local
origin (<15 km up-ice) (DiLabio and Shilts, 1979;
Krüger, 1979; Lawson, 1981a; Eyles *et al.*, 1982;
Haldorsen and Shaw, 1982; Eyles and Menzies, 1983;
Sharpe and Barnett, 1985; Shaw, 1985, 1987; Boulton, 1987b; Brodzikowski and Van Loon, 1987;
Kujansuu and Saarnisto, 1987; Rappol, 1987; Dreimanis, 1988; Eyles *et al.*, 1989; Rappol *et al.*, 1989).
It has become increasingly apparent that tills have
such a wide diversity of depositional modes and
internal structures that they can be subdivided into a
range of categories, none of which have single
diagnostic criteria for identification, but they can be
separated by observing a set of characteristics

PLATE 2.1. Large exposure (>30 m in height) of diamicton in stoss end of a drumlin, Chimney Bluffs State Park, New York State. Note
crude banding in centre of photograph, and marked diamicton colour change near upper part of exposure.

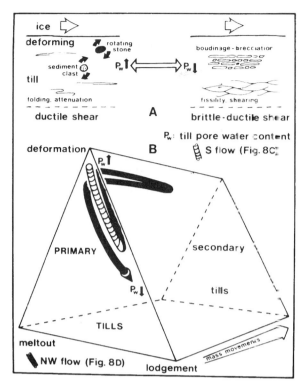

FIG. 2.11. Till 'prism' illustrating the history of till formation/deformation. (a) Model of rheologic superposition; (b) 'spaghetti' diagram depicting changing genesis and relative water contents. (After Hicock, 1992, by permission of the *Scandinavian University Press*)

(Haldorsen and Shaw, 1982; van der Meer *et al.*, 1985; Hicock, 1990).

Tills may form within several distinct glacial sub-environments such as subglacial, englacial, proglacial, and supraglacial; and may be deposited from active temperate wet-based and polar cold-based ice masses, as well as under stagnant ice conditions (Paterson, 1981; Drewry, 1986). Hicock (1990) has developed a graphic means ('Till Prism') of illustrating the complexity and interrelationships that exist within till-forming subglacial environments (Fig. 2.11). The till prism portrays the multiple pathways and associated subglacial environments through which sediment may pass before final deposition In developing this concept, Hicock utilised the concept of primary and secondary tills (Boulton and Deynoux, 1981) placed within a time frame. However, the 'prism' aptly demonstrates the inherent complexity in

till formation that has, in the past, been oversimplified and overlooked.

Although disregarded due to their comparative rarity, tills from polar ice bed conditions can develop under two states, sublimation and meltout. In the former instance, Shaw (1977, 1988) has shown that under dry polar climatic conditions, as for example today in the Dry Valleys of West Antarctica, direct sublimation of debris-charged ice can result in the formation of sublimation till that is a 'sub-species' of meltout till. Where polar ice stagnates, direct meltout tills may also form. In either instance a limited input from meltwater can be expected and thus stratification and/or sorting will be restricted. Today, with the recognition of polythermal bed conditions, it may be impossible to distinguish either till type as being directly of polar ice origin.

Under wet-based, active temperate conditions considerable input from meltwater can be expected, also thermal fluctuations and transient fluctuations in ice velocity and stress conditions will occur. Subglacial temperatures are likely to be in the $-1°$ to $-3°$ C range, with average basal ice stress levels of ≈100 kPa.

Tills formed under active temperate conditions are dominantly Lodgement and Basal Meltout Tills with a variety of local till sub-types and subglacial Mélange Diamictons. Sub-types of tills are often identified due to some peculiar characteristic of grain size, provenance and/or structural attribute (cf. Dreimanis, 1988; Raukas *et al.*, 1988). It should be borne in mind, however, that these sub-types do not reflect different glacial processes, but rather specific geological attributes, and as such should not be used in any broad 'facies' classification of tills. Both lodgement and meltout tills have been extensively described in the literature (cf. Haldorsen, 1982; Ehlers, 1983; Muller, 1983; Dreimanis, 1987, 1988; Levson and Rutter, 1987; Shaw, 1987 Fulton, 1989; Ehlers et al, 1991), whilst subglacial mélanges have only recently been regarded as separate and distinct subglacial sediments (cf. Menzies, 1989a, Hicock, 1992; Curry *et al.*, 1994).

Lodgement tills

Lodgement tills are formed subglacially under pressure melting conditions. As ice comes in contact with its bed and frictional heating occurs ice-encased

TABLE 2.2. Criteria for identification of lodgement till, melt out till and gravity flowtill (after Dreimanis, 1989)

Criterion	Lodgement till	Melt out till	Gravity flowtill
Position and Sequence in Relation to Other Glacigenic Sediments	Under advancing glaciers: lodged over older pre-advance sediments and over glacitectonites, unless they have been eroded. Under retreating glaciers: the lower-most depositional unit, if the deposits related to glacial advance have been eroded. Locally underlain by meltwater channel deposits. May be overlain by any glacigenic sediments.	Usually deposited during glacial retreat over any glacially eroded substratum or over lodgement till. Also as lenses in lodgement till. May be interbedded with lenses of englacial meltwater deposits, and locally is underlain by syndepositional subglacial meltwater sediments and subglacial flowtill.	Most commonly it is the uppermost glacial sediment in a non-aquatic facies association. Associated also locally with subglacial tills, where cavities were present under glacier ice, or where the glacier had over-ridden the ice-marginal flowtill. May be interbedded or interdigitated with glaciofluvial, glaciolacustrine or glaciomarine sediments, particularly away from its original source at the glacier ice.
Basal Contact	Since both lodgement and melt-out tills begin their formation and deposition at the glacier sole, their basal contact with the substratum (bedrock or unconsolidated sediments) is similar in large scale, being usually erosional and sharp. The glacial erosion-marks underneath the contact and the alignment of clasts immediately above the contact have the same orientation. Glacitectonic deformation structures formed by the fill-depositing glacier may occur under both tills, and they strike transverse to the direction of local glacial stress.		Variable; seldom planar over longer distances. The flows may fill shallow channels or depressions. The contact may be either concordant, or erosional, with sole marks parallel to the local direction of sediment flow. Loading structures may be present at the basal contact of waterlain flowtill and the underlying soft sediment.
	Basal contact, representing the sliding base of the glacier, is generally planar if over unconsolidated substratum, but it may be grooved. The bedrock contact is usually abraded, particularly on stoss sides of bedrock protrusions. Since the sliding base of the glacier represents a large shear plane, sheared and strongly attenuated substratum material may be deposited as a thin layer along this plane, and from place to place it is sheared up into the lodgement till. Clast pavements, both erosional and depositional, may be present along the basal contact, but they occur also higher up in lodgement till. If lodgement till becomes deformed by glacial drag shortly after its deposition, the basal contact may become involved in the deformation with tight recumbent folding, over thrusting, and shearing.	If the basal contact of glacier ice was tight with the substratum during the melting, the pre-depositional erosional marks characteristic for moving glaciers, are as well preserved as under lodgement till. However, subsole meltwater may modify the basal contact locally, and produce convex-up channel fills and various other meltwater scour features.	

debris is released into the contact zone between the ice and the bed by a process of regelation (Menzies, 1995a, Chapters 4 and 5). This debris is released at a rate of only a few centimetres per year (Mickelson, 1971, 1973; Nobles and Weertman, 1971; Boulton and Dent, 1974; Boulton et al, 1974) and is smeared along the ice/bed contact where it is lodged at the bed. Boulton (1974) suggested that under increasing

TABLE 2.2. Continued

Surface Expression, Landforms	Mainly in ground moraines and other subglacial landforms. Also along the proximal side of some end moraines are always associated with lodgement till.	In those ice-marginal landforms where glacier ice had stagnated.	Associated with most ice marginal landforms. Also, as a thin surface layer on many other direct glacial landforms.
Thickness	Typically one to a few metres; relatively constant laterally over long distances.	Single units are usually a few centimetres to a few metres thick, but they may be stacked to much greater thicknesses.	Very variable. Individual flows are usually a few decimetres to metres thick, but they may locally stack up to many metres, particularly in proglacial ice marginal moraines and some lateral moraines.
Structure, Folding, Faulting	Typically, described as massive, but on closer examination, a variety of consistently oriented macro-and micro-structures indicative of shear or thrust may be found. Folds are overturned, with anti-clines attenuated downglacier. Deformation structures are particularly noticeable, if underlying sediments are involved, or incorporated in the till, developing smudges. Subhorizontal jointing or fissility is common. Vertical joint systems, bisected by the stress direction, and transverse joints steeply dipping down-glacier, may be formed by glacier deforming its own lodgement till. The orientation of all the deformation structures is related to the stress applied by the moving glacier, and therefore it is laterally consistent for some distance.	Either massive, or with palimpsest structures partially preserved from debris stratification in basal debris-rich ice. Lenses, clasts, and pods of texturally different material preserve best, for instance soft-sediment inclusions of various sizes, and englacial channel-fills. Loss of volume with melting leads to the draping of sorted sediments over large clasts. Most large rafts or floes of substratum are associated with melt-out tills, and they may be deformed by glacial transport and by differential settlement during the melting.	The structure depend upon the type of flow and associated other mass movements, the water content and the position in the flow (see Table 2.9a, b). Either massive, or displaying a variety of flow structures, such as: (a) overturned folds with flat-lying isoclinal anticlines, (b) slump folds or flow lobes with their base usually sloping downflow, (c) roll-up structures, (d) stretched-out silt and sand clasts, (e) intraformationally sheared lenses of sediments incorporated from substratum, with their upper downflow end attenuated, if consisting of fine-grained material, or banana shaped
Grain Size Composition	Usually a Diamicton, containing clasts of various sizes. Grain-size composition depends greatly upon the lithology and grain-size composition of the substrata up-glacier and the distance and mode of transport (basal, englacial) from there. Comminution during glacial transport and lodgement has produced a multimodal particle size distribution. Most resulting subglacial tills are poorly to very poorly sorted (σ =2-5), described also as well graded, and their skewness has a nearly symmetrical distribution (Sk = -0.2 to 0.2), except for those tills that are rich in incorporated presorted materials.		Usually a Diamicton with polymodal particles size distribution. It is texturally similar to that primary till to which it is related, but with a greater variability in grain size composition, due to washing out of, or enrichment in fines, or incorporation of soft substratum sediments during the flow. Some particle size redistribution takes place during the flow. The grain size composition depends greatly upon the type of low, and the position or zone in it (see Table 2.9a, b). Sorting, inverse or normal grading may develop in some zones of flows, and parts of clasts may sink to the base of flow.

effective pressures, depending upon basal ice velocity and ice thickness, a 'critical lodgement index (L_c)' can be found which, once reached, leads to increasing lodgement of debris at the ice/bed interface i.e. as friction increases so lodgement increases. The index is an empirical relationship where:

$$V_r^1 \approx \left(\frac{N}{L_c}\right)^{\frac{1}{m}} \qquad (2.3)$$

V_r^1 is the relative forward velocity of a particle in traction at the ice/bed interface; N is the effective stress and m is an empirical constant of $\approx$ 0.3.

TABLE 2.2. Continued

	The abrasion in the zone of traction during lodgement produces particularly silt-size particles typical for lodgement tills. Most lodgement tills have a relatively consistent grain-size composition, traceable laterally for kilometres, except for the lower 0.5 to 1 m that strongly reflects the local material. Clusters or pavement of clasts are common.	The winnowing of silt- and clay-size particles in the voids during the melt-out may reduce the abundance of these particle sizes in comparison with their lodged equivalents. Some particle size variability is inherited from texturally different debris bands in ice. Extreme variations in grain size may occur over short distances in the vicinity of large rafts and other inclusions of soft sediment.	
Lithology of Clasts and Matrix	Lithologic composition tends to b e less variable than in other genetic varieties of tills; most constant is the mineralogic and geochemical composition of the till matrix. Materials of local derivation increase in abundance towards the basal contact of the tills with substratum.		The lithologic composition is generally the same as that of the source material of flowtill —a primary till or glacial debris, plus some substratum material incorporated during the flowage. Material of distant derivation dominates in the flowtills derived from supraglacial and englacial debris, but dominance of local material indicates derivation from basal debris. Soft sediment clasts derived from the substratum, or from sediment interbeds in multiple flows, are common.
		Since glacial debris of distant derivation is more common in he englacial zone than in the basal zone of a glacier and since the englacial zone has a greater possibility to be deposited as melt-out till, rather than by lodgement, materials of distant derivation may be more abundant in the melt-out till than in the lodgement till of the same till unit, particularly in supraglacial melt-out till. Great compositional variability occurs in the vicinity of incorporated 'megaclasts,' 'rafts' or 'floes' of sub-till material. Soft sediment clasts, for instance consisting of sand, may be found in melt-out till, but not in typical undeformed lodgement till.	
Clast Shapes and Their Surface Marks	Following criteria apply to lodgement till and basal melt-out till where most clasts are derived from a single cycle transport: subangular to subrounded shapes dominate, depending mainly upon the distance of transport in the basal zone of traction. Bullet-shaped ('flat-iron,' 'elongate pentagonal') clasts are more common then in other tills and nonglacial deposits, and their tapered ends usually point upglacier. Some of the elongate clasts have a keel at their base. Glacial striae are visible mainly on medium-hard fine grained rock surfaces. Elongate clasts are striated mainly parallel to their long axes, unless they have been lodged or transported by rolling.		If present, soft sediments clasts are either rounded or deformed by shear or dewatering. The more resistant rock clasts are in the same shape as they were in the source material when resedimented by the flowage. Therefore, the relative abundance of glacially abraded subangular to subrounded clasts versus completely angular clasts in Flowtills of mountain glaciers will indicate the approximate participation of basal debris versus supraglacial debris in the formation of the flowtill. Some rounded water-reworked clasts, without striations, may derive in flowtills from melt-water stream deposits.

Boulton found values of L_c in the range of 5–25. However, some debate as to the validity of this concept has since arisen (cf. Hallet, 1981; Drewry, 1986; see Menzies, 1995a, Chapter 7). The concept of the 'index' related the processes of erosion, entrainment and deposition within a single 'grand' hypothesis. However, much remains to be understood about the specific process of lodgement and its

TABLE 2.2. Continued

	The bullet-shaped and faceted clasts also crushed, sheared and stressed-out clasts are more common in lodgement till than in other tills. Lodged clasts are striated parallel to the direction of the lodging glacial movement, and they have impact marks on both the upper and lower surfaces, but in opposite orientation; on the surface the stoss end is upglacier, but on the underside--the stoss end is downglacier. Clast pavements with sets of striae parallel to the direction of the latest glacial movement over them may occur at several lodgement levels. Their top facets are either parallel with the general plane of lodgement, or they dip upglacier.	If, in an area of mountain glaciation, the source of supraglacial melt-out till is englacially or even supraglacially transported supraglacially derived debris, then the clasts are angular. Most commonly, supraglacial melt-out till in such areas also contains an admixture of glacially abraded basal debris, also englacially transported.	
Fabrics Macro-fabric (Orientation of Clasts) or Micro-fabric (Orientation of Particles in the Matrix)	Strong macro-fabrics with the long axes parallel to the local direction of glacial movement are reported from diamictons identified either as lodgement or melt-out tills. Occasionally transverse maxims have developed, associated with folding and shearing. The fabric strength may vary also, depending upon the grain-size of till, the abundance of clasts, and postdepositional modification.		Variable, and depending greatly upon the type of flow and the position in the flow. It may range from randomly oriented to strong fabric, in thin flow tills. Fabric maxima are either parallel or transverse to the local flow direction, unrelated to glacial movement; the a-b planes are either subparallel to the base of flow, or they dip up-flow. Fabric maxima may also differ laterally on short distances.
	The lodgement till fabric may be of complex origin: produced by lodgement, or by deformation of the already deposited dilated till, under the same glacier. If both stress directions coincide, a strong fabric will develop; if not - the lodgement fabric becomes weakened. Typically, the a-b planes dip slightly upglacier, if slightly up-glacier, if lodgement alone is involved. The micro-fabric is usually as strong as the macro-fabric.	In melt-out tills, fabric is inherited from glacier transport, where fabric dominates, parallel to the direction of glacial movement, unless deformation changes it to transverse fabric locally. However the melting-out process may weaken the fabric particularly the micro-fabric. Also, the dip of the inclination of clasts becomes reduced by the reduction of the volume of ice during melting.	
Consolidation, Permeability Density	Most lodgement tills, particularly the poorly melt-out tills are usually sorted matrix-supported varieties, are over-consolidated, provided there was adequate subglacial drainage. Their bulk density, penetration resistance, and seismic velocity are usually high, permeability - low, relative to other varieties of till of the region.	Supraglacially formed melt-out tills are usually less (normally to weakly) consolidated than the subglacially formed, commonly over-consolidated melt-out tills, provided there was adequate drainage of meltwater. Bulk density and penetration resistance may be lower and more variable than in related lodgement till. Also, permeability is more variable.	Primarily normally consolidated and relatively permeable. If clayey, may become over-consolidated due to post-depositional desiccation. Density lower than in primary tills.

* Modified from Goldthwait, 1988.

relationship to other subglacial processes and bed conditions.

Lodgement tills form widespread continuous till plains often associated with bedforms (drumlins, Rogen moraine, fluted moraine) These plains may be of considerable thickness (up to and > 30 m) and cover large areas of terrain. Typical characteristics of lodgement tills are illustrated in Table 2.2 (Boulton and Paul, 1976; Boulton and Deynoux, 1981; Eyles and Menzies, 1983; Dreimanis, 1988).

Structural discontinuities are common within lodgement tills. (It should be pointed out that,

although many of these structures have been described as occurring in lodgement tills, the likelihood is
that many of these tills have formed under different
sedimentological processes and are not lodgement
tills. Many lodgement tills display distinctive fissile
structures (McGown *et al.*, 1978; McKinlay et al,
1978; Krüger, 1979, 1994; Dreimanis, 1993). Fissility,
typically, appears as sub-horizontal thin partings
within fine-grained till, that are interpreted as being
indicative of stress application during deposition
(autokinetic discontinuity). The fissile structures may
be weak plate-like forms or may appear as bedding
units in the form of laminations. These laminations
are not regular and often pass around boulders
(Dreimanis, 1976). Other structures observed are sub-
horizontal foliations of similar origin to fissures; and

macro- and micro-shear joints (Plate 2.2). Occasionally, bedding planes (Virkkala, 1952) and other
structures have been thought to be caused by
dewatering and unloading (Sitler and Chapman, 1955;
Boulton *et al.*, 1974; Bjelm, 1976; McGown *et al.*,
1978; Boulton and Paul, 1979; Muller, 1983; Rappol,
1987). In general, faults and folds are uncommon
within lodgement tills but sheared inclusions, boudins, and overthrust folds do not exist. Fracture
patterns thought indicative of allokinetic loading/
unloading (Section 2.6.3.2.) due to stress relief have
been described from drumlins in Scotland (McGown
et al., 1977, 1978). Reports of increased permeabilities in drumlin lodgement tills possibly related
to the development of autokinetic (Section 2.6.3a)
discontinuities have been made (Knutsson, 1966).

PLATE 2.2. (a) Brecciation in diamicton immediately above a sand intraclast. Note scale card is 8.5 cm long. (b) Shear planes within a
diamicton, note two major planes in the centre and top left of the photograph. Several smaller planes intersect these large planes. Scale width
of photograph is 45 cm. (c) 'Bed limits' within a diamicton. Scale bar of 13.0 cm. (d) Distinct lamination within a diamicton with several small
faults and a shear plane in bottom right of photograph. Note knife for scale, 9.0 cm. All photos from Mohawk Bay, southern Ontario

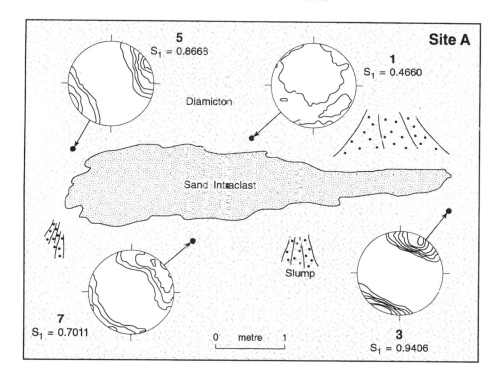

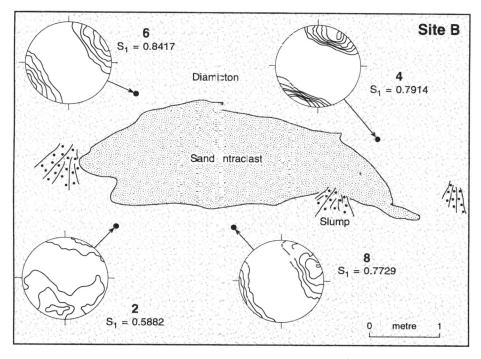

FIG. 2.12. Schmidt equal-area projections of clast fabric data at four sites around each of two sand intraclasts within a glacial mélange diamicton, Mohawk Bay, Ontario. (After Menzies, 1990c)

A characteristic of tills of all types is the presence of a discernable clast fabric (Fig. 2.12) Within most tills, it is argued that there is an *in situ* or inherited preferred clast orientation of the clasts within the sediment. Clast orientation applies to both macro- and micro-clasts (Evenson, 1971). In lodged tills a strong fabric orientation is present indicative of autokinetic stress application during deposition (Fig. 2.13). It is suggested that as the fine-grained matrix of the till is being lodged, large particles within the mobile sediment or at the sediment/ice bed interface are oriented with their long axes in the direction of the principal stress direction (on average in the down-ice direction). Since this involves ploughing during the lodging process (Alley, 1989a,b; Menzies, 1995a, Chapter 5), most clasts also exhibit a slight up-ice dip (for recent reviews of till fabrics see Dowdeswell *et al*, 1985; Dowdeswell and Sharp, 1986; Rappol, 1987;

Menzies, 1990c; Hicock, 1992). Associated with these preferred clast orientations are frequent small joints and shear planes often intersecting but of limited lateral extent.

Finally, lodgement tills may contain exotic sediments apparently rafted into or deposited within the lodged till (Shilts, 1978; Ruszczynska-Szenjach, 1983, 1987; Eyles *et al*., 1984). These inclusions often have surrounding joint sets, fractures and clay injection features. The size of these inclusions vary enormously and may be composed of fluvioglacial sands and gravels, laminated lacustrine clays and silts, re-sedimented tills, boulder beds and pavements, and weathered bedrock (Plate 2.3). Inclusions may serve as structural defects around which discontinuities develop and may act as natural anisotropic rheological components within otherwise massive till.

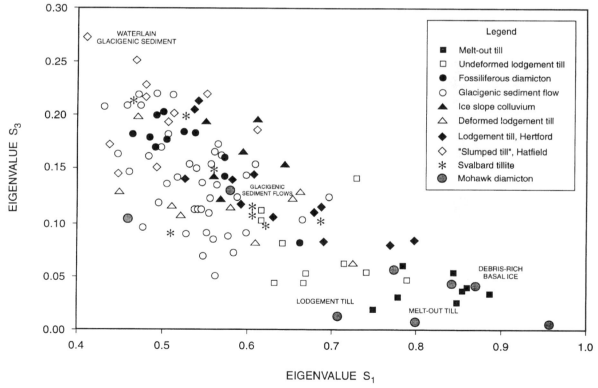

FIG. 2.13. Plot of S_1/S_2 eigenvalues for clast fabric data. (Data from Dowdeswell *et al*., 1985; Dowdeswell and Sharp, 1986 and Menzies, 1990c)

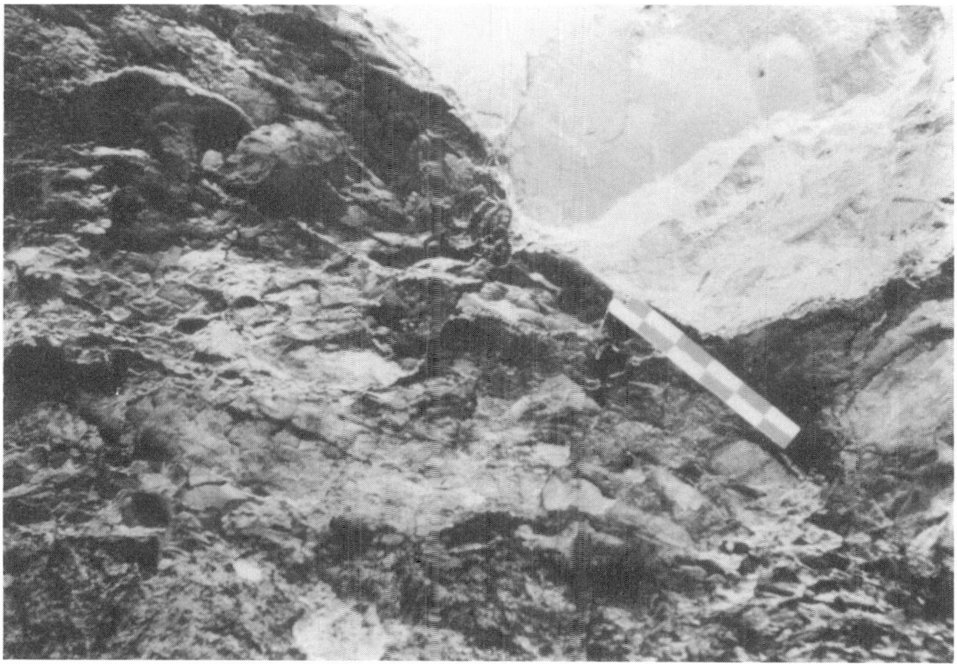

(a)

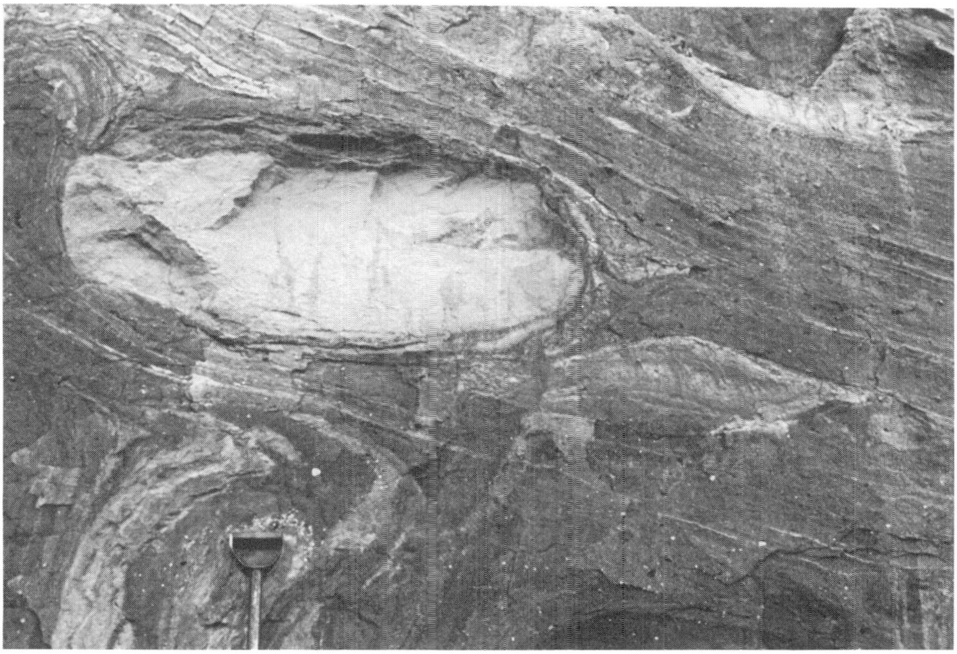

(b)

PLATE 2.3. (a) Bottom sharp contact between a sand intraclast and surrounding diamicton. Scale bar of 15 cm. (b) Sand intraclast within Cromer Till, West Runton, East Anglia, England. Note distinct lamination within diamicton, and flow structures around the intraclast and behind shovel handle. Shovel handle approximately 12.5 cm wide. (Photo courtesy of Phil Gibbard).

(c)

PLATE 2.3. (c) Contorted sand intraclasts within transition between laminated clays and Port Stanley Till, Bradtville, southern Ontario. Knife is approximately 25 cm long. (Photo courtesy of Phil Gibbard)

Lodgement tills, if fine-grained, usually exhibit a high degree of overconsolidation, are often massive in appearance, have low permeability and rarely display large-scale fracture geometries.

In terms of micromorphological examination, lodgement tills tend to exhibit a degree of foliation in thin section, a distinct unistrial fabric (Brewer, 1976) and often 'card-house' structures within massive clay units (van der Meer, 1987a; see Chapter 12). Brecciation has been observed in tills thought to be lodgement type but is possibly indicative of allokinetic effects not peculiar to the lodgement process (Derbyshire et al., 1985a; Menzies, 1990b).

Several studies of the geotechnical properties of lodgement tills have been carried out (Boulton, 1976b; Milligan, 1976; Boulton and Paul, 1979; Anderson, 1983; Forde, 1985) that attempt to correlate the depositional environment with intrinsic geotechnical characteristics. These studies illustrate the enormous variability of physical properties within this till type (Section 2.6).

Meltout tills

Meltout tills are formed as a result of massive *in situ* melting and subsequent deposition from the upper (supraglacial) and lower (subglacial) englacial zones of an ice mass (Boulton and Eyles, 1979; Haldorsen and Shaw, 1982; Sharpe and Barnett, 1985; Dreimanis, 1988; for a recent review see Paul and Eyles, 1990). The likely importance of meltout tills in the stratigraphic record is now under some doubt and their 'once thought' ubiquity is questionable (Rappol, 1987; Shaw, 1987; Paul and Eyles, 1990).

Subglacial meltout till formation is thought to occur as a result of basal ice stagnation and subsequent thawing of the encased englacial debris down onto the glacier bed (Shaw, 1982). The process of meltout till formation would seem to occur under both active and passive ice conditions (Section 2.11.1a). Meltout tills are saturated and, therefore, subject to syn- and post-depositional processes of disaggregation and re-sedimentation. The process of meltout is normally visualised as a grain by grain sedimentation process which, due to large volumes of meltwater, results in considerable winnowing and loss of fines. Local pondings, channel and pipe flow conditions occur and thus stratified sands, gravels and laminated clays and silts are often part of an entire meltout till package. The process of meltout till formation is long enough to allow porewaters to dissipate and stress adjustments be compensated for within the structural framework of the sediment. In some instances, the thaw process may last for several hundreds of years and be accompanied continuously by thaw consolidation and other stress adjustment effects (Shaw, 1982). Meltout due to sublimation (producing Sublimation Till) has also been described under arid polar conditions (Shaw, 1977a, 1988b).

Meltout tills are not found extensively over wide areas of glaciated terrains principally due to their poor preservation potential after initial deposition, and because of the restricted spatial distribution of thick englacial sediment packages from which meltout tills might form (cf. Rappol, 1987). Thick meltout sequences are restricted to marginal locations (Moran, 1971; Attig et al., 1989; Paul and Eyles, 1990). As a consequence of the style of deposition, meltout tills are usually relatively thin, coarse-grained, occurring as spatially disjunct patches within complex marginal stratigraphic packages (Chapters 3, 8 and 9). The typical characteristics of meltout tills are listed in Table 2.2 (Boulton, 1980; Haldorsen and Shaw, 1982; Shaw, 1985).

The most pertinent aspects of meltout tills recognisable in the field are; a) the presence of stratified units within a meltout till, b) the large number of slump-type structures preserved within these tills; and c) structures inherited from the englacial debris architecture.

(a) Meltout tills typically can be identified by the presence of large units of undeformed stratified/ laminated sediments intercalated within a fine, massive matrix. These units may be more or less extensive than the finer matrix and typically exhibit little or no lateral deformation. The presence of these stratified units often allows meltout tills to exhibit a massive or banded or laminated appearance (Lavrushin, 1971, 1978; Shaw, 1982; Stephan and Ehlers, 1983; Eyles et al., 1985). Within these tills are

channel-fill deposits that illustrate the past existence of englacial and subglacial meltwater conduits.

(b) Due to the influence of high porewater pressures, overloading by newly-deposited overlying meltout sediment, and associated thaw consolidation processes, a wide range of slump features, dewatering pipes and diapirs, and faults usually are found within these tills. Within fault wedges, however, undeformed stratified units may be identified with their bedding planes preserved (cf. Shaw, 1982).

(c) A commonly attributed characteristic of meltout till is the preservation, after melting, of inherited englacial structures: in particular folds, shear-stacked units and clast fabrics (cf. Lawson, 1979b) (Fig. 2.13).

Discontinuities within meltout tills abound as a result of thaw consolidation, autokinetic slumping, loading and dewatering. Fractures and normal and reverse faulting are common and where lateral slumping or consolidation has occurred shear fractures of both tensile and compressive mode are found. Observations from near Edmonton, Alberta (Shaw, 1982) revealed transcurrent faulting and folding between separate units within the till, and post-depositional intercalation of overlying units into lower subjacent sediment bodies. Discrimination between meltout till deposition and glaciotectonic allokinetic disturbance is very difficult. In comparison to lodgement tills, meltout tills probably have a higher percentage of near-vertical macro-discontinuities that in some instances extend across different sub-units within the sediment package (>1–2 m). Unlike lodgement tills, since the environment in which the meltout occurs is, by definition, at or above the freezing point, the potential effects of autokinetic cryogenic disruption can be ignored. The presence, on the other hand, of large often sub-horizontal stratified sand lenses and laterally extensive units allows for a greater degree of susceptibility toward post-depositional dewatering, cryogenic effects and seismically-induced structural disruptions. Allokinetic diapirism, the formation of sand dykes, brecciation and other inter-stratal disruptive processes are likely to be much more common and widespread in meltout tills than in lodgement tills.

A characteristic often used in the past to differentiate meltout tills from other till types is their
strongly preferred clast fabric, in some cases, retained
from prior englacial encasement (cf. Lawson, 1979b;
Dowdeswell and Sharp, 1986). However, recent
discussions have centred on the problems of fabric
replication in other subglacial stress environments
(Fig. 2.13); and the effectiveness of this diagnostic
criterion (cf. Menzies, 1990c).

The presence of undeformed inclusions within
meltout tills has been used as a diagnostic characteristic in the past but, as will be discussed below in the
context of glacial mélanges, even this criteria is
suspect.

Discussion on the characteristic geotechnical properties of meltout tills remains limited (Boulton and
Paul, 1976). In general, meltout tills do not exhibit the
same degree of consolidation as lodgement tills. In
addition, bulk permeabilities of meltout tills tend to be
higher than lodgement tills as a result of the number
of near-vertical faults and joints, and stratified
sediment inclusions.

2.5.1.2. Subglacial glaciofluvial and glaciolacustrine sediments

Many tills contain intraclasts indicative, in numerous
instances, of subglacial stratified sediment deposition
(cf. Carruthers, 1953). Today, it is commonly recognised that within subglacial cavities, glacier bed
depressions, channels and tunnels a considerable
volume of stratified sediments may be deposited (cf.
Eyles and Menzies, 1983; Eyles and Miall, 1984;
Brodzikowski and Van Loon, 1987, 1991; Jurgaitis and
Juozapavičius, 1988; Aylsworth and Shilts, 1989; see
Section 2.10.2.1.). These subglacial stratified sediments range from delicately laminated lacustrine clays
to coarse bouldery gravels with limited bedding
structures (Plate 2.4).

These sediments, deposited normally within fluvial
environments, exhibit evidence of enormous short-
term fluctuations in meltwater discharge, velocity and
depth (cf. Gurnell and Clark, 1987). This variability in
meltwater stream competency is manifest in a wide
range of sedimentary facies and internal structures, and
grain size distributions. Sedimentation in the subglacial environment is usually rapid (within a few

(a)

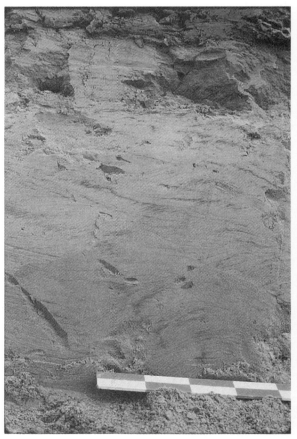

(b)

PLATE 2.4. (a) Inclusion of stratified sands within a diamicton,
Port Burwell, southern Ontario. Note coin for scale is 2.5 cm in
diameter. (b) Stratified sands containing angular clay fragments and
some cross-bedding, near Hanover, southern Ontario. Note change
in sand toward top and contact with a diamicton. Scale bar of
15 cm

TABLE 2.3. The types and origins of discontinuities within glacigenic sediments

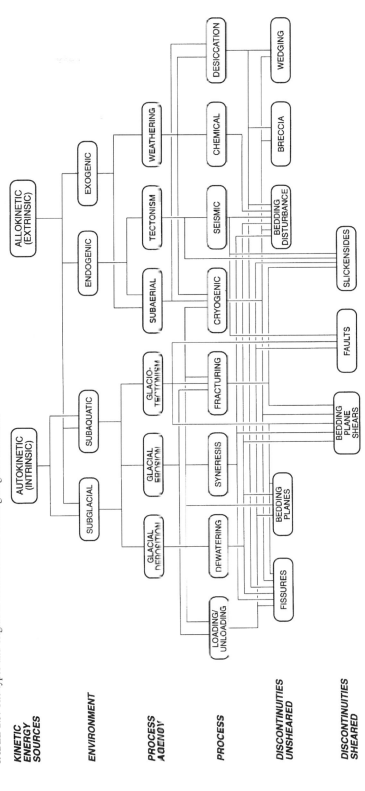

hours), causing these sediments to exhibit rapid porewater expulsion and migration features, as well as load and shear structures due to the influence of overriding ice.

Subglacial glaciofluvial sediments tend to be areally restricted; occurring as thin patches or linear ribbons (eskers) parallel or subparallel to ice direction. Subglacial and proglacial stratified sediment can be segregated on the basis that the latter are tabular with a large areal distribution, are of considerable thicknesses (>15 m) and, if preserved, may have a pitted surface pock-marked by kettles (see Chapter 3; Menzies, 1995a, Chapter 12). Subglacial stratified sediments commonly reach thicknesses <15 m, and are restricted in areal distribution.

Discontinuities are often found in stratified sediments within lodgement tills. Discontinuities include almost all forms associated with stresses applied to saturated cohesionless sediment (Table 2.3). In particular, due to overriding by glacier ice, stratified sediments may exhibit evidence of liquefaction and fluidisation, faulting and shearing. The lack of cohesion of these sediments usually precludes folding but post-depositional cryostatic conditions may support folding and buckling (Brodzikowski *et al.*, 1989; Menzies, 1990c; Brodzikowski and van Loon, 1991). In many cases these sediments, if frozen, may be 'rafted' into mélange sediments (see Section 2.5.1.3. below) or plucked from the glacier bed and incorporated into other till types (cf. Menzies, 1979, 1990c; Eyles et al, 1982; Ehlers, 1983; Ruszcznska-Szenjach, 1983, 1987) (Plate 2.5) (see also colour plates).

Within subglacial and proglacial stratified sediments, inclusions are common in the form of till balls, injections from overlying sediments and underlying diapirs, clay shards, and other exotic sediments which may have fallen into subglacial meltwater streams and cavities. Where dense clay sediment overlying saturated stratified sediment have suffered rapid loading, dish and pillar structures and, in some cases, complete ball and pillow structures may develop (Lowe and Lopiccolo, 1974; Lowe, 1975; Allen, 1982; Mills, 1983).

Subglacial glaciolacustrine sediments are often found in the lee-side of large topographic obstructions, boulders or on the floor of cavities. The

(a)

(b)

(c)

PLATE 2.5. (a) Ductile banding in diamicton mélange, Mohawk Bay, southern Ontario. Photo is approximately 1.5 m across. (b) Similar to (a) but containing a large number of angular sand units. Scale card is 8.5 cm long. (c) Faulted and laminated ductile mélange sediments. Each bar scale segment is 5 cm

presence of small units of glaciolacustrine sediment, within other subglacial sediments, is common.

2.5.1.3. Glacial diamicton mélange sediments

In the past ten years there has been increasing evidence that, under certain basal ice conditions, major ice sheets may have been underlain by areas of deformable debris layers or beds (Boulton and Jones, 1979; Menzies, 1995a, Chapter 5). These deformable beds or debris traction layers, beneath ice masses, essentially act as a lubricant resulting in high basal ice velocities and associated thin marginal ice surface profiles (Boulton and Jones, 1979; Beget, 1986, 1987; Alley et al., 1986; Boulton and Hindmarsh, 1987; Menzies, 1989; Alley, 1992). As discussed in Section 2.3.2.2. such a layer is mobilized by external applied stresses from overlying active ice (Echelmeyer and Wang, 1987; Boulton and Hindmarsh, 1987; Menzies, 1989). Such conditions are likely to occur beneath fast moving outlet glaciers and ice streams (Alley et al., 1986, 1987; Blankenship et al., 1986, 1987; Beget, 1986, 1987; Brown et al., 1987; Alley, 1989). The thickness of this layer varies according to the level of applied stress and the material's viscosity which, in turn, is controlled by its porewater content. As stress levels and/or porewater contents increase, areas of a previously immobile substrate may become increasingly mobile.

The emplacement (deposition) of a deformable bed sediment is similar to lodgement in that deposition occurs when the level of strain applied to the sediment falls below the yield strength of the sediment.

Therefore, deposition is likely to be incremental, occurring, at varying times, randomly across different parts of the glacier bed. The deposit resembles a mélange (a sediment composed of a wide variety of sub-units of differing sediment types, often from various distinctive facies environments although usually from the same sedimentary basin; see Cowan, 1982, 1985). The term mélange implies that deformation or massive ductility has occurred in the formative process. Cowan (1985) distinguished four mélange types of which Type III described as 'Block-in-Matrix' perhaps best typifies diamicton mélanges (Table 2.4).

Diamicton mélanges are, perhaps, much more widespread than hitherto recognised. Extensive areas of thick diamicton, previously considered lodgement till, may in fact be mélange diamictons (cf. Hicock, 1992). Specific subglacial bedforms and sediment suites may be part of mélange units.

As in the case of lodgement tills, the intrinsic variability of mélanges almost defies any general description (Table 2.2) of sedimentological and geotechnical characteristics. Prior to discussing the discontinuities inherent in diamicton mélanges, it must be pointed out that many characteristics of matrix, lithology, provenance, inclusions, and clast fabric are similar to those in lodgement and meltout tills (see the till 'prism', Fig. 2.11) (Hicock, 1990).

Discontinuities are common and a distinctive characteristic of diamicton mélanges (cf. Menzies and Woodward, 1993). The type of discontinuity may vary depending upon the nature of the particular sediment within the mélange (Maltman, 1987, 1988; van der

TABLE 2.4 Concept of Mélange (after Cowan, 1985)

Definition	Fragments enveloped within a fine-grained matrix; typically of obscure stratigraphy, stratal disruption and/or chaotic 'block-in-matrix' fabric
TYPE I	Stratified sequence of sands/gravels and fine-grained clays/silts. Origin probable due to disruption/fragmentation caused by layer parallel extension
TYPE II	Analagous to TYPE I with progressively disrupted sequences but of a more extreme form of extension deformation
TYPE III	'Block-in-Matrix' fine-grained 'polymict'. Exotic inclusions are typical. Formation is multi-stage by progressive highly ductile (en masse) deformation
TYPE IV	Analagous to TYPE III but typically inclusions bounded by systems of subparallel faults. Dominant deformation process is cataclasis (Structural Slicing) within a Brittle Fracture Zone

Meer, 1987a; Menzies, 1990c; Menzies and Maltman, 1992 and see Chapter 12).

There is increasing evidence that no distinctive clast fabric pattern should be expected for these mélange diamictons (cf. Rappol,1985; Menzies, 1990c; Hart, 1994). Clast fabrics, depending upon where they have been measured within the sediment package, may reveal a wide array of orientations consistent with massive deformation. Secondary fabric replication due to deformation may occur in many cases and in others primary fabrics are preserved (Fig. 2.13). Thus discrimination of these fabric artefacts would be almost impossible.

Inclusions within subglacial mélanges may consist of any sediment entrained by the mobile debris layer beneath the ice mass and, therefore, could be both glacial or non-glacial in primary origin. Some inclusions may be entrained and immediately destroyed or drastically altered while others remain, in various states of preservation, while others may be entrained in a frozen state thawing later within the deposited mélange.

2.6. SEDIMENT STRUCTURES AND RELATED SEDIMENTOLOGICAL AND GEOTECHNICAL CHARACTERISTICS WITHIN SUBGLACIAL SEDIMENTS

Sediment structures are geometric or internal architectural arrangements indicative of sediment depositional and deformational processes (cf. Allen, 1982; Collinson and Thompson, 1982; Tucker, 1982). These structures are delimited by distinct sets of discontinuities, the lineation, orientation and precise geometry of which can also be used in recognising sedimentation processes in the subglacial environment (Dreimanis, 1993).

2.6.1. Structural Discontinuities

Structures within subglacial sediments are the consequence of *in situ* contemporaneous development (Autokinetic or Intrinsic); the result of post-depositional effects (Allokinetic or Extrinsic); or a compound variety of both intrinsic and extrinsic forms (Compound). Structures may be bedding planes, faults, joints, cracks, natural cleavages, fractures, fissures, ice wedges, and breccia (Plate 2.6). Rather than distinct lineations, discontinuities may also be zones or areas of less competence. These zones may develop by a variety of processes occurring during or immediately following deposition (for example, slickensides, clay eluviation, liquefaction, fluidisation, boudinage effects and other 'tectonic' events, and dewatering processes leading to pipe channels, flame structures, and ball and pillow structures (Table 2.3) (Allen, 1982; Mills, 1983; Jones and Preston, 1987; van Wateren, 1987; Dreimanis, 1993).

Discontinuities can be viewed as structural attributes of subglacial sediments since their presence may grossly affect the bulk geotechnical nature and response of the sediment while under compressive or tensile stress (Fookes, 1965, 1978; Skempton *et al.*, 1969; Fookes and Parrish, 1969; Piteau, 1970, 1973; Deere and Patton, 1971; Chandler, 1973; Kazi and Knill, 1973; McGown *et al.*, 1978; Hencher, 1987; Feeser, 1988; Hicock, 1992; Dreimanis, 1993). The degree of their development appears to be a function of the intensity and magnitude of the processes that generated them.

Structures may be geometrically interconnected in various dendritic or arborescent branching systems, or they may be discrete and autonomous. They may influence large or only localised areas of sediment. They may be confined to a particular sediment unit. However, they can cut across sediment contact boundaries or exist solely at sediment contact boundaries between different stratigraphic units leaving the units, themselves, unaffected.

2.6.2. Forms of Structural Discontinuity within Diamictons

Before discussing individual groups of discontinuites, it should be pointed out that although some research on these structural 'partings' has occurred, all too often an incorrect interpretation of till facies has been made causing the relationship of the structure to genetic till type to be erroneous.

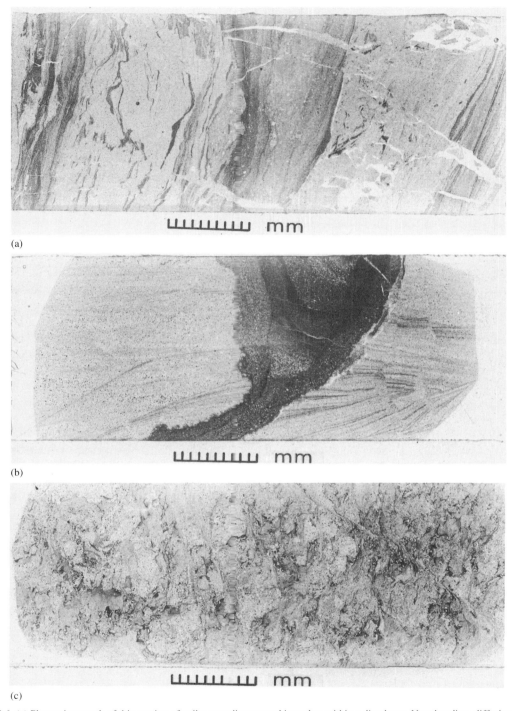

(a)

(b)

(c)

PLATE 2.6. (a) Photomicrograph of thin section of mélange sediment sand intraclast within a diamicton. Note banding, diffusion layering, micro-faulting, clay clots and balls. (b) Stratified and faulted sand intraclast with clay intrusion from below. Note banding within clay, and presence of clay balls. (c) Heavily deformed sand intraclast containing large clay content. Effect of clay is to cause the sediment to have a 'marbled' appearance. Note numerous small sub-horizontal shear planes. Large oblique shear plane at bottom right of micrograph

2.6.2.1. Autokinetic (syndepositional) or intrinsically formed or induced discontinuities

1. Bedding planes, joints and cleavage systems may form in any glacial sediment. Within diamictons (tills/mélanges) weak bedding planes are occasionally observed and the concept that tills are massive homogeneous sediments must be rejected (Dreimanis, 1993). Typically, tills have fine partings/joints that may be result of penecontemporaneous low strain occurring during deposition. These partings are usually less than a millimetre in width and often of relatively short lateral extension (<1 m) (Virkkala, 1952; Harrison, 1957; Elson, 1961; Penny and Catt, 1967; Flint, 1971; Boulton and Paul, 1976; Krüger, 1979; Menzies, 1979a; Muller, 1983; Dreimanis, 1993). These joints usually have no preferred orientation or pattern. Joints can be activated by shear or tensile stress and occasionally display surface microfluting or -slickensiding. They may also act as conduits, if well-interconnected, for post-depositional solute-rich groundwater and porewater transfer from overlying sediments. Infiltrating surface water can cause geochemical deposits to form along joint margins, while porewater may transport fine clays to form clay-skins (argillans) (Brewer, 1976). Both of these latter effects may act to seal many joints, serving to strengthen the bulk shear strength of a particular diamicton (Attewell and Farmer, 1976).

2. Where tills and mélanges have been subject to high bulk strain during the depositional process, 'fissile structures' are often observed. This fissility, sometimes termed 'platy' or 'laminar' structures, is typically near horizontal in orientation with a non-random spatial pattern, perhaps best described as resembling the leaves of a tightly shut book. Fissility is commonly observed in clay and silt-rich tills but may also occur in coarse grained tills. Debate exists as to whether these fissures are the result of stress during the depositional process or simply distinct depositional units separated by a short hiatus in sedimentation similar to normal bedding cleavage (Virkkala, 1952; Boulton and Dent, 1974; Boulton et al., 1974; Bjelm, 1976; Lavrushin, 1976; Muller, 1983; Dreimanis, 1976, 1988).

3. Evidence suggests that many tills, previously considered basal lodgement tills, may have formed via a process of basal traction. These tills, in the past referred to as deformation, comminution or shear tills (Elson, 1961; Stephan and Ehlers, 1983; Dreimanis, 1988; Parkin and Hicock, 1988; Stephan, 1988; Hart, 1990; Hicock, 1992; Hicock and Dreimanis, 1992b), are now considered subglacial mélanges (Lunkka, 1988; Hart, 1990; Menzies, 1990a,b,c). The effects of sediment translation under fairly high strain causes a very wide range of structural discontinuities to develop. These structures may evolve both during the deformation process itself or immediately following cessation and emplacement. Observations show that subglacial mélange sediments may deform pervasively or non-pervasively (Maltman, 1987, 1988).

In pervasive deformation, the complete sediment package is homogenised. This process destroys all inclusions, intraclasts and previously existing structures. Discontinuities under this regimen would essentially be at grain-to-grain boundary levels and few major structural discontinuities would exist.

Under non-pervasive deformation, widespread mobility of the sediment package occurs along discrete shear planes that act as the focus of applied strain. Small micro-shear planes take up the strain remaining, creating a myriad of structures. Between these shear planes, sediment may remain unstrained and un-disturbed. In effect there is no homogenising process and individual clasts, intercalations and intraclasts may survive intact (Lunkka, 1988; Hart et al., 1990; Menzies, 1990a, b, c).

Two groups of structures tend to develop under non-pervasive deformation: macro-shears that extend laterally, usually with a sub-horizontal disposition (1–3 m); and micro-shears extending in random directions (<25 cm) in response to intrinsic anisotropic defects within the sediment (Table 2.3).

The main structures within subglacial mélanges are fissile partings, joints, folds, faults, shear planes, kink band arrays, boudins and glacially mylonised zones. Typically, the structures have a definite preferred orientation approximately parallel to the principal plane of strain or in the case of kink bands, faults and Riedel thrusts at some angle transverse to the principal axis of deformation (rarely more than 45°). Folds, faults (cf. van der Meer, 1987b) and mylonised shear zones (Schack Pedersen, 1988; Stephan, 1988) may extend for several metres

(1–5 m) and may intersect numerous micro-struc-tures. In general, boudins and kink bands are of limited lateral extent (<1 m) and can be regarded as micro-structures. Shear planes and zones, in con-trast, exist at all scales.

Intraclasts or 'rafts' of different sediment, usually of an exotic origin, are commonplace in many subglacial tills and mélanges (Kupsch, 1962; Moran, 1971; Banham, 1975, 1977, 1988a,b; Ruszczeńska-Szenjach, 1976, 1983, 1987; Stalker, 1976; Christian-sen and Whitaker, 1976; Moran et al., 1980; Rappol, 1983, 1987; Ringberg, 1983; Stephen and Ehlers, 1983; Bluemle and Clayton, 1984; Harris and Bothamley, 1984; Attig, 1985; Stanford and Mick-elson, 1985; van der Meer et al., 1985; Clayton et al., 1987; Dreimanis et al., 1987; Dreimanis, 1988, 1993; Hicock, 1988; Lunkka, 1988; Aber et al., 1989; Menzies, 1990c). Intraclasts may be defined as sediments that were deposited within a different sub-environment but within the same depositional basin as the host sediment (till/mélange) and, following trans-portation, were fully incorporated within the till/mélange (cf. Blatt et al., 1980; Selley, 1981). Often these sediments are poorly consolidated with inher-ently high permeability values and their original structures or bedding intact (cf. Eyles et al., 1982; Ruszczeńska-Szenajch, 1987).

As a mélange is mobilized, usually, other sediments become incorporated as rafts and intraclasts. On deposition the effect of these intraclasts is to create rheological anomalies leading to stress concentrations and disparities in sediment viscosity. Where total destruction and intercalation of rafted sediment takes place 'Stratified Till' (cf. Sveg Till, Kalix Till) may be formed (Boulton, 1971, 1975, 1976a; Lundqvist, 1969; Shaw, 1979, 1983; Hicock et al., 1981; Lawson, 1981a; Dreimanis, 1982; Haldorsen and Shaw, 1982; Muller, 1983; Rappol, 1983; Levson and Rutter, 1988; Parkin and Hicock, 1989; Menzies, 1990c). Intraclasts can act as nuclei for cryogenetic processes that, in turn, can cause the surrounding sediment (Menzies, 1990b) to be brecciated (Van Vliet and Langohr, 1981; Brodzikowski and Van Loon, 1985; Derbyshire et al., 1985a, 1985b, Menzies, 1990b) or partially fractured permitting adjacent sediment injection into the intraclast (Ruszczeńska-Szenajch, 1983, 1987). Likewise, many of structures in subglacial mélange

are re-activated by post-depositional glaciotectonism (Menzies, 1995a, Chapter 10).

4. In the process of deposition most glacial sediments contain a very high water content. As the sediment 'settles' this water drains or is re-distributed. Drainage may occur under the intrinsic weight of the sediment (dewatering) or due to interparticle bonding forces and particle flocculation (syneresis) (Lachen-bruch, 1961, 1962, 1963; Allen, 1982). In either instance, if sufficient tensile stresses develop then tensile cracks form. Such crack systems can be subdivided into orthogonal and nonorthogonal and can be random or nonrandom, complete or incomplete (see Allen, 1982, fig. 13–24).

2.6.2.2. Allokinetic (epidepositional) or extrinsically formed or induced discontinuities

1. In many cases, as subglacial/subaqueous sedi-ments are being deposited or following an ice retreat and subsequent readvance, sediments are subjected to externally applied stresses due to overriding ice. This secondary stress application is generally referred to as glaciotectonism (cf. Hicock and Dreimanis, 1985; van der Meer, 1987b; Croot, 1988; Aber et al., 1989) (Menzies, 1995a, Chapter 10). Glaciotectonism usually is not restricted to single sediment units but affects whole packages of differing stratigraphic units and may pass into underlying bedrock strata (see bibliography in Aber et al., 1989).

In general, glaciotectonism may be described as sediment deformation on the 'grand scale'. However, it does not occur pervasively beneath active ice masses but tends to occur differentially under specific conditions at certain localities (Kupsch, 1962; Mackay and Mathews, 1964; Moran, 1971; Stalker, 1973; Banham, 1975; Rotnicki, 1976; Berthelsen, 1978, 1979; Sauer, 1978; Sønstegaard, 1979; Moran et al., 1980; Ruegg, 1981; Klassen, 1982; Bluemle and Clayton, 1984; Ruszczeńska-Szenajch, 1983, 1987, 1988; Stephan, 1985; Croot, 1987; Dredge and Grant, 1987; Eybergen, 1987; Van Wateren, 1987; Fernlund, 1988; Pedersen and Petersen, 1988; Eyles et al., 1989; Tsui et al., 1989; Hart, 1990; Woodworth-Lynas and Guigné, 1990).

Glaciotectonics may affect subglacial sediment and underlying bedrock as result of (1) overriding

(a)

(b)

PLATE 2.7. (a) Glaciotectonized section in Cromer Till, West Runton, East Anglia, England. Note sand block centre right is approximately 4 m long. (Photo courtesy of Phil Gibbard). (b) Glaciotecontized glaciofluvial sediments near Stouffville, Ontario. Note person on bottom right of photograph. (Photo courtesy of Jens Paterson)

horizontal drag forces, (2) through the transmission of large volumes of porewater injected into underlying sediment and by the liquefaction and/or fluidisation of saturated sediment due to overlying sediment consolidation or porewater surcharge effects, (3) pushing and/or thrusting resulting in bedrock block emplacement (Moran et al., 1980; Van Wateren, 1985; Aber et al., 1989), and (4) 'bulldozing' or loading/unloading at grounding lines beneath floating ice masses (Plate 2.7) (cf. Molnia, 1983; Dowdeswell and Scourse, 1990; Powell, 1990; Menzies, 1995a, Chapter 14).

The effects of glaciotectonics are often evident in sediments 'stacking' (Fig. 2.14) (Kupsch, 1962; Moran, 1971; Banham, 1975; Christiansen and Whitaker, 1976; Rotnicki, 1976; Babcock et al., 1978; Berthelsen, 1979; Andrews, D.E., 1980; Aber, 1982, 1985; Klassen, 1982; Bluemle and Clayton, 1984; Hicock and Dreimanis, 1985; Croot, 1987; Gijssel, 1987; Meyer, 1987; Wateren, 1987; Menzies, 1995a, Chapter 10). Where 'slices' of underlying sediment

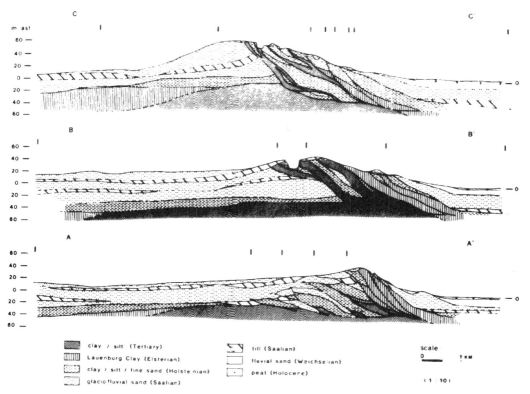

clay / silt (Tertiary)

Lauenburg Clay (Elsterian)

clay / silt / fine sand (Holstenian)

glaciofluvial sand (Saalian)

till (Saalian)

fluvial sand (Weichselian)

peat (Holocene)

scale

0   1 KM

(1 : 10)

FIG. 2.14. Schematic cross-section through the Westerburg push moraine (reprinted from van Gijssel, 1987, in: J.J.M. van der Meer (ed.) courtesy of A.A. Balkema, Rotterdam)

and/or bedrock are sheared upward into a subglacial mélange, for example, they can produce small hills transverse to ice direction that have, in the past, been mistaken for terminal moraines (Fig. 2.14). A post-depositional effect of such stacking is the disruption of groundwater flow and the formation of small perched aquifers.

Soft sediment deformation in subaquatic environments occurs when there is insufficient time for porewater drainage due to rapid sediment deposition (Attewell and Farmer, 1976; Allen, 1982; Mills, 1983). When deposition is quickly followed by ice overriding or additional sedimentation, liquefaction and/or fluidisation is likely to occur. Diapirs, faults, flame structures, cataclasis, ball and pillow structures, and dish structures form under these conditions (Allen, 1982; Allen, Collinson and Thompson, 1982; Muller, 1983; Owen, 1987).

Where an ice margin floats in a body of water, terrestrial ice is detached from its bed (Menzies, 1995a, Chapter 14). The point of detachment (the 'grounding-line') (Thomas, 1979; Molnia, 1983; Dowdeswell and Scourse, 1990), is sensitive to changes in ice thickness, water depth, wave and current movement, and tidal movement. For example, as water levels rise the grounding-line moves either in the up-ice or down-ice direction depending upon basin floor topography. Where water level rises of short duration (few hours) and small magnitude (<1–2 m) occur the position of the grounding-line tends to remain stationary, however, the ice mass rises and falls creating a cycle of localised stress loading/unloading (see Chapter 4). These localised loading/unloading cycles can be detected in sediment sequences in the form of push/thrust ridges (Washboard, De Geer, and Cross-Valley Moraines) (Moncrieff and Hambrey, 1988, 1990; Lundqvist, 1989; Powell and Molnia, 1989; Zilliacus, 1989; Boulton, 1990; Solheim et al., 1990; Woodworth-Lynas and Guigné, 1990) (Figure 2.15).

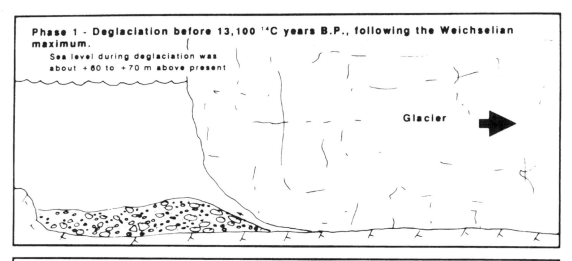

Phase 1 - Deglaciation before 13,100 ¹⁴C years B.P., following the Weichselian maximum.

Sea level during deglaciation was
about +60 to +70 m above present

Glacier

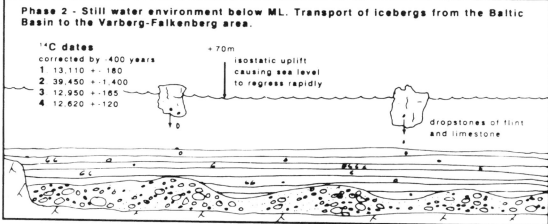

Phase 2 - Still water environment below ML. Transport of icebergs from the Baltic Basin to the Varberg-Falkenberg area.

¹⁴C dates
corrected by -400 years
1. 13,110 +/- 180
2. 39,450 +/- 1,400
3. 12,950 +/- 165
4. 12,620 +/- 120

+70m

isostatic uplift
causing sea level
to regress rapidly

dropstones of flint
and limestone

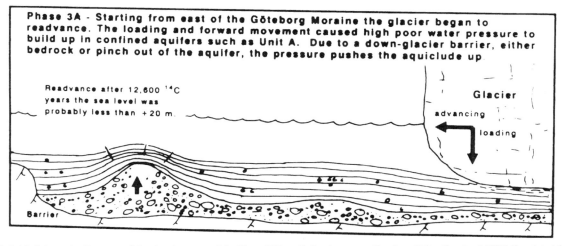

Phase 3A - Starting from east of the Göteborg Moraine the glacier began to readvance. The loading and forward movement caused high poor water pressure to build up in confined aquifers such as Unit A. Due to a down-glacier barrier, either bedrock or pinch out of the aquifer, the pressure pushes the aquiclude up.

Readvance after 12,600 ¹⁴C
years the sea level was
probably less than +20 m.

Glacier

advancing

loading

Barrier

FIG. 2.15. Schematic diagrams of the development of the Torpa Ridge, Halland, western Sweden. (After Fernlund 1987. Reprinted from Croot, (ed.), courtesy of A.A. Balkema, Rotterdam)

Phase 3B - When the aquiclude ruptured the aquifer sediments flowed up and out as clastic volcano. During this stage the soft sediment deformations occured, such as clastic intrusions, diapiric folding and mixing of sediments into diamicton.

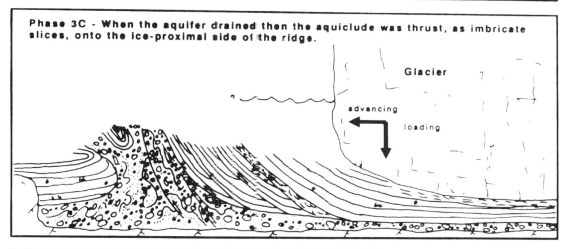

Phase 3C - When the aquifer drained then the aquiclude was thrust, as imbricate slices, onto the ice-proximal side of the ridge.

Phase 3D - The glacier subsequently advances over the ridge, resulting in the deformation of the upper 0.5 to 1 m of sediments into diamicton. The SW- ward extent of the readvance was probably several km off the present shore.

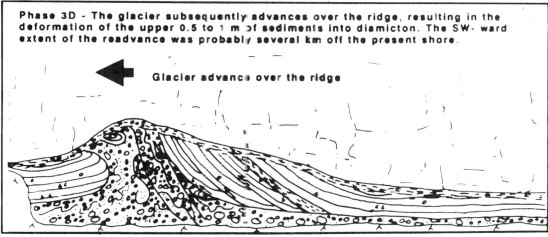

2. Subglacial sediments, unrelated to grounding-line environments, are subject to considerable loading and unloading events. These processes may be due to additions and removals of overlying sediment or time-dependent stress relief after ice sheet retreat. The effect of all of these processes, singly or in concert, appears to lead to cracks developing in many subglacial tills and mélanges (Kazi and Knill, 1969, 1973; Anderson, 1972, 1983; Fookes and Denness, 1969;; McGown et al., 1974, 1977; Banham, 1975, 1977, 1988; McGown and Radwan, 1975; McKinlay et al., 1975; Derbyshire et al., 1976; Babcock, 1977; McGown and Derbyshire, 1977; Möbus and Peterss, 1983; Derbyshire, Edge and Love, 1985b; McGown, 1985; Feeser, 1988). The origin of any joint set in till or glacial mélange is complex. Many researchers attribute joint formation to the effect of loading and unloading. The genesis of these discontinuities remains poorly studied but would appear to be a function of sediment brittle or ductile deformation behaviour (Hubert and Willis, 1957; Attewell and Farmer, 1976; Jaeger and Cook, 1976).

Work by McGown and others has demonstrated that tills frequently exhibit several distinct joint sets. Often these sets are conjugate and geometrically related to each other (McGown et al., 1977; Feeser, 1988). Usually, there exists one or two large-scale joint sets and possibly other smaller scale sets that are either random in orientation or apparently unrelated to the larger set(s).

3. Many neotectonic events lead to post-depositional soft sediment deformation within Pleistocene sediments (e.g., Hendry and Stauffer, 1975; Allen, 1982; Davenport and Ringrose, 1985, 1987; Brodzi-kowski and Van Loon, 1987). These post-depositional effects have often been unrecognised in the past or explained by subaerial weathering and pedological processes. In general, sediment deformation results from seismic waves causing liquefaction in cohesionless sediment when interbedded between finer, less permeable units or internal deformation in saturated, fine grained sediments (Seed, 1968, 1976, 1979; Sims, 1975; Seilacher, 1984; Leeder, 1987). The effect of liquefaction produces a suite of soft sediment deformation structures (dewatering pipes, diapirs, micro-faults and -joints, and small folds) that may be hard to distinguish from autokinetic structures. Seilacher (1984) has termed sediments affected by seismic-induced deformation as seismites, but their identification remains problematic. Finally, where bedrock faulting has taken place after subglacial deposition, joints and other structures can occasionally pass upwards into the sediment and create new or compound structures within the overlying sediments (cf. Brodzikowski et al., 1987a,b).

2.6.2.3. Compound structures

Although many discontinuities within subglacial sediments are of primary origin, probably most are the result of subsequent secondary process overprinting. In many instances, a primary structure is re-activated by the same or a different process resulting in its enlargement and often extension. As Figure 2.16 illustrates many structures may be of a hybrid origin. In general, where primary structures already exist the stress necessary for secondary re-activation or re-utilisation is much less.

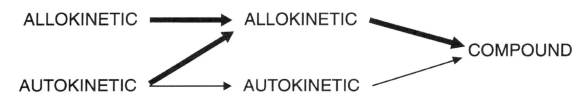

FIG. 2.16. Schematic diagram of types of discontinuities and their interrelationships

STAGES OF DEVELOPMENT

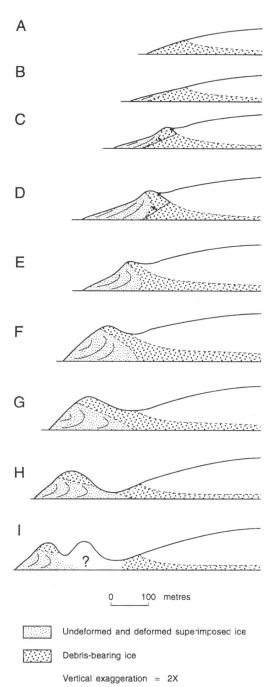

A

B

C

D

E

F

G

H

I

0   100 metres

▢ Undeformed and deformed superimposed ice

▢ Debris-bearing ice

Vertical exaggeration = 2X

FIG. 2.17. Model of evolutionary stages of an ice mass margin with progressive development of a debris-covered, ice-cored moraine. (After Hooke, 1973, reproduced by permission of the *author*)

2.7. SUBGLACIAL – PROGLACIAL TRANSITION ENVIRONMENTS

The boundary between subglacial and proglacial environments marks the site of greatest localised deposition within the glacial system (Fig. 2.1a, b). At this transition zone, large volumes of sediment derived and transported from within the subglacial system enter new sedimentological and glaciodynamic environments (Menzies, 1995a, Chapters 12 and 13). Many of the sediments discussed in this Section are considered to be dominantly proglacial, However, due to ice advance, they are often incorporated into subglacial environments. It is also not unusual to find primary flow tills deposited subglacially in tunnels and cavities. The exact delimitation between those sediments which are subglacial and those proximal proglacial, whether on land or subaqueously, is at times extremely difficult to decipher.

2.7.1. The Subglacial and Terrestrial Proglacial Environment

Subglacial debris is transported toward the margins of ice masses either along the ice/bed interface or by being sheared upward to become englacial debris close to the margin and eventually released as supraglacial

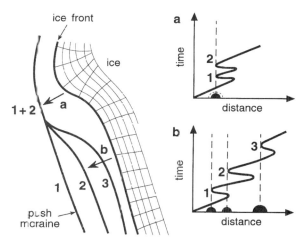

FIG. 2.18. Diagram illustrating the formation of end moraines. Diagram to the left illustrates a plan view of moraines and ice front. To the right, the two graphs chart the position of the ice front through time following advances, still stands and retreat. (Modified from Bennett and Boulton, 1993)

meltout debris (Fig. 2.17). The mechanism by which subglacial debris is sheared up into basal ice close to ice margins remains controversial (Weertman, 1961; Hooke, 1973). Debris transported to the ice front via the subglacial interface enters the proglacial system and may, at the ice margin, be squeezed and/or pushed by the ice mass (Fig. 2.15) (Menzies, 1995a, Chapter 10). In both instances, moraines may develop to a size concordant with the volumes of debris and ice, stress levels, the rheology of the debris and the length of time the ice front is stationary at a particular location

(Bennett and Boulton, 1993) (Fig. 2.18). Sediments within these morainic forms are composed of a wide range of individual lithofacies types, grain sizes and provenances. Where subglacial debris has been transported englacially and ablated as supraglacial debris, flow tills develop as the debris moves off the ice under low effective stress levels (Boulton, 1970, 1971; Lawson, 1979a, 1981a,b). Lawson, working at the edge of the Matanuska Glacier, Alaska, USA, has demonstrated that $\geq 95\%$ of the sediments in the snout area of the glacier have been influenced by, or are

TABLE 2.5a Principal characteristics of Lawson-type sediment flows (after Lawson, 1979)

	Lawson flow type			
	I	II	III	IV
Morphology	Lobate with marginal ridges, non-channelized	Lobate to channelized	Channelized	Channelized
Channel-wise profile	Body constant in thickness with planar surface, head stands above body, tail thins abruptly upslope	Body constant in thickness with ridged to planar surface, head stands above body (less than Type I), tail thins upslope	Mass thins from head to tail, irregular surface	Thin continuous 'stream', planar surface
Thickness (m)	0.01 to 2 (0.5 to typical)	0.01 to 1.4 (0.1 to 0.7 typical)	0.03 to 0.6	0.02 to 0.1
Bulk water content (wt %)	~ 8 to 14	~ 14 to 19	~ 18 to 25	725
Bulk wet density (kg m^{-3}) 2 000 – 2 600	1 900 – 2 150	1 800 – 1 950	< 1 800	
Surface flow rates (mm s^{-1})	1 – 5	2 – 50	150 – 1250	10 – 2 000
Typical length of flow (m)	10 to 300+	10 to 300+	100 to 400+	50 to 400+
Surface shear strength (MPa)	0.04 – 0.15	0.06 or less	Not measurable	Not measurable
Approx. bulk grain size (mm)	2 to 0.3	0.4 to 0.1	0.15 to 0.06	<0.06
Flow character (laminar)	Shear in thin basal zone with override at head	Rafted plug with shear in lower and marginal zones	Discontinuous plug to shear throughout	Differential shear throughout
Grain support and transport	Gross strength	Gross strength in plug; traction, local liquefaction and fluidization, grain dispersive pressures and reduced matrix strength in shear, zone	Reduced strength, traction, grain dispersive pressures; possibly liquefaction-fluidizati on, transient turbidity	Liquefaction; some traction; buoyancy (?)

+Maximum length of flow reflects boundary conditions of terminus region.

TABLE 2.5b Characteristics of sediment flow deposits in the terminus region of Matanuska Glacier, Alaska (after Lawson, 1979)

Lawson flow type	Bulk texture (1) mean (τ) (2) std dev (τ)	Internal organization		
		General	Structure	Pebble fabric
I	Gravel-sand-silt, sandy, silt (1) − 1 to 2 (2) 3 to 4.5	Clasts dispersed in fine-grained matrix	Massive	Absent to very weak; vertical clasts. $S_1 + = 0.49–0.55$
II	Gravel-sand-silt, sandy silt, silty sand (1) 2 to 3 (2) 3 to 4	Plug zone; clasts dispersed in fine-grained matrix. Shear zone; gravel zone at base, upper part may show decreased silt-clay and gravel content; overall, clasts in fine-grained matrix	Massive; intra-formational blocks. Massive; deposit may appear layered where shear and plug zones distinct in texture.	Absent to very weak; vertical clasts Absent to weak; bimodal or multimodal; vertical clasts. $S_1 = 0.50–0.65$
III	Gravelly sand to sandy silt (1) -2.5 to 2.5 (2) 3.5 to 2	Matrix to clast dominated; lack of fine-grained matrix possible; basal gravels	Massive; intraformational blocks occasionally	Moderate, multimodal to bimodal parallel and transverse to flow. $S_1 = 0.60–0.70$
IV	Sand, silty sand, sandy silt (1) >3.5 (2) <2.5	Matrix except at base where granules possible	Massive to graded (distribution, coarse-tail)	Absent

Lawson flow type	Surface forms	Contacts and basal surface features	Pene-contemporaneous deformation	Geometry* and maximum observed dimensions (length ×width, thickness, m)
I	Generally planar; also arcuate ridges, secondary rills and desiccation cracks	Nonerosional, conformable contacts; contacts sharp; load structures	Possible subflow and marginal deformation during and after deposition	Lobe: 50 ×20, 2.5
II	Arcuate ridges; flow lineations, marginal folds, mud volcanoes, braided and distributary rills on surface	Nonerosional, conformable contacts; contacts indistinct to sharp; load structures	Possible subflow and marginal deformation during and after deposition	Lobe: 30 ×20, 2.5; sheet of coalesced deposits
III	Irregular to planar; singular rill development; mud volcanoes	Nonerosional, conformable contacts; contacts indistinct to sharp	Generally absent; possible subflow defromation on liquefied sediments	Thin lobe; 20 ×10, 0.5; fan wedge; 30 ×65, 3.5; rarely, sheet of coalesced deposits
IV	Smooth, planar; mud volcanoes possible	Contacts conformable indistinct	Absent	Thin sheet; 20 ×30, 0.3; Fills surface lows of irregular size and shape

* Length and width refer to dimensions parallel and transverse to direction of movement prior to deposition.
† S_1 is a normalized eigenvalue which gives the strength of the cluster of long axes about the mean axis (see Mark, 1973).

directly the result of debris flow and resedimentation. The characteristics of these sediments are illustrated in Table 2.5 (see Chapter 3; Menzies, 1995a, Chapter 11).

2.7.1.1. Flow tills

Flow tills have been described in several differing glacial environments. For example within subglacial cavities, on the surface of ice masses within supraglacial systems and, dominantly, at the ice front. Boulton (1968, 1971, 1972b) has described in detail the flow tills of several glaciers on Svalbard (Spitsbergen), as did Lawson (1979, 1981) from the Matanuska Glacier in Alaska (cf. Dreimanis, 1988).

Flow tills moving under very low gradients (≈2°) can spread over extensive proglacial areas (Menzies, 1995a, Chapter 12). Within the proglacial environment complex stratigraphies can develop with large volumes of buried glacier ice being overlain by these tills (see Chapter 3). Flow tills are generally formed as wedge style deposits and are often stacked one flow on another. Thicknesses of stacked flow tills have been reported in excess of 10 m. Typically, stacked flow tills occur with each layer laterally offset by several metres in an on-lap→off-lap sequence (Humlum, 1981; Ingólfsson, 1988; Owen and Derbyshire, 1988).

Flow till units are usually composed of a central unsorted body or plug surrounded by heavily sheared

PLATE 2.8. Flow nose with a diamicton, Vancouver Island, British Columbia. (Photo courtesy of Steve Hicock)

and sorted outer units. Hicock *et al.* (1981) illustrate a typical flow till unit encased within subaquatic diamictons from Vancouver Island, BC, Canada (Plate 2.8).

Structures associated with flow tills are complex, exhibiting a wide range of folds, faults, riedel shears and kink band arrays. Flow noses, of great size ranges, can be occasionally detected.

Flow till clast fabrics tend to exhibit strong, uni-directional orientations (Fig. 2.13). However, similar fabrics have also been measured in sediments from other environments negating the value of clast fabric, alone, as a potential identifying characteristic (Lawson, 1979b, 1982; Gibbard, 1980; Sharp, 1982; Dowdeswell *et al.*, 1985; Rappol, 1985; Dowdeswell and Sharp, 1986; Moncrieff and Hambrey, 1988; Menzies, 1990c).

Flow tills may contain clasts from many sources. Some are angular frost-riven clasts whose source is supraglacial (in valley glaciers and ice sheets where nunataks exist), others are heavily comminuted,

striated, sub-angular to sub-rounded clasts of englacial and subglacial derivation. Frequently, small inclusions and rafted units of sediment occur within flow tills especially where flows have crossed weaker subjacent sediments. These units may be strongly contorted but often are intact.

2.7.2. The Subglacial and Subaquatic Proglacial Environments

As major ice sheets enter large bodies of water, the ice margin may float (Fig. 2.1b) (Menzies, 1995a, Chapter 5). Ice masses with a floating tongue whether as a free floating ice front (tidewater) or contiguous with an ice shelf become buoyant at the grounding-line. At this point of detachment from the bed a considerable volume of subglacial debris exits into the proglacial subaquatic environment. Various sediment types are deposited at or close to the grounding-line with characteristics derived from both subglacial and subaqueous sedimentary environments (Fig. 2.19)

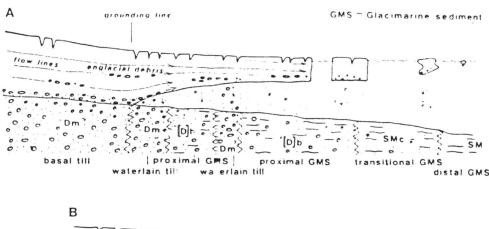

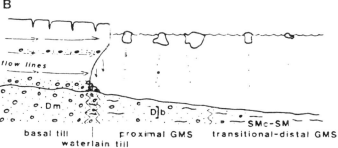

FIG. 2.19. Schematic cross-section of A, a floating ice shelf and B, a grounded ice front showing the relationship between basal till, waterlain till and glaciomarine sediments. (Reproduced with permission from the *Geological Society of London*)

(Powell, 1981a, 1984, 1990; Eyles and Eyles, 1983; Mackiewicz et al., 1984; Boulton, 1986; Talbot and Von Brunn, 1987; Albino and Dreimanis, 1988; Eyles et al., 1989; Dowdeswell and Scourse, 1990) (Menzies, 1995a, Chapter 14 and see Chapter 14).

At subglacial meltwater portals extensive linear or deltaic assemblages (subaqueous fans) of stratified sediments are deposited into the proximal subaquatic proglacial zone as meltwater flow velocities rapidly drop (Rust and Romanelli, 1975; Smith, 1982; Molnia, 1983; Visser, 1983; Moncreiff and Hambrey, 1990; Powell, 1990).

In more distal subaquatic areas a different imprint characterises the sediments and a truly subaquatic sedimentation environment develops (Fig. 2.19) (Eyles et al., 1989; Dowdeswell and Scourse, 1990; Moncreiff and Hambrey, 1988; 1990).

Tills deposited in the vicinity of a grounded floating ice front can be subdivided into: (1) Waterlain Tills (cf. Dreimanis, 1976, 1988; Evenson et al., 1977; Gibbard, 1980) and (2) Proximal Subaquatic Diamicton Mélanges (cf. Talbot and Von Brunn, 1987; Dreimanis et al., 1987; Dreimanis, 1988; Alley et al., 1987; Hicock, 1992). The distinction between these tills is made on the basis that waterlain tills are formed as a result of rain-out through a water column, while proximal subaquatic diamictons are formed as a result of ductile extrusion or gravitational deformation (both soft sediment deformation) extending from the grounding-line into the subaquatic environment. This distinction, however, is often impossible to establish, much research remains to be done.

2.7.2.1. Waterlain tills

Waterlain tills are deposited within a subaquatic environment by continuous rain-out of basal glacial debris from floating icebergs, melting under-surfaces of the main ice body, debris plumes from exiting subglacial meltwater portals, debris flowing off the main ice body and debris falling into the water from these various sources. Little or no reworking is accomplished by bottom currents and, therefore, the influence of subaquatic (lacustrine/marine) processes are minimal except for post-depositional bioturbation (Evenson et al., 1977; Kurtz and Anderson, 1979; Gibbard, 1980; Hicock et al., 1981; Eyles et al., 1983;

Gravenor et al., 1984; Morawski, 1985; Broster and Hicock, 1985). These tills have been classified under a series of different terms in recent years such as waterlaid, subaqueous, glacioaquatic, para- and aquatills; also submarine, shelf and basin moraine, and lacustrotill among many (Harland et al., 1966; Lavrushin, 1968; Dreimanis, 1969, 1976, 1988; Krygowski et al., 1969; Raukas, 1969; Francis, 1975; Morawski, 1985, 1988).

Waterlain tills can be deposited over wide areas depending upon the rate of debris rain-out and the rate of grounding-line retreat across the terrain. Waterlain tills tend to be relatively thin (<1 m) with greater thicknesses occurring in bands transverse to ice flow direction.

The characteristics of waterlain tills are determined by the rate of debris rain-out; debris grain size distribution; the style of land-ward debris transport; water column depth, temperature gradient and turbidity; the angle of lake/sea bed slope; the nature of the underlying sediment; the presence or absence of subglacial meltwater portals and meltwater discharge fluctuations; the rate of sediment deposition; the rate of grounding line retreat/advance; and the nature and style of iceberg debris addition.

Waterlain tills may be rhythmically bedded, almost laminated in appearance, but also may be massive and structureless. Apart from a crude flow-like stratification, waterlain tills usually contain abraded and striated clasts and have a preferred clast fabric (Fig. 2.13) (Domack and Lawson, 1985). Fine-grained waterlain tills may contain partially rounded clay and silt balls. Occasional deformed and/or flow lenses and sediment intraclasts can be found, the former often of considerable lateral continuity. In comparison to terrestrial tills, waterlain tills often have higher clay and silt concentrations (Elverhøi et al., 1980; Syvitski and Murray, 1981; Gilbert, 1982; Powell, 1983; Mackiewicz et al., 1984; Cowan and Powell, 1990; Stevens, 1990).

The distinction between terrestrial tills and waterlain tills remains problematic (cf. Gelinas, 1974; Dreimanis, 1976, 1988; Lawson,1988). One, often quoted, method of discrimination is the presence of 'dropstones' and other artifacts indicative of deposition through a water column, but even this attribute is far from conclusive.

Primary structures formed within waterlain tills that have not been subject to the loading/unloading stress cycles of a nearby grounding-line or lateral gravitational slumping are dominantly autokinetic related to bedding planes, and porewater transmission processes (Dreimanis, 1993).

Waterlain tills, because of their high initial water contents, are highly susceptible to allokinetic stress events. These tills tend to be relatively sensitive immediately following deposition thus dewatering, synergic, collapse, lateral shear processes and seismically-induced structures are all potential post-depositional features. Where flocculation has been a major physico-chemical process in the rain-out of debris into a water body (especially into brackish water), distinct microscopic 'card-house' packing of floccules may occur. These floccules are susceptible to even minor secondary structural stresses. On collapse, a distinctive structural geometry of discontinuities is formed with faulting and shear zones developing. Contemporaneous with particle structure collapse, porewater migration usually occurs resulting in clay particle migration, clay coating formation and piping.

In areas where sea or lake floor slope is sufficiently steep, sediment build-up may cause gravitational slumping and a series of structures related to folding and shear deformation akin to those discussed under diamicton mélange may develop. The main distinction between these deformed sediments is that in gravitationally slumped waterlain tills, the degree of stress is much lower, extrusion flow is not involved and the sediment units are much thinner (<1 m).

Where dropstones or inclusions caused by rain-out of sediment or large clasts/boulders falling through the water column and embedding into the till occurs, a distinct suite of stratal warping/buckling structures develop with associated diapiric water escape features developing around the dropstones (Chapter 4) (Dreimanis, 1976, 1979; Evenson et al., 1977; Gibbard, 1980; Hicock et al., 1981; Orheim and Elverhøi, 1981; Eyles et al., 1983; Gravenor et al., 1984; Thomas and Connel, 1985; Parkin and Hicock, 1988).

Clast fabrics within waterlain tills reveal weak principal preferred orientations, tending toward random, as can be expected with sedimentation through a water column rather than under a unidirectional shear stress (Fig. 2.13) (Dowdeswell et al., 1985; Dowdeswell and Sharp, 1986).

Geotechnically these sediments are similar to 'soft' lodgement tills, having been formed under high saturation but low stress levels. Often these tills have lower bulk densities, higher void ratios, and higher coefficients of porosity than lodgement tills (Legget, 1961; Easterbrook, 1964; Lavrushin, 1968; Boulton and Paul, 1976; Hartford, 1985).

2.7.2.2. Proximal subaquatic diamicton mélanges

These tills form as a result of two dominant processes. (a) extrusion of subglacial diamicton mélange into a water body as a 'Till Delta' (King and Fader, 1986; Alley et al., 1986, 1987a,b; Vorren et al., 1989) and (b) massive internal pervasive deformation by weight of overlying waterlain sediments to such a degree that virtually all primary structures are destroyed (cf. Talbot and Von Brunn, 1987; Dreimanis et al., 1987; Dreimanis, 1988; Moncreiff and Hambrey, 1988, 1990; Hicock, 1990). The formation of type (a) is dependant upon a subglacial diamicton mélange continuing to deform as it is extruded from beneath the ice front at the grounding-line. Evidence of this process is well documented in West Antarctica where a till delta is under investigation at present beneath Ice Stream B (Fig. 2.20) (Alley et al., 1987a,b) (Menzies, 1995a, Chapter 14). Proximal subaquatic mélanges lose the applied shear stresses from overlying ice on exiting from beneath the ice mass. Overburden stresses are likely to be higher if water column depths are greater than ice thickness but the loss in shear force will cause the deforming layer to stop moving.

In the latter case, the diamicton mélange forms under a different set of processes where hydraulic pumping occurs laterally into proximal subaqueous sequences leading to strata disruption, faulting, folding and some lateral deformation (Talbot and Von Brunn, 1987). The pumping effect is thought to be a manifestation of the grounding-line lifting and falling due to tides thereby causing lateral surges of porewater migrating into the sediment package.

The areal extent of these tills is likely to be very limited. The tills will tend to be relatively thin layers (<2 m) covering zones parallel to the grounding-line

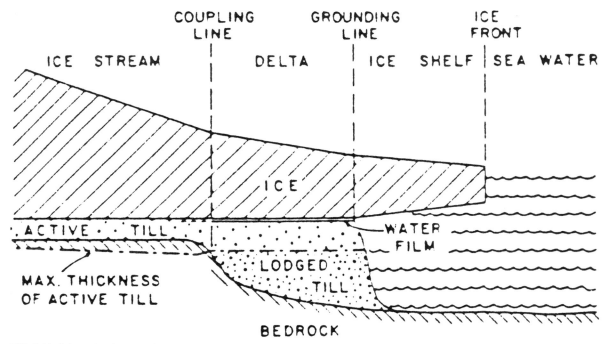

FIG. 2.20. Schematic diagram of ice stream/ice shelf model with a till 'delta' or 'tongue'. (After Alley *et al.*, 1987, copyright by the *American Geophysical Union*)

in an ice front position (Benn, 1989). Only when lengthy stillstands have occurred is there any likelihood the tills will reach substantial thicknesses.

These mélange sub-types exhibit a wide array of autokinetic structures similar to those already discussed under subglacial diamicton mélanges. The mélanges developed under lateral hydraulic pumping effects have micro-scale structures (Barnes and Reimnitz, 1979; Thomas and Connell, 1985; Fischbein, 1987; Eyles and Clark, 1988; Hodgson *et al.*, 1988; Woodworth-Lynas and Landva, 1988; Zilliacus, 1989; Woodworth-Lynas and Guigné, 1990). Within this subaquatic environment, the development of compound structures and the reactivation and re-utilisation of discontinuities will occur more easily than under terrestrial conditions due to high saturation levels, differential compaction, particle-to-particle buoyancy effects, the higher void ratio and reduced consolidation values. The size, orientation and depth of structures within these mélanges are related to the rate and degree of applied stress, and the stratigraphic nature of the sediment sequence. Many structures, due

to lateral movement, will tend to be sub-horizontal and of limited lateral extension (<1 m). Near-vertical fractures are potentially deeper with depths of around 1 m having been reported (Hodgson *et al.*, 1988).

As has been discussed elsewhere (cf. Menzies, 1990c) (Fig. 2.13), the intrinsic value of clast fabrics in these heavily deformed tills is subject to great debate. Measured clast fabrics tend to range from well oriented strong fabrics to scattered, almost, random fabrics (Hart, 1993).

The geotechnical characteristics of these subaquatic diamicton mélanges are remarkably similar to terrestrial lodgement tills and diamicton mélanges (Gelinas, 1974; Solheim *et al.*, 1990).

2.8. SUBGLACIAL LANDFORMS AND BEDFORMS

A fundamental characteristic of subglacial sedimentation is the formation of depositional or constructive landforms or bedforms (Fig. 2.21). These forms are indicative of subglacial environmental processes.

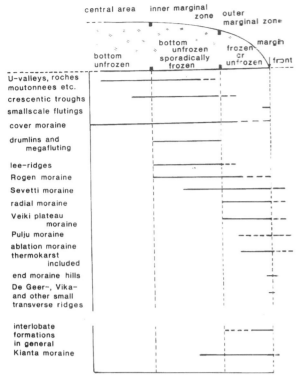

FIG. 2.21. Glacial landform/bedform types with reference to their respective basal thermal environment and position at the bed of an ice sheet. (After Aario, 1990)

Over the past century there has been a proliferation of ideas and hypotheses on the formation of these landforms (cf. Flint, 1947, 1957, 1971; Charlesworth, 1957; Embleton and King, 1968; Sugden and John, 1976; Menzies, 1984). These ideas have, almost exclusively, developed 'unique' hypotheses for individual landforms. Most of these hypotheses reflect a view based upon landform morphology or certain apparent unique characteristics. This 'morpho–sedimentological' approach which in the past has been one of the fundamental paradigms of research investigations by geologists and geomorphologists may, in isolation, be tautological and potentially sterile.

In recent years a new paradigm has been advocated that considers glaciodynamic conditions prevailing under active temperate ice masses and attempts to interrelate these subglacial conditions to subglacial sediments, facies and landforms/bedforms. This 'glacio-sedimentological' paradigm may substantiate and refine some of the above hypotheses and advance our understanding subglacial environments.

Subglacial landforms can be subdivided into two distinctive groups, bedforms and non-bedforms. The former group are integral elements of a continuum of bedforms evolving into each other as conditions alter at the subglacial interface. The latter group are landforms that are interrelated and have certain common characteristics but are not part of a spectrum of forms, but are individual reflections of and responses to varying conditions at the ice/bed interface. Within the first group are drumlins and Rogen moraines and in the second, Kalix till ridges and De Geer moraines. In general, both bedform and non-bedform landforms are lineated either parallel or transverse to major ice flow directions (Table 2.6). Before discussing individual landforms, it is important to consider the glaciated terrain on and within which these landforms are constructed or eroded. At the large scale, vast areas of glaciated terrains are covered by subglacial deposits in the form of till plains. These plains are representative of subglacial and ice mass conditions over huge areas of the glacier bed.

2.8.1. Till plains

Although many characteristic landforms of glaciation are products of active ice formed within the subglacial environment, perhaps the most widespread evidence of subglacial action is the formation of low relief, rolling till plains. These plains cover vast areas and are perhaps indicative of the more typical non-landforming or bedforming aspect of subglacial processes. Because of their somewhat monotonous appearance, research has all too often concentrated on more puzzling subglacial landforms. Till plain morphology is of particular relevance, for example, in the interior plains of North America (Moran et al., 1980, Fulton, 1989).

The sedimentology of these till plains is highly variable but tends to closely reflect the underlying bedrock composition and topography (Christiansen, 1971; Scott, 1976; Shetsen, 1984; Shilts et al., 1987, 1989; Prest and Nielsen, 1987).

Across till plains distinctive features (thrust block ridges, depressions, and push ridges) can be found

TABLE 2.6. Types of subglacial landforms/bedforms

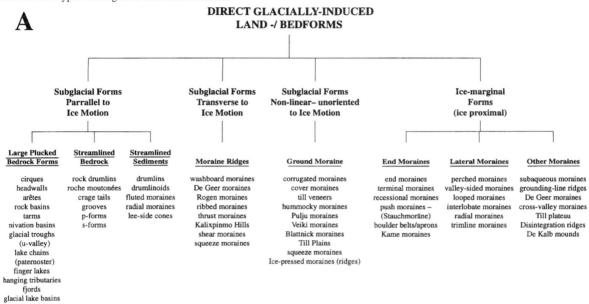

A DIRECT GLACIALLY-INDUCED LAND -/ BEDFORMS

Subglacial Forms Parallel to Ice Motion

Large Plucked Bedrock Forms	Streamlined Bedrock	Streamlined Sediments
cirques	rock drumlins	drumlins
headwalls	roche moutonées	drumlinoids
arêtes	crage tails	fluted moraines
rock basins	grooves	radial moraines
tarms	p-forms	lee-side cones
nivation basins	s-forms	
glacial troughs		
(u-valley)		
lake chains		
(paternoster)		
finger lakes		
hanging tributaries		
fjords		
glacial lake basins		

Subglacial Forms Transverse to Ice Motion

Moraine Ridges
washboard moraines
De Geer moraines
Rogen moraines
ribbed moraines
thrust moraines
Kalixpinmo Hills
shear moraines
squeeze moraines

Subglacial Forms Non-linear– unoriented to Ice Motion

Ground Moraine
corrugated moraines
cover moraines
till veneers
hummocky moraines
Pulju moraines
Veiki moraines
Blattnick moraines
Till Plains
squeeze moraines
Ice-pressed moraines (ridges)

Ice-marginal Forms (ice proximal)

End Moraines	Lateral Moraines	Other Moraines
end moraines	perched moraines	subaqueous moraines
terminal moraines	valley-sided moraines	grounding-line ridges
recessional moraines	looped moraines	De Geer moraines
push moraines –	interlobate moraines	cross-valley moraines
(Stauchmoräne)	radial moraines	Till plateau
boulder belts/aprons	trimline moraines	Disintegration ridges
Kame moraines		De Kalb mounds

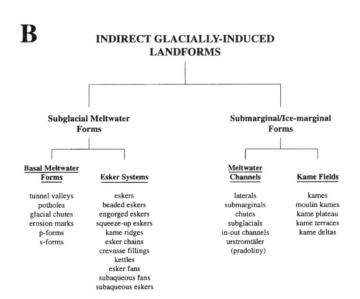

B INDIRECT GLACIALLY-INDUCED LANDFORMS

Subglacial Meltwater Forms

Basal Meltwater Forms	Esker Systems
tunnel valleys	eskers
potholes	beaded eskers
glacial chutes	engorged eskers
erosion marks	squeeze-up eskers
p-forms	kame ridges
s-forms	esker chains
	crevasse fillings
	kettles
	esker fans
	subaqueous fans
	subaqueous eskers

Submarginal/Ice-marginal Forms

Meltwater Channels	Kame Fields
laterals	kames
submarginals	moulin kames
chutes	kame plateau
subglacials	kame terraces
in-out channels	kame deltas
urstromtäler	
(pradoliny)	

Note: based upon Goldthwait, 1988; Sugden & John, 1976; Prest, 1968.

that reflect localised subglacial differential stresses, or local sediment rheology variations, or the influence of topographic control (Moran *et al.*, 1980; Aber *et al.*, 1989; Aylsworth and Shilts, 1989).

These plains have formed as a result of a widespread uniformity of subglacial depositional conditions. These conditions, however, do not necessarily signify uniformity of till type. Till plain sediments are a complex reflection of local and regional bed geology, topography, ice dynamics and subglacial bed conditions. Lodgement, meltout, flow tills, and mélanges can be found over large or restricted areas of till plains. Typically, they contain rafted units of subjacent sediments and bedrock (Stephan and Ehlers, 1983; van der Meer 1987b; Aber *et al.*, 1989) These units are often small (< 1 m³) whilst other rafts several kilometres in size have been reported (Aber *et al.*, 1989).

2.8.2. Subglacial Bedforms

Subglacial bedforms have been and remain the subject of intense study and debate (cf. Menzies, 1984; Menzies and Rose, 1987, 1989; Brodzikowski and van Loon, 1991; Clark, 1993). It has become apparent that these subglacial forms can be considered as bedform suites produced at the ice/bed interface (Lundqvist, 1969; Aario, 1977a; Boulton, 1982, 1987; Menzies, 1987, 1989a; Rose, 1989; Aylsworth and Shilts, 1989; Boulton and Clark, 1990a,b; Clark, 1993). This view is not universally held, as alternate hypotheses have also been advocated (Dardis and McCabe, 1983; Dardis *et al.*, 1984; Shaw and Kvill, 1984; Hanvey, 1987, 1989; Sharpe, 1987, 1988b; Shaw and Sharpe, 1987b; Shaw *et al.*, 1989). Both approaches to the problem of subglacial bedform development will be discussed below.

As will become apparent, the study of subglacial bedforms is the subject of strongly held, and often divergent views concerning the processes of formation. However, certain central issues emerge that, in most instances, are common to all approaches to the problem:

1. Specific subglacial environmental conditions favour the development of bedforms and non-bedforms within specific regional topographic, sedimentological and glaciodynamic settings.

2. Macroscale spatial patterns of subglacial bedforms within bedform belts, zones or fields may provide insight into the subglacial formative dynamics of individual ice masses.

3. The relationship between subglacial interface conditions of sediment rheology, thermal conditions and subglacial hydraulics is critical in bedform/non-bedform generation.

Ultimately, any subglacial bedform/non – bedform must be a reflection of the interaction of basal ice/bed stress conditions of a particular ice mass at a specific location and time.

When subglacial bedforms are observed at the very large scale (Boulton and Clark, 1990a; Clark, 1993) it is apparent that a mega-scale pattern of lineation of > 50 km in length can be seen (Fig. 2.22). At this large scale, two previously undocumented ice-moulded forms are apparent, mega-scale lineations and cross-cutting lineations. The work of Boulton and Clark gives some credence to subglacial pervasive deformation of sediments and the development of cross-cutting lineations at various scales developed in response to changing ice centres (Clark, 1993, fig.15).

2.8.2.1. Bedform continua

Before discussing individual subglacial bedforms, it is worth considering the spatial and morphological continua that appears to exist in some glaciated terrains between Rogen moraines, drumlins and fluted moraines (Fig.2.23) (Rose, 1987a). While working in the Rogen area of Sweden, Lundqvist noted a perceived spatial relationship between these bedforms over terrain in northwestern Sweden (Lundqvist, 1969; 1970, 1989; Moran *et al.*, 1980; Shilts *et al.*, 1987). The transition from Rogen moraines to drumlins has been misreported in the past (Sugden and John, 1976, Fig. 13.16; Lundqvist, 1989). In general, in areas where Rogen moraines and drumlins exist, often the Rogen moraines occur only in topographic basins and valley bottoms while drumlins occupy interfluves, there are exceptions however, to this rule (Aario, 1987). A relationship seems to exist between concave trending terrain and compressive basal ice flow where the Rogen moraines occur and convex terrain and extending basal ice flow where drumlins are found.

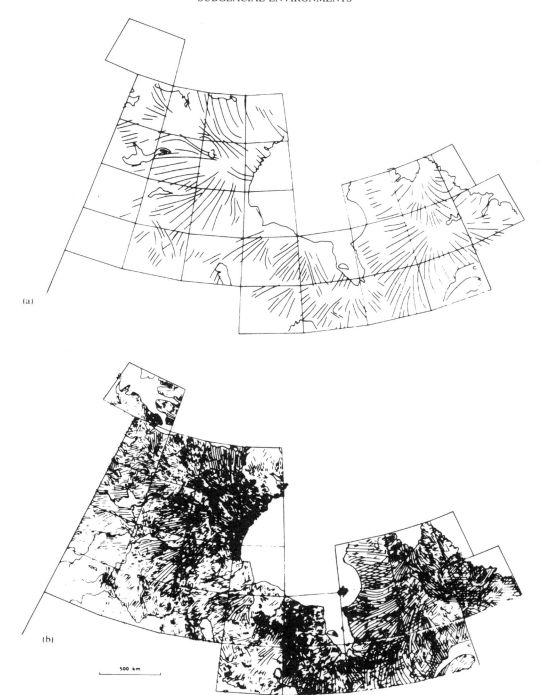

FIG. 2.22. (a) Summarized ice flowlines interpreted from the work of *Prest et al.* (1968) based upon aerial photographic interpretation. Note the two main radial flow patterns, and that there are no areas of cross-cutting or overlain flowlines. (b) Summarized ice flowlines interpreted by Clark (1993) based upon glacial lineation mapping using Landsat images. Note the extensive areas of cross-cutting patterns and, in comparison with (a), the numerous flowline patterns identified (from Clark, 1993; reproduced by permission from *Earth Surface Processes and Landforms*)

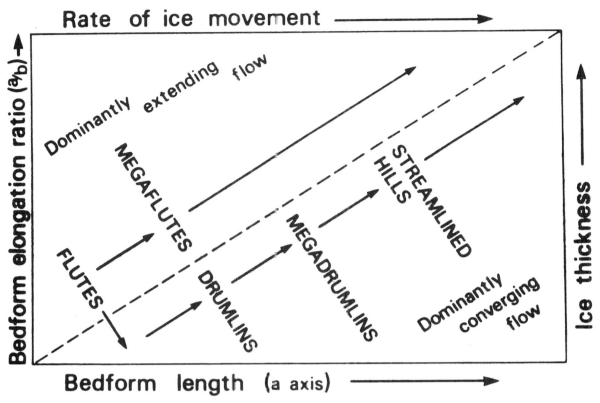

FIG. 2.23. Continuum of subglacial bedforms in relation to ice thickness and rate of ice movement. (After Rose, 1987. Reprinted from Menzies, and Rose, (eds.), courtesy of A.A Balkema, Rotterdam)

The transition from one bedform to another is almost imperceptible. Lundqvist (1969, 1970, 1989) describes three styles of bedform transition. First drumlins become incomplete with their down-ice lee ends truncated or replaced by a concave end. These incomplete drumlins begin to become aligned side by side forming ridges transverse to the main ice direction. This transition zone has been separately termed Blattnick moraine (Markgren and Lassila, 1980). The 'horns' of the incomplete drumlins always point in the down-ice direction. The 'change-over' from drumlins to Rogen moraine and *vice versa* may repeat itself several times where terrain conditions permit. Second, where drumlins begin to coalesce forming side-by-side ridges, Rogen moraines essentially 'grow' from this adjacency. Finally flutings, in some places, begin to appear on the surface of the Rogen moraine and assume increasing topographic expression in the down-ice direction to the point

where the flutes become the predominant bedform. This third transition may also occur where flutes first 'convert' to drumlins and then, into very long elongated drumlins and finally, into fluted moraine (Fig.2.24).

Within 200 km of the ice divide in Finland, Rogen moraine often appear to alter, in the up-ice direction, to Pulju moraines (Kujansu, 1967; Aario, 1990) and in the down-ice direction into Blattnick moraines (Markgren and Lassila, 1980). Pulju moraines are random winding ridges perhaps indicative of slower basal ice velocities closer to the ice divide or a freezing basal ice/bed interface state (Section 2.10.3.2.). Similar continuum transitions to those described in Sweden (Lundqvist, 1958, 1979b, 1970; Hoppe, 1968; Shaw, 1979; Markgren and Lassila, 1980) have been noted in bedforms in many parts of the world (for example, central and northern Finland (Aario, 1977a,b; 1990; Kurimo, 1977; Punkari, 1982,

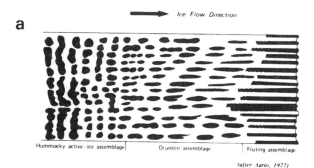

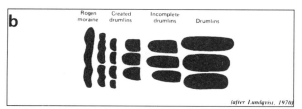

FIG. 2.24. Bedform transition patterns beneath active ice. ((a) After Aario, 1977; (b) Lundqvist, 1970). (After Menzies, 1987. Reprinted from Menzies and Rose (eds.), courtesy of A.A. Balkema, Rotterdam)

FIG. 2.25. Distribution of Rogen morainic forms in Sweden. (Courtesy of C. Hällestrand)

1984; Heikkinen and Tikkanen, 1989); central Norway (Sollid and Sørbel, 1984); Quebec (Bouchard, 1980, 1989), Keewatin and Labrador (Ignatius, 1958; Lee, 1959; Hughes, 1964; Cowan, 1968; Shilts, 1973a; Aylsworth and Shilts, 1985. 1989), Saskatchewan (Moran *et al.*, 1980) and Ontario, Canada (Terasmae, 1965; Menzies, 1987), North Dakota, USA. (Moran *et al.*, 1980) and Patagonia, Chile (Clapperton, 1989). In general, the transition from Rogen moraine to drumlins and/or fluted moraine ceases closer to ice margins where perhaps basal ice velocities are too great or basal extending flow too dominant or basal debris deformation too rapid (Boulton, 1987) (Fig.2.25). However, the relationship between the sites of Rogen moraine formation and their relationship to ice margins at the time of formation remains an interesting research problem (cf. Markgren and Lassila, 1980).

The significance of this continuum of forms demonstrates the close relationship between the ice/bed interface processes of sediment rheology, basal ice glaciodynamics and subglacial hydraulics (Fig. 2.25). Table 2.7 illustrates some possible relationships between ice/bed interface conditions and bedform/

non-bedform type development in the subglacial environment.

A final question still to be answered is whether this continuum of forms is in fact a 'natural' evolution of bedforms as ice/bed conditions transmutate or whether this continuum is simply a spatial preference of one bedform as opposed to another in particular topographic settings (Boulton and Clark, 1990a,b; Clark, 1993).

2.8.2.2. Rogen Moraine

The term Rogen moraine or ribbed moraine is applied to a series of conspicuous morainic ridges found transverse to the main ice direction occurring in such

TABLE 2.7

Interface locations	Interface conditions	Bedform type	Internal sediments and structures	Orientation of bedforms to flow of ice
Lower	1. Ice-bedrock/ immobile bed	No depositional bedforms but erosion forms	Diamicton and occasional stratified sediments, deformation structures, intraclasts, distinct herringbone clast fabrics	Linear and transverse
	2. Ice-thin mobile traction zone	Isolated forms, possibly bare bedrock between		
	3. Ice-thick mobile traction zone	Well-developed continua of forms of unstratified sediment		
	4. Ice-meltwater-bedrock	Forms but unlikely to be part of a continuum	Stratified, melt-out deformation structures, intraclsts, faulting, drag-folds	Linear and transverse but not part of continuum of forms
	5. Ice-meltwater-bedrock	Isolated forms of stratified sediments with bare bedrock	Stratified, melt-out, squeeze and cavity-infill structures	Linear and transverse perhaps part of continuum of forms
	6. Ice-meltwater-thin mobile traction zone	Isolated forms containing stratified and unstratified sediments with bare bedrock between		
	7. Ice-meltwater-thick mobile traction zone	Well-developed continua of forms of stratified and unstratified sediment		
Upper	5. Thin mobile-immobile traction zones	Small isolated forms of unstratified sediment with bare bedrock between	Diamicton dominant, deformation structures, strong clast fabric, lodgement, melt-out, flow tills present. Occasional stratified intraclasts.	Linear and transverse
	6. Thick mobile-immobile traction zones	Well-developed continua of forms of unstratified sediment		
	7. Mobile traction zone-bedrock (thin?)	Isolated well-developed forms of unstratified sediment with bare bedrock between		

large numbers to constitute a field. These ridges may reach heights of 10–20 m, are 50–100 m in width, with lateral extents of sometimes several kilometres and often with an interval spacing of 100–300 m (Plate 2.9). Where the moraines transversely cross valleys and other topographic depressions gaps are often found in some of the central parts of the ridge sets. Occasionally, the ridges exhibit a slight down-ice

arcuate pattern. Lateral ridges usually are composite interfingered ridge mosaics rather than single forms. Individual ridges are often asymmetric with a steeper lee-side giving a fish-scale appearance to the terrain (Shilts et al., 1987). Over large areas of terrain ridge crest heights are often remarkably accordant. Not all Rogen moraine are well developed and occasionally moraines are found superimposed upon drumlinoid

(a)

PLATE 2.9. (a) Rogen moraine, Uthusslön, Krattelstön, Sweden. (Photo courtesy of Jan Lunqvist). (b) Rogen moraine near Portimo, Ranua, northern Finland. (Photo courtesy of Risto Aario)

(b)

ridges (Bouchard, 1980, 1986; Cunningham, 1980; Markgren and Lassila, 1980).

At the regional scale, Rogen moraine occur at distances of ≈ 200 km from ice sheet centres and rarely within 200–300 km of the ice margin (Fig. 2.9) (Prest *et al.*, 1969; Kurimo, 1977; Shilts *et al.*, 1987; Bouchard, 1989; Lundqvist, 1989). In other words, Rogen moraine occur typically just beyond he centres of glaciation (Fig. 2.24) (cf. Aylsworth and Shilts, 1987, 1989).

At the local scale, as discussed above, Rogen moraine often occur within topographic depressions and are intrinsically associated with drumlinised terrain (Markgren and Lassila, 1980).

The composition of Rogen moraines exhibit, in general, a wide range of sediment types and structures. However, stratified sediments are often the dominant type (Cowan, 1968; Lundqvist, 1969; Shaw, 1979; Bouchard, 1980, 1986; Johansson, 1983; Aylsworth and Shilts, 1985; Fisher and Shaw, 1992). Aylsworth and Shilts (1985) noted that in some places Rogen moraines seemed to have a coarser texture than

neighbouring drumlins. Shaw (1979) studied sections in Rogen moraine from three areas in Sweden concluding that the sediments were of definite subglacial origin but of a complex lithofacies association. The till within the ridges exhibited folding and the presence of dislocated till units, with clast fabrics exhibiting random orientations closely allied to apparent autokinetic glaciotectonism. The till, termed Sveg Till, was weakly stratified with thin lenses of sorted material intercalated within the units. Bouchard (1989), in Quebec, discovered similar sediment structures with distinctive thrust shear planes manifest as 'slabs' of till stacked one on another.

Several hypotheses explaining the formation and spatial pattern of Rogen moraines have been advanced (Fig.2.26) (see reviews by Bouchard, 1980, and Lundqvist, 1989). Three main hypotheses (1) subglacial, (2) stagnant ice and (3) subglacial tectonics, have gained some support. Other origins connected with subglacial meltwater floods, crevasses fillings and marginal moraine formation have also been advanced.

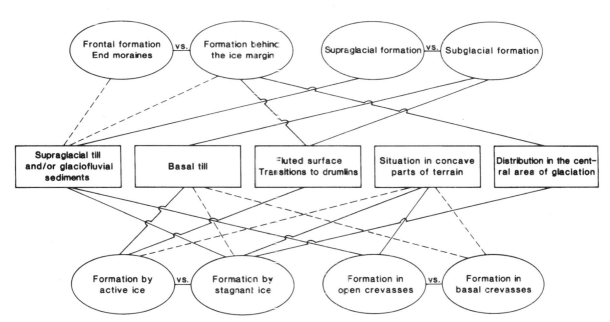

FIG. 2.26. Schematic illustration of proposed interpretations of Rogen moraine (in ovals) compared with facts observed in the field (in rectangles). The interpretation are shown in pairs, representing opposing hypotheses. The lines indicate the implications of each parameter; broken lines are less apparent relationships. (After Lundqvist, 1989, reproduced with permission from *Elsevier Science Publishers*)

A subglacial origin would appear from all the prevailing evidence to be the most likely. Several variants exist (a) meltout from basal ice layers and subsequent selective subglacial deformation causing folding and intercalation of stratified sediment (Shaw, 1979); (b) shear stacking of deformable debris layers beneath an active deformable bed-type ice mass in which an intimate relationship exists between the shear stacking process and the underlying topography (Fig. 2.27) (Bouchard, 1986, 1989; Boulton, 1987a); (c) formation within a fluctuating basal thermal regime where as colder conditions evolve, compressive flow leads to differential loading of underlying subglacial debris causing a ribbed morphology to develop (Sollid and Sørbel, 1984) and (d) subglacial meltwater floods creating transverse ripples on the underside of the ice mass sole akin to those produced on the underside of river ice (Fisher and Shaw, 1992).

However, none of these hypotheses alone would appear to satisfy all the morphological, spatial and sedimentological constraints. Lundqvist (1989) suggests, as did Mannerfelt (1945), that a multiple origin is likely. It still needs to be clarified whether Rogen moraine ridges were first formed subglacially then modified by subsequent supra- and proglacial processes, or that the ridges were formed supra- or proglacially and later altered subglacially by overrunning active ice. Finally, the relationship with drumlins and fluted moraines must be addressed (Markgren and Lassila, 1980). No hypothesis which ignores this spatial and, possibly, generic relationship can be easily accepted.

2.8.2.3. Drumlins

Drumlins are roughly ovoid shaped hills of dominantly-glacial debris that typically occur within groups or fields of several thousands. Drumlins exhibit very strong *en echelon* long axis preferred orientation paralleling the main direction of ice flow (Menzies, 1979b, 1984). Isolated drumlins are known to exist and occasional fields may contain only a few dozen drumlins. The classical shaped drumlin usually has a steeper stoss end and a tapered lee-side, however variants on this shape are perhaps more common than the classical shape itself (see Plate 2.10 in colour plates). Drumlins may range in height from 5–200 m, in width 10–100 m and in overall length from 100 m to several kilometres. An extensive statistical analyses of drumlin dimensions has been pursued from which several explanations of drumlin origin have been derived (Chorley, 1959; Heidenreich, 1964; Smalley and Unwin, 1968; Barnett and Finke, 1971; Trenhaile, 1971, 1975; Crozier, 1975; Mills, 1980, 1987; Evans, 1987). A narrow standard deviation in drumlin dimension ratios is typically encountered and a random spatial location for individual drumlins is often observed (Smalley and Unwin, 1968; Crozier, 1975; Mills, 1980, 1987; Evans, 1987). Few drumlins have been observed appearing from beneath modern day ice masses.

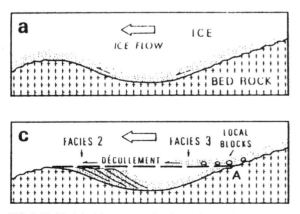
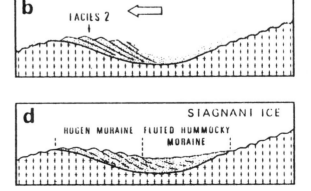

FIG. 2.27. Model of Rogen moraine formation. (After Bouchard, 1989, reproduced with permission from *Elsevier Science Publishers*)

(a)   (b)

PLATE 2.10. (a) Drumlin of the Green Bay Lobe, Wisconsin. (Photo courtesy of Donna Stetz). (b) Drumlins from Kuusamo drumlin field, eastern Finland. (Photo courtesy of Risto Aario)

PLATE 2.11. Drumlin in the proglacial zone of the Biferten Glacier, Switzerland. Photo shows drumlin width to be approximately 150 m

Drumlins on James Ross Island, Antarctica (Rabassa, 1987), in the proglacial zones of Myrdalsjökull, Iceland (Krüger, 1987) and the Bifertensgletscher, Switzerland (Meer, 1983) have been observed but none, as yet, emanating from beneath the Greenland or Antarctic Ice Sheets (see Plate 2.11 in colour plates). Vast drumlin fields, numbering in the thousands exist, for example, in Canada, Estonia, Finland, Ireland, Germany, Poland, Russia, and the USA. In almost all glaciated terrains smaller drumlin fields are to be found.

The topographic locations within which drumlins are found are many and varied (Menzies, 1979a, b). Drumlins occur in both lowland and highland terrains beneath ice sheets and valley glaciers. They occur close to terminal moraines and may, in places, appear contiguous with these moraines, while elsewhere drumlins occur on the edge of ice sheet centres (Fig. 2.9) (Shilts et al., 1987, 1989). Both Smalley and Unwin (1968) and Menzies (1981) have suggested that beneath an ice sheet certain areas might preferentially be conducive to drumlin formation (Fig. 2.28) (Boulton et al., 1977, p.243). Occasionally, a radiating pattern can be observed within a drumlin field (Fig. 2.29) (Alden, 1911; Fairchild, 1911; Wright, 1962; Glückert, 1973; Krall, 1977; Goldstein, 1989) that has been interpreted, in the past, as evidence of basal crevasse infilling due to divergent ice flow close to an ice margin. Recently, it has been suggested that drumlins, in association with Rogen moraine and fluted moraines, may be related to deformable beds beneath ice sheets and are, therefore, linked to fast basal ice (> 500 m yr^{-1}) and a preferential location within ice streams in ice sheets (Shilts et al., 1987; Dyke and Morris, 1988; Menzies, 1989a). Limited relationships appear to occur between drumlins and topographic effects although in certain fields some, perhaps causal relationships do appear to exist.

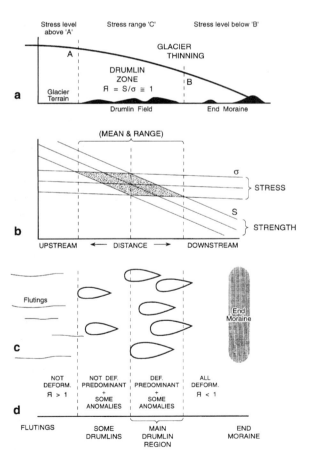

FIG. 2.28. Schematic diagram of general conditions for drumlin formation following the dilatancy theory. (Modified from Smalley, and Unwin, 1968; and Piotrowski and Smalley, 1987)

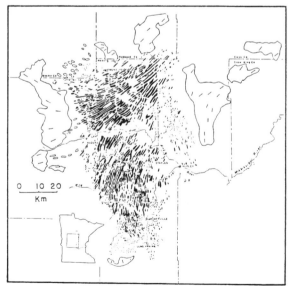

FIG. 2.29. Radial pattern of drumlin distribution in the Wadena drumlin field, Minnesota. (After Goldstein, 1989, reproduced with permission from *Elsevier Science Publishers*)

Drumlins are, such striking landforms that for many years little cognisance was taken of their internal composition. Drumlins are composed of a vast range of sediment types of varied provenance, containing an array of sediment structures and forms (see Fig. 2.30 in colour plates). In the past, drumlins were mistakenly perceived as being composed almost exclusively, of subglacial tills. Although the dominant sediment within drumlins is till, many other drumlins may contain a dominance of stratified sediment. Stratified sediment may compose the internal materials of whole drumlin fields, as in Velva, North Dakota (Lemke, 1958) or Livingstone Lake, Saskatchewan (Shaw and Kvill, 1984) with only a few isolated till intraclasts. While individual stratified drumlins may 'sit' adjacent to till drumlins as in Peterborough, Ontario (Sharpe, 1987). The range of sediment types and structures found within drumlins is illustrated in Figure 2.30 (see colour plates). Many drumlins have observable cores of bedrock, boulder dykes and other non-glacial nuclei around which subglacial debris has accreted or been emplaced by some mechanism (Barkla, 1935; Björnsson, 1953; Kupsch, 1955; Krüger, 1969; Glückert, 1973; Aario et al., 1974; Birch and Trask, 1978). In some cases, drumlin or drumlinoidal forms can be observed 'carved' from bedrock in the form of roc-drumlins (Muller, 1963; Parizek, 1964; Svensson and Frisen, 1964; Laverdière and Dionne, 1969; Finch, 1977; Menzies, 1981a). However, most drumlins do not appear to have obvious

cores around which they have been 'built' and these forms remain puzzling to explain.

Many researchers have investigated the clast fabrics found within drumlins (Hoppe, 1951; Wright, 1957, 1962; Harris, 1967; Sauter, 1967; Andrews and King, 1968; Hill, 1968, 1971, 1973; Savage, 1968; Roberts and Mark, 1970; Evenson, 1971; Johansson, 1972; Shaw and Freschauf, 1973; Walker, 1973; Aario et al., 1974; Rõuk, 1974; Kurimo, 1974; Heikkinen and Tikkanen, 1979; Aario, 1977a,b; Krüger and Thomsen, 1981; De Jong et al., 1982; Krüger, 1987; Piotrowski and Smalley, 1987; Rabassa, 1987; Goldstein, 1989; Stea and Brown, 1989; Bluemle et al., 1993). Clast fabrics appear, in some cases, to follow the outer morphology of the drumlin (Walker, 1973), while others exhibit transverse orientations (Andrews and King, 1968); or a 'herring-bone' style pattern (Shaw and Freschauf, 1973; Aario, 1977a). In many cases the complexity of internal sedimentological structures provides a random orientation. It can be questioned as to whether clast fabrics within drumlins have anything other than local significance and may indicate little concerning the origin of the drumlin form or its initiation (Menzies, 1979a, 1987).

Drumlins exhibit such a wide complexity of form and internal composition that it is almost impossible to characterise what is an 'ideal' drumlin. Many drumlins, for example, are found lying on top or obliquely across other larger drumlin forms (mega-drumlins) (Rose and Letzer, 1977). Drumlin shapes may vary enormously and may reflect formative penecontemporaneous processes or simply post-depositional subaerial mass movement (Fig.2.31) (Alden, 1918; Glückert, 1973, Shaw and Kvill, 1984; Rõuk and Raukas, 1989; Bluemle et al., 1993). Many drumlin fields progressively change as part of a continuum of bedforms thus drumlin genesis would appear tied, in those instances, to subglacial environments conducive to Rogen and fluted moraine formation.

The question of drumlin formation has attracted a vast array of research work (Menzies, 1984) (Table 2.8). Before discussing this most definitive of all 'glacial' questions, it is pertinent to state the 'conditions' that must be met by any hypotheses attempting to explain drumlin formation, assuming that a single explanation does exist for such a diverse bedform type (cf. Menzies, 1979a). Any explanation of drum-

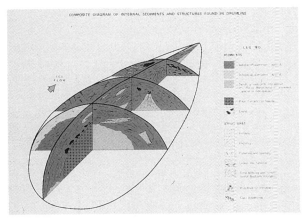

FIG. 2.30. General model of internal sediments and structures found within drumlins

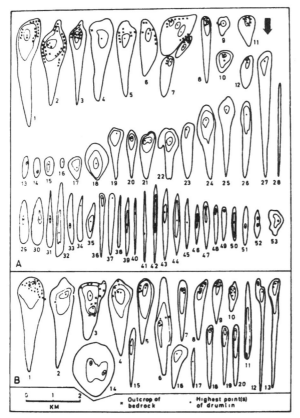

FIG. 2.31. Examples of drumlin morphology from the Pieksämäki (A) and Keitele (B) drumlin fields, Finland. (Reprinted from Glückert, 1973, reproduced by permission of the Geographical Society of Finland)

TABLE 2.8 Hypotheses of Drumlin Formation*

1.   Erosion of pre-existing glacial materials

2.   Deposition *en masse*

3.   Accretion/deposition
    (a) with a pre-existing rock/boulder/sediment core
    (b) with a 'generated' core of sediment/boulders
        kinematic fluting
        porewater dissipation
        freezing front effects
        dilatancy
        thrust-block – sediment emplacement

4.   Subglacial bed deformation

5.   Subglacial meltwater processes
        erosion mark – cavity infilling
        erosion of pre-existing sediment
        combined effects of cavity infilling and erosion of
    pre-existing sediment

* modified from Menzies, 1984

single continuous event or interrupted accretionary events; and, finally, (9) a 'trigger' mechanism(s) needs to be found that is operative in certain specific conditions yet not under others.

At present, three main groups of drumlin forming hypotheses can be identified:

(1) Formation by moulding of previously deposited material within a subglacial environment in which a limited amount of subglacial meltwater activity occurs (Q-Bed state possibly where a frozen bed transforms to a melted bed). Meltwater may influence moulding and deformational processes that are subsequently produced by acting either as a lubricating basal film at the upper ice/bed interface or as debris held porewater thereby reducing subglacial effective stresses (Whittecar and Mickelson, 1979; Krüger and Thomsen, 1984; Boulton, 1987a; Krüger, 1987). Debris is moulded by direct deformation of previously deposited sediment (both glacial and nonglacial) into drumlinoidal shapes by direct basal ice contact following some type of smearing on or sculpting process(es).

(2) Formation due to anisotropic differences in the subglacial debris (under dominant M-Bed conditions) due to: (a) dilatancy (Smalley, 1966b; Smalley and Unwin, 1968; Crozier, 1975; Carl, 1978; Whittecar and Mickelson, 1979; Markgren and Lassila, 1980;

lin formation needs to account for: (1) the varied location of drumlins and their close association in fields; (2) the diverse shape and morphology of drumlin form; (3) the enormous range in sediment type and structures within drumlins; (4) the existence of rock-cored and non-rock cored drumlins, often in proximity to each other; (5) the presence of drumlins in bedform continua in some but not all cases; (6) the relationship of drumlins to subglacial glaciodynamics and hydraulics; (7) the chronology of drumlin formation whether drumlins form simultaneously as a single field or develop into a field by repeated 'overprinting' in a single glacial phase or repetition over several glacial phases; (8) stages of drumlin development whether formed en masse or by gradual accretion in a

Smalley and Piotrowski, 1987) ; (b) porewater dissipation (Boulton, 1972a; Menzies, 1979a, 1989; Whittecar and Mickelson, 1979); (c) localised freezing (Baranowski, 1969, 1970, 1977, 1979; Menzies, 1989a); (d) localised helicoidal basal ice flow patterns (Shaw and Freschauf, 1973; Aario, 1977a,c; Car, 1978; Shaw, 1977b; Heikkinen and Tikkinen, 1979; Seret, 1979); or (e) localised subglacial debris deformation (Menzies, 1984, 1987, 1989d; Aario, 1987; Boulton, 1987a; Rose, 1987a; Alley, 1991). Within this specific group meltwater activity is of limited impact, whereas porewater in debris is considered critical in mobilizing or immobilizing local bed debris. The debris is not usually considered as being in a state of mobilized deformation except in case (e). It is suggested that changing stress field and/or stress/ strain histories due to transient basal glaciodynamics locally affecting subglacial debris rheology are the important parameters in determining whether drumlins begin to form or not. The normal anisotropy within subglacial debris may influence effective stress levels due to fluctuating debris rheology, the result of internal porewater changes. This effect, discussed in more detail below, may control both the initiation process and later moulding and morphological evolution of subsequent bedforms.

(3) Formation due to the influence of active basal meltwater (under H-Bed conditions) carving cavities beneath an ice mass and later infilling with assorted but predominantly stratified sediment or by the subglacial meltwater erosion of already deposited sediment at the upper ice/bed interface. This hypotheses stems from the presence of stratified sediments in drumlins either in part, in lee-side positions (Dardis and Mccabe, 1983, 1987; Dardis et al., 1984; Dardis 1987; Hanvey, 1987, 1989; Sharpe, 1985, 1987 1988a), or through the entire drumlin (Shaw, 1983b Shaw and Kvill, 1984; Shaw et al., 1989). or the sculpting by fluvial processes of previously deposited sediment (Shaw et al., 1989). This hypothesis demands meltwater flow at catastrophic discharges from beneath certain areas of an ice mass across the upper ice/bed interface yet permitting the overall ice mass to remain glaciodynamically stable (for detailed calculations see Shaw et al., 1989). This form of drumlin development, as with the hypothesis in (1) requires a two-stage process of initiation, beginning

first with either a pre-formed cavity or pre-existing sediment at the upper ice /bed interface. The latter stage need not be linked directly to the former stage therefore in some cases although conditions may be suitable for initiation for the first stage, the second stage may not continue toward the critical point (trigger) of drumlin development.

In all of these hypotheses the conditions at the subglacial interface(s) are the key to subsequent drumlin formation and, in the long term, to drumlin 'survival'. A complex relationship must exist between basal glaciodynamics, subglacial sediment rheology and hydraulics for any particular area of ice bed. Fluctuations in state or stress levels or meltwater production and pathways will affect all other parameters to some degree. Therefore, the nature and form of the subglacial interface(s) eventually reflects these parameter variations (Boulton and Hindmarsh, 1987; Clarke, 1987; Lingle and Brown, 1987; Alley, 1989a,b; Menzies, 1989a,b). Whatever changes that do occur may be areally widespread or limited and take place rapidly or slowly. Certain changes may cross critical thresholds that cannot be reversed, while others may exhibit varying degrees of hysteresis. The likelihood or otherwise of subglacial conditions occurring in any or all of these hypotheses remains one of the fundamental research problems of glacial geo(morpho)logy.

2.8.2.4. Fluted moraine

In certain locations it is difficult to distinguish fluted moraine from extremely elongated drumlins (Plate 2.12). A close relationship does appear to exist between these bedforms but whether such a relationship exists in all instances remains debateable. Fluted moraines, on average, are long ridges of dominantly-glacial sediment 100–500 m in length, 1–3 m in width and less than a metre to 2 m in height with a transverse spacing of 0.5–1.5 m (Table 2.9) (Dyson, 1952; Hoppe and Schytt, 1953; Schytt, 1963; McPherson and Gardner, 1969; Anderson and Sollid, 1971; Boulton, 1971; Czerwinski, 1973; Paul and Evans, 1974; Lawson, 1976 ; Morris and Morland, 1976; Åmark, 1980). Exceptions do occur where flutes may be much larger in all dimensions and stretch for several kilometres. Flute ridges usually

(a)　　　　　　　　　　　　　　　　　　　　(b)

PLATE 2.12. (a) Fluted moraine in the proglacial zone. (b) Large flute extending from a boulder (approximately 1.5 m in height). Note other flutes adjacent to central flute. Both photographs from proglacial zone of Storbreen, Norway

TABLE 2.9 Maximum Height (m) of Flutes Observed by Various Authors

Paul and Evans (1974)	Blomstrandbreen, Spitsbergen	0.2	Silt and fine sand with pockets of coarse sand and gravel
		1.0	Till
Boulton and Dent (1974)	Breiðamerkurjökull, Iceland	0.5	Till, $c = (3-8) \times 10^3$ N m^{-2}, $0 = 27°$
Morris and Morland (present study)	Breiðamerkurjökull, Iceland	0.67	Undrained till, $c = (4 + 2) \times 10^3$ N m^{-2}
Ray (1935)	Mendenhall Glacier, Alaska	0.075	Gravel
Grant and Higgins (1913)	Petrof Glacier, Alaska	0.45	Gravel
Todtmann (1952)	Bruarjökull, Iceland	2.0	
Hoppe and Schytt (1953)	Bruarjökull, Iceland	1.0	Finest material, sand and finer, on crests of ridges
	Isfallsglaciären, Kebnekajse	0.45	Unsorted material, all grades from fine clay to large boulders
Bogacki (1973)	East of river Sandgigjukvisl, Iceland	1.0	Piles of stone and rubble
Kozarski and Szupryczynski (1973)	Sidujökull, Iceland	1.3	Stone layer at top, mainly pebbles 0.2 m diameter strongly compacted moraine deposit, tabular structure thin layer very strongly pressed slate-like material, moraine with numerous cobbles and pebbles
Baranowski (1970)	Werenskioldbreen, Spitsbergen	0.3	abundance of silty material
Dyson (1952)	Grinnell Glacier/Sperry Glacier, Montana	0.9	Largest and most pronounced ridges on moraine which contains a relatively high proportion of rock flour
Shaw and Freschauf (1973)	Athabasca, Alberta	20.0	Coarse till (these features may not be flutes in our sense, however)

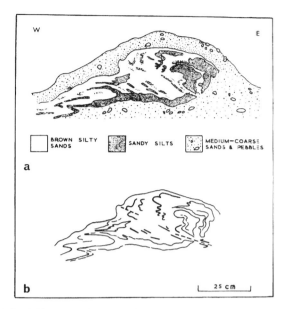

a

b

FIG. 2.32. (a) Sediments within fluted moraine from Blom-strandbreen, north-west Spitsbergen. (b) Schematic representation of structures observed in (a). (After Paul and Evans, 197–; reproduced by courtesy of the *International Glaciological Society*)

exist in groups paralleling the main direction of ice flow and may, as discussed above, be part of a continuum of bedforms. The ridges are usually symmetric in cross-section but often mass wasting alters the flanks where meltwater from the glacier snout has undercut ridge flanks. In a some places mega-flutes have been observed (Gravenor and Meneley, 1958; Shaw, 1975; Moran *et al.*, 1980; Rose, 1987a, 1989b; Shilts *et al.*, 1987).

Flutes may occur in front of a drumlin belt (Miller, 1972; Stahman, 1992), in places on top of drumlins (Rose, 1989b), occasionally on the surface of Rogen moraine (Bouchard, 1989), or as isolated bedforms in the proglacial zones of present-day glaciers (McPherson and Gardner, 1969). Typically, flutes exhibit a strong preferred uni-directional orientation but occasionally a radiating pattern may occur.

Fluted moraine may be composed of a wide variety of glacial sediments and often a structureless or massive inner core can be observed. In other places, deformed sediments, with folding and faulting, have been described (Fig. 2.32) (Paul and Evans, 1974).

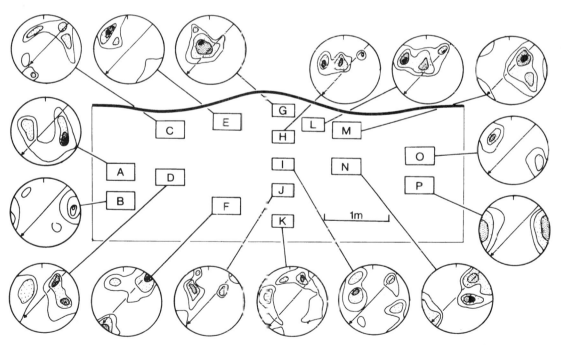

FIG. 2.33. Clast fabrics, plotted on Schmidt equal-area projection. from flute-forming till from beyond the margin of Breiðamerkurjökull, Iceland. Arrow gives the direction of ice movement and of flute crest. 2σ contour interval. (After Boulton, 1976; reproduced by courtesy of the *International Glaciological Society*)

Sediment grain size, typically, coarsens both toward the surface of individual ridges and, in present proglacial areas, in a down-valley direction (Lawson, 1976). In many cases ridge surfaces are covered with a gravel lag deposit. Inter-flute troughs may contain glaciofluvial sediments, and meltout and supraglacial flow tills.

Clast fabrics within flutes exhibit strong orientations usually paralleling the ridge long axis (Fig. 2.33) (Anderson and Sollid, 1971; Paul and Evans, 1974; Boulton, 1976c; Lawson, 1976; Åmark, 1980; Rose, 1987, 1989).

Hypotheses on the origin of fluted moraine include those related to drumlins, basal crevasse infilling, surface crevasse infilling and collapse, localised

subglacial frozen patches, subglacial tectonisation, and simple lee-side ridges in the down-ice side of large boulders or similar obstructions (Fig. 2.34) (Hoppe and Schytt, 1953; Galloway, 1956; Baranowski, 1969; McPherson and Gardner, 1969; Anderson and Sollid, 1971; Paul and Evans, 1974; Shaw, 1975; Boulton, 1976c; Morris and Morland, 1976; Aario, 1977a, c; Åmark,1980; Moran et al., 1980; Jones, 1982). Since fluted moraine is only a morphological term, the possibility of multiple origins is very likely.

Flutes would appear to be divisible into three significant groups (1) large flutings related in origin to drumlins and Rogen moraines; (2) flutes related to active subglacial processes close to the margins of ice

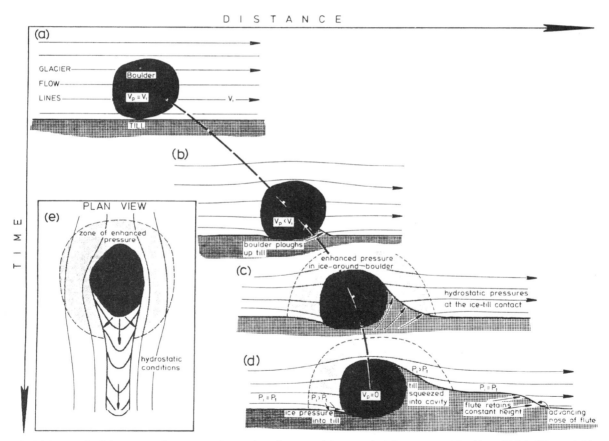

FIG. 2.34. A graph of time versus distance superimposed on diagrams of (a) an englacially transported boulder, which is (b) retarded by ploughing into the till, and which (c) subsequently develops a wedge of till on its lee side. This till can (d) be traced into a flute. Diagram (e) shows a plan view of diagram (d). (After Boulton, 1976; reproduced by courtesy of the *International Glaciological Society*)

masses; and (3) small flutes formed at the subglacial/ proglacial margin where saturated debris is squeezed into small cavities formed at the base of thin marginal ice by boulders or empty meltwater conduit.

2.8.3. Non-bedform subglacial landforms

It is generally accepted that many subglacial landform features are the products of non-bedforming processes that are related to specific topographic, geological and other non-glacial controlling variables. Landforms such as De Geer moraines and Crag and Tails are major examples of these types of subglacial forms.

2.8.3.1. De Geer moraines

These moraines form in the subglacial/subaquatic proximal environment close to either local grounding lines or basal crevasses fractures. These moraines are also described as Cross-Valley and Washboard moraines. De Geer moraine ridges are, typically, transverse to the main ice direction often with a slightly arcuate down-ice plan form with an asymmetric cross-section, the distal slope tending to be the steepest. Ridges are of highly variable height, length and width but, typically, are <5 m in height, ≈50 m wide and laterally may extend as a continuous ridge for <1 km (Elson, 1957; Hoppe, 1959; Andrews, 1963; Zilliacus, 1976, 1981, 1989; Sollid and Carlsson, 1984). Ridges usually occupy a low point in the terrain. De Geer moraines may, in many cases, be seasonal in formation. Zilliacus (1981, 1987, 1989) has reported a regular alternation between larger annual and interannual ridges.

In glaciated fjord-type terrain they are found as a series of ridges across the fjord valley floor (Fig. 2.35a) and in more open terrain may exist as contiguous ridges stretching over many kilometres paralleling successive ice marginal positions (Fig. 2.35b). De Geer (1889) was the first to recognise these ridge moraines and thought they must have formed in front of the ice. Later others described De Geer moraines as forming in 'calving bay' locations (Strömberg, 1971, 1981; Sollid and Carlsson, 1984).

Since De Geer moraines form at the boundary between the subglacial and subaquatic environments, they form in an area of sediment concentration both in transit and deposition. These moraines are, therefore, composed of a vast array of sediment types but with a dominance of stratified sediment units and often

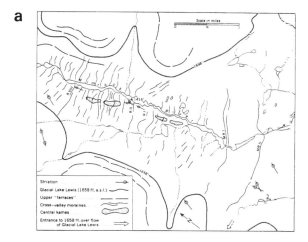

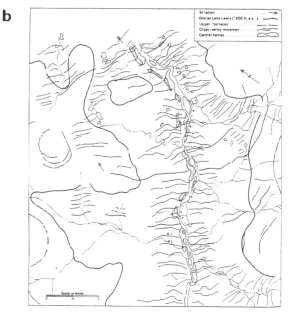

FIG. 2.35. (a) Map of cross-valley moraines in the Rimrock area, Baffin Island, Canada. (Reprinted from *Geographical Bulletin*, Vol 19, 1963) (b) Map of cross-valley moraines in the Isortoq area, Baffin Island, Canada. (Reprinted from *Geographical Bulletin*, Vol. 19, 1963)

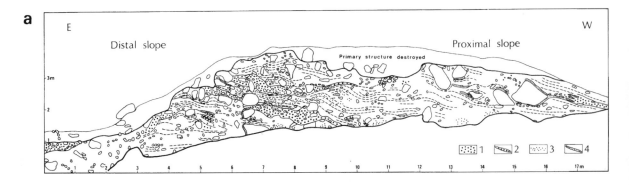

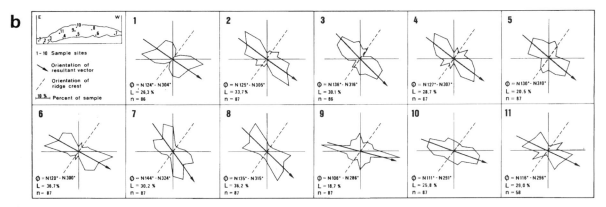

FIG. 2.36. Cross-section through a De Geer moraine showing internal structures and macrofabrics. 1, Gravelly till; 2, lenses of gravel; 3, sand/silt; 4, lenses of sand/silt. The section consists mainly of sandy till. Fabrics of pebbles (2–6.4 cm) from the cross-section. 20° class intervals, n = number of particles. Resultant vector (θ) and vector magnitude (L). (Reproduced by courtesy of the *Societas Upsaliensis pro Geologica Quaternaria*, from *Striae* 20)

close association to meltwater deposits such as eskers (Fig. 2.36). The sediments usually contain rafted intraclast units but often do not exhibit evidence of glacial tectonisation (Zilliacus, 1989). Clast fabrics within the moraines usually exhibit a strong preferred orientation normal to the ridge crest with proximal fabrics showing up-ice plunges and the distal fabrics indicating down-ice orientations (Sollid and Carlsson, 1984; Zilliacus, 1989). Zilliacus suggests that moraine fabrics developed at random revealing a slope-conformable pattern.

De Geer moraine formation would appear to occur within the subglacial/subaquatic grounding line zone only the mechanism of formation is in debate. All hypotheses of moraine formation tend to concede that deposition has occurred at the grounding-line, only the mechanism of transport to that position is in conten-

tion. It has been suggested that debris concentration is the result of englacial thrust zones (Elson, 1957), or shear planes (Andrews, 1963), or near-marginal basal crevasses (Holdsworth, 1973; Sollid and Carlsson, 1984; Zilliacus, 1987, 1989). Zilliacus (1989) has presented a model of De Geer moraine formation as illustrated in Figure 2.37. In this model, it is suggested that basal crevasses transverse to the main ice direction may be a function of basal ice fracture the result of surging conditions. The debris is then squeezed into these crevasses from several directions and thus fabrics and sediments exhibit a wide range and variety. A significant conclusion from Zilliacus' work demonstrates that De Geer moraines are not annual moraines but instead merely reflect the temporary position of a nearby dynamic grounding-line and thus have limited geochronological significance.

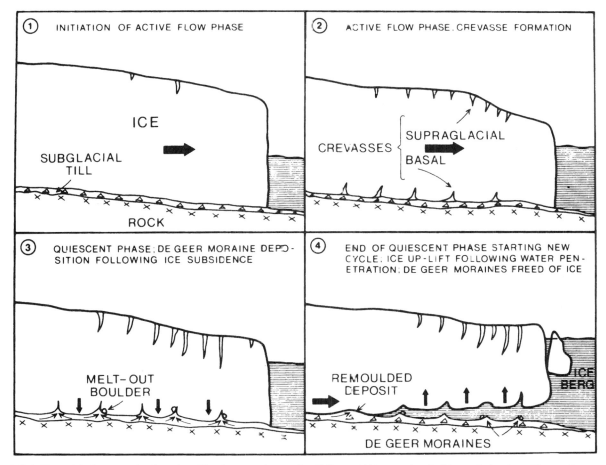

FIG. 2.37. Model of the genesis of a series of De Geer moraines (After Zilliacus, 1989, reproduced with permission from *Elsevier Science Publishers*)

2.8.3.2. Crag and tail forms

The distinction between crag and tail forms and fluted moraine forms, developed in the lee of boulders, is one possibly only of scale. The formative mechanisms of both forms would appear to be similar (Sugden and John, 1976). The characteristic crag and tail may be tens of metres high at the stoss end and may stretch for several kilometres in the down-ice direction. The classic form is well represented by the crag and tail that Edinburgh Castle and the Royal Mile in Edinburgh, Scotland sits upon (Fig. 2.38). In many cases the crag is composed of bedrock while the tail may contain a variety of dominantly-glacial sediments often of a glaciofluvial nature. However, as in the case

of the form in Edinburgh, the bedrock protrusion that represents the crag has been moulded in the down-ice direction reflecting a large bedrock-cored tail. Crag and tails are typical forms in terrain where resistant but isolated up-standing bedrock obstructions have been overridden by active ice. Therefore in areas where extinct volcanic cores, reefs or igneous intrusions such as dykes have occurred crag and tails of a range of sizes generally are to be found.

The moulding and erosive effect of active ice was thought, in the past, to be the formative mechanism involved in crag and tail development. However, usually associated with the crag is an up-ice crescentic hollow and hollows along the sides of the tail that may represent the additional impact of subglacial meltwater

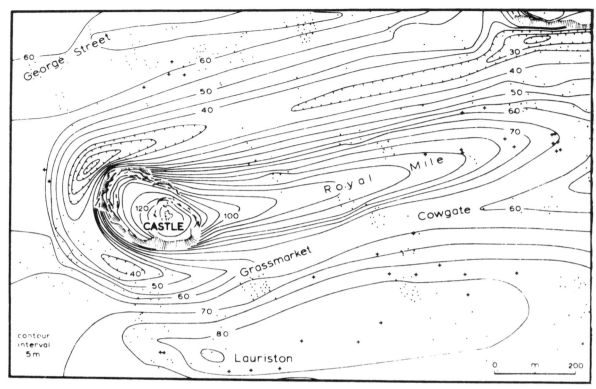

FIG. 2.38. Crag and tail shown by rockhead contours, Edinburgh, Scotland (ice flow direction from left to right). (After Sissons, 1976)

under high hydrostatic pressures (Fig. 2.51). The crag and tail form is a morphological expression of active ice motion past a large subglacial bed obstacle and, therefore, is both erosional and depositional in development. Where the tail contains substantive debris a combination of ice moulding at the stoss end and cavity infilling in the lee-side due to pressure reduction can be expected. Subglacial conditions under these circumstances possibly are typical of Q-bed states. Where the form is a bedrock flute with limited lee-side deposition, the crag and tail is perhaps more accurately explained as a erosional form developed under high subglacial hydrostatic meltwater pressures in association with H-bed conditions.

2.9. SUBGLACIAL EROSIONAL FORMS UNDER ACTIVE ICE

Subglacial erosion processes are pervasive at the active ice/bed interface. In some terrains under certain subglacial conditions, erosive processes become dominant (Menzies, 1995a, Chapter 7). Almost any description of the impact on the land surface of past glaciers invariably notes, for example, the grandeur and size of fjords and the rugged sculpted bedrock features of once glaciated terrains. However, our understanding of how these distinctive features were fashioned remains limited.

Erosional forms exist at an immense range of scales from the surface microscopic features of particles composing glacial debris (Chapter 13) to the mega-scale landforms of the District of Keewatin in Canada's North West Territories.

Regional erosional features can be subdivided, for the purposes of discussion, into regional and local forms. However, some forms, such as roche moutonées, can be found at all scales, illustrative of the pervasiveness of most erosional wear processes (Menzies, 1995a, Chapter 7). Regional erosional features occur as either areal (spatially pervasive) or

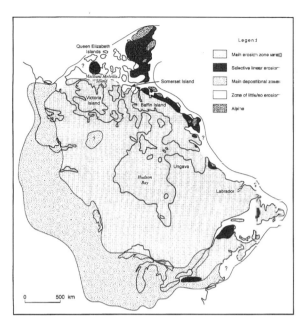

FIG. 2.39. Main glacial landscape zones, Laurentide Ice Sheet. Compiled from Landsat-1 images and topographic maps. (After Sugden, 1978; reproduced by courtesy of the *International Glaciological Society*)

2.9.1. Regional Areal Erosion

Areas of terrain where regional areal erosion has dominated the landscape, typically, exhibit low relief amplitude, limited sediment deposition and are dominated by a moulded and scoured appearance (Plate 2.13). Geological structure has often been partially exhumed in these terrains and irregular depressions and small roche moutonnées are common. Linton (1963) described these terrains as 'knock and lochan' since the landscapes of northwest Scotland are typical of this form of glacial erosion. Similar landscapes exist in shield terrains in Canada and Fenoscandia, along the edges of the Greenland and Antarctic Ice Sheets, in Patagonia and the South Island, New Zealand.

The formation of this regional erosional landscape would appear to be related to relatively slow moving ice masses under H-bed conditions with limited debris present. Zones with these characteristics within ice sheets may be present either close to ice sheet centres or in thinner marginal areas. A range of wear processes appear to have operated across a dominantly bedrock surface where occasional protuberances have led to pressure melting of the ice and discharges of meltwater often under high hydraulic pressure heads. Within this landscape, at a lower scale, crag and tails, roche moutonnées and grooves are prevalent (see below). Typically P-forms and associated forms related to rapid subglacial meltwater flow are also common. These forms are often correlated to changes in local bedrock topography and structure.

2.9.2. Regional Linear Erosion

Beneath specific zones within ice sheets, regional linear erosion appears to occur probably as a consequence of ice streaming and related fast moving, but spatially-restricted, basal ice. Under these conditions, major linear forms of glacial erosion appear to develop, for example, incised bedrock troughs, fjord valleys and tunnel valleys.

2.9.2.1. Bedrock troughs and fjord valleys

Deeply incised bedrock troughs and coastal fjords are both spectacular landscape forms indicative of intense

linear (spatially discrete) landscape types. In the former case, an areally scoured terrain develops where limited debris existed at the ice/bed interface and H-bed conditions prevailed. In contrast linear erosion processes, by definition, are differential in their impact upon terrain being confined within specific areas. Linear erosional forms are indicative of subglacial bed states in which rapid but spatially-restricted basal ice movement and/or meltwater channelling has occurred. Sugden and John (1976) illustrate this regional subdivision of landscapes with a classification for Canada (Fig. 2.39).

Alpine-type landscapes are recognised by Sugden and John (1976) as a third and separate erosional landscape. However, the degree to which subglacial erosional processes contribute to such a landscape type is restricted to valley bottoms, long-profiles, riegels and valley sides as far up as the trimlines (cf. Gurnell and Clark, 1987; Gerrard, 1990; Kleman, 1994).

PLATE 2.13. Areally scoured glaciated terrain in the Canadian Shield. Note the tracery of bedrock faults and fractures. Lac Troie area, Nouveau-Québec, Québec, Canada. (Photograph from the Government of Canada)

highland glaciation (McGee, 1883, 1894; Sugden, 1978). These, almost archetypal forms of linear erosion are typically deep, parallel-sided valleys cut in bedrock that have short straight sections and often several deep basins (Syvitski *et al.*, 1987) (Fig. 2.40). The comparison between troughs and fjords, however, although similar in many respects, differs in a few critical aspects; bedrock troughs usually have a stepped longitudinal profile (riegels) (Bakker, 1965; King, 1970; Röthlisberger and Iken, 1981) and relatively few transverse sections, while fjords may have one or several bedrock sills or thresholds (Fig. 2.41) (Gjessing, 1966; Holtedahl, 1967; Clague and Bornhold, 1980; Roberts and Rood, 1984; Nesje and Whillans, 1994) and typically many transverse valleys (Plate 2.14) (see also colour plates).

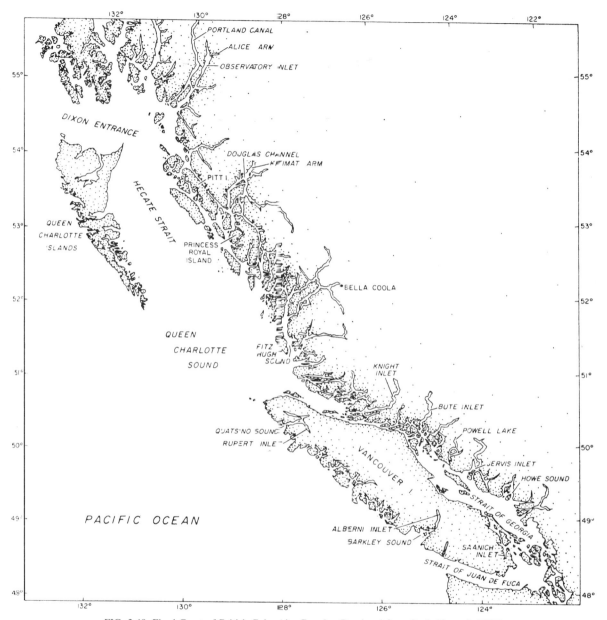

FIG. 2.40. Fjord Coast of British Columbia, Canada. (Reprinted from Syvitski *et al.*, 1987)

Both forms are indicative of intense glacial down-cutting and must be related to specific zones or lineations of fast ice movement either as outlet glaciers or ice streams within ice sheets. It has been suggested in the past that the location of both forms must in some way be related to bedrock influences and control specifically in the form of faults, shatter and mylonized zones (Gregory, 1913; Randall, 1961; Nicholson, 1963; Bird, 1967; Holtedahl, 1967; Nilsen, 1973; Roberts, 1974; England, 1985; Augustinus, 1992; Nesje and Whillans, 1994). However, what becomes apparent on comparing geologic structure

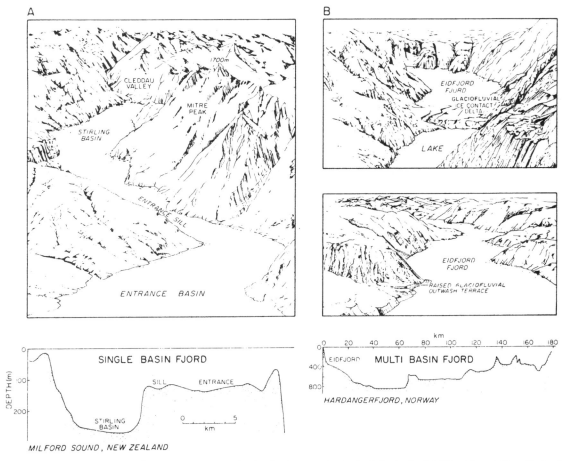

FIG. 2.41. Schematic presentation of two of the world's 'classic' fjords, (a) a simple single basin fjord, Milford Sound, New Zealand; (b) a complex multi-basin fjord, Hardangerfjord, Norway. (From Holtedahl, 1975; reproduced by permission of the Geological Survey of Norway)

and tectonic zonation is that although geological control must have a strong influence upon trough orientation there is no evidence that this influence is anything other than secondary. Rather, what has emerged is that pre-glacial geomorphic control in the form of pre-existing landscape dissection of pre-glacial topography determines to a greater degree the topographic lows that localised fast-moving ice masses may begin to follow and exploit eventually developing into troughs (Zumberge, 1955; Rudberg, 1954, 1973; Gjessing, 1966; Blake, 1978; Ryder, 1981; England, 1986). A final comment upon geologic structure and linear erosion must, however, recognise that inherent variation in rock mass strength

and 'erodibility' permits preferential removal of certain lithologies, crystalline altered forms, weathered rocks, deformed, fissured, faulted and fragmented rock types (Matthes, 1930; Lewis, 1954; Tricart and Cailleaux, 1962; Linton, 1963; Sugden, 1974; Addison, 1981; Whitehouse, 1987; Harbor et al., 1988; Augustinus and Shelby, 1990; Augustinus, 1992; Nesje and Whillans, 1994).

These trough are often described as U-shaped (Fig. 2.42). This characteristic cross-profile shape is not found in all troughs and in others sediment infill has disrupted the original cross-profile. However, it has long been held that an explanation of this typical cross-profile shape may hold a key to the under-

(a)

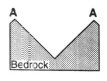

Unglaciated Valley
A-A Unglaciated Valley

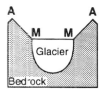

**Maximum Vertical
Ice Extent**
A-A Glacial Valley
M-M Active Glacial Channel
M-M Zone of Glacial Influence

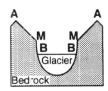

**Ice Extent
Less Than Maximum**
A-A Glacial Valley
M-M Zone of Glacial Influence
B-B Active Glacial Channel

(b)

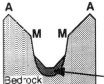

After Deglaciation
A-A Glaciated Valley
M-M Zone of Glacial Influence
Talus and Alluvium

FIG. 2.42. Schematic cross sectional evolution of a valley due to glacial erosion. (From Harbor, 1992; reproduced by permission of the *author*)

(c)

PLATE 2.14. (a) Grenville Channel (fjord), British Columbia. (Photo courtesy B.C. Ferries). (b) Princess Louise Inlet (fjord), British Columbia. (Photo courtesy of W.H. Wolferstan). (c) Princess Louise Inlet (fjord) and Jervis Inlet (top left). (Photo courtesy of W.H. Wolferstan)

standing of the formation of glacial troughs. A comparison with fluvially-eroded valley cross-profiles has been made to some degree in that it has been suggested that the volume of ice moving through a glacial trough should be directly related to the drainage or ice evacuation area up-ice of the trough (see Plate 2.15 in colour plates) (cf. Haynes, 1972; Sugden, 1978; Roberts and Rood, 1984). Roberts and Rood (1984), using data from British Columbia, have compared fjord length, width and depth with ice contributing area and found a significant relationship between fjord length and ice contributing area ($r^2 = 0.81$) (Fig. 2.43). While in Baffin Island, Sugden (1978) found a strong relationship between ice discharge and fjord cross-profile ($r^2 = 0.96$). It is generally agreed that troughs do reflect an efficient routeway for ice discharge and that the cross-profile symbolises the subglacial erosional processes involved.

PLATE 2.15 Glaciated trough, head of Glen Muick, Cairngorm Mountains, Scotland. Snow in centre foreground beyond the loch highlights a zone of hummocky moraine possibly of Loch Lomond Stadial age

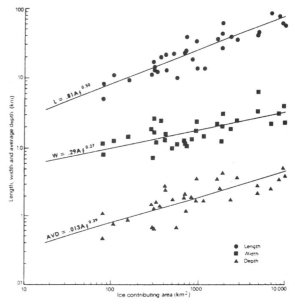

FIG. 2.43. The relationship between ice-contributing area and fjord properties of length, width and average depth. Data from British Columbia. (After Roberts and Rood, 1984; reprinted from Roberts and Rood, 1984, by permission of the *Scandinavian University Press*)

Recognition of the catenary or parabolic or U-shape has been attributed to McGee (1883, 1894) and to Davis (1916), but Svensson (1959) appears to be the first to attempt to delineate the cross-profile shape using a mathematical expression ($Y = a\ X^b$) of a parabola where Y is the vertical, and X the horizontal distances, respectively, from the midpoint of the valley bottom and a and b are constants. Subsequently many workers using variations of this expression have tested various mathematical best-fit expressions against real data (Kanasewich, 1963; Corbato, 1965; Nye, 1965; Graf, 1971; Doornkamp and King, 1971; Aniya and Welch, 1981; Hirano, 1981; Wheeler, 1984; Aniya and Naruse, 1985; Hirano and Aniya, 1988, 1989; Harbor et al., 1988; Augustinus, 1992; Harbor and Wheeler, 1992; Harbor, 1992). It is assumed (as it is generally) that a valley progresses from a V-shape ($b \approx 1.0$) to a U-shape ($b \approx 2.20$), then the value of component b is indicative of the stage of valley form development (Fig. 2.44). Harbor (1992) has shown using an iterative finite-element model of glacial trough, development from an initial V-shape to a U-shaped form that, irrespective of the initial valley form, over a period on the order of 1,000 years a steady state, quasi-parabolic cross-profile can develop.

The key element in the generation of the U-shaped profile was recognised by early workers that greater

erosion must have occurred not at the base of the valley form but some distance laterally from the central lowest point thereby reducing over time the elevation difference between the centre and adjacent areas (Fig. 2.44).

The location of glacial trough erosion seems to now be accepted but the actual mechanisms of subglacial erosion within a confined valley needs to be further explored. As Harbor (1992) has pointed out, as ice thickness increases in association with greater trough depth there is limiting point if a constant ice discharge and erosion rates are to be maintained. Since for a given ice discharge a specific surface gradient must be maintained conditions at the subglacial interface are sensitive to changes in this gradient (Menzies, 1995a, Chapter 5). Where changes in discharge and therefore surface gradient occur, it can be assumed that trough erosion rates will also vary (Harbor, 1992, fig. 11). The erosion of a glacial trough must occur due to a combination of subglacial processes of abrasion,

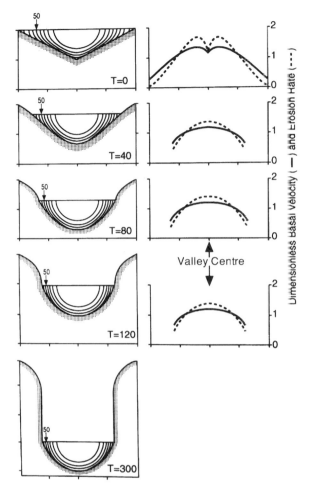

FIG. 2.44. Model results of a simulation of form development with erosion scaled to local basin velocity squared. The figures show the glacial valley cross section at different time steps (T), with velocity contours on the glacier in units of 10% of the maximum velocity for the section, and with the central contour being 90%. The plots are dimensionless and are shown with no vertical exaggeration. The graphs show corresponding cross-glacier variations in basal velocities and erosion rates, in each case scaled to an average cross section value of one. (From Harbor, 1992; reproduced by permission of the *author*)

exist in confined troughs. In the basal and side-wall environments of glacial troughs where areal changes in basal temperature, for example, cannot be dissipated or conserved and where fast basal ice motion occurs, it is likely that erosion processes are unconstrained. Under this scenario a 'run-away' style of basal erosion occurs that can only be limited where ice discharge or other external constraints upon ice velocity can reduce erosion (the back-pressure exerted by ice shelves or the buoyancy effect of a grounding-line) the effect of greater debris concentrations at the base of an ice mass (Shoemaker, 1986a,b) may retard basal sliding especially in the presence of a deformable bed, or ice mass divergence on reaching the coastline (Shoemaker, 1986a,b).

A characteristic of many glacial trough systems is the presence of transverse trough sections (Plate 2.16). These transverse troughs appear to be related to

PLATE 2.16. Landsat image of interior Rocky Mountains showing typical glaciated landscape of linear glacial troughs often linked by cross-valleys (diffluent routes). This image includes part of the continental divide between Alberta and British Columbia. The main feature in the centre of the image from right to left is part of the Rocky Mountain Trench now containing the Fraser River. This trench is a major fault zone extending from Alaska to Montana. Snow covers only the highest mountains (> 3000 m) – many carrying glaciers (image taken July, 1975). North arrow in top left corner

plucking and hydraulic erosion by debris-charged high pressure meltwater. Unlike areal erosion beneath unconstrained ice sheets (Menzies, 1995a, Chapters 3 and 5) where sensitive and transient feedback fluctuations in the basal thermal regime in close association with basal meltwater occurs, this delimiting control on effective stress levels may not

erosion by diffluent lobes of ice crossing interfluves possibly representative of pre-glacial tributary valleys.

The sill at the sea-entrance of many fjords would seem attributable in many cases to ice divergence and buoyancy effects on reaching the edge of coastal mountains and deep seawater conditions (Shoemaker, 1986a,b). In some cases subglacial diamictons are found on sill bedrock surfaces indicative of the limited erosion that may have occurred close to the final stages of glaciation in these troughs. Also many bedrock sills have a rough undulating surface that may be maintained during erosional episodes in a the self-sustaining manner discussed by Röthlisberger and Iken (1981) (Shoemaker, 1986a,b).

The presence of riegel bedrock steps and associated basins in many glacial troughs remains a difficult problem. There have been many hypotheses put forward, for example, differential bedrock resistance, faulting, valley constrictions, localised freeze/thaw at the bed beneath heavily crevassed ice, and proximity to the entrance of tributary glaciers (Cotton, 1941; Flint, 1971). The basin and step profile is not peculiar to these valleys and may simply reflect a persistence and magnification of an already existing profile by glacial erosional processes. However, such an explanation seems rather too rudimentary, unless the balance between local ice erosion and non-erosion is sufficiently delicate that at any point along the bed where relative deepening occurs a self-limiting feedback mechanism restricts further erosion until the balance between different elevations along the bed are re-established. However, how such a mechanism can be resolved at the larger scale with the existence of very deep fjord basins remains to be examined.

The origin of glacial troughs and fjords remains problematic when three-dimensional ice mass flow of variable discharge, surface gradient and bed conditions are considered in relation to bedrock properties and structures. Whether trough erosion is a self-maintaining 'run-away' process that is an abnormal manifestation of normal ice flow needs to be investigated. The fast flowing ice streams of Antarctica may supply some answers but much remains to be investigated.

2.9.2.2. Tunnel valleys

Tunnel valleys can be found in many marginal areas of once continental glaciated terrains. In particular, they are well studied in Denmark, Germany, Poland, Britain, the North Sea Basin, southern Ontario, the Canadian prairie provinces and northern states of America (Ussing, 1903; Woldstedt, 1952; Sissons, 1961a,b; Wright, 1973; Ehlers, 1981; Nilsson, 1983; Paterson, 1994). These valleys, sometimes known as rinnen, rinnentaler, tunneltaler, or tunneldalen in Europe, are often referred to as buried valleys. The term valley is perhaps a misnomer since these forms usually do not have a dendritic drainage system although tributaries and braided reaches do occur. These forms are deeply incised usually into bedrock and may better be termed as deep slits cut into the rock since they are usually only a few tens of metres in width but sometimes over a hundred metres in depth. Exceptions do occur to these dimensions but they are rare. These valleys are usually infilled with glaciofluvial sediments with occasional diamictons present either as flow sediments or rafted sediment intraclasts (Fig. 2.45) (cf. Woodland, 1970; Dingle, 1971; Wright, 1973; Whittington, 1977; Ehlers, 1981; Gaunt, 1981; Menzies, 1981; Cox, 1985; Cameron et al., 1987; Karrow, 1989; Piotrowski, 1991, 1994; Brennand and Sharpe, 1993; Booth, 1994). The effect of infilling is to bury the valley below ground surface. Additional glacial sediments may often overly the valley therefore the term 'buried valley'. The long profile of these valleys is usually undulating with deep depressions separated by short or long shallow reaches (Schou, 1949; Homci, 1974; Hinsch, 1979; Kuster and Meyer, 1979; Sissons, 1976; Grube, 1983; Cox, 1985; Cornwell and Carruthers, 1986). Tunnel valleys may extend for tens of kilometres and exhibit reverse gradients to that of the ground or bedrock surface. These gradients have, in the past, be employed to suggest that these valleys were incised under hydrostatic pressures beneath an ice mass. Valley walls are smooth with evidence of meltwater erosional scour marks and P-forms (see below) often present. In some cases the remnants of older diamictons may occur in the deepest depressions (for example the Clyde tunnel valley in west Scotland (Menzies, 1976)). An association between tunnel

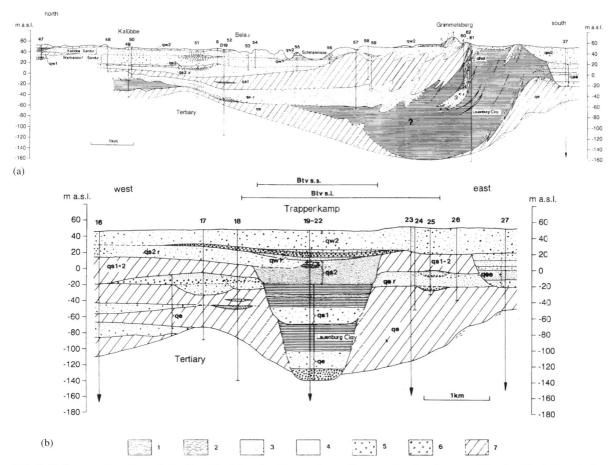

FIG. 2.45. Cross-sections through the Bornhöved tunnel valley, northwest Germany. Stratigraphy: *qe* = Elsterian Glaciation; *qhol* = Holsteinian Interglacial; *qs1, qs2* = first and second icve advances of Saalian Ice (local terminology); *qw1,qw2, qw3* = first, second and third Weichselian Ice advances; suffix *v* = advance outwash; suffix *r* =retreat outwash; *Btv ss.* = Bornhöved tunnel valley *sensu stricto*; *Btv sl.* = Bornhöved tunnel valley *sensu lato*. Lithology: 1, clay; 2, silt; 3, fine sand; 4, medium sand; 5, coarse sand; 6, gravel; 7, diamicton. Vertical exaggeration 12.5X. (After Piotrowski, 1994, reproduced with permission from *Elsevier Science Publishers*)

valleys and eskers has been noted (Wright, 1973; Grube, 1983), as has a possible link between drumlin formation, tunnel valleys and subglacial hydraulics and thermal conditions (Wright, 1973; Mooers, 1989a,b; Brennand and Sharpe, 1993).

The location of tunnel valleys is far from obvious on the present-day ground surface leading to problems in hydrogeology, in particular difficulties in waste disposal, and the undetected presence of local and/or linear and perched aquifers (Freeze and Cherry, 1979). Tunnel valleys appear to be related to subglacial ice dynamics and the evacuation of subglacial meltwater in large volumes (the Sable Island tunnel valleys cut across the Scotian Shelf off the east coast of Canada may have had volumes of ≈0.45 ×10^7 m^{3}s^{-1} when all tunnels were operational (Fig. 2.46) (Boyd et al., 1988; Wright, 1973; Sissons, 1976; Grube, 1983; Ehlers et al., 1984; Shoemaker, 1986a; Barnett, 1990; Rains et al., 1993). In some instances, as in the mid-west of the United States, tunnel valleys extend to the margins of the Late Wisconsinan Ice Sheet margins and form outwash aprons within the terminal moraine complex (Mickelson et al., 1983), however in others areas such as Indiana and Illinois such an association has not been observed (cf. Chapter 3).

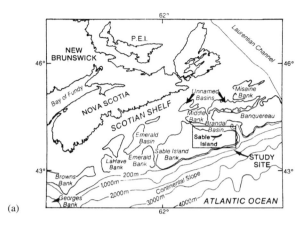

(a)

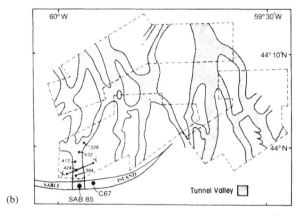

(b)

FIG. 2.46. Tunnel valleys near Sable Island, Nova Scotia. (From Boyd, Scott and Douma, 1988; reprinted with permission from Macmillan Magazines Limited)

Tunnel valleys form as a result of intense localised subglacial meltwater erosion within subglacial meltwater channels. The water in these channels is under hydrostatic pressure and is heavily charged with angular debris resulting in an efficient and high speed erosive tool. The undulating profile and the water-smoothed walls are consistent with high pressure but fluctuating meltwater discharges. The location of these channels may in part be influenced by underlying topography but is generally thought to be controlled by ice pressures. Many tunnel valleys obliquely cross slopes or are cut into the side-wall of a larger glaciated valley in both instances without regard for local gradients or topographic control (cf. Mannerfelt, 1949; Sissons, 1960a,b; Booth and Hal-

let, 1993). The difference between these valleys and subglacial channels (discussed below) is probably only one of scale and the length of time taken to cut tunnel valleys. Conditions most suitable for valley incision appear to be related, in many cases, to frozen bed environments down-ice of a temperate bed thus probably close to an ice sheet margin (Wright, 1973; Hughes, 1981; Mickelson *et al.*, 1983; Mooers, 1989a,b; Paterson, 1994). Due to the volume of water up-ice under M-bed conditions, the thinness of the marginal ice and probably He-bed conditions prevailing close to the margin, it is thought that subglacial meltwater can be evacuated toward the margin by crossing the He-bed and incising into the bed and producing tunnel valleys. Whether this scenario is correct in all instances and whether tunnel valleys are always created under He-bed conditions close to an ice margin requires further investigation. Finally, the relationship of these valleys to drumlin fields is an intriguing problem (Wright, 1973; Paterson, 1994) that needs much more research since the underlying bedrock topography under most drumlin fields is poorly known. Whether tunnel valleys and drumlins simply coincide across the same area of terrain may be more chance than causal but, it has been suggested that subglacial hydraulic conditions conducive for tunnel valley formation may equally permit drumlin development (cf. Shaw, 1983b; Rains *et al.*, 1993) (Section 2.8.1.3.). A possible scenario that might be suggested is that as tunnel valleys are being cut excess meltwater across inter-valleys areas may cause the basal ice to increasingly decouple leading to fast ice movement causing rapid sculpting at the sediment/ice bed interface generating flutes and drumlins both associated with glaciofluvial sediments and meltwater erosion and deposition. Only subsequent research can establish the validity or otherwise of this relationship.

2.9.3 Local Linear Erosional Forms

Linear erosional forms are the products of direct ice-contact erosion or channelled meltwater erosion or a combination of both. The scale of these forms varies enormously from the micro-lineations of striae to large-scale forms such as roche moutonées.

2.9.3.1. Striae and associated percussion erosional forms

Bedrock striations are perhaps one of the best indicators of glacial erosion (Plate 2.17) (see also colour plate). These scratches indicate the local vagaries of basal ice movement that occur due to variations in principal stress directions and micro-topographic bed control (Virkkala, 1960). For too long striae have been ignored as being of rudimentary significance and limited value. However, when time is taken to measure the typical cross-cutting relation-ships of superimposed striae, long axis depth varia-tions and the presence of micro-faulting, striations can be and are a useful tool in deciphering glacial movements, chronology and ice/bed interface condi-tions (Fig. 2.47) (cf. Peach and Horne, 1880; Hoppe,

(a)

PLATE 2.17. (a) Striations visible on the stoss-side of large streamlined bedrock protrusion, near Espanola, Ontario. Scale card is 8.5 cm long. (Photo courtesy of Greg Hamelin). (b) Striations of Pleistocene ice across a Precambrian diamictite of Huronian age, near Whitefish Falls, Ontario. Note coin 3.0 cm in diameter. (Photo courtesy of Greg Hamelin)

(b)

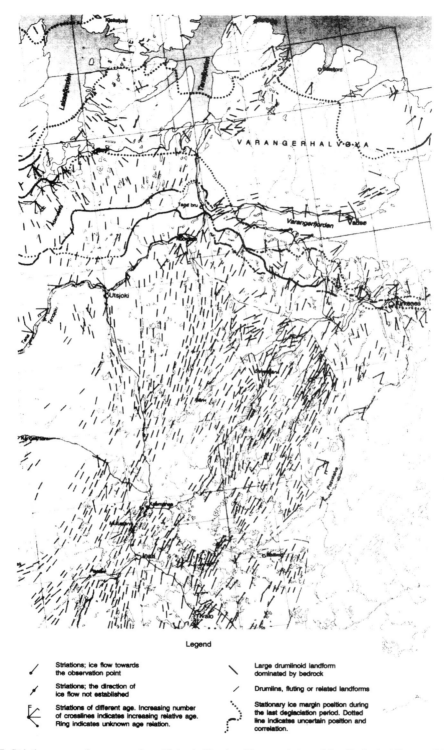

Legend

	Striations; ice flow towards the observation point		Large drumlinoid landform dominated by bedrock
	Striations; the direction of ice flow not established		Drumlins, fluting or related landforms
	Striations of different age. Increasing number of crosslines indicates increasing relative age. Ring indicates unknown age relation.		Stationary ice margin position during the last deglaciation period. Dotted line indicates uncertain position and correlation.

FIG. 2.47. Striations mapped across northern Finland. (Reprinted by permission of the *Geological Survey of Finland*)

1974; Veillette, 1989; Bouchard and Saloner, 1989; Pronk et al., 1989; Rappol, 1986, 1989 1993; Warren, 1991).

Striae exhibit a wide variation in form from straight micro-grooves of a few millimetres in depth extending for only a few centimetres to some that are almost a centimetre deep and extending for over a metre (Laverdière et al., 1985). In some cases smaller striations occur at the base of large striations. Delicate tracery of cross-cutting striae commonly occur on boulder and clast surfaces and where resistant knobs occur striae may circumvent these obstructions. Many striae have a variable depth becoming shallower in the direction of ice movement possibly indicative that the striating tool became blunt with travel down-ice.

Striae on individual clasts and bedrock surfaces are products of either direct scratching of clasts and boulders held in the basal ice or clasts within a moving basal debris layer. The depth, width and length of individual striations, therefore, reflect the combined influence of basal ice stress levels, ice velocity, meltwater presence, debris concentrations, effective stress levels, the sharpness of the individual clast indenter and the properties of both the indenter and the surface being scratched (for example the fracture toughness and penetration hardness (cf. Scott, 1979; Riley, 1982). In general, a striation is thought to be produced by plastic deformation of the rock

surface against which the tool is applied. Often small leveés are produced on either side of the main trough (Plate 2.18). In those cases where the difference in hardness between the indenter and the surface is small, a shallow striation without distinct leveés tends to be produced. In some cases the track of an indenting tool producing a striation reveals that the tool did not scratch continuously across a surface but rather moved in a percussive manner, producing a series of subparallel transverse percussion gouges (Plate 2.19). Finally, it can be theoretically shown that the angle at which the indenting tool approaches the surface also determines the depth of indentation. Since most clast tools held in the ice or in mobile debris will rotate or reorientate under the stress generated during the indentation process, it can be assumed that many striations will exhibit variations in the processes of cutting, depth and general morphology along their length. A final aspect of striation formation is that although the indenter must be sufficiently sharp and be able to hold a cutting edge for some distance, a striation can only occur if the surface of the clast or bedrock can retain the scratch mark, thus soft materials and those easily weathered do not carry striations.

Although not linear forms of glacial erosion, there exists a suite of local surface erosional forms associated with striations that can considered along

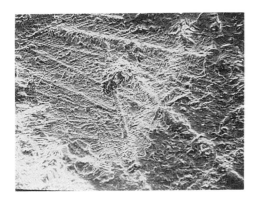

(a)

(b)

PLATE 2.18. (a) SEM photomicrograph of striations (Scale 2cm = 50μm, Magnification -200×). (b) Higher magnification SEM photomicrograph of striation showing upturned edges (Scale 2cm = 3.33μm, Magnification – 3000×)

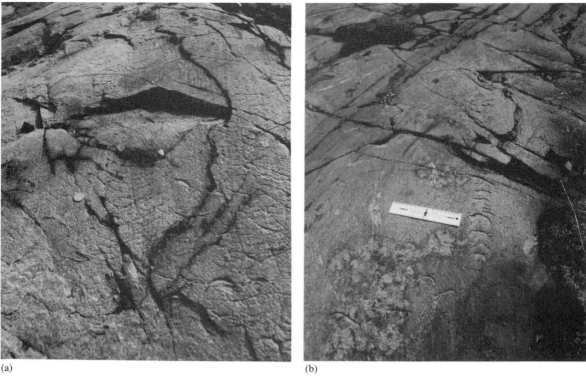

(a)   (b)

PLATE 2.19. (a) A series of chattermarks across a bedrock surface, view in the down ice direction. Note coin, centre-left, for scale (3.0 cm diameter). (Photo courtesy of Greg Hamelin). (b) Line of chattermarks across large bedrock surface, view in the up ice direction. Note ruler for scale (25 cm in length). (Photo courtesy of Greg Hamelin); both photographs from near Whitefish Falls, Ontario

with striae, these include chattermarks, crescentic gouges, lunate and conchoidal fractures and friction cracks (Gilbert, 1906; Ljungner, 1930; Harris, 1943; Embleton and King, 1968; Flint, 1971; Laverdière et al., 1985). It is critical, however, that these forms not be confused with plastically formed p-forms associated with subglacial meltwater erosion (Section 2.9.3.3.) (cf. Kor *et al.*, 1991). These forms are sharp-edged with surface evidence of conchoidal fracturing usually related to percussive action of an indenting boulder or clast. The effect of percussion tends to be localised, therefore these erosional forms occur across small areas of a bedrock surface. Gouges and cracks typically are only a few centimetres wide and millimetres in depth. These forms occur in groups or sets and can be used to estimate approximate ice direction. However, inferred ice flow direction cannot always be assumed (Dreimanis, 1953). It is likely that

cracks and gouges of this type may be the site of later subglacial meltwater flow disruption thereby acting as instigators of larger s- and p-forms.

2.9.3.2. Roche moutonnées

Roche moutonnées may be regarded as streamlined bedrock hills. Typically, these forms have a smooth stoss-end and a precipitous lee-end slope (Plate 2.20). These bedrock forms may range in scale from a few centimetres high to hills several hundreds of metres in elevation (Rudberg, 1954; Burke, 1969). The formation of a roche moutonnée would appear to be partially the result of ice erosive moulding on the upstream end of the form and glacial plucking on the lee side. In order that a lee side be affected by plucking action it is probable that no cavity of any size formed in the lee of the obstruction (Carol, 1947; Okko, 1955;

(a)

(b)

PLATE 2.20. (a) Small roche moutonées in northern Lake Euron, Ontario. Direction of ice flow from left to right. Buoy in centre approximately 1 m in height. (b) Roche moutonées in the proglacial zone of Nigardsbreen, Norway. Direction of ice flow from right to left. The roche moutonées is approximately 25 m in length

Embleton and King, 1968; Laverdière *et al.*, 1985)). As a consequence, tensile stresses and freeze-thaw processes in combination create a plucked lee face. Rastas and Seppälä (1981), in an extensive survey of roche moutonnées in southern Finland, noted the close relationship between bedrock joint systems, fissures and the size and morphology of individual forms (Fig. 2.48).

2.9.3.3. Meltwater erosional forms

As ice pressure melts on the up-ice side of bedrock and other topographic obstacles, meltwater is released which may flow as turbulent, debris-charged, meltwater streams under hydrostatic pressure. Where large subglacial lakes form beneath ice sheets, it has been suggested that vast, regional, catastrophic floods may periodically occur resulting in enormous discharges of meltwater across the subglacial bed (Shaw, 1983; Elfström, 1987; Shaw *et al.*, 1989; Shoemaker, 1991, 1992a,b). Likewise jökulhlaups may provide sudden catastrophic meltwater drainage (Menzies, 1995a, Chapters 5 and 12). Several meltwater, associated, erosional forms, under H-bed conditions, are typically found, for instance, potholes, p-forms and subglacial meltwater channels. These forms develop in association with other erosional forms such as rock drumlins, roche moutonnées and crag and tails. There are certain characteristics that these forms have in

common; . smooth, rounded bedrock surfaces, channels with fluctuating longitudinal thalwegs, limited development of channel networks and the siting of many of these forms without apparent regard for present topographic position, slope or gradient (Ljungner, 1930; Dahl, 1965; Kor *et al.*, 1991). For example, in terms of topographic siting channels may cross hillsides at angles inappropriate for 'normal' drainage pathways and may traverse topographic divides or interfluves without regard to channel slope.

(1) Subglacial meltwater erosional forms: In many glaciated areas, exposed or recently uncovered bedrock surfaces carry distinctive erosion marks that are finely detailed, often tortuous, smoothed depressions and hollows of a wide range of morphologies (Plate 2.21) (Ljungner, 1930; Dahl, 1965; Bernard, 1971a, b; Laverdière *et al.*, 1985; Kor *et al.*, 1991). These forms would appear to result from intense subglacial meltwater erosion either the result of high pressure turbulent debris-charged meltwater or, in some cases, high pressure fluidised sediment slurries. These forms were first studied in detail in Scandinavia (Ljungner, 1930; Dahl, 1965; Gjessing, 1966, 1967) where the term p-form was used to denote that these forms were 'plastically formed'. The continued validity of this term is in some doubt and perhaps 's-forms' as a more exact term representing 'sculpted forms' may be now appropriate (Kor *et al.*, 1991).

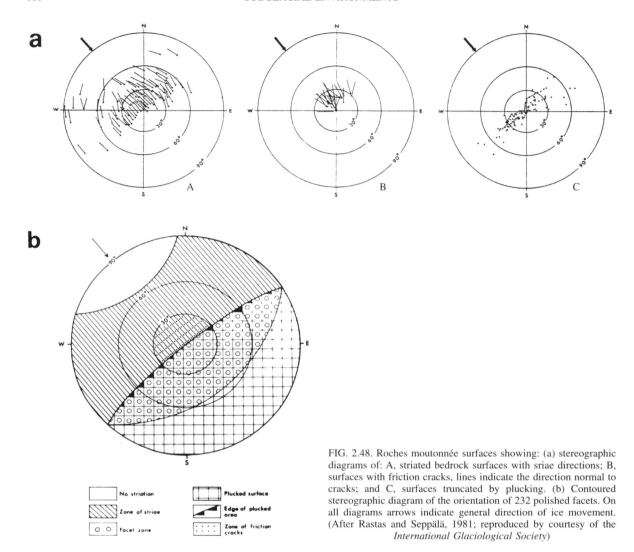

FIG. 2.48. Roches moutonnée surfaces showing: (a) stereographic diagrams of: A, striated bedrock surfaces with sriae directions; B, surfaces with friction cracks, lines indicate the direction normal to cracks; and C, surfaces truncated by plucking. (b) Contoured stereographic diagram of the orientation of 232 polished facets. On all diagrams arrows indicate general direction of ice movement. (After Rastas and Seppälä, 1981; reproduced by courtesy of the *International Glaciological Society*)

Examples of the range of forms are illustrated in Figure 2.49. In terms of scale these forms most typically are small only a few metres or less in length and width but occasionally very large scale multiple associations of these forms exist stretching over tens of metres of bedrock surfaces. Kor *et al.* (1991) suggest that, instead of just occurring in patches as has usually been assumed, s-forms may exist at the regional scale as shown in an example from north-eastern Georgian Bay, Ontario illustrative of much larger subglacial meltwater catastrophic flows (Fig. 2.50).

The debate as to whether s-forms develop due to a meltwater dominated flow or fluidised debris slurry still remains but convincing evidence of the morphology and continuity of one form with another produced under high pressure, high velocity flow conditions seems to negate most forms being produced by slurried sediments (Johnsson, 1956; Gjessing, 1966; Boulton, 1974; Allen, 1982; Kor *et al.*, 1991). These erosional forms appear to develop due to meltwater flow separations over distinct time periods related, in the first instance, to defects or bed irregularities or up-flow vortices impinging on the

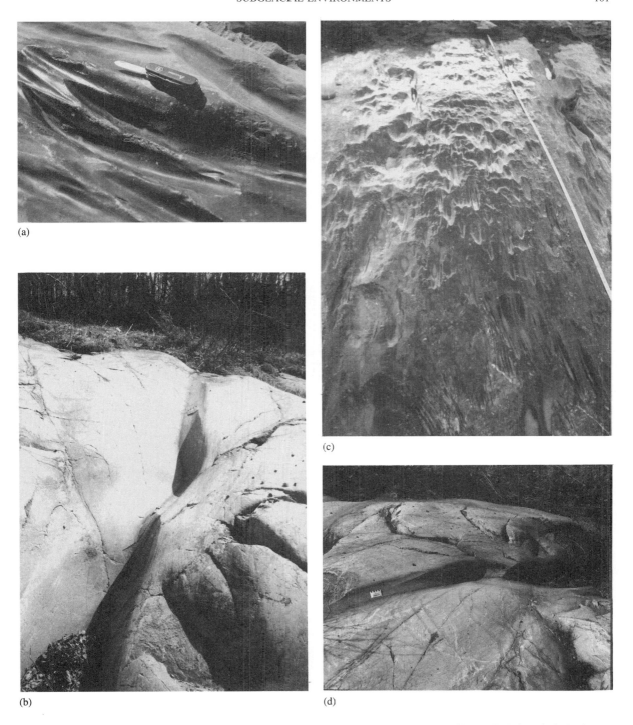

PLATE 2.21. (a) Straight small-scale flutes near Wilton Creek, Ontario. (Photo courtesy of John Shaw). (b) P-form in bedrock near Espanola, Ontario. Scale card is 8.5 cm long. (c) Scallops, subglacial meltwater erosional marks near Wilton Creek, Ontario. (Photo courtesy of John Shaw). (d) P-form cutting across a bedrock surface near Espanola, Ontario. Note conchoidal fracture in lee side (far centre-left) and iron staining (top left). Scale card is 8.5 cm long. (Photo courtesy of Greg Hamelin)

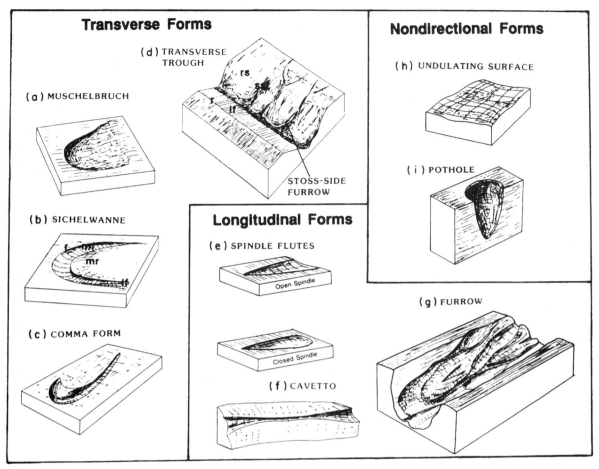

FIG. 2.49. Examples of s-forms as identified in the French River complex, Ontario. Flow direction is from left to right in all but non-directional forms. (Reproduced by permission of the *authors,* from Canadian Journal of Earth Sciences, Vol. 28)

bed (Allen, 1971, 1982; Shaw, 1988b) (Fig. 2.51). As the fluid passes across the bedrock surface, cavitation, fluid-stressing and corrasion processes act upon the rock surface as erosional wear processes, the former being likely the most important (Barnes, 1956; Dahl, 1965; Kenn and Minton, 1968). As primary s-forms develop new vortices are generated thus new erosional forms are created as secondary or tertiary forms in close proximity to the original defect or s-form. A typical association of s-forms is shown in Figure 2.52 across the stoss-side of a rock drumlin. Kor *et al.* (1991) have classified these forms into: (a) transverse s-forms (muschelbrüche, sichelwannen, comma forms and transverse troughs), (b) longitudinal s-forms (spindle forms, cavettos, stoss-side furrows and furrows), and (c) non-directional s-forms (undulating surfaces and potholes). Potholes of variable depth and diameter appear to develop from vertical meltwater vortices impinging upon a rock surface and drilling into the rock surface by a combination of corrasion and, once a hollow has formed, by the grinding action of boulders caught in the vortex of swirling meltwater. Some pot holes have near vertical sides while others are bulbous in shape with overhanging lips (Rosberg, 1925; Alexander, 1932; Hattin, 1958).

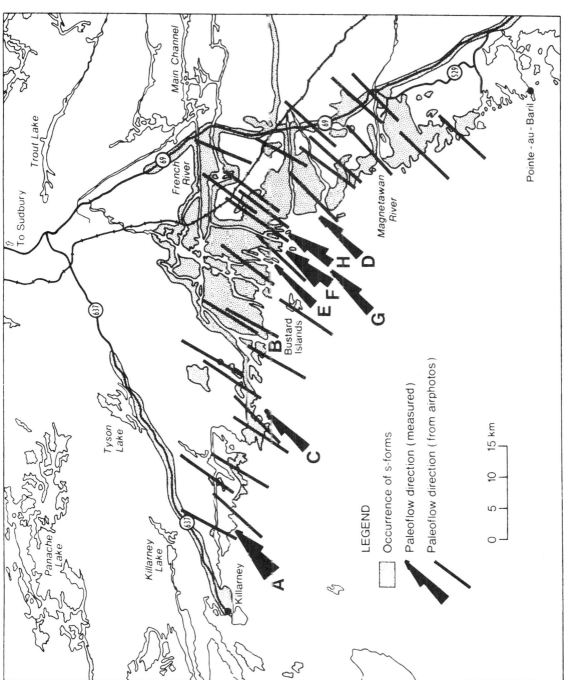

FIG. 2.50. Northeastern section of Georgian Bay, Ontario, showing observed distribution and orientation of s-forms. (Reproduced by permission of the *authors*, from Canadian Journal of Earth Sciences, Vol. 28)

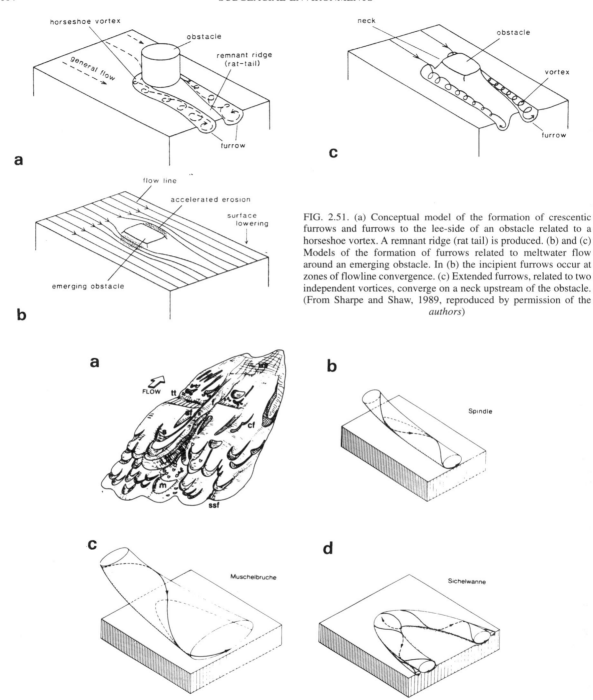

FIG. 2.51. (a) Conceptual model of the formation of crescentic furrows and furrows to the lee-side of an obstacle related to a horseshoe vortex. A remnant ridge (rat tail) is produced. (b) and (c) Models of the formation of furrows related to meltwater flow around an emerging obstacle. In (b) the incipient furrows occur at zones of flowline convergence. (c) Extended furrows, related to two independent vortices, converge on a neck upstream of the obstacle. (From Sharpe and Shaw, 1989, reproduced by permission of the *authors*)

FIG. 2.52. (a) Schematic diagram of s-forms distributed on rock drumlins. Flow away from the reader. Abbreviations: tt, transverse troughs; s, sichelwannen; m, muschelbrüche; ssf, stoss-side furrows, cf, comma form; us, undulating surface; c, cavetto; f, furrow; sf, spindle furrow. Inferred relationship between form and flow structure: (b) spindle flutes form as a result of low-angle vortex impingement; (c) muschelbrüche are formed by higher angle vortex impingement; (d) sichelwannen are formed when flow separation beyond the rim produces a roller eddy and vortices. (Reproduced by permission of the *authors*, from *Canadian Journal of Earth Sciences, Vol. 28*)

The location of s-forms appears to be related to a complex interaction between topography, bedrock structures and fissure geometry, subglacial stress fields and subglacial meltwater hydraulics and evacuation routes. Typically, sichelwannen and most s-forms exhibit an independence from topographic control. Sichelwannen often occur preferentially on the stoss-side of bedrock forms such as roche moutoneés and rock drumlins but are found in distal flanks of some larger bedrock forms and may on occasion be related to joint-specific sites (Ljungner, 1930; Dahl, 1965; Holtedahl, 1967; Gray, 1981; Rastas and Seppälä, 1981; Laverdière *et al.*, 1985; Shaw and Sharpe, 1987). In contrast, muschelbrüche and spindle forms are observed occurring on upper surface of risers (riser slope) or in distal slope localities. Potholes often occur where major breaks of slope exist across a longer bedrock surface (Menzies, 1995a, Chapter 6, Fig. 6.15). In the past it was thought that potholes might be associated with surface crevasses impinging on the bedrock below the ice but

this seems unlikely since crevasses rarely reach bedrock and if so rarely remain at one site for any length of time (Streiff-Becker, 1951; Faegri, 1952, Higgins, 1957). Lateral meltwater flow vortices alter from sub-horizontal to vertical flow elements. Potholes sidewalls reveal spiralling grooves and ridges along their walls indicative of the high velocity flow regime needed in their formation (Plate 2.22) (see also colour plates). At a smaller scale similar meltwater flow regimes result in transverse troughs and conjugate sichelwannen being generated.

(2) Subglacial meltwater channels: Beneath ice sheets and valley glaciers where meltwater flows in channels, numerous examples of channels cut in bedrock and sediment can be observed. These channels can be distinguished on the basis of several unusual characteristics that are inconsistent with subaerial channels: (a) independence of topographic control, (b) undulating long profile, (c) lack of a significant drainage basin, (d) abrupt inception and termination, (e) proximity to other channels with no

(b)

(a)

PLATE 2.22. (a) Large pot-hole, Finland. Note 'roller' stones on edge. For scale note measuring tape top centre. (Photo courtesy Geological Survey of Finland). (b) Small assymetric pot-hole cut into bedrock wall, Norway

fluvial connection, (f) cross-cutting relationship with other subglacial and subaerial channels, (g) occasional association with eskers and glaciofluvial deltas within the channel or it's terminus, (h) abrupt changes in channel direction, and (i) general lack of tributaries (Fig. 2.53) (Mannerfelt, 1945, 1949; Derbyshire, 1958, 1961; Sissons, 1958a, 1976; Price, 1963; Clapperton, 1968, 1971; Young, 1974, 1975, 1980; Pasierski, 1979; Booth and Hallet, 1993). These features are attributable to most glacial meltwater channels whether under active or passive ice.

In general, subglacial meltwater channels can be classified according to their position beneath an ice mass and in relation to specific topographic locations and association with certain glaciofluvial sediments and deposits (Table 2.10) (Sugden and John, 1976). Beneath ice sheets only ice-directed subglacial channels are likely to occur showing little or no dependence on topography and in marginal areas subglacial/proglacial channels can be found. In valley glaciers or where ice lobes of ice sheets have been confined both submarginal and marginal channels occur. The distinction between all five types is at times subtle especially since channels may have been exploited repeatedly during active and passive phases and modified or had sediments deposited within the channel itself. Mannerfelt (1945, 1949) and Sissons,

1958a,b, 1960, 1961a,b, 1963) utilised channel gradient as a key indicator of channel type since in mountainous terrain the distinction between ice sheet and valley glacier activity extending from the maximum to the final retreat phases of glaciation is, at times, difficult to unravel. Sissons (1961b) suggested that channels within dissected terrain with gradients of <1:50 were marginal types (Fig. 2.53a). This same rule, however, could not be applied with validity in lowland areas such as the vast southern edge of the Laurentide Ice Sheet.

Meltwater channels can range in size from barely a metre deep and wide to huge valleys tens of metres deep and as much as a kilometre wide. Channels may also extend for only a few metres before terminating while others extend over tens of kilometres. Channels exhibit incised meanders, bifurcations and complex stream patterns (Fig. 2.53). Channels may have multiple intakes at the same or differing elevations. Channels may have long straight reaches and discordant junctions and tributary entries. Some channels are exceptionally short in length yet very deep often seen occurring on valley spurs, drumlin crests, cols and other topographic highs. These channels have been interpreted as evidence of englacial meltwater channels touching the subglacial bed or being superimposed upon parts of the glacier bed for only short

TABLE 2.10. Meltwater channel types.

Channel type	Some typical characteristics
Closed Channel	
Subglacial	Deeply incised aligned roughly parallel to the main ice direction, limited topographic control, oblique to topographic gradient, in valleys close to valley bottom, often reverse gradient, may be associated with submarginal channels as chutes in lower reaches with engorged eskers, may cut valley-side spurs or cross-interfluves at high levels from one valley to the next
Submarginal	Normal general gradient less than terrain gradient, some topographic control, may intersect or grade into marginal channel system, or alter to subglacial chutes flowing directly down-slope, typically run parallel or near-parallel to ice mass margins (or coincident with trim lines)
Open channel	
Marginal	Normal or steep gradient, some topographic control, may intersect with proglacial channel systems, associated with kame terrace deposits
Subglacial/proglacial (land-based)	Topographic control in lower reaches, associated with glaciofluvial outwash fans
Subglacial/proglacial (subaquatic)	Topographic control in lower reaches, associated with glaciofluvial deposits especially subaquatic outwash fans and deltas.

(a)

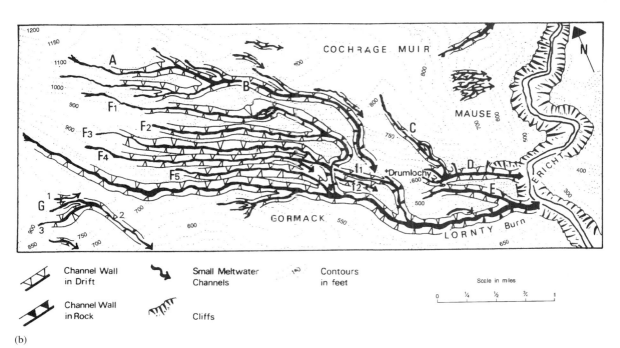

(b)

FIG. 2.53. (a) Aerial photograph of area of gradient subglacial/submarginal meltwater channels near Blairgowrie, Scotland. (b) Map of the same meltwater channels as shown in (a).

distances (Price, 1960, 1973; Clapperton, 1968, 1971).

Subglacial channels and submarginal channels classified by Sugden and John (1976) as ice-directed forms are closed systems that may be under hydro-static pressure, if far enough back from the ice margin, or, if connected to an open portal, may be only under atmospheric pressure (Hooke, 1984; Röthlisberger and Lang, 1987) (Menzies, 1995a, Chapter 6). The dimensions of these channels,

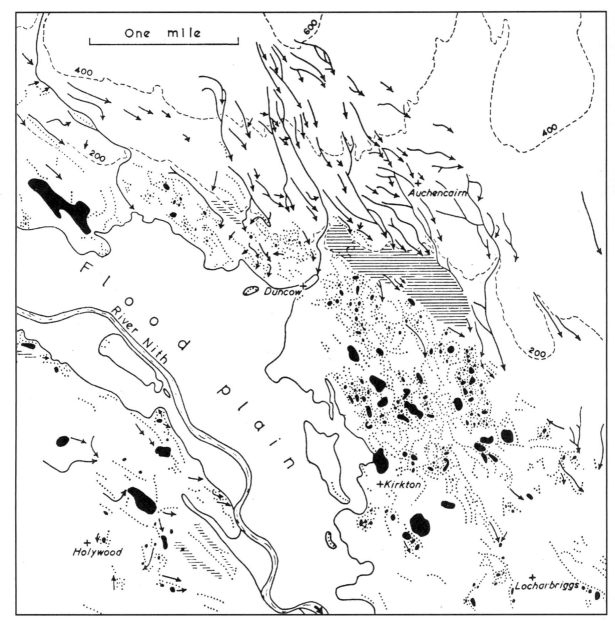

FIG. 2.53. (c) Distribution of meltwater channels and glaciofluvial deposits in Nithsdale, Scotland. (After Stone, 1959; reproduced with permission of the Royal Scottish Geographical Society.)

however, are regulated by the meltwater volume, temperature and debris content, and both ice wall closure rates and squeezing of fluidised sediment into the channel space (for a detailed discussion on channel persistence see Menzies, 1995a, Chapter 6).

Figure 2.54 illustrates the relationship between sediment (till) channels and R channels (Alley, 1989a) in terms of closure rates and effective stress levels. Where effective stress values are low, thus the driving force to close channels is small, sediment channels

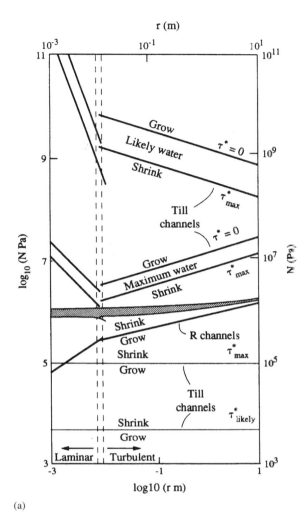

(a)

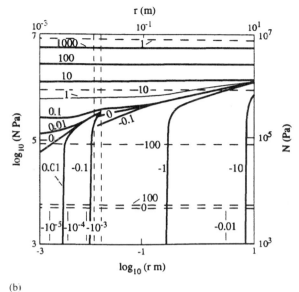

(b)

FIG. 2.54. (a) Effective confining pressure against subglacial channel radius for R and 'Till' channels. At high effective pressures 'Till' channels exhibit stability. A 'Till' channel plotted above a high effective confining pressure tends to grow and below the channel shrinks. Within the stippled zone both R and 'Till' channels shrink at equal rates. (After Alley, 1989b; reproduced by courtesy of the *International Glaciological Society*)

would appear to be potentially stable and R-channels, if they exist will be rapidly infilled with sediment. However, Alley (1989a) suggests that where sediment channels are stable, under low effective stresses, R-channels may be capable of growth. This state may be the instability suggested by Walder (1982) for hard beds where a small perturbation at the ice/bed interface once began perhaps due to an obstacle at the boundary interface self-perpetuates and triggers the growth of an interconnected cavity system (Walder, 1986).

In places where subglacial channel meltwater is under hydrostatic pressure, the characteristic undulating long profile, incised meanders and strong inde-

pendence from topographic control can occur leading to channels with reverse gradients, cross-cutting topographic slopes and passing from one valley system to another at high elevations. Under these circumstances full pipe-flow conditions prevail and active scouring of the bed is likely to occur rapidly (s-forms, tunnel valleys, potholes etc.). Pressure or discharge fluctuations in meltwater conditions are likely to lead to rapid sedimentation within the channel itself leading to glaciofluvial sedimentation and esker formation (cf. Saunderson, 1977; Saunderson and Jopling, 1980; Clark and Walder, 1994). The complex pattern of many of these channels may in part be to fluctuations in basal ice pressures and the effects of localised channel wall and ceiling collapse or slumping of debris from meltout along channels walls and ceilings temporarily blocking channel routes. It is also possible that many channel patterns are products of multiple phases of subglacial meltwater activity utilising similar routeways across the subglacial bed or are the result of englacial superimposition creating overprinting of one channel system on another leading to channel configuration and morphological complexity.

Submarginal subglacial channels appear to occur in most valley glaciers (Lliboutry, 1983; Röthlisberger and Lang, 1987) where overdeepened sections or bedrock constrictions can be bypassed at higher levels leading to channels flowing close to valley sides. These channels are similar to marginal channels in both their close proximity to these open channels and in relation to topography. Often these channels are intersected by later marginal channel suites. Submarginal channels may grade into marginal channel systems or abruptly turn down-slope as subglacial chutes. As subglacial chutes, eskers and fan deltas occasionally lie close to their abrupt termini. Rarely do these chutes grade into subglacial channel systems but appear to be connected to englacial channels. The extent down-slope of chutes may, as has been speculated, be related to the presence of the englacial piezometric water surface within the ice mass (Röthlisberger, 1972; Shreve, 1972) (Menzies, 1995a, Chapter 6). This same surface level has been termed an englacial water table (Sissons, 1976; Gray, 1991) and may be related to the upper elevation of glaciofluvial sedimentation in valleys.

Marginal and subglacial/proglacial channel types are open channel systems often with only enclosing ice walls for short distances or in the former type only along the down-slope side of the channel. Where subglacial meltwater channels enter bodies of water, jets of turbid meltwater have been reported with occasional fountains and upwelling at the ice margin illustrative of the artesian effects due to hydrostatic pressure release at the portal mouth (Menzies, 1995a, Chapter 14). The terrestrial channel types develop under open channel flow conditions with only glacial ice controlling topographic position and channel gradient. Typically, marginal channels in confined valley sites reveal the three-dimensional retreat of ice

masses both down-slope and up-valley resulting in a suite of subparallel channels being formed as each successive higher channel is abandoned in favour of a lower, new channel at the ice's edge (Fig. 2.53).

2.10. SEDIMENTS AND LANDFORMS OF PASSIVE ICE FLOW

2.10.1 Subglacial Sedimentation Processes

The processes and mechanisms that lead toward sediment deposition beneath a passive land-based ice sheet can be subdivided into those processes producing a) meltout tills (diamictons) and b) glaciofluvial and glaciolacustrine sediments. Under passive or stagnant ice conditions, an ice mass begins to decay *in situ* resulting in the production of vast quantities of meltwater. Stagnation would appear to occur under two principal conditions. First, total *in situ* decay of an ice mass, wholly or in part, may occur within confined valley systems where the ice supply to parts of the glacial system has been cut-off once the regional glacier surface descended below a certain elevation; or secondly, where, in active retreat, portions of the frontal sections of an ice mass become detached from the main ice mass due to burial under massive volumes of debris thus leading to buried ice stagnation.

2.10.1.1. Tills – diamictons

(1) Meltout tills: The fundamental characteristics of meltout tills have already been described in Section 2.5.1.1. Meltout tills can occur under both active and passive ice conditions. At present, discrimination as to within which ice conditions particular meltout tills were formed are difficult if not impossible without confirmatory evidence in the form of stratigraphic (see Chapter 8) or facies associations (see Chapter 9).

(2) Flow and waterlain tills: Under stagnating conditions, where high porewater contents prevail, flow tills are released within subglacial cavities and from channel walls (see Section 2.7.1.1.). These tills

PLATE 2.23. (a) Glaciofluvial sediments overlying a tectonised glaciolacustrine unit, near Rosenheim, Bavaria. Note rapid shifts in sediment type from coarse gravels to fine sands, visible as crude bedding. Section is approximately 20 m in height. (b) Delicate bedding structures in glaciofluvial deltaic sediments, Allen Park, Hanover, Ontario. Note load structures, dropstones and clast casts in centre, sharp contact between coarse and fine sands at base and rapid transition from fine sands to gravel at top of photo. (Photo courtesy of Greg Hamelin). (c) Fine grained planar bedded sands with smaller assymetric cross-bedding units in ice-contact delta, Allen Park, Hanover, Ontario. Note near upper part of the section a large channel infill grave unit. (Photo courtesy of Greg Hamelin)

(b)

(a)

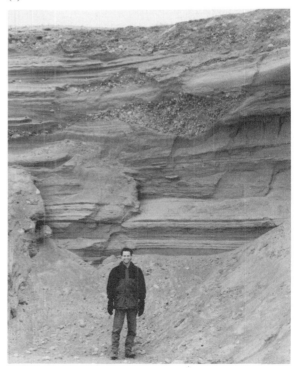

(c)

interfinger with meltout tills and may become indistinguishable from the meltout till package. Where massive stagnation occurs in subglacial/subaquatic environments at the grounding-line an increase in sediment rain-out causes large volumes of waterlain tills to be deposited (see Section 2.7.2.1.) (Menzies, 1995a, Chapter 14).

2.10.1.2. Subglacial glaciofluvial and glaciolacustrine sediments

Under conditions of passive ice decay large quantities of meltwater produce vast outpourings of stratified sediment into the subglacial and eventually the proglacial systems (Menzies, 1995a, Chapter 12). Within the subglacial system, cavities and channels are likely to become infilled and choked with stratified sediments exhibiting a wide range of grain size and depositional settings.

(1) Glaciofluvial Sediments: Stratified sediments are found in the subglacial environment in a variety of locations and forms (Section 2.10.2.1.). Sheets, lenses and small inclusions of stratified sands and gravels are found intercalated within diamicton sequences (e.g., Brodzikowski and van Loon, 1991; Menzies, 1981; Eyles *et al.*, 1982; Ehlers, 1983; Stephan and Ehlers, 1983; Harris and Donnelly, 1991; Fernlund, 1994; Meer *et al.*, 1994) (Plate 2.23). It is likely that these glaciofluvial sediments are formed intermittenly whenever the ice separates from its bed, where tunnels develop then eskers may form, but elsewhere subglacial openings, many of a temporary nature, may develop that permit meltwater flow, sediment transportation and thus sediment sorting. In most cases these glaciofluvial sediments exist as sand stringers, lenses and occasional small spreads within the more general depositional environment. In the field it is often difficult, if not impossible, to tell if stratified sediments within diamictons are necessarily of primary subglacial origin. Where large sheets of glaciofluvial sediments occur that are not related to esker deposition it is more likely that such sediments are not subglacial in origin (Menzies, 1990c). Stratified sediments appear in many instances to grade into finer grained diamictons thus indicating a gradational switch from one form of subglacial process to another, the exact mechanism of such a switch remains an

issue of considerable research significance (see Plate 2.24 in colour plates).

(2) Glaciolacustrine sediments: Subglacial glaciolacustrine sediments are common but usually of small dimensions and areal distribution (Plate 2.25). Typically, these sediments are laminated and often grade into surrounding fine grained diamictons. As in the case of glaciofluvial sediments, laminated sediments can be deposited where a small cavity forms due to ice-bed separation and a pool or small lake develops (Huddart, 1983; Morawski, 1984, 1985, 1988). In

(a)

(b)

PLATE 2.24. (a) Large sandy diamicton inclusion within a sandy, banded stratified sediment unit near Clyde, New York State. Note sharp contacts but also gradational sub-facies unit surrounding inclusion. Scale card is 8.5 cm long. (b) Gradational contact zone between a lower diamicton and overlying stratified sand. Note banding in diamicton and bedding in sand, the latter prominent due to heavy mineral preferential deposition. Several clay ball inclusions possibly intruded in a line above major contact. Pencil for scale is 15 cm long. Photo from near Niagara-on-the-Lake, Ontario

(a)

(b)

PLATE 2.25. (a) Typical winding esker ridge, Northwest Territories, Canada. Note arrow indicating person for scale. (b) Steep-sided, meandering esker in north-central District of Keewatin, Northwest Territories, Canada

many cases it is almost impossible to resolve whether the glaciolacustrine sediments are of primary origin or if they have been rafted or otherwise incorporated into a subglacial sediment (see Chapters 8 and 9).

2.10.2. Subglacial landforms

As a consequence of the dominance of meltwater action under passive ice conditions, subglacial land-forms are characteristically glaciofluvial in composi-tion with intercalated tills present. These forms can be classified into two groups (a) eskers, and (b) random ice disintegration forms such as kames, De Kalb

mounds, and Veiki and Sevetti moraine (Hoppe, 1952; Stalker, 1960; Clayton, 1964; Drozdowski, 1979; Lundqvist, 1969, 1977; Parizek, 1969; Johansson,1972, 1983; Price , 1973; Koteff, 1974; Aario, 1977a,b,c; Raukas, 1977; Minell, 1979; Moran et al., 1980; Gustavson and Boothroyd, 1981; Jurgaitis et al., 1982; Jurgaitis and Juozapavičius, 1988). Many of these forms, principally eskers and kames may be formed under slow active ice.

2.10.2.1. Eskers

Eskers are straight to sinuous ridges of glaciofluvial gravel and/or sand that vary from a few tens of metres to over 100 km in unbroken length (Plate 2.25). The ridges locally range from a few metres to over 50 m in height and from less than 50 m to hundreds of metres in width at their base. Eskers may be transitional to large ice-contact stratified drift complexes deposited by interlobate meltwater streams, and this is reflected by ambiguous references to the origin of large, linear glaciofluvial complexes in the literature (cf. Farring-ton and Synge, 1970; Price, 1973; Banerjee and MacDonald, 1975; Young, 1975; Shilts, 1984; Lind-ström, 1985; Shreve, 1985; Sutinen, 1985; Bryant, 1991; Gray, 1991; Brennand and Sharpe, 1993; Clark and Walder, 1994; Warren and Ashley, 1994).

Esker ridges generally trend parallel to the final direction of ice flow in the area where they are preserved, suggesting that although they may be formed by meltwater drainage throughout a glacia-tion, only those formed during the latest stages of glaciation are preserved. Notwithstanding the above observation, eskers can be seen to cross ice-flow features, such as drumlins, at sharp angles, suggesting that regional ice flow was, at best, sluggish during their formation (Plate 2.26).

Eskers tend to occupy the lower portions of a landscape, whether they occur in low relief terrain of the Canadian or Fennoscandian Shields, or mountain-ous terrain, such as that in the Appalachian Mountains of eastern North America. Where esker ridges pass from one depression or valley to another, they are draped, with no apparent deformation, over divides or cols. This observation has led to the concept that eskers are formed in closed subglacial tunnels, in streams flowing full and under hydrostatic head

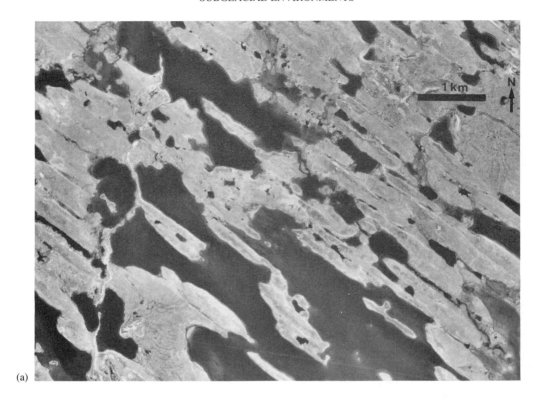

(a)

(b)

PLATE 2.26. (a) Esker superimposed at almost a right angle on drumlins, west of Rankin Inlet, southern District of Keewatin, Northwest Territories, Canada. (b) Esker draped and the nose of a drumlin in northern Manitoba, Canada

(a)

(b)

PLATE 2.27. (a) Ice-cored esker ridge melted out of glacier on Bylot Island, Canada. (Photo courtesy of C. Zdanowicz, summer 1992. (b) Close-up photograph of same esker ridge (Photo courtesy of C. Zdanowicz, summer 1992)

(Olsson, 1965; Saunderson, 1977; Saunderson and Jopling, 1980; Gale and Hoare, 1986; Gray, 1988). This model dates from the nineteenth century and is widely accepted today as the principal environment of esker formation.

Observations of modern glaciers show them to be honeycombed by tunnels, passages, crevasses, etc., conduits that pass through the ice mass in various orientations from vertical to subhorizontal (Menzies, 1995a, Chapter 6). During the melt season, in the ablation zone, these conduits are commonly full of water, and those at or near the base of the glacier are under a hydrostatic head that is proportional to the vertical distance of their outlets below their origins, either within the ice or at the glacier's surface. Meltwater in conduits that reach or pass near the base of the glacier becomes charged with mud melted out of the glacier's debris-rich basal layers. The finer portions of the mud are carried through and out the mouth of the conduit into proglacial streams or directly into lake or marine water, depending on whether the glacier is retreating in the sea, up gradient (allowing unimpeded proglacial drainage), or down gradient (blocking drainage and forming proglacial lakes). The coarser components are largely trapped in the conduit, partially or completely filling it, and are left as sandy or gravelly esker ridges when the glacier melts away (Plate 2.27a, b).

Where glacier ice is retreating in a proglacial lake or marine waters, any coarse material (silt to gravel) that exits the conduit mouth is quickly deposited and may form a fan or blanket of varying thickness over the part of the esker ridge exposed by ice recession. This blanket may not alter the morphology of the esker greatly if the recession is rapid and steady. However, where eskers were deposited in marine environments, isostatic uplift may have reworked and removed this blanket and altered the original form of the esker by wave erosion, (the same process may have affected eskers exposed in the shallower parts of proglacial lakes). Where ice retreat is interrupted by stillstands of the ice front, however, significant piles of sand and gravel are commonly dumped at the conduit mouth, forming subaqueous fans that all but obscure the underlying esker (Henderson, 1988). If the water is shallow enough, the sedimentation rate high enough, or the stillstand long enough, the subaqueous conduit-mouth deposits could build up to the water surface, forming an ice-contact delta, which likewise may obscure the esker locally. Where such interruptions of the regular retreat patterns occurred around the shrinking Keewatin (Canada) ice sheet, short eskers were formed in profusion between the larger systems for which the stillstands were represented by fans or deltas that interrupt and bury the main esker ridges.

Esker ridges may attain heights of 50 m or more and may have basal width to height ratios of 2:1 or less. Since the diamicton adjacent to eskers represents the total basal load of the glacier that deposited the esker, and since individual till sheet thicknesses rarely average more than a few metres, it is evident that some process(es) must enhance sediment accumulation in esker conduits. Since meltwater flowing through glacial conduits is generally at a temperature above the pressure melting point of the ice in which the conduits formed, and since frictional heat is generated by the fast-flowing water-sediment suspension, tunnel walls are subjected to melting, enlarging the meltwater cavity. If a tunnel becomes too large, it is liable to collapse, but because the basal ice of a glacier can deform plastically, as the tunnel walls melt, ice flows from the sides, maintaining a conduit-cross-section that allows the tunnel to stay open. Stable tunnel size is a function of the equilibrium among several factors, including rate of thermal wastage, rate of sediment accumulation, hydrostatic and ice pressure, and rate of ice flow (forward and lateral). As ice flows toward the conduit, it brings basal debris with it so that new material from lateral sources is constantly being dumped into the conduit, sorted, and deposited in or carried out of the conduit to be deposited ultimately in proglacial outwash streams, lakes, or marine waters. By this process, the esker grows to be orders of magnitude thicker than it would be if it were just sweeping out the debris contained in basal ice into which the tunnel was excavated.

As a consequence of lateral movement of ice toward conduits, eskers often contain components from sources that are located as far as several kilometres off the final location of the esker ridges. Regional striation patterns also can be deflected toward the conduit (Repo, 1954), and the area adjacent to the esker ridge may be impoverished in drift cover.

The conduits in which eskers form are regarded by many authors to exist only in a zone a few kilometres back from the retreating ice front. The water that fills the conduits is thought to be derived largely from the ice surface and is thought to reach the glacier's base through openings, such as crevasses, that are kept open by thermal melting, even in zones of compressive flow. On modern glaciers in the Canadian arctic (as elsewhere) some surface streams can be observed to disappear down crevasses or holes whereas other (the majority) flow to and over the glacier's edge. At the height of the meltwater season, the conduits and their surface connections are so full of water that rain or snow storms 'overload' the system, causing water to back up and erupt from the glacier's surface and sides as fountains of water under hydrostatic pressure. The fountains generally carry sediment-charged basal water which is conspicuously muddy compared to normal surface drainage. The internal drainage of a glacier during the melt season, particularly if it is retreating, is analogous to the sewage system of a modern city, with conduits at various levels carrying subsurface drainage under conditions of pipe-full flow. Excess water in the form of rain or snow causes the plumbing system to 'back up', just as happens during storms in cities.

Very few eskers have been observed melting out from modern glaciers, perhaps in part because those in the arctic are not particularly sediment-rich and in part because modern retreating glaciers are so active that conduits are not easily preserved in one place for long enough for lateral flow to build up significant deposits. Conduits in the snouts of glaciers on Bylot Island were observed to be abandoned after functioning for a melt season or two, leaving behind an ice cored, 2–10 m long, meandering sediment ridge in front of rapidly retreating glaciers (Plate 2.27a,b). The total sediment content of these ridges is so slight that, were the ice core to melt, a short sinuous ridge less than 1 metre high would remain.

Esker patterns: The spatial relationships of the substantial eskers left behind by the great Fenno-scandian and North American continental ice sheets fall broadly into 4 groups: (1) The outer parts of the

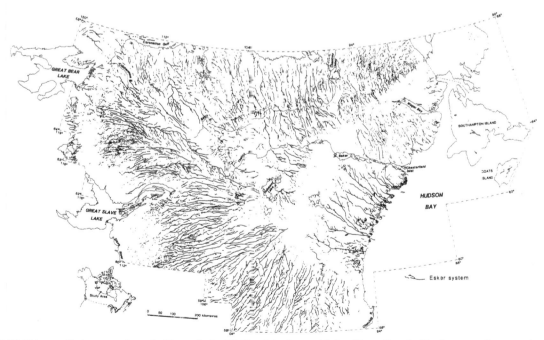

FIG. 2.55. Eskers radiating away from the Keewatin Ice Divide of the Laurentide Ice Sheet, Canada. Northeast-trending area devoid of eskers and passing through Baker Lake and west end of Wager Bay is the ice divide. (Modified from Shilts et al., 1987)

ice sheets, particularly where located on unmetamor-
phosed sedimentary rocks, have rare, isolated esker
segments, often located in tunnel valleys. Because
eskers do occur profusely on some sedimentary
terrain, their paucity in the outer reaches of continental ice sheets may have more to do with ice
dynamics than with the nature of the glacial substrate
(Clark and Walder, 1994); (2) In areas of high relief,
such as the Appalachian Mountain region of eastern
North America or the Cordilleran region of western
Canada, eskers tend to be confined to major valleys
(Shreve, 1972, 1985). Although an individual esker
may cross a divide to pass from one valley or
depression to another, these eskers generally present
an areal pattern reminiscent of the modern drainage,
but lacking in lower order (smaller) tributaries; (3) In
areas of relatively low relief around ice divides of
continental glaciers, eskers form a radiating pattern of
ridges trending generally in the main direction of late-
glacial ice flows (Fig. 2.55). Major trunk eskers are

joined by tributaries forming integrated sets representing Horton drainage systems as complex as
fourth-order (Fig. 2.56). Though individual esker
ridges tend to follow the lower part of the landscape,
as in mountainous regions, they commonly cross
divides and have numerous up-gradient reaches. In
the case of the esker systems radiating from the
Keewatin Ice Divide, the pattern of esker drainage of
the ice sheets bears virtually no resemblance to that of
the major elements of modern drainage (Fig. 2.55);
(4) A fourth esker pattern that is common in the
central sector of the Laurentide Ice Sheet is that of a
dense cluster of individual and simple dendritic esker
systems that terminate at major end moraines which
define readvance lobes or ice streams that were active
during the final stages of retreat of continental ice
sheets (Fig. 2.57). These eskers are clearly confined to
lobate areas that also are characterized by distinctive
drift compositions and landforms. For example,
eskers radiating toward the end moraines of major

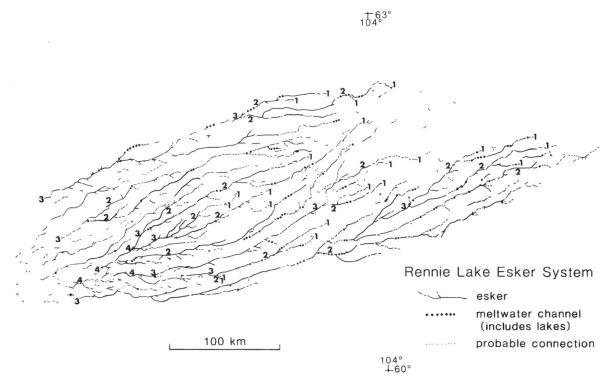

FIG. 2.56. Detail of the Rennie Lake esker system showing high order tributaries. (From Shilts *et al.*, 1987)

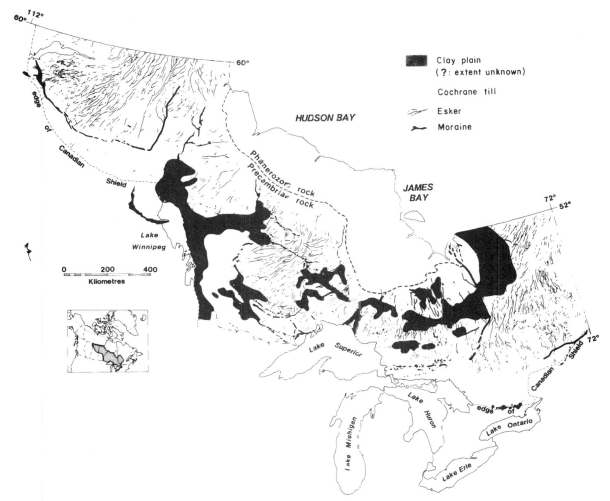

FIG. 2.57. Relationship of eskers and end moraines associated with lobate readvances, surges, or ice streams active during deglaciation in south-central Canada. Some eskers may be buried under lacustrine or marine silty clay or Cochrane Till. (From Shilts *et al.*, 1987)

lobes north of Lake Superior are intimately associated with exotic, carbonate-rich, fine-grained drift formed into drumlins and extending as dispersal trains from the Paleozoic carbonate terrain of the Hudson Bay Lowlands. The eskers appear to mark the final phase of wastage of lobes or ice streams that flowed vigorously over a short period of time, distributing large amounts of fine-grained drift. Following the period of rapid flow, they stagnated, allowing an extensive esker network to form in the relatively inactive, decaying ice.

Though numerous studies of the internal structures and textures of eskers and associated glaciofluvial sediments have been carried out, few sedimentological criteria have been advanced as being particularly diagnostic of eskers. This partially relates to the extreme natural variability of the debris that glaciers carry and which serves as the source of esker sediments. Also, the fact that debris is delivered directly into subglacial streams by ice often results in some components of the esker deposit, for example large boulders and diamicton rafts, being completely out of equilibrium with the hydraulic characteristics of the former esker streams. These almost random sedimentological characteristics make it almost as difficult to generalize about the dynamic sedimentology of eskers as it is to generalize about the sedimentological characteristics of the subglacial diamicton from which eskers derive the bulk of their components.

In lieu of definitive sedimentological clues as to esker genesis and hydrology, the form of eskers and associated sediments has been the most effective characteristic in indicating their origin. The ridge-like form of eskers indicates the presence of 'temporary' ice walls or tunnels and their tendency to be draped over topography with little change in form confirms the former existence of a closed hydrological system at the base of the ice (Plate 2.26b).

The areal pattern of esker distribution also can be used to deduce the hydrology of esker systems. If conduits in which eskers are actively deposited extend only a few kilometres or a few tens of kilometres back from the ice margin at any one time, then it is logical to assume that the dendritic patterns relate to either large-scale basal drainage or to traces of surface drainage of a dying ice sheet. Though this author favours the latter control, many authors favour the former. The definitive study of the relative importance of esker meltwater sources has yet to be made.

Models for formation of individual esker systems have existed since the nineteenth century, but with the increased access to esker-covered terrain and to air photo coverage since the 1950's, several modern models have been proposed (Shilts, 1984; Shreve, 1985; Clark and Walder, 1994; Warren and Ashley, 1994). An example of one of these models, based largely on form and aerial distribution of esker elements is given here for the Deep Rose Lake esker system in north central Canada. This and similar studies have been driven by the need to understand esker sedimentation in relation to using esker sediments as sample media for mineral prospecting, particularly for gold and diamonds.

The Deep Rose esker system is located in an area of northward flowing modern drainage and an area where glacial flow was also northward toward an ice front of the Keewatin sector of the Laurentide Ice Sheet that retreated southward. Proglacial drainage was, for the most part, unimpeded, except where major bedrock ridges crossed the esker's path, blocking small, short lived lakes.

The Deep Rose Lake Esker System – a case example

The Deep Rose Lake esker is located northwest of the Keewatin Ice Divide (KID) (Lee *et al.*, 1957; Shilts, 1980, 1984) and is tributary to a large esker system that starts near the KID. The system can be traced for more than 200 km northward to where it disappears in Queen Maud Gulf. Regional ice flow and drift transportation direction in the region was 325° ± 5°, which is clearly indicated by orientations of long axes of lakes, flutings, drumlin-like forms, and by displacement of dolomite and orthoquartzite boulders northwestward from their outcrops. Trains of Rogen moraine, consisting of hummocks and short, arcuate ridges oriented approximately at right angles to 325°, are disposed in ribbon-like patterns trending approximately 325° over much of the region. Surface meltwater drainage followed shallow surface depressions sloping generally away from the ice front.

(1) Glaciofluvial Complex: Remnants of gravel terraces were found on and adjacent to the esker ridge (Fig. 2.58). These terraces are remnants of a rather complex glaciofluvial-glaciolacustrine system typical of those that were developed during the last stages of glaciation in this part of the District of Keewatin. The Deep Rose glaciofluvial complex can be used as a model for late glaciofluvial events wherever a glacier front was retreating up-gradient, allowing meltwater to flow more or less unimpeded from the ice front.

The glaciofluvial-glaciolacustrine system comprises several distinct elements, most of which are time-transgressive because of their deposition in association with the steadily retreating ice front.

(2) Esker ridge complex: The oldest component of the late glacial meltwater system at any given point is the esker ridge or ridge complex which was formed in a tunnel at or near the base of the glacier. The tunnel more or less followed low areas in the bedrock surface. Where esker sediments were deposited near the base of the ice, and were not modified by postdepositional processes, they now form a sharp-crested ridge with side slopes in excess of 25° (Plate 2.28).

In some places the single esker ridge splits into several ridges which pass downstream back into a single ridge (Plates 2.29 and 2.30). This is thought to have occurred where sedimentation rates were so high, because of increased sediment supply, slope

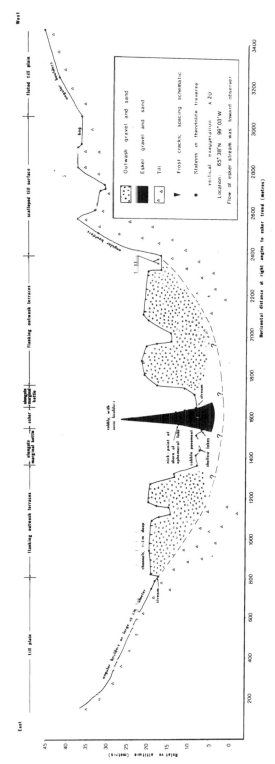

FIG. 2.58. Surveyed cross-section of kettled glaciofluvial terraces adjacent to Deep Rose esker; survey by B.C. McDonald

PLATE 2.28. Steep-sided single esker ridge of the Deep Rose esker,   (a)
District of Keewatin, Northwest Territories, Canada

(b)

PLATE 2.29. Examples of multiple esker ridges in southern District
of Keewatin and Manitoba, Canada. (a) Low altitude (< 300 m)
view of multiple ridges in southern Keewatin. (b) Bifurcating esker
ridge in northern Manitoba from ~2000 m altitude

changes, slowing of ice front retreat, or other
unknown factors, that the tunnel was blocked by
sediment, causing formation of a bypass tunnel,
which, if blocked, was bypassed again, etc.

A question of fundamental importance in developing a sedimentation model for the Deep Rose Lake
esker is, 'how much of the glaciofluvial system was
functioning as a tunnel at the base of the ice at any
one time?' The deposits of the Deep Rose Lake and
other major esker systems in the District of Keewatin
reflect what appears to have been a fully integrated
Horton system of tributaries and trunk streams,
regularly bifurcating upstream into lower order tributaries until the deposits disappear near the Keewatin
Ice Divide. This pattern may be interpreted in at least
three ways:

(a) The whole system may have functioned subglacially in sub-ice tunnels extending from the centre
of a thin, stagnant glacier to its retreating margins,
which lay at one time some 300–500 km away.
Although this model has some merit, the very size of
the ice sheet at the inception of esker deposition
would seem to argue against it, the thicker ice near the
divide being too plastic at the base to maintain open
tunnels. Also, the possibility of surface drainage
existing or being able to penetrate downward in the
thickest part of the ice sheet, which must have been
above the equilibrium line, is remote. It is hard to
imagine how the Horton system now expressed by
these eskers could have developed so fully within
such a large mass of ice; local topographic irregu-

larities at the base of the ice logically would have
exercised greater influence on esker trends than is
evident from the Horton pattern.

(b) Eskers may have been deposited by streams
flowing in short tunnels near the margin of the glacier,
continuity being maintained by up-ice migration of
the heads of the tunnels by melting (St-Onge, 1984, p.
274). Although this model is more compatible with
observed sedimentation features and probable
dynamic conditions in the retreating ice, it does not
explain well the Horton pattern of tributaries. It is
hard to imagine how a subglacial tunnel would
bifurcate regularly as it melted up ice without some
external control.

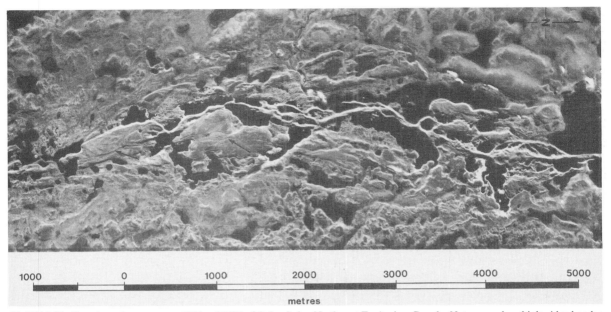

PLATE 2.30. Complex esker system ~ 100 km NNW of Baker Lake, Northwest Territories, Canada. Note several multiple ridged nodes and partial cover by outwash gravel terraces

(c) Probably the best model for englacial glacial meltwater drainage, at least in this region, is one in which an integrated system of drainage channels were developed in the ablation zone of the surface of the ice sheet. Meltwater plunging to the base of the glacier through crevasses or other conduits would have flowed in a subglacial tunnel the last few kilometres of its course before issuing from the retreating glacier front (similar to the model suggested by St-Onge, 1984, p. 273). This system would have developed quite late in the glacial cycle, when a relatively large surface of the ice sheet was below the equilibrium line. Tunnels located near the ice edge would have extended themselves headward by melting, as in the preceding model, but their headward migration would have followed roughly the traces of the surface drainage, thus accounting for the regular bifurcation upstream. This hybrid model best explains the Horton drainage pattern, the manifest evidence of subglacial origin of esker sediments and the general lack of correspondence of esker trends to local trends of modern drainage.

Determining whether hypothesis (a) or whether hypotheses (b) and (c) controlled esker deposition is absolutely fundamental to interpreting both the verti-

cal and the longitudinal changes in provenance of the clasts that make up an esker. If a basal tunnel was only a few kilometres long (as in models b and c), no clast could have been derived from bedrock, already deposited diamicton, or the basal load of the glacier from farther upstream than the average length of the tunnel. Even so, debris may have been carried considerable distances over the terrain on which the esker was eventually draped by the glacier itself before being incorporated into the esker sediment at the head of a tunnel. If the tunnel was part of an integrated system functioning at the base of the ice everywhere at once (as in **model a**), clasts at a given site could have been derived from anywhere upstream to the head of the esker system. If part of the system was supraglacial and part subglacial (as in **model c**), clasts could have been derived anywhere downstream from where the surface drainage plunged to the base of the ice, a point that would have shifted up ice with time.

In trying to assess the provenance of esker sediment, it should be borne in mind that the bulk of the sediment comprises clasts originally liberated from their source outcrops by glacier erosion and carried, in some cases tens or hundreds of kilometres, as the poorly sorted basal load in a glacier. Very little esker sediment is produced by the direct effects of fluvial action on bedrock.

In the continental ice sheet that covered the relatively flat terrain of central District of Keewatin, the volume of debris available for transportation and erosion in any particular segment of the esker tunnel comprised mainly the sediment that was available in the basal layers of the glacier and in the subglacial diamicton lying between the tunnel floor and bedrock (cf. Kleman, 1994). Supraglacial deposits in this region consist mainly of large boulders mantling the till surface, particularly in areas of Rogen moraine, suggesting that the higher parts of the glacier were relatively free of debris. If ice was perfectly rigid, once the unconsolidated basal debris was carried away by meltwater, little additional erosion would have occurred. Since this and other eskers comprise a much greater volume of coarse debris than erosion and redistribution of the till immediately beneath their tunnels can account for, additional debris must have been brought into the esker tunnel by ice flowing into it from its sides, bringing debris from lateral areas into the tunnel, as shown by Repo (1954). Ice would have flowed laterally into the tunnel at a rate determined by how fast the subglacial meltwaters could melt the tunnel walls and carry away the debris entrained in them. Thus, the meltwater that formed the Deep Rose Lake esker could have been transporting debris fluvially at 350° that had been first transported at 325° by the ice *or* first transported 325°, then transported 235° or 55° by lateral ice movement into the tunnel as shown schematically in Figure 2.59.

(3) Outwash terraces and buried ice: The Deep Rose Lake esker system, like many eskers in District of Keewatin, is flanked by one or more levels of sand and gravel terraces, fragments of which are superimposed on the valley sides and on the esker ridges, themselves. These terraces are remnants of sediment deposited in braided outwash streams, traces of the braided channels being preserved on many of the fragments (Plates 2.30 and 2.31). The outwash streams exited from esker tunnel mouths in the steadily retreating ice front (Fig. 2.60).

The only way that fluvial outwash could have been deposited in valleys with modern fluvial systems that

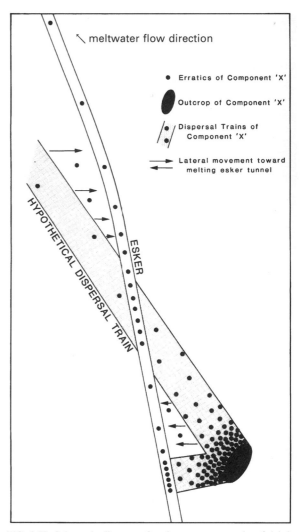

FIG. 2.59. Schematic diagram of various transport paths of debris into a conduit within which an esker is forming

are now interrupted by numerous lakes, is if the shallow, broad valleys in this region were temporarily occupied by thin masses of remnant glacier ice, which served as temporary floors for unrestricted drainage away from the ice front.

Observations of modern glaciers draining Precambrian metamorphic terrains indicate that they carry significant concentrations of debris only in a zone 10m or less above their base, the rest of the glacier generally being relatively free of debris. Ice

PLATE 2.31. Flat-topped esker knob in Deep Rose Lake esker system, District of Keewatin, Northwest Territories, Canada. Fluvial terraces occur extensively on valley sides, above kettle lakes that flank the esker ridge

front recession in these modern land-based glaciers is accomplished by downwasting; once the top of the debris-rich zone is exposed, the insulating effect of debris melting out of the ice retards melting, leaving a thin debris-covered ice 'apron' in the valley immediately in front of the glacier (Fig. 2.60b). Icings (*aufeis*) also form readily in this proglacial environment, and if covered by outwash sediment, contribute to valley-filling, buried ice. In Keewatin, deposition of fluvial or lacustrine sediments in depressions on this buried ice surface would have further retarded melting, leaving buried glacial (and icing) ice stranded in valleys and closed depressions long after adjacent uplands became ice free.

As buried ice melted in the downstream reaches of esker/outwash systems, lowering local base level, the upstream portions of meltwater streams would have adjusted gradient by cutting down into both the remnant glacial ice and into the superimposed glacio-fluvial deposits. The end result of this controlled downcutting was the formation of a series of paired terraces, some on ice, some on the valley sides or emergent esker ridge, throughout each valley now

occupied by an esker (Plates 2.30 and 2.31). Although the terraces are largely depositional, where the buried esker ridge or other deposits projected through the ice, they were sometimes planed off by fluvial erosion to the level of the depositional part of the terrace or the ice floor, leaving remarkably flat-topped esker segments in places (Plate2.31). Like the esker ridge, the terrace deposits are in part time-transgressive, being deposited by meltwater emanating from a constantly retreating source (the tunnel mouth in the ice front).

(4) Esker reaches in segments where the ice front stood in a proglacial lake: In one part of the Deep Rose Lake esker the ice front retreated south of a quartzite ridge, which blocked meltwater issuing from the esker tunnel mouth, forming a lake. A massive ice-contact delta, fed by two esker streams, was built at 200–210 m a.s.l. into this lake and stands now as a prominent landmark about 3 km south of the ridge (Fig. 2.60 and Plate 2.32).

Several clearly defined meltwater channels were cut by water overflowing northward from the proglacial lake when its surface stood below the crest of the quartzite ridge at about 190 to 210 m a.s.l. One of

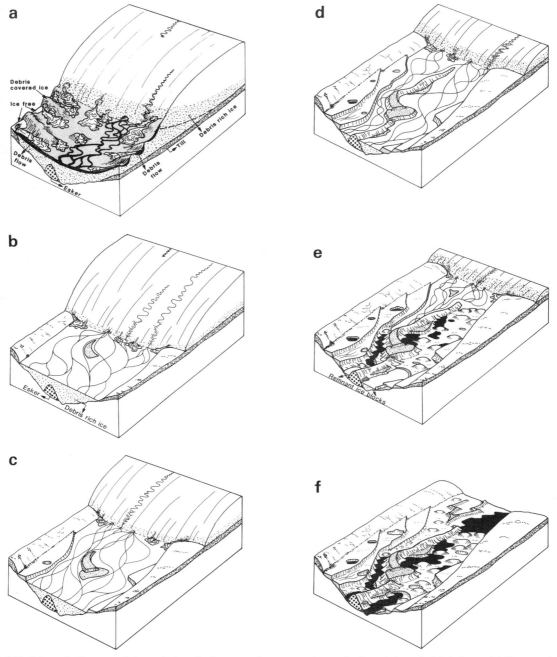

FIG. 2.60. Schematic diagrams of the evolution of esker-outwash sequences in an of unimpeded proglacial drainage. (a) Cross-section of glacier with active sedimentation in a conduit within the dirty ice zone of a retreating glacier. (b) Layers of flow till, meltout till and outwash protecting dirty ice and inactive conduit in valley in front of retreating glacier. Outwash sediments emanating from retreating active conduit tunnel mouth. (c) Melting of stagnant ice floor causes downcutting, esker emerges, is flattened in places by erosion, and terrace fragments are preserved on valley walls or esker. (d) Continued downcutting as ice wastes in lower reaches of valley. More terrace fragments creates. (e) Continued downcutting and downwasting, kettle lakes begin to emerge and intercept free drainage. (f) Final landscape of elongate kettle lakes, bisected by esker ridge, depositional and erosional terrace remnants. (Original drawings by J.A. Aylsworth)

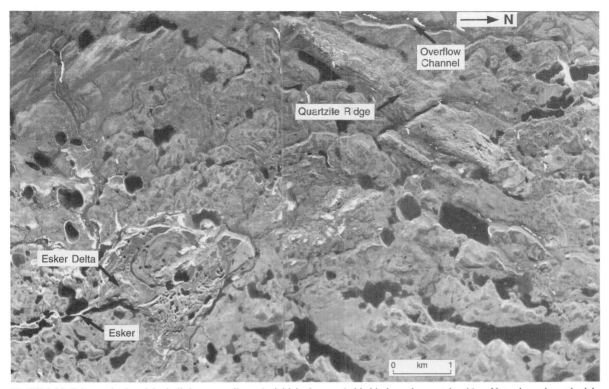

PLATE 2.32. Esker and esker delta built into a small proglacial lake impounded behind a orthoquartzite ridge. Note channels cut by lake outflow and the highly kettled surface of delta

these channels was excavated in diamicton and issued into a lake that had already formed on the ice preserved in the valley downstream from the ridge (Fig. 2.61). The sand and gravel fractions of diamicton eroded from the channel and from the quartzite ridge were redistributed into the lake as deltas which now form low terraces and subaquecus sand flats at its south end. These terraces can be traced directly into fluvial deposits in the abandoned meltwater channels and are not related in genesis or composition to the paired outwash terraces found elsewhere around the modern kettle lake of the valley occupied by the esker.

(5) Terraces and sediments relating to draining of the proglacial lake: The glacial lake ponded south of the quartzite ridge eventually drained completely. The glaciofluvial deposits south of the delta include the same types of esker tunnel sediments and ice-floored outwash trains that are found north of the quartzite

ridge. Before the lake drained completely, however, falling lake levels caused meltwater from the south to cut into the delta, forming a series of erosional terraces on it, particularly along its southwest side. Elevated fragments of meandering channels and braided stream patterns scar the west side of the delta, whereas the surface of the east side is almost wholly disrupted by kame and kettle topography, reflecting collapse of the sediment over melting buried glacier ice blocks (Plate 2.32).

(6) Conclusions: It is important to understand the origin of the various elements of the glaciofluvial system because, although they are related to each other in space, they are diachronous, that is they are not closely related in time of formation. To find the source of sediment deposited in a terrace it would be helpful to know where the ice front was at the time of its inception and how much it retreated during formation of the terrace, since controls on esker

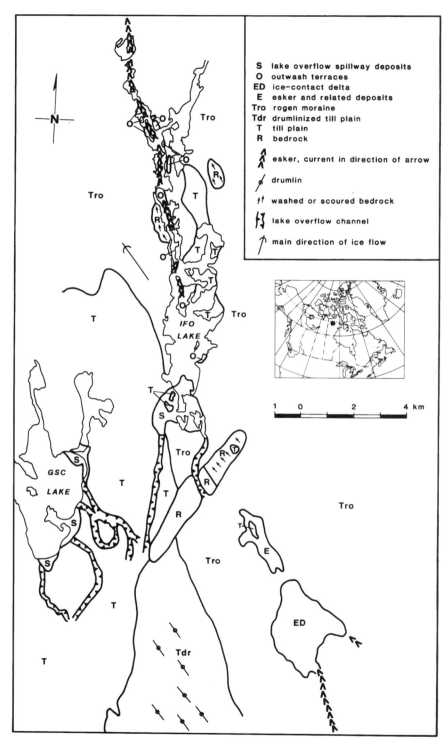

FIG. 2.61. Map of surficial deposits associated with the esker delta complex

compositions also apply to their proglacial fluvial extensions – the terraces.

Only scattered fragments of terraced glaciofluvial sediment remain along this segment of the Deep Rose Lake esker. The gradient and extent of the original terrace systems were estimated by plotting elevations of terrace remnants against distance northward down the paleoslope from the quartzite ridge (Fig.2.62). By interpolating the elevations of once contiguous terrace fragments, one can deduce the existence of a series of terrace levels, each having more or less the same gradient, and each of which can be extrapolated back approximately to the last position of the ice front before downstream melting of buried ice caused downcutting to the next lower terrace level.

The Deep Rose Lake esker system comprises three main sedimentation elements: (1) Sharp-crested single or multiple esker gravel ridges were deposited in a tunnel(s) at the base of the ice; (2) Sand and gravel terraces were built at one or more levels on and adjacent to esker ridges by meltwater streams issuing from the esker tunnel mouth in the retreating ice front. Only fragments may remain because the outwash trains were largely deposited on temporary floors of debris covered glacier ice that was preserved in shallow depressions. Initially, ice may have been preserved by the insulating properties of debris melted out of sediment-choked basal layers of the glacier, exposed during downwasting at the retreating ice front. As these remnant ice masses were melted downstream from the tunnel mouth, local base level was lowered upstream, causing downcutting and insetting of the outwash train below the higher outwash deposits. The net result of this over-ice deposition and downcutting was the preservation of terrace remnants where the sediments were deposited on solid ground or on the esker, itself in the valleys now occupied by the Deep Rose Lake esker system; (3) Glaciolacustrine deposits, including esker deltas and offshore, fine-grained sand, were deposited in a lake dammed between the retreating ice front and a

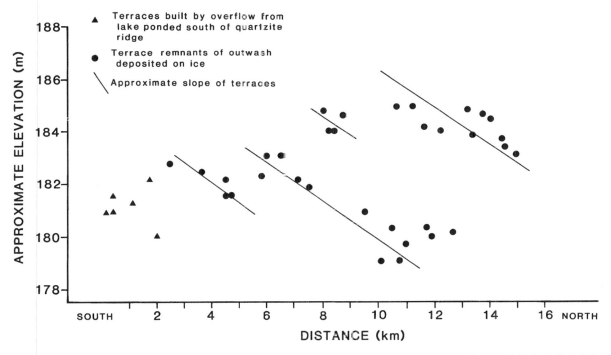

FIG. 2.62. Plots of terrace trends determined by spot altitude measurements on terrace fragments along a tributary of the Deep Rose Lake esker system

bedrock ridge lying athwart the depression now occupied by the esker.

The provenance of the sediments in the esker ridge is likely to be unrelated to that of the adjacent terrace deposits which were deposited sometime after the esker sediments were deposited. The provenance of both types of deposits depends on (1) the length of the basal tunnel functioning at one time, (2) the initial direction of glacial transport of the basal debris that was eroded, sorted and transported by the subglacial stream, and (3) the rate of ice front retreat relative to the rate of sediment deposition in the various elements of the esker system.

2.10.2.2. Non-directional subglacial forms and ice pressed forms

Within this group of subglacial forms generated in the basal zone of an ice mass are forms such as Pulju, Veiki and hummocky moraines. Considerable debate continues as to the origin of these forms and the location in which the forms are created (cf. Lagerbäck, 1988). The composition of these forms is varied. Some forms are composed of lodgement and meltout tills, and stratified sediments in distinct layers while in others evidence of thrusting, diapiric action and deformation by both active and passive ice gravitational consolidation can be found (Hoppe, 1952, 1957, 1959; Gravenor and Kupsch, 1959; Clayton and Cherry, 1967; Kujansu, 1967; Aartolahti, 1974; Daniel, 1975; Lundqvist, 1977; Minell, 1979; Johannsson, 1983; Lagerbäck, 1988; Aario, 1990).

The location of these subglacial moraines appears to correspond to two sites. First, in the central zones of ice sheets close to ice divides, and secondly close to the margins of ice sheets where large terminal moraines occur. Pulju moraines appear to coincide with the former, while Veiki and hummocky moraine often relate to the latter. Since these moraine types exhibit non-directional surface morphologies it is unlikely that their origin can be totally attributed to active ice movement but rather, although the process of instigation may have occurred under active ice, the final morphology and often chaotic distribution seems indicative of passive ice conditions following ice disintegration and collapse. Since the dominant formative processes involved in the generation of these

various moraines involves passive ice wastage, all of these forms are dealt with in Section 2.10.3.2.

2.10.3. Subglacial Erosional Forms under Passive Ice

As ice stagnates under passive conditions the volume of meltwater produced tends to lead to a dominance of meltwater erosion. However, even during passive ice, small erosional forms due to ice wear processes are generated. Most of these forms are small scale surface etchings and abrasive scratching manifesting as deflected striae. Under active ice conditions, striae tend to reflect the principal stress pathways beneath the ice, often near-paralleling the main ice flow direction. Under passive ice, striae, to a greater extent, indicate the deflecting influence and control of micro-topography and meltwater hydraulic gradients (Sollid et al., 1973; Svein et al., 1979).

The impact of meltwater erosion in the subglacial environment is possibly more intense than under active ice conditions. Although hydrostatic pressures are likely to be much less and channels and conduits are not so rapidly sealed off due to ice motion, ice overburden pressures at key locations on the bed of an ice mass will still lead to fluctuating hydraulic pressures. The effect of net ablation also leads to greater sediment input into the meltwater systems than under normal active (non-surging) ice conditions. Finally, unless the ice mass is reactivated, most passive erosional forms survive into the proglacial environment.

2.10.3.1. Meltwater erosional forms

Meltwater erosional forms such as s-forms fashioned under passive ice conditions are similar to those formed under active ice motion (Section 2.9.3.3.). Perhaps the greatest difference between forms created under these active and passive ice conditions is in the survivability, 'freshness' and degree of integrity of forms under the latter ice conditions.

Although subglacial meltwater channel development has been discussed under active ice conditions (Section 2.9.3.3.), within passive ice these channels reach there greatest level of development individually and as networks. Since Mannerfelt (1949) and later

Sissons (1967, 1972), in particular, drew upon the use of subglacial meltwater channel morphology, gradient and position within the landscape relative to past ice positions, only limited use of their significance has been made. The value of meltwater channel system morphology pattern in relation to ice disintegration landforms and sediments can be of immense service in deciphering the final stages of ice retreat (Price, 1960; Thompson, 1972; Young, 1974, 1975; Sissons, 1982; Rodhe, 1988; Gray, 1991).

2.10.3.2. Ice disintegration forms

There are several morainic forms that although initiated under active ice conditions, appear to attain their final form and morphological character under passive ice disintegration. These moraines are not commonplace but are important in several glaciated areas, for example, Finland, Canada, Alaska and Sweden.

(1) Pulju Moraines: These moraines have been best described from locations in northern Finland (Kujansu, 1967; Aario, 1990) (Plate 2.33) (see also colour plates). In plan form these moraines consist of an area of chaotic ridges that have a winding, hummocky appearance (Fig. 2.63). They seem to occur preferentially within 200km of an ice divide and although usually found in valley lowlands do occur on rising ground. The origin of these moraines remains problematic. It is observed that within tills at depth within the moraines strong preferred fabrics reflecting general ice flow direction exist.

It has been suggested that Pulju moraines are formed due to sediment being squeezed up into basal crevasses (Kujansu, 1967) or that the moraines are a result of general ice disintegration during deglaciation

(a)

(b)

(c)

PLATE 2.33. (a) Pulju moraines, northern Finland. (Photo of courtesy Risto Aario). (b) Closer view of a Pulju moraine, northern Finland. (Photo of courtesy Risto Aario). (c) Pulju moraine at the eastern end of Lake Råstojaure, north of Kiruna, Sweden. The formation is about 100 m across. (Photo courtesy of Robert Lagerbäck)

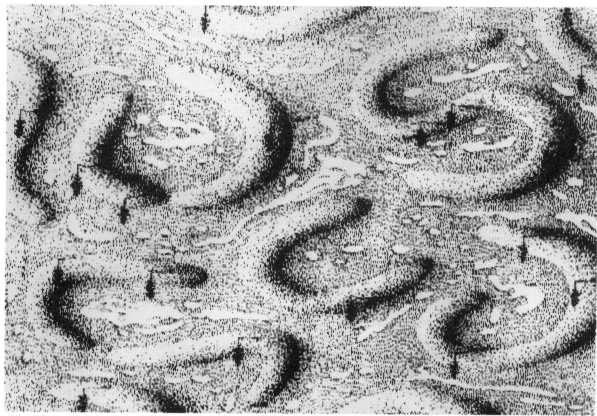

FIG. 2.63. Sketch of Pulju moraine ridges. Ridges are typically 2–5 m high, 10–15 m wide and 50–100 m in length (drawing by Kari Törmänen). (After Aario, 1990)

or lateral pressure-induced ridging, both, under passive ice conditions. However, the strong fabrics noted by Aario (1990) in these moraines appear to preclude passive ice conditions. Aario has speculated that the tills at depth may have formed under the central parts of an active ice sheet. Subsequently, as deglaciation took place, subglacial frozen bed conditions prevailed and the surficial morphology of these moraine ridges developed sediment pressure differential loading with little lateral motion due to the location beneath an ice sheet centre. This formative process remains highly speculative and much more data and research is necessary.

(2) Veiki moraines. The origin of these moraines has been the subject to a long and intense debate principally in Sweden but also in North America (Lundqvist, 1943; Hoppe, 1952, 1957; Stalker, 1960;

Clayton and Cherry, 1967; Parizek, 1969; Aartolahti, 1974; Daniel, 1975; Minell, 1979; Lundqvist, 1981; Lagerbäck, 1988). The morphology of Veiki moraines can be described as a distinctive near-circular plateau surrounded by a single or on occasion double rim or ridge (Plate 2.34). Hoppe (1952, 1957) drew attention to the proximity of Veiki moraines to drumlins suggesting that the moraines had to be deposited subglacially under active ice since often drumlins were observed superimposed upon the moraines. Lagerbäck (1988) suggests that drumlin superimposition is rare, however, Veiki moraines in Sweden, typically, have a thin carapace of till that exhibits a strong preferred fabric perhaps indicating the presence of active ice in the late stages of moraine development but not necessarily in their formative stages of instigation.

(a)   (b)

PLATE 2.34. (a) Veiki moraine, Finland. (Photo courtesy of Risto Aario). (b) Veiki moraine plateau with rim-ridge, northwest of Nattavaara, northern Sweden. (Photo courtesy of Robert Lagerbäck)

Mechanisms of Veiki moraine formation can be subdivided into: (a) subglacial origins under active or passive ice, or (2) supraglacial origins. Hoppe (1952) supported an active ice subglacial origin while others agreed that a subglacial preliminary influence must have occurred causing debris to be squeezed up into basal crevasses but under passive stagnant ice conditions and then suggested a supraglacial final set of formative processes (Gravenor and Kupsch, 1959; Parizek, 1969; Aartolahti, 1974; Daniel, 1975). Clayton and Cherry (1967), however, argued for a entirely supraglacial origin within surface ice-walled lakes prior to final collapse under deglaciation (cf. Reid, 1970; Daniel, 1975). Lagerbäck (1988) supports formation of the Veiki plateau under passive ice downwasting in which the removal of lateral support by ice wastage leads to the formation of ice-contact slopes and the formation of the Veiki moraine rim ridges. Subsequent alteration of the plateau is then thought to occur under interstadial conditions followed by readvance of the ice over the moraines and the addition of a surface carapace of till before final ice wastage. However, the relationship with drumlins and other active subglacial bedforms needs to be further investigated in glaciated terrains before a generally accepted formative hypothesis can be obtained.

(3) Hummocky moraines. The term hummocky moraine tends to be applied to a rather wide array of glacial forms and is perhaps more a morphological

than sedimentological term. Hummocky moraines are chaotic steep-sided piles of dominantly subglacial debris that lack a coherent directional pattern and are often associated with marginal areas of ice masses. However, in surface appearance hummocky moraines, seen in aerial photographs, can be easily confused with the chaotic pattern of stratified outwash aprons and dead-ice topography associated with 'kame and kettle topography.'

Considerable debate has occurred in Britain and Sweden as to the origin of these moraines. It has been argued that the moraines are formed under active ice conditions (Hoppe, 1952, 1957; Sutinen, 1985), or are glaciofluvial ice decay forms (Sugden, 1970), while others have suggested that the moraines were basal ice stagnation forms (Sissons, 1974, 1979, Gray and Lowe, 1977; Gray, 1982; Bennett and Boulton, 1993) and finally, it has been suggested that the moraines mark the position of an active but retreating ice terminus, the debris being largely derived from sidewall periglacial activity (Eyles, 1979, 1983a,b) (see Plate 2.35 in colour plates). In the past, the term ablation moraine has probably been misapplied to these moraines (Lundqvist, 1981).

Detailed sedimentological work by Hodgson (1982), however, indicates that the debris composing these moraines is often subglacially derived and was later overridden by active ice; thus an active origin may be correct. The common association of hummocky moraine with down-ice fluted moraines adds

PLATE 2.35. Hummocky moraines marking the outer limit of the Loch Lomond Stadial in Glen Turret, Tayside, Scotland. Hummocks, individually, are approximately 4 m in height

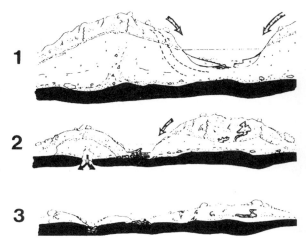

FIG. 2.64. Development of 'hummocky moraine' terrain. Ice cored sediment dumps melt under thinner covers of sediment creating topographic lows thus inverting the topography (Stages 1 →2 → 3). (After Eyles, N., 1983b).

credence to this active ice hypothesis. However, Thorp (1986, 1991) in the western Grampian Mountains of Scotland, suggests that there is often an association of hummocky moraine with the availability of rock debris from solifluced valley walls thus suggesting that supraglacial debris down-wasted and formed supraglacial diamicton dumps. In contrast, on Rannoch Moor in Scotland where no higher elevation sources of solifluced debris exist, this model (similar to that proposed by Eyles, 1983b) does not apply and instead a subglacial active ice origin seems appropriate (Fig. 2.64).

It is perhaps, as was alluded to above, a misnomer to refer to these glacial forms as all hummocky moraines since several different mechanisms of formation would appear possible. Only continued sedimentological research linked to glaciological parameters can help to unravel this complex set of forms.

(4) Stratified 'Kame and kettle' terrain. It is difficult to be certain whether this type of terrain can be attributed to subglacial passive ice conditions or supraglacial ablation and not, as is often the case, to proglacial surface collapse due to subsurface melting of stagnant ice (Menzies, 1995a, Chapter 12). Where hummocky moraines are composed dominantly of stratified sediment, a subglacial glaciofluvial origin might be plausible (cf. Clayton, 1964; McKenzie, 1969; Daniel, 1975; Sugden and John, 1976; Johansson, 1983; Sutinen, 1985; Mäkinen, 1985; Lagerbäck,

1988). In most cases, the topographic position and association with eskers and sandur plains reveal the nature of this type of topographic expression.

2.11. REPETITIVE EVENT HISTORIES IN SUBGLACIAL ENVIRONMENTS

In any sedimentary environment there is often witnessed the repetition of a sequence of sedimentological events associated with a set of processes and resultant sedimentary forms and structures at the same site over and over again (Fig. 2.10). Within glacial environments this repeated succession of events is commonplace. In particular, in subglacial environments there are two levels of repetitive event histories that can be distinguished, (1) those occurring during a single phase of glacial activity and (2) those during each successive phase of glaciation. Too often subglacial environments are discussed with a somewhat static view of sedimentological activity when in fact the subglacial environment is one of repeated dynamic activity in which multiple events occur with such enormous variations in time and space that differentiation of each event history is an intricate and, at times, almost impossible task (see Chapter 8).

During single glacial events sediments can pass through many sequences of erosion, transport and deposition. The ambiguous term – 'resedimentation' has been extensively used in attempts to convey a sense of the non-static sedimentary systems active within subglacial environments. Landforms and bedforms may be formed, or, if from a previous glacial phase or earlier within the same phase, partially or wholly destroyed or unaffected, depending upon time of formation, sediment shear strength, effective stress levels etc. (Menzies, 1989a). In some cases the change of one bedform to another may be only partially complete while in other locations a new form emerges replacing the previous form with no vestige of the earlier form visible. However, sediments, their characteristics, structures and properties may still retain the imprint of past sedimentological histories that indicate frozen conditions, rafting, thrusting etc. Overprinting, reorientation, reworking, realignment and redirection are all terms indicative of episodes of repetitive event histories.

Where deglaciation has occurred and subsequent ice readvance has occupied an area once again, a quite different set of circumstances must be considered. First, the length of time involved between separate advances. Second, the impact of climatic changes on soil development, vegetation colonization and the effect of subaerial processes in altering topographic features and slopes. Third, the presence of postglacial lakes or sea-level rise in flooding areas of terrain. Finally, whatever length of time and changed surface conditions that have taken place, previous subglacial sediments and landforms may be drastically or only slightly altered. In many instances all record of previous subglacial sediments may be totally removed and reincorporated into a new subglacial environment. It is possible that no sediment units may survive even as rafted inclusion, however, geochemical signatures may be detectable in the form of altered minerals and neoformed clay minerals. In some cases, the presence of organics in the form of woody materials, pollen or bone fragments may be indicative of past warmer climate conditions, however, all too often such evidence is lost. Some artifacts cannot be so easily destroyed (striae, chattermarks, conchoidal fractures) but instead cross-cutting and overprinting relationships

may be detected (cf. Lagerbäck, 1988; Rodhe, 1988).

It is critical that the field researchers be aware of the influence and extent of overprinting and multiple event histories. For example, in the 1960s it was commonplace to examine the distribution of drumlins using Nearest Neighbour Analysis. This statistical method relied on the dubious assumption that all drumlins in any one field were formed at roughly the same time and that adjacent drumlins were influenced by the presence of their nearest neighbour. Instead, drumlin fields may contain many elements of several formative drumlin phases and superimposition.

The overlaying effect of multiple sedimentary events within one or multiple glacial phases must be considered whenever the origin of subglacial sediments and forms are debated and when glacial stratigraphy is examined (see Chapter 8).

2.12. SUMMARY

The importance of subglacial sediments in terrigenous glaciated areas of the world cannot be undervalued. Volumetrically, other than glaciomarine sediments, these sediments cover the largest areas of the earth's glaciated surface. Their influence upon all forms of human activity cannot be underestimated from agriculture, to construction as foundations and aggregate resources, to sites for waste disposal, and as aquifers of major groundwater sources for large urban areas. Studies, therefore, of subglacial sediments remain of great significance within glacial geology/ geomorphology both in understanding processes and forms, and from an applied perspective.

There remains an enormous wealth of knowledge to be gleaned from these sediments and forms. As our understanding of subglacial processes, from both modern and past glacial environments, increases so new avenues of research emerge. Likewise, paradigms shift and mutate such that what has been accepted understanding is now in some cases in a state of flux. For example, the existence and detailed analyses of microstructures within diamictons has opened up a huge area of new research which demands new perspectives on how diamictons are deposited (see Chapter 12). Likewise, there is increasing need to grasp the intricate interplay between thermal, geological, glaciological and rheological conditions that coexist at subglacial interfaces of which subglacial sediments are the endproduct. Interpretation of sub-glacial facies associations, sub-facies units and the diagnostic criteria that might be used to discriminate environment, process and form remains a fertile research area. The need to relate sediment structures to porewater and other subglacial melwater effects requires investigation, as does the geochemical pro-cesses both during and after sediment deposition. Although a great deal is known about subglacial processes little is understood concerning rates of processes. It is likely that in many cases sediments and attendant structures may be deposited and generated over very short periods of a few hours or days while others such as meltout may take centuries.

Since subglacial sediments of past glacial environ-ments are heavily utilised today for innumerable purposes it is critical that the effects of post-depositional diagenesis be much better understood. Diagenesis appears to begin almost immediately after sediments are deposited even before ice retreat and accelerates especially in the first years of subaerial exposure. However, the extent, depth and degree of alteration caused by diagnetic processes still remains largely unknown (see Chapter 10). Diagenesis princi-pally influences the geotechnical, geochemical and structural character of the sediment, in some cases only slightly but in others totally transforming some or all of a subglacial sediment unit. As detailed analyses of subglacial sediments continues it is essential that diagenetic influences be separated from all others.

As new and controversial avenues of research and model building are developed, understanding of subglacial environments, sediments and landforms have also grown. In the past decade the importance and central role of subglacial meltwater in almost all subglacial environments and processes has become apparent, yet much remains unknown of the effects and influence of meltwater on sediments, the glacier bed and landforms.

It would fair to state that our still limited under-standing of past ice sheets, the dynamics of their growth and decay and the interrelationship of ice sheets with oceans is revealed by our as yet restricted grasp of the processes and dynamics of subglacial environments. The 'drumlin enigma' stands as a conspicuous instance of how much there is still to understand about ice sheet dynamics and the interplay of glacier motion, sediments, topography and envi-ronmental 'conditions' at the base of ice sheets and valley glaciers. Until we fully recognize the complex-ities of environments beneath ice sheets and other ice masses it is unlikely we can correctly postulate the explanantion(s) necessary for subglacial landform development, sediment deposition and other geom-orphic processes at the ice/bed interface.

FIG. 2.3. Models of subglacial thermal regimes, their spatial relationships and processes of subglacial erosion, transportation and deposition. (Reproduced from Brodzikowski and van Loon (1991) *Glacigenic Sediments*. Elsevier Science Publishers).

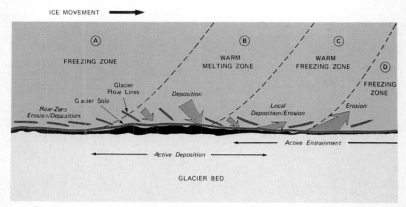

A SCHEMATIC DIAGRAM OF BASAL THERMAL ZONES AND THEIR DYNAMIC RELATIONSHIP TO AREAS OF GLACIAL DEPOSITION AND EROSION

(A) Glacier totally frozen to its bed

(B) Glacier melting at bed with isolated patches remaining frozen, larger toward zone A, smaller toward zone C

(C) Glacier beginning to freeze to bed, patches melted larger toward zone B, smaller toward down-ice edge zone D

PLATE 2.5. (a) Ductile banding in diamicton mélange. Mohawk Bay, southern Ontario. Photo is approximately 1.5 m across.

PLATE 2.5. (c) Faulted and laminated ductile mélange sediments. Each bar scale segment is 5cm.

PLATE 2.10. (a) Drumlin of the Green Bay Lobe. Wisconsin. (Photo courtsey of Donna Stetz).

PLATE 2.10. (b) Drumlins from Kuusamo drumlin field, eastern Finland. (Photo courtesy of Risto Aario).

PLATE 2.14. (b) Princess Louise Inlet (fjord). British Colombia. (Photo courtesy of W.H. Wolferstan).

PLATE 2.11. Drumlin in the proglacial zone of the Biferten Glacier. Switzerland. Photo shows drumlin width to be approximately 150 m.

PLATE 2.15. Glaciated trough, head of Glen Muick, Cairngorm Mountains, Scotland. Snow in centre foreground beyond the loch highlights a zone of hummocky moraine possibly of Loch Lomond Stadial age.

PLATE 2.17. (b) Striations of Pleistocene ice across a Precambrian diamictite of Huronian age, near Whitefish Falls, Ontario. Note coin 3.0cm in diameter. (Photo courtesy of Greg Hamelin).

(a)

(b)

▲ PLATE 2.24. (a) Large sandy diamicton inclusion within a sandy banded stratified sediment unit near Clyde, New York State. Note sharp contacts but also gradational sub-facies unit surrounding inclusion. Scale card is 8.5cm long. (b) Gradational contact zone between a lower diamicton and overlying stratified sand. Note banding in diamicton and bedding in sand, the latter prominent due to heavy mineral preferential deposition. Several clay ball inclusions possibly intruded in a line above major contact. Pencil for scale is 15cm long. Photo from near Niagara-on-the-Lake, Ontario.

PLATE 2.22. (a) Large pot-hole, Finland. Note 'roller' stones on edge. For scale note measuring tape top centre. (Photo courtesy Geological Survey of Finland). ▶

Fig. 2.30. General model of internal sediments and structures found within drumlins.

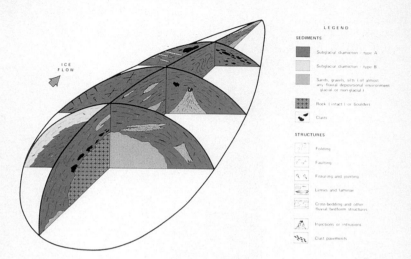

COMPOSITE DIAGRAM OF INTERNAL SEDIMENTS AND STRUCTURES FOUND IN DRUMLINS

ICE FLOW

LEGEND

SEDIMENTS

Subglacial diamicton - type A

Subglacial diamicton - type B

Sands, gravels, silts (of almost any fluvial depositional environment glacial or non-glacial)

Rock (intact) or boulders

Clasts

STRUCTURES

Folding

Faulting

Fissuring and jointing

Lenses and laminae

Cross bedding and other fluvial bedform structures

Injections or intrusions

Clast pavements

PLATE 2.33. (a) Pulju moraines, northern Finland. (Photo courtesy of Risto Aario).

PLATE 2.33. (b) Closer view of Pulju moraine, northern Finland. (Photo courtesy of Risto Aario).

PLATE 2.35. Hummocky moraines marking the outer limit of the Loch Lomond Stadial in Glen Turret, Tayside, Scotland. Hummocks, individually, are approximately 4m in height.

Chapter 3

PLEISTOCENE SUPRAGLACIAL AND ICE-MARGINAL DEPOSITS AND LANDFORMS

W.H. Johnson and J. Menzies

3.1. INTRODUCTION

Process studies in modern supraglacial and ice-marginal environments (Menzies, 1995a, Chapter 1) provide the data base that is utilized in the interpretation of Pleistocene and older supraglacial and ice-marginal sediments and landforms. Because direct observation and monitoring is possible in these modern environments, our understanding and ability to interpret ancient supraglacial and ice-marginal deposits is much better than with those that formed in the subglacial environment. Supraglacial and ice-marginal deposits and landforms are a significant and important component of the Pleistocene glacial record. These deposits commonly are the uppermost portion of the sediment sequence resulting from a glacial event, locally dominate the landscape, and their properties and characteristics have a profound influence on land use and land use decisions. In areas of multiple glacial events, the occurrence of proglacial and/or supraglacial sediment between deposits of subglacial till allow the delineation of ice-marginal fluctuations in the stratigraphic record (Chapter 8, Menzies, 1995a, Chapter 2). Supraglacial and ice-marginal landforms leave a distinctive imprint on the landscape and record the nature of glacial retreat,

whether by marginal backwasting or widespread downwasting, and the history of ice-marginal fluctuations.

Supraglacial and ice-marginal environments are transitional to and have close ties with both the subglacial and proglacial environments (Fig. 3.1). The supraglacial environment by definition occurs on the surface of glacier ice, which separates it from the subglacial environment. These two glacial subsystems come closer together as the intervening ice ablates and commonly are interconnected by englacial water passages within the ablation zone (Chapter 6). Toward the ice-margin, continuity with the proglacial environment develops as the ice thins, and the supraglacial environment, as well as the subglacial environment, give way to proglacial conditions (Menzies, 1995a, Chapter 12). In many situations it is impossible to distinguish sediments deposited in one environment from those of another in these transitional areas (Chapters 4 and 8).

From a sedimentary and geomorphic standpoint, the supraglacial environment is most important along the outer zones of glaciers and ice sheets where sediment becomes concentrated on the surface of the ice, either derived from valley and nunatak slopes that rise above the glacier or through ablation and

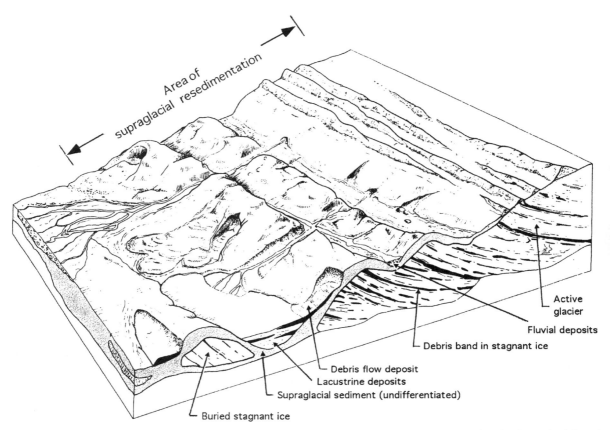

FIG. 3.1. Schematic diagram of supraglacial and ice-marginal environments; subglacial sedimentation beneath active-ice to the right and proglacial deposition to the left. Basal debris is concentrated on the glacier surface by ablation and undergoes resedimentation by mass wasting, fluvial and lacustrine processes as the glacier downwastes. (Reprinted with modifications from Edwards in Reading, (ed) 1986, by permission of Blackwell Scientific Publications)

concentration of basal and englacial debris in the ice-marginal zone. The former situation was a common occurrence in most major mountains of the world during the Pleistocene as cirque and valley glaciers and mountain ice caps formed, and large valley glaciers extended for many kilometres to lower elevations. These glaciers left a sediment and morphologic record that shows the extent of the various glacial advances and their subsequent fluctuations, and from which a chronology has been developed for the most recent portion of the Pleistocene in many glaciated valleys and mountain areas (Chapter 14). For the large ice sheets and smaller ice caps that formed during the Pleistocene, it was basal and englacial debris primarily that became concentrated

on the ice surface and was subsequently reworked in the supraglacial environment during backwasting and downwasting of the ice sheet (Fig. 3.1).

Areas where Pleistocene supraglacial and ice-marginal deposits are most common and important are the lower and lateral portions of glaciated valleys and troughs, the marginal areas of former ice sheets where the ice-margin position was maintained for some length of time, areas where the glacier flowed up an escarpment or regional slope, and in former inter-lobate positions where adjacent glacier bodies flowed toward each other or joined in the case of valley glaciers. In these situations compressive flow moved sediment toward the glacier surface where it became concentrated through ablation. These deposits are not restricted to the outer and lower zones of glaciers and ice sheets, however, and locally they can be a significant component of the glacial sequence in any former glaciated area.

Lawson (Menzies, 1995a, Chapter 11) has described the mass wasting, fluvial and lacustrine processes that are involved in the resedimentation of debris in the supraglacial environment, as well as the controls on those processes. In this chapter the end result of such processes, as preserved in the sediment and morphologic record of Pleistocene glaciation, is summarized. Somewhat similar results preserved in the pre-Pleistocene sedimentary record are described later in Chapter 7.

3.2. SEDIMENT AND SEDIMENT ASSOCIATIONS

3.2.1. Types of Sediment and Sediment Variability

Pleistocene supraglacial sediment sequences contain the entire range of clastic sediment types, from coarse gravels to fine clays to cobbly and pebbly diamictons with a wide range of matrix textures (Plate 3.1). The extreme range in sediment type reflects the different processes that were active in the environment, namely: fluvial, lacustrine and mass wasting, and the dynamic and continually changing environmental conditions in response to controls by glacier dynamics, bed topography and materials, regional and local climate, and sediment availability. The type of sediment record that developed in any one particular area and/or site depended on the interaction among these factors and the processes that were most active during resedimentation. Because both conditions and processes varied spatially and temporally, almost every site has its unique lithologic and morphologic characteristics. In spite of this variability, some general characteristics are common to most supraglacial sequences.

Although some Pleistocene supraglacial sequences are dominated by a particular type and texture of sediment, most are characterized by extreme lateral and vertical variability. Beds are relatively thin and discontinuous, and consist of well to poorly sorted lacustrine and/or fluvial deposits interbedded with

PLATE 3.1. Sediment sequence near Two Creeks, Wisconsin, USA with subglacial till in lower part and supraglacial sediment composed of variable silts, sands, and diamicton in the upper part. Handle of shovel 0.5m. (Photography by D. W. Moore)

poorly sorted mass wasting deposits of variable texture. Among the latter sediment, debris flow deposits are particularly common but they in turn vary considerably depending on sediment texture and water content of the flow when it was active. Many are massive diamictons <2 m thick that have a primary till-like appearance and are called 'flow till' in many glacigenic sediment classifications (Hartshorn, 1958; Boulton, 1971a; Dreimanis, 1988) or supraglacial subaerial mass-transport deposits by Brodzikowski and van Loon (1991) (Chapter 2). Crude stratification and internal sorting (grading) are evident in some of these deposits and the diamicton commonly contains irregular unlithified sediment

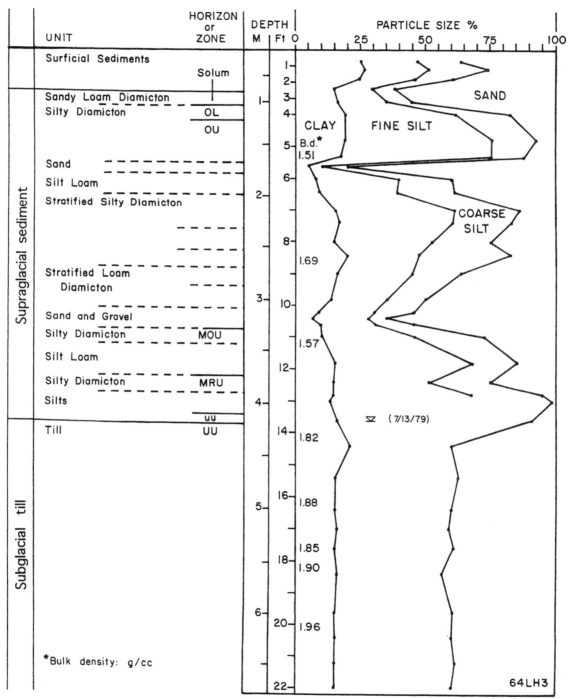

FIG. 3.2. Sediment log, weathering zones, matrix grain-size data, and bulk density for a core taken through a supraglacial and subglacial sediment sequence in Story County, Iowa, USA. Weathering symbols: OL oxidized leached, OU, oxidized, unleached, MOU mottled, oxidized, unleached, MRU, mottled, reduced, unleached, UU unoxidized, unleached. The supraglacial sequence above 4 m is texturally variable and less dense than the lower subglacial till. (After Kemmis *et al.*, 1981)

clasts. Rock fall, slump and colluvial accumulations locally are associated with the debris flow and sorted deposits. Collapse structures from melting of underlying or adjacent ice, deformational shear structures from the moving debris flows or overriding active ice, and soft sediment deformational structures from loading of saturated sediment are not uncommon. Significant lateral and vertical variability is the norm, and the end result is a complex and often chaotic sediment association.

3.2.2. Areas of Low Relief

Extensive lowland areas of northern North America, Europe and Asia were glaciated during the Pleistocene. As a case example, studies along both the eastern marginal area and the north-central portion of the Pleistocene Des Moines lobe in Iowa illustrate the variability of the supraglacial sediment sequence (Kemmis et al., 1981; Kemmis, 1991). The Des Moines lobe of the Laurentide Ice Sheet reached its maximum extent during the last glaciation about 14,000 BP. This advance and several subsequent readvances have been interpreted as surges followed by regional stagnation (Kemmis, 1991).

Along the eastern margin of the lobe, the glacial ice flowed up a regional slope. Borings and exposures in this area contain an upper zone of silts and sands interbedded with thin and variable diamictons that are interpreted to be of supraglacial origin. These deposits overlie a zone of texturally-uniform diamicton interpreted to be subglacial till (Fig. 3.2). The thickness of the two zones varies, for example, in some areas, particularly toward the margin of the glacial advance, the supraglacial sediment facies dominates and little or no subglacial till is present; in other areas such as illustrated by the boring in Fig. 3.2, both facies are well developed. Elsewhere, and particularly in the up-ice direction, the supraglacial facies is thin and completely absent in some areas. Most of the diamictons in the supraglacial facies are likely to be of debris flow origin, but some may be meltout till or ice-slope colluvium. These diamictons differ from those of subglacial till in being thin, a few centimetres to tens of centimetres thick, discontinuous, and often stacked one on top of another with diffuse to sharp contacts. Locally, vertical variations in both matrix texture and pebble size occur, as well as crude layering and deformation structures. These diamictons are less dense than the subglacial till (Fig. 3.2), and although

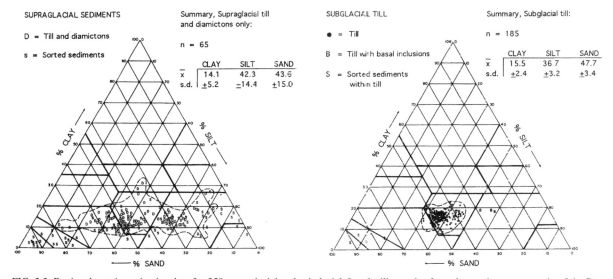

FIG. 3.3. Regional matrix grain-size data for 250 supraglacial and subglacial (basal) till samples from the southeastern margin of the Des Moines Lobe in central Iowa, USA. Although mean grain sizes are relatively similar, the large standard deviations for the supraglacial deposits indicate their textural variability. (After Kemmis et al., 1981)

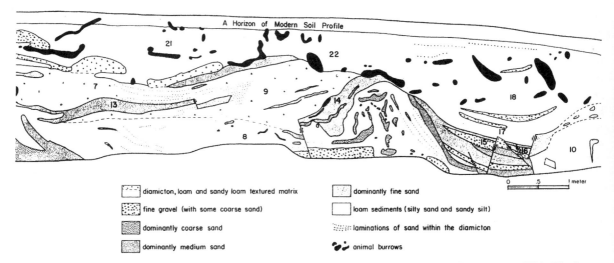

FIG. 3.4. Sketch of a road-cut exposure of supraglacial sediment from a hummock in Winnebago County, Iowa, USA. Diamictons, including the loam sediment, vary in pebble content; animal burrows indicate the friable nature of the upper sediment. Notice stacked channel deposits, faulting, and evidence for subsidence and collapse. (Reprinted from Kemmis, 1991, by permission of the author)

regionally the mean textures of the supraglacial diamictons and the subglacial diamictons (till) are about the same, the former are much more variable and have large standard deviations (Fig. 3.3).

On a local site basis in this region, Fig. 3.4 shows the complex sediment relationships in the upper portion of a hummock in an upland area of moderate relief (Kemmis, 1991). The upper part of the sediment sequence contains pebbly gravel and coarse sand associated with mostly massive silty sand and sandy silt; both are interpreted to be glaciofluvial sediment deposited in a supraglacial position. The underlying sediment, also supraglacial in origin, is composed primarily of interbedded sands and diamicton. Within the diamicton sequence, individual diamicton beds are distinguished by either subtle colour differences, reflecting variations in texture or porosity, or sand laminae at contacts. Contacts often become diffuse laterally and individual beds become indistinguishable in massive diamicton. Where distinguishable, beds commonly are 5 to 20 cm thick. Associated sorted deposits, some as distinct channel fills, have abrupt but often contorted or folded contacts with the diamicton. These deposits dip in different directions at varying angles, and are cut by high angle normal and reverse faults. Kemmis (1991) notes the similarity of the

sequence with those described by Lawson (1981b, 1982) and Boulton (1972b) at the margins of modern glaciers (Menzies, 1995a, Chapter 11). The diamicton is interpreted as consisting of debris flow deposits with thin meltwater deposits, the sand laminae, interbedded with glaciofluvial and glaciolacustrine deposits. The contorted beds and faults suggest deposition on ice with accompanying soft-sediment deformation, followed by collapse as the supporting ice melts. Although no data are available at this site, Kemmis (1991) reports generally weak pebble fabrics in diamicton of similar character and origin at other sites (cf. Rappol, 1985; Dowdeswell and Sharp, 1986); S_1 significance values for the eigenvector of maximum clustering generally were <0.59. Where pebble orientation was stronger with S_1 values about 0.7, the mean azimuth and plunge of the fabric data were in the direction of dip of the diamicton beds. Boulton (1971a) has reported relatively strong pebble fabrics from some modern glacigenic debris flows. The stronger fabrics in Iowa likely represent more fluid water-sediment mixtures and greater shear within the flow, possibly a Type III flow of Lawson (1982).

Similar sediment sequences have been described from a number of localities in Denmark (Marcussen, 1973, 1975; Nielson, 1983). There, as in Iowa, thin

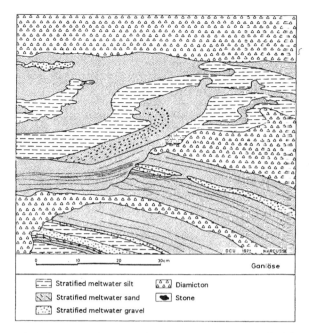

FIG. 3.5. Sketch of supraglacial sediment exposed in a ditch in Ganlose, Denmark. Stratified silts and sands and diamicton with concordant and discordant contacts. Diamicton is friable and is texturally variable; notice soft-sediment and other deformation. (Reprinted with modifications from Marcassen, 1973, by permission of Scandinavian University Press)

PLATE 3.2. Photograph (rule extended 2m) of an exposed wall (Knapatorpet site) in a hummock in an area of hummocky moraine in southern Sweden. Photograph shows lower portion of supraglacial subunit. Diamicton beds distinguished by sorted sediment between beds (arrows on photograph) and by stone concentrations in the lower portions of some units. (Reprinted from Möller, 1987)

diamicton beds are interbedded with stratified sorted sediment. The diamicton layers are discontinuous, generally <0.5 m thick, friable, vary in texture, and some have stone-rich layers in their lower part. The latter characteristic indicates that the viscosity of the debris flows, as controlled by water content and texture, was low enough so that larger clasts settled within the flows and moved through traction in the basal part of the flow, as has been observed in flows on modern glaciers (Lawson, 1982). These diamictons commonly have irregular contacts with the stratified deposits with both concordant and discordant relationships (Fig. 3.5). Slump structures, and flow deformation structures are not uncommon. Pebbles in the diamicton generally lack a strong orientation, and most fabrics are unrelated to local topography or ice-flow direction. Stratified deposits are more common in these sections and include silts, sands, and gravels. With the diamictons, they form the uppermost facies association of a glacigenic sequence, and occur in

hills (kames or hummocks) with steep ice-contact slopes.

Sedimentological studies by Möller (1987) in southern Sweden include similar supraglacial deposits. A small hummock in an area of hummocky topography contains about 4 m of interbedded diamicton and sorted sediment abruptly overlying diamicton interpreted to be primary till. The upper supraglacial sequence is composed of two subunits, the lower one containing more sorted sediment and more distinct diamicton beds (Plate 3.2). Some of the diamicton beds are distinctly graded; others are not. S_1 significance values for pebble fabrics are all <0.58 except for one with an S_1 value of 0.71. The former fabrics have a relatively large number of near-vertical clasts and the latter fabric is from the lower part of the diamicton sequence where sorted sediment is more common. Sorted sediment occurs as thin laminae within diamicton and as distinct beds of sand or fine gravel. The upper subunit is mostly massive diamicton with few laminae or lenses of sorted sediment and with weak pebble fabrics. The sequence is interpreted to represent sedimentation in a depressional area in stagnant debris-rich ice during a late stage of downwasting (Möller, 1987). More water and fluvial activity was present early in sedimentation as indicated by the sorted sediment and graded debris flow deposits, some of which have relatively strong pebble fabrics. Later, as suggested by the upper subunit, fluvial activity decreased with time, possibly as a result of decreasing exposure of ice, and the debris flows were thicker and more viscous, probably similar to Type I flows of Lawson (1982).

Although the above types of sequences are typical of supraglacial deposits in low relief areas, sorted sediment also dominates some sequences and sediment flow deposits are relatively minor. These sequences will vary depending on whether fluvial, lacustrine or marine processes were dominate. Ponded drainage was common near many Pleistocene ice margins as a result of isostatic depression of the land, regional slopes toward bedrock lowlands, and end moraine damming (Teller, 1987). These lakes locally extended onto the ice and in some cases many small to large lakes existed in the supraglacial position. Sediment deposited in these lakes generally is typical of that of ice-contact proglacial lakes with sorted gravels and sands near conduits to the lake and along the lake margin, and finer silt and clay in more distal positions. Interbedded with the sorted deposits are diamicton beds of varying character resulting from various subaqueous mass-wasting processes (Chapter 5). Supraglacial lacustrine sediment sequences differ in that faulting and flowage of material was common as a result of collapse during subsequent melting of the adjacent and subjacent ice (Clayton and Cherry,

1967). The resulting structures and deformed sediment, as well as irregular hummocky topography, serve to distinguish supraglacial lake sediment from proglacial lake deposits.

3.2.3. Moderate-relief Areas

In areas of more relief, such as portions of the shield terrane in northern Europe and North America, as well as glaciated dissected plateaus, topography played a stronger role with respect to glacial dynamics and sedimentation. Near ice margins and particularly during deglaciation, stagnant ice became concentrated in valleys and basins and eventually separated into isolated ice blocks (Flint, 1971; Kaszycki, 1987). In such situations, sedimentation was concentrated between stagnant ice and valley sides and around buried ice blocks, as well as subglacially. Sediment variability, similar to that previously described, exists within these sequences, except locally many sequences may be dominated by either ice-contact fluvial or lacustrine deposits.

Kaszycki (1987, 1989) has developed models for sedimentation in such regions based on studies of Pleistocene deposits in the shield terrane of southern Ontario (Chapter 9). Knobs and perched terraces along valley sides are characterized by sediment flow deposits of varying character interbedded with either glaciofluvial or glaciolacustrine deposits. These deposits also fill depressions on some surfaces, or overlie till or bedrock. Kaszycki (1989) delineated three facies within these deposits (Fig. 3.6).

The first facies, rhythmically interbedded diamicton and silt (A in Fig. 3.6), consists of sandy diamicton beds about 10 cm thick that contain small pebbles (1–2 cm) and pinch out laterally over distances of 5 m or more. They are interbedded with thin beds of sand and silt that are inversely graded and have gradation contacts with the diamicton. The rhythmic sequence is interpreted to be the result of a series of subaqueous debris flows in shallow ice-marginal or supraglacial ponds and lakes. Inverse grading likely resulted from dispersive pressures in a zone of basal shear within the flow and the diamicton beds represent either the plug zone of the flows or possibly cohesive flow with laminar shear throughout.

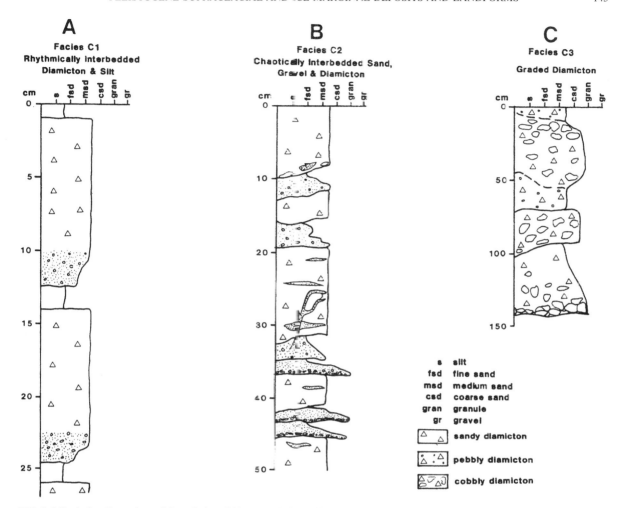

FIG. 3.6. Typical sediment logs of three facies within supraglacial and ice-contact deposits in the Haliburton region, south central Ontario, Canada. (Reprinted from Kaszycki, 1989, by permission of the author)

A second facies (B in Fig. 3.6), with extreme sediment variability, consists of chaotically interbedded sand, gravel and diamicton. Diamicton beds range in thickness from 5 to 20 cm, are matrix dominated, and contain large clasts in the cobble to boulder size. Although diamicton is not graded, cobbles locally mark the lower contact, and inclusions (clasts) of sorted sediment commonly occur near the base of diamicton where it overlies similar deposits. Pebbles have a weak to random orientation within the diamicton. Sorted sediment beds, up to 8 cm thick, commonly grade upward from granules and coarse sand to fine silty sand. Their lower contact with

diamicton is abrupt and erosional in character; internal bedding generally is deformed and chaotic with small scale normal and reverse faulting and folding. Sand dikes, about 6 mm thick, extend upward from the sand beds into diamicton for several meters. Kaszycki (1989) interprets the facies to be the result of deposition in ice-contact debris-fan environments with an abundant supply of meltwater. The diamicton beds were deposited from debris flows that likely were erosional, somewhat channelized, relatively thin, and turbulent in their lower portions, possibly similar to the Type II flows of Lawson (1982). The interbedded granules and sand with normal grading are interpreted to represent either sandy high density turbulent flows or channelized sheet flow deposits. Folds and faults within the sand beds suggest shear deformation from suprajacent and adjacent sediment flows, and the dikes and flame structure indicate rapid loading of water saturated sediment and the development of high pore water pressures and fluidization.

The third facies consists mainly of graded diamicton beds with occasional thin beds of sorted sediment (C in Fig. 3.6). Sediment variability in this facies primarily is the result of the varying concentration of large clasts in diamicton. Kaszycki (1989) recognized three textural classes of diamicton, cobbly, pebbly and sandy, with both decreasing size and concentration of large clasts in the classes. Both normal and inverse grading occur in the cobble and boulder fractions of the coarsest diamicton, which commonly grades to the finer diamicton classes. Diamicton beds range from 0.2 to 1 m thick. Inclusions of sorted sediment are not common, and the thin sorted sediment beds exhibit soft sediment deformation structures. Pebble fabrics are variable, ranging from random to some that are more strongly clustered. The latter have considerable on-site variability, and suggest pebbles were oriented parallel to local depositional slopes. Kaszycki (1989) interprets this facies to be the result of relatively low water content debris flows, similar to the Type I and II flows of Lawson (1982). Ungraded diamicton or portions of a diamicton bed reflect deposition of a non-deforming, rigid plug. Normally graded diamicton suggest the development of transient turbulence that allowed large clasts to settle through the lower portions of a flow. Inversely graded diamicton reflect dispersive pressures in the flow that affected only the larger clasts. Variation in fabric strength reflects the various flow dynamics with the weaker fabrics in the non-deforming plug and stronger, slope-related fabrics in the more sheared portions of the flows. The facies represents deposition in a variety of ice-contact supraglacial and proglacial situations.

Similar supraglacial diamictons and sorted deposits have been described from exposures and subsurface borings for highway construction in the Petteril Valley, Cumbria, England (Huddart, 1983). The Petterill Valley drains generally northward and was the site of glacial downwasting during the last major deglaciation as flow from the main ice sheet was cut off by the Pennines and other local highlands. Sediment sequences vary in the region, but commonly consist of an upper sequence of interbedded diamicton and sorted sediment. The diamicton occurs in thin beds, ranging from a few millimetres to generally less than 30 cm. The beds contain deformed clasts of clay, silt or sand, and are discontinuous laterally. Both pebble and microfabrics are remarkably strong and have an orientation generally parallel to the valley with a downvalley plunge or an orientation either toward or away from the valley side. Sorted sediment varies but is generally sand, silt or clay with some coarser sand and gravel. These deposits commonly are deformed, and locally the entire sediment sequence is highly disturbed. Huddart (1983) suggests that valley marginal, supraglacial and ice-walled lakes and drainageways came into existence with downwastage of the ice sheet. Reworking of exposed debris on the ice and exposed land areas contributed sediment to depositional basins that changed in size and character as the underlying and lateral ice support gave way through ablation. Debris flows locally were dominant and the strong fabric and thin character of the resulting diamicton beds suggests they were relatively fluid. Lacustrine deposits consist of rhythmically bedded silt and clay with thin diamicton beds, which locally grade laterally into deltaic sands and eventually to subaerially deposited sands and gravel of fluvial origin. Deformation structures are largely the result of collapse, but some resulted from soft sediment deformation during sedimentation.

3.2.4. High-relief Mountain Regions

In mountainous regions, topographic relief played a strong role in the character, nature and preservation of the Pleistocene depositional record. Drainage was concentrated within valleys, often between valley sides and the lateral ice margin, as well as downstream from ice-margins. Proglacial lakes were a common feature as a result of glacial dams across tributary valleys and in major valleys when drainage was toward the ice margin. Many such lakes extended onto the glacier and thus were in part supraglacial. Lateral and medial zones of concentrated debris within valley glaciers, basal debris, recently deposited drift, and steep valley sides provided abundant sediment for reworking in the supraglacial, ice-marginal and proglacial environments. Sedimentological studies of Pleistocene supraglacial and ice-marginal deposits in mountainous regions indicate that, except for commonly being coarser, the deposits generally are similar to those described previously in areas of continental glaciation. Mass movement deposits of varying origin are particularly common.

Levson and Rutter (1986, 1988, 1989) report and delineate a number of facies in the glacial sequence in the Jasper area of the Canadian Rocky Mountains in Alberta, Canada (Chapter 9). Among the 12 diamicton facies delineated in the region, six facies were interpreted to be supraglacial or ice marginal/proglacial in origin and, of these, one was proglacial lacustrine in origin and is not considered here. In addition to textural variations in the diamicton and character of the sorted sediment, these different facies were distinguished on (1) the relative amounts of diamicton and sorted sediment (10% to >50%), interpreted to reflect the varying amounts of meltwater fluvial activity near the ice-margin and often related to ice margin proximity; (2) the presence and type of deformation structures present, with normal faults, folds and steep dips suggesting collapse of underlying or adjacent ice, compressional folds and thrusting suggesting overriding by active ice, and small-scale soft sediment deformation structures indicating saturated sediment susceptible to loading and deformation from debris flows; and (3) composition, size, roundness, surface features and abundance of large clasts, with angular and unstriated clasts of non-local origin suggesting a valley side source and supraglacial transport; with local composition clasts suggesting an original basal or subglacial origin; and with degree of roundness and frequency of striations suggesting source and amount of reworking in the ice marginal environment. Pebble fabric data from the diamictons have low S_1 values, generally <0.6, and pebble orientation is not related to the direction of ice flow. All of these diamictons were interpreted to be deposits of varying types of debris flows interbedded with fluvial gravels, sands and silts. Facies associations, position in the sequence, and geomorphic occurrence allowed Levson and Rutter (1989) to place the respective facies in supraglacial or ice-marginal

environments, and among the latter to distinguish ice frontal sequences from lateral sequences.

Other studies in the Canadian Rockies, as well as Alaska. have emphasized the importance of mass wasting processes, often in association with ice-marginal and supraglacial lakes (e.g., Eyles, N., 1987; Eyles et al., 1987; Shaw, 1977c, 1988a). Sediment commonly was supplied to such lakes both subglacially and supraglacially, as well as from valley sides. as debris and turbidity flows. The resulting sediment associations consist of diamictons of varying character and abundance interbedded with sandy sediment and fine-grained silts and clays, often containing dropstones.

Sedimentological study of deposits in Late Pleistocene end moraines near Durango in the San Juan Mountains of southwestern Colorado indicate they are in part composed of sediment derived from the ice-marginal surface (Johnson, 1990). In addition to till, two diamicton facies. matrix-rich diamicton with abundant silt and clast-rich diamicton, were interpreted to be primarily the result of mass-wasting off the frontal ice margin. The latter facies also may consist of significant amounts of fluvially-deposited sediment. Beds in both facies dip downstream and have many of the characteristics of debris flow deposits described previously. They differ in that there is no evidence of collapse or overriding and deformation by active ice. Thus, Johnson (1990) suggests that these facies formed in a proglacial position from debris derived from a supraglacial setting.

3.3. LANDFORMS

Pleistocene supraglacial and ice-marginal landscapes include a varying and complex assemblage of landforms that reflect the interplay of local and regional topographic settings, geology, glacier dynamics, and local and regional climatic controls. These complexities are well illustrated by the differing and strongly contrasting landform assemblages that occur along the southern marginal areas of the Laurentide Ice Sheet in the United States (Mickelson et al., 1983). Glacial landscapes in the plains area of North and South Dakota are significantly different than those in western Ohio, Indiana, and Illinois, and both contrast strongly with those in New England and the plateau areas of New York, Pennsylvania and eastern Ohio. Moving north from Indiana and Illinois to Wisconsin and Michigan, other morphologic differences are apparent among ice marginal landforms. Similar morphologic contrasts exist for other sectors of the Laurentide, as well as for other Pleistocene ice sheets.

3.3.1. Ice-marginal glacial forms

3.3.1.1. End Moraines

End moraines are ridges of glacial sediment that accumulate along and at the margins of glaciers (Menzies, 1995a, Chapter 11). The term moraine was first used by peasants and farmers who observed mounds and accumulations of rocky debris around glaciers in the French Alps (Flint, 1971). The term moraine has since come to be widely used for referring to and describing many types of constructional glacial morphology, and commonly is used with a preceding adjective. When used alone, end moraine is implied and an ice-marginal origin is inferred. The term originally was used for both the morphology and the sediment making up the landform and that practice is still followed in some areas, particularly in parts of Europe. The practice of using the same term for both morphology and material is confusing, however, and the term moraine preferably should be used only with respect to morphology.

Pleistocene end moraines are commonly given a geographic name for some locality located on the moraine, e.g., the Johnstown moraine of the Green Bay lobe in Wisconsin, the Guelph-Paris moraine of the Lake Ontario Lobe in southern Ontario, or the Kirkham moraine of Scottish Readvance Ice in Cumbria, England. They commonly are complex landforms and may partially reflect bedrock topography, earlier glacial events, or multiple episodes of deposition (Mickelson *et al.*, 1983). Those composed of deposits from two or more glacial advances are called superposed end moraines; those that occur over bedrock highs are referred to as rock-cored end moraines; and those that are buried by a relatively thin drift cover of a younger glacial event are known as palimpsest end moraines. The latter term is derived from ancient manuscripts where several writings were superimposed, and the older markings were often evident through younger writing (White, 1962). Wide tracks of end morainic topography, usually with several distinct ridges, are referred to as a morainic system, and morainic topography that develops between adjacent glacial lobes is referred to as interlobate moraine. In mountainous regions, the term lateral moraine is used for the ridge or hummocky topography that result from the accumulation of debris along the lateral glacier margin, and medial moraine for ridges and topographic forms resulting from the deposition of linear debris-rich concentrations in ice, which result when two valley glaciers join and their debris-rich lateral margins merge together.

Pleistocene end moraines are widely used to interpret glacial history, particularly the fluctuations of the ice margin during recent glaciations (cf. Ehlers, 1983b; Šibrava *et al.*, 1986; Ehlers *et al.*, 1991). Correct interpretation is essential if a viable glacial history is to be reconstructed. Their formation generally is considered to mark a time when the glacier was active, the ice margin was relatively stable, fluctuating within a narrow range, and the former glacier's mass balance was more or less in steady state (i.e., they represent stillstands of the ice margin). As discussed below, this is not true for all end moraines and some likely formed during short intervals of time. Those end moraines marking the farthest advance, as well as major readvances during a glacial event, are sometimes classified as terminal moraines, and those formed during deglaciation with limited or no readvance as recessional moraines. Flint (1971) criticized this nomenclature because of the ambiguity of distinguishing end moraines marking major readvances from those that formed during recessional stillstands. However, readvances locally may be recognized by the truncation and overriding of older moraines by younger ones, significant and rather abrupt changes in the shape of successive moraines, abrupt changes in till character at the distal position of a moraine, the occurrence of incorporated proglacial sediment in till of the moraine and not in till beyond the moraine, and the initiation of heads of outwash or other drainage features at or in the moraine (Mickelson *et al.*, 1983).

Pleistocene end moraines vary greatly in both morphology and material content (Sugden and John,

1976; Brodzikowsi and Van Loon, 1991). Those related to ice sheets and ice caps commonly are arcuate in outline, and rise anywhere from 5 to 50 m or more above the surrounding drift plain. Some are short or made up of discontinuous ridges, whereas others can be traced for hundreds of kilometres with only breaks at former meltwater valleys. Some are narrow, only a kilometre or two wide, others form belts ten or more km wide. Local relief within end moraines is just as variable; some have rounded, relatively smooth slopes with relief related primarily to a developed drainage net. Others are highly irregular with many hills or hummocks and depressions or kettles, and such topography classically is what one thinks of as being typical of end moraines (Plate 3.3). It results from downwasting of the ice in the marginal area and much reworking of sediment in contact with ice, and is best described as hummocky topography.

3.3.1.2. End and ground moraine composed primarily of diamicton and sorted sediment

Materials that compose end moraines are highly variable. Many Pleistocene end moraines are composed of diamicton and sorted sediment that originated in the supraglacial environment. Although some subglacial till may occur, mass wasting deposits, such as sediment gravity flow deposits and ice-slope colluvium, and sorted sediment of fluvial and lacustrine origin are more common. Beds usually are relatively thin, discontinuous, and show deformational structures as a result of collapse and ice push and glacier overriding. These deposits, as is the topography, are the result of ice-contact resedimentation processes concurrent with downwasting and backwasting of the ice and active ice flow toward the margin. Many of the deposits described previously are of this type, and the topography of these end moraines is quite hummocky (Plate 3.3).

Other Pleistocene end moraines are dominated by diamicton that is interpreted to be mostly subglacial till, although some likely is of supraglacial melt-out origin. For example, Lundqvist et al. (1993) report that the Johnstown moraine on the western side of the Green Bay Lobe in Wisconsin (Fig. 3.7) is composed of several facies, but that the dominant

PLATE 3.3. Hummocky topography and kettle in the St. Croix moraine, western Wisconsin, USA

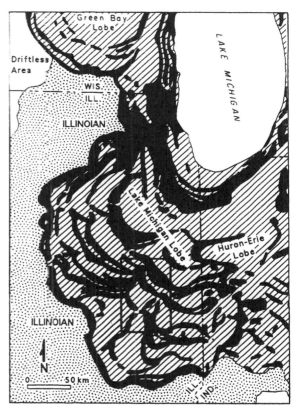
FIG. 3.7. Late Wisconsin end moraines (dark) and drift plains (ruled pattern) of the Green Bay lobe (in Wisconsin), Lake Michigan lobe and Huron-Erie lobe (in Indiana). Rectangle area shown in Plate 3.4 (After Frye and Willman, 1973)

ones are diamicton of varying character interpreted to be till of melt-out and lodgement origin (Chapter 9). The only supraglacial deposits are a thin veneer (<1 m) of bouldery loamy gravel. Although exposures are rare, many moraines of the last glaciation to the south in Illinois (Fig. 3.7; Plate 3.4) appear to be composed primarily of uniform till with varying and generally minor amounts of sorted sediment and debris flow deposits. Wickham *et al.* (1988), for example, report over 90 m of uniform diamicton in Marengo moraine in northern Illinois, and thicknesses of 20 to 60 m are not uncommon in the Bloomington and Shelbyville morainic systems of Illinois (Willman and Frye, 1970). These moraines generally lack well-developed hummocky topography, and typical supraglacial and ice-marginal deposits, although present, are relatively thin. These moraines contrast strongly with many modern end moraines where till is rare in the marginal zone, for example, at the Matanuska Glacier in Alaska where Lawson (1979a) suggests that only about 5% of the marginal deposits are primary till. Large modern end moraines composed mostly of subglacial till are not known, and the origin of these Pleistocene moraines is unclear. They apparently reflect intervals of a millennia or so when the glacier margin was relatively stable, basal or subglacial debris was continually transported to and stacked near the margin, and till slowly accumulated subglacially over a relatively long time interval (Wickham *et al.*, 1988). It is not inconceivable that they in someway may in part be the result of pervasive subglacial deformation (Menzies, 1995a, Chapter 10) (see Chapter 2).

Other moraines, particularly around the margin of the Lake Michigan basin in northeastern Illinois (Fig. 3.7), contain more variable sediment, and debris flow deposits are common (Hansel and Johnson, 1988). In these areas, the regional slope was toward the ice margin, and more basal debris became concentrated in a supraglacial position where resedimentation processes were active. This debris accumulated in the ice-marginal position and forms much of the end moraine. Successive ice-marginal positions often more or less coincided, as previous moraines served as obstacles to glacial movement, resulting in a morainic complex of the superposed type. Totten

(1969) and White (1974) have described similar moraines in northern and eastern Ohio (cf. Karrow, 1989).

In many lowland areas, such as central United States (mid west States), southern Canada (southern Ontario), northern Germany, and southern Sweden, Pleistocene end moraines are separated by lower areas with less local relief, known as ground moraine or drift (till) plain. These undulating areas usually have relief less than 10 m, and often less than 5 m (Chapter 2). Although exceptions occur, the supraglacial sediment association is thin and often discontinuous. It resulted from resedimentation processes in the ice-marginal zone during active ice-margin backwasting and downwasting of the glacier surface. Subglacial deposits commonly occur at or near the surface and subglacial bedforms often are conspicuous and may dominate the landscape (Chapter 2). Locally, minor transverse moraines may occur and, if of ice-marginal origin, represent short-lived ice-marginal positions during deglaciation. Some intermoraine areas were the sites of proglacial lakes, as drainage was dammed behind end moraines and the retreating ice margin. Most or all glacial morphology may be masked by lacustrine deposits, and in such cases these areas are mapped and classified as glacial lake plains, rather than ground moraine.

3.3.1.3. End Moraines Primarily of Glaciotectonic Origin

Glaciotectonic deformation near ice margins commonly results in ridges known as push moraines (Menzies, 1995a, Chapter 10). They are composed primarily of large thrust masses of sediment and bedrock, often imbricated, that have been derived from beneath the glacier. They are particularly well known from the Saalian glaciation of Germany and the Netherlands (Meyer, 1987; Berg and Beets, 1987; Wateren, 1987, 1992), but are not uncommon elsewhere (e.g. in England (Hart, 1990) and in the Plains area of Canada and the United States (Clayton and Moran, 1974; Christiansen and Whitaker, 1976; Moran *et al.*, 1980; Bluemle and Clayton, 1984)). Because such landforms can originate back from the ice-margin (Moran, 1971), only those glaciotectonic forms that originate at ice margins should be

PLATE 3.4. Satellite view of end and ground moraines in east-central Illinois (see rectangle area in FIG. 3.7; Champaign-Urbana, Illinois located in lower center part of scene). Note arcuate end moraines (light tone), reentrant between two glacial sublobes in central and west-central part of area; relict braided pattern on fluvial surface, west-central part of area; parabolic sand dunes on glacial lake plain in north-east part of area, and local small scale fluting and disintegration features

considered as end moraines. The processes involved and resulting morphological forms are described further in Chapter 10 (Menzies, 1995a).

3.3.1.4. End moraines composed primarily of stratified drift

Although the term moraine commonly implies diamicton or 'till', many end moraines are composed primarily of stratified and sorted deposits. In fact, Shaw (1988b) suggests that most end moraines are composed primarily of such material. Although probably an overstatement, the occurrence of fluvial and deltaic deposits in ice-marginal ridges is common, and as long as the term end moraine is used in a morphological sense, these ice-marginal ridges should be classified and mapped as end moraines. Such ridges exist in a terrestrial setting, where they are the proximal parts of outwash or sandur fans and plains, or in a lacustrine or marine setting, where they commonly are the result of ice-contact deltaic or subaqueous fan sedimentation. Diamicton occurs within these moraines, most commonly in the proximal zone, and most is of subaerial or subaqueous sediment gravity flow origin.

Some of the best known end moraines of this type occur in Canada and Scandinavia where the retreating Late Pleistocene ice margin was in contact with water, either lake or sea. The Late Pleistocene Ra and younger moraines in Norway, the Levene, Skovde, and Billingen moraines of Sweden, and the Salpausselkä moraines of Finland are particularly well known (Fyfe, 1990; Lundqvist, 1990). They were deposited in contact with the sea to the west in Norway and in contact with Baltic Ice Lake to the east. Most were associated with the Younger Dryas event, a time of glacier expansion during the general interval of Late Pleistocene deglaciation.

The moraines vary in character laterally but most contain stratified gravels, sands and silts with some interbedded diamicton (Lundqvist, 1990). Deltaic foreset beds are not uncommon, and the moraines often are wider and more deltaic in character near zones where meltwater discharge was discrete and eskers occur in the up-ice direction. In Sweden, the ridges of sediment are less continuous, and in some

area it is difficult to map the ice-marginal position by morphology alone.

The morphology and sediment character of the moraines is controlled in part by water depth and the nature of the subglacial drainage system, as illustrated by the Salpausselkä I moraine in Finland (Fyfe, 1990). Where the glacier margin was slightly above water level, Gilbert-type deltas formed with topset and foreset beds in the distal part and diamicton and coarse gravels in the proximal portion of the moraine (A in Fig. 3.8). The moraine complex is commonly several km wide with about 35 m of relief. The proximal slope is steep and of ice-contact origin with kettleholes common; the distal slope is less steep and has distinct upper fan and delta front surfaces. These fan delta plateaus commonly connect with eskers and

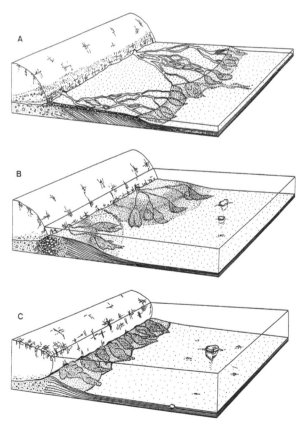

FIG. 3.8. Schematic diagrams suggesting the origin of different portions of the Salpausselkä I moraine in Finland. (Reprinted from Fyfe, 1990, by permission of Scandinavian University Press)

were the result of sedimentation from a distinct subglacial conduit drainage system. Closely spaced conduits resulted in overlapping fan deltas and a continuous end morainic ridge.

Where the ice margin was grounded at a relatively shallow water depth, sedimentation resulted in a moraine that is narrower, has more relief, and both proximal and distal slopes are steep (B in Fig. 3.8). Under these conditions the moraine consists primarily of foreset deposits. In areas where the water depth was greater, in the order of 60 m or so, sedimentation continued to be restricted, and a narrow, but lower ridge resulted. It is made up of a wedge of cross-bedded sands and some gravels that grade distally to fine sand and silt. In even deeper water, the ice-marginal ridge looses much of its morphology and is several kilometres wide with about 10 m of relief (C in Fig. 3. 11). These latter subaqueous moraines generally are not connected with eskers in the up ice direction. Fyfe (1990) interprets the changing morphology to be a reflection of the nature of sediment delivery to the ice-margin. As water depth increased, the basal shear stress of the ice sheet near the margin was reduced, which led a lowering of the ice sheet profile and a destabilization of large conduit sub-glacial drainage. The subaqueous ice margin had a more distributed drainage system, either of linked cavities or small closely spaced conduits, and there was more even distribution of sediment to the marginal area.

Similar moraines have been described from Canada and the United States. Sharpe and Cowan (1990) report long, arcuate moraines about 10–40 m high and 1/2 to 3 km wide in Ontario. The moraines have a steep proximal slope and a more gentle distal slope, and are composed primarily of stratified sand and gravel, with large boulders on the surface. They formed during backwasting of the Laurentide Ice Sheet when it was in contact with either proglacial lakes or the sea. The Vermilion moraine in Minnesota locally is similar (Lehr and Hobbs, 1992; Lehr and Matsch, 1987). It is narrow and deltaic in origin with well developed foreset beds. The distal side of the moraine is distinctly scalloped as a result of spacing between conduits, which were discharging into a proglacial lake. Somewhat similar ridges and forms in Maine, called 'stratified end moraines' by Borns

(1973), primarily are the result of marine fan sedimentation (Ashley et al., 1991) (Menzies, 1995a, Chapter 13). An ice-marginal origin is indicated by an up-ice association with esker deposits, and more diamicton and considerable active ice deformation in the proximal deposits.

Within bodies of water where an ice mass has temporally grounded, moraine ridges variously termed De Geer, cross-valley, ribbed or washboard moraines can form.

Other Pleistocene moraines are similar to these, except instead of forming in a water body, they were land-based and are composed of fluvially deposited sand and gravel. In essence, these moraines represent 'heads of outwash' situations where closely spaced drainageways from the glacier coalesced to form a more or less continuous ridge, which has a steep ice-contact proximal side and a more gentle distal side. Kettles near the former ice margin give the ridge a morphology that is irregular and distinctly morainic in character. Several moraines and moraine complexes of this character have been described in Michigan (Rieck, 1979; Blewett and Reick, 1987).

3.3.1.5. Lateral, End and Medial Moraines of Mountain Glaciation

Pleistocene glacial depositional landforms in mountainous regions commonly are quite distinct and contribute to the spectacular scenery of these glaciated regions. Most are ice-marginal in origin and consist of end and lateral moraines, as well as medial moraines. The latter are the longitudinal ridges along valley bottoms that formed downice from the junction of two tributary glaciers. Because drainage and fluvial activity was concentrated in valley bottoms, end and medial moraines commonly were not preserved as part of the geomorphic record. Nonetheless, there are many mountain valleys where late Pleistocene forms are extremely well preserved, and the former position of the ice margin easily can be reconstructed (Plate 3.5).

Lateral and end moraines form a continuum around the former valley glacier margin. They vary from distinct, narrow ridges with steep slopes, which loop around the former margin (Plate 3.5), to broader tracts of hummocky morainal topography marking the

PLATE 3.5. Lateral and end moraines of Bloody and Sawmill Canyons, Mono Basin area, California, USA. Moraines represent several major glaciations; Walker Lake dammed behind youngest moraine. (Photograph by Peter Birkeland)

PLATE 3.6. Broad area of Eightmile end moraines, Yellowstone valley, Montana, USA. Moraines mark the limit of the last glaciation; region beyond the dashed line not glaciated. (Photograph by W. B. Hall, from Pierce, 1979)

terminal zone (Plate 3.6). As with ice sheets, the character of the moraine depends on glacial dynamics, including mass balance, and the nature and style of deglaciation. Although these end moraines commonly are cut by one or more drainageways, many have not been dissected and form dams encircling picturesque lakes of varying size and shape. Such lakes are common in the lower portions of valleys in many mountain ranges that were glaciated during the

Pleistocene. The end moraines have characteristics suggesting they formed primarily by dumping of debris at the ice margin. Resedimentation processes, particularly mass wasting, were active and the moraine forms consist of a complex assemblage of deposits as previously described. Because smaller

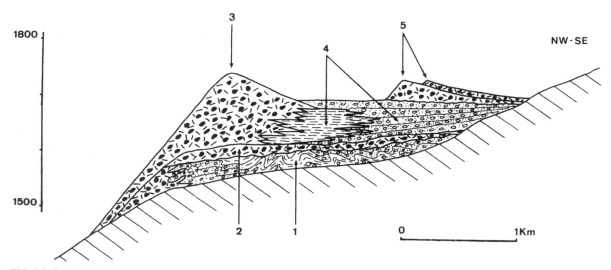

FIG. 3.9. Composite cross-section the Cerler lateral moraine complex, Esera valley, central southern Pynenees, Spain. Diamicton of lateral moraine, (3), overlies older deposits, (1 and 2), and interfingers with lacustrine deposits, (4), that formed as a result of glacial damming of a tributary valley. A later glacier advance in the tributary valley deposited younger end morainic ridges, (5). All deposits related to the last glaciation. (Reprinted from Bordonau 1993, by permission Pergamon Press Ltd)

tributary valleys were often dammed by major valley glaciers, lateral moraine deposits often interfinger with marginal fluvial and lacustrine deposits in the dammed valley (Fig. 3.9).

3.3.2. Ice-marginal fluvial forms

3.3.2.1. Ice-marginal meltwater channels and systems

Along the margins of Pleistocene ice masses, considerable volumes of meltwater, especially during the ablation season, carved channels of various sizes and/ or deposited an assortment of stratified sediment. Such channels were most common in areas of moderate to high relief where the drainage was confined between the glacier and valleys sides or in situations where the glacier flowed up a regional slope and drainage again was confined between the ice and land areas.

Many of the marginal channels reflect a compromise between the slope of the topography and the ice margin itself, thus in many cases channels appear to disregard the normal slope gradient cutting instead across hillslopes at low angles (Mannerfelt, 1945, 1949; Sissons, 1961) (Chapter 2). Such channels or erosional benches vary in character, both in cross-section and longitudinally, depending on whether drainage was completely on land, on or in ice, or some combination of the these. As a result, the erosional forms tend to be discontinuous longitudinally and many channel cross-sections are not complete, with only the portion cut in rock or sediment preserved. In many instances, marginal channels intersected with sub-marginal and subglacial channels forming an intricate network of open and closed channel systems (Chapter 2). Where debris was limited or where very large meltwater discharges occurred meltwater erosional p- or s-forms are typically found on the channel surfaces cut in rock (Chapter 2).

3.3.2.2. Kame terraces and kame deposits

Where glaciofluvial deposition was the dominant process along former ice margins, various landforms of stratified drift developed. Along valley sides, and in some cases between separate but adjacent ice lobes, large volumes of glaciofluvial sediment were deposited. The former situation was particularly common and the resulting landforms, kame terraces, dominate many valleys (Plate 3.7). They commonly have an irregular upper surface as the result of kettle development, and a relatively steep ice-contact side facing the central valley. Their gradient usually reflects the ice-margin slope in the down-ice direction (Sissons, 1958b; Young, 1974, 1975, 1980; Gray, 1991). Kame terraces formed at the same time but on opposite sides of a valley commonly are not at the same elevation because of variations in microclimate related to the aspect of the valley. This relationship, as well as their morphology and spatial relationship to other glacial features, serves to help distinguish them from alluvial terraces composed of proglacial fluvial sediment (Gray, 1991). In many valleys, as the ice downwasted and the margin retreated, a sequence of kame terraces formed along the valley sides (Jahns, 1941; Sissons, 1958a). These terraces are composed of a variety of stratified and sorted sediment with a wide range of particle sizes, as well as masses of diamicton of sediment flow origin. It was from the occurrence of diamicton in such kame terraces that the term flow till was originally coined (Hartshorn, 1958). Sediment in the ice-contact slope typically exhibits faulting and slump structures indicative of post-depositional movement during and following ice retreat. The

PLATE 3.7 Kame terrace surface with kettle to right, White River valley, Vermont, USA. Sand and gravel locally is mined from this landform.

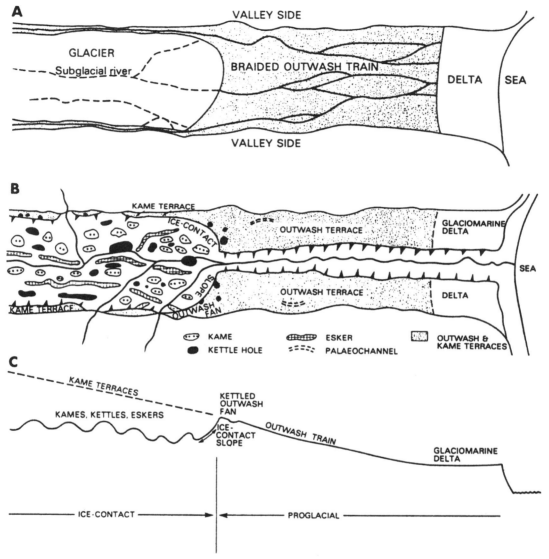

FIG. 3.10. Schematic maps (A), during glaciation; (B) after glaciation, and (C) longitudinal profile of B; showing inferred origin of glaciofluvial morphosequence composed of ice-contact and proglacial deposits. (Reprinted from Gray in Ehlers, Gibbard and Rose (eds.) 1991, courtesy of A.A. Balkema Publishers, Rotterdam)

surface of the terraces are often pitted and contain kettle holes where abandoned ice blocks were left to slowly ablate while sedimentation occurred around them. Also the terrace surface may have small meltwater marginal channels that may dissect the terraces into a large number of accordant segments. In many instances kame terraces merge down-stream into proglacial deltaic terraces, valley train or sandur plains or lacustrine plains. (Gray, 1991 (Fig. 3.10)).

Such a continuum of meltwater deposits originating on, in, along, and below the ice, and extending to and beyond the stagnant ice margin form a genetic sequence that has been termed a morphosequence (Jahns, 1941; Koteff, 1974; Koteff and Pessl, 1981) based on studies in New England. Each sequence is inferred to have formed during an interval when the base level for regional drainage in the valley was relatively stable and a relatively distinct zone of stagnant ice existed for some distance back from the ice margin to the region of active ice. Different types of sequences have been distinguished depending on the environments of deposition, fluvial, lacustrine or marine, and on whether the sequence begins in contact with the stagnant-ice margin or not. A typical fluvial ice-contact sequence might consist of a head of outwash, kame terrace, associated kames and eskers, and outwash plain (Fig. 3.10) or a delta plain. if drainage terminated into a lake. Koteff (1974) emphasized debris in the basal ice as the main sediment source, and suggested a 'dirt machine' existed whereby shearing and compressive flow at the active ice margin would continually bring basal debris to the surface. More recently, subglacial fluvial erosion has been recognized as an important and probably dominant contributor of sediment, and Gustavson and Boothroyd (1987) suggested that the Malaspina Glacier in Alaska is an appropriate modern analog for these landforms and sediment sequences from the Pleistocene of northeastern United States.

Kames are roughly circular mounds that are composed of stratified glaciofluvial sediment. They form in contact with ice and usually occur in association with other forms of glaciofluvial deposition. Kames originate in a variety of ways, sometimes from sediment being deposited in cavities within or below ice, or in supraglacial depressions, with the resultant mound forming as the sediment is let down onto the ground surface. Highly conical and circular kames often form below moulins, as surface drainage, upon reaching the ground surface, spreads laterally and deposits its load in a cone-shaped deposit. As in the case of kame terraces, sediments are of a wide range in particle size and sorting, and due to the removal of ice walls, sediment in kames commonly exhibits faulting and slumping. In some instances, as in Sweden and Scotland or locally in the Kettle Interlobate moraine in Wisconsin, valley lowlands or plains are filled with kame fields, the result of *in situ* downwasting of ice where diffluent ice lobes have been stranded (Black, 1970; Ehlers, 1983b; Ehlers *et al.*, 1991).

3.3.4. Reconstruction of Glacier Form From Ice-marginal Landforms

In valleys and lowlands, and in other areas where nunataks protruded through the glacier, ice-marginal landforms, such as end and lateral moraines, kame terraces and erosional benches, have been used not only to map the extent of a glaciation, but also to reconstruct Pleistocene glacier profiles and surface form (Mathews, 1974; Thompson, 1972; Pierce, 1979, Thorp, 1986, 1991; Sutherland, 1984; Beget, 1987; Klassen and Fisher, 1988; Clark, 1992). Deposits and landforms along the lateral glacier margin usually reflect the glacier slope. From these data, as well as elevation data from marginal erratics and trimlines on former nunataks, the surface form of former glaciers has been reconstructed. Ice thickness can be inferred from the reconstruction, and with the ice-surface slope, former basal shear stresses can be calculated. Such data have been important in evaluating the dynamics of Pleistocene ice sheets, particularly with respect to the strength of bed materials and the possibility of pervasive subglacial deformation (Chapter 2; Menzies, 1995a, Chapter 10).

3.3.4. Supraglacial forms and landscapes

3.3.4.1. Disintegration features — Hummocky moraine

Large areas glaciated during the Pleistocene are characterized by distinctive topography, which resulted from stagnation of the marginal zone and downwasting of the glacier (cf. Hoppe, 1952; Sissons, 1967; Aartolahti, 1974; Hodgson, 1982; Eyles, 1983b; Attig and Clayton, 1993). Areas of stagnant ice developed either after an extensive advance, possibly a surge (e.g. Sharpe 1988b; Kemmis, 1991), or incrementally as the marginal zone became insulated by supraglacial debris, eventually stagnated, and the region of stagnation grew progressively larger in the up-ice direction (Clayton *et al.*, 1980). The latter situation locally was common and eventually large regions were covered by debris-covered dead ice. Stagnant ice landforms are particularly well developed where the glacier flowed up a regional slope and large quantities of basal debris moved toward the surface as a result of intense compressive flow. After melt out, this debris was reworked during the melt-down process and the stagnant ice landforms are composed in large part of typical supraglacial sediment – mass wasting, lacustrine and fluvial deposits. Diamicton, mainly of sediment flow origin, commonly is the dominant material, but locally sorted sediment may be widespread. The landscape is irregular and chaotic in appearance, with many small hills and depressions with steep to gentle slopes. Most hills reach about the same elevation and lack any consistent trend (Plate 3.8). These areas generally lack end moraines, and the areas could be considered as 'high relief' ground moraine.

Portions of the plains area of western Canada and north-central United States are characterized by terrain of this type, and much of the descriptive nomenclature used for it originated there (cf. Clayton, 1967; Moran *et al.*, 1980; Aber *et al.*, 1989). The lack of trees in the region make features particularly evident, and they are quite distinct on aerial photographs (Plate 3.9). Gravenor and Kupsch (1959) introduced the term 'disintegration' to refer to the separation of ice blocks during the wastage and decay of the ice sheet, and suggested that the term could also be used to describe individual landforms that resulted from the process. Disintegration features are considered controlled if the landforms maintain some regular and repetitive pattern that resulted from ice control, such as a crevasse pattern or differential debris distribution in the ice as a result of shearing or folding; if the forms are randomly distributed on the landscape, they are described as uncontrolled. Features can then be described by adjectives and nouns, (e.g. 'circular disintegration ridges' or 'linear disintegration troughs'). The overall landscape is best described as hummocky moraine, and it can be further subdivided into high- (>10 m), intermediate-, and low-relief (<3 m) areas for mapping and descriptive purposes. Genetic terminology commonly refers to such areas as stagnant ice or dead-ice moraine, or collapse moraine in order to emphasize the process (Clayton *et al.*, 1980). It should be realized, however, that all hummocky topography is not of supraglacial origin, but can form as a result of subglacial processes

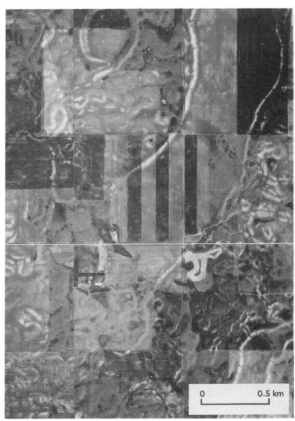

PLATE 3.9 Aerial photograph of hummocky moraine with small circular disintegration ridges and larger perched lake plains with raised rims (see Fig. 3.12). The larger ice-walled lake extended onto buried ice to the lower left and lake sediment subsequently collapsed resulting in hummocky topography. Mountrail County, North Dakota, USA. (US Department of Agriculture Photograph)

PLATE 3.8 Hummocky ground moraine resulting from down-wasting of stagnant ice concurrent with reworking and collapse of supraglacial sediment. Mountrail County, North Dakota, USA. (Photograph by Lee Clayton)

(Chapter 2). In fact, the features described here have been interpreted by some to be the result of squeezing of debris in subglacial crevasses in the marginal zone (Hoppe, 1952, Stalker, 1960). In addition, Attig and Clayton (1993) described multiple origins of hummocky topography along a former ice margin in northern Wisconsin.

A supraglacial landscape usually reflects the final stages of wastage of the ice, which commonly is the last of a sequence of topographic inversions as sediment moves from high areas on the glacier to low areas, only to be reworked later as a result of differential melt rates resulting from differential insulation of ice by supraglacial debris (Menzies, 1995a, Chapter 11). Thus, high areas of the landscape today reflect low areas immediately following glacier wastage, and low areas of the landscape reflect higher ice-cored areas in the past. Two of the most common forms are circular hummocks and circular disintegration ridges or 'doughnuts', the latter being hummocks with a central depression (Gravenor, 1955; Parizek, 1969; Aartolahti, 1974). Clayton (1967) and Clayton and Moran (1974) interpret these forms to be related to 'sinkholes' or large funnel-shaped moulins on the glacier surface (Fig. 3.11). Sediment that accumulates in the depression is let down to form the hummock as the surrounding ice melts. If the hummock late in its formation has an ice core, a depression forms in the central area resulting in a circular disintegration ridge (Fig. 3.11). Clayton and Moran (1974) suggest that sinkholes on glaciers have a maximum diameter of about 200 m, which is the size of many Pleistocene hummocks, lending support to the collapse origin of these forms. They also suggest that the amount of relief in hummocky moraine is directly related to the thickness of supraglacial sediment. Many of the hills, ridges and depressions in stagnant-ice moraine are irregular in shape, and reflect the complex and continually changing supraglacial topography during downwasting of the glacier surface.

As the supraglacial debris cover became thicker, glacier ice was completely buried and the glacier surface eventually stabilized as vegetation became established. Melting rates were reduced and the supraglacial environment was much less dynamic. Under such conditions, thick deposits of melt-out till formed below the supraglacial sediment cover. Some of the hummocky moraine areas in Iowa (Kemmis, 1991) are characterized by such sediment sequences, with kettle depressions reflecting the former position of englacial and subglacial drainageways through the stagnant ice (Price, 1969, 1973).

3.3.4.2. Raised plateaus and perched lake plains

Many hummocky moraine areas also contain small to large irregularly-shaped plains or plateaus that usually stand higher than the tops of the hummocks. These features are relatively flat, may or may not have a rim around their margin, and generally lack evidence of collapse topography (Plate 3.9) (Chapter 2). They are composed of a variety of sediment types; many contain mostly lacustrine deposits, but others consist of interbedded mass-wasting, fluvial and lacustrine deposits. The lack of collapse topography indicates that deposition took place on firm ground, not on ice, but the depositional basin must have had walls of stagnant ice (cf. Clayton and Cherry, 1967; Markren and Lassila, 1980). Such raised features in North America have been called moraine plateaus (Gravenor and Kupsch, 1959), ice-walled lake plains (Clayton and Cherry, 1967), and moraine-lake plateaus

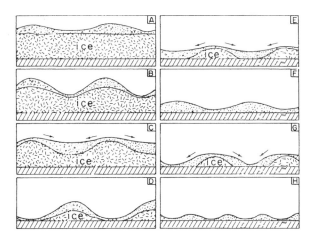

FIG. 3.11. Schematic diagrams showing the formation of hummocks and circular disintegration ridges during the redistribution of supraglacial sediment as stagnant ice downwastes. A, B, C, D, Inversions of topography as a result of nonuniformity of debris and variable melt rates; E, F, formation of hummocks; G, H, formation of circular disintegration ridges. (From Clayton and Moran in Coates (ed), 1974, by permission of SUNY Binghamton)

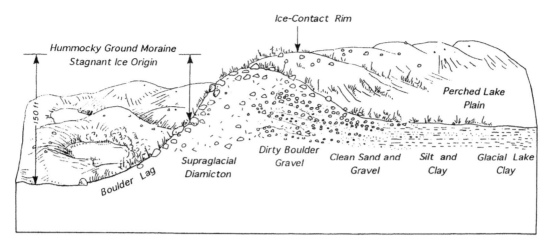

FIG. 3.12. Schematic cross-section showing sediment relationships among rim, perched lake plain, and hummocky ground moraine (see Plate 3.9). (Reprinted from Parizek, 1959, permission of the author).

(Parizek, 1969). The rims (Fig. 3.12) reflect sediment that accumulated near the margin of the depression as sediment was transported to or flowed down the slopes of the basin, and are more common where ice was exposed around the margin of the lake or depression (Clayton and Cherry, 1967). These forms are similar to the Blattnick moraine of Scandinavia (Markgren and Lassila, 1980) and may have some relationship to forms of the Veikki moraine as described in Finland (Minell, 1979; Aario, 1990) (Chapter 2).

3.4. SUMMARY

Supraglacial and ice-marginal deposits and landforms are among the more varied, complex, and distinctive of the glacial system. Those of the Pleistocene have been studied for over a century, and much of the terminology that is used for them have developed from studies of this ancient record. In turn, interpretations of the historical record of glaciations during the Pleistocene locally are based primarily on determining the relative age of these deposits and landforms. In ideal situations they have been used to reconstruct the profiles and thickness of the vanished glaciers that were responsible for them, and to infer their dynamics.

Many of the sorted deposits are an important resource as an aggregate material, and where buried, locally serve as major sources of groundwater (Melhorn and Kempton, 1991). The varied lithologic and textural character of supraglacial deposits and their discontinuous nature and varied morphology make them an important consideration in land-use and construction site evaluations. Thus, Pleistocene supraglacial and ice-marginal deposits and landforms have and will continue to play an important role in both academic and applied aspects of glacial geo(morpho)logy.

Chapter 4

ICE SCOUR AS AN INDICATOR OF GLACIOLACUSTRINE ENVIRONMENTS

C.M.T. Woodworth-Lynas

4.1. INTRODUCTION

A considerable body of knowledge exists concerning the hydraulic and depositional environments of glacial lakes and the nature of laminated sediments that are commonly deposited in them. For a thorough review of modern glacial lake depositional environments, processes and sediments the reader is referred to Chapter 13 in Menzies, 1995a (Church and Gilbert, 1975; Eyles and Miall, 1984; Drewry, 1986: Allen and Collinson, 1986). An extensive literature on the Quaternary continental glacial lakes of North America is found in Teller and Clayton (1983) and Karrow and Calkin (1985). However, there are few descriptions of Pre-Pleistocene glacigenic sediments that are unequivocally of glaciolacustrine origin. This is largely because of the absence of distinctive fossil fauna or flora in ancient glaciolacustrine and glaciomarine sediments, and because of some general similarities in facies. For example, Hambrey and Harland (1981a), in their compilation of the earth's Pre-Pleistocene record, do not attempt to distinguish glaciomarine from glaciolacustrine deposits, and present three key criteria that may be used as evidence for identifying either facies, namely, (1) the presence of dropstones, (2) finely-graded stratification, such as laminated clay or siltstones, and (3) association of tillites with resedimented deposits.

A survey of the papers in Hambrey and Harland (1981a) reveals many authors who describe glacial-aquatic deposits but refrain from giving a definite interpretation of glaciolacustrine or glaciomarine deposition, preferring to acknowledge the possibility of either environment as viable hypotheses. However, some authors do make explicit or implicit interpretations of glaciolacustrine conditions. Most of these interpretations are founded on the existence of varvites/laminites with dropstones associated with facies consistent with continental glaciation. But, as pointed out above, laminated sediments with dropstones can be associated with either glaciolacustrine or glaciomarine environments. Unfortunately, this ambiguity in possible interpretations does little to assist in the recognition of ancient glaciolacustrine rocks. Consequently, there is a need to develop specific criteria that distinguish ancient glaciolacustrine from glaciomarine sediments. These criteria must be based on detailed knowledge of facies associations from Pleistocene and Recent glacial lakes (see Menzies, 1995a, Chapter 13).

This chapter specifically focuses on recognizing the effects of ice scour as a possible environmental indicator. Ice scour is a post-depositional mechanical process that affects both glaciolacustrine and glaciomarine sediments. As icebergs and ice pressure ridges drift in glacial lakes or oceans they periodically contact unconsolidated lakefloor or seabed sediment when their draft exceeds water depth (Gipp, 1993). Once in contact, if a floating ice mass continues to be driven forward by currents and winds, its keel will plough through the surficial sediments. This action creates a characteristically curvilinear scour mark that consists normally of two raised embankments (berms) of sediment on either side of a central trough which has been excavated by the moving keel. Sediment beneath the scouring keel responds to loading by compaction and faulting.

Ice scour is a common modern phenomenon in both the northern and southern polar ocean regions and has been recognized as an important marine geological process during the Quaternary period. Surficial marine sediments tend to be exposed to the effects of ice scour for long periods of time (on the order of tens of thousands of years) allowing multiple scour events to destroy internal sedimentary structure and rework previous scour events such that individual scour marks are rarely preserved. The result is a reworked deposit called an ice keel turbate (Barnes and Lien, 1988).

In contrast, ice scour in glaciolacustrine sediments rarely has been recognized. It has been argued that surficial sediments in glacial lakes probably were exposed to fewer scour events, in relative terms, than glaciomarine sediments due (a) to rapid sedimentation rates burying and preserving single scour marks, and (b) the ephemeral nature of glacial lakes. Therefore, although ice keel turbates may be created it is more likely that discrete scour marks will be preserved in glaciolacustrine sediments. As a consequence it is useful to be able to recognize ice scour marks in the rock record because they might assist in distinguishing between glaciolacustrine and glaciomarine facies. There is little information on ice scour in glacial lakes, thus the following discussion draws extensively on knowledge of ice scour in the ocean.

The process of ice scour in glacial lakes and in the ocean is identical. This is confirmed by the work from Lake Superior where Berkson and Clay (1973) collected sidescan sonographs of relict iceberg scour marks having surface features directly comparable with similar oceanic scour marks (Geonautics Limited, 1989, for Canadian eastern continental shelf; Gilbert and Pedersen, 1986, for the Beaufort Sea).

4.1.1. Previous Work

Modern-day scour by seasonal ice and icebergs in the earth's polar and sub-polar oceans is well known (Geonautics Limited, 1985; Goodwin et al., 1985; Hodgson et al., 1988; Lewis and Woodworth-Lynas, 1990; Woodworth-Lynas et al., 1991). Present-day scouring by seasonally-occurring ice generally takes place during the spring breakup in large polar to temperate freshwater bodies such as Great Slave Lake and Lake Erie, Canada (Weber, 1958; Grass, 1984, 1985). The planform pattern of relict scour marks visible on the present-day land surface has been reported, for example, from large areas of Canada and the northern United States that were formerly occupied by Quaternary glacial lakes, such as Lake Agassiz (Horberg, 1951; Clayton et al., 1965; Dredge, 1982; Mollard, 1983) Lake Ojibway (Dionne, 1977) and Lake Iroquois (Gilbert et al., 1992). Similar patterns, created by icebergs during deglaciation, also have been identified on the modern lake floor of Lake Superior (Berkson and Clay, 1973).

However, complete cross-sections of scour marks have been described from only four localities worldwide. Three of these localities comprise glaciolacustrine environments, and one (in Norway) records a glacial outburst flood that inundated partially exposed marine sediments (Longva and Thoresen, 1991). In Scotland, Thomas and Connell (1985) described a 10 m long grounding structure, at least 2 m deep, from Devensian age (Wisconsinan/Weichselian) glaciolacustrine sediments. Small scale reverse faults were observed below the inner margins of the trough and 'downfolded' strata were seen extending to at least 1.3 m below the trough (Fig. 4.1). Thomas and Connell interpreted the structures to be the result of ice/sediment interaction during a grounding event caused by a slow lowering of water level, with no horizontal keel movement: processes that are not typical of most ice scour marks. This feature is analogous to modern 'scour pockets' created by grounding icebergs during annual breakout floods (Fahnestock and Bradley, 1973).

Eyles and Clark (1988) described a well-preserved scour mark, approximately 9 m wide and 2.5 m deep, at Scarborough Bluffs, Ontario. They interpreted the scour mark, incised into delta foreset sediments, to

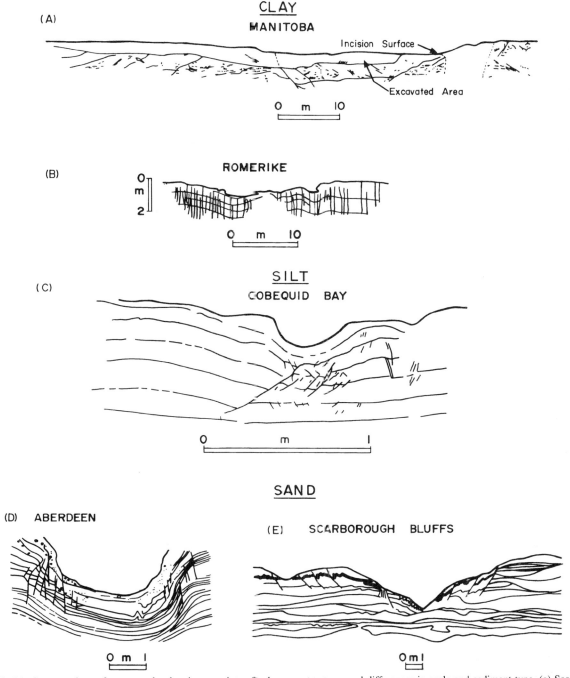

FIG. 4.1. Cross-sections of scour marks showing a variety of sub-scour structures and differences in scale and sediment type. (a) Scour mark (approximately 9,900 years old) from glacial Lake Agassiz (from Woodworth-Lynas, 1990); (b) Scour mark (approximately 9,200 years old) from Romerike, southern Norway formed during a glacial outburst flood (from Longva and Bakkejord, 1990); (c) scour mark (approximately two months old) from modern tidal flat, Cobequid Bay, Nova Scotia (unpublished data); (d) Scour mark of Wisconsin age, in glaciolacustrine sediments (from Thomas and Connell, 1985); (e) Scour mark (approximately 60,000 years old) in delta front glaciolacustrine sediments, Scarborough Bluffs, Ontario (from Eyles and Clark, 1988)

have been made by a pressure ridge keel in water depths of 20 m about 60,000 years ago. They described thrust and normal faults, load casts and folds below and on either side of the scour mark trough (Fig. 4.1), and suggested that sediments had been affected by shearing up to 2 m below the scour mark trough during the scouring event.

In Norway, Longva and Bakkejord (1990) reported on excavations of two iceberg scour marks and an iceberg pit that were formed during a Holocene glacial outburst flood in the Romerike area about 9,200 years ago. Excavated sections across the scour marks revealed evidence of folding, faulting and sediment liquefaction (Fig. 4.1). Deformation in sub-scour sediments occurred to approximately three times the depth of scour mark incision (Longva, *pers. commun.* 1986).

In an analysis of scour marks from glacial Lake Agassiz, Woodworth-Lynas and Guigné (1990) described major normal faults with dip-slip displacements of at least 3.5 m beneath a scour mark trough (Fig. 4.1). These faults are likely the result of bearing capacity failure of the clay sediment beneath the scouring keel (Poorooshasb *et al.*, 1989). The faults propagate to depths beyond 5.5 m below the deepest part of the scour mark trough (Woodworth-Lynas, 1990). Sub-horizontal thrust faults occur beneath the scour mark berms (Woodworth-Lynas and Guigné, 1990), and at depths up to 4 m below the central scour mark trough.

4.2. SCOURING IN THE GEOLOGICAL RECORD

Prolonged reworking of submerged sediments by the action of grounding and scouring iceberg keels will produce facies called iceberg turbates (Vorren *et al.*, 1983). However, the general term ice keel turbate (Barnes and Lien, 1988) is more useful because (a) it describes the effect of scouring by all types of floating ice, and (b) it may not be possible to distinguish between turbates created by scouring icebergs, seasonal ice or some combination of both. Ice keel turbates have been recognized from marine acoustic profiles in unconsolidated offshore Quaternary glaciomarine surficial sediments (Vorren *et al.*, 1983; Josenhans *et al.*, 1986; Barnes *et al.*, 1987).

Until recently neither ice scour marks nor ice keel turbates had been documented in the rock record (Woodworth-Lynas, 1988). There are two elementary reasons for this: (1) criteria for the visual recognition of scour marks in rocks have not been established; and (2) ice keel turbates consist of multiple, superimposed ice scour marks, and because scour marks have not been recognized, criteria for the visual recognition of ice keel turbates also have not been established. The only exceptions are descriptions of Permian age scour marks in Brazil (Rocha-Campos *et al.*, 1990) and in Australia (Powell and Gostin, 1990) (see discussion below).

Development of criteria for the recognition of discrete ice scour marks is essential before establishing some common elements that define ice keel turbates. As noted above, glaciolacustrine sediments offer several advantages in studying the detailed morphology and structure of ice scour marks namely, (1) the relatively short life of glacial lakes restricts the amount of time, and hence the number of scour events, during which ice scour marks may be created; (2) there is usually a seasonally high rate of sediment accumulation derived from turbid meltwater originating from subglacial outlets at the grounding line, and from proglacial rivers and streams (distal deposits are often well stratified, providing ideal marker horizons that will register and display deformation structures associated with scouring); and (3) changes in lake level effectively segregate scour mark populations either by exposure above lake level or by removing lakebed sediments from the effects of scour as lake depths increase. These three factors tend to favour the creation and preservation of discrete ice scour marks and mitigate against the creation of ice keel turbates.

In comparison, modern glaciomarine sediments affected by ice scour are located on continental shelf areas. Glaciomarine shelf areas tend to be sediment-starved (Piper, 1991), and sea levels may fluctuate as in glacial lakes but at generally slower rates. Additionally, scouring by icebergs may continue long after glaciation has ceased on the adjacent continental area because of the drift of icebergs supplied by distant calving margins in higher polar regions. These factors tend to mitigate against the preservation of discrete scour marks and favour the production of ice keel turbates. Field investigations of marine scour marks are considerably more difficult, requiring expensive logistics and remote sensing tools that, no matter how technologically advanced, cannot match the level of detail acquired by onland studies.

4.2.1. Fossil Scour Marks and Ice Keel Turbates

The recognition of fossil ice scours and ice keel turbates is important first, as a practical aid to the location and safety of offshore pipelines. For example, the Canadian and United States oil industry reappraised engineering design for the protection of offshore oil and gas pipelines in ice-scoured continental shelf areas as a direct consequence of the definition of sub-scour deformation from beneath large scale relict iceberg scour marks in glacial Lake Agassiz (Woodworth-Lynas and Guigné, 1990; Woodworth-Lynas, 1990; Clark et al., 1990). However, from an engineering standpoint, the observations from Lake Agassiz scour marks must be used with caution because the deformation structures are developed in relatively well sorted clays which are uncommon beneath northern latitude continental shelves. In addition, the Manitoba clays had a history of subaerial exposure and re-submergence prior to scouring (Woodworth-Lynas and Guigné, 1990) which may have affected their response to ice keel loading; in Canada most continental shelf surficial sediments presently affected by scouring icebergs have never been exposed, except for parts of the Grand Banks of Newfoundland and the Beaufort Sea. The key issue arising from acceptance of the Lake Agassiz study results by offshore engineers concerns the effects of variation of grain size on the response of sediment to ice scour. In geological terms this concern can be translated into the question: What are the types and depths to which deformation structures occur in different sediment types?

A second reason for studying fossil ice scour marks and ice keel turbates is their potential environmental impact. Scouring of sediments by floating ice keels is a global phenomenon, occurring seasonally over vast areas of high latitude and temperate continental shelves and lakebeds (Fig. 4.2) and probably reworking large volumes of sediment. For example, a conservative estimate of the volume of sediments

reworked by scouring icebergs on the Labrador continental shelf each year is 0.03 km^3 (Lewis et al., 1989), or as much as 300 km^3 in the last 10,000 year period. The latter figure must be regarded as a very conservative estimate because it is based upon modern iceberg flux rates which are probably far less than the Holocene average. Unfortunately there are no data documenting scouring rates or volume of sediment reworked in contemporary glacial lakes or glaciolacustrine facies of any age.

Prolonged exposure to scouring by many keels can modify sediments to the point of almost complete homogenization of original sedimentary structures (Vorren et al., 1983; Josenhans et al., 1986). In Lake Agassiz, for example, the period of scouring probably lasted for as little as ~1,200 years (9900–8700 BP) and in places resulted in severe reworking of original strata. If the ice scour process can be demonstrated in ancient glaciolacustrine and glaciomarine rock sequences it will provide a new, powerful diagnostic indicator of environment that, in conjunction with other sedimentological and paleontological environmental information, has implications for the presence of proximal or distal calving margins and/or iceberg and seasonal pack ice extent, water depth, paleoocurrent and paleowind direction (Gilbert et al., 1992). The effects of scouring may be recorded in marine sediments far removed from direct glacial action (Eyles et al., 1985). As a result, scour by seasonal pack ice and by far-drifted icebergs not associated with glaciation, and thus ice keel turbates, may be prevalent in the 'interglacial' record (present-day Beaufort Sea, Labrador Sea and Grand Banks).

Despite the wide geographic occurrence of Pleistocene ice scour marks on land, the range of grain sizes in which they are preserved is restricted to clays and silts. However, tillites and other glaciolacustrine and glaciomarine sediments in the rock record in which the effects of ice scour may be preserved generally offer a wider range of grain sizes.

4.2.2. Soft-sediment Striated Surfaces versus Scour Marks

Most Proterozoic and Phanerozoic features described and tentatively reinterpreted as scour marks are recognized largely based on soft-sediment striations

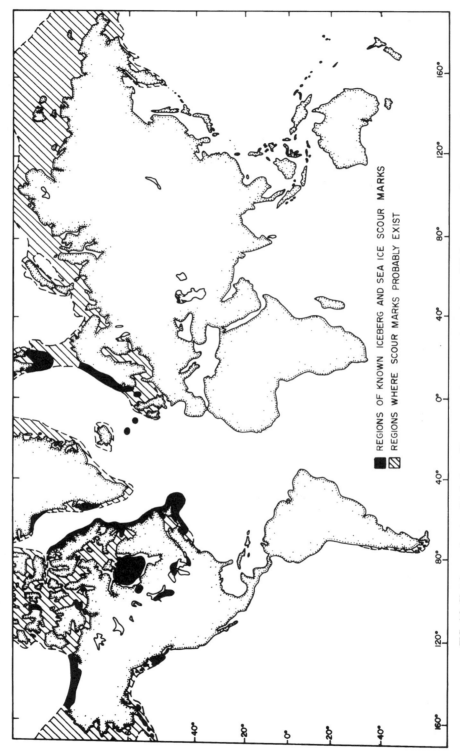

FIG. 4.2. Sketch map showing distribution of known oceanic scour marks and areas of likely occurrence (not yet surveyed)

on bedding plane exposures. Soft-sediment striations are perhaps the single most compelling diagnostic feature on which to base an initial interpretation of the presence of scour marks. Although there are other striating mechanisms, reported soft-sediment striated surfaces are interpreted to have been formed by the mechanical action of grounded or nearly neutrally buoyant, floating ice sheets (Beuf *et al.*, 1971; Eyles, 1988b; Visser, 1989; Eyles and Lagoe, 1990).

4.2.2.1. Subglacial flutes

Striating and fluting mechanisms can operate at the sole of a glacier (Dyson, 1952; Hoppe and Schytt, 1953; Baranowski, 1970; Paul and Evans, 1974) but there are notable differences between the morphologies and genesis of glacially- and free-floating ice mass-produced features. Glacial flutes often originate subglacially on the lee sides of rigid obstructions, frequently boulders, that project at least 0.3–0.5 m above a lodged till surface (Boulton, 1976c). Flutes, which may extend up to 1 km in length, may be created when deformable subglacial sediment is intruded into cavities which open up in the base of the ice and propagate from the lee sides of obstructions (Boulton, 1987a) or by deposition of debris from the glacier sole into cavities on the lee side of rigid obstructions (Boulton, 1982). In contrast, ridge-and-groove microtopography in a scour mark trough is produced by flow of seabed material into irregularities at the trailing edge of a scouring keel, and although boulders have been observed at the start points of ridges, such obstructions are not a pre-requisite for ridge formation (Woodworth-Lynas *et al.*, 1991). The crest lines of glacial flutes generally intersect the lee sides of initiating obstructions in an upward-concave till wedge so that the downstream height of the flute is often less than the height of the obstruction. Also, flute width may be slightly less than boulder width (Boulton, 1976c). These characteristics differ from scour mark ridge crest lines which are at the same elevation as the top of a boulder from the point of initiation onward, and ridge width is the same as boulder width. Also ridge shape may mirror the cross-sectional shape of the initiating boulder (Woodworth-Lynas *et al.*, 1991). Glacial flutes are often deflected around boulders in their path as a

result of disturbance of ice flow lines. Such deflections are not seen in scour mark ridges-and-grooves, and indeed are not expected since, unlike most glacial fluting, they are an extrusion phenomenon generated as the sediment passes from beneath the scouring keel.

Glacially-fluted surfaces are not always developed at the glacial sole. Bråsvellbreen, Svalbard is an example of a glacier with a marine terminus where, on the seafloor exposed between the surge moraine and the present marine terminus, flutes are absent (Solheim, 1991). The seafloor features were generated during the 1936–38 surge, and consist of a rhombohedral pattern of ridges, up to 5 m high, that have orientations sub-parallel and sub-perpendicular to the present ice margin, and also discontinuous arcuate ridges that are of similar height trending sub-parallel to the ice margin. The rhombohedral ridge pattern was probably formed by flow of seabed sediment into crevasses at the base of the glacier, and the arcuate ridges are small, annual push moraines (Solheim, 1991). None of these features resemble the products of ice scouring.

4.2.2.2. Ice shelf striations

A seabed- or lakebed-grazing buoyant ice sheet may produce ridges and grooves similar to those developed in scour mark troughs because irregularities in the ice sheet base will allow processes to function similar to those operating at the trailing edge of a scouring iceberg keel. However, floating ice sheets (ice shelves) are now thought to be an anomalous rather than normal outcome of seaward-flowing tidewater glaciers (Syvitski, *in press*). This is because ice shelves require a number of special conditions to form, including limitation to deep water due to buoyancy considerations, high ice velocity, low calving rates, and lateral pinning points (Syvitski, 1991a). Thus, if floating ice sheets are rare phenomena their frequent invocation to explain the origin of striated surfaces described from the rock record by several authors may be largely untenable.

In glaciomarine environments, sediments deposited from rain-out of basal material will characterize facies seaward of the grounding line of an ice shelf (Drewry and Cooper, 1981; Hambrey *et al.*, 1989). Striated surfaces may be developed in these polymict sediments if the grounding line position fluctuates. Orientation of striae in the grounding zone will be regional in extent. On the other hand, scour mark striations may be widely variable in orientation, and individual sets of striae will be confined between parallel berms or margins.

4.2.3. Recognizing Ice Scour Marks in Lithified Sediments

In order to recognize the effect of ice scour in the Pre-Quaternary glacial record a definitive set of diagnostic features has to be established. Scour mark surface morphology and sub-scour structure (Woodworth-Lynas et al., 1991) can be used as a guide to identification of scour-affected lithified sediments (Table 4.1).

The key to successful identification of sub-scour deformation structures in modern terrestrial settings has been initially to positively identify ice scour marks as they appear on the present ground surface above water (Woodworth-Lynas *et al.*, 1986; Thomas and Connell, 1985; Woodworth-Lynas and Guigné, 1990; Longva and Bakkejord, 1990). It is then necessary to establish correlations between the surface expression of scour marks and sub-surface structures. However, because of the nature of rock outcrops, it is not always possible to observe an association between features developed on bedding surfaces and sub-scour deformation structures in pre-Quaternary sediments. This is largely because cross-sections are more common than are expansive bedding plane outcrops. However, initially it is best to examine outcrops where bedding surfaces are exposed in order to define diagnostic surface features (see Woodworth-Lynas *et al.*, 1991) and to link these with diagnostic cross-sectional features (Thomas and Connell, 1985; Eyles and Clark, 1988; Woodworth-Lynas and Guigné, 1990; Longva and Bakkejord, 1990). A synthesis of diagnostic attributes from modern features provides criteria for identifying single scour marks. Caution is advised when using Table 4.1 since, to paraphrase Hambrey and Harland (1979, p. 274), 'Few of the criteria in themselves demonstrate . . . origin, . . . but normally, to be certain, a combination of these factors on a sufficient scale is necessary'.

4.2.4. Glacigenic Sediments that may Contain Fossil Ice Scour Marks

4.2.4.1. Pre-Quaternary sediments

Six to nine Phanerozoic and Proterozoic glacial 'ages' have been recognized, including the Quaternary (Harland and Herod, 1975; Hambrey and Harland, 1979) (Menzies, 1995a, Chapter 2) (Fig. 7.1). There follows a selective discussion of some of the 'tillites' that define these ice ages with respect to soft-sediment striated surfaces and various other features. These tillites are worth re-examining in the context of criteria for identifying the effects of ice scouring.

(A) *Proterozoic.* Short (< 1 m), randomly oriented furrows occur on bedding surfaces of the Kuibis Series quartzite within the Nama System of Namibia (Plates 4.1 and 4.2). These features, along with meandering low-amplitude (< 1 m) soft-sediment ridges, have been attributed to the grounding and scouring action of ice floes (Martin, 1965; Kröner and Rankama, 1972), but other occurrences and the significance of these small-scale phenomena were not discussed.

PLATE 4.1. Small-scale, randomly-oriented furrows on a bedding plane surface

TABLE 4.1. Criteria for identifying ice scour marks in the rock record

---

Morphological features (bedding surface) (after *Woodworth-Lynas et al. 1991*)

- Flat or dish-shaped troughs with defined margins (usually positive relief). Trough width ≈10 s cm to 100 s m. Trough depth ≈ cm – m.
- Ridge-and-groove microtopography and striations parallel to margins. Relief ≈ several metres.
- Ridges-and-grooved on the inner berm may be at an angle to main set in trough centre.
  - A cobble or boulder may mark beginning of a ridge with same X-section area as initiating boulder.
  - Width and relief of ridges may alter significantly over a few metres.
- Occasional ice dissolution voids between scour mark margins truncate ridge-and-groove microtopography. Coarse may fill void bottom.
- Irregularly-shaped flat-topped mounds in trough. Mounds may have ridge-and-groove microtopography on top surface. Inter-mound space may have sediment-infill.
- Berms. Width varies in proportion to trough width ranging from cm to 10–20 m
  - Blocky berms: tension cracks on inner flank. Berm crest of discrete subangular blocks. Outer berm of disaggregated sediment possibly incorporating rafts of blocky material in matrix.
  - Round berms: may be original (if coarse sediment) or due to winnowing. Winnowed berms may be armoured with coarse lag deposits if sediment poorly sorted.
- Surcharge. Undulose pile of disaggregated material at leading edge of scour marl. Truncates ridge-and-groove microtopography in rear.

Other features
- Trough and associated grooves have disjunct orientations with respect to regional glacial flow patterns and to other scour marks.
- Scour marks may have cross-cutting relationships on bedding surface.
- Transverse ridges may be present in trough reflecting keel oscillations (e.g. Reimnitz *et al.* 1973; Lien, 1981).
- Trough may be filled with well-sorted sediment from winnowing of material in suspension due to the scouring keel, or reworking.

Sub-scour structures (cross-section) (after *Woodworth-Lynas and Guigné, 1990, Woodworth-Lynas, 1990; unpubl. material*).
- Flat or dish-shaped trough with defined margins (margins usually positive relief). Trough width ≈ 10 s cm to 100 s m. Depth ≈ cm – m.
- Conjugate normal faults. Faults typically in short segments, may/may not intersect, may be folded. Possibly single major fault set below trough centre defining a passive wedge of downward-displaced sediment. Later fault reactivation, due to overburden pressures, may cause normal offset of trough surface (cm) and penetrate a short distance into overlying unit.
- Sub-horizontal faults beneath berms. Possibly a large continuous fault below trough centre offset/connected to conjugate faults.
- Fracture cleavage in small scale fold hinges in fine-grained sediment of large scour marks, unrelated to regional tectonism.
- Berms. Width varies in proportion with trough width, ranging from a few cm to 10–20 m.
  - Berm blocks defined by subvertical cracks (?1–2 m+ deep) filled by superseding unit, may drape over berm. Collapsed outer berm resting in disaggregated matrix.
  - Round berms, may be defined by lag deposit on upper surface.
- Downward deflection of layering. Maximum deflection and bed thickness reduction beneath central trough.
- Possible zone of upwarping on either side of and immediately adjacent to trough.

Other features
- Ridge-and-groove microtopography may show as undulations or more angular crenulations relief ≈ 30 cm.
- Trough may be filled with well-sorted sediment from winnowing of material in suspension due to scouring keel. If laminated, sediment unit infilling trough may onlap towards inner berm flank.
- Local ice dissolution voids with steep walls may truncate the incision surface in the trough. Pit bottom may be filled with coarse material from underlying, scoured unit, and clasts may exhibit evidence of surface abrasion.
- Flat-topped mounds may appear similar to dissolution voids but trough floors between mounds may be characterized by flat surface of undisturbed seabed and absence of coarse material resting on it.

Associated environmental indicators
Sediments with above features contained in or bounced by facies indicative of lacustrine or marine conditions.

PLATE 4.2. Meandering, low-amplitude soft-sediment ridge. From the Kuibis Series quartzite, Proterozoic of South Africa. (Photographs from Martin, 1965)

Montes *et al.* (1985) described a soft-sediment striated surface on a disconformity surface in Brazil that separates the glacigenic dropstone-bearing Bebedouro Formation from underlying deltaic deposits of the Morro do Chapéu Formation. The striated surface is characterised by grooves with relief between 1–5 cm and striae that curve and intersect and have abrupt terminations. Dropstones in this transgressive sequence imply the presence of free-floating ice, although the interpreted formation of the curvilinear grooves by scouring ice keels was not made.

In a reinterpretation of soft-sediment deformation structures from the Port Askaig Formation exposed on the Garvellach Islands of western Scotland, Eyles and Clark (1985) noted the restricted development of a polygonal network of sandstone wedges on the upper surface of diamictite 22 (Plate 4.4) which they interpreted as 'suspended polygon' structures formed subaqueously by earthquake shock-induced loading of sands into underlying muddy diamictites. Spencer (1971), in contrast, interpreted the wedges as perma-

PLATE 4.3 Soft-sediment ridge, similar to the one in Fig. 4.1b, in unconsolidated sediments from the trough of a new 50-m wide iceberg scour mark, Makkovik Bank, Labrador Sea. Water depth 150 m. (Photograph taken from a submersible – field of view about 3 m across)

(A)

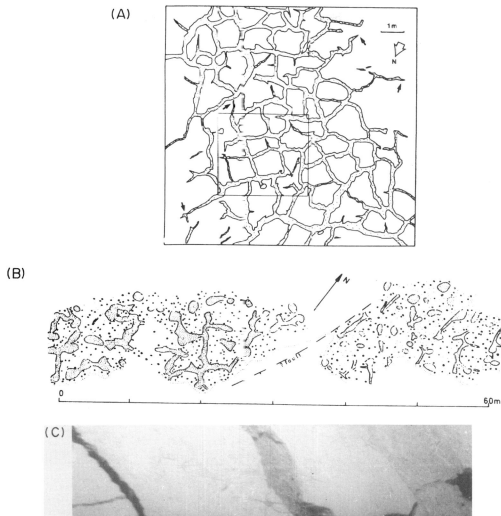

(B)

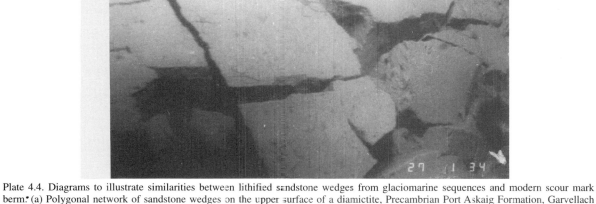

(C)

Plate 4.4. Diagrams to illustrate similarities between lithified sandstone wedges from glaciomarine sequences and modern scour mark berm. (a) Polygonal network of sandstone wedges on the upper surface of a diamictite, Precambrian Port Askaig Formation, Garvellach Islands, Scotland (from Eyles and Clark, 1985); (b) polygonal network of sandstone wedges on a bedding plane surface within the Ulveso Formation, east Greenland (from Hambrey and Spencer, 1987); (c) polygonal network of open cracks from the inner berm flank of a fresh 50 m wide iceberg scour mark, Saglek Bank, Labrador Sea. Water depth approximately 120 m. (Photograph taken from a submersible – field of view about 3 m across)

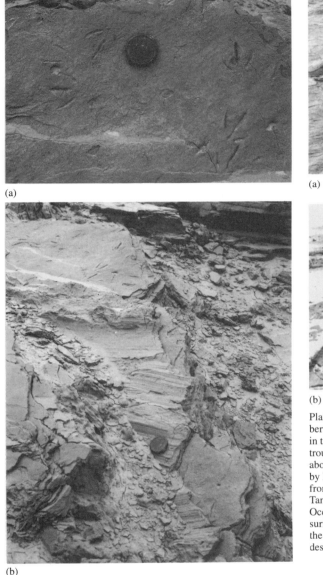

(a)

(b)

Plate 4.5. (a) Close-up view of top portion of (b) showing ice crystal pseudomorphs and bird footprints associated with: (b) striated surfaces (lens cap) of small scale ice scour marks on bedding surface of modern tidal flat sediments, Cobequid Bay, Nova Scotia. Note the different striation orientations

(a)

(b)

Plate 4.6. (a) Double-sided feature with small, raised, parallel berms. Small surcharge pile (tape measure) truncates fine striations in trough behind. Note ridge in background obliquely truncating the trough. This feature occurs on the Silurian transgression surface, above the Tamadjert Formation, Algerian Sahara, and is interpreted by Beuf *et al.* (1971) to have been formed by littoral sea ice. (Taken from Beuf *et al.*, 1971). (b) Soft-sediment slumped ridges, Tamadjert Formation, Algerian Sahara (from Beuf *et al.* 1971). Occurrence of slumping implies the existence of a free sediment surface at the time of ridge formation. These ridges are similar to the ones illustrated in Plates 4.2 and 4.3, and also to ridges described by Savage (1972) from the Late Carboniferous Dwyka Formation (see Plate 4.8)

frost contraction cracks that developed subaerially. Hambrey and Spencer (1987) also used a subaerial, periglacial tension crack theory to explain similar polygonal sandstone wedges in the Vendian Tillite Group of east Greenland (Chapter 5). However, these polygonal sandstone wedges also are similar in morphology and extent to the network of tension fractures observed along the inner berms of some modern scour marks (see Woodworth-Lynas *et al.*, 1991). The continental shelf-type water depths interpreted for the Port Askaig Formation by Eyles (1988a), and the presence of substantial ice-rafted

(a)

(b)

Plate 4.7 (a) Soft-sediment striations, Ordovician Tamadjert Formation, Algerian Sahara. Note unstriated margin on left side and gradual amplitude change and lateral migration of ridges across adjacent grooves – features characteristic of ridges and grooves formed at the trailing edge of scouring ice keels. Liftoff of the graving tools is implied where striations are developed on either side of but not within small (few cm wide) depressions along the left margin (from Beuf *et al*. 1971). (b) Ridge-and-groove microtopography with striations in the trough of a new 50-m wide iceberg scour mark, Saglek Bank, Labrador Sea. Water depth approximately 120 m (photograph taken from a submersible – field of view about 3 m across)

material, including dropstones and sandy diamiet 'clots', suggests the presence of icebergs or sea ice and implies the possibility of scouring by ice keels

(B) *Ordovician/Silurian*. Beuf *et al*. (1971) ascribed 1–2 m wide sinuous features (one with a surcharge pile) exposed on a marine transgression surface from the Silurian of the Sahara Desert to the action of sea ice keels (Plate 4.6a). Beuf *et al*. noted several finely striated flat bedding plane surfaces exposed in over 1200 km of outcrop from the Upper Ordovician Tamadjert Formation but interpreted these as being generated by clastic material embedded in the sole of a moving glacier. However, they did note that curved striations may indicate scouring by floating ice. Plate 4.6b illustrates from Beuf *et al*. the flanks of two prominent ridges that have slumped and appear to truncate adjacent fine striations. These

features bear striking similarity to the feature shown in Plate 4.2 from the Precambrian Kuibis Series quartzite of South Africa. Plate 4.7a shows ridges and grooves that change shape and transgress longitudinally, reflecting dynamic change of irregularities in the bottom topography of the ice (see Woodworth-Lynas *et al*., 1991). Also, an unstriated, slightly raised margin is apparent on the left side. These elements, plus variations in striation orientation that Beuf *et al*. interpreted as possibly the result of local changes in glacial drainage patterns, although not conclusive by themselves, may permit a reinterpretation of these features as possible ice scour marks.

(C) *Carboniferous/Permian*. Seven glacial and two interglacial units have been defined in the Dwyka Formation of South Africa. Subaqueous outwash deposits indicate water depths ranged from 40 m to in

(a)

(b)

(c)

Plate 4.8. Soft-sediment striations on a sandstone bedding surface, Late Carboniferous Dwyka Formation, South Africa. (a) General view of the striated surface; (b) striations partially masked by slumping of adjacent ridge (arrows); (c) striations above hammer handle partially masked by adjacent material (Reproduced from Savage, 1972)

excess of 250 m within the Karoo Basin (Visser, 1989). Units 2 and 5 have been interpreted as debris rain-out deposits from possible 'iceberg zones'. However, soft-sediment grooved and striated pavements found throughout the Dwyka Formation have been typically interpreted as having formed beneath an ice sheet.

Savage (1972) described a soft-sediment striated surface in Dwyka Formation pebbly quartz sandstones (Plate 4.8). He noted slumping of ridges across adjacent grooves in a manner similar to that described by Beuf *et al.* (1971), indicating soft-sediment deformation immediately after the groove was formed. Savage acknowledged difficulty in explaining how the marks would be preserved if they had formed beneath an ice sheet. Visser (1990) described the same surface from a nearby location, and suggested that associated erosional troughs and channels could have been formed by scouring icebergs although he preferred an origin due to subglacial meltwater flow. Elsewhere, Visser and Hall (1984), commenting on a soft-sediment striated surface interpreted to have formed at the sole of a glacier, noted that the preservation of these features indicated rapid localised pressure relief. Near Kenhardt, current ripples are oriented normal to soft-sediment sandstone grooves that have a disjunct orientation with respect to regional glacier flow (Visser, 1985). Such associations may be evidence of current-driven drift of scouring ice keels.

(D) *Permian.* Iceberg scour marks have been interpreted on bedding plane surfaces from the glaciolacustrine Rio do Sul Formation (Itararé Subgroup) in Brazil (Rocha-Campos *et al.*, 1990; Rocha-Campos, *pers. commun.* 1988, 1989, 1991, 1992; Paulo Santos, *pers. commun.* 1991). These authors described sinuous, sub-parallel furrows with 'V'-shaped or flat bottomed striated troughs up to 50 cm wide, 20 cm deep and at least 80 m long. Cross-sectional views show downward-displaced varved bedding (Plate 4.9) cut by small faults dipping towards the centre of the structures, that bear striking similarity to modern scour marks found on the tidal flats of Cobequid Bay, Nova Scotia (Paulo Santos, *pers. commun.* 1991) (compare Plate 4.9 with Fig. 4.1c). Small scale low angle reverse faults and recumbent folding occur on either side of the Rio do Sul trough features.

Plate 4.9 Downward-displaced varved bedding beneath a scour mark trough in the Permian Rio do Sul Formation, Parana Basin, Brazil. Note upwarping on either side. (Photograph kindly provided by A.C. Rocha-Campos, Universidade de Sao Paulo.) Compare similarity with scour mark from Cobequid Bay illustrated in Fig. 4.1c

Small-scale iceberg scour marks (< 1 m wide) have been described from bioturbated diamictites in the Pebbley Beach Formation, Australia (Powell and Gostin, 1990) (Plate 4.10). The a/b planes of boulders immediately beyond the scour mark margins dip towards the scour mark trough (R. Powell, *pers. commun.* 1991), suggesting sub-scour compaction during scour events was sufficient to locally re-orient clasts.

Soft-sediment striated surfaces marking sharp sedimentary contacts within structureless diamictites have been described from the Pagoda Formation, Transantarctic Mountains (Miller, 1989). One of these surfaces is developed above a winnowed diamictite (cf. Miller, 1989, Fig. 5c). Glaciolacustrine conditions are interpreted to have existed during a series of glacial retreats.

'Striated and grooved pavements . . . clearly of direct glacial origin' are common in the lower part of the Buckeye Formation, also in the Transantarctic Mountains (Aitchison *et al.*, 1988, p. 101). However,

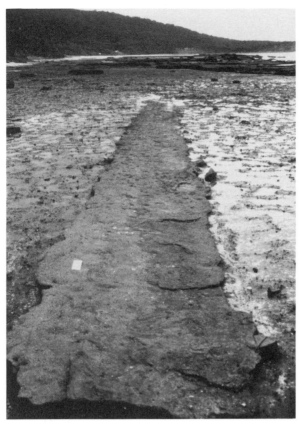

Plate 4.10. Small-scale ice scour mark (approximately 1 m wide) on bedding plane surface of a bioturbated glaciomarine diamictite, Permian Pebbley Beach Formation, Australia. The trough is filled with sandstone thus obscuring any features, such as ridges and grooves, that may occur on the incision surface. The contrast in grain size and sorting between the diamictite and sandstone in the trough is similar to observations from large-scale, Pleistocene scour marks exposed on King William Island, Northwest Territories, Canada described by Woodworth-Lynas et al. (1986). (Photograph kindly provided by R. Powell, Northern Illinois University)

water with sediments affected by storm wave action. These are all environmental indicators compatible with the presence of icebergs and sea ice.

4.2.4.2. Ice and iceberg scour

In the rock record, a distinction between scour marks made by seasonally-occurring ice and by icebergs is not possible using the criteria for identification presented above. However, it may be possible to discriminate between the two types of ice using the following criteria:

(1) Generally, scouring by pressure ridge keels (and small icebergs) is restricted to about <60 m water depth in present-day oceans (Reimnitz et al., 1984); beyond this depth, scouring is by glacial ice masses only. Therefore, if water depth of an ancient example is accurately known, it may be possible to select the most probable scouring agent. Due to a lack of data on the drafts of pressure ridge and iceberg keels in freshwater bodies, the limiting value of 60 m, derived entirely from marine studies, is assumed to be approximately valid for deep glacial lakes. For example, icebergs (with drafts of 110 m) are inferred to have created the Lake Agassiz scour marks (Woodworth-Lynas and Guigné, 1990). Maximum iceberg draft is about 550 m on the east Greenland continental shelf (Dowdeswell et al., 1993), about 230 m on the Canadian eastern continental shelf (Hotzel and Miller, 1983) and may be as great as 500 m in the Antarctic (Barnes and Lien, 1988).

(2) In some cases, the presence of either icebergs or pack ice (but not both) may be known. For example, in an ice contact lake, scouring by lake ice and by icebergs may occur. Conversely, in a distal glacial lake, scouring can only be by the action of lake ice. An example of an ancient distal environment is glacial Lake Iroquois where scouring was by seasonal pack ice only (Gilbert et al., 1992). Modern examples of scour by pack ice only are known in Lake Erie, Great Slave Lake, the Beaufort Sea and northern Caspian Sea (Grass, 1984, 1985; Weber, 1958; Lewis and Blasco, 1990; Koshechkin, 1958).

(3) The character of the coarse sand fraction of associated ice-rafted debris may point to glacial ice-rafting (Von Huene et al., 1973).

the one illustration of such a striated bedding surface shows an abrupt margin beyond which no striations are evident (Aitchison et al., 1988, Fig. 5). Aitchison et al. grant the possibility that soft-sediment striated surfaces may have originated from 'the grounding of floating ice such as bottom scraping berg ice' (p. 103). Dropstones occur in diamictite units with which the pavements are always associated. Hummocky stratification noted in parts of the section indicates the presence of relatively shallow open

(4) Striations and facets on particles, if present, are strong evidence of glacial (and thus iceberg) origin (Huggett and Kidd, 1984; Gilbert, 1990).

4.2.4.3. Ice-rafting

Ice-rafted sediment generally originates from glacial, colluvial, aeolian and littoral sources (Gilbert, 1990). A fifth source of sediment is derived from the post-scour rafting of sediment that has been incorporated in the keel of a scouring ice mass. Such mechanically incorporated debris consists of a mix of 'normal' lacustrine or marine sediment deposited from suspension or traction currents, ice-rafted sediment (from the other four sources) and previously scoured-and-rafted sediment. One consequence of prolonged influence of this form of rafting action is the generation of sediments containing a homogeneous mix of clasts from a variety of different original sources. It is difficult to identify the proportion of a sediment that has been affected by the action of scouring-and-rafting.

Scouring-and-rafting is probably the single most important factor affecting sediments on the Canadian eastern continental shelf. The seabed of this shelf is annually exposed to iceberg-scour turbation and therefore to scouring-and-rafting. The latter process probably contributes to a net removal of sediment on the shelf as material is rafted and redeposited on the continental slope and beyond. Additional removal may occur during the scouring process as fine-grained sediment is lifted into suspension and then winnowed by ocean currents (Woodworth-Lynas et al., 1986). Scouring-and-rafting may assist in removal of sediment from glacial lakes through the mechanical action of sediment being lifted into suspension during scour, and from post-scour rainout of material from the debris-impregnated keel. These suspended sediments may be flushed from the lake via outlet rivers before the grains can be redeposited.

4.2.5. Scour Marks, Ice Keel Turbates and Tillites

From the preceding discussion it is apparent that the most easily recognized diagnostic features of scour marks are those exposed on bedding plane surfaces. If scour marks can be identified from bedding plane structures then sediment deformations that occurs below the striated surface can more confidently be interpreted as scour-related, rather than as products of other types of soft-sediment or tectonic deformation. Once the association of sub-scour deformation structures with striated surfaces is established the recognition of scour marks in cross-section alone is possible

Identification of ice keel turbates is more problematical because the physical appearance of structures in sediments resulting from multiple scour events is not known. One approach is to examine a bedrock sediment sequence that contains discrete, recognizable scour marks, and then to ascend through the sequence into beds that contain increasing numbers of scour marks that have been overprinted by subsequent scour events. Such an approach necessitates that scouring intensity must have been increasing with time, or that there must have been decreasing sediment deposition to ensure overprinting of early by later events. This approach may not be feasible when a gradation between discrete scour marks and an ice keel turbate does not occur, such as where there is an abrupt transition into massive tillite from sediments that may not contain any evidence of scouring. In this latter situation the tillite can be examined for scour mark attributes that, even if not in discrete associations, may collectively help to define the unit as an ice keel turbate. It is possible that the lower, and more likely the upper, surface of an ice keel turbate unit may preserve a relic topography created by the first or last scouring ice keels. This surface may preserve a closer association of discrete scour mark elements than turbated sediments below such as the shapes of flat or dish-shaped troughs with defined margins, ridge-and-groove microtopography and striations, round berms defined by a lag deposit on their upper surfaces, berm blocks defined by sub-vertical cracks, ice dissolution voids and flat-topped mounds.

4.3. DISCUSSION

The keels of icebergs and of ice pressure ridges typically create curvilinear scour marks when they touch, penetrate and move through unlithified sediments. The most obvious effects of scour are seen on a lake bed or seabed which, for each scouring event, usually preserves a furrow with a berm of expelled material on either side (Woodworth-Lynas *et al.*, 1991). Obscured from view in modern offshore glaciomarine or glaciolacustrine areas are the effects of scour on sediments beneath the trough. Sub-scour deformation structures can be seen in excavated or naturally exposed cross sections of ancient scour marks exposed in glaciolacustrine sediments on land. The limited availability and accessibility of Pleistocene scour marks preserved in unlithified sediments above sea and lake levels makes a more rigorous search of ancient, lithified glaciomarine and glaciolacustrine sequences a useful avenue of investigation. Although ice keel turbates can be formed in glaciomarine and glaciolacustrine sediments, they are probably more common in the former, single scour marks being more likely in glaciolacustrine sediments.

4.3.1. Ice Scour in the Rock Record: Engineering and Stratigraphy

At present, little is known of the effect of grain size on the style of sub-scour deformation. It is likely that coarser-grained sediments and bimodal and poly-modal sediments will behave differently during scour, developing and retaining different types of sub-scour deformation structures (Thomas and Connell, 1985; Eyles and Clark, 1988) (Fig. 4.1). Suitable outcrops of scour marks in such heterolithic, unconsolidated sediment types have not been reported. Alternatively, a useful approach in quantifying variations in deformation style may be to examine inferred ice scour marks in lithified glaciolacustrine sediments that contain a range of grain sizes. Such an analysis may produce quantifiable data on sub-scour deformation that have direct engineering applications to modern-day seabed engineering problems in areas affected by ice scour. An examination of glaciolacustrine rocks is thus appropriate: (1) to recognize ice scour marks and ice keel turbate facies in a variety of glaciolacustrine (and glaciomarine sediment) types, and (2) to advance understanding in seabed engineering. Engineering applications of such a study include defining the zone of sediment deformation beneath scour marks from a variety of lithologies, and defining the types of deformation mechanisms that operate during the scouring process (such as bulk strain vs. discrete failure, bearing capacity failure, horizontal shear etc.) and quantifying their effects (such as the amount of offset along faults, degree of compaction etc.).

Chapter 5

GLACIOMARINE ENVIRONMENTS 'ANCIENT GLACIOMARINE SEDIMENTS'

A. Elverhøi and R. Henrich

5.1. INTRODUCTION

The origin and genesis of ancient (Pre-Holocene) glaciomarine sediments is still a subject of controversy (for definition of glaciomarine sediments see Menzies, 1995a, Chapter 14). Until recently, knowledge of modern glaciomarine sediments was limited in comparison to that of terrestrial glacial deposits. Consequently, terrestrial glacial sediments were overemphasized in studies of Cenozoic as well as pre-Cenozoic sediments. However, as information on modern environments has progressed, reinterpretations of ancient sequences have demonstrated that glaciomarine sediments are widespread (Anderson, 1983). Glaciomarine sediments are now well documented as an important constituent of all the major glacial episodes in the Early Proterozoic, Late Proterozoic (Late Pre-Cambrian), Paleozoic and Cenozoic (Hambrey and Harland, 1981a) (Chapter 7; Menzies, 1995a, Chapter 2).

A basic problem in interpreting ancient glaciomarine sediments has been, and still is, the lack of reliable criteria for differentiating between marine and terrestrial glacial sediments. During the last two decades understanding of the glaciomarine environment has progressed significantly, in particular, with regard to the proximal ice regime, where detailed studies have been conducted (Powell, 1984; Dowdeswell and Scourse, 1990; Anderson and Ashley, 1991). For open marine conditions, and especially for glaciomarine sedimentation in the deep ocean, knowledge is still limited. However, new marine sampling techniques have provided a large number of undisturbed deep sea sediment cores. Hence, it has been possible to combine conventional methods of glaciomarine facies analysis with micropalaeontological and geochemical data analysis to interpret palaeo-oceanographic, ecologic and glaciomarine processes. Furthermore, a high degree of stratigraphic resolution has become possible through the use of oxygen isotope stratigraphy and AMS ^{14}C dating (Chapter 14).

It is now possible to identify the basic elements of an interglacial–glacial cycle within shelf, slope, and deep sea sediments (e.g. Henrich, 1990). In this chapter the glaciomarine sedimentary environment will be illustrated by data from Norwegian–Greenland Sea and the northern Norwegian slope and shelf, including the western Barents Sea (Fig. 5.1). In addition, two well-known Early and Late Pre-Cambrian sections will be described and discussed with regard to classical conflicts in interpreting ancient glaciomarine sediments.

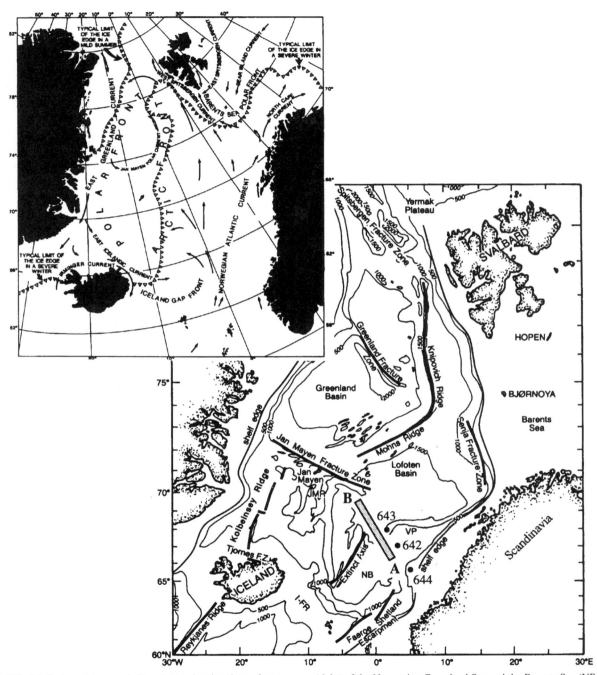

FIG. 5.1. Bathymetric map (left) and map showing the surface currents (right) of the Norwegian-Greenland Sea and the Barents Sea (NB = Norwegian Basin, VP = Vøring Plateau, I-FR = Iceland Faroe Ridge, JMR = Jan Mayen Ridge). The coring location of ODP Leg 104 is also shown. A-B marks the location of the transect shown in Fig. 5. 21

5.2. GLACIOMARINE SEDIMENTS: CLASSIFICATION AND IDENTIFICATION

Traditionally, deep sea sediments have not been referred to in discussions of the glaciomarine environment (Menzies, 1995a, Chapter 14). However, progress in geological marine research has demonstrated that deep sea sediments represent an important source of information concerning the long-term glacial record. In high latitude areas, the deposition of the deep sea sediments is largely controlled by the glacial conditions onshore and on the continental shelf (e.g. ice-shelf regimes, grounded ice-margins, tidewater glaciers). It is crucial to understand the origin of these sediments in the Cenozoic record in order to relate them to the correct glacial regime.

In the analysis of deep sea glaciomarine sediments, it is important to realize that many glaciomarine processes have daily and/or seasonal fluctuations and the progradation or regression of glacial environments can occur on a scale of years to tens of years. In addition, shifts in pelagic sedimentation may occur, but on a much longer time-scale. Although seasonal pulses in pelagic particle flux within polar oceans are the rule rather than the exception, the average pelagic accumulation rates are more continuous over longer periods (e.g. centuries or thousands of years); while glaciomarine sedimentation is generally characterized by fluctuating conditions with episodic peaks of deposition. As a consequence, flux calculations using a linear sedimentation rate for a mixed pelagic/ glaciomarine lithology tend to overestimate the pelagic fluxes and underestimate the glaciomarine fluxes (Henrich et al., 1989a). Another potential problem is that stratigraphic data points may be displaced by bioturbation, especially in sedimentary sections that contain alternating glaciomarine-dominated and pelagic-dominated lithologies (Henrich et al., 1989a).

Problems in dating are highly relevant when using the deep sea record for deciphering glacial histories (Chapter 14). Varying lithologies and facies must be well-dated if glacial events are to be correctly identified. Within the Cenozoic, a set of inter-calibrated chronostratigraphic methods are available, in addition to the combination of bio- and magnetostratigraphy. For example, oxygen isotope stratigraphy

(Imbrie et al., 1984; Martinson et al., 1987) permits a stratigraphic resolution of 3 to 5 kyr for the past 800 ka. The time-scale of oxygen isotope stratigraphy has recently been extended to the past 3.5 Ma (Sarnthein and Tiedemann, 1989). Radiocarbon datings show an even better resolution, on the order of hundreds of years for the past 20 to 40 ka (but note correction of the ^{14}C scale (i.e. Bard et al., 1990). Dating of the predominantly terrigenous glaciomarine sediments has always been problematic due to a generally very low content of pelagic carbonates. Application of the tandem accelerator mass spectrometer to ^{14}C dating has revitalized chronologic efforts to date glaciomarine sediments in polar regions (Andrews, 1985; Domack et al., 1989). New approaches for measuring the decay of thorium (Eisenhauer et al., 1990) and amino acid-stratigraphy (e.g. Sejrup et al., 1984) have recently been developed and show promising preliminary results. Thus, the existing stratigraphic methods enable correlation of individual glaciomarine units deposited during several hundreds of years in sections that cover the past 20 to 40 ka, whereas units spanning thousands of years can be correlated in sections covering the past 800 ka.

The following parameters have frequently been used for interpreting ancient glaciomarine sediments: dropstones, grain-size distribution, particle shape, surface texture, clast fabric, mineralogy/geochemistry, stratification/lamination, thickness and lateral extent, stratigraphic and facies relationships, fossil content, and geomorphological features (e.g. ice ploughing) (Hambrey and Harland, 1981a; Anderson, 1983) (Chapter 4). In reconstructing former environments, interglacial/glacial cycles need to be identified. As illustrated below, glacial–interglacial variations play a major role in shaping the deep sea glaciomarine environment due to changes in the circulation system, formation of dense bottom water, and distribution of sea ice. In recording the composition and distribution of deep sea glaciomarine sediments, which are strongly related to the oceanic circulation, parameters reflecting changes in the water masses should be included. Essential parameters include carbonate production and dissolution, organic carbon and amorphous silica content, the oxygen/carbon isotope signal and sedimentation rate or flux.

5.2.1. Ice-rafted Detritus

Deposits of pebbles that disrupt basal laminations, while being draped by overlying undisturbed laminae, are prime indicators of a glaciomarine environment, reflecting ice-rafting of coarser materials in combination with fall-out from suspended particles in the water column (Flint, 1971, pp. 195–197). In the study of deep sea glaciomarine sediments, the proportion of coarser materials (ice-rafted debris, IRD) present is a meaningful parameter. Variations in IRD concentration are commonly related to changes in the glacial regime of the adjacent shelves. Well-dated cores from the last glacial period document that the IRD peaks at the early phase of deglaciation are associated with extensive iceberg influx (Fig. 5.2). Compared to the oxygen isotope record, which reflects global ice volume, the IRD parameter reflects more localized glacial events. However, icebergs derived from an ice-shelf generally have a very low content of debris, which is concentrated in a thin basal layer, whereas icebergs from outlet glaciers seem to be the dominant source of debris-rich icebergs (Menzies, 1995a, Chapter 14). Thus, a shift from high to low IRD content may reflect a change from an outlet glacier to an ice-shelf regime, as well as a change from calving outlet glaciers to a dominant source from ice margins entirely on land (Chapter 4).

A fundamental problem in interpreting ice-rafted material is in distinguishing between clasts rafted by sea ice and those rafted by icebergs (Gilbert, 1990). Ice-rafted material is often classified as particles >63 μm in the open ocean. Icebergs are traditionally regarded as the major rafting agent. However, recent studies have documented the incorporation of coarse silt- and sand-sized material in sea ice (Barnes *et al.*, 1982a; Barrett *et al.*, 1983; Pfirman *et al.*, 1990). Difficulties arise from the lack of reliable criteria for differentiating between the two modes of rafting. In particular, it is difficult to discriminate iceberg versus sea ice-rafted sand-sized material. Ice-rafted material should therefore be interpreted with caution (Chapter 4; Menzies, 1995a, Chapter 14).

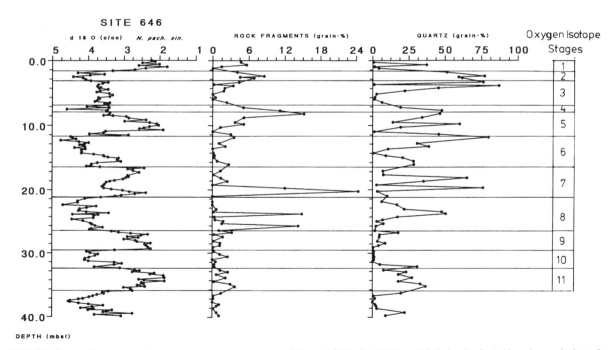

FIG. 5.2. Oxygen isotope, rock fragments and quartz content variations at ODP site 646 (Leg 105, Labrador Sea). Note that peaks in rock fragments and quartz grains correspond to high $\delta^{18}O$ values, e.g. glacial periods and cool phases during interglacials (modified from Wolf and Thiede, 1991)

5.2.2. Grain-size Distribution and Sedimentation Rate

Glaciomarine sediments may range from coarse-grained diamictites to fine-grained deep sea muds with only a small percentage of sand. The grain-size distribution of a glacial diamictite is similar to that frequently found in gravity flow deposits (Menzies, 1995a, Chapter 8). As illustrated by Anderson *et al.* (1980b, 1983), basal till, debris flow and glaciomarine sediments from the continental shelf of Antarctica were found to be indistinguishable on the basis of grain-size distribution. In the deep ocean (e.g. Arctic Ocean and Norwegian–Greenland Sea) the grain-size distribution varies in a cyclic pattern, apparently corresponding to interglacial–glacial changes (Morris *et al.*, 1985; Henrich, 1989). The most coarse-grained units (diamictons) seem to correspond to periods of deglaciation and maximum extension of the ice sheets across the continental shelves. Fine-grained mud is characteristic of interglacial periods (Section 5.4.5).

The sedimentation rate, also, reflects interglacial-glacial changes. In deep sea sediments, beyond formerly glaciated margins (e.g. Svalbard/Barents Sea, Scandinavia), interglacial sedimentation rates are in the range of $1–2\,cm\,kyr^{-1}$, while during glacials, rates increase to several tens of $cm\,kyr^{-1}$ (Jones and Keigwin, 1988; Henrich *et al.*, 1989a). In open shelf regions (distal), typical interglacial rates are in the range $1–5\,cm\,kyr^{-1}$ (Elverhøi, 1984), while during glacials, rates are more variable, with $>10\,cm\,ky^{-1}$ reported (Vorren *et al.*, 1984). During glacial periods, the maximum sediment input peaks at early phases of deglaciation, when sediment discharge and flux of icebergs are at a maximum. During interglacials, sea ice may represent an important sediment source, as seen in the northern Barents Sea, which is covered by sea ice 8–10 months of the year. Here, the sedimentation rate is in the range $3–5\,cm\,kyr^{-1}$ ($40–65\,t/km^2$) and sea ice-rafted sediments (fine-grained mud) have been calculated to account for $\frac{1}{3}$ of the annual flux (Elverhøi *et al.*, 1989). The remaining sediment is derived from land and sedimentation by currents, while iceberg-rafting is of minor volumetric significance.

In summary, sedimentation rates reach their maximum during glacials, and diamictons may form in deep sea environments, due to iceberg-rafting. In contrast, during interglacials, sedimentation rates decline, and in Arctic regions, rafting of fine-grained materials from sea ice may become a major sediment source, both in shelf areas and in deep sea environments (Chapter 4).

5.2.3. Deep Water Formation

During interglacial periods, deep water formation in high latitude areas plays a significant role in deep water circulation. Deep water is formed in polar regions in two ways: (1) dense brines are derived from adjacent shelves as a result of salt rejection during sea ice formation and super cooling beneath the ice shelves; and (2) as a result of the convection of open ocean deep currents in regions with low density stratification.

Dense shelf-water formation is also important for sediment erosion and transport on polar continental shelves and slopes. In Antarctica, the cold water forms coast-parallel currents that flow underneath large ice shelves, such as the Filcher and Ross Ice Shelves. During its sub-ice-shelf flow, the water temperature decreases ($<-1.9°C$). In the Weddell Sea, the outflow follows the western slope of Crary Trough (Fig. 5.3), with a water discharge rate calculated to be 10 times that of the Amazon River (Foldvik, *pers. commun.*, 1991). Current velocities $>100\,cm\,s^{-1}$ have been measured at a depth of 2000 m on the slope in the Weddell Sea (Foldvik and Gammelsrød, 1988). This flow strongly influences sedimentation, resulting in erosional features tens of meters in incision depth on the continental slope and rise (Fig. 5.4). During glacials, Antarctic glacier ice most likely extended to the shelf break (Denton and Hughes, 1981b), and the flow of such cold water stopped. Thus, a major change in flow regime occurred on the shelf and slope from interglacial to glacial periods. In the Barents Sea, dense shelf water forms during the winter season in response to sea ice formation (e.g. Midttun, 1985). This cold water episodically cascades into the adjacent deep sea regions. Sediment trap studies in the Norwegian–Greenland Sea show short, intensive episodes of sediment discharge into the deep ocean during the winter months, thought to be associated with such influxes of cold water (Honjo *et al.*, 1988).

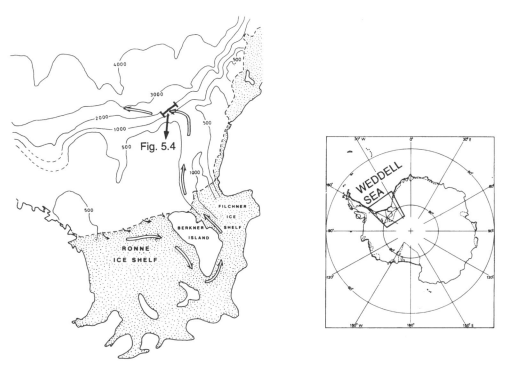

FIG. 5.3. Bathymetric map of the Weddell Sea showing the possible flow of the cold shelf water (ISW = ice shelf water). ISW flows along the bottom. The discharge is on average approximately 10 times greater than that of the Amazon River (from Foldvik and Gammelsrød, 1988 reproduced with permission from Elsevier Science Publishers) (The location of the seismic section shown in Fig. 5.4 is shown by a bar)

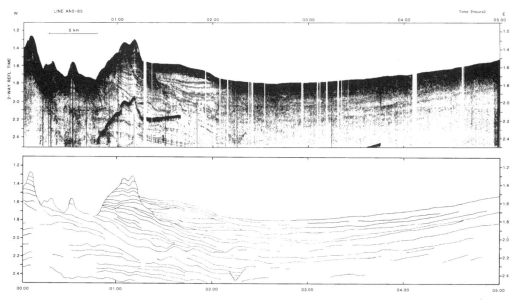

FIG. 5.4. Seismic section (sparker) along the upper continental slope of the Weddell Sea (see Fig. 5. 3 for location), running perpendicular to the ice shelf water. The area is characterized by a 'ridge and trough terrain' where individual sediment layers are cut and seen in the walls. The erosion is thought to have been caused by the downflowing ice shelf waters (Solheim and Elverhøi, unpublished data)

This episodic flow may flush the shelf regions of fine-grained material and be important in eroding the upper continental slope. Similarly, as in Antarctica, the Barents Sea has been glaciated repeatedly during the Late Cenozoic (Solheim and Kristoffersen, 1984; Vorren et al., 1988). Consequently, during glacial maxima, this cold water flow may not have existed on the western Barents slope.

5.2.4. Sea Ice Cover

During glacials, the perennial sea ice cover expanded towards lower latitudes. An important aspect of a perennial or seasonal ice cover is the reduced impact of wind-induced current activity, especially in shallow shelf areas. Of more regional importance is the thermal capping effect of the ice cover. Ice cover effects are well demonstrated in the modern Arctic Ocean, where the ice cover hinders thermal convection, and, during the summer, melting forms a relatively low salinity layer which also prevents convection. The result is a well-stratified upper water column which strongly reduces convection and renewal of the deep water masses.

Glacial–interglacial changes may also influence the transport of debris in sea ice in Arctic regions. At present, the wide and shallow shelf areas represent important regions for debris entrainment into the sea ice (Menzies, 1995a, Chapter 14). However, during glacial periods, these shallow shelf regions become either subaerially exposed or covered by grounded ice. Accordingly, the conditions would be comparable to present-day Antarctica, where the sea ice is almost 'clean'. Under such conditions the sediment discharge by sea ice of typical fine-grained material to the deep ocean diminishes.

5.2.5. Circulation System/Current Regimes

The interglacial transport of warm, saline water (and air) masses into the Norwegian–Greenland Sea and Arctic Ocean (Fig. 5.1), typical of the present day, was opposed during glacials by a general southward movement of cold air and water masses. However, during glacial periods, a northward flow of surface water (e.g. Atlantic water or northeast currents

parallel to the ice margin) along the eastern margins of the Norwegian–Greenland Sea was maintained for long periods. The current velocity was strongly reduced, causing limited erosion of the shelf sediments and upper slope, as it is typical today.

5.2.6. Oxygen and Carbon Isotopes in Glaciomarine Settings

The oxygen isotope values measured on pelagic and benthic calcareous tests record a mixed signal of (1) the overall characteristics of the water masses (e.g. temperature and salinity) and (2) global ice volume. An increase in the $\delta^{18}O$ record corresponds with either a decrease in temperature or an increase in global ice volume, or a combination of both. However, if we combine the 'global ice volume signal' of $\delta^{18}O$ with the IRD content in a deep sea core, a more local ice record can be deduced (Wolf and Thiede, 1991). As seen in Fig. 5.2, an increased content of IRD corresponds with intervals of high $\delta^{18}O$ isotope content, that is, with glacials.

In the polar regions of the Norwegian–Greenland Sea, changes in bottom water temperatures (ΔT) and salinity (ΔS) undergo only small-scale variations between glacials and interglacials ($\Delta T = 1-2$ °C and $\Delta S = <1\%_o$). Therefore, the global ice volume effect can be estimated from the benthic foraminifer $\delta^{18}O$ signals by a simple subtraction of the $\delta^{18}O$ interglacial value from the total $\delta^{18}O$ value. Subsequent subtraction of the global ice volume $\delta^{18}O$ from the total values measured in planktonic foraminifer tests will then indicate the surface water salinity/temperature $\delta^{18}O$ signal. If a reasonable assumption can be made about the salinity of the surface water, its temperature can be reconstructed (Duplessy, 1978). On the other hand, if regional comparisons allow distinctions to be made between areas with pronounced salinity changes and areas with constant salinity, it might be possible to calculate the respective local temperature and salinity profiles, and thus to identify meltwater peaks in the oxygen isotope record (Hald and Vorren, 1987; Jones and Keigwin, 1988; Sarnthein et al., 1992). Nevertheless, the distinction between effects of salinity and temperature is not unequivocal and such interpretations should be scrutinized with caution.

The $\delta^{13}C$ signal of benthic and planktonic organisms is a reflection of much more complex processes. A high $\delta^{13}C$ signal in planktonic organisms records a rapid exchange between CO_2 of the surface waters and the atmosphere, while low $\delta^{13}C$ values may indicate a more stratified water column, for example, caused by high meltwater introduction. In addition, a high productivity of marine organic matter preferentially extracts ^{12}C from sea water, which results in even more positive $\delta^{13}C$ values of the calcite precipitated in equilibrium with the ^{12}C-depleted sea water. Production of young deep water in the Norwegian and Greenland Seas is documented by positive $\delta^{13}C$ values of benthic foraminifers (Jansen and Erlenkeuser, 1985). 'Aging' of deep waters is recorded by negative $\delta^{13}C$ values of benthic foraminifers, which results from the incorporation of ^{13}C-depleted CO_2 derived from oxidation of organic matter at the sea floor.

In summary, oxygen and carbon isotopes of planktonic and benthic tests provide a useful tool for the recognition of water mass characteristics. In glaciomarine environments, interpretations should always consider the problems concerned with the distinction between effects of salinity and temperature on the planktonic oxygen isotope signal.

5.2.7. Carbonate Sedimentation in Glaciomarine Environmental Settings

A classical controversy in the interpretation of ancient glacial sediments is the association of diamictites and carbonates (Chapter 7). The association of carbonates and diamictites is particularly common in Late-Precambrian sediments (Spencer, 1971), where it seems to be the rule rather than the exception (Anderson, 1983). Carbonates are found as interbedded layers as well as forming the base or top of a glacial sequence. However, the facies association of carbonate-diamictite has also been interpreted as evidence against a glacial origin of the diamictites (Schermerhorn, 1974) because carbonates are traditionally thought to reflect warmer climates. Recent studies have shown that skeletal carbonate (from molluscs, red algae and barnacles) may well accumulate in cold climates (Bjørlykke et al., 1978; Domack, 1988; Freiwald et al., 1991) (Menzies, 1995a, Chapter

14). The frequent and extensive accumulations of carbonate on Arctic shelves were formed in response to the Holocene transgression. Such carbonates cap the underlying Weichselian/Wisconsinan glacial sediments as far north as 80° N. In these very high latitudes, 75–80° N, carbonate-bearing glaciomarine sediments are being deposited at the present day in an ice distal setting, and carbonates can therefore be regarded as an integral component of glaciomarine sedimentation (Elverhøi et al., 1989). In Antarctica, carbonate and, in particular, siliceous ooze, form the main constituents of shelf sediments in a number of locations (Domack, 1988; Dunbar et al., 1989). Similarly, as in the high Arctic, the biogenic sediments of Antarctica form a component part of the present-day glaciomarine sedimentation and are found as ice-proximal deposits. There is therefore no environmental conflict between carbonate accumulation and clastic glaciomarine sedimentation. The Holocene high latitude shelves can be seen as providing suitable conditions for carbonate accumulation: low clastic sediment supply (due to sediment-starved shelves) and high nutrient input (from upwelling).

However, some ancient carbonate accumulations developed differently from those in Holocene deposits. The Late-Precambrian carbonate accumulations have often been attributed to the formation of stromatolites. Additionally, a completely different chemical composition of sea water in the Precambrian oceans (e.g. the early 'soda ocean') (Kempe and Degens, 1985; Kempe et al., 1989) would enable direct precipitation of carbonate from sea water over a wide temperature range. Some of the Late Precambrian carbonates have also been shown to be of clastic origin (e.g. reworked from underlying or adjacent areas) (Fairchild, 1983, Fairchild and Spiro, 1990). In Permian sequences, the carbonate-diamicton association is common, and a purely biogenic origin, formed under cold water conditions, has been proposed (Rao, 1981; Domack, 1988).

In deep sea sediments, the carbonate content has been found to vary in a cyclic pattern corresponding to interglacial–glacial changes. In low- and mid-latitudes of the Pacific, the carbonate cycle seems to be caused by dissolution: during interglacials, the CCD-depth (carbonate compensation depth) rose, while during glacials, it receded to a lower water depth (Berger, 1973). However, in high latitude areas, the conditions are much more complex, and the observed content of CaCO₃ is a function of the variation between biogenic production relative to influx of non-biogenic (detrital) carbonate and dissolution (Ramm, 1989). The general trend is of a high influx of biogenic carbonates during interglacials, with minimum values corresponding to glacials. In particular, the introduction of meltwater plumes into the pelagic system has the effect of strongly reducing the biogenic production in surface waters (Menzies, 1995a, Chapter 2).

In principle, the contrast between glacial and interglacial carbonate production in high latitude open sea environments can be illustrated by the modern carbonate production in the Norwegian–Greenland Sea (Figs 5.5 and 5.6a,b). Atlantic water with no ice cover characterizes the eastern areas (Norwegian Current), while cold Arctic waters and permanent or semi permanent sea ice (i.e. glacial conditions) describe the environments to the north and west (Fig. 5.1). Recent data from sediment traps in the Norwegian Sea (Fig. 5.1), (Honjo *et al.*, 1988; Samtleben

and Bickert, 1990) show decreasing pelagic carbonate production within the Norwegian Current along its path northwards. This gradient records the cooling of the Norwegian Current. Further to the north and west, in the Fram Strait and Greenland Sea areas (Fig. 5.1) (where there are 'glacial' conditions), the carbonate flux decreases to half to one third of the total flux determined for the Norwegian Current.

The main biogenic carbonate content of surface sediments under open ocean cold water conditions is formed by a monospecific assemblage of the left-coiling foraminifera, *Neogloboquadrina pachyderma*. High reproduction rates of this species have been reported from Arctic and Antarctic sea ice environments (Spindler, 1990). Sediment trap measurements have shown coccoliths as an important carbonate source. Areas with almost permanent sea ice cover are barren of coccoliths, while areas with some open water have regular summer coccolith blooms, dominated by *Coccolithus pelagicus*. However, most of the coccolith carbonate is generally rapidly dissolved in the water column and at the sediment surface. Introduction of meltwater plumes into the pelagic system has the effect of strongly reducing both

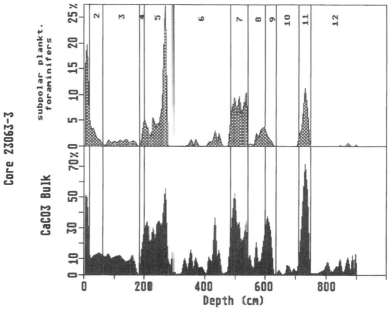

FIG. 5.5. Vertical variations in the carbonate content of Quaternary sediments from the Norwegian-Greenland Sea. From the vertical section, carbonate peaks and high contents of subpolar planktonic foraminifers are found in interglacial periods (1,5,7 and 11).

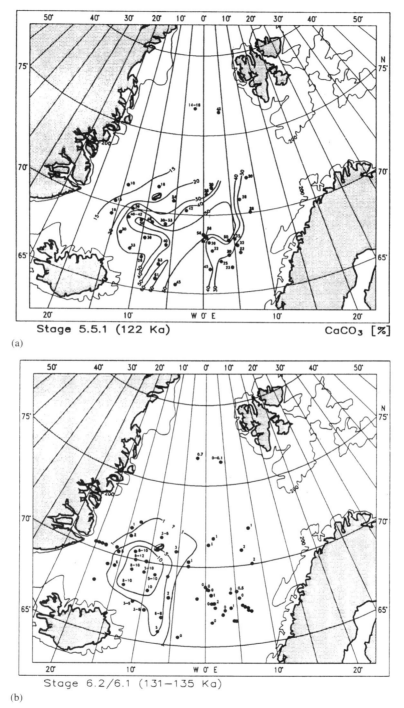

(a)

(b)

FIG. 5.6a,b. The horizontal distribution maps reveal that the highest carbonate contents can be seen on the eastern side of the basin, deposited under the warm Atlantic surface waters (Fig. 5.6b, Eemian interglacial, stage 5). During glacial periods (Fig. 5.6c, stage 6.2), generally much lower carbonate contents are recorded with a much narrower extension along the western and central regions, i.e. areas which have been less affected by ice-rafting and meltwater dispersion

foraminifer and coccolith abundance in surface waters, due primarily to the resultant loss of light caused by particle-rich glacial runoff.

Thus, carbonate flux calculations, compositional analysis and species abundances provide useful tools for the recognition of water mass characteristics in glaciomarine environments, and permit reconstruction of carbonate production during glacial and interglacial climatic shifts (Gard, 1988; Henrich et al., 1989a; Ramm, 1989; Gard and Backman, 1990).

5.2.8. Biogenic Opal Production and Preservation in Glaciomarine Environmental Settings

Maps of the modern distribution of biogenic opal in surface sediments (Leinen et al., 1986) indicate a clear correlation between opal-rich sediments and zones of higher primary productivity in the equatorial divergences, coastal upwelling areas and circum Antarctic and Arctic diatom ooze belt. In addition, phytoplankton production in the Arctic and Subarctic (Sakshaug and Holm-Hansen, 1984) is related to spring blooms in the vicinity of the ice edge. During spring melting of sea ice, nutrient-rich water develops at the surface, inducing a vigorous phytoplankton bloom in the wake of the retreating ice edge. In general, biogenic opal is rapidly dissolved within the water column and, in particular, at and/or below the sediment water interface, as a result of the under-saturation of sea water with respect to opaline silica (Broecker and Peng, 1982; Calvert, 1983). The fraction of opal preserved in the sediments varies sharply. In general, the amount of biogenic opal and its degree of preservation increases in areas with high silica fluxes (Broecker and Peng, 1982). On a global scale, the amount of opal buried in the sediments is estimated to be <5–10% of the total production. Consequently, specific conditions are required to preserve biogenic opal blooms in the sediments (e.g. high production of siliceous tests, high bulk sedimentation rates and a low level of bioturbation). Despite the fragmentary record of opal production in glaciomarine environments, specific opal occurrences have been used successfully in glaciomarine palaeo-oceanographic reconstructions.

A time-transgressive diatom maximum in subarctic sediments during the last deglaciation has been interpreted to indicate the northward passage of the oceanic polar front from the southern tip of Norway to the Fram Strait (Fig. 5.1) (Stabell, 1986). Grobe (1986, 1987) and Grobe et al. (1990) have interpreted the cyclic occurrence of biogenic opal in glacial/interglacial sections from the Weddell Sea continental margin in Antarctica (Fig. 5.3) to be due to siliceous blooms. Highest opal fluxes are recorded during peak warm interglacial stages, while reduced production occurs during glacial stages. Triggering mechanisms of opal production in the Antarctic are complex and controversial (Mortlock et al., 1991). Nutrient availability obviously is not a limiting factor (van Bennekom et al., 1988). Instead, temperature, light penetration, and the amount of dissolved iron in surface waters (Martin, 1990; Mortlock et al., 1991) have been identified as the major controls of opal production.

5.2.9. Carbonate Dissolution Records and Early Diagenetic Reactions as a Response to Changing Bottom Water Properties in Glaciomarine Settings

Bottom water properties and circulation patterns in deep sea basins with glaciomarine ice margins on their surrounding shelves are strongly influenced by glaciomarine processes. Dense brine formation during seasonal sea ice growth and open ocean deep convection episodically injects a young oxygen-rich and dense deep water into the basins, whereas the introduction of high quantities of meltwater into the open ocean can stabilize the water column and thus inhibit deep water circulation and exchange. Hence, it is important to record changes in bottom water properties and to evaluate the effect which glaciomarine processes contribute to these changes. Changes in the O_2 and CO_2 content of bottom waters can be deduced from studies of carbonate preservation. In addition, carbonate dissolution may strongly modify carbonate fluxes, especially in glaciomarine settings where carbonate production rates may be low. In addition to changes in bottom water chemistry, carbonate dissolution may indicate early diagenetic reactions. Hence, dissolution studies encounter two principal problems: first, the development of reliable methods which allow the quantification of dissolution; and second, distinguishing between dissolution in bottom water and in porewater. Most parameters commonly used for dissolution studies in low- and mid-latitude oceans (e.g. the planktonic/ benthic ratio, quantitative determination of the insoluble residue of pelagic carbonates, and the percentage of dissolution-sensitive taxa) cannot be applied to polar and subpolar sediments due to large-scale variations between ecological conditions and sedimentary inputs of sources from glacial to interglacial stages (Henrich et al., 1989a).

Conventional fragmentation indices of left-coiling *Neogloboquadrina pachyderma*, the only planktonic foraminifer species occurring in abundance in both glacial and interglacial sediments, and the more sensitive SEM-based dissolution indices (Henrich, 1986) of this species, allow estimation of dissolution on a semi-quantitative scale. If these investigations are accomplished by geochemical studies that allow identification of the early diagenetic evolution of the sediment, differentiation might be possible between dissolution at the sea floor and within the sediment (Henrich et al., 1989a). These diagenetic studies include metal enrichment in oxic, suboxic and anoxic diagenetic environments (Berner, 1981; Wilson et al., 1986) in relation to sedimentation rate, supply and quality of organic matter and changes in bottom water properties. Early diagenetic oxidation of organic matter in oxic and suboxic porewater environments provides a potentially effective mechanism of carbonate dissolution, which may be reflected in a gradual decrease in the organic carbon content of the sediments. But, if strongly dissolved calcareous tests can also be detected in the diagenetic zones below these levels, the dissolution obviously should be explained as a sea floor phenomenon indicating increased PCO_2 in bottom waters. In addition, the identification of changes in the benthic organism assemblages could also contribute further essential information. An example that uses such a combined set of data to identify changing properties of bottom waters and related early diagenetic reactions is discussed below (Section 5.4.5).

5.2.10. Ecologic Conditions and Environmental Reconstructions

When reconstructing glaciomarine environments, the use of fossils and knowledge of their ecological requirements represent important parameters. Distinctions between paleoenvironments must occur on smaller as well as larger scales, including surface versus deeper waters and glacial/interglacial climatic shifts. Examples of changing conditions include: (1) advection and mixture of warm and cold, seasonally pack-ice-covered surface waters, (2) introduction of meltwater at tidewater ice fronts and melting sea ice/ icebergs, (3) the production of cold, dense brines on the shelves, and (4) deep convection in subpolar and polar seas. Micropaleontological methods, including statistical analyses of specific planktonic and benthic foraminifer assemblages, have now been developed in order to estimate physical properties such as the temperature and salinity of surface and deep waters; these are based on planktonic and benthic assemblages, respectively (Haake and Pflaumann, 1989). Unfortunately, the resolution of this method in polar waters is limited because (1) the assemblage is dominated by left-coiling *Neogloboquadrina pachyderma*, and (2) interpretational problems arise from salinity effects in surface waters. Diatom assemblages have been successfully used to trace changes in surface water mass conditions during postglacial and Holocene times (Koc Karpuz and Jansen, 1992). However, due to low sedimentation rates and strong diagenetic solution of biogenic silica, this approach can not be applied in most older sequences. Quantitative analysis of radiolarian assemblages depict characteristic patterns indicative of cold surface and shallow subsurface waters (Morley, 1983; Jansen and Bjørklund, 1985). Benthic foraminiferal assemblages reveal pronounced glacial/interglacial shifts related to changes in bottom water conditions (Streeter et al., 1982). Important information on glacial/interglacial variations in deep water circulation can also be obtained from measuring the Cd/Ca ratios of benthic foraminifers (Fig. 5.7) (Boyle, 1988). It has been found that the distribution of cadmium matches the content of phosphates and nitrate nutrients in today's oceans. This work has shown that the cadmium ion may substitute for calcium in foraminiferal shells, and

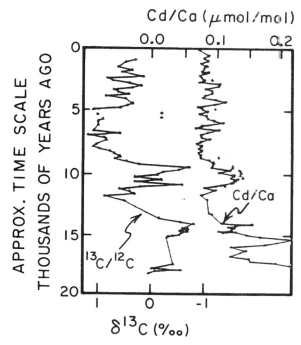

FIG. 5.7. Variations in the Cd/Ca ratio of benthic foraminifers in the North Atlantic. Relatively high Cd/Ca values correspond to glacial periods, e.g., lack of cold deep water formed from surface cooling (and nutrient depletion from the surface biomass) during these intervals (From Boyle, 1988; Broecker and Denton, 1989)

by analyzing cadmium, information on the nitrate and phosphate content of the water masses can be obtained. Accordingly, low Cd/Ca values in benthic foraminiferal shells indicate nutrient-depleted water masses (e.g. water masses formerly exposed to the surface). This type of analysis has recently been used for tracing the formation of cold water in the North Atlantic during the last glacial /interglacial period. From Fig. 5.7 it is evident that the Cd/Ca ratio was higher during glacials, which indicates that the present-day process of cold deep water formation from chilling of surface waters in the Norwegian–Greenland Sea was less efficient during these periods (Boyle, 1988; Broecker and Denton, 1989).

On the shelves, typical stratigraphic successions of macro- and microfaunal assemblages outline changes in benthic habitat conditions, which are related to shifts in the sedimentary environment during the waxing and waning of continental ice sheets (Vorren *et al.*, 1989; Thomsen and Vorren, 1986).

5.3. EXAMPLES OF PRE-CENOZOIC GLACIOMARINE SEQUENCES

Commonly, interpretations of ancient sequences can be characterized by conflicting points of view. In the following, two well-known Pre-Cenozoic sequences are described and various interpretations discussed. The main objective is to show how the interpretation of diamictites has changed radically, from one of being ascribed largely to terrestrial deposits, to the present-day view of essentially being of a glaciomarine origin (see Chapter 7).

5.3.1. Gowganda Formation

The Gowganda Formation of Ontario, Canada, is an Early Proterozoic deposit, with a probable age of 2.7–2.3 10^9 BP (Young, 1981). The formation is primarily exposed in three locations close to eastern Lake Superior in Canada. The sediments are characterized by various types of diamictites (tillite deposits) and varved argillites with abundant dropstones. The coarser diamictites are normally structureless, while laminations are often seen in the mudstones (Young, 1981). Sedimentary structures are generally well preserved, but folded and metamorphosed rocks are found locally (up to amphibolite facies). Typically, the individual members of the Gowganda Formation are discontinuous on a scale of kilometres. However, some members can be followed for tens of kilometres. The thickness of the formation normally varies between 300 to about 1000 m, but thicknesses of up to 3000 m have been reported. In general, the diamictite members are massive, while the interbedded argillites are laminated and contain IRD (Fig. 5.8). These laminated units also contain mudstone pellets. Previously, the formation was interpreted as a lodgement tillite with interbedded lacustrine varved argillites (Lindsey, 1969; Young, 1981). Glaciomarine sediments were thought to be present only in a restricted area. As depicted in the schematic stratigraphic section (Fig. 5.8), the diamictites represent ice advances, with the interbedded argillites as lacustrine ice-recessional deposits. However, in more recent studies, a subaqueous origin with predominantly glaciomarine deposition has been suggested, but there are still clearly different opinions on the depositional

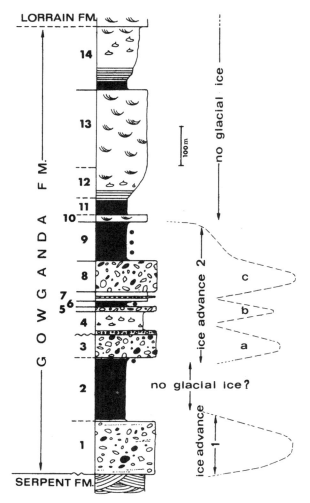

FIG. 5.8. Schematic stratigraphic section through the Gowganda Formation. Two major ice advances are postulated (member 1 and 3 to 9), while members 10 to 14 are considered to be non-glacial and consist of two deltaic cycles. Members 2, 6, 9 and 11 represent laminated argillite. Black dots at the right side of the column represent dropstones in argillite sediments (from Young, 1981, and Young and Nesbitt, 1985)

indicated by debris flow, sandy fluidized flow and Bouma-turbidites, indicating a marine environment; (2) the massive- to faintly-bedded diamictites tens of metres in thickness are interpreted as ice-rafted debris and distal mud, while some thinner units may represent debris flows; (3) no convincing evidence of any lodgement till formation has been found. In addition, long 'fingers' of the Gowganda Formation extending northward on the Archean basement were earlier thought to represent paleo-valleys or deep water lacustrine environments. However, the lack of coarse debris along the margins excludes the valley hypothesis, and, due to the large scale (400 × 500 km) and thickness, a lacustrine formation is also challenged.

According to Eyles *et al.* (1985), the Gowganda Formation was formed as an open marine deposit, and the lower section in the Elliot Lake region (southwestern exposure) is interpreted as the continental slope and submarine channel-fill depositional system of a glaciomarine environment.

5.3.1.2. Resedimentation of glacially-transported debris

According to Young and Nesbitt (1985), the lower diamictite (Fig. 5.8) results from resedimentation of glacially transported debris in an actively subsiding region. The sediments were supplied from outlet glaciers entering the sea, probably as ice tongues. Subsequent glacial recession was followed by deep water sedimentation of fine-grained materials; however, clasts in its lower and upper units reflect nearby glaciers. The second ice-advance (Fig. 5.8) was responsible for deposition of the upper diamictite complex. Here, conglomerates and turbidites are widespread, and are thought to be related to ice advances and active tectonic subsidence.

5.3.1.3. Ice proximal regime

Mustard and Donaldson (1987) suggested a submarine ice proximal depositional regime, at least for the lower part of the Gowganda Formation. They separate the lower member into the following four units: (1) a 'basal diamictite' unit, which is overlain by (2) a coarsening-upward rhythmite/ sandstone/clast-sup-

environment (Eyles *et al.*, 1985; Young and Nesbitt, 1985; Mustard and Donaldson, 1987). Three varying interpretations are summarized below.

5.3.1.1. Continental slope/channel fill

Eyles *et al.* (1985) suggested the following depositional regime: (1) sediment gravity flow deposits are

ported diamictite, which may grade laterally into (3a) either a stratified or (3b) massive matrix-supported diamictite ('Fan' or 'Interfan' association, respectively). The top unit (4) consists of a complex series of diamictites (Upper diamictite), including matrix-supported as well as clast-supported diamictites, with additional fine-grained noncyclic, poorly-laminated

mudstones, containing IRD. Subglacial erosional features such as plucking and glacial lee-side quarrying at the top of the Archean–Proterozoic unconformity surface demonstrate the presence of grounded ice. Consequently, the 'Basal Diamictite', 1–12 m thick, is interpreted as a basal tillite, while stratified interbeds within this diamictite-zone are explained as

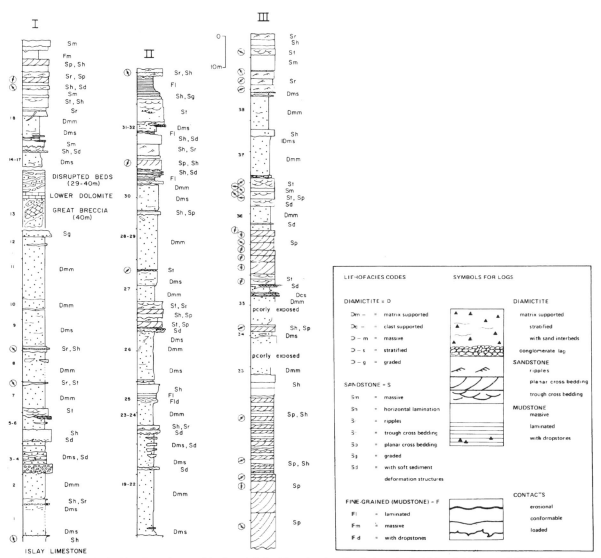

FIG. 5.9. Sedimentological log through part of the Port Askaig tillite, showing diamictite units 1–38 (total thickness of 540 m). Diamictites (D) are the main lithofacies, while the interbedded sand layers (S) become more important towards the top. Paleocurrent directions are shown by arrows and all have been corrected for tectonic tilt. The three sections are from Garbh Eileach in the Garvellach Islands, SW Scotland. (From Eyles, 1988, reproduced with permission from Elsevier Science Publishers)

deposits in subglacial meltwater channels. The two overlying 'Fan' and 'Interfan' associations are interpreted as: subaqueous outwash deposits ('Fan' association) and a more complex regime of ice-marginal debris/morainal bank/debris flow/high density sediment gravity flows ('Interfan'). Scattered dropstones reflect rafting from an adjacent calving glacier. The 'Upper Diamictite' is thought to represent deposition under the readvance of a floating or partly floating ice sheet. The lack of glaciotectonic structures or features reflecting erosion or incorporation of the underlying sediments suggests an ice-shelf depositional regime rather than an advancing grounded ice sheet.

5.3.1.4. Gowganda Formation – concluding remarks

The study of the Gowganda Formation illustrates the recent paradigm shift in means of interpretation, from a terrestrial assignment to an inferred glaciomarine depositional environment; the massive diamictites being interpreted as glaciomarine deposits, possibly resedimented. As corroborative evidence along the margins of the Norwegian–Greenland Sea, thick diamictites combined with large-scale slump and slide scars are found as a typical facies association for a glaciated margin.

5.3.2. Port Askaig Formation

The Port Askaig Formation in Scotland, belonging to the Dalradian Subgroup, is another deposit characterized by conflicting views as to its origin (Spencer, 1975, 1981; Eyles and Eyles, 1983b; Eyles, 1988a). This formation, with a maximum thickness of up to 850 m, is of Late Proterozoic Age (approximately 670 Ma), includes 47 diamictite horizons (Fig. 5.9). It is under- and overlain by carbonates, limestones and dolomite. The lower diamictite horizons have a dolomitic silt matrix, now shown to have been reworked from the underlying formation. The diamictite horizons are characterized by massive or stratified units containing discontinuous interbeds of sand. In general, the individual diamictite beds have a sharp, conformable lower contact. At a number of the diamictite horizons, sandstone wedges (< 12 m), interpreted as periglacial permafrost (Spencer, 1981), are a common phenomenon.

5.3.2.1. Subglacial origin

The Port Askaig Formation has been studied extensively. According to Spencer (1981), the diamictites are formed mainly as subglacial deposits (Fig. 5.10). This is based on the follows aspects: (1) The 'Great Breccia' (Fig. 5.9) and its underlying sediments show signs of glaciotectonism, glacial erosional features and recycling of earlier mixtites. (2) The 'clean' sandstone and conglomerate units interbedded with the mixtites are unlikely to have developed in response to winnowing of glaciomarine sediments in a marine environment. Instead, these deposits have formed in sub- and englacial tunnels. (3) The horizons containing ice-rafted pebbles – clasts penetrating and deflecting laminae – are clearly different from the homogeneous/lenticularly stratified mixtites, which were formed subglacially.

5.3.2.2. Glaciomarine formation

The Port Askaig Formation was recently reinterpreted to be of glaciomarine origin (Fig. 5.10), including tidally-influenced marine sedimentation (Eyles et al., 1985; Eyles, 1988a). The revised interpretation can be

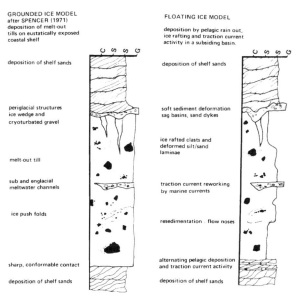

FIG. 5.10. Illustration summarizing the two contrasting models of deposition of the Port Askaig diamictites. (From Eyles and Eyles, 1983, reproduced by permission of the authors)

summarized as follows: (1) The massive diamictites were formed by the settling of fine-grained suspended matter in combination with ice-rafted debris in water depths less than 250 m; (2) the sand layers, suggested by Spencer (1981) to be sub- or englacial tunnel deposits, are interpreted as formed on a tidal y-influenced shallow marine shelf. The sand is thought to have been derived from deltaic sources on the basin margins and transported onto the shelf during storms; (3) the conglomerate layers were formed in response to increased current activity and winnowing of underlying sand and diamictite; (4) the distinct diamictite layers such as the Great Breccia and the Disrupted Bed are interpreted as products of down-slope resedimentation under periods of increased tectonic activity.

In general, the Port Askaig Formation consists of a number of coarsening-upwards sequences and each sequence consists of a diamictite at its base, followed by sandstone, and, finally, capped by a thin conglomerate layer. As discussed by Eyles (1988a) and as seen from the Barents Sea, such coarsening-upwards sequences are typical for Late Quaternary high latitude shelf sediments. In the Late Quaternary, such sequences have developed in response to the glacio-marine sedimentation during the Late Weichselian/Wisconsinan, which was followed by glacio-isostatic rebound and sea shallowing, with increasing current activity and winnowing during the Holocene.

5.3.2.3. Port Askaig Formation – concluding remarks

The interpretations of a terrestrial origin have been challenged by the emergence of the concept of glaciomarine deposition. As in the case of the Gowganda Formation, the association of thick diamictite units combined with sand/conglomerates and carbonates in the Port Askaig Formation may have an analogue in Quaternary glaciomarine sediments.

5.3.3. Pre-Cenozoic Glaciomarine Sediments – General Comments

When analyzing ancient, Pre-Cenozoic glacigenic sediments, the problem of preservation potential must be considered. In contrast to Cenozoic glaciomarine

sediments, which are essentially laid down along a passive margin, the preserved Pre-Cenozoic glacigenic sediments are mostly from intercratonic or rift basins (Bjørlykke, 1985; Nystuen, 1985). The ancient margin sediments have been involved in orogenic events and metamorphosed, while the intercratonic basins have a higher likelihood of preservation. However, thick Pre-Cenozoic sequences (e.g. the Sturtian succession in South Australia [Late Proterozoic]) may reach thicknesses of up to 3–5 km (Young and Gostin, 1989a). Glaciomarine sedimentation, combined with sediment gravity flow deposits, has been suggested as the principal mechanism of deposition for these latter successions. As demonstrated below, the same processes are also the most important for the Late Cenozoic passive margin deposits, even though the tectonic settings may be different between the Cenozoic and Pre-Cenozoic deposits.

5.4. GLACIOMARINE SEDIMENTS AND THE SEDIMENTARY ENVIRONMENT OF A PASSIVE MARGIN AND ITS ADJACENT OCEAN – EXEMPLIFIED BY THE NORWEGIAN–GREENLAND SEA AND THE NORWEGIAN/BARENTS SEA MARGIN

5.4.1. Cenozoic Glaciations – General Background

On a global scale, the change from an Early Cenozoic temperate, non-glacial to a Middle and Late Cenozoic glacial/interglacial mode of sedimentation is documented by an increase in benthic foraminiferal $\delta^{18}O$ values (Miller et al., 1987). Glacial onset in the southern (42 Ma) and northern hemispheres (10–5 Ma) was a response to significant changes in global ocean circulation. In the sediments, the onset of a cold and glacial climate is reflected by the introduction of diamictons on the continental shelves and the appearance of ice-rafted fragments (>63 μm) in deep sea sediments in and adjacent to polar regions (Domack and Domack, 1991) (Menzies, 1995a, Chapter 2).

In the southern hemisphere, glaciomarine sediments are widely distributed, and dropstones are observed as far north as 50°S (Antarctic Convergence). From recent ODP drilling and drilling in McMurdo Sound, glaciomarine sediments as old as Middle Eocene age (Prytz Bay) and Early Oligocene (McMurdo Sound) have been reported, indicating a more or less continuous glacial coverage of the East Antarctic region during the last 40 million years (Domack and Domack, 1991; Barrett *et al.*, 1989). In more distal areas, dropstones are largely confined to the Late Cenozoic, while individual layers of dropstones have also been found in Oligocene sediments (ODP Leg 120, site 748: Kerguelen Plateau). For further reading of the literature on the Cenozoic glaciomarine sedimentation in Antarctica, see Barrett *et al.* (1989), Barker *et al.* (1989), Barron *et al.* (1991), Domack and Domack (1991) and references therein.

In the northern hemisphere, deep sea sediments containing probable ice-rafted fragments have been traced back to: (1) ≈8.5 Ma in the Baffin Bay (Wolf and Thiede (1991); (2) ≈5.3 Ma in the Norwegian Sea (Jansen *et al.*, 1990; Jansen and Sjøholm, 1991); and (3) probably >4 Ma in the Arctic Ocean (Aksu and Mudie, 1985). It is not yet clear if the IRD represents iceberg or sea ice rafting, or a combination of both. Glaciomarine sediments of Pliocene age are also recognized in the Yakataga Formation of Alaska (Plafker and Addicott, 1976).

At 2.6–2.5 Ma there is a major global change in the sedimentary record, including that of the North Atlantic and Norwegian–Greenland Sea (Fig. 5.1). Well-defined glacial/interglacial cycles appear, and the input of IRD increases significantly, resulting in the deposition of deep sea diamictons (Henrich, 1989). These events have been correlated with the onset of northern hemisphere continental glaciations. Another threshold in the development of northern hemisphere glaciations occurred at ~1 Ma, as indicated by a significant increase in both IRD and finer sediment supplied to the northern oceans.

In the southern ocean, the cooling and build up of the Antarctic Ice Sheet is attributed to the isolation of the Antarctic continent and development of the circum-polar current regime, which dramatically changed the sedimentary regime around Antarctica. In the northern hemisphere, Cenozoic plate tectonics and the oceanic gateway configuration (e.g. key locations

of oceanic circulation features, such as submarine ridges, influencing deep water exchange between oceanic basins) are much more complex. The opening and development of fundamental gateways, such as Fram Strait and the Greenland–Scotland Ridge, are among the most important submarine topographic developments in global oceanographic circulation.

The sedimentary record of the Norwegian–Greenland Sea and its adjacent margins offers a good prospect for gaining insight into the development of glacially-influenced passive margins. Four major ice sheets terminate or have terminated on these margins: the British, Scandinavian, Greenland and Svalbard-Barents Sea Ice Sheets. Recent deep sea drilling (ODP Leg 104), combined with extensive sediment coring and shallow seismic profiling, enable reconstruction of the sedimentary record and environment of the past 2.6 Ma and postulation of different scenarios for the glaciated continental margins. The basic tools are now available to allow tracing of specific glacial events from their sources on land over the shelf and slope into the deep sea.

5.4.2. General Aspects of the Late Cenozoic Glaciomarine Sedimentation of the Norwegian–Greenland Sea

One of the most interesting results from ODP Leg 104 was the observation of a rhythmic occurrence of dark diamictons frequently intercalated within the glacial/interglacial section of the past 2.6 Ma (Eldholm *et al.*, 1987; Henrich *et al.*, 1989b). This was the first time such layers had been discovered in deep sea sections. Glacial/interglacial sediments in the Norwegian–Greenland Sea reveal cyclic and rhythmic variations (Section 5.4.5) in biogenic and terrigenous composition, indicating corresponding changes in surface and bottom water properties. Interglacial sediments are characterized by high pelagic carbonate shell production (e.g. planktonic foraminifers and coccoliths), with high percentages of warm subpolar species reflecting inflow of warm Atlantic water. Glacial sediments indicate much lower carbonate shell production, dominated by colder water species. In addition, variable inputs of IRD and bulk terrigenous sediment supply are observed during glacial stages, with rhythmic deposition of dark diamictons. In general, the peak deposition of IRD material occurs during early periods of deglaciation. Shifts in benthic faunal associations, especially benthic foraminifers, indicate variations in bottom water ventilation and circulation. Highest production rates of dense, young, oxygen-rich deep waters occurred during interglacial stages.

From the sedimentary record in the Norwegian–Greenland Sea, a number of environmental shifts (glacial/interglacial) are observed that may reflect changing conditions on land and in the sea. Interglacial deposits do not appear prior to 1.2 Ma at sites located

offshore on the edge and slope of the outer Vøring Plateau (Fig. 5.1), a marginal plateau at the Norwegian mid-continental margin. They are restricted to a few occurrences at a high sedimentation rate site on the inner plateau close to the shelf. Persistent strong carbonate dissolution is indicated throughout the section from 2.57 to 1.2 Ma, followed by a long transitional period with a gradual decrease in dissolution from 1.2 Ma to about 0.6 Ma (Henrich, 1989). Low amplitude oscillations in IRD input with peak supplies in the dark diamictons are indicated throughout the

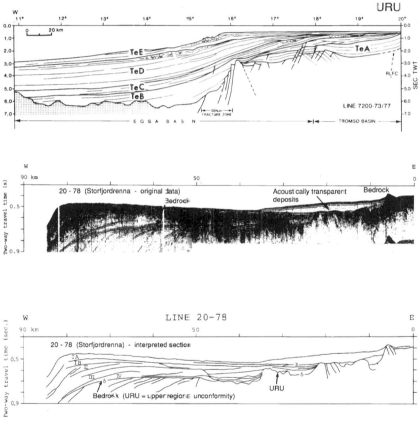

FIG. 5.11. Interpreted seismic sections from the Bear Island Trough (a) and Storfjordrenna (b) and original data from the latter area (c). (For location, see Fig. 5.11). (a) Schematic profile showing the main structure of the fan complex outside the Bear Island Trough (From Vorren et al., 1991). Unit TeE and TeD is characterized by chaotic seismic reflection pattern. According to Vorren et al. (1991) unit TeE and TeD represent Late Cenozoic glacigenic sediments, while Eidvin and Riis (1989) also include unit TeC into the Late Cenozoic deposits. Unit TeB and TeA represent probably Middle and Early Tertiary deposits (b) The glacial nature of the sediments is clearly seen from the large-scale ridge features forming up to 50 m local highs, and a number of erosional unconformities, which reflect periods of glacial advance. Four regionally correlative reflectors have been identified: a, b, γ and d, while I, II, III and IV define four seismic sequences. Acoustically transparent deposits in front of a bedrock threshold are interpreted to represent glaciomarine sediments deposited in front of a glacier which had its front pinned at the threshold (from Solheim and Kristoffersen, 1984, reproduced by permission of the Norsk Polarinstitutt).

period since 2.57 Ma. Considering all environmental parameters, the section from 2.57 to 1.2 Ma is characterized by high frequency and low amplitude oscillations of IRD input, a lowered salinity, and, most probably, a decreased carbonate shell production in surface waters as well as a very shallow lysocline. During this period, the northern continents were covered by ice caps of smaller dimensions than those of the last million years and the oceanic polar front was maintained within the northeastern North Atlantic and the Norwegian Sea. However, the overall cool climate in the Norwegian Sea was episodically interrupted by weak intrusions of warmer Atlantic waters along the eastern margin. The last 1 Ma is characterized by a higher supply of IRD and bulk sediment. Large amplitude variations characterize the IRD and carbonate records, indicating growth and decay of huge ice masses in the northern hemisphere.

5.4.3. Barents Sea/Northern Norwegian Shelf

The glaciomarine shelf record may be incomplete or represent disturbed sequences due to repetitive glacial advances and iceberg ploughing. In the Barents Sea, only a thin cover (< 15 m) of sediment is found in intermediate and shallower areas (50–300 m water depth). On the deeper parts of the shelf and towards the slope, thick prograding glaciomarine sediment wedges which were formed during successive glacial advances terminate at or close the shelf edge (Fig. 5.11).

The morphology of the Barents Sea (Fig. 5.12), which is an epicontinental sea characterized by northwest trending basins (300–500 m water depth) and shallow banks (30–150 m deep), was strongly influenced by Late Cenozoic glaciations (up to 1000 m of glacial erosion has been estimated) (Eidvin and Riis, 1989). The entire shelf was probably covered by grounded ice during the last glaciation (Vorren et al., 1988) and at present (and during the Holocene) the northern and central parts of the area are influenced by sea ice and iceberg sedimentation (Elverhøi et al., 1989).

The southwestern Barents shelf and slope reveal a variety of glacigenic units typical for passive polar margins. On the outer shelf up to 300 m of stratiform glacigenic sediments overlie an upper regional uncon-

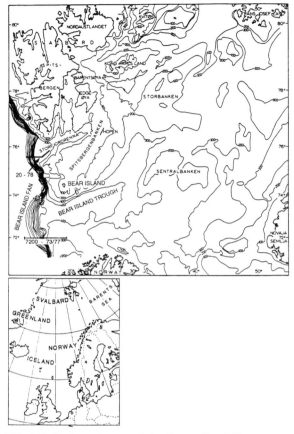

FIG. 5.12. Bathymetric map of the Barents Sea (100 m contour interval), including location of seismic profiles

formity (URU) with a glacially-eroded morphology (Fig. 5.11). In the central and shallower regions (300–100 m water depth), the sediment thickness is generally less than 15 m. However, local sediment accumulations of 30–50 m in thickness are found as moraine ridges and ice proximal sediments or acoustically transparent deposits (Profile 20–78 in Fig. 5.11). The ice proximal sediments, which may cover up to 2000 km² are found in the inner part of major embayments commonly at a water depth of 300 m (present day). These accumulations may have formed from settling of fine-grained materials from turbid surface plumes along the ice margin (Elverhøi et al., 1990). A modern analogue is probably represented by Bråsvellbreen (part of the 8000 km² large Austfonna Ice Cap at Nordaustlandet) in the northern Barents

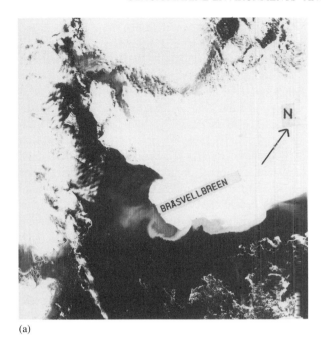

(a)

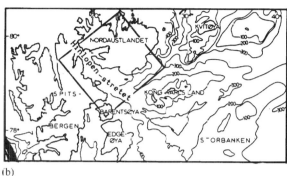

(b)

PLATE 5.1. Satellite image (Landsat, July 23, 1976) of Nordaustlandet, Austfonna Ice cap, showing the meltwater plumes and offshore flow pattern (from Elverhøi et al., 1989)

Sea (Plate 5.1). Here, the glacial sediment discharge is concentrated in a few melt water plumes and transported along the ice margin by the coast parallel current. The main deposition from the plume occurs within 20 km of the ice front and sedimentation from proximal parts (< 2 km) of these plumes has been estimated to be on the order of 10–15 cm year^{-1} (Pfirman and Solheim, 1989). The ice proximal deposits, at 300 m water depth, appear to represent a major halt in the recession of the Barents Sea Ice Sheet ~12–15 ka (Elverhøi et al., 1990).

At the outer shelf, the sediments are comprised of up to 150 m-thick seismic units which are separated by erosional surfaces (Fig. 5.11). Various depositional environments have been inferred, based on the seismic signatures (Fig. 5.13). These include glaciomarine sediments (= a semitransparent signature), submarine outwash fans (= a diverging stratified signature), glaciomarine/marine trough fills (= a stratified onlap pattern) and till (= a chaotic reflection). On the upper slope and at the shelf margin, thicknesses of ≥1000 m are reported, with a thickening towards deeper water (Vorren et al., 1989, 1991). Slope sediments are composed of prograding sequences with a complex sigmoidal-oblique character. Glacial periods are characterized by progradation and

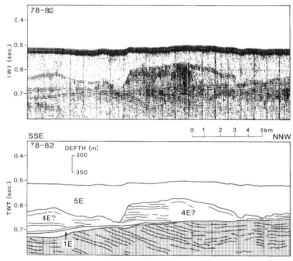

FIG. 5.13. Examples of interpretations of acoustic character and sediment type from glacigenic sediments in the outer part of the Bear Island Trough. 5E, acoustically-transparent/semitransparent, e.g. glaciomarine sediments. 4E, chaotic reflection pattern, e.g till deposits (from Vorren et al., 1989)

build-up, while during interglacials, sediments were removed from the continental shelf and transported into the deep ocean, bypassing and eroding the upper slope sediments (Fig. 5.14) (Vorren et al., 1989).

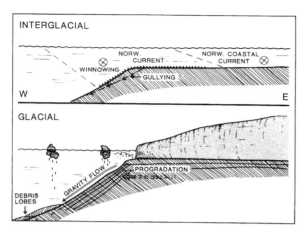

FIG. 5.14. Generalized model for glacial – interglacial sedimentary regimes on a passive margin. Glacials are characterized by wedge progradation while interglacial periods are associated with current-winnowing and gullying (from Vorren *et al.*, 1989, reproduced with permission from Elsevier Science Publishers)

Due to glacio-isostatic rebound, a shallowing of 60–100 m has taken place in the central and northern Barents Sea during the last 10,000 years. The sedimentary section on the shallow Spitsbergenbanken (Fig. 5.12) (30–80 m water depth) reflects these conditions in a typical regressive facies succession (Fig. 5.15a). This sequence is comprised of Late Weichselian till, overlain by glaciomarine sediments, capped by a residual diamicton, with a final layer of approximately 0.5 m polygenetic carbonate-rich sediments. In the Barents Sea, the carbonates are interpreted as post-glacial in development. The carbonates change from an initial soft bottom fauna (dominated by the mollusc *Mya truncata*) to a hard bottom-dominated fauna consisting mainly of the barnacle, *Balanides* (Bjørlykke *et al.*, 1978). The carbonates accumulated under conditions typical for carbonate deposition: low clastic sediment discharge and high nutrient supply due to strong currents and upwelling. The change from a soft to hard bottom is related to shallower water depths and a probable increase in bottom currents. Thus, excavation of mollusc shells and ice-rafted pebbles from the underlying diamicton produced a current-reworked, shell-rich, residual diamicton which formed the substrate for the more recent (< 4000 years) *Balanus* barnacle fauna. Icebergs and also sea ice pass over

this region, and coarser IRD has been dropped and mixed into the post-glacial sediments.

In intermediate and deeper parts of the central and northern Barents Sea (200–300 m water depth), the sediments have not been affected by the post glacial erosion, and the Late Weichselian glaciomarine sediments also grade gradually into fine-grained Holocene mud (Fig. 5.15b). Winnowing and re-suspension of the shallower banks represent a major source of the Holocene sediments, in addition to melt-out from sea ice and minor amounts of iceberg-rafted material (Elverhøi *et al.*, 1989).

In the deeper troughs, such as the Bear Island Trough (300–500 m water depth), the 50–100 m thick sediment sequence consists of relatively homogeneous diamicton of more than one depositional event (Fig. 5.15c) (Hald *et al.*, 1990). The sediments are overconsolidated and a till origin has been proposed; however, the presence of foraminifers, typical of a glaciomarine environment, show that parts of the till deposits are derived from glaciomarine sediments. The glacially-reworked nature of these sediments is seen also from shearing, folding and fissile textures.

The glaciomarine sequence on the shelf off northern Norway demonstrates the various facies formed during deglaciation of a shelf region and their distribution (Figs 5.16 and 5.17). Here, it is seen clearly how the shelf topography has controlled the facies distribution. Compared to the sequence from the central Barents Sea, the sequence off northern Norway shows a minor glacial readvance (Figs 5.16 and 5.17), which results in a more varied lithostratigraphy than that found to the north. Additionally, the Holocene – or post-glacial – sequence off northern Norway differs from the time-equivalent deposits in the central and northern Barents Sea by not being formed in a glaciomarine environment. The most complete and thickest sections are found in the troughs, while bank profiles have reduced thicknesses, erosional features, and are affected by iceberg turbation. Based on the sediment and microfossil composition, the following scenario has been constructed (Fig. 5.17):

(1) Basal till/overconsolidated diamicton, including scattered reworked shell fragments, indicates lodgement and/or deposition close to a tidewater glacier front (stage A);

A Shallow banks (Spitsbergenbanken, 30-80m water depth)

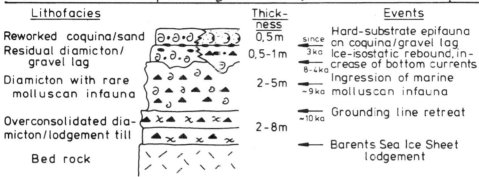

B Deep banks (200-300 m water depth)

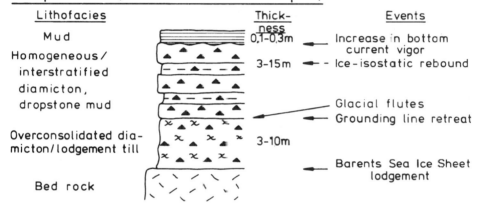

C Deep troughs (>300m water depth)

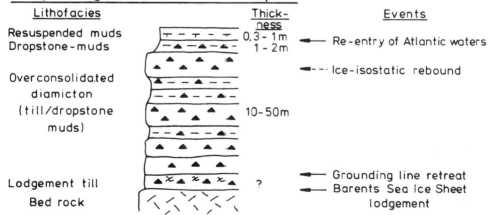

FIG. 5.15. Regressive glaciomarine sediments formed in response to glacioisostatic rebound in the central Barents Sea (from Bjørlykke *et al.*, 1978; Elverhøi *et. al.*, 1990; Hald *et al.*, 1990; Henrich, 1990)

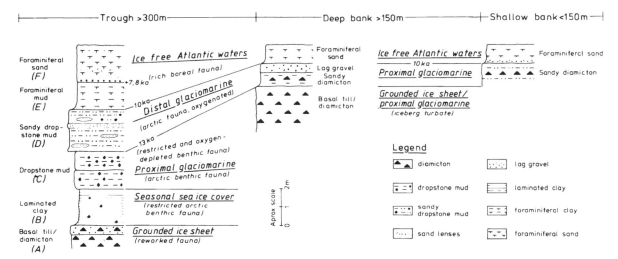

FIG. 5.16. Sediments during the last deglaciation on the continental shelf off northern Norway (from Vorren *et al.*, 1984; Henrich, 1990)

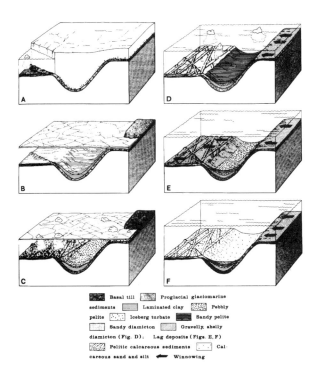

FIG. 5.17. Sedimentary environments during the last deglaciation on the continental shelf off northern Norway (from Vorren *et al.*, 1984; Henrich, 1990)

(2) Laminated clay with a restricted Arctic benthic fauna and scattered dropstones indicates seasonal sea ice cover during a stillstand in the ice recession (stage B);

(3) Dropstone-rich mud with a rich bottom fauna indicates readvance of the ice front, with frequent iceberg rafting. During this period the sediment on bank areas was intensively reworked by iceberg ploughing (Gipp, 1993) (Chapter 4), forming typical iceberg turbates (accumulations of sediments reworked by icebergs) (stage C). A more restricted fauna towards the upper part of the dropstone-rich mud may indicate less saline and less oxygenated water masses;

(4) Sandy dropstone mud, including sand lenses both on banks and in the troughs, demonstrates an open marine environment with icebergs and intensified current activity – indicating the intrusion of Atlantic water and withdrawal of polar water masses to higher latitudes (stages D, E);

(5) Similarly, as in the Barents Sea, carbonate-rich sediments cap the diamictons in the bank areas (Freiwald *et al.*, 1991). However, in contrast to the accumulations on the Spitsbergenbanken, Barents

Sea, the carbonate accumulations off northern Norway are not part of a glaciomarine environment. There are no sea ice or icebergs off northern Norway and the water masses all have temperatures well above 0°C (stage F).

5.4.4. The Continental Slope

The slope system is the main depositional centre for glacially-eroded sediments. Two to three km-thick sequences have been reported on the margins off Svalbard and the Barents Sea (Myhre and Eldholm, 1988; Vorren et al., 1991) (Fig. 5.11). Compared to lower latitude slopes, the polar slopes have been less thoroughly studied, and any understanding of the sediments and the sedimentary regime is still fragmentary. However, typical glacial/interglacial sediment cycles and settings have been identified (Piper et al., 1990). Along the margins of the Norwegian–Greenland Sea, the slope outside the Barents Sea Trough has been the most studied (Vorren et al., 1989). Characteristic features of the upper slope are: (1) slide scars and (2) gullies (Fig. 5.18). A major slide affecting approximately 1000 km³ has been observed, with a width of up to 40 km and a depth of about 300 m. The slide is located just below the shelf break (400 m water depth) and extends down the slope to a water depth of at least 1500 m. In addition to this major slide, there are a number of minor slides on the upper slope. Typically, the scars have been refilled with acoustically fine-grained sediments.

The gullies, which are up to 150 m deep and 1 km wide, may also form significant features of the upper slope. The gullies commonly start at the shelf break, about 400 m water depth, and extend to a water depth of ~1000 m. Lack of sediment infill in the gullies and the finding of shelf foraminifers, indicate that these features presently act as active sediment conduits from the shelf to the lower part of the fan. It has been suggested that the down-flow of cold saline shelf water is essential for erosion and down slope transport. Buried gullies are also seen in the sediment records.

On the lower slope, the most pronounced features are debris lobes. Sediment lenses up to 15 km wide and 30–150 m thick have been identified from seismic records. The origin of these features is not fully explained; they may represent deposition from debris flows originating on the upper slope.

Additional information on the slope regime of a glacially influenced margin can be gained from examination of the Scotian shelf, eastern Canada (Piper et al., 1990). Here, the Laurentian Fan, south of Grand Banks, acts as an important centre of deposition from the Laurentide Ice Sheet. Similarly, as on the Bear Island Fan, the upper slope is dissected by gullies and valleys, which appear to originate at about 500 m water depth. Towards the deeper part of the upper slope, the gullies and valleys coalesce, and only a few major valleys continue to the continental rise.

Based on seismic profiling, three groups of facies have been identified: valley, intervalley and chaotic facies. The valley facies, with an internal reflection pattern of cut and fill sequences, is comprised of muds and variably-sorted coarser beds. Cores from the intervalley facies are composed of mud and the continuous sub-parallel internal reflection pattern shows a typical mound topography. The composition and origin of the chaotic facies varies. In the uppermost part of the slope, the chaotic facies is caused by unsorted sediments (diamicton), while in deeper waters it is due to redistributed materials.

A key problem in interpreting the slope sediments is the relationship of facies with interglacial/glacial changes. Based on data from the Bear Island Fan and the Scotian slope, the following concepts have been proposed (Vorren et al., 1989; Piper et al., 1990): During glacial maxima, the ice edge may expand to the shelf edge and till is deposited on the shelf edge/upper slope. Fine-grained, as well as coarse-grained

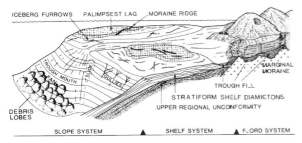

FIG. 5.18. Morphology, centres of deposition, and characteristic sedimentary features on the Bear Island Fan (FIG. 5. 12) (from Vorren et al., 1989, reproduced with permission from Elsevier Science Publishers)

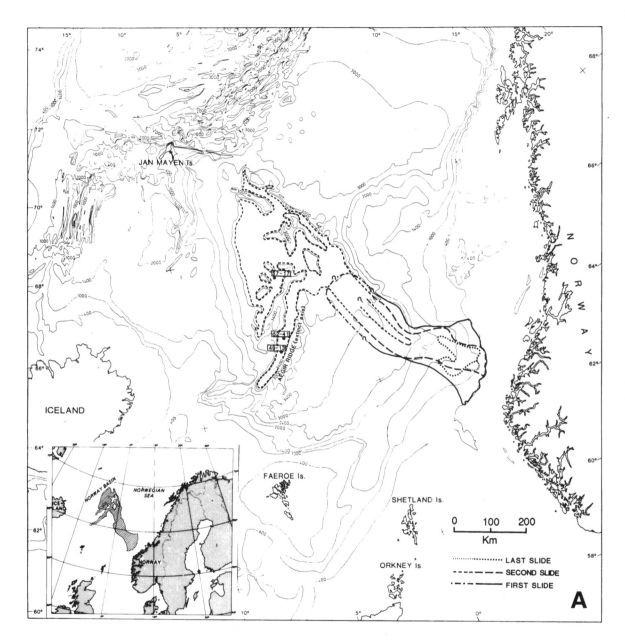

FIG. 5.19. (a) Bathymetric map of the Norwegian Sea showing the areal distribution of the Storegga slide (a) and a longitudinal section through the slide scar (b) and pre-slide reconstructions (c). The samples 49–21 and 49–23 (18B) contain debris flow deposits with lumps of Eocene-Oligocene sediments. Q, base of Pleistocene; P, base of Pliocene; MO, mid-Oligocene; Pal, Paleocene-Eocene; B, Tertiary basalt. Note that the slide deposits are shaded. In the cores, D, debris flow deposits,; T, turbidite; Pe, pelagic post slide deposits. (Modified from Jansen *et al.*, 1987; Bugge *et al.*, 1988)

sediments, are delivered to the upper slope at relatively high rates. Part of the sediments are transported directly to the deeper areas, causing gulley erosion. Rapid mud deposition on the upper slope, probably sedimented at the highest rate during recession of the ice, leads to unstable sediments being deposited. Failure of these deposits generates thick, mud-rich turbidity currents. (The debris lobes on the lower part of Bear Island slope may have formed under such conditions.) Interglacials are characterized by low clastic sediment supply, reduced turbidity current transport and hemipelagic sedimentation. The most pronounced cases of down slope transport, particularly during interglacial periods, may be caused by the cascading of cold, saline shelf water (Midttun, 1985). (The uppermost part of the Bear Island slope is eroded also by the relatively warm Atlantic Water in the Norwegian Current, which flows along the margin.)

Sediment failure and slides seem not to be directly correlated with interglacial/glacial cycles. However, as the main body of the sediments displaced during the slide is of glacigenic origin, such processes are of importance in understanding the facies/sediment distribution of glacially-influenced margins. The Grand Banks slide of 1929 (Heezen and Ewing, 1952) was triggered by an earthquake, and a similar mechanism was proposed for one of the world's largest submarine slides, the Storegga Slide on the mid-Norwegian margin (Jansen et al., 1987; Bugge et al., 1988). On the basis of coring and acoustic profiling, it was possible, in the latter case, to identify three main slide events, covering 34000 km^2 (Fig. 5.19). The total volume of sediment involved is ~5500 km^3. The average gradient of the whole slide scar is about 0.6°. The oldest and largest slide (~4000 km^3), probably of a Middle Weichselian age, consisted mainly of a debris flow of uncompacted clay of Plio-Pleistocene

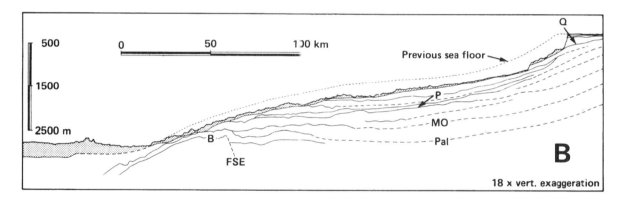

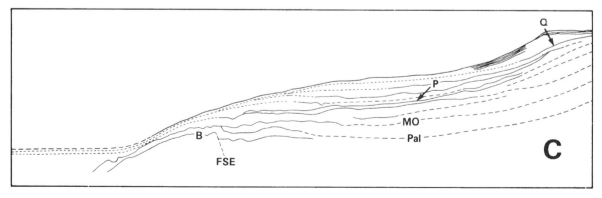

FIG 5.19 (b) and (c)

age. The second and third slides, both from Holocene time, also involved Neogene and Palaeogene sediments. Huge sediment slabs, 200 m thick and 10–30 km wide, were transported up to 200 km downslope to 2000–2500 m water depth. The sliding sediments probably flowed downslope on a liquified layer.

5.4.5. The Deep Sea Environment

5.4.5.1. General background

Rhythmically-alternating continuous seismic reflection patterns characterize wide areas of the sediment cover on the sea floor of the Norwegian–Greenland Sea. Undisturbed sediment cores retrieved from that region reveal typical glacial/interglacial sedimenta-

tion patterns that can be correlated over distances of several hundred kilometres. Individual glacial units can be traced continuously from the central sector of the Norwegian–Greenland Sea to the adjacent continental margins, where most expand considerably in thickness. As a result, a thick pile of glacial/ interglacial sediments (on the order of 250–500 m) was deposited at the mid-Norwegian margin (e.g. Vøring Plateau) during the past 2.6 Ma (Eldholm *et al.*, 1987). This wedge decreases in thickness to about 50–70 m within less than 150 km. This general pattern of pelagic and glaciomarine sedimentation may be considerably modified in areas of strong bottom current activity. Areas of almost permanent deep sea erosion, and of accumulation with deposition of large wave-like sediment deposits (drifts), have been discovered in the Greenland Sea and the Fram Strait.

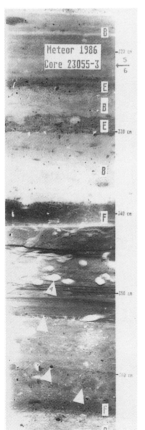

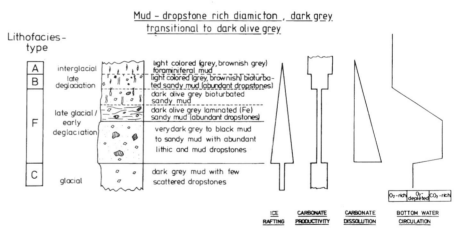

FIG. 5.20. Characteristic facies succession formed in the Norwegian–Greenland Sea during major deglaciation events (from Henrich, 1990, reproduced by permission of the author). The sediment sequence is also illustrated by radiographs

5.4.5.2. Facies succession

The sediments deposited predominantly from the interactions of pelagic sedimentation and glaciomarine processes record a spectrum of glacial and interglacial sediment facies. Each facies is characterized by a set of sedimentological, geochemical and micropaleontological features. Rhythmic and cyclic repetitions of sediment facies types dominate the stratigraphic record illustrative of glacial/interglacial climatic shifts (Figs 5.20 and 5.21). A noticeable feature of these cyclic variations is the pronounced colour changes between light and dark layers (Eldholm *et al.*, 1987). The layers are of greatly varying thicknesses (e.g. on metre scale for light layers and on the centimetres to tens of centimetres scale for dark layers). The light sediment facies are comprised of two prominent

lithologies (Fig. 5.21). Brownish, intensively bioturbated foraminiferal muds and foramnanno oozes (e.g. Facies A; Henrich *et al.*, 1989a) have been deposited during interglacials at low sedimentation rates ($1-2\,\mathrm{cm\,kyr^{-1}}$). A highly diverse benthic fauna (benthic foraminifers, echinoderms, sponges) with good carbonate preservation indicate well-oxygenated bottom water environments. The second lithology consists of brownish (Facies B) and grey (Facies C), moderately bioturbated, sandy to silty muds, with intermediate to low carbonate content and variable mixtures of coarse terrigenous debris and reworked organic carbon, deposited at moderately high sedimentation rates ($2-5\,\mathrm{cm\,kyr^{-1}}$). The scattered occurrence of dropstones and enrichment of sand-sized IRD along discrete layers point to episodic input of coarse terrigenous particles from icebergs or sea ice.

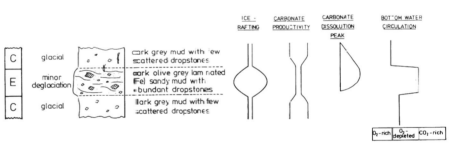

FIG. 5.21. Characteristic facies succession formed in the Norwegian-Greenland Sea during glacial periods, with the continental ice sheet close to the shelf break (from Henrich, 1990, reproduced by permission of the author). The sediment sequence is also illustrated by radiographs

Three types of dark diamictons (Facies D, E and F; Henrich *et al.*, 1989a) are intercalated with the light-coloured glacial sediment packages. These dark diamictons occur at thicknesses of centimetres to tens of centimetres. Most layers reveal a sharp base and gradational top contact often modified by bioturbation. A chaotic fabric of variably-sized dropstones and scattered mud clasts in a sandy to silty mud matrix is the predominant internal structure. Occasionally, enrichment by coarse debris is observed along discrete bands. In addition, these dark diamictons have been influenced by early diagenetic processes occurring probably in response to changing bottom water conditions.

Most spectacular with respect to diagenetic overprint is the diamicton facies type F. An example of diagenetic overprint is sulphate reduction processes in porewaters during stages of incipient burial followed by downward protrusion of a secondary oxidation front during later stages, with an approximate time lag of a few thousand years. It forms a complex layer consisting of a basal very dark grey diamicton grading upwards into a dark olive grey diamicton. Higher parts of Facies F commonly reveal brownish diagenetic laminations that cut sedimentary structures, but are themselves truncated by burrows filled with brownish sediment from above.

All three dark diamictons reveal very low carbonate contents (most frequently, ≥0.3% CaCO$_3$), strong dissolution features on the planktonic foraminifera, high organic carbon values, and high content of sand-sized terrigenous debris, densely scattered dropstones and abundant sediment pellets. Rock fragments and large dropstones consist of various igneous, metamorphic and sedimentary lithologies. Most interesting with respect to provenance are coal fragments and chalk, in addition to Cretaceous foraminifers and *Inoceramus* prisms as residue of chalk in the coarse fraction (Spiegler, 1989). Dark diamictons may contain both ice-rafted chalk and coal particles. Since chalk is only exposed to glacial erosion along shallow subcrops of the southern Norwegian shelf (Bugge *et al.*, 1984), or in the North Sea region and its adjacent continental margins in southern Scandinavia and Britain, a northward drift of icebergs is indicated for these specific deposits. This is possibly associated with advection of Atlantic water or strong longshore currents parallel to the ice margin. Coal is probably also derived from southern source regions, but high quantities of ice-rafted coal particles have been found in the Fram Strait, where exposures on the Arctic shelves and surrounding continents are the most likely source region (Bischof *et al.*, 1990).

Mapping of facies distributions along an East-West transect across the Norwegian Sea reveals typical facies successions that correspond to glacial/interglacial climatic shifts (Fig. 5.22). Dark diamictons occur in greatest thicknesses and highest frequency close to the continental margin, decreasing in number and thickness westward, and grading far offshore into oxic sediments of Facies B. From this pattern, the major source of diamictons would appear to have been the continental margin. Based on sediment flux calculations, Henrich *et al.* (1989a) assume that the diamictons were deposited rapidly.

5.4.5.3. Environmental interpretations

Environmental parameters associated with the rhythmic facies successions are shown in Fig. 5.23. The most prominent rhythm is the facies succession C–F–B–A, which represents a complete shift from full glacial to interglacial conditions. The environment during deposition of Facies C was characterized by a low carbonate shell production and some ice-rafting in the surface waters. Bottom waters were oxygen-depleted, but far from anoxic, as is indicated by a low diversity benthic foraminiferal fauna, moderate bioturbation, and good carbonate preservation. Facies C was deposited during an advance of the continental ice sheets onto the shelf. The offshore situation was characterized by a seasonally varying sea ice pack interspersed with a few drifting icebergs.

Deposition of Facies F resulted in a drastic increase in ice-rafting and a strong decrease in carbonate shell production in surface waters. Highly corrosive bottom waters, indicated by the strong dissolution of carbonate tests and an interval barren of benthic organisms, suggest bottom water conditions unfavourable for the growth of fauna. Facies F documents the maximum extension of the ice sheet close to, at, or even below the shelf edge, in addition to an early period of ice-sheet retreat. During this period, large-scale calving or frequent surges along the tidewater front of the

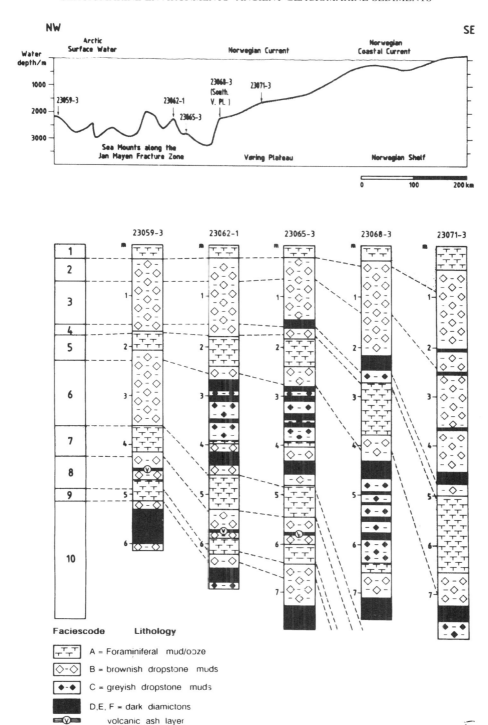

FIG. 5.22. Facies successions from isotope stages 1 to 10 in the Norwegian Sea. Note the E-W transitions of dark diamictons (Facies F, D and E) to dropstone muds (Facies B and C) (from Henrich, 1990; see Fig. 5.1 for location of the transect). (Reproduced by permission of the author)

a) Facies C: Advance of continental ice out to
   the shelf

c) Facies B: Rapid deglaciation on the shelf
   and intrusion of Atlantic water

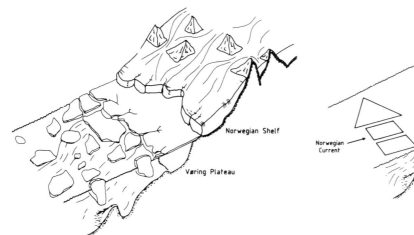

b) Facies F: Ice sheet close to the shelf
   break during glacial maximum
   and its early retreat late
   glacial / early deglacial time

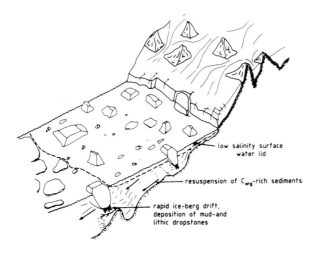

FIG. 5. 23. Paleoenvironmental model of the Norwegian Sea and adjacent shelf. (a) Extensive glaciation, with the ice margin located at the shelf break, associated with offshore pack-ice drift. (b) Early or late glacial period, with intensive iceberg calving from grounded outlet glaciers. (c) intrusion of the Norwegian Current and reduced iceberg drift, combined with increased carbonate production in open sea areas. The ice margin is now located within the present day coastal areas (from Henrich, 1990, reproduced by permission of the author)

grounded ice sheet delivered huge amounts of debris-laden icebergs into the open sea. In addition, large meltwater discharges may have formed a low-salinity lid on the surface and deposited fine-grained sediment from sediment-laden plumes. A decrease in bottom water renewal and the oxidation of organic matter at

this time would have resulted in highly corrosive bottom waters.

Facies B is characterized by a gradual decrease in ice-rafting and an increase in carbonate shell production (Fig. 5.23). Bottom waters became progressively less corrosive, more oxygenated, and dissolution decreased, reflected by a change in the redox potential of the sediment and the reappearance of benthic fauna. Facies B represents the retreat of the ice sheet on the shelf and a major deglaciation, characterized by a rapid retreat of the low-salinity surface water lid towards the coast. A rapid rise in sea level would have probably caused a sudden disintegration of the marine-based parts of the continental ice sheets.

The uppermost facies, Facies. A, records a high interglacial carbonate productivity and completely oxygenated bottom waters (= a rich benthic fauna).

Other rhythmic facies successions commonly observed within certain glacial stages (e.g. C–D–C, C–E–C or C–F–B–C) indicate similar environmental changes but without a return to or initiation of full interglacial conditions. Such a situation might occur at any time when a grounded ice sheet on the shelf became unstable. The release of enormous masses of icebergs might also be triggered by glacio-isostatic processes (e.g. the strong subsidence of the shelf in response to increased ice-loading on the continent might have destabilized grounded ice on the shelf).

The paleo-oceanographic model outlined above might prove to be a useful tool for offshore paleo-oceanographic reconstructions. It incorporates glacial stages with major advances of continental ice sheets on the shelves and allows for correlation between open ocean records and records from the shelf.

The glaciomarine section older than 1 Ma in the Leg 104 drill sites reveals significant changes in most of the isotopic, micropalaeontological and sedimentological parameters (e.g. a drastic decrease in overall sediment supply [suggested by low bulk accumulation rates], generally low carbonate shell production in surface waters, combined with a shift to ^{18}O-depleted isotope values in planktonic foraminifer tests and persistent strong dissolution). These changes have been interpreted as indicating a different palaeo-oceanographic pattern, with generally lowered surface water salinities, a more stabilized water column and long-term persistence of corrosive bottom waters (Jansen et al., 1988; Henrich, 1989). Under such conditions, dark diamictons would have formed over longer periods, since overall sediment supply to the basins may have been reduced due to the existence of more stable, smaller ice caps on the surrounding continents.

5.5. CONCLUSIONS

The results and models of Cenozoic sedimentation in Arctic and Antarctic glaciomarine shelf, slope and deep sea settings show that the slope and adjacent deep sea have acted as major centres of glaciomarine deposits in the past. In addition, the pelagic sections, intercalated in the deep sea record and on the slope, can be accurately dated by various high resolution stratigraphic methods. Hence, they permit a correlation between specific glaciomarine units and events on the shelf and back toward the ice margin. It is recommended that future studies on glaciomarine sections concentrate on the distal glaciomarine settings on the slope and in the deep sea, rather than try to identify ice margins.

In the analysis of Pre-Cenozoic glacigenic sediments, preservation potential remains a major problem. In contrast to the Cenozoic glaciomarine sediments, essentially laid down along passive margins, the Pre-Cenozoic glacigenic sediments that have been described are mostly from intercratonic or rift basins (Bjørlykke, 1985; Nystuen, 1985). Even though models from passive margins cannot be directly applied to intercratonic basins, the data from Cenozoic margins demonstrate certain principles for the overall distribution of glacial sediments. Results from the Cenozoic polar margins also demonstrate that deep sea glaciomarine and intercalated pelagic sediments might provide essential supplementary information pertaining to the glacial history of adjacent land areas. Furthermore, knowledge of the three-dimensional geometry of glaciomarine units might allow reconstruction of similar deep sea, shelf-slope settings in ancient glaciomarine sections.

Carbonate accumulation on the shelves proximal to glaciomarine environments has now been well documented, and does not contradict a glaciomarine origin of diamictites. In stratigraphic sections, the repetitive occurrence of diamictons and carbonates might correspond to glacial and interglacial climatic oscillations, respectively.

Chapter 6

GLACIOAEOLIAN PROCESSES, SEDIMENTS AND LANDFORMS

E. Derbyshire and L. A. Owen

6.1. INTRODUCTION

The frequently strong association of aeolian processes with present and former glaciation has been recognised for over sixty years from field observations (Högbom, 1923). The distribution of geological evidence such as coversands, sand dunes and loess mantles in western Europe and North America was used to develop the hypothesis of a glacial anticyclone associated with the European ice sheet (Hobbs, 1942, 1943a, b), a framework used by a number of authors to characterise the extraglacial environment of the last glacial maximum in North America and Europe (e.g. Schultz and Frye, 1968; Dylik, 1969; Demek and Kukla, 1969; Galon, 1959; Velichko and Morozova, 1969; Tricart, 1970; Maarleveld, 1960; Dylikowa, 1969; Krajewski, 1977; Poser, 1932, 1948, 1950; Reiter, 1961; Lill and Smalley, 1978; Williams, R.G.B. 1975).

For over a century, glaciers have been regarded as the greatest single source of aeolian silt, and the nature and distribution of much of the world's loess (aeolian silt) has been explained in terms of the former presence of large ice sheets on the northern continents at various times during the Quaternary. A number of authors have regarded glacial grinding of rock as the only earth-surface process capable of producing the huge volumes of silt making up the world's loess lands. Although some recent experimental evidence has pointed to a larger contribution to the world's silt production by desert-related processes, notably salt weathering, it is fair to say that the prevailing view still favours an origin in cold climate conditions with increased aridity in which periglacial and montane environments are added to the strictly glacial.

On a more local scale, the relationship between glacial deposition and aeolian reworking of sediments has been illustrated from a number of glacial forelands, an early example being the observations in Greenland by Poser (1932). Information on sediment generation and sources, the operation of the aeolian processes, facies discrimination and paleoenvironmental reconstruction is, not surprisingly, widely scattered through the literature on sedimentology, glacial and periglacial geomorphology, and Quaternary stratigraphy. Glacial aeolian sedimentation is, however, often disregarded or only briefly dealt with in most textbooks (Flint, 1971; Embleton and King, 1975; French, 1976; Washburn, 1979). This chapter draws on a selection of this literature in order to examine the relationship between glaciers and aeolian sediments and landforms.

6.2. SEDIMENT PRODUCTION AND SOURCES

Comminution of rock to produce fragments with a particle size susceptible to movement by aeolian suspension, saltation and traction is effected on a large scale by several mechanisms including glacial grinding, weathering (notably by crystal growth) and fluvial and aeolian abrasion of particles in transport. More localized and less well documented processes include hydration effects, chemical and biological weathering and mass slope failures in rock and soil including those associated with earthquake shock.

Most of the quartz in sedimentary rocks is derived initially from igneous and metamorphic types and is predominantly of sand size. Most of the sand size particles in the world's sedimentary system consist of quartz derived from massive plutonic rocks and gneiss, with a mean grain size of 720 microns, sources such as schist, with a mean quartz grain size of 440 microns, bringing the initial average particle size down to about 600 microns (Blatt, 1987). The ancient granitic rocks such as those of the heavily-glaciated Laurentian and Baltic Shields contain quartz in a stressed state which facilitates comminution (Smalley, 1966). In contrast, the quartz in the Quaternary sands and silts of High Asia derives less from highly stressed granite and more from thick, recycled sedimentary units in the Karakoram–Himalayan–Tibetan zone of crustal thickening. Break up and release of quartz particles is effected by the considerable energy derived from tectonic and crystal-growth processes in this rapidly-rising, cold and rather dry region. The observation by Blatt (1987) that the $\delta^{18}O$ values of quartz from plutonic igneous rocks are consistently lower than those found in quartz from many sedimentary and metamorphic sources is used by Smalley (1990) as a further distinction between loess derived from the Pleistocene ice sheet glaciations (glacial loess) and that produced in High Asia by tectonic and weathering processes (mountain loess).

The production of large amounts of silt requires comminution of sand grade material and there has been considerable discussion of the relative importance of the glacial and non-glacial processes involved. Attrition in blowing and saltating sand grains produces some silt-size material but the experimental results of Kuenen (1960) suggested that the volume is small and this led Smalley and Vita-Finzi (1968) to reject this origin for the great loess sheets. Instead, production of silt by the grinding action at the bed of glaciers and ice sheets has led some writers to express the view that this is the only mechanism capable of producing such large volumes of silt (e.g. Smalley and Cabrera, 1970; Boulton, 1978; Collinson, 1979). Other evidence, notably particle shape and surface texture, and the mineralogy of the clay grade (<0.002 mm) has been invoked to strengthen this view. Smalley and Cabrera (1970) interpreted the predominantly angular shape of loess grains as consistent with glacial breakage, and fracture surfaces were attributed to glacial processing of silt grains in some European loess (Smalley et al., 1973), as well as those of the Mississippi valley and New Zealand (Smalley, 1966, 1971).

There is accumulating evidence that weathering processes are important producers of silt. Mechanical breakage exploits the inherent planar microfractures (Moss, 1966; Riezebos and van der Waals, 1974; Moss and Green, 1975) generated in sand-size quartz grains by stress gradients during metamorphism and as a result of contraction or cooling from the molten state. This can result from freeze–thaw stressing (Zeuner, 1959; Minervin, 1984), and impact during fluvial transport (Moss et al., 1973; Whalley, 1979; Palmer, 1982) as well as impact-breakage during aeolian transport (Smalley and Vita-Vinzi, 1968; Whalley et al., 1982, 1987). The products of such splitting are predominantly blade-shaped fragments (Smalley, 1966), a view consistent with the observed dominance of face-to-face fabrics in the younger, little-modified loesses of China and Western Europe (Derbyshire and Mellors, 1988; Derbyshire et al., 1988). The evidence for splitting of quartz sand grains along discontinuities by dissolution in tropical climates with marked seasonality has been summarised by Nahon and Trompette (1982) including the release of silt-size quartz from ferricretes which degrade to produce mantles of quartz silt up to 1 m thick. A process is also described from semi-arid areas in which calcite replacement reduces the quartz in calcretes to silt grade debris which is eventually released with dissolution of the calcitic cement. Of particular interest is the laboratory and field evidence of silt production by the combined processes of hydration and salt crystal growth (Goudie *et al.*, 1970; Goudie, 1974; Goudie *et al.*, 1979; Sperling and Cooke, 1985). Salt weathering is a significant cause of rock taffoni and silt production in a range of dry environments including intertropical desert plains and piedmonts (Cooke, 1981; Goudie and Day, 1981; Bradley *et al.*, 1978), polar deserts (Prebble, 1967), and dry mountain slopes (Goudie, 1983). Quartz sand grains treated with saturated sulphate solutions in a hot desert environment simulated in a climatic cabinet showed abundant fracturing and comminution to produce angular silt-sized particles after only 40 cycles of temperature and humidity oscillation (Goudie *et al.*, 1979). Scanning electron microscope examination of the fragments resulting from the experiments of Sperling and Cooke (1985), for example, revealed conchoidal and stepped fracture surfaces which are indistinguishable from those attributed by some authors specifically to glacial crushing. Activity of this kind may be high where salts become concentrated as in saturated sites such as arroyos, alluvial fans and playa basins: locations in which the finer-grained sediments also tend to coat the surface. A combination of salt enrichment and contraction cracking on desiccation in wetting and drying cycles yields silt-sized bundles of clay-grade particles (Yaalon and Ganor, 1973; Dare-Edwards, 1984) which are removed by the wind at threshold velocities significantly lower than those required to lift the grains making up such aggregates (Gillette *et al.*, 1980). It thus appears that the cold but semi-arid to arid environments that cover so much of central Asia are capable of producing abundant silt grade materials which, following concentration in wadis, fans and plains, are readily susceptible to deflation (Yaalon, 1969; Bruins and Yaalon, 1979; Smalley and Smalley, 1983).

The intimate association of till matrices of glacially-comminuted rock flour, periglacial river sediments and the European loess sheets has sustained a strong body of opinion favouring the glacial-aeolian origin of loess (e.g. Hobbs, 1933, Smalley, 1966). However, the extension of this explanation to the Loess Plateau of China (Smalley, 1968) was dependent on derivation of glacial silt from the mountains of subtropical China to the southeast which, according to

prevailing opinion at that time as summarized in the map of Sun and Yang (1961), underwent at least three glaciations during the Pleistocene. Recent re-evaluation has concluded that there is no unequivocal evidence of Pleistocene glaciation south of the Yangtze River (Derbyshire, 1983, 1992). A further attempt to explain the main body of the Chinese loess in glacial-aeolian terms postulates derivation from the Qinghai–Xizang (Tibet) Plateau as a result of a glacial anticyclone induced by widespread ice sheets (Smalley and Krinsley, 1978). The older literature on which this interpretation was based (e.g. Trinkler, 1930; Sinitsin, 1958) has been revived in the form of radiation-balance modelling using scattered and, in some cases, equivocal geomorphic evidence without reference to diagnostic sedimentary data (e.g. Kuhle, 1987). All of this work is considered by the Chinese field teams to greatly over-estimate the degree of Pleistocene ice cover (Shi *et al.*, 1986; Derbyshire, 1987; Burbank and Kang, 1991; Derbyshire *et al.*, 1991). Uplift throughout the Pleistocene rendered the plateau progressively less favourable to the maintenance of large dynamic glacial systems after the maximum (piedmont and mountain ice cap) glaciation of the middle Pleistocene (Zheng, 1989), but much more susceptible to severe periglacial and salt weathering, as well as tectonic shattering of rocks (Smalley, 1990).

6.3. WIND ACTION AROUND GLACIERS

Wind action in and adjacent to glacially-covered terrain is expressed directly in the erosion of bedrock surfaces and unlithified deposits, and transportation of the detached particles to be deposited as a variety of aeolian sediments and landforms. Wind also acts in a number of ways that influence landforms indirectly, most notably in its effect upon snow distribution on glacial forelands, but consideration of aeolian processes will be limited here to direct effects.

Sources of the abrasive particles essential to wind erosion of bedrock surfaces have been outlined above. To these should be added the hard granular snow of the high polar regions where grain hardness reaches a Moh value of 6 in snow at a temperature of −50°C (Bird, 1967). Wind abrasion produces a suite of distinctive landforms including asymmetrically-

PLATE 6.1. Wind eroded sandstones in Canon Fiord, Ellesmere Island

PLATE 6.2. Polished dikes and veins of microgranite cutting frost-shattered granites in Taylor Valley, Antarctica

smoothed pebbles and cobbles (ventifacts), stream-lined ridges (yardangs) and grooves, polished pave-ments and a variety of cavernous forms, under the general term taffoni, in which both weathering and aeolian abrasion appear to play a part (Samuelson, 1926; Fristrup, 1952; Pissart, 1966; Cailleux, 1968; Bogacki, 1970; Akerman, 1980; McKenna-Neuman and Gilbert, 1986; Koster and Dijkmans, 1988).

The distribution of wind-eroded surfaces is very uneven, however, and there are extensive areas in, for example, northern Canada almost devoid of such features (Bird, 1967), localized effects being observed only in mechanically-weak lithologies (Saint-Onge, 1965). The dry forelands of polar and continental glaciers and ice sheets, on the other hand, display abundant and well-developed forms (Plate 6.1). To the early reports from northeast Greenland (Fristrup, 1952) have been added many observations from the ice-free enclaves (Sekyra, 1969) and the dry valleys of Antarctica (Nichols, 1966). Ventifacts have been observed in many lithologies including coarse-grained granites, but the commonest abrasion-pol-ished surfaces on both pebbles and exposed bedrock occur in the finer-grained basic intrusives. Frost-shattered coarse-grained granite contrast with wind-polished dikes and veins of microgranite in Taylor Valley, Antarctica (Plate 6.2). Ventifact orientation leaves little doubt that glacially-induced katabatic winds move very dense, very cold air at high velocities in winter from the Antarctic ice sheet and its outlet (valley) glaciers along the deep troughs known as 'dry valleys'. The wind pattern is not always so simple, however, as can be seen in the aeolian abrasion of both sandstone and gneissic bedrock around the cirque containing Sandy Glacier which lies almost 2000 m above the floor of Wright dry valley (Plates 6.3a, b). That the very strong winds blow across as well as along the axis of this dry valley is attested by the composition of the englacial sediments of the cirque glacier (Dort et al., 1969).

The cavernous weathering forms seen in the Antarctic dry valleys have also been explained, in part, by wind erosion. There appears to be rather general agreement that weathering processes weaken rock surfaces which are then etched by wind abrasion. Periodic localized melting by direct insolation of thin accumulations of snow around rock stacks and boulders provides a brief wetting phase used by Cailleux (1968) to explain the mobilization of salts (notably mirabilite) to form veins and efflorescences which effect substantial localized granular disag-gregation of the coarser-grained intrusive rock sur-faces and old glacial boulders. Such features are also well known from Greenland (Washburn, 1969) and the glaciated desert valleys of the Karakoram in northern Pakistan and China. Thus weakened, rock surfaces are much more susceptible to wind abrasion than in the unweathered state. However, the observed asymmetry of the taffoni does not simply reflect present-day dominant wind directions. Case hard-ening, by which the surface of a boulder becomes relatively stronger than the taffoni by evaporation of solutions leaving a thin cemented surface layer, certainly plays a role in both Antarctic and high mountain deserts. Such rock varnish may develop through a number of stages until it, too, is removed in part by aeolian abrasion (Plate 6.4) (Derbyshire et al., 1984). Although a preferred orientation has been recorded in some groups of taffoni, the orientations do not appear to match current dominant wind directions in Antarctica (Cailleux and Calkin, 1963) or in the Karakoram where residual snowpatch localisation (a combined product of insolation and an indirect effect of wind action) appears to be the essential cause.

The erosive action of the wind in the form of deflation from the surfaces of unlithified deposits such as till, glaciofluvial and glaciolacustrine facies is probably far more significant in terms of volume of material moved per unit of time, than is the case of abrasion of even severely weathered bedrock. The winnowing of fines (sands and silts) from freshly-deposited proglacial sediments may be sufficiently rapid to produce a stone pavement within just a few years. Blowing sand and silt has been observed on many glacial outwash plains including, for example, those in the Yukon (Nickling, 1978), Baffin Island (Church, 1972; McKenna-Neuman and Gilbert, 1986), Iceland (Bogacki, 1970), west Spitsbergen (Riezebos et al., 1986), and in the Karakoram glacial valleys (Plate 6.5).

The relationships between particle size and fluid and impact threshold velocities are summarized in Fig. 6.1 (Mabbutt, 1977). Particles <0.1 mm in diameter are moved by suspension, particles >0.1 mm in diameter

(a)

(b)

PLATE 6.3. (a) Aeolian deposition on an alpine glacier in the Wright Dry Valley, eastern Antarctica. Note wind-rippled sand. (b) Wind-cut grooves in the sandstone slopes of an alpine glacier in the Wright Dry Valley, eastern Antarctica

PLATE 6.4. Removal of rock varnish by aeolian abrasion in the Karakoram Mountains

PLATE 6.5. Aeolian sands deposited within depressions between moraine ridges on the Ghulkin Glacier, Hunza valley, Pakistan

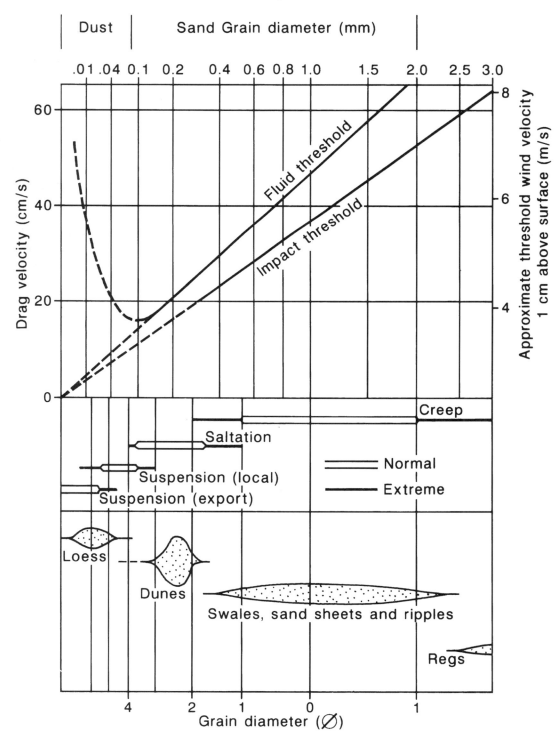

FIG. 6.1. The relationship between particle size, fluid and impact thresholds, dominant mode of aeolian transport and aeolian landforms (after Mabbutt, 1977 and Cooke and Doornkamp, 1990)

are moved by saltation, while those >0.6 mm move by creep. Threshold velocities for entrainment of particles, in traction, saltation and suspension loads, are reached with higher frequency in winter on snow-free surfaces in high-latitude glacial forelands. Wind velocities and air densities may be sufficiently high in the Antarctic winter to move particles of gravel grade by traction (Selby *et al.*, 1974).

Particle size has a strong influence not only on the mechanism of transport but also on the sedimentary characteristics and landforms (Fig. 6.1). Aeolian sand is deposited as dunes and sheets within short distances of its source in glacial outwash plains, lake basins and proglacial coastal flats. In contrast, aeolian silts are transported predominantly in suspension and may be carried hundreds of kilometres or more from their proglacial source locations, contributing a fertile and readily-worked fraction to the soils of extensive regions remote from the glaciers.

6.4. GLACIOAEOLIAN SEDIMENTS AND LANDFORMS

The wind acts as an agent of deposition on glacier surfaces, on the morainic and outwash areas adjacent to glaciers, and in more distant extraglacial terrain.

Given that the wastage of glaciers tends to destroy supraglacial aeolian landforms and to re-work aeolian deposits, it is not surprising that much aeolian sedimentation on glacier surfaces has gone unreported. The dry conditions required tend to restrict aeolian entrainment and transport to the winter season. Nevertheless, wind-winnowing of the supraglacial load of some Alpine glaciers also takes place locally in summer when the debris dries out and eventually collapses as the subjacent glacier ice melts down, the sand and coarser grades being redeposited by sliding and rolling while the silts are blown off. This is an important process of grain size fractionation in the terminal zone of glaciers with hot, dry summers as in the Karakoram, Himalaya, and some valleys on the eastern slope of the Southern Alps of New Zealand (Plate 6.6). Aeolian deposition of sediments on glacier surfaces appears to be substantial only in the case of hyper-arid polar and continental glaciers such as the Qilian Shan of northern Tibet and the Alpine glaciers in the dry

PLATE 6.5. Grain size fractionation in the terminal zone of glaciers in the Southern Alps of New Zealand

valley region of eastern Antarctica. Most glaciers in the latter region show a finely dispersed silt component along the englacial foliation derived from aeolian deposition of dust and snow in the accumulation zone. On some Antarctic glaciers, the supraglacial load is predominantly of wind-rippled sand deposited by winds blowing across the glacier from adjacent debris-strewn desert surfaces. The intercalation of such sand with snow in the glacier's accumulation zone is sometimes preserved in the englacial foliation. In the case of Sandy Glacier in Wright Valley, Antarctica, the consistency of the alternation of thin layers of sand and clean ice in the snout area has been used to support the view that movement in this cirque glacier has been by simple rotational sliding with little deformation of the original sedimentary structure

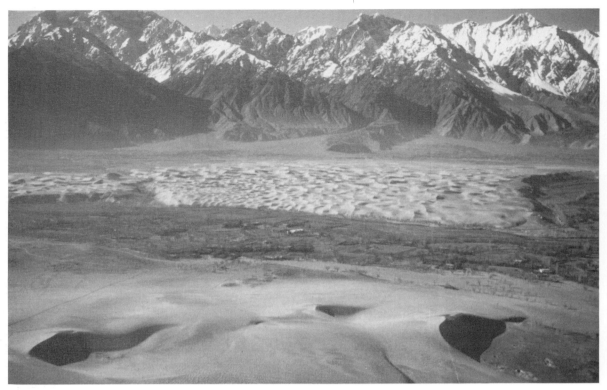

PLATE 6.7. Field of barchan dunes adjacent to the upper Indus River (to right of photograph) in the Skardu Basin, Karakoram Mountains, Northern Pakistan

derived from aeolian deposition above the equilibrium line (Dort, 1967; Dort *et al.*, 1969).

The sands and silts blown across the proglacial plains and beyond are deposited as dunes and a number of sheet-like or planar bodies. Sand dunes derived from glacial outwash plains are known from many parts of the world including Antarctica (Webb and McKelvey, 1959), Alaska (Koster and Dijkmans, 1988) and the Karakoram Mountains (Plate 6.7). Dunes of barchan, transverse and whaleback form have been described from the Antarctic dry valleys where they characteristically consist of interstratified sand and snow-derived ice which acts as a cement (Calkin and Rutford, 1974). Deposition of sand with snow in this way is predominantly a winter process. Although, outside the polar regions, such niveo-aeolian deposits (Rochette and Cailleux, 1974) tend to be destroyed by seasonal melting, some structures have been recognized in dunes (Steidtmann, 1973; Ahlbrandt and Andrews, 1978; Ballantyne and Whit-

tington, 1987; Bélanger and Filion, 1991). Ahlbrandt and Andrews (1978), for example, recognize cold-climate deposits with distinctive tensional, compressional and dissipation sedimentary structures in Colorado (Cailleux, 1968, 1972, 1974) (Fig. 6.2). Small-scale sand accumulations in the lee of vegetation tussocks ('sand shadows'), and in the lee of moraines, gravel braids in outwash plains and beach ridges, are common on glacial forelands. Processes of deflation may also contribute to dune morphologies such as the classic blow-out dunes in Iceland (Bogacki, 1970). Parabolic dunes are particularly common in cold climate regions, often forming extensive complexes (Ahlbrandt and Andrews, 1978; Koster, 1988; Bélanger and Filion, 1991) (Fig. 6.3). The formation and movement of dunes has been observed with good examples being described in northwest Alaska (Koster and Dijkmans, 1988) and on the sub-arctic eastern coast of Hudson Bay (Bélanger and Filion, 1991). Reconstructions of formation and

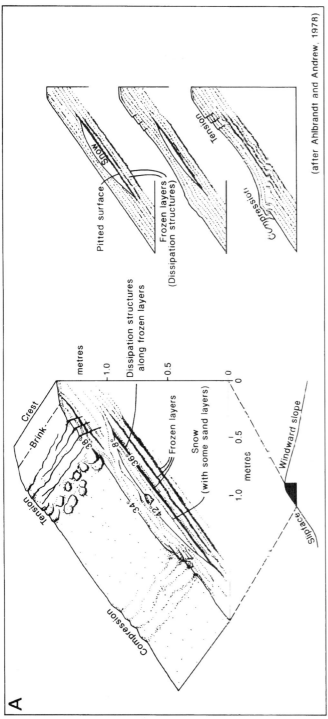

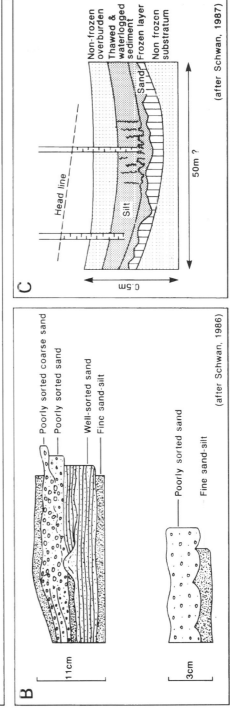

FIG. 6.2. Sedimentary structures in niveo-aeolian deposits. (a) Characteristics and formations of tensional, compressional and dissipation structures in sand dunes, North Park, Colorado (after Ahlbrandt and Andrews, 1978); (b) Loading structures in Weiselian aeolian sands (after Schwan, 1986); (c) Fluidization of sands and interbedded silts during springtime in periglacial conditions (after Schwan, 1987, reproduced by permission of Elsevier Science Publishers)

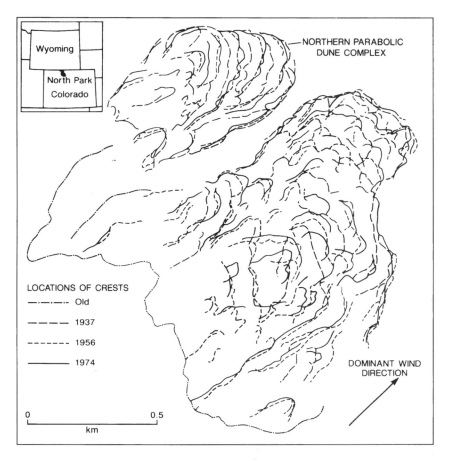

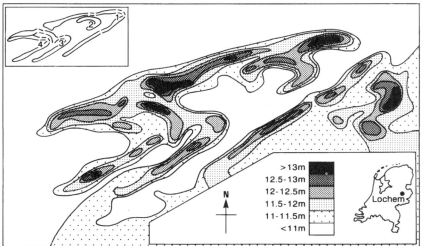

FIG. 6.3. Morphologies of cold climate parabolic dunes. The upper map illustrates the migration of recent parabolic dunes in North Park, Colorado (after Ahlbrandt and Andrews, 1978). The lower map shows parabolic cover sand ridges of Younger Coversand II (Younger Dryas Stadial), near Lochem in the Netherlands. Note four stages of formation of this ridge pattern (after Koster, 1988, reproduced by permission of Elsevier Science Publishers)

movements of dunes include work in Colorado (Ahlbrandt and Andrews, 1978), western Nebraska (Smith, 1965) and The Netherlands (Koster, 1988) (Fig. 6.3).

Sand, with varying amounts of silt, occurs as sheets in down-wind locations adjacent to outwash plains and temporary, dried-out lakes or ponds. Sand sheets, often with distinctive sub-horizontal planar bedding, are best known from the Pleistocene record of The Netherlands where they have been termed 'coversands' (Maarleveld, 1960). Figure 6.4 shows the distribution of the coversand belts in Europe.

This, along with evidence such as the distribution of ventifacts and loess, the percentage of frosted coarse sand grains in aeolian sediments (Fig. 6.4b), sedimentary structures indicating dominant palaeowind directions, as well as evidence from other glacial and paraglacial sediments, are important in reconstructing Pleistocene paleoenvironments (Chapter 13; Menzies, 1995a, Chapter 15). This has facilitated reconstruction of ice margins (Fig. 6.4) and the configuration of the past atmospheric pressure system during the last glaciation (Fig. 6.4c). Cailleux (1942), for example, believed that areas with higher

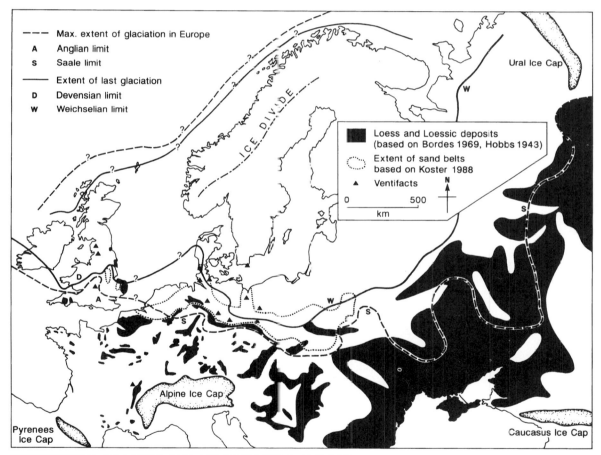

DISTRIBUTION OF LOESS AND ICE EXTENTS IN EUROPE

FIG. 6.4. Maps of northern Europe comparing: (a) the distribution of loess and cover sand, and reconstruction of former ice extents (after Bordes, 1969; Hobbs, 1943; and Koster, 1988); (b) the percentage frosted coarse sand grains (after Cailleux, 1942, reproduced by permission of the Geological Society of France); (c) reconstruction of summer atmospheric pressure and wind directions in Europe during Late Glacial time (modified after Poser, 1948, Reiter, 1961 and 1963; Washburn, 1973)

percentages of frosted grains were regions that had experienced the most severe periglacial activity and strongest winds (Fig. 6.4c). Increases in aeolian activity and the formation of coversands are thought to have coincided with increased aridity during periods of colder climate. In the European Lowlands the deposition of coversands during late glacial times occurred during the Early Dryas Stadial (ca. 12,000–11,800 BP) and the Late Dryas Stadial (ca. 11,000–10,000 BP), with older coversands deposited in the Earliest Dryas Stadial (ca. 14,000–13,000 BP) and the Pleniglacial (older than 29,000–22,500 BP : Kukla, 1975; Koster, 1988). Deposition of sands younger than 8000 BP may have been initiated by human activity. Sarntheim's (1978) computer sim-

ulation of the distribution of moisture during the last glacial and the present interglacial, and evidence for the distribution and extent of past deserts show that increased aridity during the glacial maximum (~18,000 BP) enhanced aeolian activity, the forma- tion of dunes and the extension of deserts, while during the climatic optimum (~6,000 BP) deserts and dune formation were at a minimum.

The down-wind fining of aeolian deposits from sand to sandy silt and to silt has been demonstrated in a number of currently-glaciated regions including Spitsbergen (Riezebos et al., 1986; Van Vliet-Lanoë and Hequette, 1987). In the case of the Holm- strömbreen area of Spitsbergen comminution of mica also increases rapidly down-wind from the source

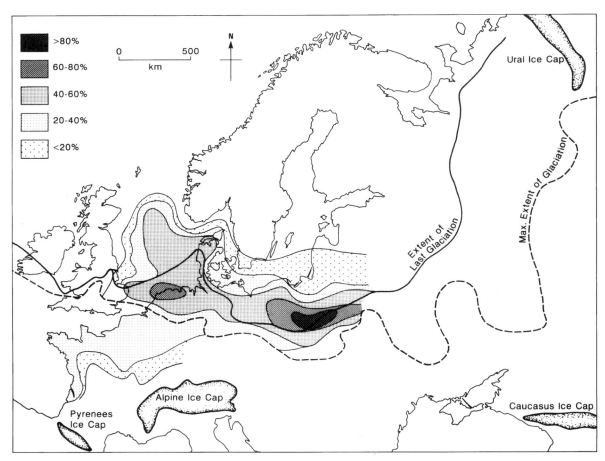

FIG. 6.4. (b)

sites, a characteristic noted in the Pleistocene cov-ersands of The Netherlands (Riezebos *et al.*, 1936). Periodic wetting and drying are important, even in very dry polar regions where vertical cracks arising from desiccation, thermal contraction, or both, may become filled with aeolian sands to form sand wedges (Washburn, 1980) (Plate 6.8). Other indicators of periodic moistening of the surface laminae of wind-deposited sediments are adhesion structures (the so-called adhesion ripples and warts: Kocurek and Fielder, 1982) known from both modern and Pleisto-cene proglacial aeolian sand sheets (Ruegg, 1983).

Wind-deposited silts (predominantly in the grain size range 0.01–0.05 mm) are known as loess (Gill and Smalley, 1978). These are unstratified, relatively porous, dry, and yellow buff to brown yellow in colour. Figures 6.5, 6.6, and 6.7, and Table 6.1 show the particle size characteristics of aeolian sediments from selected areas of the world. Note that the particle size of loess is dominantly medium silt (Fig. 6.6) while the coversands have a much wider particle size variability and are often loamy (Fig. 6.5). The mineralogy of the aeolian sediments is strongly influenced by source area although carbonates, fre-quently occurring as cements, are common in many loesses (Table 6.2). Yellow laminated silts in the floor deposits of Lake Constance, Switzerland, occurring as inter-laminae within glaciolacustrine varves, have been interpreted as loess laid down in late winter and spring during the Older Dryas (Niessen *et al.*, 1992),

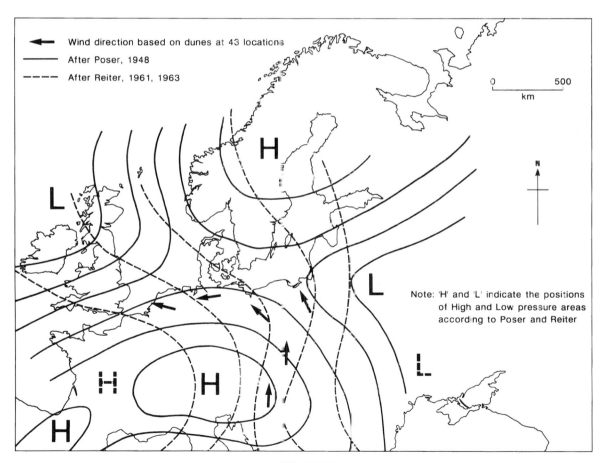

FIG. 6.4. (c)

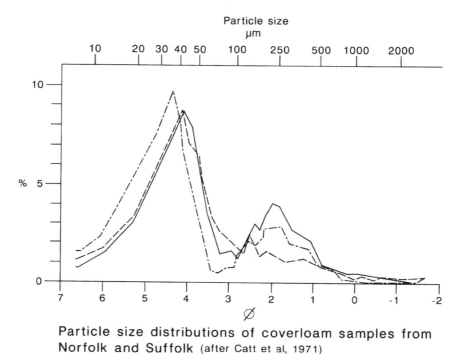

Particle size distributions of coverloam samples from
Norfolk and Suffolk (after Catt et al, 1971)

FIG. 6.5. Particle size distributions of coverloam from Norfolk and Suffolk (after Catt *et al.*, 1971, reproduced by permission of Blackwell Science Publishers, Oxford). Note the bimodal distribution

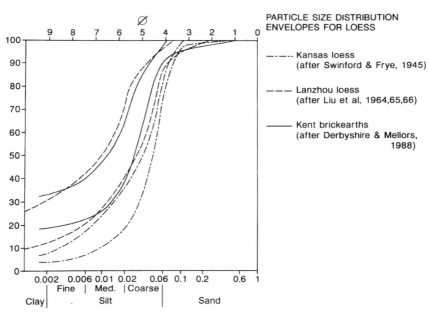

FIG. 6.6. Particle size distribution envelopes for loess from Kansas, USA (after Swinford and Frye, 1945), Lanzhou, China (after Lui *et al.*, 1964, 1965 and 1966) and Kent, UK (after Derbyshire and Mellors, 1988). Note the 50–80% of the sediments is medium to coarse silt grade. (Reproduced by permission of Elsevier Science Publishers)

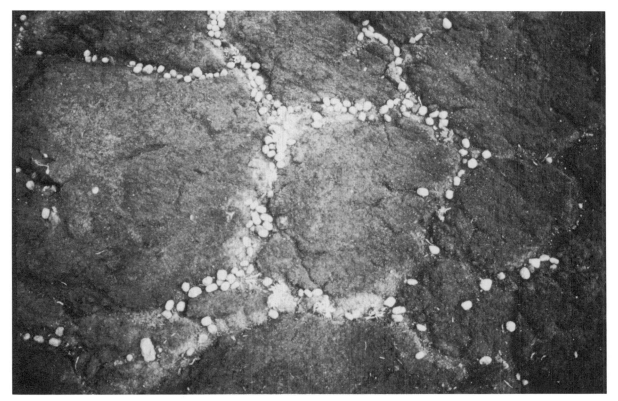

PLATE 6.8. Contraction crack polygons with sand and gravel fillings. Ellesmere Island, N.W.T., Canada

a time of extreme aridity. Loess deposition here appears to have been a winter phenomenon rather than a summer one as in most present-day proglacial zones.

6.5. FACIES

The facies progression coversands–sandy loess–loess is well documented from the proglacial forelands around the northern hemisphere Pleistocene ice sheets (Ruegg, 1981a, 1983; Schwan, 1986, 1987; Koster, 1988) (Figs 6.8 and 6.9). Figure 6.8 is a schematic diagram showing the possible landform and sediment associations in a glacioaeolian-dominated region with lithofacies variations shown in section. The scale has been omitted because such facies variations can change over short distances (several kilometres), as is the case in the European Alps or Iceland, as well as over many hundreds or thousands of miles as is the

case in Europe, North America and Asia. Schwan (1986) (Fig. 6.9) showed that facies may vary seasonally. Over large areas particular lithofacies associations can be shown to be geographically governed by their proximity to the ice margins: that is, the dominant sediment source area. Much, although by no means all, loess is derived from the silt produced by subglacial grinding. At the present time, loess is not accumulating in great quantities even adjacent to the polar deserts, although thick deposits of Holocene age are known from the Yukon and central Alaska (Péwé, 1955). The proglacial loess of the last ice-sheet maximum in Europe contains sedimentary indicators of a complex environmental history. Ruegg (1981a, 1983), Schwan (1986, 1987) and Koster (1988) all recognize a broad succession of facies associations that vary through time from fluvial to aeolian–lacustrine to aeolian in northwest Europe, representing climatic amelioration (Figs 6.10 and

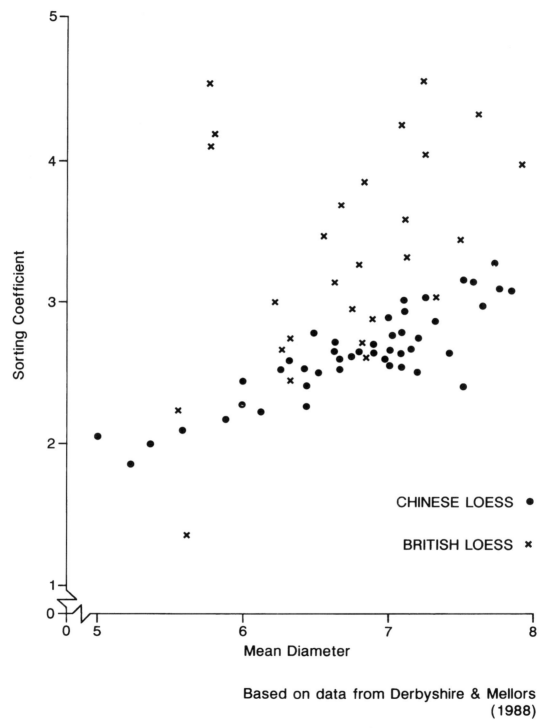

FIG. 6.7. Co-plot of sorting coefficient against mean particle diameter comparing Chinese and British loess. Note the higher degree of sorting in Chinese loess (data from Derbyshire and Mellors, 1988, reproduced by permission of Elsevier Science Publishers)

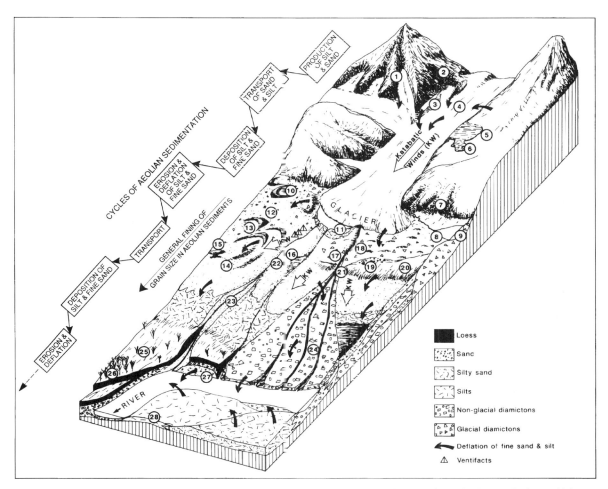

FIG. 6.8. Landform-sediment associations in the glacioaeolian environment. 1, Silt and fine sand formation by weathering on high steep valley slopes; 2, Silts and sands produced and supplied to the glacier by rockfall and other mass movement processes; 3, Silts deposited in a supraglacial lake; 4, Silts, sands and other rock debris falling into crevasses and incorporated into the ice; 5, Silts and sands washed into a small lake; 6, Terraces comprising lacustrine silts and sands produced as lake dries up or drains; 7, Fine sediments produced by overland flow and deposited at base of slope; 8, Meltwater stream feeding a proglacial lake. 9, Proglacial lake into which lacustrine silts and sands are deposited; 10, Parabolic dune; 11, Ice-contact lake; 12, 13, Barchan dune; 14, Longitudinal dunes; 15, Deflation hollow; 16, Deflation of lacustrine silts and sands in a dried up proglacial lake; 17, Rock strewn surface, reg-like; 18, Hummocky moraine; 19, Aeolian sand infilling depressions within till ridges; 20, End moraine; 21, Meltwater stream dissecting the end moraine; 22, Cover sands; 23, Floodplain sands and gravels; 24, Alluvial fan; 25, River terraces capped by loess; 26, Vegetated surface with formation of palaeools; 27, Fines being deflated from floodplain sediments; 28, Loess hills.

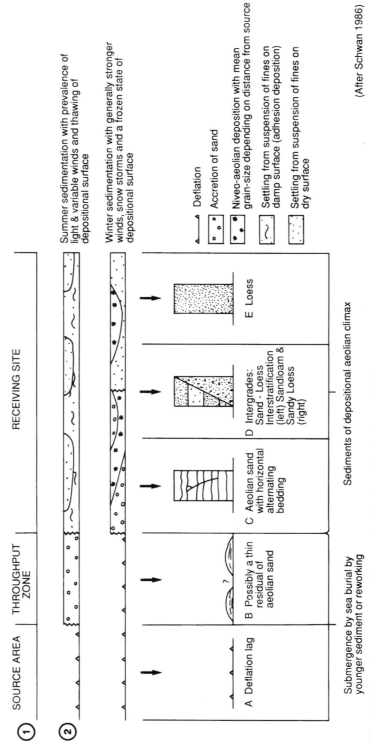

FIG. 6.9. Schematic representation of seasonal aeolian deposition in the periglacial environment (after Schwan, 1986, reproduced by permission of Elsevier Science Publishers

TABLE 6.1. Range in the median size of loess reported in the literature. (Methods of pretreatment and analysis vary so data are not exactly comparable). After Pye (1983)

Area	Range in median size	Average	Sample size	Authors
Nebraska	4.42–4.56	4.47	4	Swineford and Frye, 1951
Kansas	4.29–5.59	5.00	43	Swineford and Frye, 1951
France	4.85–6.55	5.59	4	Swineford and Frye, 1951
Germany	4.95–5.70	5.29	8	Swineford and Frye, 1951
Argentina	4.00–5.10	4.38	12	Teruggi, 1957
Yakutia	4.18–6.88	5.10	26	Pewe and Journaux, 1983
New Zealand	4.40–6.35	5.51	33	Young, 1964
Mississippi	4.99–5.93	5.76	42	Snowden and Priddy, 1968
Tajikistan	5.70–6.93	6.31	11	Goudie et al., in press
China	5.10–5.15	6.36	12	Derbyshire, 1983

TABLE 6.2. Bulk mineralogical composition of loess samples from several different regions of the world

Location	Kaolinite	Illite	Smectite	Chlorite/ Chlorite-Vermiculite	Vermiculite	Quartz	Feldspar
China							
Luochuan, Shaanxi[1]	+	+	tr	tr	tr	+	+
Jiuzhoutai Mtn.[2]	−	+	tr	−	+	+	+
UK[2]							
Pegwell Bay	+	+	+	−	+	+	+
Pine Farm Quarry	+	+	−	−	+	+	+
Northfleet	+	+	−	−	+	+	+
West Germany[3]							
Wallertheim	−	+	−	−	−	+	+
Karhlich	+	+	−	−	−	+	+
France[4]							
Saint Romain	+	+	−	−	−	+	+
Saint Pierre-les-Elbeuf	+	+	−	−	−	+	+
Primarette	+	+	−	+	+	+	+
Italy[4]							
Copreno	+	+	−	+	+	+	+
Vauda Grande	+	+	−	−	+	+	+
Vauda di Nole	+	+	−	−	+	+	+
USA[3]							
Vicksbury, Mississippi	+	+	−	−	−	+	+
Richardson, Alaska	−	+	−	+	−	+	+
Poland[3]							
Tyszowce	+	+	−	−	−	+	+
New Zealand							
Timaru	−	+	−	−	−	+	+
CIS							
Nurek, Tajikistan	+	+	+	+	−	+	+
Mingtepe, Uzbekistan	+	+	+	+	−	+	+

[1]after Liu et al., 1985. [2]after Derbyshire and Mellors, 1988. [3]after Pye, 1983. [4]after Derbyshire et al., 1988.
+ present
tr trace
− absent

TABLE 6.3. Principal facies of Weichselian periglacial aeolian sand in Western Europe. Based on Koster (1982), Pyritz (1972), Ruegg (1983) in Schwan, 1986

Facies	Characteristic sedimentary structures	Modal range of grain-size distribution in µm	Shape of sediment body	Morphology	Dominant period of deposition
1	Dune-foreset cross-bedding	150–300+	Dunes (aeolian dune sands)	River dunes and inland fields	Weichselian late glacial
2	Even horizontal or slightly inclined parallel laminations, low-angled cross-lamination and occasional strings of granules or small pebbles	105–210			
3	Horizontal alternating bedding and irregular waviness of layers	Bimodal with peaks 105–150 and ieither 50–75 or 16–63	Sand sheets (aeolian cover sands)	Gently undulating mantle blanketing pre-existing topography	Weichselian upper pleniglacial

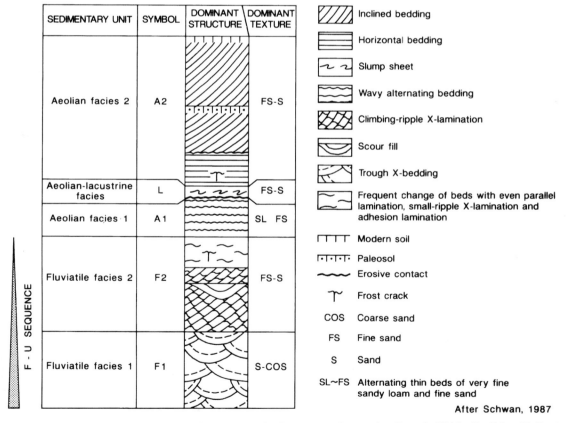

FIG. 6.10. Lithofacies association of a typical section in the Weichselian cover sand succession. From the Weichselian Talsand in Emsland, Germany showing the transition from fluvial to aeolian sedimentation (from Schwan, 1987, reproduced by permission of Elsevier Science Publishers)

FACIES		DEPOSITIONAL ENVIRONMENT	STRUCTURES
6 DUNE SAND FACIES (older inland dunes and river dunes)		dry aeolian	dune-foreset cross-bedding, (sub)horizontal lamination
5 SAND SHEET FACIES A (younger cover sands)		deflation surface, desert pavement	evenly laminated/even horizontal or slightly inclined parallel lamination, rarely cross-bedded/ low-angle cross-lamination, granule and adhesion ripples/occasional strings of small pebbles, deflation levels, small frost cracks and cryogenic deformations
		dry aeolian (seasonal frost?)	
4 SAND SHEET FACIES B (older cover sands)		moist aeolian (permafrost?)	evenly laminated ('layer-cake')/horizontal alternating bedding, silty laminae or silt layers, adhesion ripples, frost wedges and major cryogenic deformations, 'vertical-platy' cracks
		wet aeolian (permafrost?)	
3 (AEOLIAN-) LACUSTRINE FACIES		shallow pools, aeolian supplied material	evenly or wave-ripple laminated, silt and gyttja layers, adhesion lamination
2 (LOCAL) FLOWING-WATER 1 OR FLUVIATILE FACIES A OR B		water current velocity low	fining-upward sequence of
		low energy (ripple phase) fast running water	climbing-ripple cross-lamination, scour troughs, horizontal lamination, adhesion lamination
		low energy (dune phase) fast running water	on large-scale trough cross-bedding, sand with scattered granules, no cryogenic deformations
		high energy very fast running water	

(After Koster 1988)

FIG. 6.11. Facies associations of Weicheselian dune sands and sand sheets in Northwestern Europe. The facies are in sequential order; however, facies type 6 and 5 may be synchronous and facies 4 and 3 can be in reverse order (after Koster, 1988 according to Ruegg, 1981, 1983; Schwan, 1986, 1987, reproduced by permission of Elsevier Science Publishers)

6.11). There is also abundant evidence of syndeposi-tional permafrost development in the form of contraction cracks, sand wedges and pseudomorphs of ice veins and ice wedges (Harry and Gozdzik, 1988; Svensson, 1988), as well as indications of widespread reworking by a number of processes including frost creep, solifluction, rainbeat, rainwash and snow meltwater flow (Maarleveld, 1964; Mücher and Vreeken, 1981; Vreeken and Mücher, 1981).

In the glaciated valleys of some of the world's greatest mountain ranges, the considerable volumes of silt produced by subglacial abrasion occur in a number of 'storages' which persist for varying lengths of time depending on the climatic conditions affecting, in particular, the density of the vegetation cover, and also on the dynamism of the proglacial landscape as influenced by climate, topography and tectonics. Silt initially stored as the matrix in the till

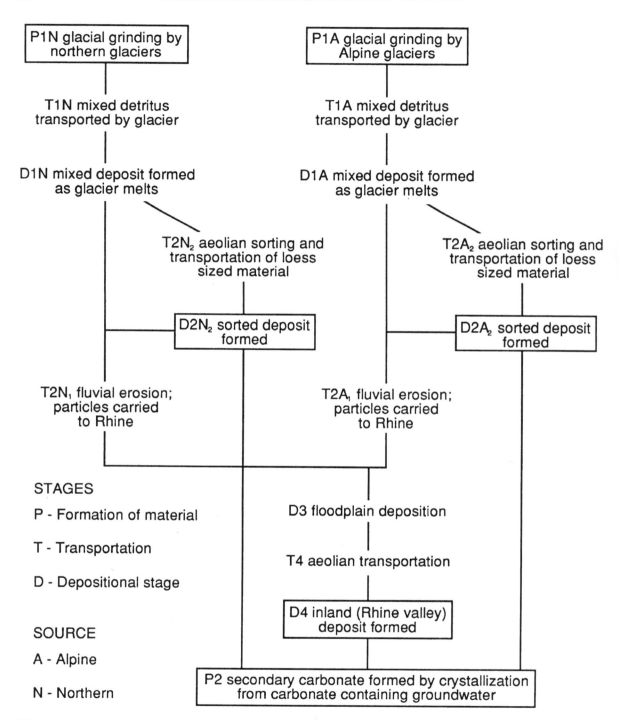

FIG. 6.12. Sequences of events leading to the formation of western European loess deposits. For example, loess deposits in the Munich area would undergo events P1A-T1A-D1A-T2A$_2$-D2A$_2$ while loess deposited in the North European belt would undergo events P1N-T1N-D1N-T2N$_2$-D2N$_2$ (after Smalley *et al.*, 1973, reproduced by permission of Elsevier Science Publishers)

of large moraines is reworked as mudflows, is incorporated in outwash rivers as suspended load and is held in enclosed depressions as lake beds (Owen, 1988a). The volumes of such stored silts are considerable in the Karakoram and Himalaya (Li et al., 1984; Fort et al., 1989; Owen, 1988a), but the occurrence within such mountain regions of silts as aeolian accumulations tends to be limited to topographically favourable locations where thicknesses rarely exceed a metre or two. As with the Pleistocene loess of western Europe, there is much evidence of colluvial reworking. Most silt-grade material is blown considerable distances from the immediate proglacial source sites in the high mountains, and the thick loess deposits of regions such as the Potwar Plateau in northern Pakistan are probably derived mainly from glacial silts that have been concentrated into thick alluvial sequences by the major rivers along the Karakoram–Himalayan piedmont. Smalley et al. (1973) suggested that much of the silt that occurs in the North European Plain and hill lands, including that in the Munich area. This originated from glaciers, was then fluvially transported, deposited as floodplain sediments and subsequently deflated and deposited by aeolian processes to form loess (Fig. 6.12). These examples help demonstrate the often polygenic origin and complex dynamics of the glacioaeolian system. Part of this complexity is illustrated in Fig. 6.8.

6.6. CONCLUSION

Studies of contemporary glacioaeolian processes and sedimentation are few compared to the research undertaken on other aspects of glacial geology. Such studies, however, are essential to the interpretation of the vast thicknesses of Quaternary glacioaeolian sediments found throughout the world. These sediments have great potential for use as palaeoenvironmental indicators for continental regions, especially with regard to reconstructing palaeoclimatic change. This is particularly true of the thick loess sequences in northwestern Europe, midwestern USA and central China which contain paleosols. Unfortunately, relatively little is known about the process and timing of paleosol formation and associated aeolian sediments within such sequences, yet they hold important information that has still to be comprehensively collected and evaluated (Chapter 10).

The study, interpretation and mapping of glacioaeolian sand and silt have important economic aspects. Not only are some of the richest agricultural regions of the world located on glacioaeolian sediments, but these regions yield some of the best sand deposits for industrial use, such as for aggregates, concrete and manufacturing which includes heat resistant bricks, ceramics and glass. An understanding of the sediment geometries and facies variability is therefore important for the exploration and exploitation of these resources.

Chapter 7

GLACIAL ENVIRONMENTS OF PRE-PLEISTOCENE AGE

G.M. Young

7.1. INTRODUCTION

There is a large body of literature on the identification of ancient glacial deposits (Hambrey and Harland, 1981a). Among the various diagnostic physical criteria, perhaps the most important are the presence of a striated basement (Plate 7.1) or intraformational surface (Plate 7.2), dropstones (Plates 7.3 and 7.4), particularly when they are abundant and of varied provenance, striated stones (Plate 7.5) and the presence of widespread diamictites. Other criteria, more commonly found in a continental setting, include fossil eskers or other subglacial channel fills, ancient U-shaped valleys, roche moutonées, step fractures, patterned ground, fossil ice wedges and other criteria indicative of permafrost, varved sediments and till pellets. In some cases structures such as small thrust faults, drag folds and clast fabric have been used as evidence of ice contact processes. Some of these features can, however, be developed in non-glacial settings (Schermerhorn, 1974; Oberbeck et a., 1993).

In this chapter emphasis will be placed on marine glacigenic deposits since most of the preserved record is marine. Exceptions include the terrestrial glacial facies of both Late Proterozoic and Late Ordovician age in the western Sahara (Deynoux, 1985) and

evidence of the existence of permafrost conditions (Williams, 1986).

There are severe limitations on the interpretation of ancient glacigenic deposits. In the Pleistocene record glacial sediments are usually unconsolidated and can be excavated to examine clast fabric, for example. Clasts can be extracted to study shape and surface textures. Most older glacial deposits, however, are consolidated and commonly metamorphosed. Re-orientation of clasts may have taken place in response to tectonic squeezing and new fabrics may be superimposed on those related to depositional processes. Small striations are commonly developed on clast surfaces due to movement of the matrix around clasts during tectonic flattening. In many cases clasts cannot be removed from the matrix of diamictites; they commonly reveal tectonic surface textures and, in extreme cases, tectonic shape modification. Using scanning electron microscopy (SEM), Krinsley and Doornkamp (1973) and others studied surface textures of sand grains in an attempt to differentiate depositional environments. The characteristic attribute is the presence of sub-microscopic chattermarks on hard minerals such as zircons (Chapter 13). In ancient deposits, however, such textures can be modified by diagenetic and other processes. Glacio-tectonic features, commonly used to identify ice-

PLATE 7.1 Cast of striated surface developed on volcanic rocks beneath the Late Proterozoic (Sturtian) glacigenic succession in the northern part of the Adelaide Geosyncline, Australia. Pen for scale

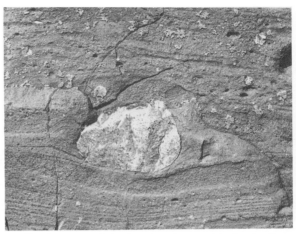

PLATE 7.3 Dropstone in laminated sandstones of the Sturtian (Late Proterozoic) glacigenic succession, South Australia. Note 'splash up' structure to left of clast. Apparent long dimension of clast about 6 cm

PLATE 7.2 Intraformational striated surface developed on mudstones of the Early Proterozoic Gowganda Formation, Ontario, Canada. Note that striated surface is covered by a diamictite forming party of the Gowganda Formation. Coin is 2.5 cm in diameter

contact deposits in the Pleistocene, are difficult to differentiate from other tectonic features in many ancient deposits. The usefulness of radiocarbon dating is limited to relatively recent deposits and therefore other dating techniques such as whole rock Rb-Sr dating have been applied (with varying success) to older glacigenic rocks. In most cases glacial deposits cannot be directly dated and geochronological studies

PLATE 7.4 Dropstone in laminated (varved?) mudstone of the Gowganda Formation, Ontario. Note large size of clast relative to thickness of laminations of enclosing sediment

PLATE 7.5 Faceted and striated quartzite clast from diamictites of the Late Proterozoic (Sturtian) succession, South Australia

7.2. DISTRIBUTION OF PRE-PLEISTOCENE GLACIATIONS IN TIME AND SPACE

Glacigenic deposits are known from the Archean to the present. As noted by Crowell (1982), the distribution of such deposits throughout geologic time and space is irregular. The temporal distribution is shown in Fig. 7.1 and can be summarized as follows:

(a) There is scant evidence of glaciation in the Archean. Possible glacigenic diamictites are locally present in the Witwatersrand succession of South Africa and possible dropstones have been described in Archean sedimentary rocks in close proximity to the Stillwater Complex in the United States;

(b) Glacigenic rocks are more widespread in the Early Proterozoic and are documented in Canada from the Huronian of Ontario, the Hurwitz Group in the Northwest Territories and the Chibougamau Formation in Quebec. In the United States glacial deposits of this age are known from Michigan, southeastern Wyoming and neighbouring states. Outside of North America Early Proterozoic glacigenic rocks have been reported from Finland, South Africa and Western Australia, and are possibly present in India. Descriptions of most of these occurrences can be found in Hambrey and Harland (1981a);

(c) There is a global dearth of glacial deposits in the long time interval between approximately 2.0 Ga and 1.0 Ga. In some ways this long 'interglacial' is more difficult to explain than the occurrences of glacigenic rocks;

(d) Late Proterozoic successions contain the most abundant and widespread evidence of glaciation on earth. Such deposits are now known from all continents. The wide distribution of these glacial deposits has raised questions concerning their origin. Were they all formed contemporaneously or was there sequential glaciation of the continents as they moved into high paleolatitudes during sea floor spreading? The problem is compounded by the discovery that many of these deposits appear to have formed in low paleolatitudes. Many authors have also noted the common association with apparently incongruous rock types such as dolostones and iron-formations (Menzies, 1995a, Chapter 2) (Section 7.3.4.2);

merely provide a broad time range within which the ancient glacial deposits can be placed. New techniques involving U-Pb dating of single zircon crystals may potentially provide useful data on the age of provenance terrains (Rainbird et al., 1992). Metazoan fossils, which are extremely useful in differentiating marine from terrestrial environments in the Phanerozoic, did not exist during most of the Precambrian. In descending the geologic column, the number of definitive attributes decreases.

The Pre-Pleistocene record is valuable in providing a much better window on events occurring over long time periods. This is in stark contrast to the short time period of the Pleistocene. Many studies of Pleistocene deposits have insufficient relevance to the preserved glacial record because the majority of Pleistocene workers are concerned with terrestrial deposits which are mostly ephemeral. The 'Pleistocene' glaciation may not yet be over, thus attempts to understand it may be premature.

Understanding is sometimes facilitated by a distant perspective. Ancient deposits provide such a perspective from which to view the stratigraphic context, including deposits formed both before and after glaciations. A further advantage in studying ancient glacigenic deposits is that they provide a window through which environments that are very poorly known, particularly modern marine glacial environments that are largely inaccessible, can be viewed.

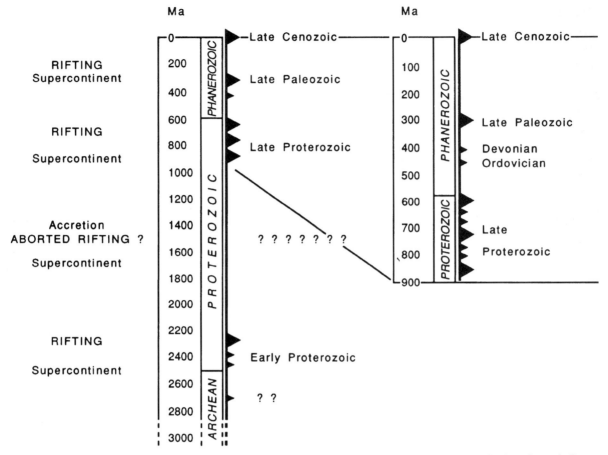

FIG. 7.1 The distribution of glacigenic deposits throughout geologic time (black triangles). Open triangle for Late-Cenozoic-Recent glaciation signifies that this glacial phase may be incomplete. Note that most preserved major glacial deposits can be correlated with rifting of supercontinents. Some, such as the Devonian (?) and Late Cenozoic-Recent, may be related to mature ocean stages (phase C of FIG. 7.3). The lack of glacial deposits throughout most of the Proterozoic is puzzling but may be linked to peculiar plate tectonic conditions during much of that period. In North America, and possibly elsewhere, widespread intrusion of anorthosites and rapakivi-type granites has been interpreted as a period of aborted rifting, and the southern half of North America appears to have undergone a spectacular accretionary phase at this time

(e) Significant glaciations in Phanerozoic time took place in Permo-Carboniferous times (mainly in Gondwana) and are recorded in the Tertiary to Pleistocene of Antarctica and Alaska. There is also local evidence of glaciation in the Cambrian of China, the Late Ordovician (mainly in Africa and South America), the Devonian of Brazil and in the Cretaceous.

7.3. LITHOFACIES AND SEDIMENTOLOGY OF PRE-PLEISTOCENE GLACIGENIC ROCKS – A BRIEF REVIEW

7.3.1. Diamictites

Diamictites (Plate 7.6) are the most enigmatic and equivocal glacial deposits. Features such as striated and faceted clasts, a polymictic clast population and

PLATE 7.6 Sandy diamciton from the Late Proterozoic (Sturtian) succession, South Australia. Note the wide range of grain sizes and rock types. Knife is 10 cm long

possibly geochemistry, might permit a glacial origin to be ascertained but do not resolve the question of precise depositional environment. Facies associations are therefore important in interpretation (Chapter 9). Some are true tillites, in the sense of having formed more or less directly from glacial ice (as opposed to those formed by rain-out from floating ice or deposits that have been 're-sedimented' as sediment gravity flows). Criteria such as small scale glaciotectonic thrusting and preferred orientation of clasts may be used to identify tillites but many diamictites have undergone tectonic compression, making interpretation of such features equivocal. Some diamictites are associated with appropriate structures (striated basement) and paleolandforms such as roche moutonées (Deynoux, 1985) or drumlins but these are rare (Rocha-Campos et al., 1968). It is probable that the majority of preserved diamictites did not form in continental settings, but rather in marine basins. This bias is simply related to preservation potential. The stratigraphic record and simple logic indicate that sediments, in general, are much more likely to be preserved in basinal than terrestrial settings. Glacial sediments are no exception.

Attempts to interpret marine diamictites have relied heavily on models derived from studies of recent glacial deposition (Chapter 5; Menzies, 1995a, Chapter 14). Perhaps the most influential of these was the

now largely abandoned dry- and wet-based glacier model of Carey and Ahmad (1961). Parts of this model were rejected because of the realization that very little sedimentation of glacially-transported debris is taking place beneath ice shelves in Antarctica at the present day (Drewry and Cooper, 1981; Anderson et al., 1983) (Chapter 5). Most deposition of this kind is now inferred to take place at, or close to, the grounding line. Ice shelves appear to form only under special conditions, where there is a protective coastal embayment or where there are sufficient submarine 'pinning points' to stabilize the ice mass. One of the most striking aspects of deposition in the polar glacial setting, typified by most of the Antarctic, is the dearth of meltwater deposits. This is in marked contrast to the situation in 'temperate' glaciated regions such as the Gulf of Alaska, where, at the present time, there is virtually no direct contribution from ice to the sediments of the Gulf region (Molnia, 1983). Sedimentation on the marine coastal region is dominated by fine grained materials carried in suspension across outwash plains by meltwater streams (Menzies, 1995a, Chapter 12).

In the absence of critical indicators such as glaciotectonic features, striated pavements and till fabric, diamictites are among the most difficult sedimentary rocks to interpret (Fig. 7.2) for they can form directly from ice, or in a more distal glacio-marine or -lacustrine setting by 'rain-out' from floating ice or by mass flow processes (Eyles et al., 1985). The discussion between Christie-Blick (1982) and Ojakangas and Matsch (1980, 1982) serves to underline the difficulties involved in interpretation of these rocks.

In some cases, associated sediments provide essential clues but, even with this additional information, interpretation can be difficult. Incorporated fossils can be used as evidence of marine deposition and can provide clues regarding depth of accumulation (Plafker and Addicott, 1976; Birkenmajer, 1982). Diamictites containing lenticular cross-bedded sandstone bodies (Young and Gostin, 1988a), formed in meltwater streams, have been interpreted as ice contact diamictites or tillites (Dreimanis, 1976; Dreimanis and Schlüchter, 1985). Many massive diamictons and diamictites have been interpreted in this way (Williams and King, 1979; Edwards, 1976).

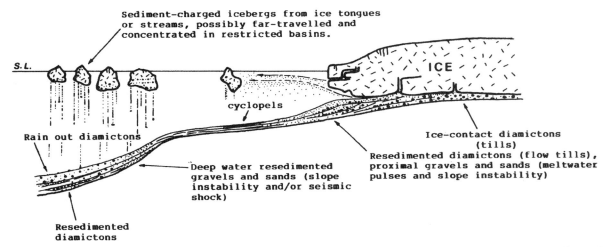

FIG. 7.2 Simplified sketch to show the problems in interpretation of homofacial associations (diamictons and resedimented orthoconglomerates and sandstones) in ice-proximal and ice-distal settings in a glaciomarine environments. 'Cyclopels' are fine grained rhythmic beds formed by interaction between tidal currents and sediment-charges meltwater driven currents (Mackiewicz *et al.*, 1984, reproduced with permission from Elsevier Science Publishers). Figure not to scale

Another prevalent interpretation for such diamictites is that they developed under an ice shelf (Reading and Walker, 1966; Miall, 1983). This interpretation was used to explain the absence of other interbedded lithologies. Eyles *et al.* (1985) suggested that some massive diamictites could have formed by rain-out processes. They argued that depositional processes active at the grounding line (Drewry and Cooper, 1981), including meltwater processes, together with commonly cited back and forth migration of the grounding line, should lead to a complex interdigitation of a variety of rock types. Thick successions of stratified diamictites, which are common in the geologic record (Plate 7.7), are more likely to have formed by rain-out in a distal glaciomarine setting. Some caution is probably warranted in using information from modern glacial environments to interpret ancient occurrences. The apparent formation of some ancient glacial deposits at low paleolatitudes makes it difficult to interpret them in terms of 'temperate' or 'polar' settings. Many Early and Late Proterozoic glacial successions appear to be preserved in rift basins. This tectonic setting could have played an important role in controlling the nature of preserved glacial sediments. For example, the semi-restricted nature of such basins perhaps led to massive

iceberg jams which, in turn, could have caused deposition of much more abundant 'rain out' diamictons than in present day environments.

Another viable depositional mechanism for diamictites is by mass flow processes in both glacial and non-glacial settings. Such diamictites will normally be stratified, with flow structures such as wispy laminations and flow noses and commonly contain

PLATE 7.7 Stratified diamictites in the Late Proterozoic (Sturtian) succession, South Australia. Cliff is about 40 m high

rip-up clasts and rafts. In glacial settings this process has traditionally been associated, by Pleistocene geologists, with slumping of debris from the ice margin (Evenson *et al.*, 1977; Dreimanis, 1979) and is accordingly (like the diamictites discussed above) assigned to an ice-proximal environment. Study of the nature and distribution of associated lithofacies in many cases in the geologic record provides no evidence of proximity to an ice margin (Young and Gostin, 1991). Such diamictites are more likely to have formed in 'deeper' water environments due to oversteepened depositional slopes, steep topography or seismic shock. Schermerhorn (1974) argued that since many Late Proterozoic diamictites are associated with turbidites and other, coarser products of sediment gravity flow processes (re-sedimented facies), that they were not of glacial origin. Subsequent work (Wright and Anderson, 1982; Anderson, 1983) has shown that re-sedimentation processes are extremely common on the present day Antarctic shelves (Chapter 5). Re-evaluation of many of these Late Proterozoic diamictites (Hambrey and Harland, 1981a) has provided support for a glacial origin.

The interpretation of extensive massive diamictites remains problematic. The modern protagonists favour two settings. One school, concerned mainly with recent sedimentation in Antarctica, invokes subglacial deposition close to the grounding line of marine ice sheets. Others, more involved with interpretation of the geologic record, favour a more distal setting and an origin by rain out, under conditions that may have differed somewhat from those of the present-day Antarctic shelves. Clearly all lines of evidence should be considered in an interpretation but perhaps the most powerful, at least in the geological record, remains the stratigraphic context – what are the associated lithofacies? Another possible way to discriminate between these different origins is to study the petrography/geochemistry of diamictites. Materials derived from grounded ice might reflect a relatively local provenance with distinct petrographic provinces (Anderson *et al.*, 1984; Mazzullo and Anderson, 1987), whereas ice-rafted material could be very far travelled and homogeneous over large areas. This approach has not received much attention in the study of glacigenic diamictites but promises to be useful.

Even where there is agreement about the depositional mechanism (mass flow) there is still a dichotomy of opinion concerning depositional setting. Those working in the Recent and Pleistocene generally favour ice-proximal settings because they can see evidence of these processes; those working in the ancient feel that they are seeing evidence of mass flow in more distal glaciomarine settings (Fig. 7.2). Both interpretations are probably correct and the depositional setting can only be identified by studying associated rock types.

7.3.2. Orthoconglomerates and Sandstones

As with diamictites, these rock types can occur in both proximal and distal glacial settings. In proximal environments they have been attributed to deposition in subaqueous fans where glacial meltwater tunnels debouch from the submerged but grounded ice front (Rust and Romanelli, 1975; Cheel and Rust, 1982; Mustard and Donaldson, 1987). In proximal parts of such fans, coarse gravels can accumulate in channels. Some of these are deposited as sediment gravity flows (Plate 7.8) and display characteristic features such as massive beds and inverse grading (McCabe and Eyles, 1988). Such flows could have originated as meltwater pulses (Cheel and Rust, 1982) or by slumping related to steep depositional slopes. These coarse-grained rocks are commonly associated with finer-grained cross bedded gravels and sands that accumulated in the inter-channel areas. Diamictites may also be intimately associated with these better sorted facies in areas where traction currents were episodic.

A thick assemblage of rocks of this kind is preserved on the eastern margin of the Late Proterozoic North Flinders Basin, South Australia. These rocks have been interpreted as coarse glacial outwash materials that were subjected to re-sedimentation as sediment gravity flows triggered in part by contemporaneous fault activity on the margins of a rift basin (Young and Gostin, 1988b). These rocks probably represent a deeper depositional setting than the Pleistocene examples cited above. As with subaquatic flow tills, there has been a tendency for those studying Pleistocene deposits to invoke ice-proximal settings, whereas deeper, more distal envir-

PLATE 7.8 Inverse grading in polymictic conglomerate overlying parallel-bedded sandstone and granule conglomerates. Such deposits are sediment gravity flows but the depositional setting can be problematic (ice-proximal or distal?)

onments are commonly cited for ancient deposits. Wright and Anderson (1982), however, suggested that in the present Antarctic continental shelf, relatively clean sands could be generated over down-current distances of about 10 km by downslope winnowing of sediment gravity flows derived from diamictons (Chapter 5). Ice-proximal gravels and sands should be associated with cross bedded sands (McCabe and Eyles, 1988) whereas those in deeper water settings would be associated with finer-grained laminated muddy sediments and turbiditic sands.

The Serle conglomerate at the top of the Sturtian glacial succession in the Flinders Ranges of South Australia is an example of resedimentation of glacial materials in a submarine fan channel system (Young and Gostin, 1989). Materials up to cobble grade form lenticular bodies interpreted as channel-filling conglomerates. Associated finer-grained rocks are turbi-

dites and deep water shales. In this case deposition of such coarse-grained materials in a relatively deep water setting may have been a response to isostatic uplift following withdrawal of the ice from a neighbouring land mass.

In a very different tectonic setting – that of a convergent plate margin – Eyles (1987) described a succession of channelized graded gravels and diamictites forming part of the Late Cenozoic Yakataga Formation on Middleton Island in the Gulf of Alaska. These were interpreted as having formed in a series of broad channels cutting the continental shelf edge. The coarse material was inferred to have been deposited as sediment gravity flows, derived from subglacial meltwater streams.

7.3.3. Mudstones

7.3.3.1. Loessite

About 10% of the land surface of the earth is covered with loess deposits of Pleistocene age (Chapter 6). These windblown deposits are generally considered to have formed from materials crushed during glacial transport but others have cited a desert origin. Ancient loess deposits are relatively rare, probably because of poor preservation potential. Some Late Proterozoic mudstones associated with glacigenic sediments in Norway (Edwards, 1979) have been interpreted as 'loessites', or indurated loess. The identification is based mainly on textural attributes and grain fabric, which is purported to reflect wind direction. Wood (1973) inferred that parts of the Early Proterozoic Gordon Lake Formation were tidal flat deposits, derived largely from loess, blown into the depositional basin. He also suggested that glacial grinding may have played an important role in production of the fine-grained siliceous material. Taylor and McLennan (1985, p. 42) have used the composition of loess to obtain estimates of upper crustal composition. This interpretation is based on the idea that the fine-grained material making up loess represents rock flour derived by glacial grinding. Such materials should have undergone little chemical weathering and should therefore be representative of crustal composition of the area of continental crust sampled by the glaciers.

7.3.3.2. Varved mudstones

Muddy sediments can accumulate in a number of depositional settings related to glaciation. Perhaps the most well known are the varved clays that are typical of many glacial lakes. Rhythmic bedding of this kind has been described from a number of ancient examples, including the Early Proterozoic Gowganda Formation (Plate 7.9) (Jackson, 1965). Dropstones (Plate 7.4), 'till pellets' and pebble nests (Ovenshine, 1970) are common in such sediments and are interpreted to have formed from floating ice derived from glaciers. G.E. Williams (1985, 1989) also suggested that laminae in fine-grained sediments from Late Proterozoic rocks of the Adelaide Geosyncline in South Australia were annual, and described a cyclicity that he linked to sun-spot activity. Subsequently, however, Williams (1989) suggested that the larger scale cycle might be annual and interpreted the fine laminations as being due to variable velocity and range of ebb tides. He used these data to make inferences about the paleorotation of the earth and the dynamics of the Earth-Moon system in the Late Proterozoic. These studies show the problems inherent in interpretation of laminations in fine-grained sediments and the difficulty in identifying annual rhythms or varves (Menzies, 1995a, Chapter 14).

PLATE 7.9 Finely laminated (varved?) mudstones in the Early Proterozoic Gowganda Formation, Ontario. These fine grained rocks contain large rafted clasts (see Plate 7.4)

7.3.3.3. Glaciomarine mudstones

Some authors (Powell, 1981a; Osterman and Andrews, 1983) have suggested that, in a proximal marine glacial environment, a zone of muddy sediment will develop adjacent to subaqueous outwash fans in an area from which icebergs and other sediment-rich ice are removed by meltwater currents. Some muddy sediments formed in this sort of setting, by the interaction of tidal currents and suspended material in the upper part of the water column, have a rhythmic or laminated character and have been called 'cyclopels' by Mackiewicz et al. (1984) (Fig. 7.2). Muddy sediments are also typical of distal glaciomarine environments, where there is a widespread (up to thousands of square kilometres) and abundant supply of suspended sediment. In some cases such sediment may also contain an ice-rafted component from far-travelled icebergs (Chapter 5). In the absence of such ice-rafted debris and in non-fossiliferous deposits, these muddy sediments would be difficult to differentiate from normal marine muds. One possible criterion is the use of major element geochemistry to determine the degree of weathering that the material has undergone (Section 7.3.4). Such sediments can be massive under conditions of rapid or steady sediment supply but could be subjected to a variety of currents such as the contour currents around the Antarctic and can become involved in downslope movements resulting in stratified sediment gravity flow deposits (Kurtz and Anderson, 1979; Wright and Anderson, 1982).

7.3.4. Associated Lithofacies

7.3.4.1. Dolostones

Experimental and theoretical studies and observations of present and past occurrences of dolomite all suggest that formation of the mineral is facilitated by relatively warm conditions. The common association of diamictites and dolostones in Late Proterozoic successions is therefore puzzling and was one of the reasons why Schermerhorn (1974) questioned the glacial origin of many these rocks. Dolostones take on a variety of forms in these Proterozoic successions. Dolomite occurs as matrix and clasts in many diamictites. It also takes the form of laminated dolomicrites, some of which contain dropstones (Spencer, 1971). Flat chip conglomerates composed of rip-up clasts of dolomicrite also occur. Some stromatolitic dolostones are associated with glacigenic diamictites (Young and Gostin, 1988a). In some cases lenses of dolostone are present in the matrix of diamictites.

Bjorlykke *et al.* (1978) suggested analogies with modern cold water carbonate deposits in high latitudes but suitable skeletal organisms were not present in Late Proterozoic seas. Walter and Bauld (1983) drew attention to occurrences of evaporites and stromatolitic sediments in lakes in the Dry Valleys region of Antarctica. They suggested that such environments might provide an analogue for Late Proterozoic occurrences but it seems probable that the depositional setting of many of the Proterozoic examples was rather different. Deynoux (1985) suggested that carbonate in Late Proterozoic diamictites of North Africa was derived both as clastic particles and by dissolution of the substrate at the time of deposition. Re-distribution within diamictites was brought about by solution and precipitation processes, related in part to concentration of salts by freezing (Menzies and Maltman, 1992). Rapid climatic changes were invoked by Fairchild and Hambrey (1984) to explain the association of diamictites and dolostones in the Late Proterozoic succession of Spitsbergen (cf. Spencer, 1971). Williams (1975) invoked a strong seasonality, brought about by increased obliquity of the ecliptic (Menzies, 1995a, Chapter 2). This hypothesis is in keeping with some paleomagnetic data favouring the deposition of many Late Proterozoic glacial sequences in low paleolatitudes (Embleton and Williams, 1986). Continuing work on the stable isotopic composition of the carbonates (Williams, 1979; Lambert *et al.*, 1980) may help to shed light on the problem.

Dolostones associated with Late Proterozoic (Sturtian) diamictites in the northern part of the Adelaide Geosyncline are probably mostly detrital deposits derived by erosion of dolostones from the underlying Burra Group. Stromatolitic dolostones obviously had some organic influence, although this also may have been a passive one if the stromatolites formed by the trapping of carbonate detritus (see Fairchild, 1991, for a detailed discussion of this problem). Lenses of dolostone in diamictite may have formed by diagenetic remobilization of carbonate. The presence, in such lenses, of silica blebs which are possibly replacements of evaporite minerals, suggests that concentration of solutes took place, possibly by freezing shortly after deposition (cf. Deynoux, 1985).

It is likely, however, that a high percentage of Late Proterozoic dolostones within glacigenic successions are of detrital origin. The common association with glacigenic deposits is because many Late Proterozoic glacigenic deposits overlie, and were largely derived from dolostone-rich platformal assemblages. Many Late Proterozoic glacigenic formations are also overlain by dolomitic units, suggesting rather dramatic climatic fluctuations.

7.3.4.2. Iron-Formations

Iron-formations are traditionally considered as part of the Archean or Early Proterozoic stratigraphic package but some very significant iron-formations occur in intimate association with glacigenic deposits in a number of Late Proterozoic basins. This unique association has recently been explained (Yeo, 1981; Young, 1988) as reflecting deposition of glacigenic sediments in a rift-basin setting where concomitant hydrothermal activity produced Fe-rich (and in some cases, Mn-rich) brines. Overturn related to movement of cold glacial meltwaters resulted in displacement of the metal-charged brines which, by dilution (Seyfried and Bischoff, 1977) precipitated out. Precipitation took place together with both diamictons and laminated muds, resulting in a very intimate interdigitation of clastic glacigenic deposits and chemically precipitated iron (usually oxides) and chert.

The best known occurrences of this intriguing combination of rock types are in the northern Cordillera of North America in the Rapitan Group (Yeo, 1981), in Alaska (Allison *et al.*, 1981; Young, 1982;), in the Damara orogen of South Africa (Breitkopf, 1988) and in the Adelaide Geosyncline in South Australia (Link and Gostin, 1981; Preiss, 1987). Other examples include those in Brazil (Dorr, 1945) and in south China (Tang *et al.*, 1987). All of these examples are from Late Proterozoic successions, and may reflect the special tectonic setting attendant on fragmentation of a Late Proterozoic supercontinent (Section 7.5).

7.4. GEOCHEMICAL ASPECTS

Weathering processes are accelerated in warm wet climates. Conversely, sediments formed in a frigid climatic regime should be relatively unaltered. Thus, by studying the major element geochemistry of sediments or sedimentary rocks it should be possible to identify products of glacial deposition. Nesbitt and Young (1982, 1984) developed a weathering index which they called the Chemical Index of Alteration (CIA) and attempted to apply it to Early Proterozoic rocks in the Lake Huron area. Using molar proportions, the CIA is calculated as follows:–

$$CIA = [Al_2O_3/(Al_2O_3 + CaO^* + Na_2O + K_2O)] \times 100 \qquad (7.1)$$

where CaO* is CaO in silicates (as opposed to carbonates or phosphates). The index provides a measure of the relative amounts of labile versus stable oxides in the sediment, which gives some idea of the degree of weathering. The CIA gives low values (~50) for unweathered materials and high values (to a maximum of 100) for chemically weathered material. This technique has proved useful in the study of Early Proterozoic successions in Canada (Nesbitt and Young, 1982) and elsewhere (Marmo and Ojakangas, 1984). Since mineralogical (and chemical) composition of sediments and sedimentary rocks is clearly related to grain size, it is important that comparisons be made among samples of similar grain size. Another limitation in the use of the CIA is that of recycling; if a glacier over-rides a terrain composed of rocks produced from materials that were weathered in an earlier depositional cycle, then the diamictons (diamictites) will display high CIA values inherited from the source rocks.

Frakes (1975) and Sumartojo and Gostin (1976) used trace element geochemistry in an attempt to differentiate marine and terrestrial glacial deposits. Frakes (1985) suggested that marine glacial deposits should be depleted in Fe as a result of deposition in more reducing environments but care should be taken in extending this criterion back to Early Proterozoic

or Archean times when iron-enrichment in marine basins seems to have been common (Eriksson, 1983; Young, 1988).

7.5. TECTONIC SETTING of GLACIAL DEPOSITS

The dearth of glacial deposits in Archean times (Fig. 7.1) seems anomalous in view of the inferred lower solar luminosity during that period of earth history (Sagan and Mullen, 1972; Owen et al., 1979). Warm conditions could have been maintained on the Archean earth because of an enhanced greenhouse effect due to high concentrations of gases such as CO_2 in the early atmosphere. Alternatively the thinner Archean crust could have been largely submerged (Hargraves, 1976) and such a condition may have led to more equable climates (Henderson-Sellers and Henderson-Sellers, 1988). With the diachronous thickening and stabilization of a large portion of the continental crust during the Archean/Proterozoic transition (Taylor and McLennan, 1985), plate tectonic processes on a scale and of a type comparable to those of the Phanerozoic appear to have taken place (Hoffman, 1988). At the same time a cycle of formation and fragmentation of supercontinents was initiated (Nance et al., 1986, 1988). These processes may have been primary controls on glaciation throughout geological history.

Many authors have hunted for rhythms on various scales in the seemingly erratic distribution of glacial deposits throughout geologic time (Williams, 1975; Steiner and Grillmair, 1973). Studies of this kind have been directed towards finding a mechanism to produce glaciations but none appears to provide a satisfactory explanation for all the glacial deposits preserved on earth. Young (1988) drew attention to the common spatial and temporal association between continental rifting and glaciation in many Early and Late Proterozoic basins. A more detailed account of these relationships is given in Menzies (1994a, Chapter 2). Recent publications by Fischer (1982), Nance et al. (1986, 1988) and Worsley et al. (1986) have emphasized the possible influence of supercontinent creation and destruction on earth climate. Their model involves a 400 to 500 Ma cycle of separation and amalgamation of the earth's con-

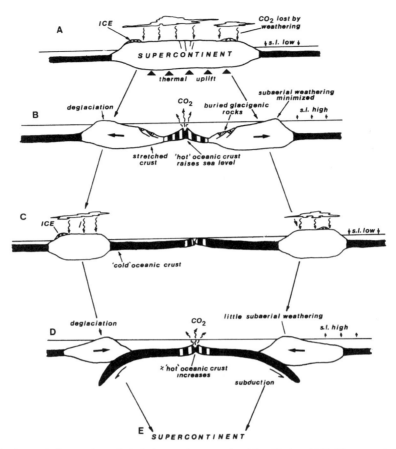

FIG. 7.3 Possible mechanisms relating continental glaciation to plate tectonics (after Fischer, 1982; Nance *et al.*, 1986, 1988; Worsley *et al.*, 1986)

Stage A: Sea level is low because of thermal uplift of the continental crust, assembled into a supercontinent. CO_2 is lost from the atmosphere in the weathering of exposed continental crust. The reduction in atmospheric CO_2 initiates a cooling trend, leading to continental glaciation.

Stage B: The supercontinent is fragmented and rifting takes place. Relative sea level rises because of production of 'high standing' hot young oceanic crust, stretching of the rifted continental crust (decreasing the area of the oceans) and isostatic and thermal subsidence of the thinned continental crust as it moves away from the 'hot' zone. This process favours preservation of glacigenic rocks formed in early rifts at the trailing margins of the newly formed continental blocks. Atmospheric CO_2 levels begin to build up because of production at the newly formed spreading centres and because of a reduction of CO_2 depletion by weathering/precipitation reactions. Greenhouse effects lead to deglaciation, which in turn contributes to high sea levels.

Stage C: The 'Atlantic'-type ocean reaches maturity. Sea level has lowered again due mainly to the high percentage of old dense oceanic crust. Exposure of the continents again leads to reduction of atmospheric CO_2 and eventually, an 'ice-house' condition, as in Stage A. Glaciation sets in. This stage is comparable to the present condition on Earth.

Stage D: The 'Atlantic'-style ocean begins to subduct. The proportion of young oceanic crust is therefore increased so that sea level rises. As the continents are drowned, less CO_2 is consumed in weathering reactions and deglaciation begins, enhancing the trend towards rising sea levels. Glacial deposits formed between Stages C and D are less likely to be preserved than those formed during breakup of a supercontinent because those at the leading edges of the continents are susceptible to tectonic destruction during the subduction phase. Slower subsidence at the trailing edge (no tectonic crustal thinning as in Stage B) would also increase the possibility of loss by erosion.

Stage E: Consumption of the ocean and continental collision result in production of a supercontinent and a return to the situation depicted in Stage A.

Note: According to this model, glaciation would be likely between Stages A and B and between Stages C and D but the preservation potential of those formed between A and B would be much greater because of higher subsidence rates related to crustal thinning and removal of the depositional site (rift grabens) from the spreading centre

tinents. Initiation of the cycle is conjectured to be brought about by the blanketing effect of the super-continent. Heat from the interior of the earth dissipates less efficiently through continental than oceanic crust. A large continental mass is less easily cooled by 'edge effects' related to subduction and will tend to heat up and become thermally buoyed, causing a relative sea level drop and eventually leading to rifting and fragmentation.

These authors suggested that glaciation takes place due to an 'ice-house' effect, during periods of low relative sea level when weathering of the high-standing continents and precipitation of marine carbonates depletes the CO_2 content of the atmosphere. At such times, albedo would also be high. According to the model, there are two periods when the oceans should be at a relatively low level. These are: (1) when a supercontinent has been assembled, and (2) when an 'Atlantic'-type ocean has reached a mature stage and is about to begin subducting to initiate a phase of closure (Fig. 7.3). The main reason for a low sea-stand in situation (2) is the large proportion of 'old', dense ocean crust, which subsides, leading to withdrawal of the oceans from continental regions. Condition (2) is close to what exists at the present time. Sea level is low (compared to most of Phanerozoic time) and many believe that the earth is currently in an interstadial or interglacial phase of the Pleistocene glaciation (although, for the first time human activities, such as the burning of fossil fuels, animal husbandry, cultivation of rice paddies and destruction of rain forests, may be causing significant modification of the earth's climatic pattern).

Major glaciations on earth such as those of the Early and Late Proterozoic may be attributed, at least in part, to the existence of elevated continents. The glacigenic rocks are preserved by virtue of development of rift basins and mioclines at the onset of breakup of the supercontinents (Young et al., 1979; Eisbacher, 1985; Young, 1988). As pointed out by Worsley et al. (1986) most Phanerozoic glaciations can be accommodated in such a theory. The Late Proterozoic and Permo-Carboniferous glaciations occurred at times of particularly low sea levels and the Pleistocene glaciation took place (and continues?) during a low sea level stand related to the mature nature of much of the Atlantic oceanic crust. The

glacial deposits of the Early Proterozoic appear to have formed just prior to a time of continental (supercontinental?) fragmentation and a low sea-stand can be inferred. Preservation of glacial deposits may have been brought about by deposition in rift basins (Young, 1988) which were buried by newly-developing mioclines on the flooded trailing continental margins (Fig. 7.3).

The cycle of fragmentation and amalgamation of supercontinents is driven by convection in the asthenosphere. It may provide a plausible first order causative mechanism for global glaciation. This is not to say that other factors, such as the length of the galactic year and the distribution of continents on the surface of the earth, have not been important. Rocks belonging to the long period between 2.0 and 1.0 Ga show little evidence of glaciation. During the first part of this long period, it has been suggested (Hoffman, 1988) that the continental crust may have been enlarged by lateral additions (accretionary tectonics). Aborted attempts at rifting may be indicated by widespread development of anorthosites and rapakivi-type granites about 1.5 Ga. These unusual tectonic regimes and attendant release of large amounts of CO_2 into the atmosphere may have interrupted the 'supercontinental' tectonic cycle and inhibited production and/or preservation of glacial deposits during this period. Latitudinal distribution of continents may also have been a factor in the presence or absence of glaciation on earth at certain times. In particular, the presence of continental masses in high latitudes appears to have been a key factor in production of Phanerozoic glacial deposits. Milankovitch-type controls, related to orbital perturbations have also been important, but probably as second order phenomena, causing advance-retreat phases during glacial periods, rather than initiating them (Menzies, 1995a, Chapter 2). Changes in oceanic circulation can also have a profound effect on climate because of the importance of ocean currents in heat transfer (Broecker and Denton, 1990b).

7.6. SUMMARY

Ancient glacial deposits play a special role in unravelling the climatic history of the earth for, in spite of some of the problems of interpretation outlined above, they provide some of the least ambiguous evidence concerning surface conditions on the planet in times past. Astrophysics suggests that in Archean times the 'faint' young sun should have resulted in a perennially frozen surface on earth. The geological record, however, shows that the first widespread glaciation did not occur until the time of the Archean/Proterozoic transition, when emergent crust first became widespread. The dearth of Archean glaciation is attributed to a 'super greenhouse effect', most probably due to a high CO_2 content in the early atmosphere. Glaciations are therefore most likely related to changes in atmospheric composition (especially CO_2), which may, in turn, be linked to plate tectonic phenomena. During periods of emergent continental crust, such as episodes of supercontinentality with attendant mountain building, enhanced weathering may have caused a drawdown of CO_2, resulting in glaciation. Suggested causes of glaciation are many, however, and caution is warranted in applying a single mechanism, particularly when it is realised that earth environments undergo secular change. For example, it is likely that the earth's interior has cooled throughout geologic time, affecting rates and style of plate tectonic processes. The atmosphere, hydrosphere, biosphere and lithosphere have all evolved; these changes must also be reflected in surface processes, including glaciation. In these days of heightened awareness about environmental problems such as global warming and its potential hazard for the human species it is particularly relevant to restudy the preserved record of ancient climatic change, for the stratigraphic record is the only long term climatic record.

CHAPTER 8

GLACIAL STRATIGRAPHY

J. Rose and J. Menzies

'Diluvium is chaos.' (Lossen, 1875)

8.1 INTRODUCTION

Stratigraphy is the discipline by which we seek to understand 'the history of how, when and why glacial deposits . . . came to be what and where they are today' (Salvador, 1994, p xvii). In other words, glacial stratigraphy provides the methods and procedures by which we can reconstruct the history and patterns of past glaciations and associated environments. For many earth scientists, stratigraphy is the ultimate aim of their subject (cf. Harland, 1992). For the science of glacial geology it should include all the information about processes, geographical distributions, timing and environment of past glaciers and glaciation.

Glacial stratigraphy is very closely related to Quaternary stratigraphy which is concerned with the history of the whole environment over the past 2.6 Ma, even though glaciations have occurred on many other occasions in Earth history (Deynoux *et al.*, 1994) (Fig. 1.1). Quaternary stratigraphy includes the study of glaciations of the past. Many features of Quaternary stratigraphy such as 'glacial' and 'interglacial' subdivisions (Flint, 1971; Rose 1989a) are derived from early studies of glaciation and from the fact that glaciers are such critical driving forces in

bringing about changes on the Earth's surface. However, Quaternary stratigraphy also includes the study of other physical, chemical and biological processes and their interaction. Indeed it is through Quaternary stratigraphy that it is possible to reconstruct former landscapes at any scale and to understand what these landscapes were like, when they occurred and the rates and processes by which they changed. In an area such as, Ontario Canada, or northern England these changes include glaciation, but in areas such as much of the south-western United States or southern England, glaciation plays no direct part in this history and the striking changes involve variations in river activity, vegetation cover or faunal types. The study of such a wide range of mutually interacting processes gives immense robustness to the scientific procedure and generates an inherent excitement in understanding the different processes, landscape patterns, landscape sensitivities, and rates of change that may have occurred in a given area over a given period of time.

Nevertheless, this chapter is concerned primarily with stratigraphic processes and procedures associated with glaciers and closely related processes, and reference to Quaternary stratigraphy will only be made when necessary. It is however exceedingly important to remember that glaciation and glacial processes drive so many other activities, such as the behaviour of meltwater rivers, and the rates of dust production and deposition, the presence of lakes and the level of the sea, deformation of the Earth's crust, the extent of permafrost, plants, and animals, and even global atmospheric circulation. Consequently links with Quaternary stratigraphy are intimate, intricate and many, and this chapter must inevitably overlap to some extent with the wider stratigraphic discipline. Indeed it is worth considering glacial stratigraphy as part of an intricate mosaic of earth and ecological systems, all of which are most likely to be in a highly unstable condition both in time and space, and which exist as a series of patterns that are very rarely repeated.

Yet, despite the obvious importance and excitement of glacial stratigraphy, many of the core aims have, in the past, been obscured by unnecessary detail. The result has been that the subject has, in many cases, become an interminable catalogue of till formations associated with glacial lobe advances and retreats, stillstands, periods of ice stagnation, and of sedimentary units caused by lowering and rising sea and lake levels, all bound up in complex network diagrams (Mickelson *et al.*, 1983; Fulton, 1984; Šibrava *et al.*, 1986). These complex diagrams constructed, at times, upon questionable dates or poorly understood glacial sedimentology or geomorphology form the basis of many glacial reconstructions and histories of glacial development. Traditionally these schemes have acquired an immutable status that has been defended at 'all costs', or have been modified reluctantly retaining an outmoded structure. The result is that too often revisions, instead of bringing greater lucidity and understanding, have wrought even more confusion and further mystification (Rose and Schlüchter, 1989).

This chapter will review the principles behind glacial stratigraphy. It will outline the appropriate stratigraphic methods available and current geochronological techniques used for correlation and dating. Emphasis will be placed on the methods and the associated complexities and stratigraphic problems. Examples will be given of stratigraphic schemes at all scales (Bowen, 1991).

8.2 RATIONALE

Stratigraphy attempts to determine the concise chronological sequence of geological events over a wide area, as manifested in periods with a relatively distinctive or characteristic geological property. For glacial stratigraphy these properties typically relate to glaciogenic sediments or landforms. Historically the subject was dominated by morphostratigraphy that emphasised landform assemblage units and their spatial arrangement and diversity (cf. Sissons, 1967, 1976; Lowe and Walker, 1984), but has recently changed towards a more balanced approach based around glacio-sedimentology emphasising lithofacies types and associations, landform/ sediment assemblages and strictly controlled geochronometry (cf. Worsley, 1967, Rose and Allen, 1977; Evenson *et al.*, 1977; Madgett and Catt, 1978; Gibbard, 1980; Dreimanis, 1982; Eyles *et al.*, 1983; Rose, 1985; Eyles and McCabe, 1989).

Stratigraphic relationships are considered within the context of Walther's Law (Middleton, 1973). Essentially, the law states that in a conformable succession of sediments only those lithofacies that exist in a vertical sequence occur adjacent to each other in nature. In other words, spatially contiguous lithofacies environments occurring across the Earth's surface can be found in the same vertical succession as stacked lithofacies units. Individual depositional systems can be, therefore, separated within a stratigraphic package and are bounded by unconformities or sharp facies transitions into adjacent (spatial or temporal) unlike systems (Edwards, 1986; Miall, 1984; Brodzikowski and Van Loon, 1991). Within a terrestrial glaciogenic sequence of sediments facies, transitions and unconformities are very common over short distances and over very short time spans, and glaciotectonic processes may transpose sedimentary units. It is these spatio-temporal variations, perhaps above all else, that complicate glacial stratigraphic interpretations. In glaciomarine sequences many of the problems inherent in land-based glacial stratigraphic differentiation are less acute. However, much of the stratigraphical evidence in the marine environment is circumstantial and rarely provides direct evidence of glacial action (cf. Bowen, 1991).

An additional problem in glacial stratigraphy is the tendency for only peripheral, marginal and distally-linked sediments to record past evidence of glacial sedimentological systems. The preservation potential of glacial sediments tends to be limited, providing spatially fragmentary and disparate stratigraphic sequences at best (Francis, 1981; Rose, 1989a). Much evidence of activity within the global glacial system is recorded in the form of proxy evidence such as isotopic signals in ocean sediments (Shackleton, 1987). There are no single lengthy records on land of direct glacial activity. This is in contrast to ocean records of almost uninterrupted successions of glacial presence on the globe (cf. Sarnthein *et al.*, 1984; Jansen and Sjøholm, 1991) (Fig. 8.1) (Menzies, 1995a, Chapter 2) (see Chapter 7). Towards the centres of past ice sheets the stratigraphic record becomes increasingly limited, although whether this is a function of persistent glaciation or effective glacial erosion is not clear (Hirvas *et al.*, 1988; Garcia Ambrosiani and Robertsson, 1992).

The preservation potential of glacial sediments apparently reaches a maximum in low lying terrains on the peripheries of major ice sheets where possibly erosional activity by ice has been minimised perhaps due to low marginal ice profiles and higher deposi-tional rates associated with highly deformable trans-ported sediment (Rose, 1989a). The other likely preservation sites are zones of minimal erosion close to low velocity ice at ice-shed locations and chance locations where pockets of glacial sediment have been injected into or infill deep depressions such as weathering hollows or joint fissures. Those areas where saprolites have been detected in, for example, the eastern townships of Quebec and upper New Brunswick, or in the north-east of Scotland are likely locations for well-preserved glacial stratigraphies. However even in these locations, the sedimentary units are discontinuous and often difficult to place within a stratigraphic framework.

Where organic remains occur, biostratigraphy may be used to determine the Quaternary history of a region, or geochronometric methods such as radio-carbon or U-series dating may be applied to derive a timescale (see Chapter 14). But neither biostratig-raphy nor the geochronometry may tell anything directly about glacial processes or the associated glacial environment. These methods simply provide a biostratigraphic, geochronometric or chronostrati-graphic framework into which the glacial events may be slotted. Despite the obvious unsatisfactory nature of this approach it is these methods that have, hitherto

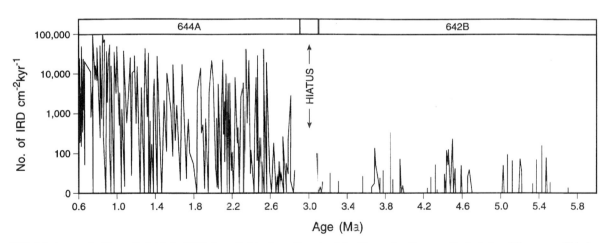

FIG. 8.1. A record of ice-rafted debris (IRD) deposited in the sediments of the Norwegian Sea over the last six million years. This is a continuous lithostratigraphic record of glaciation of Greenland over the period concerned. The results are taken from ocean cores 644A and 642B located respectively at 66° 40.7' N, 4° 34.6' E and 67° 13.5'N, 2° 55.7' E. Note that the y-axis is expressed as a logarithmic scale. The onset of intensive glacial activity around 2.6 Ma is very clear, but it is also apparent that Greenland was glaciated and IRD was discharged into the surrounding oceans in the preceding 3 Ma. (From Jansen and Sjøholm, 1991; reprinted with permission from Macmillan Magazines Limited)

formed the framework for the history of glaciations in regions like Britain (Mitchell *et al.*, 1973; Bowen *et al.*, 1986; Rose, 1989a) and Switzerland (Schlüchter, 1992).

In order that interpretations upon which glacial stratigraphies are based should have some credibility, it is essential that individual glacial processes must be understood. It is also essential that stratigraphic subdivisions and classifications should be based on *observed* properties and not *inferred* genesis. Thus it is the body of sediment, or the shape of the landform that is the critical factor, not the process by which it was formed. However, in reality this is impractical as description alone without a genetic evaluation would be overwhelming and inhibit communication and scientific progress. The result is that an iterative

process exists whereby observation is followed by interpretation which is then followed by a classification and subdivision of the observed evidence based on knowledge of the process. Thus for instance, a succession of diamicton, sand and gravel, and laminated silts followed by diamicton, would be analyzed and interpreted, and, depending upon the outcome of the interpretation, the stratigraphic succession could represent either: 1) one glacial sequence, with the lower diamicton being a till, the upper diamicton being a debris flow derived from the lower till separated by meltwater, and proglacial lake sedimentation (in which the units may be members of one formation), or: 2) two glacial sequences in which the two diamictons represent two glacial episodes separated by non-glacial conditions (in which case the

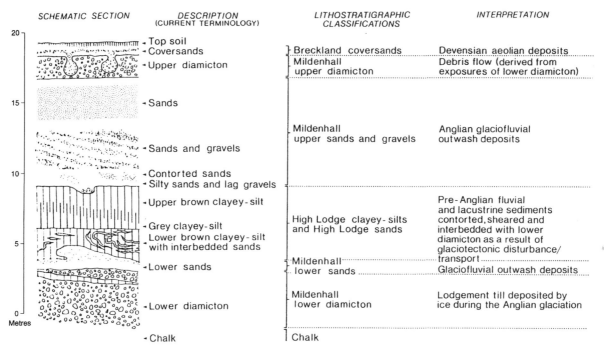

FIG. 8.2. A example of lithostratigraphic subdivision, High Lodge, eastern England. The site includes Middle Pleistocene glacial and non-glacial sediments, and is affected by glaciotectonism. Two formations are represented: The High Lodge Formation is the earlier, and is composed of 'pre-glacial' river sediments with a lithological assemblage determined by the rock within the river catchment. The Mildenhall Formation is the later and represents glacial, glaciofluvial and debris flow sediments with a lithological assemble including material transported to the site by glaciers from regions well beyond the 'pre-glacial' river catchment. Because of glaciotectonism the normal order of super-position has been greatly complicated and a kineto-stratigraphic interpretation is also appropriate with the units below the Mildenhall upper sand and gravel member being glaciotectonised, whereas this unit, along with the Mildenhall upper diamicton member, is unaffected by glaciotectonism, having been formed during and following ice wastage. It should be noted that the lithostratigraphy does not fully coincide with the kineto-stratigraphy. (Reproduced from Lewis, 1992, in Ashton, Cook, Lewis and Rose (Eds), *British Museum Press*)

lower till, sand and gravel and laminated silts would be members of one formation, and the upper diamicton would represent a separate formation) (cf. Ashton *et al.*, 1992; Lewis, 1992) (Fig. 8.2). Even in this simple case judgement is required, and alternative subdivisions may be proposed. This emphasises the critical importance of accurate description and critical understanding of the processes of formation.

8.3 STRATIGRAPHY WITHIN GLACIAL ENVIRONMENTS

The glacial environment is perhaps one of the most complex, dynamic and least well understood sedimentological environments. Glacial events can occur with devastating impact, as in the case of surges or jökulhlaups which may cause profound deformation of underlying rocks and sediments, and transport tonnes of sediments in a matter of a few hours or days. Conversely, other glacial processes such as distal lake or marine sedimentation, or subglacial erosion involve much slower rates and their effects can be enduring. Consequently, within any glacial environment a vast range of sediments, depositional mechanisms and sedimentation rates may be found juxtaposed. Imposed upon glacial sediments are the effects of subaerial non-glacial processes including pedogenesis and diagenesis (see Chapter 10). These processes may modify extensive or very restricted areas of glaciogenic sediments and landforms, over periods of time that range from hours to centuries and millennia. Subaerial processes will occur under differing climatic, vegetational and pedological regimes causing a wide variety of effects that can range from barely perceptible modification to total alteration or removal. The effects of diagenesis on glacial sediments are poorly understood and much research needs to be conducted (Chapter 10). Without adequate knowledge of the extent of erosion, subaerial post-glacial sedimentation, pedogenesis or diagenesis, stratigraphic discrimination can be extremely difficult.

A further problem for glacial stratigraphy is that many glaciated landscapes along the margins, or in lowland central areas of Pleistocene ice sheets have been submerged by either freshwater or seawater over varying periods of time. In some places, such as the

area submerged beneath Glacial Lake Agassiz in North America (Teller and Clayton, 1983), the lowlands around the Gulf of Bothnia in Scandinavia, or the lowlands around Hudson Bay in Canada, sufficiently long periods of inundation have resulted in extensive and thick glaciolacustrine and glaciomarine sedimentation, or shoreline erosion, so that the original evidence of glaciation is obscured or removed.

A persistent problem in understanding glacial stratigraphic relationships are the difficulties of correlating glacial events, both over small distances within the area of one glacial system, or over long distances between one glaciated area and another. Primarily these problems arise because of the heterogeneity of most glaciogenic sediments and the inherent lithological independence of any individual glacial system They also originate because different ice masses, and even different parts of the same ice mass respond to climate at different rates and with dissimilar geographical expressions. Indeed, glacial processes rarely operate with any degree of synchroneity even within temporal boundaries of millennia (Menzies, 1995a, Chapters 2, 3 and 4).

The problems of correlation within a single, or a number of closely linked glacial systems are well illustrated in the Great Lakes area of North America where an extensive and elaborate stratigraphic framework, originally based upon separate till sheet recognition has been developed over the past 50 years (Mickelson *et al.*, 1983; Karrow, 1984a) (Fig. 8.3). Several methods of correlation have been made. In many cases the till sheets were recognised stratigraphically on a 'layer cake' stratigraphic position. It being assumed that the lowest till unit was stratigraphically equivalent over several thousands of km^2, and thenceforth in ascending order. Other tills were correlated on the basis of lithologic composition or the presence of distinctive suites of indicator lithologies that could be traced to particular sources. Thus, in extreme southern Ontario, the Catfish Creek till (Gibbard, 1980) was recognised on the basis of its stratigraphical position immediately overlying bedrock, its typically coarse grained content and yellow coloration. However, as the complexity and vicissitudes of till depositional mechanics become better known, it has became understood that 'counting-up'

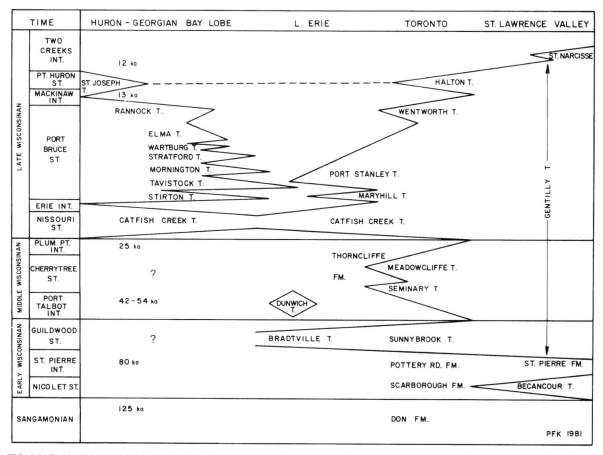

FIG. 8.3. Glacial lithostratigraphic units (T, tills) and ice fluctuations in the region of the southern Laurentide ice sheet from Lake Huron to the St. Lawrence valley for the Last Interglacial (Sangamonian) and Last Glaciation (Wisconsinan). Note that the timescale is variable, and in radiocarbon years where possible. Note also the diachoneity of the lithostratigraphic boundaries based on the presence of tills. These are classical examples of diachronic units. (Reproduced from Karrow, 1984)

or simple lithostratigraphical methods of stratigraphic correlation are inherently weak. It is now understood that many of the correlations are invalid and the fluctuations of the many ice lobes at the southern margin of the Laurentide ice sheet are essentially asynchronous.

Similarly, in the western part of the Midland Valley of Scotland, till colour has been used to suggest that two different glacial episodes are represented stratigraphically (cf. Menzies, 1981). A similar method, but including till provenance indicators and particle size distribution was used in separating three tills suggesting three glacial advances near Arthur, Ontario (Cowan et al., 1978) (Fig. 8.4). None of the above

criteria can be used with any confidence unless there is a clear scientific rationale for the development of different glacial lithostratigraphic units, such as a reason for the introduction of new lithologies into separate till units. This situation may occur if an event such as a marine transgression occurs within a region and results in the deposition of lithologically distinctive sediments in the interval between two glaciations. In this case the upper till unit could incorporate these marine sediments, the lower could not.

A marine transgression occurred during the interstadial before the Younger Dryas (c. 13,500–10,500 ^{14}C yrs BP) in the Loch Lomond basin, western central Scotland. Here the till units of the Dimlington

A

B

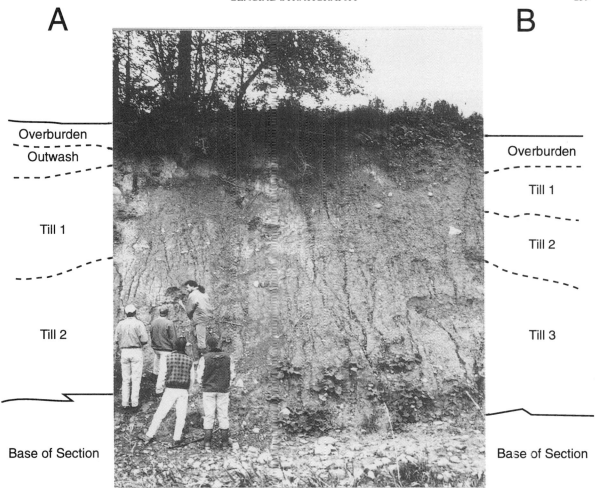

Overburden

Outwash

Till 1

Till 2

Base of Section

Overburden

Till 1

Till 2

Till 3

Base of Section

FIG. 8.4. Diamicton facies and associated sediments in the area around Arthur, Ontario. (Adapted from Cowan *et al.*, 1978)

glaciation and the Loch Lomond glaciation can be clearly defined and distinguished on lithostratigraphic criteria (Rose, 1989a) (Fig. 8.5, Table 8.4b). Similarly, other independent corroborating evidence such as radiometric dates, or luminescence ages on glacioaqueous sediments (Hutt and Jungner, 1992), may give confidence to stratigraphic interpretation (cf. Karrow, 1992) (see Chapter 14).

The problem of depositional hiatus is of critical importance. Hiatuses are not restricted to glacial environments but are particularly common within them. For example, within the subglacial terrestrial environment, periods of limited deposition, periods of equal deposition and erosion, or periods of net erosion

may result in limited to zero sedimentation for varying periods of time over differing areas of the glacier bed. The recognition of localised or widespread depositional hiatuses remains difficult to substantiate without some surface tool marks, a distinct weathering rind, a shallow ground surface imprint, or a paleosol immediately below the suspected hiatus (see Chapter 10).

Sediment reworking and re-sedimentation is a difficult and common problem of glacial stratigraphy. The term 're-sedimentation' tries to convey the concept of sediment being eroded from a primary depositional site, transported some varying distance from this initial source and again deposited (cf.

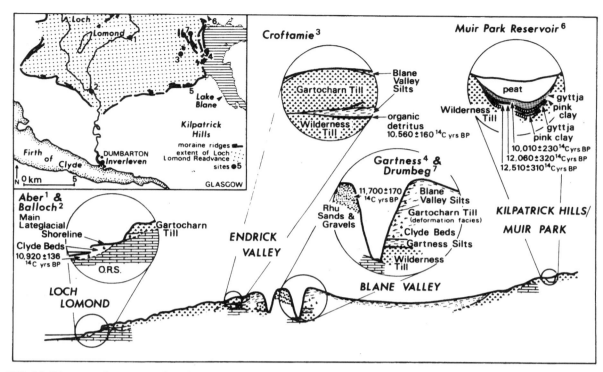

FIG. 8.5. Diagrammatic representation of the stratigraphic evidence for the Loch Lomond glaciation in the southern part of the Loch Lomond basin. The inset map shows the location of the lithostratigraphic sites and the position of the Loch Lomond Readvance moraine which is a morphostratigraphic unit. The stippled area shows the extent of ice cover at the maximum of the glacial event and the horizontal lines show the extent of the proglacial Lake Blane that formed at the same time. (Reproduced from Rose *et al.*, 1988)

Ashton *et al.*, 1992; Lewis, 1992) (Fig. 8.2). The distinction between primary and secondary deposition remains vague but has been used to attempt stratigraphic division between different stages of deposition (Boulton and Deynoux, 1981), or to reduce the number of stages of deposition when individual diamicton units have been traditionally considered evidence for separate glacial events. The concept of re-sedimentation is recognised as a common ongoing process in glacial environments that further confounds stratigraphic segregation and order.

It is a logical consequence of the above problems that even at a scale of a few kilometres, sediments from the same ice mass may defy stratigraphic correlation unless there is continuity between sites. Processes of cut-and-fill, sediment reworking and glaciotectonic deformation all create highly complex stratigraphic sequences and arrangements. The result is that the relatively simple stratigraphic procedures

of non-glacial stratigraphy are not applicable and stratigraphic correlation based upon single units must be approached with caution (Karrow, 1984b). In these circumstances correlation should be based upon lithofacies units or, more typically, lithofacies associations indicative of specific sedimentary environmental conditions at any given time.

The problem of correlation between different glacier systems on different parts of the globe is equally complex, but for different reasons. An example of this problem is the attempts to correlate the glacial response to the Younger Dryas climatic deterioration at the Last Glacial/present Interglacial transition (Bard and Broecker, 1992; Peteet, 1993; Lowe *et al.*, 1994). In north-western Europe the evidence for glacier expansion at this time is well represented by glaciogenic sediments in the region close to the North Atlantic, whereas in more continental parts of Europe evidence is mainly found in the form of proxy evidence such as lithological variations or oxygen isotope signals in lake sediments. In North America, evidence for glacier expansion is very variable. This range of evidence reflects the scale of climatic change at that time and the proximity of the region concerned to the energy source that amplified the change. In the example cited the main driving force for climatic cooling was the North Atlantic, so that the glacial responses were greatest where the heat-source was cut-off but a moisture source remained (Menzies, 1995a, Chapter 2). In all cases however, dating and correlation of the Younger Dryas glacial episode, some 11,000–10,000 ^{14}C yrs ago, is not derived directly from the glacial evidence, but is derived from non-glacial evidence, located above and below the glacial sediments, that is suitable for radiometric dating or biostratigraphic correlation (Rose et al., 1988) (Fig. 8.5). In other words it is Quaternary stratigraphic methods rather than glacial stratigraphy that provides the basis of correlation and dating. This illustration provides possibly the best stratigraphic case for correlating glaciation globally. For earlier glacial episodes the glaciogenic evidence is often badly altered by subaerial erosion or pedogenesis and dating methods have either poorer resolution or are not available. For later (Neoglacial and Little Ice Age) glacial episodes the climatic driving force was less powerful and the effects of glaciation are less extensive, and often complicated by a multitude of glacial events that reached similar limits.

It is evident that in attempting to develop glacial stratigraphies there are many major problems that have no simple solution. It is apparent that only with a clearer understanding of glacial processes and environments and resulting lithofacies units and associations, will it be possible to set up reasonable stratigraphic approximations. These schemes may subsequently be altered, revised or scrapped if new information appears.

8.4 STRATIGRAPHIC NOMENCLATURE

A problem that distinguishes glacial sediments from other sediments is their geologically short-time span and their enormous lithologic variability. Fossil assemblages are virtually absent from the glacial environment and biostratigraphy is not part of glacial stratigraphy per se., although the presence of organic remains in adjacent non-glacial or proglacial sediments may provide proxy evidence of glacial conditions (Penny et al., 1969), and is generally the basis of a chronostratigraphic structure (Mitchell, et al., 1973) (Figs. 8.3 and 8.5). Landform stratigraphy may play a very important rôle when subaerial modification of the glacial terrain is minimal. Likewise, soil stratigraphy may provide a range of evidence for reconstructing past environments, and correlation of glacial episodes. In addition to the above stratigraphic methods, correlation is also provided by a range of geochronometric techniques (see Chapter 14). Only a few, such as luminescence dating of subaerially and subaqueously deposited sediments provide direct dating, and many other methods such as weathering rinds, Schmidt hammer impact measurements, surface texture roughness, lichenometry, biomass quantity, and species abundance provide correlation through an estimate of the time since ice melted from a glaciated surface. Widely used dating methods such as radiocarbon, and amino acid geochronology may provide maximum and minimum ages for glaciation based on dates from underlying, entrained and overlying materials.

The highly complex nature of glacial stratigraphy has meant that the use of the International Stratigraphic Guide (Salvador, 1994) and the North American Stratigraphic Code (1983) (henceforth NASC) are of restricted value. Rather the Guide and Code provide only the conceptual basis of stratigraphy and stratigraphic methods and, in the main, a review of conventional stratigraphic methods appropriate to simpler geological environments (cf. Harland, 1992). The result has been that Quaternary stratigraphy has adopted, as a matter of course, a much wider range of stratigraphic methods and a much more refined, rigorous and sensitive stratigraphic scheme (e.g., Morrison, 1965; Luttig et al., 1969; Mitchell et al., 1973; Bowen, 1978; Šibrava et al., 1986; Fulton, 1989). A scheme that deals in much more precise timescales with higher levels of stratigraphic resolution than is possible in any other part of the geological column Likewise, the need to deal with frequent hiatuses in sedimentary sequences has meant that much thought has had to be given to stratigraphic judgements rather than adopting a casual approach such as is commonplace in evaluation of other, simpler stratigraphic systems. Hitherto glacial stratigraphy has avoided the constraints of committee bureaucracy similar to that implied in the publication of the International Commission on Stratigraphy (Cowie *et al.*, 1986; Rose 1989a).

Lithostratigraphic units form the critical building blocks of glacial stratigraphy. Where appropriate these units are combined with morphostratigraphic units. These are the main sources of evidence for deriving glacial stratigraphic schemes. These units may be supported by proxy evidence of glaciation which can be provided by chemical stratigraphic units as indicators of enhanced (glacial) erosion and sedimentation, and isotopic stratigraphic units as indicators of ice volume. Similarly some biostratigraphic units (especially based on insect content) may provide proxy evidence for glacial conditions (Penny *et al.*, 1969). A wide range of other lithostratigraphic evidence derived from loess, marine, and lake sediments may also act as proxy indicators of glacial activity.

These fundamental stratigraphic units have been used in glacial stratigraphy to provide evidence for glacial activity in the form of glacial advance, standstills, and retreats and essentially this, along with dating, is the critical aim of glacial stratigraphy. However, it is at this stage that glacial stratigraphy becomes intimately associated with Quaternary stratigraphy which is based essentially on *climatostratigraphy* defined in terms of glacials and interglacials. In Quaternary stratigraphy these climatostratigraphic units form the main elements of the chronostratigraphic subdivision (Stages), and are used irrespective of whether the evidence suggests glacier cover, or simply a significant change of climate without any evidence for the presence or absence of ice. The use of geochronometry to place ages on Quaternary events provides the basis of a timescale by which events may be dated (see Chapter 14) and by which it is possible to determine the rates at which processes operate.

For the purpose of glacial stratigraphy, a *glacial stage* is recognised as a major expansion of glaciers of long duration, and glacials are divided into stadials and interstadials. *Stadials* are defined as a limited

expansion of glaciers or a subdivision of a glacial characterised by a relative deterioration of climate, and *interstadials* are defined as a period of relative improvement of climate but of lesser duration or lesser vegetational development than the present interglacial (Holocene) (Rose, 1989a p. 47). Glacials are separated by *interglacials* which have been defined in terms of the vegetational development (cf. West, 1977; Karrow, 1989; Klassen, 1989). By this definition it is required that the vegetation of an interglacial should be equivalent to that which developed in a region at the thermal optimum of the present (Holocene) interglacial. These terms may apply in areas that are not glacierized and may never have been glacierized. These climatostratigraphic terms are the equivalent of 'event stratigraphy' used in other parts of the geological column (Salvador, 1994 p 117).

Altogether these terms create considerable confusion. First, although the terms are defined in terms of climate, some of the elements, such as interglacials, are based on climatic proxy such as vegetational development represented by pollen assemblage biozones that may be unreliable indicators of climate. For instance, the vegetation will not only reflect climate but the competition between the available species, soil productivity and time needed for colonisation of different species. As an example it is known that summer temperatures in Britain about 13,000 ^{14}C years ago were warmer than now, but Britain was virtually treeless at the time because trees had not yet reached Britain from the refuges in which they survived the very harsh climate of the Last Glacial Maximum (Walker *et al.*, 1993). Secondly, some of these terms imply duration, which is of course not climatic. By definition, a stadial is shorter than a glacial of which it is part, and it is expected that an interstadial will be shorter than an interglacial (Lowe and Walker, 1984, p. 8). However, this is clearly not the case with the Upton Warren Interstadial Complex in Britain which covers a period from about 45,000–26,000 ^{14}C years BP, roughly twice as long as the present interglacial. Indeed any interstadial recognised for the interval equivalent to Oxygen Isotope (OI) Stage 3 is likely to be longer than Interglacial OI stage 5e. Thirdly, climatostratigraphic terms are used as the basis of chronostratigraphy (stages) yet, as they are climatically defined it is highly improbable that the evidence for boundaries upon which these stages are defined represents quite different periods of time and is asynchronous (Lowe and Grey 1980) (Fig. 8.3). For example, the sediments representing the beginning of the Last Glacial Maximum (the Dimlington Stadial, or Dimlington Chronozone – a sub-stage of the British Devensian Stage) is considered to have begun deposition about 26,000 14C years ago (Rose, 1985). Clearly in areas where glaciation persisted throughout the Last Glaciation the evidence for the beginning of this climatic subdivision is likely to be much earlier if it can be distinguished, whereas in areas which were not so sensitive to climate deterioration, such as south-east France, the evidence for climate cooling would be later (Serat *et al.*, 1990; Pons *et al.*, 1992). Although a chronostratigraphic

boundary can be defined at the stratotype (Figs. 8.3 and 8.6), the likelihood of obtaining climatostratigraphic evidence from different climatic or geomorphic provinces to represent a synchronous boundary is very difficult. This is a problem for all chronostratigraphy that is not controlled by geochronometry, especially that which is defined by pre-Quaternary biostratigraphy. However, the level of resolution at these earlier periods is poor and the problem is less acute. In Quaternary glacial stratigraphy a very high resolution is expected and this problem may have serious ramifications for understanding glacial activity.

These problems were discussed in Watson and Wright (1980) and it is for the above reason that the NASC (1983) introduced the concept of diachronic units. A *diachronic unit* is defined as 'a unit which

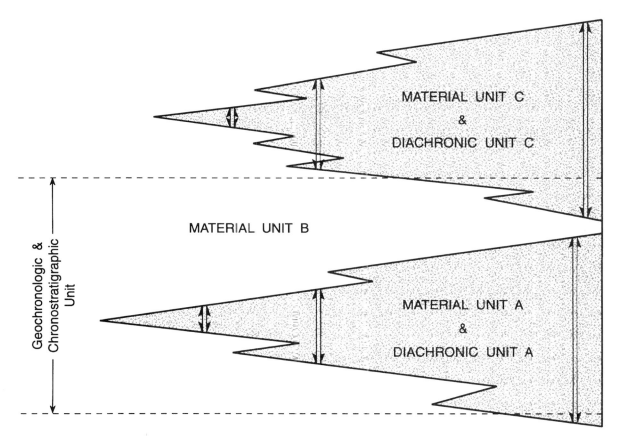

FIG. 8.6. Schematic comparison of diachronic units with geochronological and chronostratigraphic units. (Reproduced from North American Stratigraphic Code, 1983)

comprises unequal spans of time represented either by a specific lithostratigraphic ... or pedostratigraphic unit, or by an assemblage of such units' NASC, 1983, p. 870). It is clear that these units are particularly appropriate to glacial stratigraphy. The purposes of diachronic units are clearly set out in the NASC (1983 p. 870–871). 'Diachronic classification provides: i) a means of comparing the spans of time represented by stratigraphic units with diachronic boundaries at different localities; ii) a basis for broadly establishing in time the beginning and ending of deposition of diachronous stratigraphic units at different sites; iii) a basis for inferring the rate of change in extent of depositional processes; iv) a means of determining and comparing rates and durations of deposition at different localities, and v) a means of comparing temporal and spatial relationships of diachronous stratigraphic units.' The boundaries of the diachronic units are the times recorded by the beginning and end of deposition of the material evidence, and one or both of the boundaries may be time-transgressive.

'A *diachron* is the fundamental ... unit and if a hierarchy of diachronic units is needed the terms episode, phase, span and cline, in order of decreasing rank, are recommended. The rank of the unit is determined by the scope of the unit ... and not by the timespan represented by the unit at a particular place' ... 'An *episode* is the unit of highest rank and greatest scope in hierarchical classification. If the 'Wisconsinan Age' were to be redefined as a diachronic unit, it would have the rank of episode' (NASC, 1983, p. 870–871).

The concept and practical exercise of establishing stratotypes has been introduced in order that the attribute of stratigraphic units can be defined unambiguously, and in order to facilitate communication between sites and between scientists. Stratotypes only apply to stratigraphic units that have been named formally (Salvador, 1994 p. 14), and a *stratotype* is defined as 'a specific interval or point in a specific sequence of rock strata and constitutes the standard for the definition and characterization of the stratigraphic unit or boundary being defined' (Salvador,

1994 p. 26). There are many varieties of stratotype (Salvador, 1994 p. 28) defined as follows:

holostratotype – the original stratotype;

parastratotype – supplementary stratotype supplying additional information, especially about diversity of the unit;

lectostratotype – used in the absence of an adequately designated holostratotype;

neostratotype – a replacement for the original holostratotype which has been destroyed, covered or otherwise made inaccessible;

hypostratotype – an additional stratotype used to extend knowledge of the holostratotype but it is always subsidiary to the holostratotype.

8.5 GLACIAL STRATIGRAPHIC PROCEDURES AND METHODS

8.5.1 Lithostratigraphy

In glacial geology lithostratigraphy provides the fundamental approach to any stratigraphic framework. It includes within it a number of stratigraphic variants that are concerned with other aspects of sediment lithology, but have become sufficiently distinctive or important to be known by a separate term. Included amongst these of direct or indirect relevance to glacial stratigraphy are kineto-stratigraphy, chemistratigraphy, isotope stratigraphy, and magnetostratigraphy.

Lithostratigraphy 'deals with the description and systematic organisation of ... rocks .... into distinctive named units based on the lithological character of the rocks and their stratigraphic relations' (Salvador, 1994, p. 31). A lithostratigraphic unit is a 'body of rocks that is defined and recognised on the basis of its observable and distinctive lithological properties, or combination of properties and its stratigraphic relations' (Salvador, 1994, p. 31–2). A lithostratigraphic unit should be capable of being mapped and typically be a tabular entity. However, this is often difficult, but certainly not impossible, with glacial sediments because they are often of indeterminate lateral and vertical extent and may suffer intense deformation

TABLE 8.1. Lithostratigraphic terminology and facies units

Group	Two or more formations	Facies group
Formation	Primary unit of lithostratigraphy	Facies association
Member	Distinctive lithologic entity within a formation	Facies unit
Bed	Distinctive lithologic layer within a member	Subfacies unit

and intercalation (Bowen, 1978). Lithostratigraphic units are organised within a hierarchical system where the formation is the primary unit (Salvador. 1994, p. 33), and a group is a number of formations. Formations include members, which are composed of beds. These are outlined in Table 8.1 and reference is given to the appropriate lithofacies unit.

A formation or facies association is far from easy to define, but should form a body of sediment or sediment facies that has distinctive lithological properties and represents a distinctive sedimentological environment that can be segregated from others on the basis of all or a combination of the following observable properties: particle size, particle shape and surface texture, sedimentary structures, bedding characteristics, geotechnical properties, lithology and mineralogy. Major unconformities would not be expected to occur within a single formation but smaller unconformities and disconformities are typical.

Identification and designation of formations in glacial sediments is difficult because of rapid and complex lateral and vertical gradation. For example, a diamicton association may interfinger with a sand and gravel association. Separately, these associations may have sufficient importance in distinctiveness, thickness, extent, and geomorphological significance to justify the designation as formations, but because of the interdigitation, it may be difficult to separate them in the field or on maps and it is apparent that they were associated during deposition. In this case is it more sensible to bring the two facies associations together and call them a single formation. This is the case with the Lowestoft Formation in eastern England which represents a lithologically distinctive set of facies associations formed by glaciers and meltwater during the Anglian glaciation (Rose and Allen, 1977) (Plate 8.1). In this case the Lowestoft Formation

includes tills, outwash sands and gravels, and proglacial lake sediments (Allen et al., 1991).

A member is a 'part of a formation. It is recognised ... because it possesses lithologic properties distinguishing it from adjacent parts of the formation. No fixed standard is required for the extent or thickness of a member' (Salvador, 1994, p. 34). Because of the highly complex nature of the glacial sedimentary environment, in comparison with the relatively simple sedimentary environments associated with other processes, members or facies units have become the most crucial elements in understanding glacial sedimentological environments and thus the development of a stratigraphic framework. It is at this level of stratigraphic nomenclature and decision making that most problems occur. If a member is mistakenly identified then any particular formation within which the

PLATE 8.1. Example of glacial facies association in the Lowestoft Formation of eastern England. This sections shows examples of different till and sand and gravel facies determined by glacial transport path, transport position within, on, or below the ice and depostional process. The site is from the Chelmsford area of East Anglia

member resides is potentially mis-identified and an inaccurate stratigraphic framework may be erected.

The recognition of individual members or facies units may not be an easy task. There are no diagnostic characteristics that can specify whether a specific unit should be designated a member. Likewise, the status of a facies unit may change as new exposures become available, and the extent of particular lithological elements changes, or new relationships are found between the initial unit and newly discovered units. Consequently it is critical that all characteristic features, parameters and properties of a unit should be identified and tabulated. Thus, even if the initial interpretation is proven to be mistaken, the data can be later re-interpreted. In order to achieve this, a set of objective descriptive methods have been adopted (cf. Rust, 1978; Collinson and Thompson, 1982; Tucker, 1986; Eyles et al., 1983; Eyles and Miall, 1984). In addition, no single section or site should ever be designated part of a stratigraphic framework unless all other contextual information (other sites, landforms) is mapped and appraised. Despite this, an evaluation of the stratigraphic status and process of formation of members requires individual judgement and this may lead to controversy and discussion, although such debate inevitably strengthens any final decision (Shaw, 1987).

An example of this problem can be seen from studies of the large sections of the Scarborough Bluffs on the north shore of Lake Ontario, Canada. Here, the facies units have been described at different levels and explained by several widely different interpretations, with the result that are number of different members and formations have been proposed (Karrow, 1967, 1974; Eyles et al., 1983; Dreimanis, 1984a). Similarly, the facies units revealed along the East Anglian coast of Norfolk, England have led to long and detailed debates over a period of more than 100 years and a succession of different processes of formation have been suggested (cf. Reid, 1882; Ransom, 1967; Banham, 1975; Perrin et al., 1979; Boulton et al., 1984; Lunkka, 1988; Eyles et al., 1989; Hart and Boulton, 1991). However, in this later case, although members have been re-interpreted on several occasions according to process of formation, description has been of a relatively high quality and the main observed properties have been sufficiently distinctive

that the deposits continue to be allocated to the units that are known as the Lowestoft Formation (as described above) and the North Sea Drift Formation (Hart and Boulton, 1991). In Chapter 5, Elverhøi and Heinrich discuss similar interpretative problems with reference to the Precambrian rocks of the Port Askaig Formation in western Scotland.

Figure 8.7 illustrates the types of problems that assail the stratigrapher in the field. If only Core I is examined a conclusion may be reached which is different from that which would be reached when Cores I, II and III are examined together. Core I shows a section immediately above bedrock in which a dense clay-rich diamicton unit, exhibiting structures indicative of massive deformation, is conformably overlain by a dense clay-rich matrix-supported diamicton containing a well-oriented clast fabric; that is in turn conformably overlain by a coarser grained diamicton containing evidence of occasional dropstones and structures associated with sediment rain-out within a subaqueous environment. Core II, 78 m east of Core I, has a similar vertical profile except in the upper unit where a conformable but interfingering sandy unit of fine to medium massively bedded sands with occasional diamicton balls is seen to exist. Core III, 320 m east of Core II, reveals the presence of a deformed diamicton and matrix-supported diamicton overlain by well-stratified fine to medium sands that have evidence of being foreset and bottom-set beds of deltaic origin.

If not taken in context such a sequence might be interpreted in at least two quite different ways: (1) Core I alone might have been seen as evidence of land-based ice sheet with subglacial sediments replaced at the top of the section by conformable sediments deposited in a localised small pond or lake possibly itself existing in a subglacial cavern. (2) If the cores are taken together the evidence would suggest ice marginal subaqueous sedimentation occurring in a large proglacial lake. In this case the deformed diamicton is interpreted as being extruded into the lake or deposited just up-ice of a grounding line prior to lake water incursion. In either case the diamicton would be conformable with a debris flow deposit which emanated from the grounding line into a body of water, and that this debris flow was, in turn, buried by typical waterlain diamicton sediments. Such

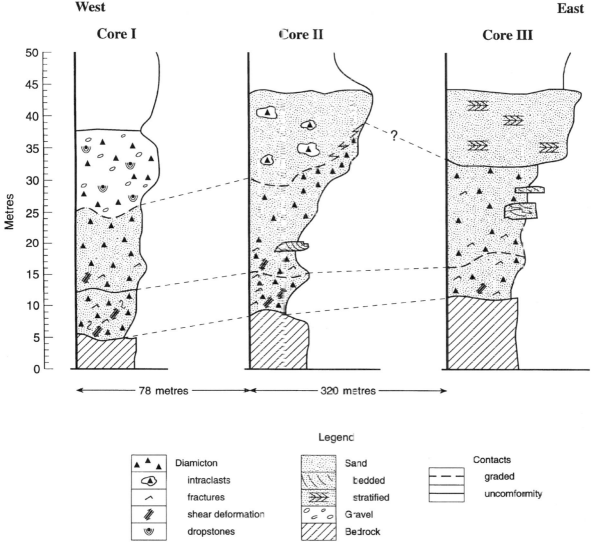

West   East

Core I   **Core II**   **Core III**

FIG. 8.7. Example of lithofacies and lithostratigraphic interpretation. See text for interpretation

a scenario would be confirmed by the presence of deltaic deposits in the upper part of Cores II and III. All the main units would have sufficient status to be defined as members, although representing quite different processes, but in either case all the members would be part of a single formation.

A bed or subfacies unit is the smallest distinct lithostratigraphic unit. 'It is a unit layer in a stratified sequence of rocks which is lithologically distinguishable from other layers above and below from which it is separated by more or less well defined bedding surfaces' (Salvador, 1994, p. 34). Usually a bed is indicative of a transient sedimentological environment. Intercalated units, tongues and lenses of sediment may be included, whereas rafts and intraclasts are exotics not indicative of the particular *in situ* sedimentological environment. Where distinctive visible changes occur in a member unit, each separate sub-unit can be defined as a bed, but only if a sedimentological basis exists for such segregation.

Beds are exceptionally common in glacial environments reflecting the complexity of this system. They may vary from a multitude of units in a sand and gravel member formed by a highly variable ice-proximal braided river system, to individual beds of laminated silt and clay, or sands enclosed in massive diamictons formed by glaciogenic or debris flow processes.

The term Group is applied to a sequence of two or more contiguous or associated formations with significant and diagnostic lithological properties in common. Hitherto, the term Group has not been widely used in glacial stratigraphy. However, as the need to define specific glacial environments increases it is likely that the value of such a designation will increase. However, due to the fragmentary nature of glacial sedimentological processes the value of a group designation beyond local and regional levels will be limited. The term has been applied in southern England to the Kesgrave Group (Whiteman and Rose, 1992), which is a body of outwash sands and gravels deposited in the catchment of the ancestral river

Thames over a period of about 1.2 Ma through much of the Early and Middle Pleistocene. Previously, the Kesgrave sands and gravels had been given formation status (Rose and Allen, 1977), but additional work demonstrated the greater extent of the deposit, the greater complexity of the facies units, the presence of two distinctive facies associations, and the fact that these facies associations were deposited during nearly 50 per cent of Quaternary time. The lithostratigraphy of the Kesgrave sands and gravels is now based on the Sudbury Formation and Colchester Formation which represent the two distinctive facies associations.

8.5.2 Variants of lithostratigraphy relevant to glacial stratigraphy

8.5.2.1 Kineto-stratigraphy

Kineto-stratigraphy stems from the work of Berthelsen (1973, 1978) on the glaciotectonically deformed sediments of Denmark. This stratigraphic approach is applied where conventional lithostrati-

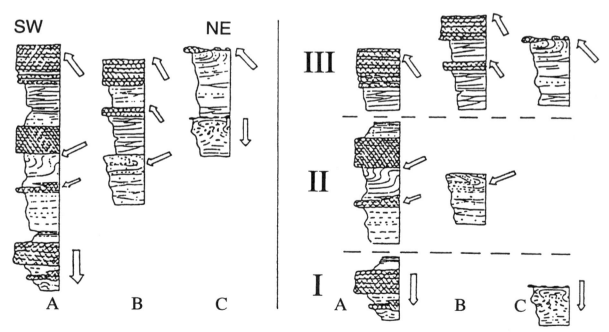

FIG. 8.8. Example of kineto-stratigraphic classification. At the left-hand side of the figure the stratigraphic successions are from three localities A,B, and C arranged along a SW-NE line. Directional elements derived from glaciotectonic deformation are indicated with the north arrow pointing upwards. The right-hand side of the figure shows the same deposits grouped into kineto-stratigraphic units I, II, and III. (Reproduced from Berthelsen, 1978)

graphic methods fail due to intense sediment defor-mation, translation and penetrative movement into adjacent sediment due to active and subsequent passive stress application (Hirvas et al., 1976; Ban-ham, 1977; Aber et al., 1989; van Wateren, 1992; Warren and Croot, 1994). Kineto-stratigraphy attempts to differentiate sediments into units that have similar deformational characteristics and in the case of glacial stratigraphy these reflect glacial deforma-tional histories. The particular attributes that con-stitute kineto-stratigraphy are the forms of glacially-induced geometric and directional indicators (Fig. 8.8). This method is especially appropriate when overthrust, folded and faulted sediments occur such that the Law of Superposition does not apply (cf. van Wateren, 1992; Lewis, 1992) (Fig. 8.2). In many cases subjacent pre-existing sediment has been so intensely altered that structurally the underlying sediment body becomes mixed with the overlying deformed sedi-ment creating a single kineto-stratigraphic unit out of what were once two distinct lithofacies units (Ras-mussen, 1975) (Fig. 8.9).

Sediment deformation can occur during the process of deposition (syngenetic), immediately following deposition but within the same lithofacies environ-ment (cogenetic), or some time after deposition and unrelated to the lithofacies environment that is generating the deformation process (epigenetic) (Menzies, 1995a, Chapter 10). Syngenetic deforma-tion can be viewed as autokinetic, that is deformation occurs within the stratigraphic unit due to internal stress conditions whereas cogenetic and epigenetic are allokinetic deformation processes related to exter-nally applied stress conditions (Menzies and Suppiah, in press). A final form of deformation occurs when various forms of deformed sediment are re-deformed (compound deformation) and this can be built into kineto-stratigraphic schemes provided the individual deformational events have each produced structures that can be discriminated in the geological analysis (cf. Ruszczynska-Szenajch, 1976, 1987; Menzies, 1990b).

Kineto-stratigraphy provides a means by which it is possible to piece together various elements of a

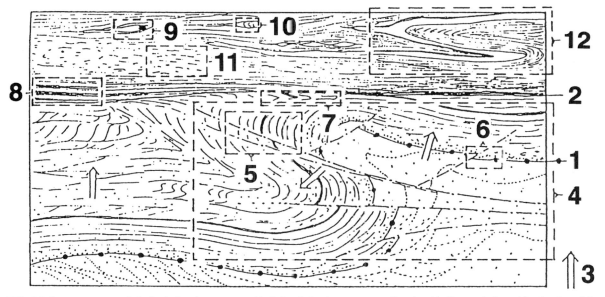

FIG. 8.9. Important types of glacier-induced structures with their relative importance as directional indicators indicated by the size of the dashed frame. 1, Boundary between an upper kineto-stratigraphic unit with domainal deformation and subjacent strata with extra-domainal deformation; 2 base of lodgement till, 3 way-up in glaciofluvial sediments; 4 field of combined glaciotectonic studies; 5 overthrusts – note how they converge in the direction opposite to the ice movement; 6 conjugate thrusts with opposite sense of movement; 7 sub-sole drag – shearing is intense in this zone; 8 striations in mylonitic till; 9 torpedo structure; 10 intrafolial folds of cm size; 11 site of till fabric analysis; 12 glaciotectonic structures of m size. Other structures of less directional value are also shown, but not framed, for instance boudinage structures. (Reproduced from Berthelsen, 1978)

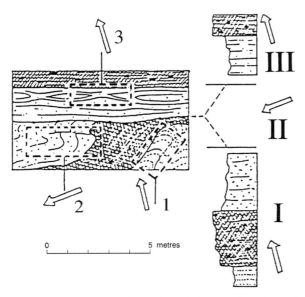

FIG. 8.10. Diagrammatic illustration of the use of an 'empty' locality. In the profile on the left there are deposits from only two kineto-stratigraphic units (below and above the unconformity), but due to the extra-domainal deformation (2) related to the 'missing' unit, the previous existence of the latter can be inferred as unit II on the right-hand profile. (Reproduced from Berthelsen, 1978)

typically fragmented glacial sedimentary record, and identify, or accommodate depositional hiatuses of incalculable time periods (Berthelsen's 'empty' locality) thus permitting a composite stratigraphic record to be developed (Fig. 8.10). Where an epigenetic imprint on a subjacent sediment has occurred followed by a depositional hiatus, it is possible that elsewhere a sediment package with the same kinetic indicators can be found that can be then 'added' to form a synthesised stratigraphy. This method is far more exact and intellectually demanding than that usually required by the procedures of International Stratigraphic Guide and provides information that is critical to the understanding of the glacial stratigraphic record.

8.5.2.2 Isotope stratigraphy

Isotope stratigraphy, as relevant to glacial stratigraphy, is derived primarily from the study of the oxygen isotope signal stored in marine organisms (foraminifera) in ocean sediments. The shells of these microfossils preserve a record of the relative abundance of oxygen isotopes 18 and 16 in sea water, which have a ratio that is determined primarily by the volume of ice on the globe (Shackleton and Opdyke, 1973; Shackleton, 1987). This is due to the fact that the light isotope of oxygen (^{16}O) is evaporated preferentially from the ocean and is stored in precipitation that contributes to the formation of glaciers and ice sheets. Thus relatively high δ^{18}O values indicate a relative abundance of ice cover on the globe, whereas relatively low δ^{18}O values indicate conditions similar to the present, with even lower values if the Greenland and Antarctic ice sheets were to disappear. These figures only apply to the global ice volume values, and tell nothing about particular ice sheets. Also the relationship between the δ^{18}O value and ice volume is not constant as a growing ice sheet is less isotopically light than a steady state ice sheet (Mix and Ruddiman, 1984). Nevertheless variations in the δ^{18}O values give an estimate of ice cover on the globe through time, and because these records are collected from long cores from continuous ocean bottom sediments they provide a record of changes of ice cover over time and are therefore an excellent, if general, proxy of the history of glaciation.

A large number of ocean cores have now been analyzed and the results have been published for large parts and even the whole of the Quaternary and earlier and provide an indication of the relative volume of ice at different times (Fig. 8.11a). These results show the inherent correlation between ice volume and orbital forcing parameters (Imbrie and Imbrie, 1979) calibrated at 65°N. Traditionally, ocean cores were dated by palaeomagnetism with particular reference to the Brunhes/Matuyama and Matuyama/Gauss magnetic reversals, assuming a constant rate of deposition between these chronological markers. Initially this provided a reasonable chronology for the record of glaciation, and it was within this framework that the oxygen isotope record was initially subdivided stratigraphically (Shackleton and Opdyke, 1973).

Stratigraphic subdivision of the oxygen isotope record has been based on climatostratigraphic principles, with climatostratigraphic units defined by numbers from the most recent downwards. Each climatostratigraphic unit was in turn equated with a chronostratigraphic unit and defined as an oxygen

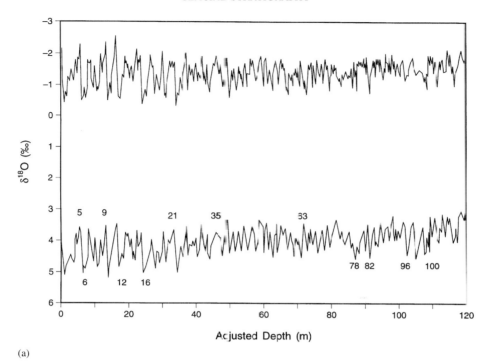

(a)

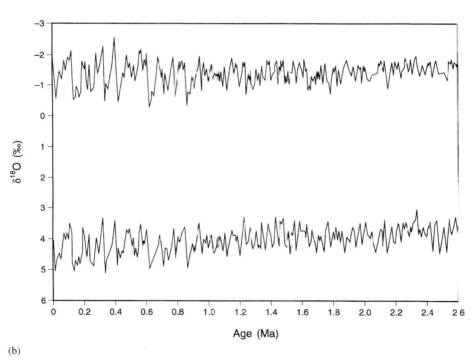

(b)

FIG. 8.11. (a) Oxygen isotope stratigraphy derived from ODP cores 677A and B down to 120 m, based on the planktonic (above) and benthonic (below) foraminifera. Particular OI stages are labelled. (b) As above. but tuned to the orbital timescale of Shackleton et al., (1990).

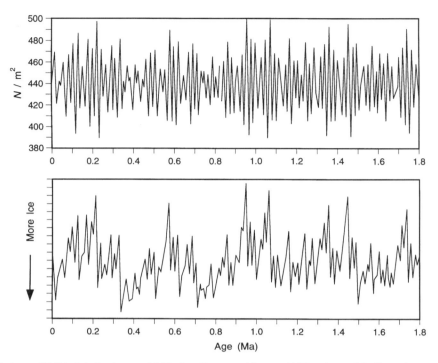

FIG 8.11. (c) Insolation at 65° N in July for the past 1.8 Ma from Berger and Loutre (1988) and a model of ice volume run using the model of Imbrie and Imbrie (1980). (Reproduced from Shackleton *et al.*, 1990)

isotope (OI) Stage. Warm episodes, similar to or almost as warm as the present, were given odd numbers with the present interglacial being designated OI Stage 1. This has been applied consistently, so that all interglacials have an odd number, except for the main interstadial of the Last Glaciation which is designated OI Stage 3. Cold episodes, equivalent to a glaciation were given even numbers, except for the Last Glaciation which comprises OI Stages 2 and 4. The climatic deterioration of the Last Glacial Maximum is designated OI Stage 2. Using this scheme, stages can be subdivided to represent climatic oscillations and sub-stages are defined by a lower case letter. Thus, the Last Interglacial as represented in the oceans, is subdivided into Sub-stages 5a, 5c and 5e which are periods of relative warmth, and Sub-stages 5b and 5d which are periods with relative glacier expansion (Shackleton, 1987; Bowen, 1994) (Fig. 8.12).

With additional samples and the replication of evidence, confidence in the meaning and quality of

the results has increased. The isotopic variations have been analyzed mathematically in terms of the probable timing of orbitally forced climatic oscillations at latitude 65°N. Essentially, these variations occur at a predictable scale so that the timing of any change or climatic signal can be expressed with mathematical precision. Using this control the observed isotopic variations have been correlated with the predicted variation and a timescale has been derived. This process is known as orbital tuning (Martinsen *et al.*, 1987; Shackleton *et al.*, 1990; Berger and Loutre, 1988). In general, these results confirm those derived by palaeomagnetism and sedimentation rates, but add a level of refinement and confidence that has been, hitherto, not possible (Fig. 8.12b). Orbital tuning places the beginning of the Quaternary/Pleistocene at 2.60 Ma, and the Brunhes/Matuyama boundary at 0.78 Ma (Shackleton *et al.*, 1990).

The results for these studies show some very interesting patterns in the history of global Quaternary glaciation (Fig. 8.12) (Menzies, 1995a, Chapter 2). At

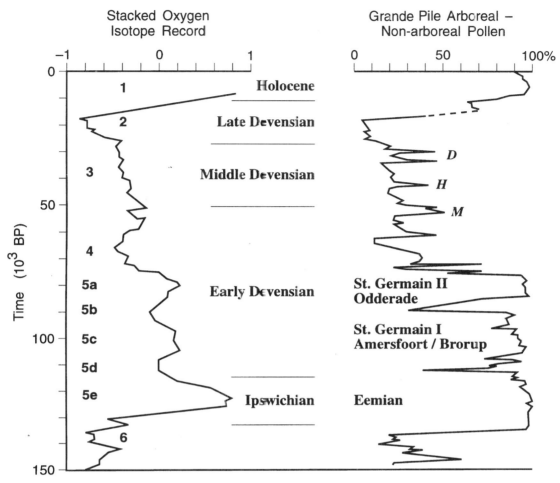

FIG. 8.12. Oxygen isotope record for the last 130 ka giving OI stages and sub-stages compared with the arboreal pollen record from Grande Pile south east France which is defined in terms of the Eemian Interglacial, the St. Germain I/ Amersfoort/ Brorup, St. Germain II/ Odderade, Moershoofd (M), Hengelo (H) and Denekamp (D) Interstadials. (Reproduced from Bowen, 1994)

the general scale, it can be seen that in the Early Pleistocene the pattern of climatic variation shows frequent oscillations of low magnitude (controlled by a 21,000 year periodicity determined by precession of the equinoxes) (Imbrie and Imbrie, 1979), and it seems that in the cold stages at this time glaciation was already established in Greenland and Antarctica, but was not important in temperate latitudes. In the period between about 1.6 Ma and 0.9 Ma, the pattern of climatic variation shows moderate amplitude variations with moderate magnitude (controlled by a 42,000 year periodicity forced predominantly by the variation in the axial tilt of the Earth) (Ruddiman and

Reymo, 1988). It has been suggested that in the cold stages at this time glaciation continued in Greenland and Antarctica but became an important process in the high altitude regions of the temperate latitudes and glaciogenic sediments were transported, for the first time, to the temperate lowlands (Rose, 1988a,b; Whiteman and Rose, 1992). In the final 0.9 Ma of the later Middle Pleistocene and Late Pleistocene the pattern of climatic variation displays high amplitude oscillations with long duration (determined by a 96,000 year periodicity forced predominantly by the eccentricity of the Earth's orbit), and it is this pattern that appears to account for the build-up of large-scale continental ice sheets that extended onto the lowland temperate regions of North America, Europe and Patagonia. There were 10 or 11 major expansions of glaciers over this period, and it was during OI Stages 12 and 16 that glaciers reached their greatest extent (Fig. 8.12).

Isotope stratigraphy can also be derived from ice cores, although the $\delta^{18}O$ values are the inverse of those for ocean sediments. However ice cores have a reliable record that goes back only as far as the Last Interglacial (Johnsen *et al.*, 1992; Dansgaard *et al.*,

1993) (Fig. 8.13). Nevertheless, these records give much higher levels of precision and provide detailed evidence of climatic variation during a glaciation, providing a record of stadial and interstadial oscillations (Johnsen *et al.*, 1993). This data, also, provides a link between the dynamics of the major ice sheets, such as Greenland, and the oceans (Bond *et al.*, 1993). Such data give evidence that is crucial to the understanding of how the Earth system works and the interactions between the hydrosphere, cryosphere and atmosphere.

8.5.3 Morphostratigraphy

Although morphostratigraphy is not unique to glacial stratigraphy it has played an important role in the development of concepts and models of glacial history and the dynamics of former glaciers. This is largely a function of the power of glaciers and glacial meltwater to form highly distinctive landforms, such as moraine ridges, kame terraces, and meltwater channels over very short periods of time (see Chapter 2). These landforms can, also, represent the positions of glacier margins as ice bodies expanded and retreated or down-wasted in response to changes in climate. Glacial landforms therefore can provide an indication of changes in glacial extent and consequently are a proxy for climate and a basis for climatostratigraphy.

'A morphostratigraphic unit is defined as comprising a body of rock that is identified primarily from the surface form it displays; it may or may not be distinctive in lithology from contiguous units; it may or may not transgress time throughout its extent' (Frye and Willman, 1962). This definition was developed as a result of mapping glaciogenic sediments in Illinois, mid-west USA, where lithofacies are exceedingly complicated and often stratigraphically undiagnostic in terms of the contemporary sedimentological theory (Chapters 3 and 9). Conversely, the landforms could be resolved into a series of moraine ridges that were interpreted as forming at a glacier margin, and therefore gave a clear expression of the changing glacier configuration.

Moraine ridges are the most characteristic morphostratigraphic unit in glacial geomorphology and

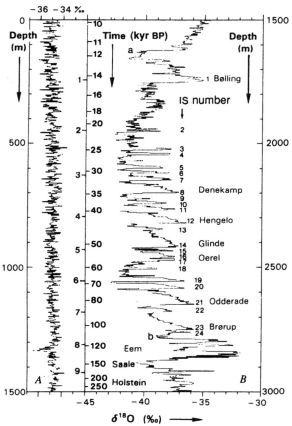

FIG. 8.13. $\delta^{18}O$ record from the GRIP ice core taken at the summit of Greenland. (a) Shows the variation from 0 to 1,500 m depth which covers the last 10 ka. (b) Shows the variation from 1500 to 3,000 m depth which extends from 10 ka to c. 250 ka. (b) Clearly shows the $\delta^{18}O$ variations between stadial and interstadial. The interstadials are numbered 1 to 24, and names are given where appropriate. The timescale is obtained by counting annual layers back to 14.5 ka BP and beyond that by ice-flow modelling. (Reproduced from Dansgaard *et al.*, 1993, reprinted with permission from Macmillan Magazines Limited)

geology. They are indicators of glacier margin position. They are recognised by their form and can, when not modified by postglacial subaerial mass movement, be mapped with great precision. As emphasised by Frye and Willman (1962), these units should be identified independently of their internal composition, and indeed moraine ridges can be formed of a great variety of materials with a great diversity of forms. For example, the Younger Dryas moraines in Scandinavia range from ice-contact deltas in western Norway and southern Finland, to push moraines formed of glaciomarine sediments in north western Norway, and boulder ridges in the Scandi-

navian north. Similarly, but at a smaller scale, the Loch Lomond Readvance moraine around the southern end of the Loch Lomond basin (Fig. 8.5, Table 8.3) ranges in composition from a till ridge on the hillside slopes in the west and south, a sand and gravel ridge in the valley bottoms of the west and south, an ice-contact sandur in part of the main outlet valley and an ice-contact delta in parts of ice-dammed valleys, and, in the south east, where the ice margin grounded in a large proglacial lake (Lake Blane), the moraine forms a low, broad ridge of glaciotectonised glaciolacustrine silts (Rose, 1981). Although the sediments are so variable and would be difficult to correlate on many lithostratigraphic criteria, this landform indicates a roughly synchronous position for the maximal extent of the Loch Lomond glacier sometime just after 10560 ^{14}C yrs BP (Rose, et al., 1988). However, despite this lithological variability, it is essential to emphasize that there are elements of inherent sedimentological uniformity to the extent that all these sediments show evidence of ice-contact on the ice-proximal side of the ridge. This evidence takes the form of glaciotectonic structures in sorted sediments and a distinctive clast fabric in tills. This case emphasises the need to link landform and sediment evidence whenever possible, and to integrate, as many kinds of stratigraphic information as possible.

Other examples of morphostratigraphic evidence are shown in Figures 8.14 and 8.16, (Sissons, 1977; Ballantyne, 1989) which include landforms such as glacial meltwater channels that are interpreted as forming at the ice margin, and kame terraces which are built-up against an ice margin. Although not strictly stratigraphical the distribution of hummocky moraine, and the distribution and arrangement of flutes and drumlins has been taken as evidence of glacier extent and behaviour, with the ice-distal limit of these features giving a minimum position for the extent of ice cover. In a similar fashion the extent of ice cover is also given by the ice-proximal limit of the distribution of periglacial soils and slope landforms, on the assumption that these could not form beneath ice or would have been destroyed if the ice had covered the area. This does, of course, assume that these periglacial features did not form after the ice retreated.

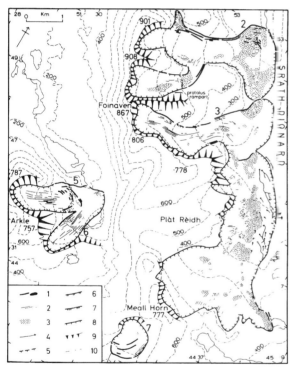

FIG. 8.14. An example of geomorphological mapping to derive a morphostratigraphy of Loch Lomond Stadial glaciers in the region of Foinaven, north west Scotland. This is a simple scheme used to identify the limit of the glacier at the maximum of the Loch Lomond Readvance. The landform evidence used consists of: 1, moraine ridges; 2, fluted till; 3, hummocky moraine; 4, meltwater channels; 5, ice-distal limit of a 'boulder spread'; 6, ice-proximal limit of periglacial features; 7, ice-distal limit of hummocky moraine or 'drift' sheet. 8 is the interpolated ice limit and 9 is very steep slopes of the glacier source area; 10 contours at 100 m intervals. (Reproduced from Sissons, 1977, in Gray and Lowe (eds), with kind permission from Pergamon Press Ltd

Innovative work by Boulton *et al.* (1985) has used the pattern of glacier (flutes, drumlins, streamlined hills) to reconstruct the activity of former glaciers and ice sheets across northern Britain, Scandinavia and North America. This technique uses this pattern of lineations to determine former ice-flow paths. Glacier marginal configuration is then reconstructed with a trend that is normal to the ice-flow vector. Recogni-tion of superimposed patterns of lineations (Rose and Letzer, 1977; Rose, 1987b; Boulton and Clark, 1990; Clark, 1994) has added a longer temporal dimension to this method of analysis and it has been possible to determine more than one stage of glacier development with different flow directions and ice dynamics.

The great strength of morphostratigraphy is that the evidence is on the land surface and may, if it is well

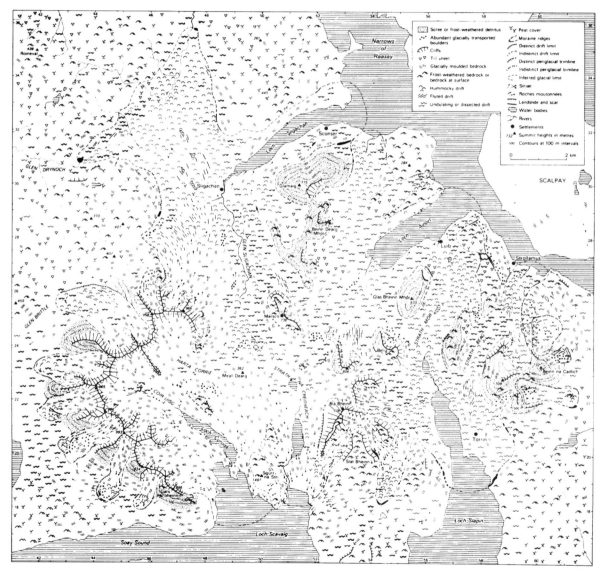

FIG. 8.15. (a) An example of a detailed geomorphological mapping to derive a morphostratigraphy of Loch Lomond Stadial glaciers in the Isle of Skye, Scotland. The evidence used is shown in the key.

developed, be traced continuously across whole regions with a level of precision that is not possible with any other stratigraphic method. In addition the spatial extent of landforms means that the relationship between one landform and another, and hence one stratigraphic unit and another can be investigated in detail. Using the example of moraine ridges, geomorphological mapping, such as illustrated in Figures 8.14 and 8.15, has the potential to demonstrate merging or overlapping ridges indicating different ice-streams within a glacier system (Mickelson *et al.*, 1983). The ability to establish such temporal and spatial relationships is known as *connectivity* and can have a high level of stratigraphic precision equivalent to the superposition of stacked sediments. Examples could include graded sandur surfaces indicating

FIG. 8.15. (b) A reconstruction of the glaciers based on the above evidence. (Reproduced from Ballantyne, 1989)

contemporaneity of the meltwater deposition, or cross-cutting moraine ridges which indicate that later ice extended across an earlier marginal position. This is illustrated on Figure 8.15 where glacier 3 has extended, at a later time, across the area of glaciers 2 and 4. On the other hand, on this same figure, glaciers 5 and 6 are shown to reach their limit contemporaneously.

Another advantage of continuity of surface expression is that glacial landforms can be readily dated by methods such as lichenometry or Schmidt hammer

impact measurements (Matthews, 1992; McCarroll, 1989). Studies of this kind allow the reconstruction of detailed morphostratigraphy with strict geochronology (Fig. 8.16). The level of precision and resolution provided by these methods across such wide areas of the Earth's surface is rarely equalled by any other stratigraphic method.

There are, of course, many problems associated with morphostratigraphy, not least the difficulty of measuring and analyzing features of immense size and, in North America, over vast areas. Geomorpho-

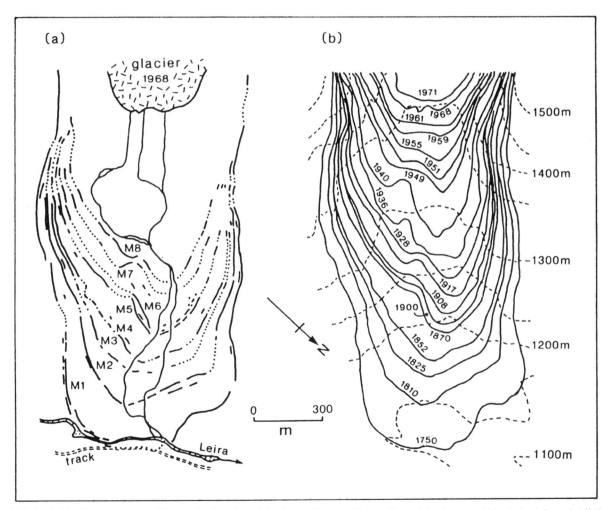

FIG. 8.16. Morphostratigraphy of the foreland region of Storbreen Norway. The position of the ice margin is derived from detailed geomorphological mapping, and the geochronology has been constructed from a variety of techniques including lichenometry and air photo analysis. (Reproduced from Matthews, 1992)

logical mapping requires enormous time, great skill, effort and experience, and many of the problems of glacial geomorphology, in the middle part of this century, stem from the failure to apply these requirements and attributes especially prior to the work of J.B. Sissons in Scotland (Ballantyne and Gray, 1984). Air photography has greatly helped in the process of geomorphological analysis, but air photographs are no substitute for field mapping, as the detailed form and internal composition of the landforms cannot be determined.

A ubiquitous problem is that once morphogenesis has taken place, the form of glacial landforms such as moraine ridges or meltwater channels is altered by subaerial processes and the original detailed form is lost. With this loss there is rarely any possibility of regaining the initial stratigraphic precision. At this stage morphostratigraphy fails to be of value as a stratigraphic technique. The use of this property, often known as the 'freshness' of landforms, has been used as a stratigraphic method. However, it is of little use in this rôle other than as a very unreliable 'first order' estimation, and this approach has created far more problems than it has solved. Hitherto, no satisfactory morphometric method has been devised to quantify this approach, or to overcome the intrinsic variation caused by the lithological variability of glaciogenic sediments and facies associations. The concept of 'older' and 'newer' drift is possibly as far as this approach can be taken (Wright, 1937) and even this approach can be shown to be fundamentally flawed, depending upon the original form of the glaciogenic landscape (Morgan, 1973). In mid-latitude regions, only glaciogenic landforms formed during or since the Last Glacial Maximum can be studied by morphostratigraphic methods. Older features are likely to have been subject to considerable geomorphological modification making application of the technique unproductive and futile (Straw, 1960; Sparks and West, 1964).

8.5.4 Stratotypes

Stratigraphic convention requires that in order to establish a stratigraphic sequence for a region it is necessary to designate a type section (stratotype) or set of type sections representing a type locality. A stratotype acts as an objective example or standard for a particular stratigraphic sequence and as a criterion against which other sections can be compared. Ideally, a stratotype should be a continuous record and contain the contact with the underlying stratigraphic unit. However, as noted by Bowen (1978) and Rose (1989) such stratotypes are rare indeed within glacial sediments, and in Britain, for instance, only for the Anglian and Devensian glacial stages have stratotypes been adequately defined and these are unsatisfactory with respect to the definition of their lower boundaries (Rose, 1989).

The existence of the lower boundary in a stratotype is critical as it contains the change from one set of sedimentary characteristics, and hence from one depositional system or subsystem to another. In glacial geology it could identify the onset of glaciogenic deposition at a site. At the stratotype for the Dimlington Stadial (Last Glacial Maximum) in Britain the lower boundary is identified by the base of a laminated silt member that was deposited by snow melt or glacial meltwater in close proximity to glacial ice (Catt and Penny, 1966; Penny et al., 1969; Rose, 1985). This facies unit is succeeded by diamictons that represent the presence of glaciers within the region. The upper boundary is not required as it can be defined by the lower boundary of the succeeding unit (West 1989) and the lack of requirement of a definition for the upper boundary means that new units can be inserted into a stratigraphic scheme without requiring re-definition of the upper boundary. For the Dimlington Stadial, for example, the upper boundary is defined by the base of the lacustrine clays that represent the onset of the Windermere Interstadial in Britain (Rose, 1985). It is known, for instance, that the transition between the stadial and interstadial conditions are missing in the type area of the Dimlington Stadial (Walker et al., 1993) and this can be accommodated by reference to the Interstadial typesite without the need to redefine the stratotype for the Dimlington glaciation.

These type sections, therefore, provide a record of sedimentological change that represent activity of importance to glaciation, such as, an ice advance or retreat. The type section may also include evidence for the approach of an ice front as indicated by increased coarsening of sediment or structures indicating progressively higher flow regimes, or by the initial influx of glacially derived sand-size aeolian sediment (Rose *et al.*, 1985)

Stratotypes may also be defined on the basis of morphostratigraphy, but this is rare because morphostratigraphic evidence looses its precision as a consequence of non-glacial subaerial denudational processes and morphological boundaries are then very difficult to recognize and define. However, in the type area for the Loch Lomond Stadial in Britain, the Loch Lomond moraine ridge is included as part of the evidence for the stratotype (Rose, 1981, 1989) (Fig. 8.5, Table 8.3).

8.5.5 Chronostratigraphy, Geochronology and Geochronometry

Chronostratigraphy is 'the organisation of rocks on the basis of their age or time of formation' (Salvador, 1994, p. 113) and *chronostratigraphic units* are 'a body of rocks that includes all rocks formed during a specific interval of geological time' (Salvador, 1994, p. 113). *Geochronology* is 'the science of dating and determining the time sequence of events in the history of the earth' (Salvador, 1994, p. 120) and *Geochronological units* are units of geological time (Salvador, 1994, p. 120). *Geochronometry* and *geochronometric units* deal with the quantification of geological time, that is, putting actual ages on the events or time periods. Reference to Salvador (1994, p. 120) for a definition of geochronometric terms shows an expected resolution of thousands or millions cf years. This emphasizes the much higher level of precision required by glacial stratigraphic science where resolution may be in decades, years or even seasons. There has been much confusion in the geological literature about the use of chronostratigraphy and geochronol-

ogy. This is amplified in Salvador (1994, p. 9–10) where it is pointed out that each chronostratigraphic unit is a rock body which has a corresponding interval of geological time (geochronological unit). This means that the chronostratigraphic units are the sediments, landforms and soils formed over a certain interval of time and that geochronological units are the interval of time during which these sediments are deposited, soils are formed or landforms are constructed. Geochronometric units also provide a quantitative value for the period of time which allows calculation of the rates at which processes operate. The hierarchy of chronostratigraphic and geochronological units is given in Table 8.2.

Bearing in mind that lithostratigraphy and morphostratigraphy remain the building blocks of glacial stratigraphy, the aims of this science are most effectively expressed by using diachronic units with a climatostratigraphic basis. These methods reflect the climatic forcing of systematic glacier expansion, standstill and retreat, and the inherent variation that exists between sites as a function of local climates, variations in glacier catchment shape, size, and steepness and bed-material (Fig. 8.3). Nevertheless, chronostratigraphy remains the aim of most glacial stratigraphic studies, if only because it provides the most useful method of communication between scientists and with the layman. Table 8.3 provides an example of a local scale stratigraphy, based on the type area for the Loch Lomond Stadial, in the south eastern part of the Loch Lomond Basin, west central Scotland. This scheme is based on the evidence summarised in Figure 8.5. The chronostratigraphy provides a basis for comparison with other sites in the British Isles, and indeed elsewhere. The geochronometry from the site provides dates that only approximate with the formally accepted ages (Mangerud *et*

TABLE 8.2. Chronostratigraphic and geochronological hierarchical terminology

Chronostratigraphic unit	Geochronologic unit	Example
Eonothem	Eon	Phanerozoic
Erathem	Era	Cenozoic
System	Period	Quaternary
Series	Epoch	Pleistocene
Stage	Age	Devensian/Wisconsinan
Chronozone	Chron	Dimlington Stadial/Younger Dryas

TABLE 8.3. Stratigraphic subdivisions and terminology at the type area for the Loch Lomond Stadial, south east Loch Lomondside, western central Scotland (based on Rose, 1989)

Lithostratigraphic & Morphostratigraphic units:		Climatostratigraphy	Chronostratigraphy	Geochronology	
Within limit of Loch Lomond Readvance (Croftamie, Aber, Gartness, Carnock Burn, Loch Lomond ①)	Beyond limit of Loch Lomond Readvance (Muir Park Reservoir ②) ③	Stadial/Interstadial	Stage	Radiocarbon Years ④	Calendar Years ⑤
Brown silty clay (freshwater)	Peat		Flandrian Interglacial Stage		
Dark organic silty clay (marine)	Gyttja				
Grey & brown silty clay (freshwater)	Clay-gyttja			10,000	11,500 ⑥
Blane Valley Silts / Rhu Gravels / Gartocharn Till/Loch Lomond Moraine / Blane Valley Silts / Main Lateglacial Shoreline	Pink clay	Loch Lomond Stadial	Late Devensian Glacial substage		
Clyde Beds	Gyttja	Windermere Interstadial		11,000	12,800 ⑤
Gartness Gravels, Sands and Clays / Wilderness Till	Silty clay Wilderness Till	Dimlington Stadial		13,000	14,620 ⑤
				26,000	c.22,000 (Max) ⑥

① Dickson et al., 1978; Rose, 1980; Rose et al., 1988.
② Donner, 1957; Vasari and Vasari, 1968; Vasari, 1977.
③ Site shown on Fig. 8.5.
④ Rose, 1985.
⑤ Alley et al., 1993.
⑥ Bard et al., 1991.

TABLE 8.4. Chronostratigraphy of the British Isles and north western Europe (based on Bowen, 1994)

OI Stages	Age ka	British Isles	NW Europe
		Chronostratigraphy	
1		HOLOCENE	
	11.5		
2		Younger Dryas	
		Bølling-Allerød	
		Late Devensian	Late Weichselian
	25		
3		Middle Devensian	
	50		
4			
	70		
5a		Upton Warren	Odderade
5b			
5c		Chelford	Amersfoort-Brorup
5d			
5e		Ipswichian	Eemian
	130		
6		(Ridgeacre)	Saale (Warthe)
	186		
7		Stanton Harcourt	Bantega/Hoogeven
	245		
8			Saale (Drenthe)
	303		
9		Hoxnian	Domnitz/Wacken
	339		
10			Fuhne
11		Swanscombe	Holsteinian
	423		
12		Anglian	Elster II
	478		
13		Cromerian	Cromerian IV
14			Glacial C
15		Waverley Wood	Cromerian III
16			Glacial B
17			Cromerian I
18	790		Glacial A
19			Cromerian I
20			Dorst (g)
21			Leerdam (ig)
22			Linge (g)
23	900		Bavel (ig)
			Menapian
			Waalian (A,B,C)
		Beestonian	Eburonian
	1.6 Ma		
		Pastonian to	Tiglian
		Ludhamian	(C5 to A)
		Pre-Ludhamian	Praetiglian
	2.3 Ma		

Note: (g)–glacial, (ig)–interglacial

al., 1974) for the boundaries between the stages and sub-stages, although local dates do contribute to the understanding of the earth system dynamics in the region. However, this site is accepted as a stratotype because of precedent (Simpson, 1933), the wide range of lithostratigraphic, morphostratigraphic and geo-chronometric evidence, and the persistence and access-ibility of the sections (Rose, 1981, 1989). For glacial behaviour specifically, the site also gives evidence for the behaviour and timing of the glacier that advanced down the Loch Lomond basin from the western Scottish Highlands during the Younger Dryas (Loch Lomond Stadial) climatic deterioration (Sissons, 1967).

Finally, Table 8.4 provides an example of a currently accepted chronostratigraphy for the Pleisto-cene of the British Isles and north western Europe (Bowen, 1994). At this level the scheme is subdivided into glacial and interglacial stages, and glaciation *per se* is subsumed under this climatic scheme, so that the actual status of glacier behaviour hardly contributes to its derivation. Indeed the general extent of glaciation in some of the 'glacial' stages is not known. The timescale is based primarily on orbital tuning. The essential aim of this scheme is to provide a framework for a highly complex period of Earth history and a basis for communication between scientists from different major regions on the globe.

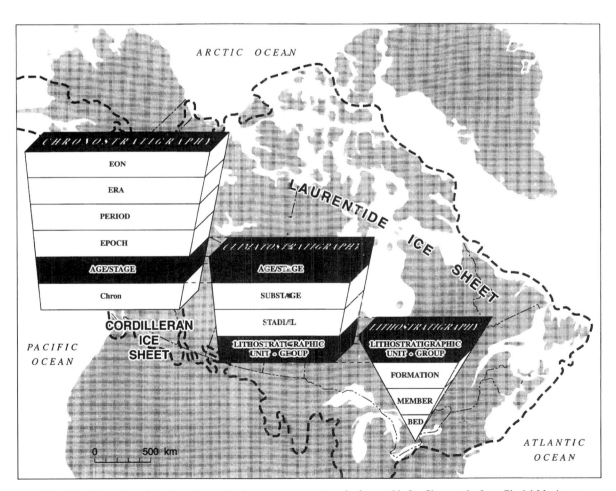

FIG. 8.17. Illustration of stratigraphic terminology, superimposed on the Laurentide Ice Sheet at the Last Glacial Maximum

8.6 CONCLUSION

Much of glacial stratigraphy has a solid scientific foundation, using an extended, and hence more powerful version, of the International Stratigraphic Code (Salvador, 1994). However the complex nature of the glacial system means that interpretation is often far more complex than for the other depositional systems that dominate the pre-Middle Quaternary parts of the Geological column. In these circumstances, and bearing in mind the immense difficulties of dealing with spatially extensive morphostratigraphic evidence, it is little wonder that in some instances stratigraphic frameworks that were constructed some decades ago now appear naive.

Currently many past stratigraphies are subject to revision, as new descriptive methods and greater understanding of glacial processes are applied. It is essential, however that any new stratigraphy must not be cast in dogma as has been the case in the past (Rose, 1987a, 1989, 1991; Gibbard and Turner, 1988) and is subject to modification along with the discovery of new evidence and understanding.

For a working framework, glacial stratigraphy can be comprehended at a range of scales from global to site. At the global, hemispheric, continental, and continental/regional levels chronostratigraphy is the most appropriate level for assimilating and communicating stratigraphic information. At the continental/regional, regional, regional/local and local levels climatostratigraphy is most effective, whereas as the local and site levels lithostratigraphy is most appropriate (Fig. 8.17).

There remain several continuing issues in glacial stratigraphy; in particular the need to establish better inter-regional and -continental stratigraphic correlations; to develop multiple dating systems applied to specific problem sites; to reconcile and validate the stratigraphic record between terrestrial and oceanic stratigraphies; to institute and expand objective methods of stratigraphic definition; finally, glacial stratigraphy must be utilised as a tool in obtaining insight into the processes and rates of operation of the glacial system, the climatic, geographical and geological conditions pertinent to and existing during ice sheet formation and decay.

Chapter 9

LITHOFACIES ASSOCIATIONS FOR TERRESTRIAL GLACIGENIC SUCCESSIONS

T.J. Kemmis

9.1. INTRODUCTION

9.1.1. Facies Models of Ancient Glacigenic Successions

Facies successions deposited in subglacial, supraglacial and proglacial environments are often used as facies models for ancient glacigenic successions (e.g. Boulton and Paul, 1976; Eyles *et al.*, 1983; Brodzikowski and Van Loon, 1991). In this chapter, examples of sedimentary sequences from ancient glacial successions are presented. In most cases, these sedimentary successions represent changing depositional environments over long time periods (i.e. changing environments from proglacial to subglacial to supraglacial, for example) at a location during one or more advance-retreat cycles of a single glacial period (such as the Wisconsinan Stage), as well as the complexities added by post-glacial modification. Hence, facies models for most ancient successions are not simple, and seldom correspond to models of a single depositional environment (such as the subglacial or the supraglacial environment).

9.1.2. Problems in Formulating Facies Models for Ancient Glacigenic Sequences

There are several reasons why it is difficult to formulate useful, predictive facies models for ancient glacigenic successions and landsystems at this time. First, comprehensive sedimentologic study of modern analogs started little more than 20 years ago. As a consequence, it has only been recently that studies of modern glacigenic sequences have provided an adequate basis for comparison and, as yet, few studies of ancient successions have been made using these sedimentologic criteria.

Second, while areas of past glaciation potentially preserve a significantly long record of glacial depositional conditions (and possibly changing environments with time), very few areas have been studied in adequate detail, rarely is the entire sequence (and, hence, record) exposed and few ancient sites have been put within the context of the larger scale glacial system of which they were a part (Chapter 8).

Third, as new research tools have been developed (such as aerial photography, satellite images, and large-scale topographic maps), formerly glaciated areas are often found to have a much more complicated landform assemblage than previously recognized (e.g. Kemmis *et al.*, 1981).

Fourth, modern analogs often have limited correspondence to ancient glaciers. For example, modern glaciers are in many respects poor analogs for former continental ice sheets because the former occur on a different (smaller) scale, in different climatic settings, and/or on different bed materials and topography. Further, observations of modern glaciers have been limited primarily to study of short-term processes occurring in the immediate vicinity of the glacier terminus. Study of environments other than the ice-marginal environment, particularly the subglacial environment and the (variable) nature of glacial hydrologic system, are largely theoretical and speculative, and still subject to much controversy (Menzies, 1995a, Chapters 5 and 6). Short-term studies of ice-marginal areas have not been completely successful in assessing either longer-term landform development (or in other terms, development of facies assemblages) or the factors affecting longer-term sediment preservation (the deposits actually retained in the sedimentary record).

Finally, despite the need and desire for simple facies models, glacigenic systems are extremely complex. Such complexity should be expected because of the large number of variables including glacial dynamics, climate (including not just annual budgets, but the magnitude of seasonal changes), bed materials and topography, system scale and various local factors affecting sediment preservation (Table 9.1). Virtually every site has its own unique depositional sequence because the combination of possible variables is immense; nonetheless, some generalizations can be made for several glacial depositional settings and these will be briefly presented in this chapter.

9.1.3. Facies of Ancient Glacigenic Sequences — Some Examples

In this chapter, presentation will be made of sedimentary successions from several different terrestrial glacigenic settings, including:

TABLE 9.1. A partial listing of variables influencing glacial depositional environments (not all of these variables are independent variables)

Glacier size (scale)
Climate
Glacier mass balance
Basal thermal regime
Ice velocity
Equilibrium vs. surging conditions
General topography (especially in mountainous areas where valley sides may be debris sources)
Bed topography
Proglacial area characteristics (slope direction, local relief, etc.)
Bed materials (rock, sediment or water)
  Rock properties
  – primary properties – hardness, cleavage, bedding, sedimentary structures, etc.
  – secondary properties – jointing, weathering, structural features, etc.
  Sediment properties
  – primary properties – grain-size distribution, mineralogy, hydraulic conductivity, sedimentary structures, etc.
  – secondary properties – weathering features, jointing, etc.
Time of year
Meltwater production and drainage, as well as how it affects pore pressures, basal shear stresses, etc.
Transport distances
Proximity to glacier margin
Active vs. stagnant ice margin
Glacier margin advance/retreat
Preservation potential

(1) Low-relief areas glaciated by continental ice sheets. (a) Underlain by pre-existing glacial deposits or relatively soft sedimentary bedrock. (b) Underlain by relatively hard igneous and metamorphic bedrock ('shield' areas).

(2) High-relief areas glaciated by continental ice sheets.

(3) Ice caps and valley glaciers of mountainous areas.

(4) Polar ice caps.

The purpose of these examples is not to erect facies models for glacial landforms into which sedimentary sequences can be forced to correspond, but to illustrate the complexity of glacigenic systems and the dissimilarity between different terrestrial glacigenic environments. Facies models for glacigenic systems, like those for fluvial systems (Miall, 1984), should be expanded rather than forced to fit into a limited number of facies models. Likewise, the reasons for differences between glacial landform/sediment assemblages should be investigated.

The examples selected in this chapter were chosen because the sedimentology of the sites is well documented, facies relationships are clearly presented, and the sites' positions in the respective glacial systems are reasonably well known. Specific details of facies characteristics, facies relationships, site stratigraphy, and inferred glacial history and ice dynamics have been deliberately omitted or incompletely discussed for reasons of brevity; the reader is encouraged to refer to the original studies for the necessary details. It must also be emphasized that these landform/sediment assemblages are not the only ones deposited by these particular glacial advances (Brodzikowski and Van Loon, 1991).

In this chapter there is no discussion of glacio-aquatic systems but the variability of glacio-aquatic systems is at least as great as that for terrestrial glacigenic systems because of variations in system scale, position with respect to the grounding line, water-circulation systems (in oceans and lakes), and the various factors influencing glacial dynamics. The reader is referred to Chapter 5 and to recent summaries by Gravenor et al. (1984), Eyles et al. (1985), Borns and Matsch (1989), Dowdeswell and Scourse (1990), Anderson and Ashley (1991) and Menzies (1995a, Chapters 13 and 14).

9.2. EXAMPLES of SEDIMENTARY ASSEMBLAGES

9.2.1. Low-relief Areas Glaciated by Continental Ice Sheets Underlain by Pre-existing Glacial Deposits or Relatively Soft Sedimentary Bedrock

9.2.1.1. End moraine of the Green Bay Lobe of the Laurentide Ice Sheet

The Johnstown moraine is the terminal moraine of the Late Wisconsinan Green Bay Lobe of the Laurentide Ice Sheet. This end moraine is a well defined ridge 10 to 20 m high and a few hundred meters wide. Lundqvist et al. (1993) studied the sedimentology of four sites along the southwestern portion of the moraine in south-central Wisconsin, USA, and a generalized vertical succession of five major facies units, informally designated from A to E, was identified at each site (Fig. 9.1).

Facies A, not exposed at the site shown in Fig. 9.1, is a generally massive, gravelly sandy loam diamicton interpreted to be a subglacially deposited till. It interfingers laterally with bouldery coarse gravel,

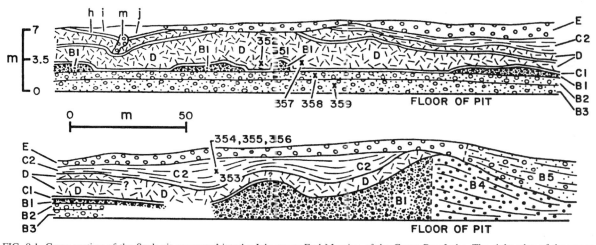

FIG. 9.1. Cross-section of the Sauk pit excavated into the Johnstown End Moraine of the Green Bay Lobe. The right edge of the upper diagram connects to the left edge of the lower diagram. Most of the section is parallel to and just behind the crest of the moraine. The right end of the lower section curves and crosses the crest at right angles (Lundqvist, Clayton and Mickelson, 1993 with kind permission from Pergamon Press Ltd)

Facies subunit B1, that occurs in convex upward, tube-like bodies a few tens of metres wide and a few metres high. These esker-like gravel bodies are generally oriented perpendicular to the axis of the moraine ridge, parallel to glacier-flow. The gravels of unit B1 locally overlie or grade to proglacial fluvial and lacustrine deposits (Facies subunits B2 and B3) and grade distally to proglacial outwash (subunit B4) (see Menzies, 1995a, Chapter 12).

Overlying Facies B1 and A is Facies C, a layered diamicton. The layering consists of diamicton beds, each up to several centimetres in thickness. The beds were lithologically uniform, but compositionally different. The beds are folded locally either from settling around melting ice bodies or from glaciotectonic deformation. Facies C is interpreted to have been deposited subglacially, but it is uncertain whether by melt-out, lodgement or a combination of both processes.

Facies D is the thickest and laterally most extensive diamicton unit in the end-moraine sequence. It is a massive, gravelly sandy loam with a well oriented pebble fabric dipping in the up-glacier direction. It is interpreted to have been deposited subglacially, probably by a combination of both melt-out and lodgement processes. Capping the end moraine is a thin veneer of very bouldery loamy gravel, Facies E. In many places, the modern soil profile is developed in this sediment, hindering genetic interpretation, but the unit is interpreted to be either a proglacial or supraglacial deposit sorted during short meltwater transport and partly slumped during or after deposition.

The Johnstown end moraine is interpreted to have formed when the Green Bay Lobe was at its terminus and remained there for a significant time period. Subglacial tunnels formed while till of Facies A was being deposited. Outwash from these tunnels was deposited proglacially (Facies subunit B4), and coarse, bouldery outwash (Facies subunit B1) was deposited in the tunnels. Layered diamicton of Facies C was discontinuously deposited over units A and B1, followed by deposition of massive, uniform till of Facies D. As ice withdrew from the moraine, a thin veneer of poorly sorted, bouldery loamy gravel was deposited in either proglacial or supraglacial environments.

It is interesting to compare this facies assemblage with those of modern Alaskan valley glaciers. The Johnstown End Moraine is composed of a relatively high proportion of lodgement and melt-out till (Facies A, C and D) with a very low percentage of resedimented deposits (Facies E). This contrasts with deposits at the terminus of the Matanuska Glacier where Lawson (1979a) estimates that 95% of the glacigenic sediment has been resedimented by various secondary mass-wasting processes, and the Maclaren and Gulkana glaciers where Evenson and Clinch (1987) found little basal and englacial deposition of debris.

9.2.1.2. Groundmoraine of the Lake Michigan Lobe of the Laurentide Ice Sheet

The terrain of the Lake Michigan Lobe of the Laurentide Ice Sheet in northeastern Illinois, USA, consists of low-relief groundmoraine and/or lake plain and a series of arcuate end-moraine ridges (Leverett, 1899; Willman and Frye, 1970; Johnson and Hansel, 1990). The studied stratigraphic succession is in a groundmoraine area near Wedron, Illinois where stratigraphic and geomorphic relationships have long been interpreted to indicate repeated advance and retreat of a grounded glacial lobe (Leverett, 1899; Willman and Payne, 1942; Willman and Frye, 1970). Detailed stratigraphic and sedimentologic study by Johnson and Hansel (1990) confirms that the stratigraphic succession results from a fluctuating glacial lobe, although the number of advance-retreat sequences are interpreted as three instead of six. The three sequences each record the preservation of successive proglacial, subglacial, ice-marginal and subaerial deposits.

The Wedron sequence was deposited from approximately 25,000 to 17,200 BP (Johnson and Hansel, 1990) (Chapter 3). Twenty-four basic lithofacies types (Table 9.2) have been grouped into a generalized vertical succession of seven facies assemblages, informally designated from A to G (Table 9.3), deposited during each glacial advance (sequence).

The succession of facies assemblages (A to G) and the lithofacies of the three advance–retreat sequences (I to III) are shown on Fig. 9.2. The most complex sequence, Sequence I, began as the advancing glacial

TABLE 9.2. Lithofacies codes and sedimentary characteristics for facies exposed in the Wedron, Illinois area (from Johnson and Hansel, 1990)

Code	Lithofacies type	Description
Dm	Diamicton, massive	Homogeneous, matrix-supported; very poorly sorted clay/silt/sand/gravel admixture; dispersed striated clasts and striated boulder pavements common; erosional lower contacts
Dm(s)	Diamicton, massive, sheared	Dm with evidence for glacier shear, e.g. bedding plane shears, basal grooves, rafting and deformation of substrate materials, and smudging and shear attenuation of softer lithologies
Dm(r)	Diamicton, massive, redeposited	Dm with evidence for resedimentation; thin to thick discontinuous beds and lenses interbedded with sorted sediment; subhorizontal to inclided beds; textural variability and fine structures (e.g. rafted silt or clay stringers with small flow noses) common
Dm(d)	Diamicton, massive, deformed	Dm with evidence for soft-sediment deformation; loaded contacts with softer sediment and diapiric structures common
Dm(p)	Diamicton, massive, pedogenically altered	Dm with evidence for pedogenesis, e.g. soil horizonation and mineral alteration
Ds	Diamicton, stratified	Matrix-supported; pronounced stratification; conspicuous textural differentiation or structure within diamicton; pronounced winnowing
Dg	Diamicton, graded	Matrix-supported; clast content generally normally graded
Fm	Clay and silt, massive	Clast poor
Fmd	Clay and silt, massive, diamictic	Fm with clasts dispersed
Fmo	Clay and silt, massive, organic	Fm with conspicuous organic matter
Fl	Clay and silt, laminated	Alternating clay, silt, and fine sand laminae
Fl(d)	Clay and silt, stratified, deformed	Fl with evidence for soft-sediment deformation, e.g. wavy lamination, folded and overturned bundles of laminae, ball-and-pillow structures
Fmo(p)	Clay and silt, massive, organic, pedogenically altered	Fmo with evidence for pedogenesis, e.g. soil horizonation and mineral alteration
Fm(s) or Fl(s)	Clay and sand, massive or laminated, sheared	Fm or Fl with evidence for glacier shear, e.g. erosional upper contact, bedding plane shears, glacier grooves, slickensides, shear attenuation
Sh	Sand, horizontal lamination	Stratified, generally normally bedded
Sx	Sand, crossbedded	Planar of trough cross-stratified; medium to coarse sand, may be pebbly
Sr	Sand, rippled	Ripple-drift cross-lamination
Sm	Sand, massive	Silty sand
S(s)	Sand, sheared	S with evidence for shearing, e.g. erosional contact(s), and disturbance or lack of primary structures
Sm(p)	Sand, massive pedogenically altered	Sm with evidence for pedogenesis; e.g. soil horizonation and mineral alteration
Sh(d)	Sand, horizontally laminated, deformed	Sh with evidence for soft-sediment deformation; e.g. loaded contacts, water-escape and ball-and-pillow structures
G	Gravel	Massive; matrix-supported
Gh	Gravel, crudely bedded	Crude horizontal beds; matrix-supported
Gg	Gravel, graded	Generally coarsening-upward sequences

lobe locally dammed the pre-existing drainage system. Lacustrine sediment (Facies A1, Lithofacies Fl) was deposited initially in valleys on the bedrock surface. As the Ice Sheet drew nearer, proglacial fluvial sediment (Facies B, Lithofacies Sx) was deposited, and continued glacial advance resulted in deformation of the top of this facies (Lithofacies S(s))

as ice overrode the site. Continued ice movement resulted in deformation (folding, shear) and lodgement of a thin layer of diamicton and intercalated sorted sediment generally <1 m thick (Facies C, Lithofacies Dm(s) and S(s)) followed by lodgement of generally massive clay loam and loam till, Subfacies D1 and D2. Facies D locally includes

TABLE 9.3. Major and minor lithofacies and inferred origin of facies (from Johnson and hansel, 1990; lithofacies explained in the text)

Facies	Major lithofacies	Minor lithofacies	Inferred environment and dominant process
A	Fl, Fm, Flo, Fmd	Fmd, Fm(s), Dm(r)	Proglacial lacustrine
B	Sx, Sh, Gh	S(s), Sr, Gg, Dm(r)	Proglacial fluvial
C	Dm(s), S(s)	G(s), Dm, Sx	Subglacial deformation
D	Dm, Dm(s)	Sh(s) (d)	Subglacial lodgement
E	Dm(r)	Fm, Fl(d), Sh, Gh, Dm(r) (d)	Ice-marginal resedimentation
F	Fl, Fl(d)	Sh, Sr, Fm, Fl(s) (d)	Paraglacial lacustrine
G	Fm, Fm(p)		Aeolian

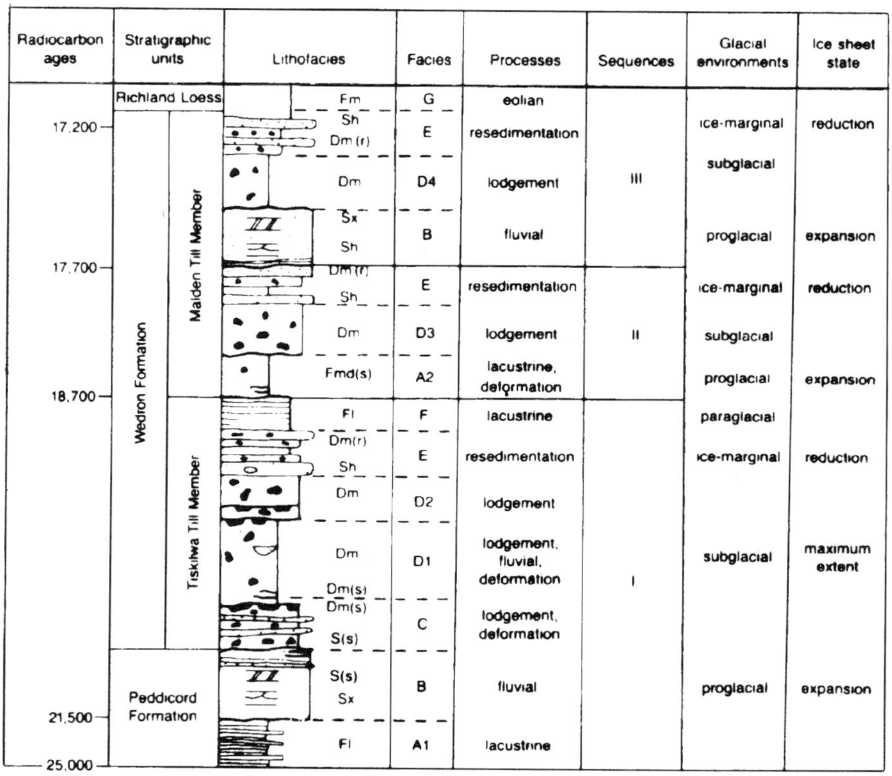

FIG. 9.2. Composite lithofacies and facies associations at Wedron, and inferred processes, sequences, glacial environments, and ice sheet state (Johnson and Hansel, 1990, reproduced by permission of SEPM)

bulletstones, boulder pavements, smudges, shear planes and truncated sand bodies indicating deposition by lodgement. Locally incised into the till of Facies D are sorted sediments deposited in subglacial channels. Although the tops of these channel fills are often truncated, there is no other evidence of deformation, and the channels formed in a stable, nondeforming bed. As ice withdrew during Sequence I, Facies E (Lithofacies Dm(r) and Sh) was deposited in an ice-marginal environment. Lacustrine sediments

(Facies F, Lithofacies Fl) were deposited as ice withdrew from the immediate site.

Sequences II and III broadly resemble Sequence I. In both, subaerial sorted deposits (Facies A2 and B, respectively, Fig. 9.2) occur at the base of the sequences. These sorted deposits are overlain by subglacial deposits, lodgement tills (Subfacies D3 and D4, respectively) which are in turn overlain by deposits resedimented in ice-marginal environments (Facies E, Lithofacies Dm(r) and Sh) as ice withdrew at the end of each advance. The clay loam, silty clay loam and silty clay lodgement tills of Subfacies D3 and D4 are compositionally and texturally different from those of D1 and D2. Not present in either Sequence II or III is Facies C, deformed diamicton and intercalated sediment – deformation till.

9.2.1.3. Drumlins of the Peterborough drumlin field of the Laurentide Ice Sheet

Drumlins display a (probable) continuum of forms and possess variable internal geometries and a wide range of facies types (e.g., Menzies, 1979b, 1987; Shaw, 1980, 1983b; Rose, 1987a; McCabe and Dardis, 1989). Unfortunately, drumlin study has suffered from (1) numerous theoretical studies giving limited regard to actual drumlin stratigraphy and structure, (2) many morphometric analyses that also ignore drumlin structure and stratigraphy, (3) incomplete field investigations where detailed facies descriptions and relationships have not been made, and (4) study restricted to the drumlin forms while ignoring inter-drumlin areas.

Recent detailed sedimentologic studies of drumlins in Canada (Ontario, Saskatchewan and the Northwest Territories) and various drumlin fields in Ireland indicate that drumlin stratigraphy is complex and often undeformed (Shaw, 1983b; Dardis and McCabe, 1983, 1987; Shaw and Kvill, 1984; Dardis et al., 1984; Dardis, 1985, 1987; Sharpe, 1987; Shaw and Sharpe, 1987; Hanvey, 1987; McCabe and Dardis, 1989). Although there are differences in facies, facies relationships and interpreted processes between these studies, they do not support the concept of drumlin formation by pervasive subglacial deformation and lodgement. For this review facies examples are taken from Sharpe (1987). Proceedings of two recent

symposia (Menzies and Rose, 1987, 1989) and references therein provide additional examples detailing drumlin stratigraphy and structure.

Exposures of several drumlins in the Peterborough field reveal similar internal stratigraphies (Sharpe, 1987). Lithofacies profiles of the Hoekstra Pit (Fig. 9.3) were made at 10 to 25 m intervals. Details of the facies and their geometry are given in Sharpe (1987). The sequence within the drumlin is conformable; contacts between facies are transitional and often interbedded, indicating continuous sedimentation, although energy conditions and sediment influx rates varied with time. The gravel and sand facies are interpreted to have been deposited by high-energy

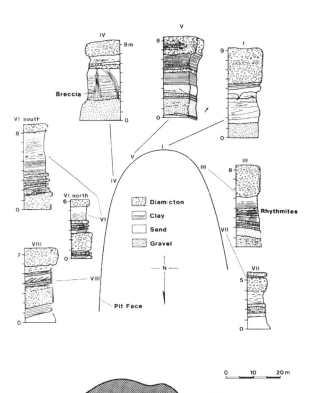

FIG. 9.3. Summary of lithofacies in measured sections at the Hoekstra pit, Peterborough drumlin field. The cross-section shows the location of the pit with respect to the drumlin. Facies are described in detail in the original text (Sharpe, 1987, in Menzies and Rose (eds), courtesy of A.A. Balkema, Rotterdam)

influx into standing water in a subglacial cavity. The sands grade upward to rhythmically bedded silt and clay with occasional thin interbeds of gravel and diamicton. The rhythmically bedded silt and clay represent low-energy sedimentation in the subglacial cavity; thin gravel and diamicton interbeds resulted from the episodic influx of sediment flows. Diamictons at the top of the sequence vary from massive to interbedded with silt and sand lenses. The diamictons are interpreted to have been deposited by subglacial melt-out, slump and sediment flow.

The sedimentary geometry of this sequence appears conformable to the drumlin shape and glaciotectonic deformation structures are absent. Sharpe (1987) interpreted the sequence to have formed as fill in a subglacial cavity. However, within this same drumlin field there are many drumlins with internal compositions of massive diamictons with little or no evidence of stratified sediments present. In some cases (close to the town of Hastings) drumlins of both types are found adjacent to each other.

Certainly much research remains before the variability in drumlin morphology and facies composition is understood. However, sedimentologic study, integrated with regional geologic history, promises to give us a better understanding of drumlin formation and to provide more realistic input for modelling drumlin processes (McCabe and Dardis, 1989). It is clear from several recent sedimentologic studies that lodgement till may not be present in all drumlins and that internal sedimentary sequences may not have been pervasively deformed (cf. Boulton, 1987a; Menzies, 1990c; Menzies and Maltman, 1992).

9.2.2. Low-relief Areas Glaciated by Continental Ice Sheets Underlain by Relatively Hard Igneous and Metamorphic Bedrock (Shield Areas)

9.2.2.1. Shield terrain of the Laurentide Ice Sheet in the Haliburton region, southern Ontario (Kaszycki, 1987)

The Haliburton area is typical of large portions of the Canadian Shield: relief ranges between 50 and 160 m (averaging ~70 m), bedrock lithologies include a variety of igneous and metamorphic rock types relatively resistant to glacial erosion, and topography is partly controlled by bedrock structure. In such an area, the erodibility and roughness of the glacier bed influence the amount of debris produced, its dispersal within the ice, and ice flow dynamics, particularly during the final phases of deglaciation (Kaszycki, 1987). Again, complex facies relationships occur and, in this setting, result from local conditions and changing depositional environments with time.

Sedimentary successions can be subdivided into upland sequences dominated by glacigenic diamicton facies deposited in subglacial and ice-marginal environments and valley fill sequences dominated by proglacial fluvial and lacustrine sedimentation in ice-marginal and proglacial environments (Chapter 3; Menzies, 1995a, Chapters 11 and 12).

Upland areas have rugged, irregular topography with local relief <50 m. The sediment cover is usually thin and discontinuous. A depositional model for upland shield areas (Fig. 9.4) depicts temporal changes from active-ice phases, with continuous glacial cover, to stagnation and the resulting local separation of ice blocks, during deglaciation.

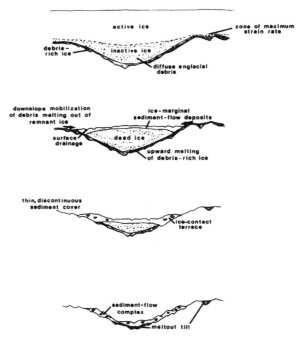

FIG. 9.4. Generalized model for subglacial and ice-contact sedimentation in upland shield terrain of the Haliburton area (Kaszycki, 1987, reproduced by permission of the National Research Council of Canada)

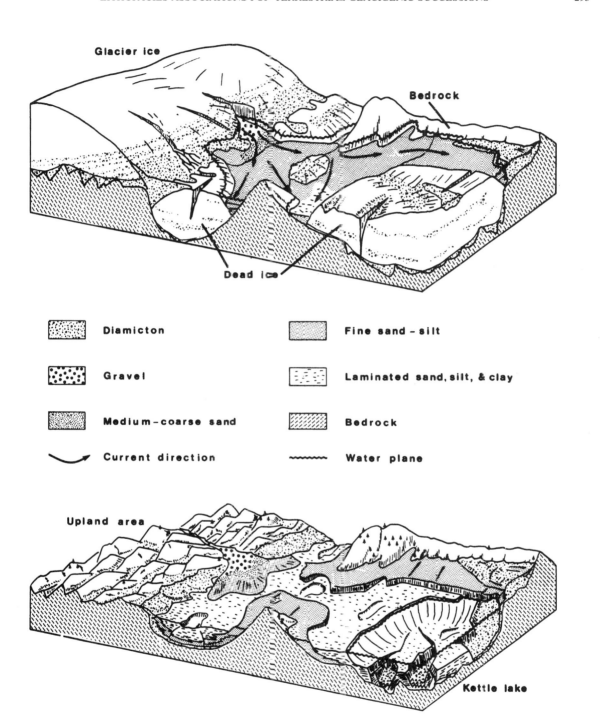

FIG. 9.5. Generalized model for proglacial basin-fill sedimentation in shield terrain of the Haliburton area (Kaszycki, 1987, reproduced by permission of the National Research Council of Canada)

TABLE 9.4.

Facies		Main Characteristics and Sedimentary Structures		Total Percent Sorted Sediment	Other Diamicton Properties				Basal Contact
No.	Abbreviated Title	Diamicton	Sorted Sediments within diamicton		Matrix Texture	Clast shape and Size Distribution	Frequency of Striations	Dominant Clast Derivation	
1	Massive diamicton	Massive, unsorted, matrix supported, dense; Underlying materials often compressively deformed, eroded and incorporated in base of diamicton as matrix or intraclasts; matrix rich in silt and clay.	Sand and silt lenses absent.	Nil	Sandy-mud (50-70% silt & clay with silt generally > clay)	Mainly subangular to subrounded Some facets; Dominantly pebbles to cobbles	Abundant (about 35 to 50% of all soft clasts, mainly shale & limestone, striated)	Local	Sharp and planar; gradational when underlain by facies 2 to 6
2	Diamicton with steeply dipping sand lenses	Same as Facies 1	Rare, thin, lenses and beds of sand and/or gravel with strong (10 - 30°) downvalley dips.	Less than 5	As above	As above	As above	Local	As above
3	Diamicton with plano-convex sand lenses	Similar to Facies 1 except more variable matrix textures	Rare, horizontal, sand and gravel lenses with strongly to slightly convex upper surfaces and planar to slightly trough-shaped lower surfaces (plano-convex); often with undisturbed cross-stratification conformable to the lenses' lower boundaries)	Less than 5	As above	As above	As above	Local	As above
4	Diamicton with abundant abraded and embedded clasts	Same as facies 1 except: clasts at base of diamicton commonly partially buried (embedded) in compressively deformed underlying sediments; all soft clasts are heavily striated parallel to a-axis; faceting on upper clast surfaces common; may be clast supported.	Absent or as in Facies 3	Less than 5	As above	"Barrel and Bullet" shapes abundant; Fractures and facets common; Pebble to boulder sizes	Very abundant (about 70-90% of all soft clasts heavily striated)	Local	Sharp and planar (erosional)
5	Stratified diamicton	Horizontal diamicton strata (commonly 5 - 20 cm thick and 10's of meters wide) distinct mainly due to color and texture differences; strata unsorted, drape boulders & have diffuse boundaries	Absent or as in Facies 3	Less than 5	As above but with clay generally > silt	Similar to facies 1; low total clast content	Common (about 20 -50% of all soft clasts striated)	Local	gradational
6	Diamicton with circular sand lenses	Similar to facies 1; large clasts often protrude into lenses from encasing diamicton	Sand and gravel lenses approximately circular (in cross-section); stratification in lenses deformed often normal faulted; sand and gravel beds exhibit large textural variation, often poorly sorted and contain diamicton inclusions	Less than 10	As above but with similar silt & clay contents	Same as facies 5	Same as facies 5	Mainly local	gradational
7	Diamicton with normal faulted sand lenses	Cobbles and boulders abundant; Massive to weakly stratified due mainly to variations in clast size between beds; generally matrix supported, intermediate matrix textures; poorly compacted; clasts from high elevations common.	Irregular layers and trough-shaped lenses of normal faulted sands and gravels occur sporadically; stratification in lenses rare; poorly sorted; crude horizontal layering in diamicton accentuated by sorted layers.	Less than 20 (usually less than 10)	Sandy-mud to Muddy-sand (35-60% silt and clay)	Commonly angular to subangular; Cobbles and boulders unusually abundant	Striations rare	Non-local (often very far travelled and from high elevations)	Usually gradational but locally is clear and planar (erosional)
8	Diamicton with silt and clay beds	Massive, matrix supported, fine-grained, occasional folded and convoluted laminations; only 10 to 20% total clast content; diamicton complexly intercalated with silt and clay	Horizontally laminated silt and clay beds a few mm's to 10's of cm's thick; individual laminae usually laterally traceable for several meters; silts and clays contain scattered pebbles; large clasts in diamicton may protrude into underlying silts and clays and deform bedding.	5 to 50	Sandy-mud to Muddy-sand (45-65% silt and clay)	Subangular to rounded; Mainly pebbles to small cobbles	Common	Local or non-local (usually local)	Planar and sharp with small scale irregularities and load structures
9	Diamicton with compressively deformed sand lenses	Matrix to clast supported, diamicton beds (1 to 2 m thick) dominate (often more than 80%); stratification accentuated by vertical variability in clast size distribution in the diamicton; angular inclusions of underlying materials common in lower parts of diamicton;	Indistinct lenses and thin diffuse beds of sorted sediment result in poorly developed stratification; trough-shaped lenses common; sorted beds dip gently or are horizontal and sometimes traceable for 10's of meters; sands and gravels commonly poorly sorted and often compressively deformed (faulted, folded and/or sheared), deformation greatest in upper part; load structures common; laminated silts and clays locally present; some large clasts deform bedding.	10 to 20	Muddy-sand to Silty-sand (40 - 50 % silt and clay)	Same as facies 8	Common	Usually local	Sharp to gradational (often scoured)
10	Stratified diamicton with trough-shaped sand lenses	Diamicton less abundant and sandier than facies 9 (usually comprises 50 - 80% of the deposit); other diamicton characteristics similar to facies 9	Regularly shaped, well defined, plano-concave sand and gravel lenses (1 - 5 m wide and 0.5 - 1 m thick); little or no deformation of lenses; trough cross-beds conform to lower boundary; small scale horizontal laminae and planar cross-beds present in undisturbed form; sands and gravels moderately to well sorted.	20 to 50	Silty-sand to Muddy-sand (30 - 45 % silt and clay)	Same as facies 8	Common	Same as facies 8	Erosive with "scour-troughs" similar in dimension to lenses
11	Sandy diamicton with sand and gravel beds	Diamicton less abundant than sorted materials; matrix supported; unsorted to very poorly sorted; sandy; little or no stratification	Sands and gravels exhibit large and sudden changes in grain-size distribution, sorting, and bed orientation (dips up to 40°); frequently folded and faulted; load and fluid injection structures common; angular inclusions often present	More than 50	Muddy-sand (25 - 35 % silt and clay)	Subangular to subrounded with some angular clasts; pebble to boulder sizes	Common to rare	Local to non-local	Highly irregular & often deformed; generally erosional
12	Gravelly diamicton with sand and gravel beds	Gravelly, generally clast supported, grading into moderately sorted gravels; higher clast contents and coarser matrix textures than facies 11	Sands and gravels generally horizontally bedded, well to very well sorted, weakly to moderately horizontally stratified and planar & trough cross-stratified; some ripple bedding; bedding commonly dips to valley sides; clasts in gravels sometimes imbricated and rounded to well rounded	More than 50	Muddy-sand to Sand (15 to 30 % silt and clay)	Subangular to rounded; mainly pebbles & cobbles, few boulders	Rare	Local to non-local	Usually sharp and planar

*   Pebble fabrics are 3-D representations of the a-axis orientations of blade and prolate shaped, medium to large pebbles.
**   S1 is a normalized representation of the strength of clustering around the principle eigenvector (V1).
*** See text for explanation.

TABLE 9.4. Continued

Frequency of Occurrence and Lateral Extent	Pebble Fabric *	S 1** Value	Facies Associations	Characteristic Trends			Facies		Interpretation***	
				Sorting	Clasts		Facies Number	Type of deposit and depositional processes	Position of Formation	Position of Deposition
					Fabric	Glacial Markings				
Very common (more than 50 localities); in thicknesses of 10 m or more, laterally continuous for 10's of meters	Strong, unimodal, V1* parallel to valley, shallow upvalley plunge	0.78 0.74	Vertically grades into facies 2 to 6				1	Basally derived till of indeterminate origin (probably mainly lodgement and meltout)	Mainly Subglacial	SUBGLACIAL / BASAL
Uncommon (less than 10 localities); occurs downvalley of bedrock obstructions usually a few meters thick as defined by associated lenses	Same as Facies 1 except a-axes exhibit steep downvalley plunge (5 - 15°) and fabric may be slightly bimodal	0.67	Vertically grades into facies 1 and 3 to 6.				2	Lee-side till deposited by sub-glacial flow, melt-out and/or lodgement processes		
Uncommon, usually only a few meters thick as defined by associated lenses	Same as Facies 1	0.70 0.67 0.67	Occurs at the base of massive diamictons (facies 1 or 2)				3	Meltout till (and possibly also lodgement till in some localities) deposited at the glacier base		
Occurs sporadically; usually thin (less than 2 m) and discontinuous; may be represented only by one or more embedded clasts (10-15 localities)	Moderately to very strong, unimodal, transverse or parallel, horizontal or slight upvalley plunge	0.64	Usually underlies facies 1, 2 or 3 & unconformably overlies sorted deposits (commonly sands and gravels) or facies 8 to 12				4	Lodgement till	Subglacial	
Uncommon, obscure, 5 to 20 meters thick, individual bands laterally traceable for 10's of meters	Unimodal, moderately strong, V1 parallel to valley, slight upvalley plunge	0.57 0.55 0.55	Associated with facies 6 & often conformably overlies facies 1, 2, 3, and/or 4.				5	Meltout till deposited at the glacier base	Englacial? Englacial	
Rare (<5 localities); thickness and lateral extent defined by associated lenses	Moderate to weak, unimodal to multimodal, V1 usually parallel to valley	0.50	Associated with facies 5 and often conformably overlies facies 1, 2, 3, and/or 4.				6	Meltout till formed at positions within ice probably near the glacier base		
Common on surface (20-30 localities); average about 5 m's thick; laterally extensive for 100's of meters; locally thin or absent	Moderate to weak; unimodal, bimodal, or multimodal; V1 parallel or oblique to valley; high a-axis dips	0.56 0.55 0.47	Commonly forms the surface deposit and is closely associated with facies 10; often overlies facies 6				7	Flow tills deposited by supraglacial mass movements with intermittent fluvial activity		SUPRAGLACIAL
Locally common; usually < 5 m thick (diamicton beds often < 1 m thick); laterally extensive (diamicton often occurs as lenses several m wide)	No data	No data	Often associated with thick sequences of silts and clays				8	Sub-aquatic till flows interbedded with glacial lacustrine sediments		
Common; thickness variable, generally less than 10 m's; laterally extensive for 10's of meters.	Weakly developed; V1 randomly oriented; multi-modal; dip of a-axis commonly low.	0.50 0.49	Usually overlain by facies 1, 2, 3 or 4; and often conformably underlain by facies 10				9	Proximal mass movement tills with intermittent fluvial sediments	Mainly Frontal (Proglacial)	ICE-MARGINAL
Common (especially on the surface); thickness and extent as in facies 9	Weak; multi-modal; V1 often parallel to valley; high, upvalley dips common on a-axes	0.49	Associated with facies 7 on the surface; in the sub-surface underlain by facies 8, 11, or 12 or sorted deposits & overlain by facies 9.				10	Subaerial till flows interbedded with stream channel deposits		
Common (especially near the surface); 10's of m thick (diamicton generally 1 - 5 m thick); laterally extensive	No data	No data	Laterally gradational with facies 10 and 12				11	Subaerial till flow and debris flow deposits interbedded with proximal glacial fluvial outwash	Frontal and Lateral	
Same as facies 11 but more common in the subsurface	No data	No data	Closely associated with thick sand and gravel deposits and facies 10 and 11.				12	Subaerial debris flow deposits interbedded with proximal outwash		

Vertical trend arrows in the "Characteristic Trends" section:

- Increasing proportion of sorted materials →
- Coarser mean grain size of diamicton matrix and greater sorting of associated sands and gravels →
- Increasing pebble fabric strength (S1 value) →
- Higher numbers of striated and faceted clasts and greater compaction of diamicton →

Upland areas are discontinuously veneered by massive and stratified melt-out till <1 m thick. Clast fabrics in the melt-out till are consistently oriented and stratification consists of few to abundant sub-horizontal stringers of sand as well as occasional clasts of contorted fine sand and silt. Locally flanking or filling narrow troughs in the bedrock surface are thicker till sequences (>5 m thick). This thicker till is an extremely compact, massive to fissile diamicton with well oriented clast fabrics, striated bullet-shaped boulders and laminae of sand and granules along the fissile partings. Occasionally present within this diamicton are meltwater sediments that were deposited in subglacial conduits. The thick, compact diamicton is interpreted to have formed by shear stacking of debris-rich basal ice in bedrock basins and subsequent melt-out beneath active ice. Ice-contact terraces also flank bedrock knobs and are perched along the valley sides. The lower part of the terrace sequence consists of subaqueously deposited sediment-flow diamictons interbedded with either glacio-fluvial or glaciolacustrine deposits, depending on local drainage conditions. This assemblage is overlain by subaerial sediment flows deposited after free drainage (breaching of stagnant ice blocks) was established. In places, the upper part of the ice-contact terrace sequence consists predominantly of massive, matrix-dominated diamicton deposited by sediment flows of low water content.

The thickest deposits occur in valley-fill sequences where over 70 m of sediment may be preserved. Complex sedimentary assemblages developed because of location with respect to retreating and stagnant ice and changing depositional environments with time (Fig. 9.5). Initially, the surface of stagnant ice within the valleys was above the level of proglacial lakes, and glaciofluvial sediment was deposited as ice-contact terraces and valley-train deposits (Menzies, 1995a, Chapter 13). Proglacially, density underflows of sediment-charged meltwater from subglacial drainage systems were funnelled between rock knobs and blocks of stagnant ice, producing fining-upward subaqueous outwash sequences later left as ice-contact terraces. These fining-upward sequences commonly consist of diffusely bedded coarse sand and ripple-drift cross-laminated to normally graded sand, capped by laminated sand, silt and clay. In large valleys with

continued ice wastage, kettle lakes formed. Prolonged sedimentation resulted in a series of lacustrine terraces which outline the configuration of the remnant ice blocks.

9.2.3. Ice Caps and Valley Glaciers of Mountainous Areas

9.2.3.1. Montane glacigenic diamictons of the Jasper region, Alberta, Canada

Montane, valley-glacier settings are also distinguished by complex facies successions representing deposition in changing depositional environments through time. Twelve major facies representing deposition in proglacial, ice-marginal, subglacial, englacial and supraglacial environments were described in deposits of the Jasper region by Levson and Rutter (1988) and given informal numbers 1 to 12. Properties, relationships, associations, and interpretations of these facies are summarized in Table 9.4, although the reader is strongly encouraged to refer to the original manuscript.

The interpretations made by Levson and Rutter are based on a variety of properties, including diamicton structure, texture, clast fabric and facies associations, and the facies designations rely heavily on the type, geometry, bed contacts and deformation structures of sorted sediments intimately associated with the diamictons.

Facies 1 to 6 comprise a sedimentary association formed in subglacial or basal environments. Facies 1 to 4 commonly grade laterally and vertically into one another and were deposited subglacially. Diamicton in Facies 1 is massive, and its origin is indeterminate; facies relationships suggest it was deposited by subglacial lodgement and melt-out processes. Facies 2, consisting of massive diamicton with sporadic, steeply dipping sand lenses, occurs on the lee side of bedrock knobs. Diamicton comprising this facies is interpreted to result from subglacial flow, melt-out and possibly lodgement on the lee-side of bedrock obstacles. Massive diamicton in Facies 3 contains sorted deposits in 'plano-convex lenses' (Shaw, 1982; Levson and Rutter, 1988). Presence of the plano-convex lenses suggests the diamictons were deposited by subglacial melt-out. Massive diamicton in Facies 4

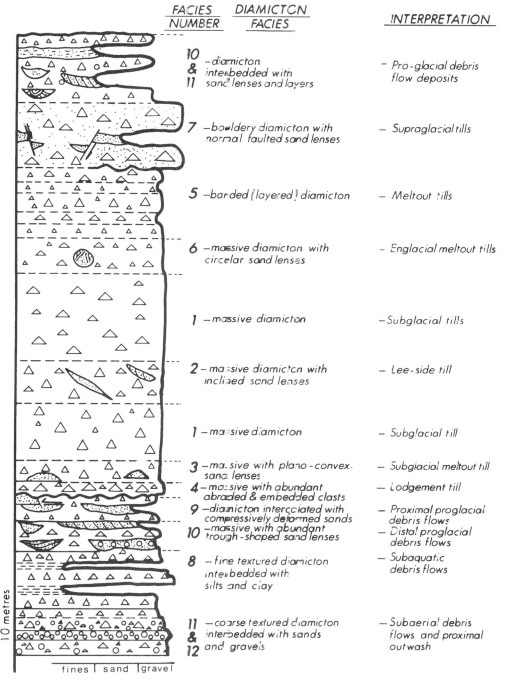

FACIES NUMBER	DIAMICTON FACIES	INTERPRETATION

10 & 11 – diamicton interbedded with sand lenses and layers — Pro-glacial debris flow deposits

7 – bouldery diamicton with normal faulted sand lenses — Supraglacial tills

5 – banded (layered) diamicton — Meltout tills

6 – massive diamicton with circular sand lenses — Englacial meltout tills

1 – massive diamicton — Subglacial tills

2 – massive diamicton with inclined sand lenses — Lee-side till

1 – massive diamicton — Subglacial till

3 – massive with plano-convex sand lenses — Subglacial meltout till

4 – massive with abundant abraded & embedded clasts — Lodgement till

9 – diamicton intercalated with compressively deformed sands — Proximal proglacial debris flows

10 – massive with abundant trough-shaped sand lenses — Distal proglacial debris flows

8 – fine textured diamicton interbedded with silts and clay — Subaquatic debris flows

11 & 12 – coarse textured diamicton interbedded with sands and gravels — Subaerial debris flows and proximal outwash

10 metres

fines | sand | gravel

FIG. 9.6. Hypothetical sequence of glacial sediments deposited during a single glacial advance-retreat cycle in a mountain area. Facies 8, 11, and 12 are conformably overlain by facies 10, which grades upward into facies 9. A sharp erosional unconformity separates all the preceding deposits from facies 1 to 4, which occur as a complexly interbedded association conformably overlain by facies 5 and 6. Facies 7 commonly occurs at the tip of the sequence, but locally grades laterally or vertically into facies 10 and/or 11. The sequence consists of proglacial deposits overlain successively by subglacial and supraglacial tills and proglacial debris flows (Levson and Rutter, 1989, courtesy of A.A. Balkema, Rotterdam)

contains abundant striated and faceted clasts. Clasts at the base of this facies are partially embedded in the underlying substrate, and indicate this diamicton was deposited by lodgement.

Facies 5 and 6 conformably overlie Facies 1 to 4 and are interpreted to have been formed by subglacial melt-out. Facies 5 consists of layered diamictons in which the layers differ in composition and/or clast percentage. The diamicton strata are probably inherited from the compositional differences in basal ice debris layers. Clast fabrics of this facies are consistently oriented. Facies 6 is composed of massive diamicton containing disturbed, tubular lenses of sand and gravel. The tubular lenses of sand and gravel are interpreted to be meltwater sediments that filled circular englacial tunnels near the glacier base, while the associated diamictons were probably deposited by melt-out of basal and/or englacial debris in the surrounding ice.

Facies 7 to 12 form a second facies association deposited in ice-marginal and proglacial environments. Consequently, Facies 7 to 12 may be found both underlying and overlying Facies 1 to 6. A supraglacial origin for Facies 7 is indicated by (1) massive to stratified diamicton with high clast content, (2) clasts whose compositional and surface properties indicate supraglacial transport, and (3) association with sorted deposits with normal faults caused by ice melt-out collapse. Facies 8 consists of massive diamicton beds in laminated clays and silts. These diamicton beds are interpreted to be subaquatic debris flows into proglacial lakes (see Menzies, 1995a, Chapter 13). Facies 9 consists of intercalated diamicton, sand and gravel with compressive deformation structures, and is interpreted to be proglacial deposits overridden by the glacier. Facies 10, characterized by diamicton with abundant trough-shaped lenses of sorted sediments, is interpreted to have formed in an ice-marginal environment where debris flows were periodically incised by meltwater streams. Facies 11 consists of sandy diamicton interbedded with sands and gravels, and is interpreted to consist of debris flows into proximal outwash. Facies 11 grades to Facies 12, poorly sorted gravelly diamictons occurring as uniformly thick, horizontal beds in horizontally stratified and cross-stratified sand and gravel. This facies is also interpreted to consist of

subaerial debris flows interbedded with proximal outwash deposits.

Fig. 9.6 shows a hypothetical vertical sequence that could form during an advance-retreat cycle in a montane glacial environment. Oscillating advances of the ice margin could result in repetition of some or all of the facies shown for a single cycle.

9.2.4. Polar Ice Caps

The limited study of depositional sequences in polar regions has usually been concentrated at specific sites along modern glacial margins in Antarctica. Fitzsimons (1990) suggests that even here climatic differences are important and that depositional sequences differ between arid polar and polar maritime environments. In both settings, basal ice at the terminus is below the pressure melting point; there is no flow by slip over the bed, and movement occurs slowly by internal deformation of the ice (cf. Echelmeyer and Wang, 1987). Since the glacier is frozen to the bed, there is little subglacial erosion and debris concentrations in the basal ice are low. In arid polar environments, there is little surface melting and sublimation occurs (Shaw, 1977a, 1988b). In polar maritime environments, melting occurs, if only for short periods during the year. Low debris concentrations and the availability of meltwater commonly result in resedimentation of surficial deposits by various mass-wasting processes (Lundqvist, 1987; Fitzsimons, 1990). Melt-out till is preserved only where debris is melted out either onto a low, stable ice slope or where the debris is released under confining conditions and cannot be mobilized (Fitzsimons, 1990).

A hypothetical vertical sequence related to a glacial advance-retreat cycle for arid polar glaciers (Shaw, 1977a,b, 1988b) is presented in Fig. 9.7. As ice advances, proglacial fluvial sediment becomes overlain by sediment flows and stratified sediment deposited in proximal ice-marginal environments. After ice advance, poorly attenuated and highly attenuated sublimation tills (which preserve intricate, glacier-derived foliation) form beneath a thin sediment cover. The sediment cover includes a complex of deposits formed in the supraglacial environment

A   Advancing glacier

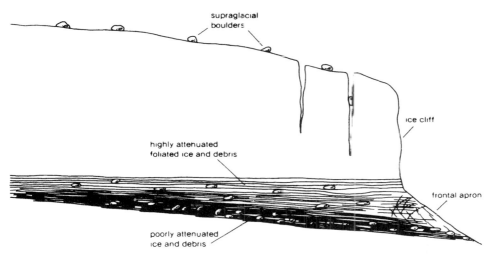

B   Retreating glacier

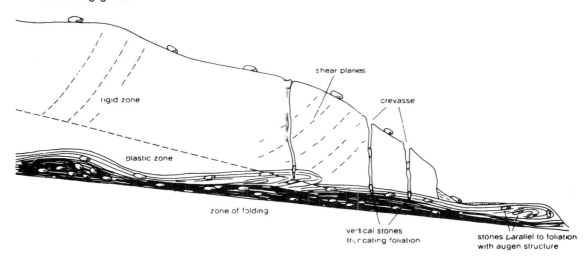

C   Depositional sequence

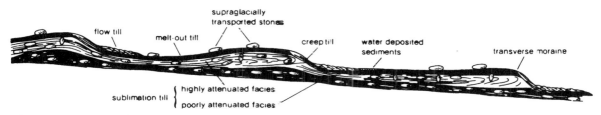

FIG. 9.7. A general model for glacial transport and deposition in arid polar environments (from Shaw, 1977a, reproduced by permission of the author)

either by let down of supraglacial debris or various mass-wasting processes. With ice retreat, resedimented materials and water-deposited sediments, formed in ice-marginal and proglacial environments, are added to the sediment sequence.

In polar maritime environments, the dominant glacigenic deposits are sediment flows and other secondary sediments which form in association with lacustrine and fluvial sediments (Lundqvist, 1987; Fitzsimons, 1990). Melt-out till is preserved only when this debris cover is consolidated and slows the melting of debris-rich ice or constrains the newly melted-out debris so that it is not mobilized directly after its release. The presence of sublimation till in this environment is problematic (Fitzsimons, 1990). Crag-and-tail forms (Lundqvist, 1987) and lodgement till smeared on erratic boulders (Fitzsimons, 1990) occur in very limited areas. Both the forms and sediments are interpreted to have formed under conditions different than at present.

9.3. CONCLUSIONS

The examples presented in this chapter illustrate the complex and differing facies associations that can be deposited in various terrigenous glacial environments. This complexity results from changing depositional environments through time, the particular ice dynamics of the ice mass involved and various local factors. It has been suggested that basal thermal regime is the most important control on depositional processes of glaciers (Boulton, 1972a,b; Eyles et al., 1983; Eyles and Menzies, 1983). However, it is unclear from examples in this chapter how or if successive changes in depositional environments are related to basal thermal regime.

At present, even though the tools to justifiably evaluate former glacial successions are available, too few studies have been made to make useful predictive facies models. Studies of modern glaciers have only recently provided the necessary criteria to make prudent genetic interpretations. These criteria require detailed documentation of lateral and vertical facies relationships, as well as documentation of deposit properties, including diamicton structure, texture and clast fabric, and the type, geometry, bed contacts and deformation structures of sorted sediments intimately associated with these diamictons. If predictable facies models are to be made, more studies must be conducted using these criteria, and they must be integrated into a regional setting. Only then will it be possible to accurately predict facies sequences and reconstruct past ice dynamics and glacial environments.

The studies cited here are merely examples of the differing sedimentary successions that develop in formerly glaciated areas. Many others could have been given. Additional complexity not presented includes discussion of various Scandinavian landforms (Rogen moraine, Veiki moraine, etc.) (Chapter 2) and widespread glacio-aquatic sequences (Chapters 4 and 5) (Menzies, 1995a, Chapters 13 and 14).

Recently, pervasive deformation of subglacial sediment beds has been shown to occur under many glaciers and ice streams (Boulton and Hindmarsh, 1987; Alley et al., 1987a,b). Recognizable facies associations would confirm this form of ice movement (Menzies, 1989a; Alley, 1991; Boyce and Eyles, 1991). At present an interesting controversy has developed over the presence or absence of (e.g. Clayton et al., 1989; Clark and Hansel, 1989; Johnson and Hansel, 1990) evidence supportive of pervasive subglacial deformation (Beget, 1986; Menzies, 1990c; Clark, 1991; Alley, 1991; Hicock, 1991, 1992; Hicock and Dreimanis, 1992) (Chapter 2). If areas of subglacial deformable bed conditions are found to be widespread, there will be even greater facies complexity than recognized in this chapter.

Chapter 10

PALEOSOLS

J. Boardman

10.1. SOILS AND GLACIAL DEPOSITS

Soils are important in the study of glacial sediments because of their stratigraphic importance: they represent periods of landscape stability prior to, within or post-dating glacial episodes. Soils are of value in estimating the time interval between glacial units. It is also essential to recognise the effects of weathering and soil formation and not to confuse these transformations with lithological variations in a glacial deposit.

Soil-forming processes begin in the glacial environment as soon as ice, meltwater or wind has ceased to disturb the surface of the sediment and weathering and biogenic processes have begun to operate. The beginning of soil formation in recently deglaciated environments is the subject of several studies. Boulton and Dent (1974) show that some changes are surprisingly quick and precede weathering and organic accumulation. In recently deposited till in south-east Iceland textural changes begin immediately after ice retreat. The changes are due to silt translocation and wind ablation, the latter resulting in a surface stone layer and loess accumulation where vegetation traps silt and fine sand (Fig. 10.1).

In contrast, soil formation in a geomorphologically active glacial environment of hummocky disintegration moraines is discussed by Acton and Fehrenbacher (1976). Figure 10.2 illustrates the interaction between mass movement processes, redistribution of sediment

and soil formation. Soils develop on slopes that are temporarily stable but the organic-rich horizon is in part stripped and redeposited as buried soil material. This contrasts with *in situ* buried soils which represent phases of stability and soil formation in a zone of episodic sediment accumulation. The distinction may be important in terms of environmental reconstruction since the transported buried soil may represent a single episode of erosion and deposition and, an *in situ* soil, several years of stability.

The process of plant succession will begin in the recently deglaciated zone if the land surface remains stable and the climate sufficiently temperate. Plant life is responsible for the build-up of organic matter which is both the first clearly visible index of soil development and is the necessary condition for further colonisation by plants and animals including micro-organisms. The effects of weathering processes such as oxidation and decalcification are likely to become visible over time scales of 10^2 to 10^3 years. In the recognition of these and other soil-forming processes such as podzolisation and gleying, much emphasis is placed on change of colour affecting the unweathered parent material and on the development of horizons. Soils progress through time from being 'poorly developed' in their lack of textural modification and horizonation to being 'well developed' with textural changes due to weathering and translocation of fine materials and clear horizonation. Laboratory analysis of soil properties is frequently used to

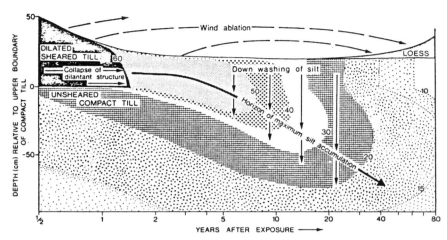

FIG. 10.1. Schematic diagram showing the change in soil structure and texture with time in the lodgement till stratum (Boulton and Dent, 1974, by permission of the *Scandinavian University Press*)

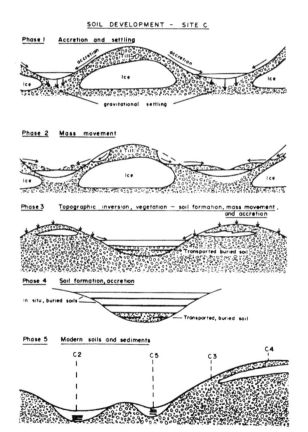

FIG. 10.2. Hypothetical scheme for the origin of deposits and the development of soils in hummocky disintegration moraines (Acton and Fehrenbacher, 1976, reproduced by permission of the *Royal Society of Canada*)

confirm what is evident in the field although the investigation of soil chemistry and mineral transformation necessitates the use of techniques such as XRD and thin section examination (Bullock, 1985; Jenkins, 1985).

In very cold environments such as Antarctica, soils on glacial deposits show signs of textural modification due to weathering but do not contain organic matter. The differentiation of soils is based on surface rock characteristics, soil colour and depth, horizon development and the presence of salts. Campbell and Claridge (1987) propose a series of soil weathering stages based on these criteria (Table 10.1).

Examination of soils that are forming at the present day in glacial environments may assist in the interpretation of soils which are encountered in geological investigations of older glacial deposits. For example, the study of features resulting from ice lensing, freezing and thawing and frost creep in periglacial areas may be used to interpret soils developed in the Weichselian loess of northwestern Europe (Van Vliet-Lanöe, 1990).

10.2. PALEOSOLS

Catt (1987, p. 487) suggests that a paleosol may be loosely defined as, 'a soil formed in past periods on old land surfaces'. Attempts at precise definition run

TABLE 10.1. Morphological development of soils with increasing weathering in the Antarctic

Weathering stage	Surface rock characteristics	Soil colour	Horizon development	Soil salts	Soil depth	Other
1	Fresh, unstained, coarse and angular	Pale olive to light grey (5Y 6/3–7/2	Nil	Absent	Very shallow, underlain by ice	Moderate patterned ground development
2	Light staining, slight rounding, some disintegration	Pale brown to light brownish grey (10YR 6/3–2.5Y 6/2	Weak	Few flecks	Shallow, underlain by ice	Moderate patterned ground development
3	Distinct polish, staining and rounding, some cavernous weathering, some ventifacts	Light yellowish brown (10YR 5/3–2.5Y 6/4	Distinct	Many salt flecks in upper party of profile and beneath stones	Moderately deep	Some disintegration of boulders in the soil, slight increase in fine fraction
4	Boulders much reduced by rounding, crumbling and ventifaction, strongly developed carvernous weathering; staining and polish well developed; some desert varnish	Yellowish brown (10YR 5/4) in upper horizons, paler in lower horizons	Very distinct	In discontinuous or continuous horizon beneath surface	Deep	(As for stage 3)
5	Few boulders, many pebbles forming pavement, extensive crumbling, staining, rounding, pitting and polish	Dark yellowish brown to yellowish red (10YR 4/4–5YR 5/8)	Very distinct	In horizon 20–30 cm from surface and scattered throughout profile	Deep	(As for stage 3)
6	Weathered and crumbled bedrock, very strongly stained	Strong brown to yellowish red and dark red (7.5YR 5/6–5YR 4/8 or 2.5YR 3/6)	Very distinct	(As for stage 5)	Shallow to deep	Bedrock sometimes crumbled to 50 cm depth

into several problems but it is retained as a convenient, widely-used term.

The value and significance of paleosols in studies of glacial sediments relates to a basic concept in soil science: that soils represent zones of weathering at and below a stable land surface. The recognition of a soil therefore implies a period of stability with relatively low rates of erosion or deposition. Many soils also contain organic material although the organic-rich A horizon may be unrecognisable or has been lost by erosion in some paleosols (Plate 10.1).

PLATE 10.1. An arctic brown soil developed on an earlier argillic soil in sand and gravel parent material, at Broomfield, eastern England. Developed within the arctic brown soil are wedges associated with a more extreme phase of the arctic climate. The soil is a buried paleosol overlain by a brown clay till. The arctic brown soil and the sand wedges are a development of the Barham Soil, which is an early Anglian (Middle Pleistocene) soil stratigraphic unit beneath the Anglian glacial deposits (Lowestoft Till, probably OI Stage 12) in eastern England. The argillic soil, (Valley Farm Soil) which can be recognised by the yellow brown material at the lower part of the section formed in warm and seasonally moist temperate climates during a long period of the 'pre-glacial' Middle Pleistocene of eastern England. The sand and gravel parent material is braided channel sediments of the River Thames (Kesgrave Sands and Gravels) and the surface upon which the soil formed is a former terrace of that river, which is now buried beneath glacial deposits

PLATE 10.2. A cold water gley soil at Kirkhill, northeastern Scotland. This soil is developed on glaciofluvial sands and gravels and is buried beneath slightly humic sands and gelifluctate. The buried paleosol shows a compressed black A horizon above a well developed white Ea horizon, overlying in turn a weakly oxidised Bs horizon. The lower glaciofluvial deposits are attributed to the Anglian (Middle Pleistocene, OI Stage 12) glaciation, the soil is thought to have formed during a subsequent Middle Pleistocene temperate stage, with the profile shown here representing development in the cold climate at the end of the temperate stage. The overlying periglacial deposits are interpreted as forming during a later Middle Pleistocene glaciation. All the sediments shown here are buried by till formed during the Last Glacial maximum

Recognition of paleosols is not always easy. In vertical section, glacial deposits may show abrupt or gradual lithological changes: these may appear to be the result of weathering. Alternatively, zones of weathering may be mistaken for discrete lithological units. For example, in East Yorkshire, England, Madgett and Catt (1978) show that what had previously been regarded as a sequence of three tills is in fact only two. The third, the 'Hesle Till', is a postglacial weathering profile on whichever of the two tills is at the ground surface. The characteristic reddish brown colour of the 'Hesle Till' is due to oxidation of pyrite and siderite to depths of approximately 5 m.

The distinction is frequently made between 'buried' and 'relict' paleosols. A paleosol is buried if it is isolated from soil-forming processes due to burial beneath a later sediment. Isolation is sometimes difficult to prove and shallow burial (0.5–1 m) frequently results in the superimposition of later pedogenic features onto earlier ones. This is the case in Fig. 10.3 where the lower part of the Farmdale Soil profile is developed in the Sangamon Soil. A relict soil is one containing features that have formed under conditions that no longer exist at the site. Relict soils may become buried soils if they are traced laterally, for example, where they have been covered by landslipped material.

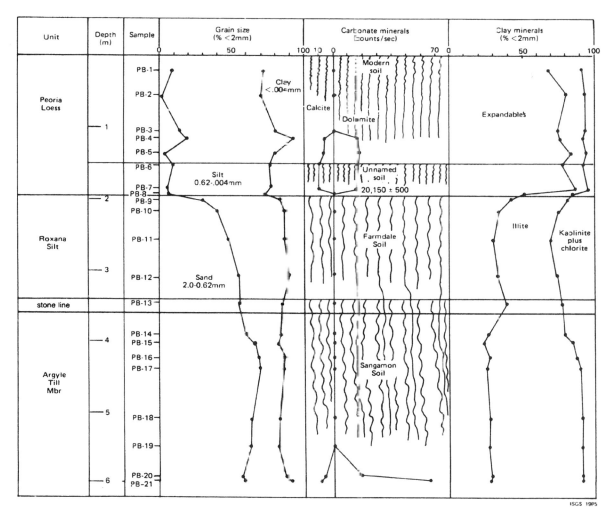

FIG. 10.3. Grain-size distribution, clay mineral composition, and carbonate data for the Wempleton southeast section, Illinois (McKenna and Follmer, 1990, reproduced by permission of the *Illinois State Geological Survey*). Vertical lines represent the extent of soil development for four named soils. Note that the lower part of the Farmdale Soil profile is superimposed upon the Sangamon Soil profile

Most soils will contain features that have formed under conditions that no longer exist at the site. These are referred to as 'paleosolic features'. The site may have become better drained, the vegetation cover may have changed resulting in a more acidic litter, or the regional climate may have become warmer; all these changes singly or in concert will affect soil properties. Many recent environmental changes that are reflected in soils are the result of site disturbance by humans.

There is no general agreement on the proportion of paleosolic features in a profile before it is regarded as a paleosol (i.e. a soil developed in a past period). Nor is it always easy to recognise features that are in disequilibrium with present conditions. There is some evidence that clay translocation in soils in Britain occurred primarily in the early Flandrian prior to 5000 BP (Weir *et al.*, 1971), in which case, clay-enriched horizons common in present day soils, are relict features. However, clay translocation is still occurring albeit at low rates in soils of northwest Europe (Burnham, 1964).

Paleosols should not be treated as isolated profiles but as part of a landscape (Valentine and Dalrymlple, 1975). However, this may be difficult because of a lack of widespread exposure. Significant progress was made by Ruhe (1969) in his studies of large areas with good stratigraphic control. Attempts to examine catenary relationships of soils developed in past landscapes that are now buried are rare, the notable exception being Follmer's (1982) detailed reconstruction of the Sangamonian surface in south-central Illinois. Much of this reconstruction has, of necessity, been based on borehole records. More rarely, quarry exposures allow soils to be examined in their paleo-hillslope position (Canfield et al., 1984).

One of the difficulties of reliance on single profile exposures is that the position in the former landscape may be unknown. Studies of soil catenas of different ages are needed. An admirable example of this type of work is provided by Swanson (1985) who examined catenas on Pinedale and Bull Lake age moraines in Wyoming. The best developed soils occur on the concave downslope position and the least well developed on the convex shoulder of the slope. This suggests that, even in the absence of landsliding and overland flow processes in these dry climates, soil creep and splash continue to modify the soils on slopes. It is also of interest that the differentiation between soils on the basis of slope position is greater on the older moraines, those of Bull Lake age. In a development of this study, Birkeland and Burke (1988) noted that on moraines of approximately 100,000 years of age in California, Bt horizons are not always found in soils on moraine crests but they do occur in footslope positions. This suggests that erosion has removed soils from sites often regarded as stable and that these processes may be ongoing rather than representing rapid adjustment to post glacial conditions. The history of soils and the geomorphology of the site are intimately linked and this is likely to be the case particularly on sloping ground such as glacial moraines.

In recent years elucidation of the pedogenic history of glacial deposits has used increasingly detailed analytical techniques. Complex histories are reflected in soil features and these have been related to environmental events known through other evidence. Micromorphological studies, however, are providing information that is unique to this mode of investigation (e.g. Fedoroff and Goldberg, 1982; Kemp, 1987).

10.3. PARENT MATERIALS

In the glacial environment the primary parent materials on which soil development takes place are till, glaciofluvial sands and gravels (outwash) and aeolian sediments, especially loess. To the soil scientist these offer great contrasts in their influence on soil development. Texturally, till may vary from a coarse-grained, well-drained sandy loam to a clayey, poorly-drained material on which soils will tend to be waterlogged. The mineralogical composition of glacial materials reflects their origin but may also provide opportunities to study the weathering history of the soil. This method requires detailed analysis of changes in the frequency of mineral species down profile and into the unweathered parent material (Weir et al., 1971; Madgett and Catt, 1978).

The study of soils and paleosols in glacial deposits is greatly simplified if the parent material is relatively uniform over long distances. The attraction of loess to the pedologist is that this condition can be met. The relative homogeneity of loess has allowed comparisons to be made of paleosols from different Quaternary temperate episodes (Catt, 1988a) (Chapter 6). With outwash deposits uniformity of parent material is less likely and with till, most unlikely. Some of the problems are illustrated with reference to outwash materials which form river terraces.

Outwash materials offer certain advantages for soil studies. Level topography and the possibility of similar parent materials allow comparisons to be made of soils on terraces which may be associated with particular glacial events (e.g. Porter, 1976). Higher terraces will normally represent older events and therefore relative age on the basis of geomorphic position may be assumed. This is not the case with, for example, glacial moraines. The possibility of dating organic material, ash or artefacts in terrace deposits adds to their attractiveness for this type of study.

In the uplands of Europe and North America glacial outwash deposits frequently form the upper members of terrace sequences that, in their lower part, are non-glacial. In lowland sequences the glacial influence is established on the basis of erratic presence and high paleo-discharges. Soils may be developed in fine-grained overbank or reworked loessic deposits which overlie coarser outwash materials.

Parent material of similar texture may occur on terraces of different age within the same river system (Robertson-Rintoul, 1986). In the uplands, sand and gravel is usually the dominant material but areas of fine-grained backwater and waning limb flood deposits may also occur (Smith and Boardman, 1989).

Terraces do not necessarily have a one-to-one relationship with glacial events. Terrace sequences of considerable altitudinal range may prove to be the result of a single glacial episode. Also, it may not be possible to separate glacial episodes on the basis of soil characteristics in associated terrace sequences. Parsons et al. (1981) examined soils on five terraces resulting from the catastrophic Spokane Floods that drained Glacial Lake Missoula. The terraces, with a combined relief of 141 m, could have formed within a period of 2000 years or less. Similar soils occur on the five terraces.

The age of soils within a terrace may not correspond to the age of the terrace surface. Some terrace sequences in upland Britain contain the lower part of soil profiles which have been truncated during recent flood episodes; young soils are developed in the upper gravelly deposits (Smith and Boardman, 1989). Pleistocene outwash sequences are frequently capped by last glaciation loess (~20,000 BP) which overlies or is mixed into older soil material.

10.4. DISTURBANCE AND SURVIVAL

Ideal conditions for the study of soils and paleosols on glacial materials are rarely met in practice. One problem frequently encountered concerns the loss of all or part of the soil profile. It is the exceptional site which has remained geomorphologically stable through long periods of Quaternary time and where the soil record reflects changing environmental conditions. This is particularly the case in glacial and periglacial environments.

In western Europe, non-glacial cold episodes have been characterized by either loess deposition, periglacial disturbance of land surfaces including mass movements of various types (Harris, 1987), or both. Mass movement has resulted in the stripping of previously formed soils. Soil formation has therefore begun again on relatively unweathered parent material at the close of periglacial episodes. This can be illustrated with reference to soils on Devensian and

TABLE 10.2. Decalsification depth and clay translocation in soils on reddish tills of Devensian and Wolstonian age from Britain (based on published Soil Survey data)

	Depth to carbonate (cm)				Minimum % illuvial clay			
	On Devensian tills		On Wolstonian tills		On Devensian tills		On Wolstonian tills	
Series	Range	Mean of finite values	Range	Mean of finite values	Range	Mean	Range	Mean
Salop	50->137	72	46->150	56	3–15	6.4	5–14	9.0
Clifton	80->144	80	–	–	1–18	7.6	5–14	–
Salwick	100->110	100	54->140	54	4–12	8.2	4–10	5.7
Cottam	80->120	103	50–100	70	2–20	8.0	4–17	9.3
Flint	43->90	52	46–62	52	1–10	5.5	3–14	8.8
Crewe	47->120	53	62–78	70	4–8	6.0	9	9.0
Wooton	77->100	78	–	–	1–7	3.8	–	–
Lea	77->100	78	–	–	–	–	–	–

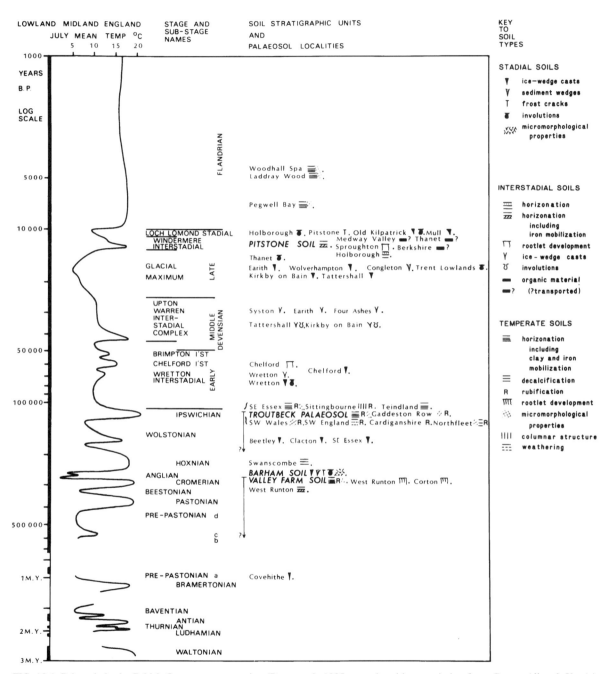

FIG. 10.4. Paleosols in the British Quaternary succession (Rose *et al.*, 1985, reproduced by permission from *George Allen & Unwin*)

Wolstonian tills in the English Midlands. Catt (1979) explains the similar values for clay accumulation and depth of carbonate removal in these soils as a result of their similar age: soils in both parent materials have developed in the last 10 or 15,000 years since Late Devensian episodes of mass movement and loess incorporation (Table 10.2). There are, however, exceptional sites where truncated soils on older glacial materials are preserved beneath later additions (e.g. Sturdy et al., 1979).

In North America the problem of periglacial disruption of interglacial soils appears to be less common. For example, in an early study, Lessig (1961) showed clear differences between soil development on outwash terraces of Wisconsinan and Illinoian age in Ohio. In the drier and flatter areas of the mid-West the survival of paleosols such as that of Sangamonian age suggests limited periglacial mass movement. Similarly, the dry climate of the eastern Rockies presumably accounts for the survival of Ball

Lake moraines and their soils in relatively pristine condition. The periglacial influence on soils is clearly much greater in the Appalachians, as reviewed by Clark and Ciolkosz (1988).

10.5. THE VALUE OF PALEOSOLS

10.5.1. Stratigraphic Markers

Paleosols have long been recognized as marker horizons that separate one glacial deposit from another. They have, therefore, a stratigraphic value that is now firmly established in Quaternary geology. In the British Isles several paleosols have been formally named and related to the British Quaternary stratigraphy (Rose et al., 1985a) (Fig. 10.4) (Chapter 8). In some areas of the world soils were accorded early recognition, notably in Illinois where Leverett's work laid the foundations for a detailed Quaternary stratigraphy including tills, outwash and loess (Foll-

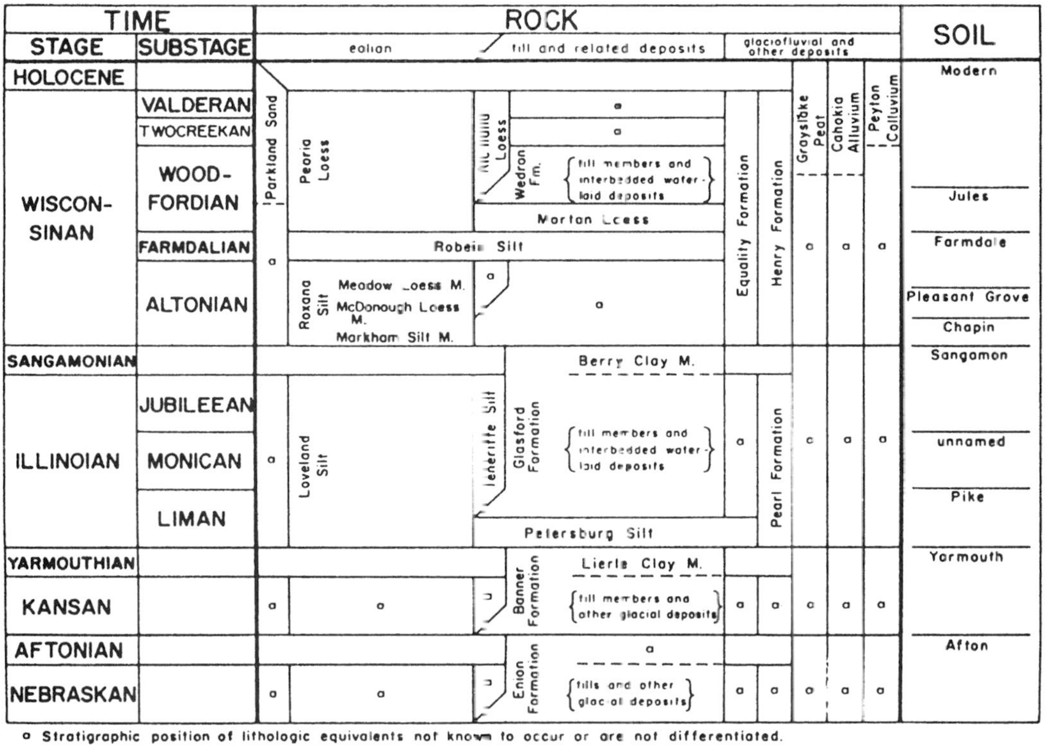

TIME		ROCK			SOIL
STAGE	SUBSTAGE	eolian	till and related deposits	glaciofluvial and other deposits	SOIL
HOLOCENE		Parkland Sand / Peoria Loess			Modern
WISCON-SINAN	VALDERAN		a	Graylake Peat / Cohokia Alluvium / Peyton Colluvium	
	TWOCREEKAN		a		
	WOOD-FORDIAN	Richland Loess / Wedron Fm.	{till members and interbedded water-laid deposits} / Morton Loess	Equality Formation / Henry Formation	Jules
	FARMDALIAN	Robein Silt		a	Farmdale
	ALTONIAN	Roxana Silt: Meadow Loess M. / McDonough Loess M. / Markham Silt M.	a		Pleasant Grove / Chapin
SANGAMONIAN			Berry Clay M.		Sangamon
ILLINOIAN	JUBILEEAN	Loveland Silt	Teneriffe Silt / Glasford Formation {till members and interbedded water-laid deposits}	a / Pearl Formation / a / a / a	unnamed
	MONICAN	a			
	LIMAN		Petersburg Silt		Pike
YARMOUTHIAN			Lierle Clay M.		Yarmouth
KANSAN		a	a	Banner Formation {till members and other glacial deposits} / a / a / a / a / a	
AFTONIAN			a		Afton
NEBRASKAN		a	a	Enion Formation {tills and other glacial deposits} / a / a / a / a / a	

a Stratigraphic position of lithologic equivalents not known to occur or are not differentiated.

FIG. 10.5. Quaternary stratigraphy of central Illinois. Modified from Willman and Frye (1970) (Follmer, 1978)

mer, 1978). The development of the concept of the Sangamon Soil, as the weathering zone between the deposits of the last and the penultimate glaciation, holds a central position in this aspect of scientific history. The relationship of soil stratigraphic units to rock stratigraphic units of largely glacial origin is shown in Fig. 10.5.

10.5.2. Age Indicators

Paleosols have proved useful as a means of dating glacial deposits. This is because they may contain datable material or their degree of development may be used to estimate the time elapsed since deposition of the parent material.

Dating methods for paleosols have been reviewed by Evans (1982). The following section emphasises those methods appropriate to paleosols in glacial deposits (Chapter 14).

10.5.2.1. Absolute age

The maximum age of recent glacial deposits has been established by the $_{14}C$ dating of buried A horizons from soils (Matthews, 1985). However, organic materials beneath glacial deposits are the exception rather than the rule. Thermoluminescence (TL) has been used for dating paleosols developed in loess (Wintle et al., 1984), and in dating ash layers incorporated into buried soils within glacial moraines in the northwestern United States (Osborn and Karlstrom, 1989). In the Antarctic the age of glacial deposits has been established by a combination of radiometric dating, especially potassium-argon dating of volcanic material, and the degree of weathering of soils and surface rocks (Campbell and Claridge, 1987). This approach has shown that soils on older glacial deposits are of Late Tertiary age.

10.5.2.2. Relative dating

In the absence of datable material, soil evidence may be used to estimate the relative age of deposits and also the time elapsed since deposition. In many areas, particularly where widely dispersed moraines occur, the number and age of glacial events is often unclear. Examination of the character of soils on these

deposits using a range of 'relative dating techniques' has proved to be a valuable line of enquiry (Birkeland, 1984). Some relative dating techniques use soils or soil properties; others examine surface morphology, clast weathering and lichen size and abundance.

Studies of this nature are based on Jenny's concept of the soil forming factors (Birkeland, 1984, chapter 6):

$$S \text{ or } s = f(cl, o, r, p, t, \ldots) \qquad (10.1)$$

In this empirical qualitative, open-ended equation, S represents the soil or s any soil property, cl the climatic factor, o the biotic factor, r the topographic factor, p the parent material and t the time factor. By holding other factors constant, the equation may be 'solved' for the time factor. Thus, for example, in a sequence of moraines, if it can be assumed that climate, vegetation, parent material and topography have remained similar, or not varied significantly, then it can be inferred that differing soil properties are a function of the length of time over which they have developed.

This approach has now been used widely in areas where there is good chronological control on the age of glacial materials, and in such circumstances it has been shown to be effective although not all soil properties display clear trends with age (Mellor, 1985; Birkeland, 1982). Many studies have tested several properties and selected those that seem to give reliable and useful results (i.e. trends with assumed age (Birkeland et al., 1980; Gellatly, 1985; Colman and Pierce, 1986)). In an investigation of four soils of estimated ages between 115,000 and 8000 years old, Evans and Cameron (1979) show that the property with the clearest age-related trend is the depth of the solum (A+B horizons). The solum increases in depth from 11 cm in the youngest to 77 cm in the oldest soil: an average rate of solum development of 1.0 cm/1000 years. The depth of solum has the advantage of being a property that is easily measured in the field; however, micro-topographical influences must be taken into account, for example, by sampling from similar geomorphological positions in the landscape (Swanson, 1985). Other investigators have insisted that a range of relative-dating techniques be used and that reliance on single criteria should be avoided

(Burke and Birkeland, 1979). In the Antarctic, relative dating of soils has played a major part in elucidating soil age and glacial history. A range of morphological criteria is used by Campbell and Claridge (1987) in defining the weathering stage of a soil (Table 10.1), that may be found on glacial deposits of different ages (Fig. 10.6).

Another approach has been to attempt to quantify profile data that reflect soil age. Indices are used that assign values to specific properties in comparison to the C horizon or parent material (Bilzi and Ciolkosz, 1977; Harden, 1982; Levine and Ciolkosz, 1983). In order to obtain absolute ages, the indices for a group of soils must be calibrated using radiometric dates. Karlstrom (1988) estimated the lengths of soil-forming intervals for paleosols on glacial tills in northern Montana and southern Alberta using this procedure. In Pennsylvania, soils in tills of Wood-fordian, Altonian and Pre-Wisconinan age are clearly distinguished by use of the Clay Accumulation Index (Levine and Ciolkosz, 1983).

Table 10.3 lists some of the soil properties that have been used to establish relative age. At many sites it will not be possible to use some of these techniques.

Soils on glacial deposits may not contain appropriate stone lithologies for weathering-rind studies. More frequently the constraint will be the age range of the soils under examination. The build-up of organic carbon in the A horizon will differentiate soils on moraines of less than 250 years age (Mellor, 1985), whereas the amount of translocated clay in the B horizon is a more appropriate method for investigating soils that span thousands of years in age. The lack of sensitivity of many of these techniques is a problem. For example, there would be little likelihood of differentiating soils of 80,000 and those of 90,000 years age using the build-up of clay. Rates of clay migration may be very sensitive to site factors which can rarely be assumed to have been constant: small differences in clay amounts may therefore be the result of factors other than time.

In using soils to date glacial materials considerable emphasis has been placed on the development of textural B horizons. It is argued that weathering of minerals in the A and E horizon produces clay and to a lesser extent silt that is translocated down the profile to form a clay-enriched B horizon (argillic horizon); seasonal climates seem to promote this process (Soil

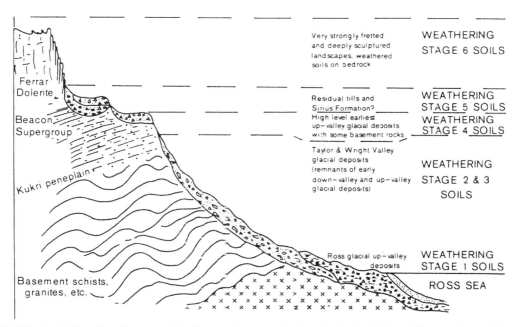

FIG. 10.6. Soils, glacial deposits and weathered landforms in the Antarctic (Campbell and Claridge, 1987, reproduced with permission from Elsevier Science Publishers)

TABLE 10.3. Relative dating of soils on glacial deposits: selected studies

Technique	Area	Reference
Weathering-rind thickness	New Zealand Idaho, USA Cascades, USA	Birkeland (1982) Coleman and Pierce (1986) Porter (1975)
Phosphate retention	New Zealand	Gallatly (1985)
Profile development	Baffin Island Cascades, USA Norway Alaska Baffin Island	Birkeland (1978) Porter (1976) Mellor (1985) Ugolini (1968) Evans & Cameron (1979)
Clay minerals	Sierra Nevada Yukon Colorado	Birkeland and Janda (1971) Foscolos et al., (1977) Shroba and Birkeland (1983)
Mica weathering	Colorado	Shroba and Birkeland (1983)
Hornblende etching	Baffin Island	Locke (1979)
Iron & aluminium	Sierra Nevada Norway Pennsylvania, USA Arctic Himalaya New Zealand	Shroba and Birkeland (1983) Mellor (1985) Levine and Ciolkosz (1989)
Organic carbon	Norway Canada	Mellor (1985) Jacobson and Birks (1980)

Survey Staff, 1975). Rates of translocation are known to be slow. However, while this mechanism certainly operates, there is also field evidence that shows fine-grained material may be blown onto soil surfaces prior to translocation. This seems to be particularly important in dry climates and on coarse-grained soils (Locke, 1986). This latter aspect has implications for the reliability of clay accumulation as a dating method in areas and at times where aeolian processes have been dominant. Birkeland and Shroba (1974) suggest that aeolian-influx rather than soil-forming intervals should be considered more pertinent.

10.5.3. Indicators of Environmental Conditions

Soil properties developed on glacial deposits have been used to reconstruct both interglacial and glacial climates. The properties include clay mineral assemblages, the presence of large amounts of translocated clay, the colour of the soil and plant pollen contained in soil horizons (Smith et al., 1986; Felix-Henningsen

and Urban, 1982). However, although climatic inferences continue to be made on the basis of soil features (Karlstrom, 1988; Butler, 1987), this is a viewpoint where circumspection is required (Boardman, 1985a). For example, considerable use has been made of the red colour of soils as an indicator of former temperatures but the relationship of soil colour and temperature in present-day soils is poorly documented. Schwertmann et al. (1982) note that on glacial deposits in the foothills of the European Alps, redness due to the presence of hematite decreases in an easterly direction as temperature decreases and rainfall increases. However, since this relationship is observed on gravelly soils and not silty ones, it is suggested that it is the pedoclimate that is critical with well-drained, warmer soils being more susceptible to hematite formation.

The value of paleosols both as stratigraphic markers and as repositories of information concerning environmental conditions is recognised in the reconstruction of the Quaternary history of western Europe.

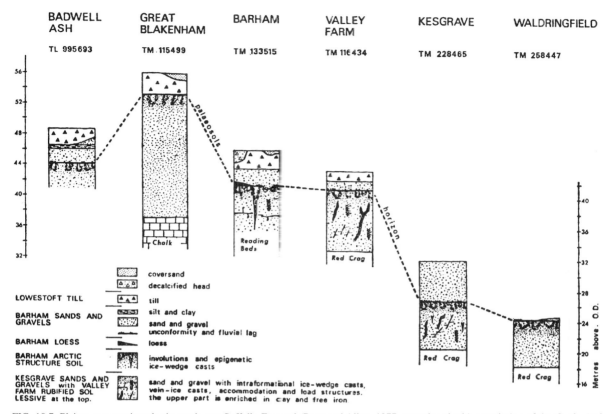

FIG. 10.7. Pleistocene stratigraphy in south-east Suffolk, England (Rose and Allen, 1977, reproduced with permission of the *Geological Society of London*). See text for explanation.

In East Anglia, the presence of a major regional paleosol has proved crucial in the subdivision of previously undivided river terrace deposits. This is illustrated in Fig. 10.7. At many sites across southern East Anglia a deep clay-enriched and reddened soil, the Valley Farm Soil, is developed in the Kesgrave Sands and Gravels of fluvial, non-glacial origin. The soil represents a long period of temperate conditions in the Early and Middle Pleistocene. Superimposed onto the Valley Farm Soil is the Barham Soil representing periglacial conditions due to the onset of the Anglian glacial stage. This is overlain by till, outwash, coversand or loess. The separation of the Kesgrave fluvial deposits from the Barham Sands and Gravels of glaciofluvial origin, on the basis of the recognition of a paleosol, was a major step forward in elucidating the stratigraphy of the area (Rose and Allen, 1977).

Recent advances have been made in describing the effects of cold climates on soils in glacial deposits. Both the type of soil profile and the features produced by different temperature and humidity regimes are now better understood (Van-Vliet Lanöe, 1985, 1986). Recognition of the meaning and significance of cold climate features in paleosols has enhanced their value as stratigraphic markers and climatic indicators, for example, the Barham Soil in East Anglia (Rose *et al.*, 1985).

The refining of regional stratigraphies using micromorphological examination of paleosols has become an important pedological tool in western Europe. In East Anglia, more details of phases of glacial/interglacial conditions in the Middle Pleistocene have been provided based on the recognition of cold-climate disruption of soil features assumed to have formed under temperate conditions (Kemp, 1987). A

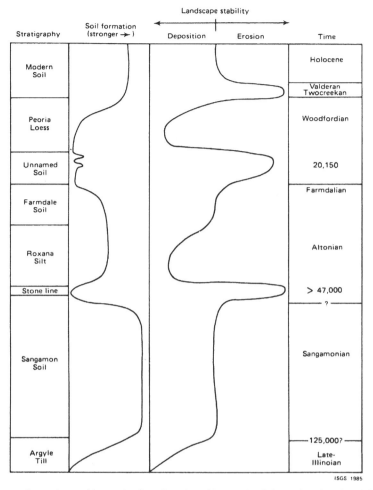

FIG. 10.8. Duration and approximate dates of intervals of erosion, deposition, and soil formation since late Illinoian time in the area of the Wempleton southeast section, Illinois (McKenna and Follmer, 1990, reproduced by permission of the *Illinois State Geological Survey)*

similar approach was developed by Bullock and Murphy (1979) and is also used by Fedoroff and Goldberg (1982).

10.6. CONCLUSION

The value of paleosols in glacial studies relates to their stratigraphic importance in the sub-division of geological sequences and in the environmental information that they provide. Inferences regarding past climates have to be made with circumspection: some

dramatic claims have been made based on soil data (Foscolos *et al.*, 1977; Karlstrom, 1988). The contribution of paleosol studies to an understanding of recent geological history, including glacial events, is best illustrated with reference to the American Mid-West. Figure 10.8 shows the relationship between soil formation and landscape stability: the importance of an understanding of soils in this region is clear from the recognition they are given in the local stratigraphy. Finally, in many upland areas the use of relative dating techniques on soils, has enabled glacial events to be differentiated and dated.

Chapter 11

GLACIO-ISOSTASY, GLACIO-EUSTASY AND RELATIVE SEA-LEVEL CHANGE

J.M. Gray

11.1. INTRODUCTION

Relative sea-level — the level of the sea relative to the land at any point on the coast is a major influence on, and is strongly influenced by, glacially-related processes.

On the other hand, the level of the sea is a primary control on sedimentary processes. Above it, glacial, glaciofluvial and glaciolacustrine processes predominate and sea-level acts as the ultimate base-level for these processes. Below it is the glaciomarine environment with its own complexity of processes related to distance from the ice margin, water depths, circulation, etc. (Boulton, 1990). However, the level of the sea relative to the land has not been stable during the Quaternary, amplitudes of change of several hundred metres having been experienced in formerly glaciated areas (Shackleton, 1987; Dott, 1992). Thus, sedimentary environments near the present coastline of such areas have invariably experienced a complex series of changes. Glaciomarine sediments, 'raised' beaches and marine deltas are frequently encountered above present sea-level in, for example, Canada, Scandinavia and Scotland, whereas the partially submerged drumlins of Boston Harbour, Massachusetts or Strangford Lough, Northern Ireland were

probably formed by glacial processes when relative sea-level was considerably below its present level.

These changes in relative sea-level are themselves largely brought about by changes in the world's ice masses. First, changing volumes of ice affect the volume of ocean water and hence global sea-levels. Secondly, in glaciated areas the lithosphere is loaded and unloaded as the ice masses come and go, and major vertical movements of the land surface occur that affect relative sea-level in coastal areas. These two effects, known as glacio-eustasy and glacio-isostasy respectively, are the major, though not the only, causes of sea-level change on a Late Quaternary timescale (Fairbridge, 1983).

As Fairbridge points out, both words are partly derived from the Greek word 'stasis' meaning standing. This is somewhat ironic considering the amplitude of Quaternary changes in land and sea, but it refers to the fact that both are constantly seeking an equilibrium position. Sea-level seeks equilibrium with the geoid, the equigeopotential surface controlled by terrestrial gravity. The land, on the other hand, tends towards hydrostatic equilibrium since the lithosphere essentially 'floats' on the asthenosphere.

In the first part of this chapter, the principles underlying glacio-eustasy and glacio-isostasy and the theoretical implications for relative sea-level change will be introduced. In the second part, some examples from previously glaciated coastal environments will be examined to determine whether the field evidence matches the theory. The chapter is not intended to be a comprehensive, advanced review of global sea-level change; instead it attempts to explain the principles involved in the complex spatial and temporal interplay between changing land and sea-levels and changing ice margins. There is a general lack of introduction to these complex relationships in earth science literature.

As outlined above, formerly glaciated coastal areas have been environments of change during the Quaternary. In simple terms, the land has been depressed and uplifted, the sea has been rising and falling and the ice has been advancing and receding across the coastal zone. The end product has been a frequently changing suite of glacial, paraglacial and non-glacial processes that is reflected in highly complex series and associations of landforms and sediments. This chapter attempts to unravel some of this considerable complexity.

11.2. GLACIO-EUSTASY

During the major global glaciations of the Middle–Late Quaternary, around three times the present volume of glacial ice developed on land with the Laurentide and Fennoscandian Ice Sheets accounting for much of this increase (Flint 1971; Denton and Hughes, 1981a; Anon., 1985). Since there is only a finite volume of water circulating in the hydrological cycle, it is clear from Table 11.1 that the only possible source for this additional ice is the oceans (roughly

similar volumes of water or ice are probably retained in other parts of the cycle during a glaciation). Although there are several complications, the general pattern is clearly established — during a major glaciation world sea-level falls but rises again as the major ice sheets melt. The amplitude of these changes is ca. 120 m.

One complication involves the timing of ice sheet growth and wastage. If different ice sheets behave differently, which they do, then for any one glaciated region eustatic sea-level may be changing very differently from the way the ice mass is changing. This is particularly true of small ice masses. For example, in theory it would be possible to have a large British Ice Sheet in existence while eustatic sea-level is still relatively high, or a small ice sheet while eustatic sea-level is low. This is because the primary control on eustatic sea-level is the behaviour of the Laurentide and Fennoscandian Ice Sheets, and the British Ice Sheet makes only a small contribution to the volume of sea-water (Denton and Hughes, 1981b). Boulton (1990) considers in detail the implications of these 'leads' and 'lags' of glaciation relative to glacio-eustasy (Dott, 1992).

Other significant complications include the gravitational attraction of water to ice-masses (Clark *et al.*, 1978), hydro-isostatic loading and unloading of the continental shelves (Bloom, 1967), and, in the longer term, changes in the geoid (Mörner, 1976) and tectonic movements that alter the capacity of the ocean basins.

Traditionally, attempts have been made to construct eustatic sea-level curves by radiocarbon dating of geological evidence of sea-level change in coastal areas. Many such curves covering changes over the

TABLE 11.1. Present day distribution of water in the hydrological cycle (after Flint, 1971)

Residence	Volume (km³ ×10⁶)	Percentage
World Oceans	1350	97.6
Rivers, lakes and groundwater	8.6	0.6
Glaciers (water equivalent)	24	1.7
Atmosphere	0.013	trace
Total	1382.613	>99.9

last 12,000–15,000 years have been produced since 1960, and the quality has increased greatly since the early ones which drew on evidence of variable reliability from widely dispersed areas (Fairbridge, 1961). However, unless the land has been stable (a very rare occurrence given problems of hydro-isostatic loading of nearshore areas, sediment compaction, etc.), these curves are relative, not eustatic sea-level curves. This may partly explain the differences between them. However, most show a general rise from before 12,000 BP, recovering to about present levels by 5000 BP with subsequent variations of the amplitude of 1–2 m. Fairbanks (1989) recently published a 'glacio-eustatic curve' obtained by radiocarbon dating reef-crest corals in Barbados and correcting for assumed land uplift (Fig. 11.1). His curve extends the record back to the maximum of the last glaciation 18,000 years ago, when sea-level was at −120 m. The rapid rise between 18,000 and 5000 BP was interrupted by a reduced rate of rise during the Younger Dryas.

Another approach for reconstructing eustatic sea-level change has been to use oxygen isotope curves

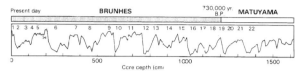

FIG. 11.2. Shackleton and Opdyke's (1973) oxygen isotope curve from deep-sea core V28-238. A 'first approximation' of glacio-eustatic sea-level change during the last million years.

derived from analysis of micro-fauna in ocean-bed cores (Shackleton and Opdyke, 1973) (Fig. 11.2). Variations in the oxygen isotopes of marine micro-fauna are partly due to changes in the temperature of sea water in which the organisms grew but, more importantly, are due to variations in the isotopic composition of the sea water. In turn, the latter are related to the global volumes of ice, and hence to the volume of sea-water and hence to sea-level (Menzies, 1995a, Chapter 2). None of these linkages is perfect, however. The isotopic composition of ice varies and, again, there are long-term changes in the capacity of the ocean basins that distort the final link between volume of sea water and eustatic sea-level, and they do so increasingly towards the Early Quaternary. Thus, the oxygen isotope record is only a 'first approximation' of global continental ice volume and hence of glacio-eustatic sea-level. Closer approximations are being attempted (Shackleton, 1987).

It may be useful to view glacially-induced eustatic changes as potentially having an amplitude of at least 200 m (Fig. 11.3). At the lower end is the situation corresponding to the large ice sheets of the Middle-Late Quaternary; 200 m higher is the projected situation with no ice sheets (for example Table 15.1 in Barry, 1985). The Late–Middle Quaternary inter-glacial world occupies a position close to three-fifths of the way up this scale. Recent interest in environmental hazards has aroused concern over the possibility of rises in sea-level caused by either displacement of water following surging of part of the Antarctic ice-sheet (an idea that dates from Wilson, 1964), or from melting of ice due to the greenhouse effect (Titus, 1987). Figure 11.3 shows the maximum and minimum predictions from the latter effect up to 2050 A.D. (ignoring the more sensationalist projections), and it illustrates the point that the current concern over coastal flooding, evacuation of low-lying lands, salt

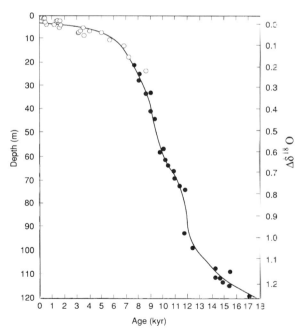

FIG. 11.1. Fairbanks' (1989) 'glacio-eustatic sea-level curve' for the last 17,000 years

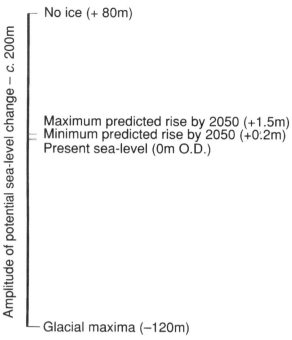

FIG. 11.3. The amplitude of potential glacially-induced sea-level change

11.3. GLACIO-ISOSTASY

'*It occurs to me that the enormous weight of ice thrown upon the land may have had something to do with this depression . . . and then the melting of the ice would account for the rising of the land, which seems to have followed upon the decrease of the glaciers*'. In these words T.F. Jamieson (1865, p. 178) first captured the idea of what is now called glacio-isostasy. Holmes (1965) illustrated the concept by means of blocks of wood of different heights floating at different levels in a tank of water (Fig. 11.4). If a block of ice is added to the top of one of the blocks such that it fits exactly with its top surface, the block of wood will be depressed by an amount (D_w) determined by the thickness of the ice (h_i), the density of the ice (ρ_i) and the density of the water (ρ_{wt}), such that:

$$D_w = h_i \frac{\rho_i}{\rho_{wt}} \qquad (11.1)$$

Thus D_w is independent of the density of the wood and, in this case, since the density of ice is close to that of water, the amount of depression is approximately equal to the thickness of the ice. If the ice block melts, the block of wood will return to its original position, staying in hydrostatic equilibrium throughout because of the low viscosity of water. If the wood and water in Fig. 11.4 are replaced by

water intrusion, increased coastal erosion, etc. stems from relatively minor variations in the volume of global ice. The world's coastal communities and ports are very dependent on an almost perfectly stable relative sea-level.

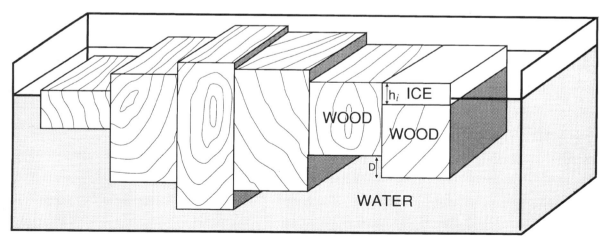

FIG. 11.4. Isostasy illustrated by blocks of wood floating in water (after, Holmes, 1965)

lithosphere and asthenosphere respectively, the expression becomes:

$$D_w = h_i \frac{\rho_i}{\rho_{as}} \qquad (11.2)$$

for glacio-isostatic depression, where ρ_{as} is the density of the asthenosphere. Since the average density of glacial ice is 0.9–1.0 Mg m^{-3}, and that of the asthenosphere is 3.0–3.3 Mg m^{-3}, Eqn. (11.2) gives isostatic depression as between one third and one quarter of the thickness of the ice (i.e. a ~300 m thick ice sheet {typical of Laurentide and Fenno-scandian Ice Sheets in the Middle–Late Quaternary} potentially could depress the lithosphere by 750–1000 m). Thus glacio-isostasy involves major vertical land movements.

However, the lithosphere does not behave as simply as a block of wood floating in water. Although the surface layers of the earth are structurally very complex, they are often viewed as a thin elastic lithosphere overlying a viscous asthenosphere. For loads applied on a short time span (<10,000 years) there is an elastic response, (i.e. all the deformation is recoverable following unloading). However loads applied over long time spans (>100,000 years) are able to displace material at depth and the surface may not return completely to its original position upon removal of the load. Since glacial loading and unloading often operate on intermediate timespans, both elastic and viscous behaviour are probably involved. Unlike wood floating in water, the high viscosity of the asthenosphere means that glacio-isostatic uplift or rebound is generally unable to keep pace with the removal of ice loads. Thus the depressed crust is frequently out of isostatic equilibrium as it rebounds, a phenomenon reflected in gravity anomalies in deglaciated areas.

A further complication involves the shape of ice sheets, which do not take the form of the giant ice-cubes shown in Fig. 11.4! In fact the exponential surface profile of an ice sheet on a non-deformable bed is well established from the present-day Greenland and Antarctic Ice sheets. As ice thickness varies, it follows that ice-loading must also be differential. Furthermore, the crust has considerable lateral strength such that depression will take place beyond

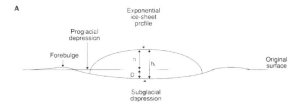

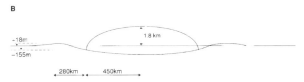

FIG. 11.5. (a) General model of glacio-isostatic displacement by an ice-sheet.(b) Walcott's (1970) model of glacio-isostatic displacement by a 900 km diameter ice-sheet

the ice margin (proglacial depression) rather than decreasing to zero at this point. Walcott (1970) produced one of the first simple models for the form of this depressed surface if a number of assumptions are made, (for example, a regular, circular ice sheet acting on a flat, non-deformable bed) (Fig. 11.5).

All surface profile models of ice sheets yield values for the surface elevation of the ice sheet (h_{se}) rather than ice thickness (h_i) as used in Eqn. (11.2). However, $h_i = D_w + h_{se}$ (Fig. 11.5a) and Eqn. (11.2) becomes:

$$D_w = \frac{h_{se} \rho_i}{\rho_{as} - \rho_i} \qquad (11.3)$$

If valid, this would allow calculation of glacio-isostatic depression using values for surface elevation of an ice sheet and density values for ice and asthenosphere. In fact, Eqn. (11.3) is normally assumed to apply under the centres of large ice sheets and, using the density values given above, Eqn. (11.4) indicates that maximum depression under a large ice sheet amounts to just under two-fifths of the surface elevation. Towards the periphery, however, crustal rigidity increasingly distorts the relationship. However, Walcott (1970), using geophysical data to estimate crustal rigidity, calculated that depression at the ice margin (w_d) can be expressed as:

$$w_d = \frac{2H \rho_i}{10 (\rho_{as} - \rho_i)} \qquad (11.4)$$

where H is maximum surface elevation for a 900 km diameter ice sheet with an exponential surface profile. With $\rho_i = 1.0\,Mg\,m^{-3}$ and $\rho_{as} = 3.3\,Mg\,m^{-3}$, Eqn. (11.4) becomes:

$$w_d = \frac{H}{11.5} \qquad (11.5)$$

For a 900 km diameter ice sheet with an exponential surface profile, H = 1800 m, ($D_w = -783$ m), and from Eqn. (11.5) the depression at the ice margin is −155 m. Walcott calculated that proglacial depression would extend 180 km beyond the ice margin, and that beyond this would be a forebulge reaching a maximum of +18 m, 280 km from the ice margin (Fig. 11.5b).

One of the most important points to appreciate is that ice thickness varies across an ice sheet, so that glacio-isostatic depression is differential, being greatest where the ice is thickest and decreasing towards and beyond the periphery. Thus, when the ice sheet begins to melt, the crust will rebound differentially, rising by the greatest amount and at the greatest rate where the ice was thickest. Any shoreline formed

during isostatic rebound will be subsequently updomed and will display a concentric isobase pattern.

Figure 11.6 is an attempt to illustrate the consequences of differential rebound for shoreline deformation. In order to portray the vertical movements as a cross-section through the centre of uplift, it uses the unlikely case of two semi-circular islands separated by a narrow sea strait (Fig. 11.6a). A number of simplifying assumptions are made including no changes in eustatic sea-level, no depression beyond the ice margin and instantaneous ice sheet melting. It is assumed that Fig. 11.6a represents the interglacial case in which Shoreline A is formed along both sides of the sea strait. If it now assumed that during a glacial period an ice sheet builds up to completely cover the two islands so that the maximum thickness is over the centre of the sea strait (Fig. 11.6b), then Shoreline A will be downwarped most at this point (assuming that no erosion or deposition occurs to remove the shoreline). If, then, it is assumed that instantaneous melting of the ice sheet occurs, the situation shown in Fig. 11.6c will take place, with a differentially downwarped crust and a new shoreline

PLAN VIEWS

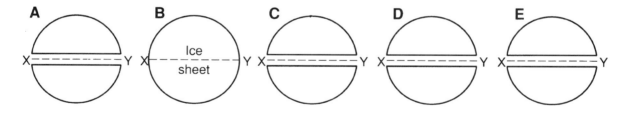

CROSS SECTIONS ALONG X – Y

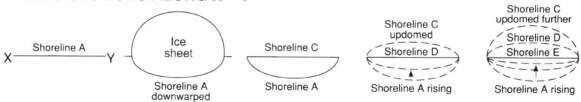

FIG. 11.6. Generalized model of glacio-isostatic depression and rebound of two semi-circular islands separated by a sea strait. For explanation see text. Many assumptions made

formed (Shoreline C). However, the crust will begin to rebound differentially and after a period of time (Fig. 11.6d), Shoreline C will have been updomed, a new shoreline (Shoreline D) will have formed at sea-level, and Shoreline A will have risen towards its original position. Figure 11.6e takes the isostatic rebound a stage further, and illustrates the important point that a series of shorelines tilted at different gradients will be formed along the sides of the strait. Those formed soon after ice-melt will be tilted most steeply, whereas those of recent origin have had little time to be updomed and thus will have low gradients. If the dome of uplift is symmetrical, shorelines of similar gradient on either side of the dome will be of similar age. Shoreline A will eventually return to its original position (i.e. interglacial shorelines will be approximately horizontal) if rebound is complete, though complications may arise if material at depth has been permanently displaced (see above) or if glacial erosion removes sufficient rock/sediment to cause additional isostatic rebound.

11.4. INTERPLAY OF GLACIO-EUSTASY AND GLACIO-ISOSTASY

11.4.1. Spatial Patterns

As stated above, Fig. 11.6 is grossly oversimplified, but it allows the basic glacio-isostatic behaviour to be viewed in isolation. Two of the assumptions made in Fig. 11.6 were: no changes in eustatic sea-level and no depression beyond the ice margin. Figure 11.7 is an attempt to replace both assumptions with more realistic situations. It illustrates a cross-section from the centre of isostatic rebound to an area beyond the reach of glacio-isostatic movements.

As already explained, at the end of a glaciation eustatic sea-level will rise due to the melting ice sheets. In Fig. 11.7 this is depicted by the rise on the left-hand side of the diagram. Isostatic uplift is also occurring but this is differential, being greatest towards the former ice sheet centre. If a shoreline is formed during isostatic recovery, it will only survive

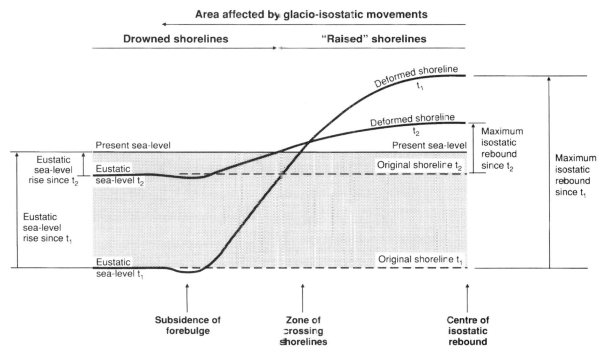

FIG. 11.7. Cross section from the centre of rebound (right) to an area beyond the influence of glacio-isostasy (left) showing the interplay between eustatic and isostatic movements

above sea-level if the land rises more quickly than the sea subsequent to its formation. This is most likely towards the centre of the former ice sheet, whereas, towards the periphery, the sea will rise more quickly than the land, with the result that any shorelines formed during rebound will be drowned subsequently. If the land rises by more than the sea after a particular shoreline is formed, then there is a net or relative sea-level fall at that locality.

An important point illustrated by Fig. 11.7 is that shorelines can be expected to cross because successive Shorelines 1 and 2 are related to progressively higher glacio-eustatic sea-levels. Furthermore, this idea is relevant in comparing maximum shoreline altitudes in isostatically uplifted areas. It follows from the above that shoreline altitude above present sea-level does not indicate the amount of isostatic rebound since it was formed because allowance must also be made for eustatic sea-level rise during this period. Thus, two areas with similar sized ice sheets will have very different maximum shoreline altitudes if eustatic sea-levels were different when rebound began. This is illustrated in Fig. 11.8 where Areas 1 and 2 have undergone the same amount of isostatic rebound but begun at different times. Area 2 was deglaciated later and hence began rebounding later

than Area 1, by which time eustatic sea-level was higher. Hence, relative to present sea-level, it has higher shorelines than Area 1.

Another assumption incorporated in Fig. 11.6 was instantaneous ice sheet melting. However, a major ice sheet will take thousands of years to disappear and glacio-isostatic rebound will begin as soon as the ice sheet begins to lose volume. Thus, in most isostatic areas, ice wastage and sea-level change are interlinked. A common occurrence in isostatically uplifted areas is that shorelines disappear towards the centre of rebound (i.e. not as shown in Figs 11.6d and e). In many places there are sudden drops in the marine limit (highest level of marine action on a coast) and in most cases this is due to the presence of glacial ice. Figure 11.9 illustrates the two types of shoreline disappearance due to ice. In some cases (Type A, Fig. 11.9) the absence of a particular shoreline is the result of ice occupying the area at the time the shoreline was formed. If so, at the locality where the shoreline disappears, it will often be related to glaciomarine deltas or outwash plains/trains and ice-contact glacio-fluvial landforms (Gray, 1991). Alternatively, the disappearance of a shoreline towards the rebound centre may be due to its subsequent destruction by an ice advance (Type B, Fig. 11.9).

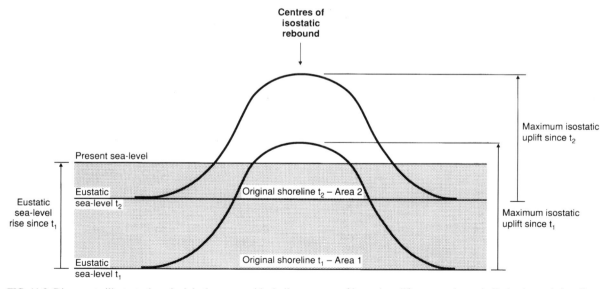

FIG. 11.8. Diagram to illustrate the principle that areas with similar amounts of isostatic uplift may not have similarly elevated shorelines if deglaciated at different times

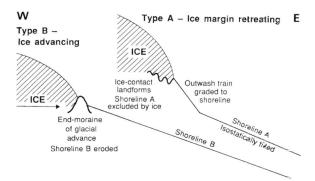

FIG. 11.9. Shorelines may not be continuous when traced towards the centre of isostatic uplift, (a) due to contemporaneous presence of a wasting ice mass, or (b) due to erosion by subsequent advance

Of course, deglaciation may proceed in a significantly different pattern from ice sheet growth, and this could result in changes in rebound pattern with time, rather than as the simple, symmetrical ice sheet growth and decay portrayed above.

11.4.2. Temporal Patterns

So far, the discussion has mainly concentrated on spatial aspects of glacio-isostasy, but the temporal sequence may also be examined. Figure 11.10 is an idealized symmetrical curve of isostatic movement at a particular locality consequent upon the simple

growth and decay of an ice sheet. The locality is between the rebound centre and the eventual periphery of the ice sheet. The sequence extends the curve of Andrews (1970, Figs 1–6) by incorporating the period of isostatic depression. The assumptions include isostatic equilibrium both initially and at the glacial maximum.

The curve may be divided into 6 sections:

(1) forebulge uplift (U_{fb}) which may be expected to extend like a wave ahead of the advancing ice sheet causing an initial slight rise in the surface;

(2) proglacial depression (D_{pro}) which follows the initial uplift as crustal rigidity causes depression before the ice sheet reaches the locality (Fig. 11.5 and discussion above). It is a period of accelerating depression;

(3) subglacial depression (D_{sub}) which is the further depression taking place between the times of ice advance over the locality and equilibrium from maximum ice loading;

(4) restrained rebound (U_r) which occurs between the start of ice sheet thinning and deglaciation of the locality. It is a period of accelerating uplift, and maximum uplift rates are experienced at about the time of deglaciation;

(5) postglacial rebound (U_p) which occurs between deglaciation of the locality and the present-day. This is generally a period of exponentially decelerating rebound;

(6) residual rebound (U_{rr}) which is the amount of

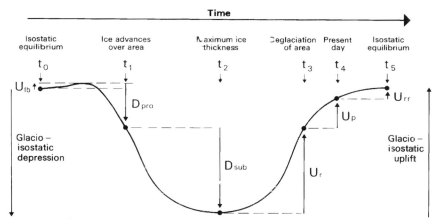

FIG. 11.10. The six stages of glacio-isostatic movement experienced by an area during the simple advance and decay of an ice-sheet (modified from Andrews, 1970)

ISOSTATIC AND EUSTATIC CURVES

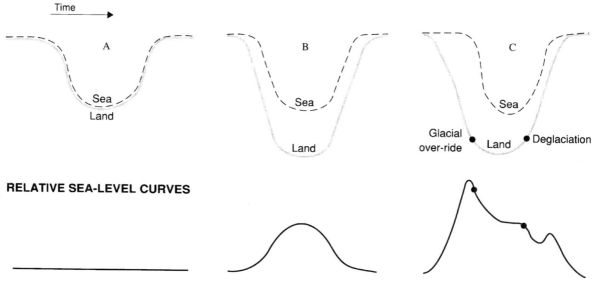

FIG.11.11. Simple glacio-eustatic and glacio-isostatic curves, and the resultant relative sea-level curves

rebound still to occur before recovery is complete and equilibrium is re-established.

Complications may occur, however, particularly if glacial advance or retreat are interrupted. For example, glacial readvances during rebound might retard or even reverse uplift. Note, also, that because of asthenosphere viscosity, maximum depression will lag behind maximum loading. Furthermore, depression and uplift are unlikely to occur symmetrically.

The influence of eustatic sea-level may now be incorporated into this temporal model. Figure 11.11a shows eustatic sea-level movement exactly in phase with, and exactly matched by, isostatic movement. Hence, relative sea-level has been stationary throughout. If, on the other hand, the two are in phase but the amplitude of the glacio-isostatic cycle greatly exceeds that of the eustatic cycle, relative sea-level will show a simple rise and fall (Fig. 11.11b). A third possibility is exemplified in Fig. 11.11c. Here the two are out of phase and of different amplitudes and, in addition, the glacio-isostatic cycle is completed more quickly than the eustatic cycle. The end result is a complex relative sea-level curve. Note in particular that relative sea-level immediately prior to glaciation is much higher than immediately following deglaciation.

11.4.3. Spatial/Temporal Models

Boulton (1990) has recently attempted to combine these spatial and temporal changes in diagrams depicting ice sheet growth and decay together with eustatic fall and rise during the glaciation on the vertical axis (Fig. 11.12). The lines on the diagrams indicate relative sea-level position at any place and time, and the three different diagrams reflect different leads and lags. Figure 11.12a has the ice sheet closely in phase with glacio-eustasy (e.g. the Wisconsinan Laurentide Ice Sheet), whereas Fig. 11.12b and 11.12c illustrate the ice sheet lagging and leading the glacio-eustatic cycle, respectively, as exemplified by the Weichselian West Spitsbergen Ice Sheet and Devensian British Ice Sheet. Note that, within the glaciated areas, relative sea level is predicted to fall well into the future, reflecting residual rebound.

11.5. ILLUSTRATION OF THE THEORETICAL PRINCIPLES

So far this chapter has dealt almost exclusively with theoretical considerations. In the remainder of this chapter, several examples are given demonstrating

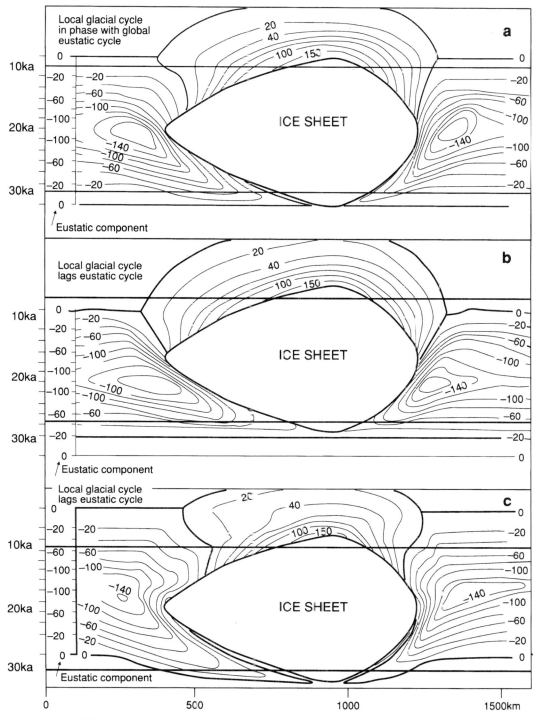

FIG. 11.12. Boulton's (1990) space-time models of glaciation and sea-level change for: (a) glaciation and glacio-eustasy in phase; (b) glaciation lags behind glacio-eustasy; (c) glaciation leads glacio-eustasy. Contours show relative sea-level. (Reproduced with permission from the *Geological Society of London*)

that many of the principles described above are reflected in the field evidence. However, the discussion also reveals several instances where further complexities are involved.

11.5.1. Spatial Patterns

In all formerly glaciated areas, the principle of differential depression and rebound has been firmly established from concentric isobase patterns. In Scandinavia, various shorelines display this pattern, indicating that the last ice sheet was finally centred over the Gulf of Bothnia/Swedish coast. Figure 11.13a, for example, shows Eronen's (1983) reconstruction of the Littorina Shoreline of 7500–7000 BP Earlier Baltic shorelines illustrate the principle of shoreline absence due to ice (Fig. 11.9). For example the Yoldia Shoreline of 10,200 BP does not extend into the Gulf of Bothnia due to the presence of a rapidly shrinking ice sheet (Fig. 11.13b). Comparison of the isobase spacing for these two shorelines (Figs 11.13a and b) indicates the much steeper gradients on the Yoldia Shoreline. This is clearer when shown in profile perpendicular to the isobases (e.g. across south west Finland) (Fig. 11.14). Several of the principles outlined above are clear on this diagram:

(1) the principle of decreasing shoreline gradients with time (Fig. 11.6),
(2) the principle of shoreline disappearance towards the rebound centre due to the presence of ice (e.g. Fig. 11.9, the 10,200 BP Yoldia Stage),
(3) the principle of crossing shorelines (Fig. 11.7).

All these points are also clear on the equidistant shoreline diagram for south east Scotland (Sissons, 1983) shown in Fig. 11.15. The arrows indicate the sequence of shoreline formation and there is clearly a decrease in shoreline gradient with time. All the early shorelines terminate at an ice margin when traced towards the rebound centre, indicating that the sea was flooding up the estuaries of the rivers Forth and Tay as the last ice sheet retreated. Finally, the Main Perth Shoreline (Lateglacial) clearly crosses the Main Postglacial Shoreline (Holocene).

This last point may be widened with reference to Fig. 11.16a which gives a more generalized view of Late Devensian and Holocene shoreline formation in western Britain (Fig. 11.7). The Main Perth Shoreline was formed ~13,500 BP when glacio-eustatic sea level was at ≈ -100 m. Only in Scotland has subsequent glacio-isostatic rebound been greater than the subsequent glacio-eustatic rise in sea-level. Thus, in England and Wales, this shoreline was drowned after

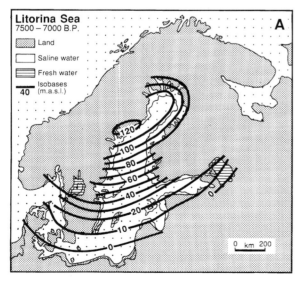

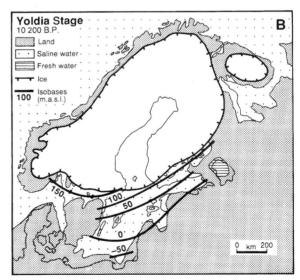

FIG. 11.13. Isobases for (a) the Littorina Sea (7,500–7,000 BP) and (b) the Yoldia stage (10,200–10,000 BP) in Fennoscandia (after Eronen, 1983, reproduced by permission of the *Institute of British Geographers*)

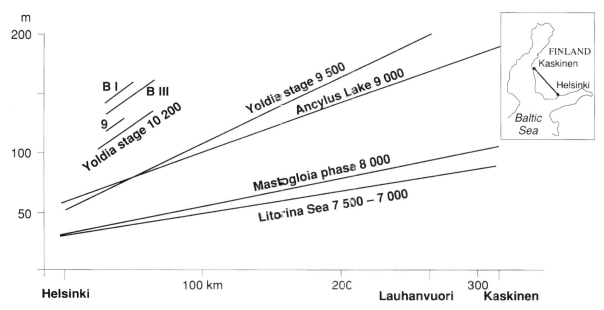

FIG. 11.14. Equidistant shoreline diagram for S.W. Finland (after Eronen, 1983, reproduced by permission of the *Institute of British Geographers*)

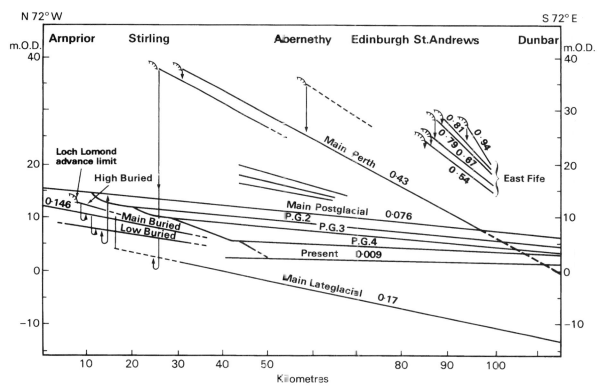

FIG. 11.15. Equidistant shoreline diagram for S.E. Scotland (after Sissons, 1983, reproduced by permission of the *Institute of British Geographers*)

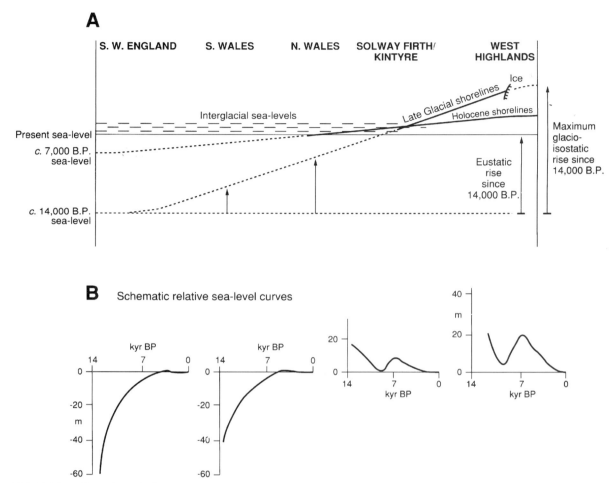

FIG. 11.16. (a) General model of shoreline development on a N-S transect through western Britain. (b) Four schematic relative sea-level curves along this N-S transect

its formation. The Main Postglacial Shoreline was formed ~7000 BP by which time eustatic sea level had risen to ≈ – 10 m O.D. to form a shoreline that cut across those formed earlier.

In North America, similar relationships exist. For example, isobase patterns have indicated that the last ice sheet had at least two ice domes centred over Labrador and Keewatin (Andrews, 1970). It is worth stressing at this point that isobase patterns provide valuable evidence in reconstructing the geometry of former ice masses. This is nowhere more true than in Canada where the shape of the last ice sheet is a subject of continuing debate (Peltier and Andrews, 1983).

Several studies have identified complexities in the pattern of glacio-isostatic rebound. Forsström *et al.* (1988), for example, recognized differences in rebound pattern between Eemian and Flandrian interglacials in eastern Fennoscandia that they ascribe to differences in deglaciation history and changes in the geographical distribution of ice load during the melting phases. Similar factors may account for the eastward shift in the centre of Scottish isostatic rebound in the early Holocene identified by Gray (1983) and Firth and Haggart (1989).

There is also increasing recognition of non-uniform uplift and faulting during rebound. In Iceland, Sigmundsson (1991) has found evidence that the low asthenosphere viscosity associated with the Mid-Atlantic constructive plate boundary influenced the rate of rebound. In Scotland, several studies have identified shoreline faulting and bending over distances of a few hundred metres to a few kilometres, indicating that isostatic uplift is itself a complex process rather than a simple tilting (Sissons, 1972; Gray, 1974; Sissons and Cornish, 1982). Recent research in Norway and Canada indicates active neotectonics though it is difficult to be certain that the movements are related to glacio-isostatic uplift (Andrews, 1987b; Anundsen, 1989). Eyles and McCabe (1989) propose 'piano-key' tectonics related to pre-existing structures. Although some will see this as a convenient means of accounting for almost any altitude anomalies, the general complexity of crustal response to glacio-isostatic loading and unloading is unquestionable.

11.5.2. Temporal Patterns

Many relative sea level curves have been produced for the period since decay of the last ice sheets. The exact shapes of these curves reflect differences in their location relative to the former ice-margin, the amount of depression and the timing of deglaciation (Carter, 1992). Clark *et al.*, (1978) produced six global zones of relative sea level change, and, at a more local scale, Shennan (1987) has demonstrated significant differences in the relative sea level curves around North Sea coasts.

In Scandinavia and North America, relative sea level curves towards the rebound centres demonstrate rapid falls of sea level during the Flandrian. One example is Hafsten's (1983) curve from the Oslo area of Norway demonstrating a uni-directional relative sea level fall of 220 m in the last 10,000 years (Fig. 11.17a). This may be compared with, and is similar to, the situation in Fig. 11.11b. Since eustatic sea level was rising strongly throughout this period, it is clear that glacio-isostatic uplift must have been very rapid, particularly in the early Holocene. This simply reflects the maximum uplift rates around the time of deglaciation, and subsequent deceleration (U_p, Fig. 11.10).

However, towards peripheral areas of both Scandinavia and North America, where glacio-isostatic uplift

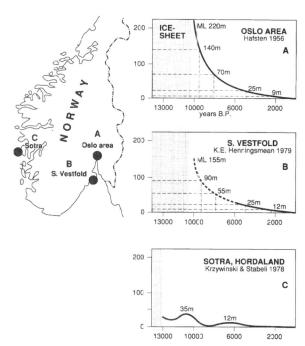

FIG. 11.17. Relative sea-level curves from southern Norway (after Hafsten, 1983, reproduced by permission of the *Institute of British Geographers*)

has been more limited, the curves become more complex and relative sea level transgressions are recorded. Examples occur in southwest Norway (Hafsten, 1983; Fig. 11.17c). In Maine, U.S.A. (Fig. 11.18), upon ice retreat at ~13,000 BP glaciomarine deltas were formed at ≈70 m but rapid uplift led to a rapid relative sea level drop to levels below present sea level prior to 11,000 BP (Belknap *et al.*, 1986). The subsequent rise in relative sea level must reflect the strong rise in eustatic sea levels consequent upon the final disintegration of the Laurentide and Fennoscandian Ice Sheets.

In Scotland an even more complex pattern emerges reflecting the thinner former ice sheet, though the rapid fall immediately following deglaciation is again evident. Subsequently, however, with rates of uplift often matched or exceeded by the eustatic rise, transgressions and regressions were numerous. Figure 11.19 shows relative sea level trends for inner Scottish coasts in the Lateglacial and Holocene and is speculatively extended back to the last

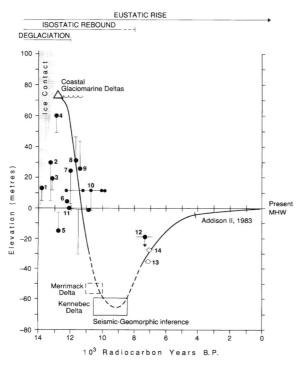

FIG. 11.18. Relative sea-level curve for Maine (after Belknap *et al.*, 1986, reproduced by permission of *Academic Press*)

interglacial (Gray, 1985). The curve incorporates Sutherland's (1981) idea that the so-called high level marine shell beds of Scotland were formed by proglacial depression (D_{pro}) by a Scottish Ice Sheet advancing while eustatic sea level was still high. He argued that the maritime situation and small size of the Scottish Ice Sheet would enable it to grow more rapidly than the Laurentide and Scandinavian Ice Sheets, and thus would be advancing and depressing the crust at a time when eustatic sea level was still relatively high (Fig. 11.11c). It is interesting to compare Fig. 11.19 with Fig. 11.11c. The similarity in trends is obvious, which is interesting considering the simplicity of the curves on which Fig. 11.11c is based. Figure 11.16b shows schematic relative sea level curves for four localities down the west coast of Britain. The northerly two reflect the dominance of glacio-isostatic rebound, while the southerly two are dominated by glacio-eustatic sea level rise.

Finally, it is now clear that whereas Scotland has virtually completed its isostatic rebound (<0.2 m/century; Shennan, 1989), there are still significant amounts of residual rebound (U_{rr}) remaining in

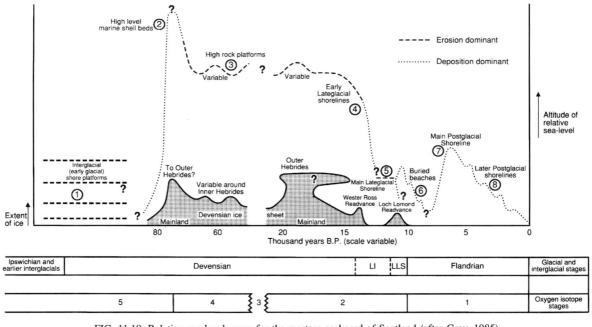

FIG. 11.19. Relative sea-level curve for the western seaboard of Scotland (after Gray, 1985)

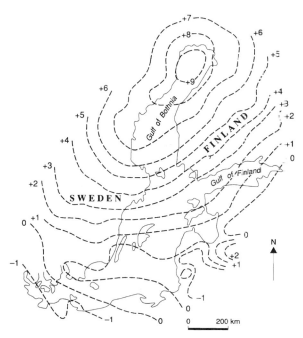

FIG. 11.20. Rates of residual rebound in Fennoscandia (after Eronen, 1983, reproduced by permission of the *Institute of British Geographers*)

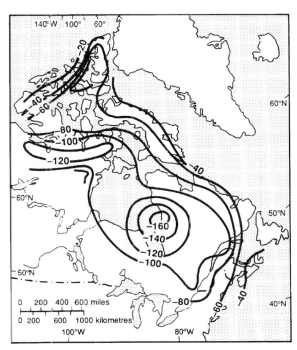

FIG. 11.22. Amounts of residual rebound in N. America (after Andrews, 1970, reproduced by permission of the *Institute of British Geographers*)

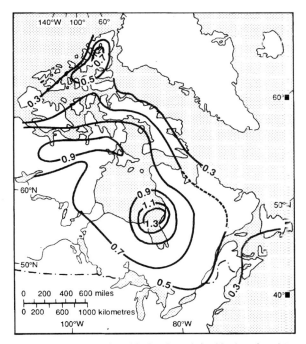

FIG. 11.21. Rates of residual rebound in N. America (after Andrews, 1970, reproduced by permission of the *Institute of British Geographers*)

North America and Fennoscandia. Figure 11.20 shows present rates of uplift in Fennoscandia which reach maximum values of 1.0 m/century in the north of the Gulf of Bothnia (Eronen, 1983). In North America, present uplift values are even higher (up to 1.4 m/century over southern Hudson Bay, Fig. 11.21) according to Andrews (1970). He also presents a map (Fig. 11.22) showing amounts of residual rebound still to occur in North America, amounting to >160 m on the east shore of Hudson Bay, and even the eastern seaboard of the U.S.A. and Canada north of New York has 30–60 m still to rise. These changes will occur on a longer timescale than greenhouse gas-induced sea level rise described above, but, where they occur, they will have a significant impact on coastlines and coastal communities. If you must buy a beach house, choose a location where the rate of greenhouse gas-induced sea level rise will be exactly matched by rates of residual glacio-isostatic rebound!

11.6. IMPLICATIONS FOR GLACIOMARINE FACIES ARCHITECTURE

Boulton (1990) has recently suggested a glaciomarine facies model relating sedimentation rate, sediment stability, depositional processes and faunal content to distance from the glacier, water circulation and bathymetry. The dominant characteristic of the model is the depth fining of the sediments on fiord sides with coarse littoral sediments overlying deeper-water glaciomarine muds. More importantly Boulton has

suggested a spatial/temporal cycle of sedimentation controlled by the glacial / eustatic / isostatic cycles described above. Figure 11.23 shows this cycle extending from one interglacial (A), through transgressive (B) and regressive (C) phases, into the next interglacial (D), reflecting the simple pattern of relative sea level change in Fig. 11.11b. The transgressive phase may coincide with glacial advance over the site and the formation of tills, including deformation tills or glacially-transported marine deposits.

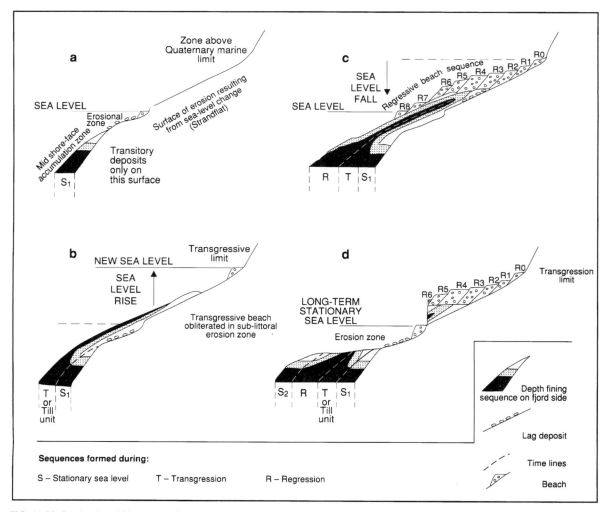

FIG. 11.23. Boulton's (1990) model of sedimentary sequences in the nearshore zone of a glacio-isostatically dominated sea-level cycle: (a) Facies distribution during last interglacial; (b) Consequence of transgression resulting from dominant glacio-isostatic depression; (c) Consequences of late-glacial regression resulting from dominant glacio-isostatic uplift; (d) Erosional notch produced by present interglacial sea-level. (Reproduced with permission from the *Geological Society of London*)

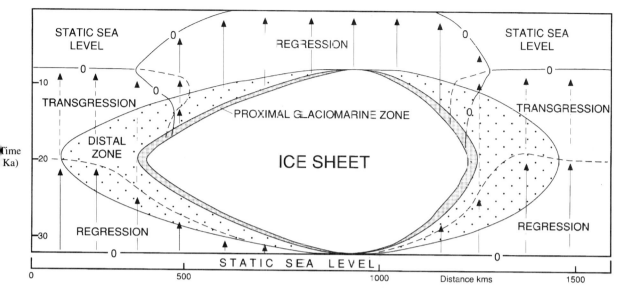

FIG. 11.24. Boulton's (1990) space-time model of changing glaciomarine environments during a glacial cycle. (Reproduced with permission from the *Geological Society of London*)

Sequences that reflect these changes can be found in many glaciated regions. In Spitsbergen, Boulton *et al.* (1982) demonstrated that following ice wastage the sea inundated the newly deglaciated areas depositing glaciomarine muds in up to 70 m water depths. However, rapid postglacial uplift caused regression and the muds began to suffer strong erosion in the sub-littoral zone and are overlain by beach sediments. Figure 11.24 illustrates schematically how the cyclic expansion and contraction of proximal and distal glaciomarine influence varies during a glacial cycle (Fig. 11.12).

11.7. TOWARDS AN INTEGRATED MODEL OF GLOBAL RELATIVE SEA LEVEL CHANGE

It should be clear from the above discussion that the interplay of land, sea and ice is highly complex. This complexity has encouraged the development of global relative sea level models similar in principle to global circulation models. The aim of these models is to be able to produce relative sea level curves for any part of the world that match those developed independently from field evidence. The factors involved in any such model will include the three dimensional pattern and timing of deglaciation for all ice sheets and the rheological properties of the lithosphere and asthenosphere. Feedback between model predictions and field data are capable of pin-pointing places where model assumptions or the field data are likely to be in error.

Developments along these lines were first reported by Peltier and Andrews (1976) with their ICE-1 model, and subsequent refinements are reported in Wu and Peltier (1983), Peltier and Andrews (1983) and Peltier (1987a, b, 1991). Nakada and Lambeck (1988) and Lambeck (1990, 1991a) have modified the ICE models to produce their ARC series of models and have demonstrated varied success in matching predicted to observed relative sea level changes from various locations around the world. Detailed studies from Britain indicate a particularly close agreement (Lambeck, 1991a, b).

The end products of these studies should allow a better understanding of the factors involved in the relative sea level curves of any part of the world. It should then be possible to predict future changes in relative sea level caused by changes in ice volume or other factors. These models are likely to be of crucial importance for both future coastal planning and the study of past sea level changes.

Chapter 12

MICROMORPHOLOGY

J.J.M. van der Meer

12.1. INTRODUCTION

Many times when studying sediments – whether glacial or other – it would be extremely valuable to see more details than can be seen by the naked eye or even with the aid of a handlense. If it were hardrock no one would hesitate to obtain a specimen for thin sectioning, in loose sediments this is not yet an obvious procedure. The main reasons are that not many laboratories have the equipment to impregnate soft sediment samples of adequate size and that not enough earth scientists are aware of the possibilities of this technique. In a number of pedology laboratories micromorphological studies have been common practice for many years already. In palecpedological studies it is seen as a must; Catt (1986) stated that ' . . . it is probably unwise to identify any buried soil without examining it first in thin section' (p. 180). The obvious advantage of thin sections is that t enables (in undisturbed samples) the direct observation and the study of the interrelation of (soil / sediment) components. Since the genetic interpretation of sediments often uses interrelations (i.e. fabric, structure) of the constituent particles (Dreimanis, 1988), the advantages of thin sections become manifest. In many cases macroscopic studies do not lead to a precise mode of deposition of a particular glacial unit being established (e.g. van der Meer *et al*, 1985; Dreimanis, 1988), thus all the more reason for applying micromorphology, as developed in pedology, to glacial sediments.

12.2. SAMPLING and SAMPLE PREPARATION

Thin section samples can be obtained in different ways, the first is to take a loose block and, after noting the orientation and the top or bottom of the sample, it should be wrapped tightly or (preferably) stored in some sort of container. This method applies especially to stony and/or dried-out compact deposits, as well as to subsamples from boreholes.

The second method makes use of steel or brass containers (15 $\times$ 8 $\times$ 5 cm or 8 $\times$ 6 $\times$ 4 cm), consisting of a rectangular box with two loose lids; the box has a cutting edge on one side. The cutting edge is pressed against the pit face and slowly cut in with a knife, thus inserting the box into the sediment (Plate 12.1). After cutting away the surrounding parent sediment, the orientation and top of the box is noted and once extracted, the lids are placed on the box. This way a rectangular block is obtained within the container. Samples are then transported with as little disturbance as possible; hard shocks should be especially avoided.

Many different kinds of sediments can be sampled with these methods; problems are mainly caused by very loose, sandy sediments and presently, water-

PLATE 12.1. Collecting a thin section sample of till. The metal container (15 × 8 cm) has been cut into the till face and part of the host sediment removed. By cutting behind the container the block can be removed. The example shows the sampling of a textural/colour boundary (arrows)

logged deposits. In both cases it may be impossible to obtain an undisturbed sample. But even if one manages to do so inevitable vibrations and/or shocks during transport and handling will cause disturbance of the original structure resulting in the sample becoming valueless. For both types of sediment no effective (nondisturbing) sampling method has been found as yet.

As each thin section only represents a small, and possibly unrepresentative part of the sampled unit or exposure, it is recommended to collect more than one sample. Samples should be taken not only of each specific litho-unit but as in pedology, also across the boundaries between units. Structural elements such as folds are obvious targets for sampling. In order to obtain a representative sample attempts should be made to try to get samples as large as possible (see sizes indicated above). For reasons given below it is strongly recommended that representative bulk samples of the studied section be obtained.

A number of procedures exist for manufacturing thin-sections. These have most recently been compiled by Murphy (1986). For the examples described below, the following method has been used: The samples (in their containers, top-lid removed) are air-dried in the laboratory. After removal from the tin, the samples are impregnated with synolite (an unsaturated polyester resin), mixed with monostyrene as thinner, cobaltoctate as accelerator and cyclonox as

catalyser (for quantities see Murphy, 1986). The actual impregnation occurs in a vacuum chamber; even then, dense, clayey sediments may partly remain unimpregnated, while salt-containing samples may cause similar problems. After impregnation samples must be left to harden. This may take as long as six weeks and is absolutely vital if broken or distorted samples are to be avoided. The next step is the cutting of each sample and mounting on glass plates. Finally, the samples are ground to a thickness of ± 20 μm.

12.3. STUDYING THIN SECTIONS, NOMENCLATURE

Thin sections should be studied under a petrographic microscope using magnifications between 2 and 10 times. Larger magnifications easily lead to study of single grains, which is not very fruitful. When comparing different thin sections, care should be taken to use the same magnification and illumination. Up to now the number of people who have used this method is limited, as shown by the papers dealing exclusively with this subject (Sitler and Chapman, 1955; Ostry and Deane, 1963; Korina and Faustova, 1964; Wisniewski, 1965; Sitler, 1968; Evenson, 1970; Romans et al., 1980; Johnson, 1983; van der Meer et al., 1983, 1992; Love and Derbyshire, 1985; Tang Yongyi, 1987; van der Meer, 1987a). Also the number of studies in which thin sections have been used is not very extensive (e.g. Gravenor and Meneley, 1958; Boulton, 1970a; Lavrushin, 1976; Evenson et al., 1977; Johnson, 1980, 1983; May, 1980; van der Meer, 1982, 1987b, 1993; Rappol, 1983; Seret, 1983; Kulig, 1985; van der Meer et al., 1985, 1992, 1994a,b; Menzies, 1986; Feeser, 1988; Owen and Derbyshire, 1988; Tsui et al., 1988; Menzies and Habbe, 1992; Menzies and Maltman, 1992; Menzies and Woodward, 1993). As a result no standard nomenclature for the description of thin sections of (glacial) sediments has been developed. Since different authors described the same feature with various different terms, van der Meer (1987a) suggested that the plasmic fabric nomenclature developed by pedologists (Brewer, 1976) should be used. Despite discussions amongst pedologists for a renewal of this terminology (Bullock et al., 1985) this still provides the best reference as it has been widely used and is applicable to sediments as well as to soils.

Micromorphological features can be grouped under several headings (see below). Of these, the plasmic fabric, which is the arrangement of the plasma, is only visible under crossed polarizers or in circularly polarized light, because of the high birefringence of (re)oriented domains. However, whether this is visible depends not only on the strength of the (re)orientation, but also on the clay and carbonate content. In the first place a minimum amount of clay is necessary for the birefringence to show, while generally the development increases with the clay content. As the strength of the (re)orientation not only depends on clay content, but also on the stresses applied to the sediment, no general figure for the necessary, minimum amount of clay can be given. In the second place fine-grained carbonates disturb the view as they scatter the polarized light again. Thus, highly calcareous tills will not show much birefringence, no matter the strength of the (re)oriented domains. Removal of fine-grained carbonates *in situ* could be tried by etching of the thin section after mounting on glass and before putting up the top glass. Also the clay mineralogy should be taken into account as swell-and-shrink processes may result in plasmic fabrics which are similar to those produced by certain glacially induced processes like rotation (Lafeber, 1964). Of course all clays will swell when wetted, but it is better to be informed about the presence and amount of expandable clays.

12.4. MICROMORPHOLOGICAL FEATURES OF GLACIAL SEDIMENTS

As first mentioned by Sitler (1968), micromorphological observations can be grouped under certain headings: texture, structure, plasmic fabric, diagenetic features, characteristics of mineral and other particles and, finally, silt and sand (micro)fabric. Here only the first four will be discussed; for characteristics of particles see FitzPatrick (1984) and for microfabrics Evenson (1970, 1971). The examples given below have been selected from samples taken from The Netherlands, Germany, Sweden, Spitsbergen, Scotland, Ireland, Switzerland and Argentina and can be taken as being representative for a large number of observations. For convenience the several headings will be treated separately although it is obvious that

the same example could be treated under several headings. A till that shows distinct banding on a centimetre-scale can be taken as an example of textural differences, but as this banding may well be the result of repeated folding it could also be treated as an example of structure.

12.4.1. Texture

Usually tills are considered to be homogeneous as far as their texture is concerned. When studying tills in the field, this homogeneity is far from clear as textural differences like changes in grain size in a profile as well as soft sediment inclusions of widely different shape and size are widespread. The latter points to reworking of older, often directly underlying deposits. If reworking of older material is macroscopically visible in the sense that soft sediment inclusions are visible, the composition of these inclusions may be difficult to determine, especially if they are of small size. Bulk samples will not provide enough insight in the nature of these inclusions, but thin sections will through direct observation of this material (van der Meer et al., 1983, 1985). Direct observation of the constituent particles of both till and inclusions will tell us more, not only about their size and shape, but also about their orientation and distribution as well as their relation to other particles (see Example 1, below). In quite a number of samples it has been observed that skeleton grains are not randomly distributed. They may be concentrated in certain bands or zones, while sometimes a polygonal pattern is suggested (van Ginkel, 1991). However, in most cases no clear pattern can be discerned.

Also, the shape of these intraclasts provides information on the genesis of the whole deposit. Deformation of such clasts by flow produces shapes (and also associated plasmic fabrics, see below) distinctly different from those produced by shear (see Section 12.5). Examples like these and also the presence of fine-grained casings around larger grains (coarse sand, fine gravel) have been presented by van der Meer (1987a). This appears to be quite common, albeit not often as clear as the example shown there. Small skeleton grains around large particles are seldom in direct contact with this large particle, although the small grains may be in direct contact with each other.

Usually there is a thin zone of fine-grained material in between small and larger particles. All this shows that even macroscopically homogeneous tills are not so on a microscopic scale. Although the origin of such small scale differences in texture is not self-evident and no explanation can be offered as yet, their occurrence shows that care should be taken in the interpretation of grain size analyses of bulk samples.

12.4.1.1. Example 1 – Banded till from Lunteren, The Netherlands

This sample has been collected from the base of the till just above a shear zone (Fig. 27 in van der Meer *et al.*, 1985), to study microscopic effects of macroscopically visible shearing. Orientation of the sample (left to right) is N–S, ice movement NW–SE (van der Meer *et al.*, 1985).

The sand–silt–clay ratio was 51–23–26% based upon a mean of two samples from the same site, the whole profile being decalcified. No clay mineralogical analyses was done of samples from this site; however, two samples from the same lithostratigraphic unit in the same pit were found to consist of: smectite 28–26%, illite 43–43%, kaolinite 19–20%, chlorite 5–2% and vermiculite 5–9% (Haldorsen *et al.*, 1989). In view of this clay content and composition, combined with the fact that the site was located some 14 m below the surface (and hence more or less constant moisture conditions can be assumed), the micromorphological features to be seen can all be ascribed to glacial processes.

The thin section (Plate 12.2a, b) shows a number of (sub-) horizontal bands, some of which are not continuous across the sample. In the upper part the bands consist of two types of diamicton, one dom-

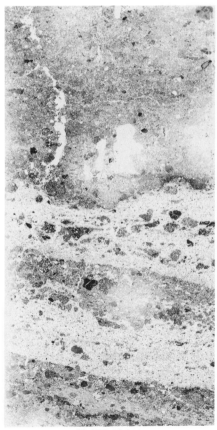

PLATE 12.2. Example 1. Textural differentiation in a banded till from Lunteren, The Netherlands. See section 12.4.1 for explanation and discussion. (a) Thin section R.656. (b) Textural division of thin section R.656, voids in white and scale in cm. (c) Distribution of silt/clay and till pebbles in the textural bands of thin section R.656. Numbered rectangles indicate position of micrographs. Scale in cm. (d) Detail of one of the dominantly sand-supported till bands; the darkrimmed rounded spots are air bubbles. Bar indicates 2 mm. Plane light. (e) Detail of one of the dominantly matrix supported till bands. Bar indicates 2 mm. Plane light. (f) Detail of one of the sand bands. Bar indicates 2 mm. Plane light. (g) Example of a silt and a clay pebble, both elongated and showing small cracks. Bar indicates 1 mm. Plane light. (h) The same pebbles seen under x nic., notice strong birefringence in both pebbles (see also section 12.4.3). (i) Example of a well-rounded till pebble. Bar indicates 1 mm. Plane light.

PLATE 12.2(a)

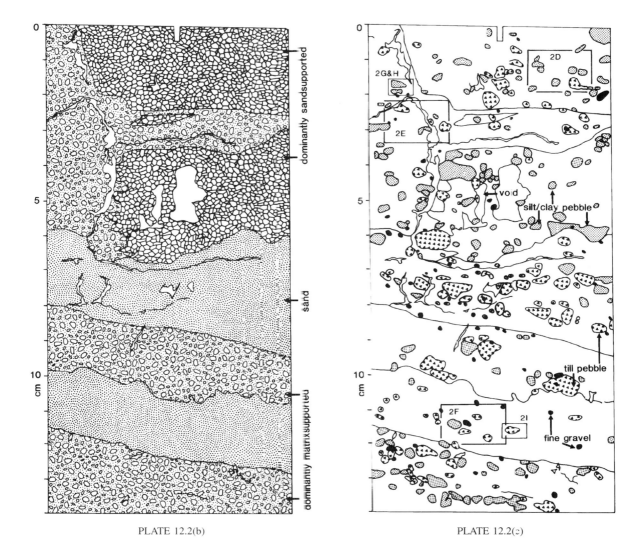

PLATE 12.2(b)   PLATE 12.2(c)

inantly sand-supported (Plate 12.2b, d), (see also colour plates) the other dominantly matrix-supported (Plate 12.2b, e) (see also colour plates). There is no obvious difference in grain size or shape between these bands. In the lower half, sand bands (Plate 12.2b, f) (see also colour plates) alternate with dominantly matrix-supported diamicton bands.

The boundaries between the bands have different characteristics. The upper boundaries of the sand bands are rather irregular as compared to their lower boundaries.

The thin section further shows a great number of rounded to elongate pebbles. The long axis of the elongate pebbles is parallel to the banding of the sample. Two types of pebbles can be recognized (van der Meer et al., 1983). The first type consists of either pure clay or silt mixed with some coarser quartz grains (see Plate 12.2g in colour plates). Sometimes, layering in the pebbles (see Plate 12.2f in colour plates) shows their derivation from the underlying rhythmites (van der Meer et al., 1983, 1985). The second type of pebble consists of till (see Plate 12.2i

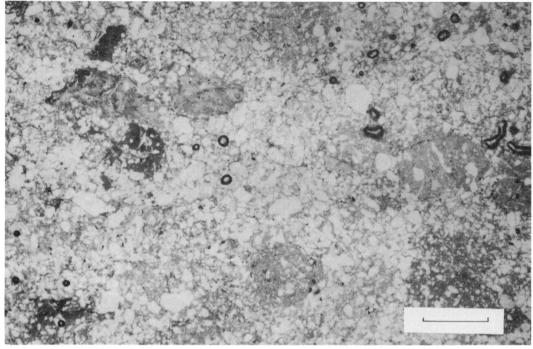

PLATE 12.2(d)

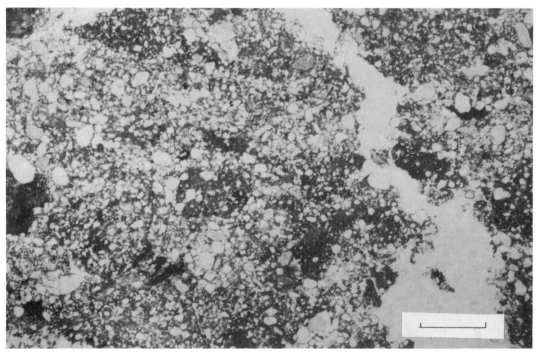

PLATE 12.2(e)

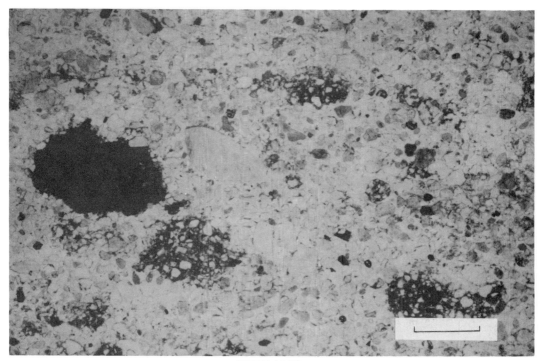

PLATE 12.2(f)

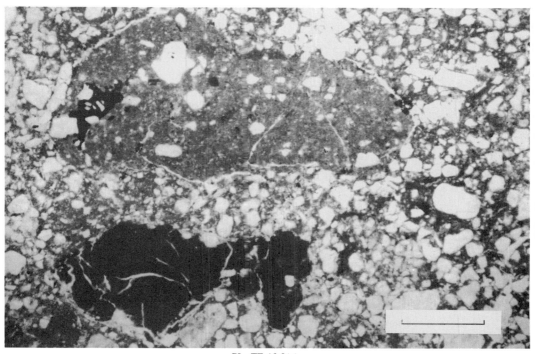

PLATE 12.2(g)

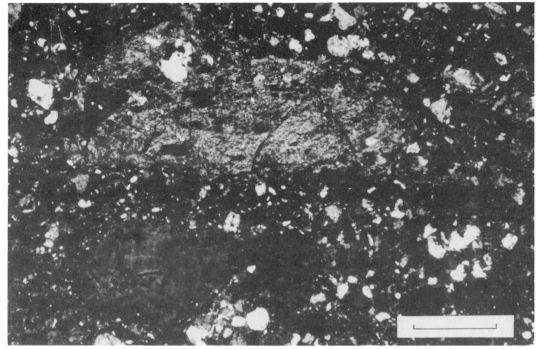

PLATE 12.2(h)

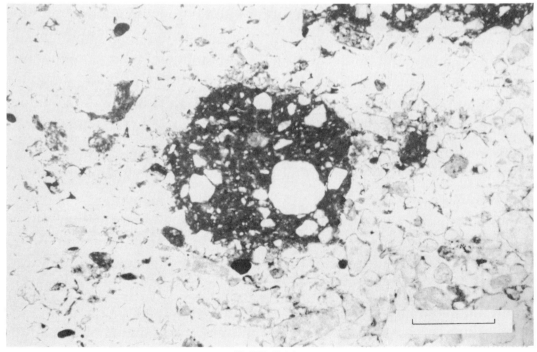

PLATE 12.2(i)

in colour plates), which differs from the main till mass by a finer texture. Many of the pebbles show small cracks, which demonstrates that they were being reduced in size, when the subglacial formation of this till sequence stopped. The distribution of the pebbles is not consistent within or between bands. Some bands contain more pebbles, while also the type of pebbles differs per band (i.e. the sand bands mainly contain till pebbles, while some of the diamicton bands mainly contain silt/clay pebbles) (Plate 12.2c). The different types of pebbles as well as their amount show that reworking has been a major process in the formation of this till. As described before (van der Meer et al., 1983, 1985) a lobe of the Saale Ice Sheet overrode an ice-pushed ridge and in doing so picked up clay/silt rhythmites which had been formed in front of the glacier. Seasonal variation in the position of the ice front is considered to be responsible for the entrainment of till. Both the clay/silt rhythmites and the till were deformed into rounded pebbles because of rotational movements. Conceivably, some of the pebbles, (i.e. the till pebbles in the sand bands) may have formed through erosion by meltwater and deposited as such in the sandy material. The rotational movements are also evidenced by the plasmic fabrics (see below) described from this site. Hence, the banding of this till can also, on micromorphological grounds, be ascribed to deformation and not to sedimentation. Consequently, the till can be considered to be rather immature (a mature till as produced by subglacial deformation] being one that is completely homogenous). Continuation of the deformation would have further attenuated the different bands, while also the pebbles would have disappeared through comminution.

12.4.2. Structure

The definition of soil structure ('the aggregation of primary soil particles into compound particles, or clusters of primary particles, which are separated from adjoining aggregates by planes of weakness') as given by the Soil Survey Staff (1951) can also be applied to the study of (glacial) sediments. Apart from thin sections, further information on structure can be obtained by making X-rays of remaining slices of the original thin section sample. Planes of weakness –

like fissures – clearly show in X-rays (Calvert and Veevers, 1962; van der Meer et al., 1983; van der Meer, 1987a) thus enabling easy study of the shape, size and orientation of aggregates.

In the field the most obvious example of primary particle aggregation is fissility, often highlighted by the presence of iron and/or manganese precipitates. It has been shown (van der Meer, 1987a, p. 83) that in the examples studied the fissures (the actual planes of weakness) are not related to differences in grain size or orientation of the primary particles. In small scale fissility the fissures themselves are not even continuous. What this may indicate about the origin of the fissility is not yet clear.

Structures that are often not visible in the field are related to smaller scale aggregations of primary particles, like the till pebbles described in Example 2, below. Smaller till and clay pebbles have been observed in several samples from different localities (Rappol et al., 1989). This latter type of pebble is best observed under crossed polarizers as they consist of (sub-) rounded aggregates of strong unidirectional birefringence, while the planes of weakness consist of thin zones of birefringence encircling the aggregates (skelsepic plasmic fabric). The rounded shape of the pebbles, together with the skelsepic fabric, points to deformation through rotation.

Another macroscopically visible structure may consist of a continuous sequence of two or more till beds. Although differences in grain size are not always present, different suites of erratics or different macrofabrics in the till beds have often been interpreted as pointing to more than one glacial episode. In the field the exact nature of a boundary (i.e. whether it represents a hiatus) is often difficult to establish. Thin sections of such boundaries, by providing detailed information on weathering but also on continuity of structures, can help to solve the question as to whether such a boundary really represents a hiatus or merely a change in direction or provenance during a continuous glacial episode (Rappol et al., 1989).

Also, small faults and folds in intraclasts (van der Meer and Laban, 1990; van der Wateren, 1992) or in interbedded waterlain sediments (Figs. 16 and 17 in van der Meer, 1987a) can best be studied in thin section (cf. Bordonau and van der Meer, 1994).

Detailed analysis of such structures in oriented thin sections leads to a better understanding of the stress-field which acted on the deposit (Chapter 10).

12.4.2.1. Example 2 – Samples from a profile at Wijnjewoude, The Netherlands (Rappol, 1983)

Four samples from different zones of the profile have been thin sectioned (Plate 12.3a). The samples are oriented (left to right) SW–NE, with ice movement NE–SW (Rappol, 1983). Grain size for the samples (sand–silt–clay) displays 65–17–18% for the decalcified part and 62–23–14% for the calcareous part. Carbonate content in the lower part of the profile is low and never amounts to more than 2–3%. The clay mineralogy of the decalcified (Plate 12.3a) and the calcareous (second figure) parts of the profile differ slightly, most likely reflecting weathering: smectite 48–30%, illite 36–42%, kaolinite 15–18%, chlorite 1–8% and vermiculite 0–0% (Haldorsen et al., 1989).

Plate 12.3a shows structures within the four samples as traced from the thin sections. The structure, as recorded in these tracings, demonstrates that the macroscopically homogenous till consists in fact of separate structural elements or pebbles (van

der Meer, 1987a, Figs. 15 and 31). It is obvious that the shape of these pebbles changes with depth. High in the profile they are well-rounded, while lower down this shape becomes more and more flattened and angular. Details of pebbles at different levels in the profile are presented in Plates 12.3b, c, d, e. These details show that many of the planes of weakness (voids) are not continuous or of equal width. The micrographs also show that the texture in all four samples is rather uniform and that, hence, the location of the voids may partly be the result of drying during sample preparation, but the diversity of these features shows that the location of the voids has been predetermined by glacial action.

The origin of this structure can be envisaged as follows (van der Meer, 1987a, p. 83): A first stage is marked by the grain-by-grain accumulation of the till mass. After a change in subglacial conditions this till mass deformed, first by brecciation (Plates 12.3d, e), followed by rotation and rounding of the aggregates (Plates 12.3b, c). This latter stage resembles movement as in a bed of marbles. However, such a bed requires a lot of pore space, and as the plates demonstrate, such space is not present. The continuous interaction between different pebbles explains the lack of continuous voids and the intergrown nature

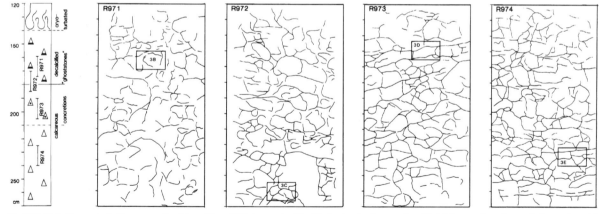

PLATE 12.3. Example 2. Structures in till profile at Wijnjewoude, The Netherlands. See section 12.4.2 for explanation and discussion. (a) Schematic till profile and tracings of the voids in four thin sections from different levels in the profile. These voids delineate the structural units (till pebbles). Numbered rectangles indicate the position of micrographs. Scale of thin sections in cm. (b) Detail of thin section R.971 showing a well-rounded pebble, mainly delineated by small voids. Bar indicates 2 mm. Plane light. (c) Detail of thin section R.972 showing a pebble complex. Bar indicates 2 mm. Plane light. (d) Detail of thin section R.973 showing the flattened outline of pebbles lower down in the profile. Bar indicates 2 mm. Plane light. (e) Detail of thin section R.974 showing the flattened and angular outline of pebbles lower down in the profile. Bar indicates 2 mm. Plane light.

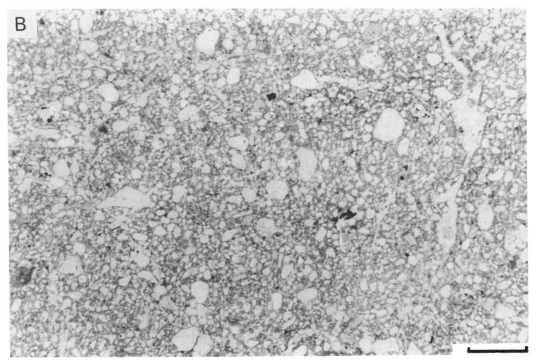

PLATE 12.3(b)

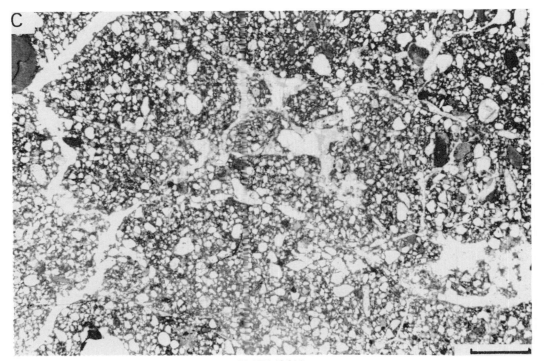

PLATE 12.3(c)

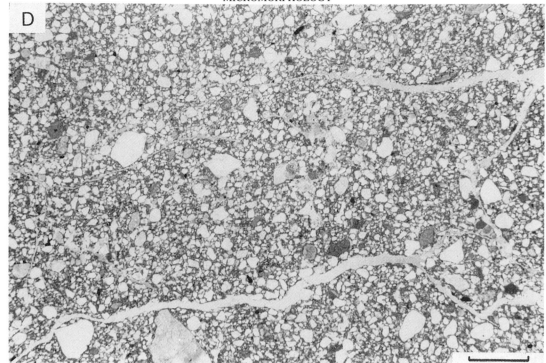

PLATE 12.3(d)

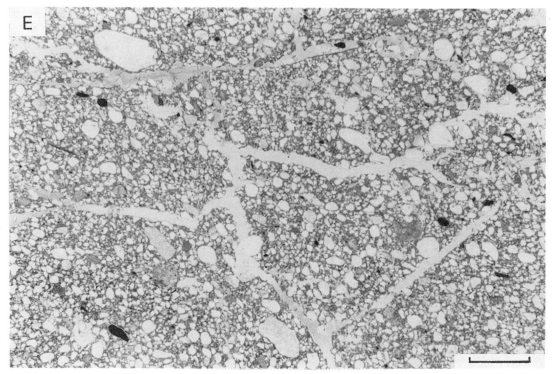

PLATE 12.3(e)

of some of the contacts between pebbles. The flattening-with-depth is thought to reflect the decrease in velocity which is known to exist in deforming glacier beds (Boulton and Hindmarsh, 1987).

12.4.3. Plasmic Fabric

Plasmic fabric is the description of the arrangement of the plasma, which is all material <2 μm; thus, it may consist of clay minerals, oxides and hydroxides of iron, for example; soluble salts, etc. Because similarly-oriented domains show up by a high birefringence, the arrangement of the plasma can be studied microscopically under crossed polarizers or in circularly polarized light and classified according to the pattern of these domains. It should be clear that this plasmic fabric is more than just the orientation of silt and fine sand grains, and that it is in fact a third type of fabric besides micro- and macrofabric. As mentioned above, the nomenclature developed by pedologists (Brewer, 1976) will be used to describe the different types of plasmic fabrics that have been recognised in glacial deposits so far.

The presence of birefringent plasma domains may be the result of sedimentation, but is usually the result of stresses applied to sediments, whereby fine-grained material is reoriented in accordance with the stress field. In tills the stresses may be the direct result of the overriding glacier, but post-depositional processes like slope movement cannot be ruled out a priori. The possibility of slope movements having affected the till to be studied should be established in the field while sampling. From the material studied so far it has become clear that a plasmic fabric is not always present. Tills which are macroscopically structured (i.e. showing well developed joints, or a strong fissility) have been found to show no or only a very weak plasmic fabric (Lagerlund and van der Meer, 1990). Similarly, in a thin section taken across a major thrust in an ice-pushed ridge in West Germany, although Tertiary loam had been thrust over lower Quaternary gravelly sands (Kluiving, 1989), the loam only shows a very weak skelsepic fabric (see below). It is thought that this lack of microscopic evidence of deformation is the result of high porewater pressures during the actual deformation. Instead of the sediments, pressurised porewaters carried the load of the overlying/overriding ice or sediment masses.

As stated before the plasmic fabric can be classified according to its pattern or the lack of a pattern. The latter is known as a silasepic or argillasepic fabric (depending on the amount of clay) and appears to be typical for flow tills (van der Meer, 1987a) and apparently also for glaciomarine deposits in as far as these have not been influenced by iceberg scour (Lagerlund and van der Meer, 1990).

Strial fabrics like (bi)masepic and unistrial fabrics (Brewer, 1976; van der Meer, 1987a) point to shear, as does kinking of clay-rich bands. This latter fabric was not recognized by Brewer (1976), but has been included by FitzPatrick (1984). In tills it has been recognized several times (van der Meer, 1982, 1987a). Also, brecciation of clay bands points to deformation by shear. As the brecciation is usually visible as short shear planes cutting each other obliquely or perpendicularly (sometimes three directions can be recognized), it shows that rotation is part of the deformation process.

Rotation appears to be the most common process in the (de)formation of glacier beds as evidenced by the widespread occurrence of the skelsepic plasmic fabric in tills. In this type of plasmic fabric the oriented domains are oriented parallel to the surface of skeleton grains (Plates 12.2d, f) (see also colour plates) (Jim, 1990). Often, it is also possible to see two more or less perpendicular orientations in relation to the skeleton grains, a fabric which is known as lattisepic. The result of rotational movements is sometimes shown by the orientation of fine sand and silt grains parallel to the surface of large grains (Plate 12.4g). As will be shown elsewhere establishing the presence of, for example, rotational elements, enables the recognition of deformable beds, and hence facilitates the reconstruction of the dynamics of Pleistocene glaciers and ice-caps.

12.4.3.1. Example 3 – Sample collected at Urk, The Netherlands (Rappol *et al.*, 1989)

The sample was taken with the purpose of studying the transition between 'red' till floes and the underlying 'normal' till (van der Meer, 1987b; Rappol *et al.*, 1989). The sample was oriented (left to right)

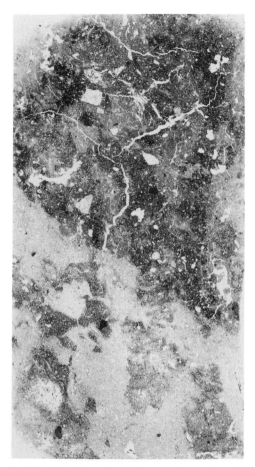

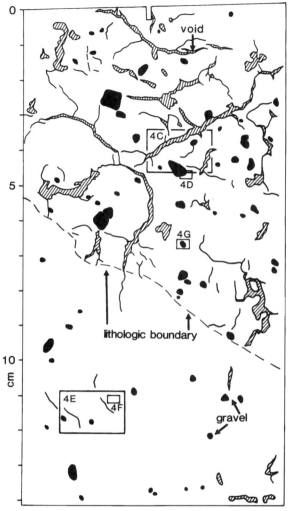

PLATE 12.4(b)

PLATE 12.4. Example 3. Plasmic fabrics in red till floe and underlying 'normal' till at Urk, The Netherlands. See section 12.4.3 for explanation and discussion. (a) Thin section O.817 which has been collected across the boundary between till floe and underlying till. (b) Sketch of thin section O.817. Numbered rectangles indicate the position of micrographs. Scale in cm. (c) Example of plasmic fabric in a clayey part of the red till. Bar indicates 2 mm. × Nic. (d) detail of plasmic fabric around gravel particle just below Plate 4C. Bar indicates 0.5 mm. × Nic. (e) Example of the plasmic fabric in the normal till. Bar indicates 2 mm. × Nic. (f) Detail of plasmic fabric in the normal till. Bar indicates 0.5 mm. × Nic. (g) Sketch showing the orientation of small particles parallel to the surface of a larger particle and in accordance with the plasmic fabric. Bar indicates 0.5 mm.

SSW–NNE, most likely with an ice movement NE–SW (Rappol, 1987). Grain size (sand–silt–clay) was 60–21–19% (mean of three samples) for the normal till, and 32–42–26% (mean of four samples) for the red till (Rappol, *unpublished*). The whole profile contains carbonates, although the amounts are quite different: 6.2% in the normal till and 25.9 % in the red till (Rappol, *unpublished*). The red till also shows extensive deposition of secondary carbonates. It is obvious that these two tills are quite different in composition, which is also reflected by the clay mineralogy (see Plate 12.4e, normal till and, Plate 12.4c red till in colour plates): smectite 33–0%, illite 41–67%, kaolinite 17–11%, chlorite 7–7%, and vermiculite 2–15% (Haldorsen *et al.*, 1989).

The abrupt transition between the two till beds is quite distinctive in Plate 12.4a, which shows the whole thin section. The higher clay and silt content of the

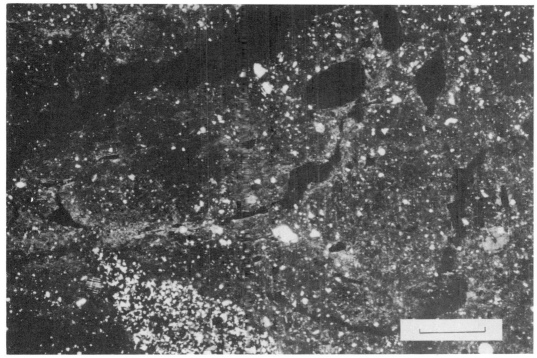

PLATE 12.4(c)

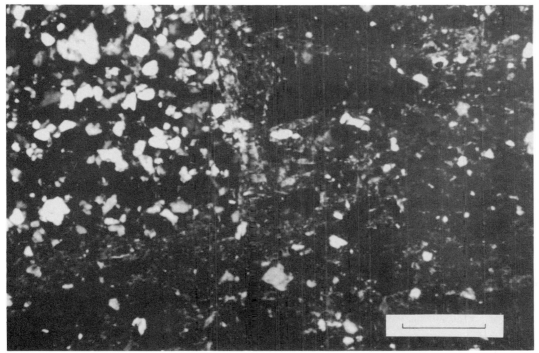

PLATE 12.4(d)

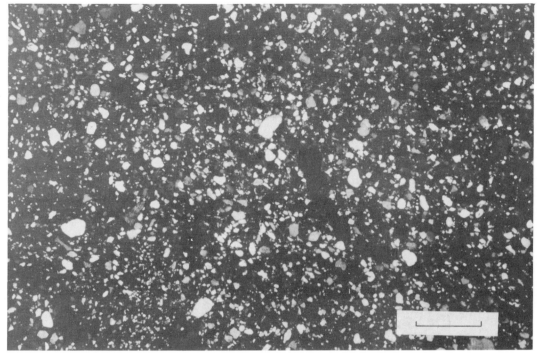

PLATE 12.4(e)

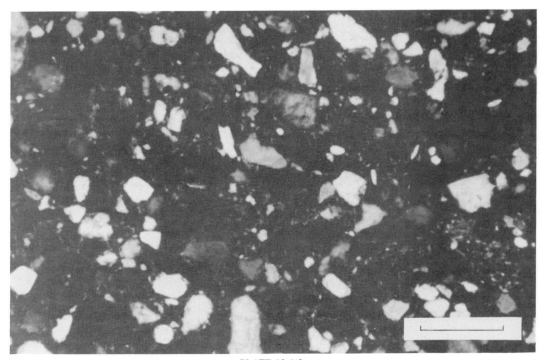

PLATE 12.4(f)

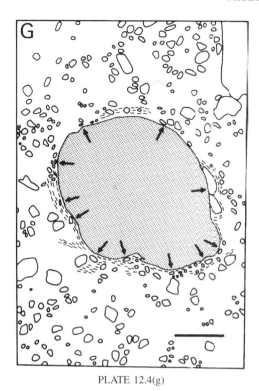

PLATE 12.4(g)

12.4.4. Post-depositional Changes

Like all sediments, glacial deposits are prone to post-depositional changes. As described in van der Meer (1987a) these include the translocation of clay, silt, (hydr)oxides of Fe and Mn and carbonates. Especially precipitates of the latter compound may lead to (sometimes strong) cementation. All of these translocations as well as initial weathering can be studied in thin sections. The results of translocation are usually visible as some form of cutan, the name of which is derived from the composition, (e.g. argillan for clay or calcitan for carbonates). Apart from strong cementation, disturbance of glacial deposits by frost action or periglacial conditions, in general, can often be easily observed in the field. However, the absence of cryoturbation does not necessarily mean that a deposit has not been disturbed by frost. Nor does the depth to which cryoturbation is present indicate the maximum depth of frost influence. Within tills in The Netherlands that date from the penultimate glaciation, it is quite common to find the top 2 to $2\frac{1}{2}$ m disturbed by cryoturbation and/or frost wedges related to the last glaciation (Rappol, 1987).

12.4.4.1. Example 4 – Profile collected at ter Idzard, The Netherlands

The profile was studied in order to relate the till on the Drente Plateau to the till in a large roadcut through ice-pushed tills nearby (Rappol *et al.*, 1989). Seven samples of different size and from different zones of the profile have been thin-sectioned (Plate 12.5a). Orientation of samples (left to right) is SSW–NNE, with an ice movement NE–SW (Rappol *et al.*, 1989). Grain size ratios (sand–silt–clay) are 65–16–19% (mean of three samples; Rappol, *unpublished*). The whole profile is decalcified. The clay mineralogy of this site has not been studied, but can be considered to be comparable to the composition of tills with smectite presented in the other examples.

Plate 12.5a shows the profile studied at ter Idzard. In the field it was obvious that cryoturbations reach to a depth of well over $1\frac{1}{2}$ m. Below that there was no macroscopic evidence of frost action. The micrographs presented in Plates 12.5d, e, f clearly demonstrate the existence of a silt droplet structure (Fig. 25

upper, red till is highlighted by the large number of cracks which developed in it during preparation of the thin section. Despite the high carbonate content of the red till the plasmic fabric is clearly visible; undoubtedly this has been caused by translocation of the carbonates which left certain zones with much lower carbonate content or none at all. Plates 12.4c, d, f, g (see also colour plates) show that the oriented domains are mainly parallel to the surface of skeleton grains (skelsepic fabric). Plate 12.4c shows that in clay-rich areas the plasmic fabric can be described as bimasepic. The oriented domains consist of short light streaks combining into longer lineations with two more or less perpendicular orientations. This fabric points to concentrated shear. Comparison of Plates 12.4c and 4e demonstrates that the strength of the plasmic fabric is not the same in the two till beds: it is obvious in Plate 12.4c (the red till) and hardly visible in Plate 12.4e (the normal till). Larger magnification (Plates 12.4d, f) shows that the skelsepic fabric is present in the lower, normal till, but that it is much stronger in the upper, more clayey red till.

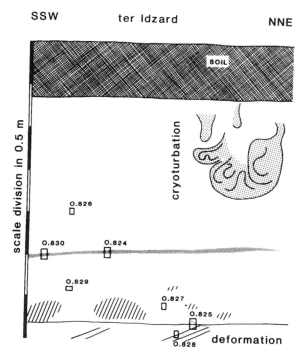

PLATE 12.5(b)

PLATE 12.5. Example 4. Postdepositional changes in a till profile at ter Idzard, The Netherlands. See section 12.4.4 for explanation and discussion. (a) Schematic representation of the till profile. Numbered rectangles indicate the position of thin sections. Hatchings represent textural and colour differentiation within the profile; after Rappol *et al.*, 1989. (b) Thin section O.826 from about 2 m depth. (c) Tracing of voids in thin section O.826. These voids delineate the silt droplet structure. Numbered rectangle indicates the position of a micrograph. Scale in cm. (d) Example of the well-developed silt droplet structure in thin section O.826. Bar indicates 2 mm. Plane light. (e) Example of weakly developed silt droplet structure near the base of thin section O.824. Bar indicates 2 mm. Plane light. (f) Example of the very weakly developed silt droplet structure in the upper part of thin section O.827. Bar indicates 2 mm. Plane light.

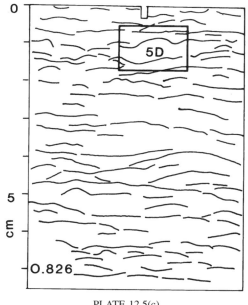

PLATE 12.5(c)

in van der Meer, 1987a) down to a depth of just over three metres. Plates 12.5b, c, d show the strong development of this silt droplet structure at a depth of about 2 m in the till profile. This structure can best be described as consisting of slightly curved elements, the upper surface of which is enriched in fines (slightly darker in the micrographs). Comparison of Plates 12.3a and 12.5c reveals the completely different nature of these secondary structures. The silt droplet structure is not related to glacial action, but to the postdepositional formation of ground-ice lenses

under periglacial conditions (Coutard and Mücher, 1985; Van Vliet-Lanoë, 1985). Plates 12.5e and 12.5f demonstrate the diminishing strength of the silt droplet development with depth. Thus it is obvious that the influence of frost reaches to far greater depth

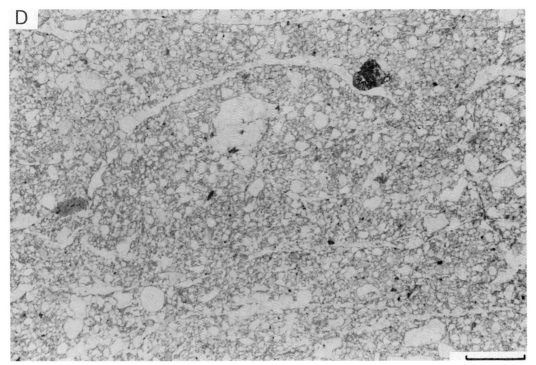

PLATE 12.5(d)

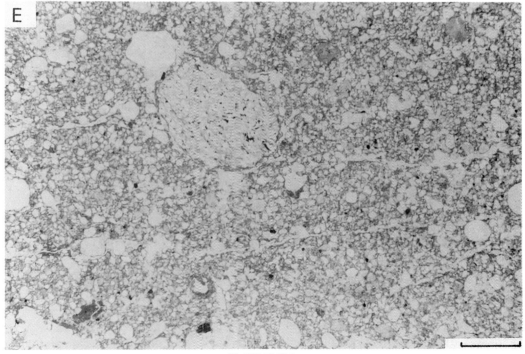

PLATE 12.5(e)

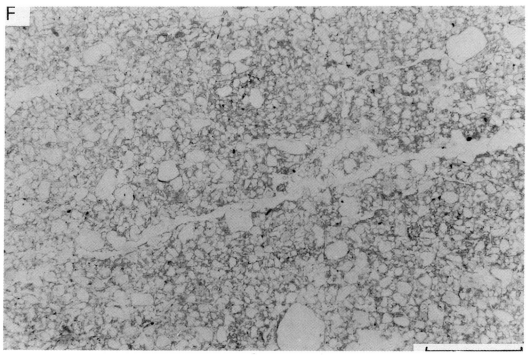

PLATE 12.5(f)

than the level of cryoturbation. It also shows that the frost influence does not affect the linear nature of the clayey band at about 2.5 m depth.

Besides the silt droplets, the ter Idzard profile also shows evidence of clay translocation in the form of ferri-argillans (clay skins containing iron compounds). The age of this clay translocation is difficult to establish. Menzies (1986) described clay skins thought to be formed subglacially, but the skins at ter Idzard may also be related to Eemian (last interglacial) or Holocene soil formation.

12.5. DISCUSSION

The description of texture, structure, plasmic fabric and post-depositional changes presented above demonstrates that much information can be gained from studying thin sections of glacial sediments. The interpretation of the information thus obtained is a different matter. Up to now mainly Pleistocene deposits have been studied, but the value of these observations can only be assessed by correlating them with tills of known origin (van der Meer, 1987a, p. 88). The main problem in such a correlation programme is to find good equivalents to Pleistocene deposits. Most (accessible) glaciers are found in alpine areas and are of a relatively small size. The length of the transport path of glacial debris in these glaciers is incomparable to the distance debris in Pleistocene icecaps travelled. Besides, Alpine glaciers often mainly overly bedrock, while the ice sheets of the past also covered vast stretches of unconsolidated sediments. In other words, the tills produced by Alpine glaciers are not directly comparable to Pleistocene tills: quite often the amount of clay in Alpine tills is not high enough to show clear plasmic fabrics.

Despite these problems it is possible to gain information on the genesis of glacial deposits by studying thin sections. Usually it is a combination of several observations (microscopic as well as macroscopic) which provides such information: the presence of textural banding can be the result of sedimentation or of strongly attenuated folding (combination of textural banding and a well developed plasmic fabric points to the latter origin) (Example 1). Similarly, the shape of soft sediment inclusions point at the most likely origin of the deposit: flow produces wavy, irregular clasts, while shear quite often produces elongated clasts with a strong (uni-) directional plasmic fabric. These examples clearly show that it is necessary to study all aspects of the thin sections As stated before, the combination of micromorphological analyses with field evidence leads to better understanding of the deposit (van der Meer *et al.*, 1985; Rappol *et al.*, 1989). Especially when field observations are lacking, as in material collected from boreholes – either on land or at sea – thin sections are about the only method for obtaining structural information on the genesis of the deposit (van der Meer and Laban, 1990).

12.6. FUTURE WORK

As indicated above it is absolutely necessary to compare observations on Pleistocene tills with those on tills of known origin. Apart from the problems sketched above, there are also other problems in obtaining such samples. This does not hold for flow tills; these are by far the easiest to observe and sample (when they are not too wet). The difficulty is mainly in the criteria to establish lodgement from melt-out, also in the active glacial environment. When ice is overlying till, how does one decide on the origin of the till? If the surface of the till is fluted, should this then be taken as evidence of lodgement or of deformation? One way of avoiding this problem is by sampling debris-rich ice and letting it melt under field conditions (i.e. from the base upwards). This is not only a very time-consuming procedure, but at the same time one has to decide on free drainage or water ponding. After all, one of the most likely criteria for melt-cut is the microscopic work of meltwater, although its presence and hence its influence underneath temperate glaciers cannot be ignored. It may well be that by far the easiest way to obtain samples of known origin is by producing them experimentally. The production of debris-rich ice following the method of Cegla *et al.* (1976) may be a good start for such experiments (see also Eyles *et al.*, 1987b).

If it is possible to establish the micromorphological criteria for melt-out tills and flow tills, lodgement tills will form the remaining group. Similarly, it should be possible to collect samples of present-day glaciomarine environments, both undisturbed and deformed by iceberg scouring, to establish their characteristics as well as the way in which they differ from on-land glacial deposits.

This will certainly facilitate the interpretation of glacigenic deposits in near-coastal settings where isostatic rebound has occurred (Lagerlund and van der Meer, 1990; van der Meer *et al.*, 1992). It will also facilitate the interpretation of Pre-Pleistocene glacigenic deposits (Visser *et al.*, 1984), because although relatively little is known of the micromorphology of Pleistocene and Recent deposits the use of thin sections in tillite research is quite common (e.g. Fairchild and Hambrey, 1984).

Chapter 13

SCANNING ELECTRON MICROSCOPY

W. B. Whalley

13.1. INTRODUCTION

Perhaps no other expensive piece of equipment has become as widely used as the scanning electron microscope (SEM) for routine analyses of sediments. Its ability to provide a range of magnifications with large depth of focus, even at high magnifications, is unrivalled. Indeed, this is the basis for using the SEM to determine surface textures of individual grains, and the identification of sedimentary transport mechanisms was one of the earliest, and still important, uses of the SEM in geology. Other techniques have been applied (Warnke and Gram, 1969) but none has gained the popularity of the SEM. Since its introduction in the early 1960s, the SEM has become progressively easier to operate. It is now possible for even casual users to attain a high degree of proficiency in its use at a basic level as a 'super microscope'. Along with this simplicity, however, there has been an increasing sophistication of instrument capability and the way it can be used. In this chapter the elements of SEM operation are covered with an overview of the uses to which it has been put in the study of glacial deposits.

Because of the complexity of some instrumentation, techniques and the range of applications, this chapter can only provide some guidelines and in no way attempts a 'how to do it' approach. Texts that cover operation of the SEM and ancillary equipment include: Smart and Tovey (1981, 1982) and Goodhew

and Humphreys (1988). Starting points for general investigations related to environmental reconstructions using surface textures of grains in sediments are the bibliography by Bull *et al.* (1986), Bull's (1981) overview and the collections of papers edited by Whalley (1978a) and Marshall (1987). Whalley (1985) provides a general introduction on work concerned with the sedimentological characterisation in soil studies which has relevance to glacial sediments.

Two main areas of application are covered in this chapter: (1) the use of the SEM to determine surface characteristics (surface textures) of sand and silt grains as a means of distinguishing provenance or geomorphic agency and (2) the characterisation and investigation of bulk materials such as tills and loess.

13.1.2. Principles of Operation

An electron gun provides a stream of electrons that are accelerated down an evacuated column. This beam is collimated by electromagnetic lens coils and finally brought to a focus on the specimen. To provide illumination over a surface, therefore, the beam needs to be scanned over a portion of the surface much in the same way as a television screen is scanned by the electron beam. This 'raster' is formed by the scanning coils in the column (Fig. 13.1). The smaller the area scanned the higher the magnification obtained.

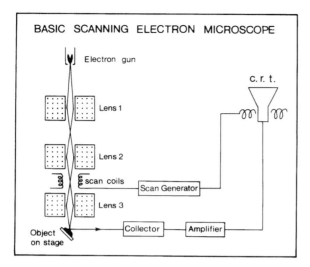

FIG. 13.1. Schematic configuration of the scanning electron microscope

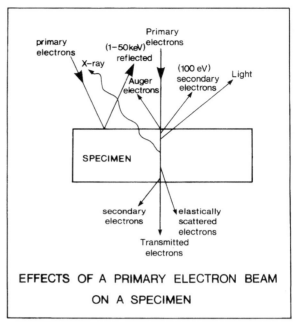

FIG. 13.2. Effects of primary electron beam on a specimen. This shows the main effects; those collected by the SEM under normal conditions are the secondary and some reflected (back-scattered) electrons. Note that the X-rays can come from within the specimen as well as from the surface

The specimen surface interacts with the beam electrons in several ways (Fig. 13.2). Two main interactions are collected in a conventional or normal mode SEM (i.e. secondary electrons together with some reflected {back-scattered} electrons). These are collected by a special device (the Everhart–Thornley detector) which converts the electrons into light that is detected by a photomultiplier. The signal from the detector is linked to the raster electronics to produce the picture seen in the normal viewing screen. Contrast, brightness and magnification controls are provided and 'split screen' and other viewing modes may be possible according to sophistication of the instrument. Modern machines now use automatically-adjusting scale bars which are precise enough for most purposes and make it unnecessary to record and convert magnifications.

13.1.3. Additional Instrumentation

Several methods of obtaining both different types of image and beam interaction information are possible and are frequently found on modern SEMs. These include, as a peak of sophistication, the analytical SEM and the electron microprobe (microanalyzer) or EPMA. These will not be discussed here except in general terms but reference should be made to

Goldstein *et al.* (1981) and Newbury *et al.* (1986). Bisdom (1981) provides a useful set of papers dealing with various forms of advanced analytical techniques that, although primarily concerned with soil analyses, may be of interest to those dealing with glacial sediments.

Back-scattered electron imagers (BSE or BSI) fitted even to simple microscopes as compact, solid state detectors have become reliable. The electrons are collected in line with the incident beam near the pole piece and are viewed in the screen. The energy of the back-scattered electron signal, unlike the secondary mode, contains atomic number (Z) information. Thus, if suitably detected, such a signal can be used to provide a simple chemical analytical facility without the use of expensive energy dispersive X-ray detectors (e.g. Dilkes and Graham, 1985). However, the interpretation of these images requires some care and calibration and the use, ideally, of flat and polished specimens. Other examples of the use of BSE in geological context are provided by Robinson and

Nickel (1979), Hall and Lloyd (1981) and Pye and Krinsley (1983).

There are two main types of detector which give 'true' elemental information: energy dispersive X-ray spectrometers (EDS or EDX) and wavelength dispersive spectrometer (WDS) systems. The latter will not be described here as they are generally found on microprobes and occasionally on analytical SEMs; furthermore, as EDS systems improve in flexibility, they are used less often at a routine level (Goldstein et al., 1981).

Energy dispersive X-ray instruments use solid-state detectors to collect emitted X-rays for, most usually, $Z > 11$ (sodium) but recent developments will allow detection to $Z = 5$ (boron). A complete spectrum of X-ray energy emitted from the sample by the excitation of the incident electron beam is collected. Software can provide operations such as separating possibly overlapping peaks and various corrections for atomic number (Z), absorbance (A) and fluorescence (F). These ZAF corrections are really only applicable to flat polished specimens but special software to analyze 'rough' specimens is available (Goldstein et al., 1981; Goodhew and Humphreys, 1988). At the simplest level, EDS can be used to check the mineralogy of a grain (e.g. quartz or feldspar). It is not uncommon to find photographs of spectra (direct from a television monitor or from a computer printer) showing elements found in a specimen. However, interpretation of such raw spectra must be treated with care; relative peak heights do not necessarily indicate relative abundance of elements. The various corrections just described need to be applied (Goldstein et al., 1981; Newbury et al., 1986).

Cathodoluminescence (CL) is also a surface emission characteristic (where light is emitted when the specimen is excited by an electron beam) that has been examined with the SEM as well as the light microscope ('Luminoscope') (Krinsley and Tovey, 1978; Grant, 1978) but the technique has been little used in recent work on glacial specimens. Recently, Kearsley and Wright (1988) have shown how modern detectors and image analysis techniques can extend the usefulness, but markedly increase the cost, and it may be that there will be renewed interest for example, in tillite, carbonate and diagenesis studies.

13.1.4. Sample Preparation and Examination

For the preparation of individual mineral grains for surface texture analysis, it may be sufficient to wash them in distilled water by swirling and decanting off finer material before drying. If there is much adhering material, then gentle boiling in a mixture of hydrochloric acid and stannous chloride or with sodium hexametaphosphate will often suffice. Although ultrasonic treatment has been discouraged by some workers, it may be suitable unless the grains are very fragile (e.g. multi-grain grus). Sieving should be avoided as this may affect edge textures. Cremeens et al. (1987) provide a good review of pretreatments and

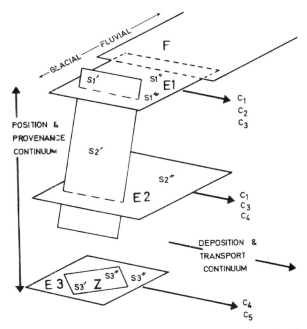

FIG. 13.3. Schematic illustration of provenance and mixing of environments and the complex problems involved under certain circumstances. E1, E2 and E3 are distinctly delimited topographic environments (e.g. supra-, en and subglacial positions) although there may be no differences between the mechanisms involved in transport. Samples taken within these environments (S1', S1" etc) need to be compared between them (e.g. S1" to S2"). The observed characteristics of each specimen (not necessarily on each grain) are C1, C2 etc. Thus, one environment sampled in the transport continuum may contain a mixture of textures C1, C3, C4 and C5. If one of these is a critical texture, enough grains must be examined in order to detect it. Z is a sample from different positions within environment E3; differences in mixing and transport rates must be assessed carefully within the transport sample space to minimize bias.

their possible effects on grain surface textures. Figure 13.3 shows the stages in grain selection for texture analysis; it should be noted just how few grains can effectively be examined with cost constraints and thus representativeness can be a problem.

A solid specimen can be mounted by gluing it onto a specimen stub. Individual grains can be mounted on double-stick tape onto the stub surface. There are several papers which provide useful information and hints on mounting and preparation (Smart and Tovey, 1982; De Nee, 1978; King and Banholzer, 1978; Walker, 1978; Murphy, 1982).

It is clear from the outline of the principles of operation that sample preparation is crucial for certain types of analysis, specifically analytical EDS. Here the sample must be polished (as a thin 'probe' slide or as a thick specimen) and coated with a film of carbon. This coating is necessary to provide a path for the incident electron beam to leak away, otherwise the charge build-up ('charging') degrades the image and analysis. This conductive coating is important where the examined material is a good insulator. For 'normal' analysis a coating of gold (by 'sputtering') is used in preference to carbon as it gives a better image and higher resolution due to its capacity to produce a high yield of secondary electrons.

13.1.5. Image Capture

The simplest way to preserve an image is to take a photograph of the viewing screen, but a special high resolution screen and built-in camera is more usual. A more recent development is the capture of the image on video-tape (external) or a solid state memory or 'frame store' (internal). This can be useful if many specimens need to be examined, thus images can be collected quickly (saving expensive 'beam time') and viewed later (saving photographic costs). However, if photographs are needed for publication they should be obtained 'first hand' as the photographed video image quality may not be as good. It is judicious practice to take both low and high magnification photographs of points of interest. This makes any subsequent reference easier. Despite time and expense, it may be more economical in the long run to take 'location' shots than to have to relocate an area (often not an easy task) and re-photograph.

13.1.6. Image Processing and Image Analysis

Sometimes both processing and analysis of images are combined but strictly, image processing deals with correcting, 'tidying' or improving the image. To some extent, this can be done on the SEM itself (e.g. mixing a BSI signal with a topographic one). Alternatively, it can be done as 'post-processing', tidying as one might a remotely sensed image (which is what an SEM micrograph is!) by applying filters etc. Analysis relates to quantifying aspects of the image. Sometimes this may be no more than calculating areas, maximum diameters etc. This is frequently done on an image analysis computer, which may, but need not, be attached to the SEM itself (Moss, 1988). Such systems have been most extensively used for quantification of grain outlines (Section 13.2.3). Because of the complexities of micrographs showing surface textures few advances have been made in quantifying complete grain surfaces. Greater progress has been made when either pore spaces or lineaments need to be characterized. Optical methods (e.g. Kaye and Naylor, 1972; Tovey and Smart, 1986) can be used for the latter. Pore spaces etc., also need complex treatment (Section 13.4.3).

13.2. SEM AND SEDIMENTS

13.2.1. SEM Characterisation of Particles and Size Analysis

If we consider energy input into geomorphic systems, then a component of energy may be used to do geomorphic work (i.e. erosion and transport). The manifestation of this energy within any geomorphic system may lead to: weathering of bedrock; transport of sediments; erosion of bedrock; abrasion of transported sediments; grading of sediments during and after transport; sedimentary structures produced upon deposition; and diagenesis of the sediments.

Several other geotechnical properties of the materials may also be reflected in one or more of the above. Although size analysis is the traditional manner of investigating relationships between energy input and sediment/geomorphic change, the SEM can be of use in all of these facets to examine the form of particles involved (Syvitski, 1991b). The effects of abrasion on

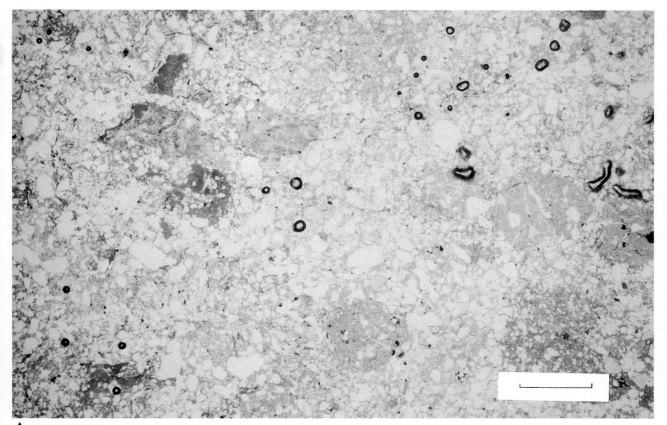

▲ PLATE 12.2(d)

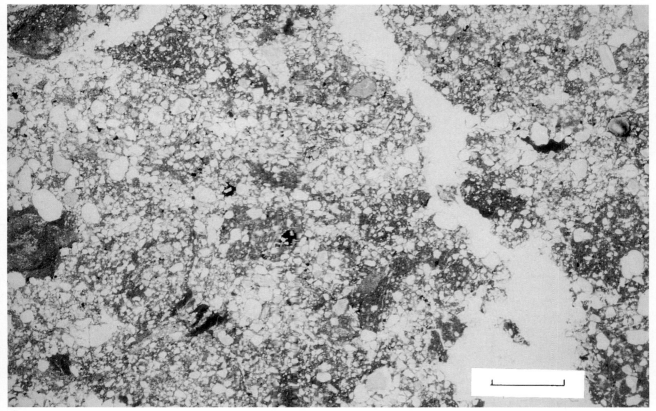

▲ PLATE 12.2(e)

PLATE 12.2.(f) ▶

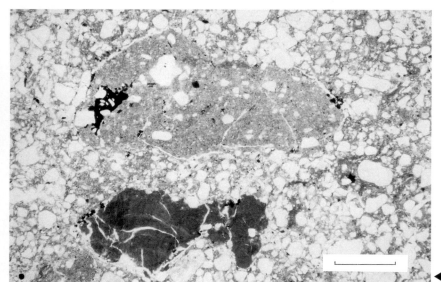

◀ PLATE 12.2.(g)

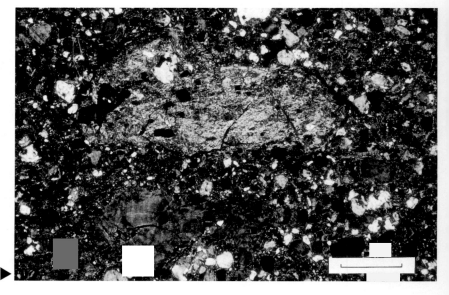

PLATE 12.2.(h) ▶

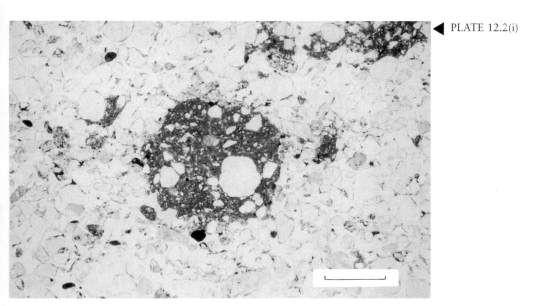

◀ PLATE 12.2(i)

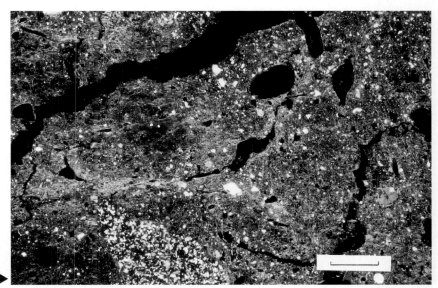

PLATE 12.4(c) ▶

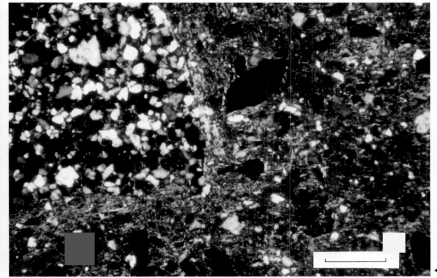

◀ PLATE 12.4(d)

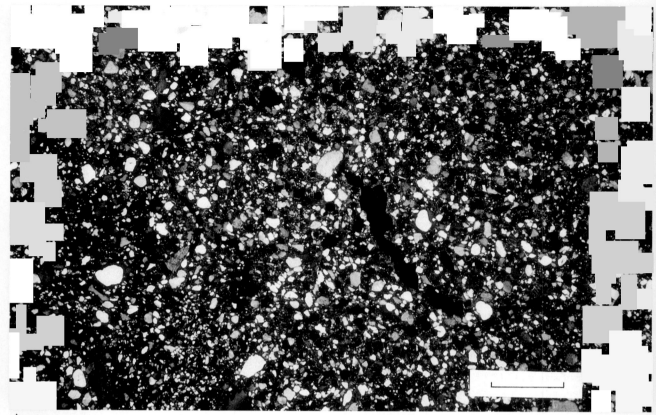

▲ PLATE 12.4(e)

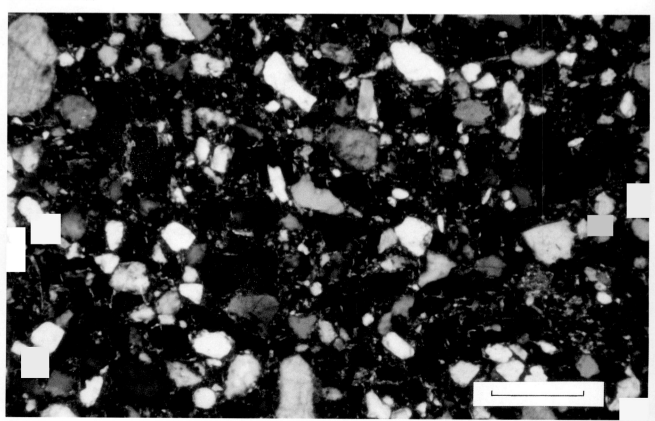

▲ PLATE 12.4(f)

edges of grains, striations and rounding for example, have long been recognized as a factor in examination of glacially-related mechanisms but gross measures of shape change are usually measured. Where small energy changes produce microscopic effects it is possible to use SEM examination to reveal such changes in surface texture. Because of the small size of sand grains, the large depth of field available makes the SEM a necessary tool to examine these surface textures which are often characteristic of particular geological environments. Thus, the SEM is a useful tool as small energy changes leaving an imprint on individual grains may be detected but could be missed by granulometric methods. The size of grains used for surface texture analysis affects, to some extent, the disposition and frequency of superimposed features. There is only limited evaluation of this at present (e.g. Nordstrom and Margolis, 1972).

Sediment 'size' is often intimately linked to particle form (Orford and Whalley, 1991). A thorough treatment of size and SEM surface characterization is still needed, although Eyles (1978) provides a useful start on this problem.

Krinsley, with various co-workers, provided details of surface texture characterization with an electron microscope (Bull *et al.*, 1986). Margolis and Krinsley (1974) set out the methodology and techniques of analysis involved in SEM. The technique depends upon the assumption that some specific, superimposed features on grain surfaces correspond to a particular formative mechanism and, thus, to a geomorphological process or sedimentary environment. These textures are generally at a smaller scale than outline angularity or roundness (Krinsley and Marshall, 1987).

In work by Krinsley, Cailleux (e.g. Cailleux and Schneider, 1968) and co-workers in the late 1960s, material sampled from known environments was examined and compared to laboratory simulations of specific mechanisms (Bond and Fernandes, 1974). More detailed and sophisticated experiments have continued since then. Photographic examples were summarized by Krinsley and Doornkamp (1973) describing the main diagnostic textures seen in various environments. To this list of textures others have been added (Bull, 1981; Culver *et al.*, 1983; Bull *et al.*, 1987). However, there has, as yet, been no up to

date compendium of photographs to replace Krinsley and Doornkamp's 'Atlas'.

Related work by Le Ribault has provided a more detailed set of subdivisions than most other workers although not all would agree that such fine detail of environmental discrimination can be achieved consistently (Le Ribault, 1975a,b, 1977, 1978).

Quartz is generally used as the mineral investigated for surface texture determination as it tends to be ubiquitous. However, heavy minerals have also been used in such studies (e.g. Stieglitz, 1969). In a glacial context, garnet has been studied in the light of the interpretation of chattermarks (see Section 13.2.4.4.) but other minerals might well be used.

Before reporting some actual results, it is necessary to preview certain problems and methods relevant to determinations of sediment origin with the SEM.

13.2.2. Sampling Problems

It is rarely economically feasible to photograph, or even examine without photographing, more than a few tens of grains per sample which means that there are definite statistical problems in sampling representativeness (Fig. 13.3). Several authors have addressed the problem of the number of grains acceptable for SEM characterization (Tovey and Wong, 1973, 1978a; Baker, 1976; Tovey *et al.*, 1978). A suitable minimum number seems to be thirty. Vincent (1976), however, used 200 grains per sample in a limited study of some periglacial features.

Sampling problems arise where the sediment origin is unknown. It is clear that mixing of distinct origins makes elucidating a mixed provenance difficult. Under these circumstances, thirty grains may be far from an adequate number and what is sufficient will depend upon the proportions of the mixed components. It is also advisable to take additional samples from known environments and to determine the range of textures where possible. Glacial sediment mixing provides a good example of this problem (Fig. 13.4). Despite this, it should be noted that any discriminatory technique (such as size analysis) suffers from similar difficulties. As well as mixing, it may be that there are, or have been, queues in a system (Fig. 13.5). This queuing might be in a moraine or esker for example. On being released into another transport

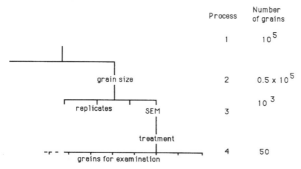

FIG. 13.4. Selection of grains for SEM analysis showing the extreme degree of selection and the need to avoid operator bias, taking the complexities shown in Figure 13.3 into account.

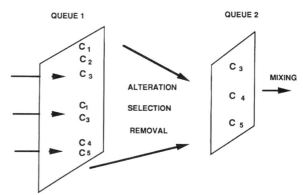

FIG. 13.5. Queuing and mixing of environments. An example could be: Queue 1 – supraglacial moraine (static or slopw moving) or lateral moraine (static); Queue 2 – Outwash, esker, river terrace, flow till. The surface texture characteristics typical in a sample may change due to material being held in two queues, mixing after Q2 may prohibit environment discrimination.

path there may be some textures which are preferentially selected for analysis unless an adequate number of grains are examined. Little work has been done on this mixing/queuing problem to date although warnings have been given, for instance, by Rogerson and Hudson (1983) who pointed out that the existence of a particular surface texture does not necessarily indicate the last mode of transport.

13.2.3. Quantitative Estimation of Particle Outlines and Surface Textures

Given that the usual assessment of surface textures is (at present) visual, quantitative analysis presents considerable difficulties. Despite some advances in quantification techniques it is fair to say that there is still no numerical and objective method which can be used for particle characterization. The best image processor here remains the eye–brain–memory combination. However, there are several ways in which some measure of numerical evaluation is possible. Two basic methods are possible: quantification of grain outlines and nominal/ordinal estimation of surface textures.

13.2.3.1. Form, shape and angularity-roundness determinations

Quantitative form determination is an important and critical area of interest (Barrett, 1980; Orford and Whalley, 1991). Whilst the Zingg biaxial ratio shape groupings are often used, they are not easily treated

numerically and the Sneed and Folk (1958) method is to be preferred when three orthogonal axes values can be obtained. For SEM analysis, of course, the third dimension is difficult to obtain. With smaller grains, reliance is placed upon two-dimensional measures and this applies also to micrographs. The problem is that the grains may not necessarily be examined in the (optimal) maximum projection plane. Whalley and Orford (1982) show some of the gross shape changes with varying viewing angle. Some previous experiments using a universal stage and small grains, however, suggested that there was little shape change as (rounded) grains were tilted (Tilmann, 1973). Nevertheless, ordinal scale measurement can be applied by way of angularity-roundness charts in the traditional manner. The use of Fourier analysis to characterize grain outlines has been used for some time in the United States (e.g. Ehrlich et al., 1987). Fractal analysis (which uses stepped-off lengths around the grain perimeter to evaluate measures of outline form) is also being used (Beddow, 1980; Kaye, 1981; Orford and Whalley, 1987, 1991; Whalley and Orford, 1989).

The need to use a custom-built image analysis instrument may have deterred much development although suitable equipment linked to micro-computers is now an inexpensive option (e.g. Telford et al., 1987). Nevertheless, software has to be written specifically. The digitization of micrographs is the means by which quantification methods are linked to SEM observations (Czarnecka and Gillott, 1980; Dowdeswell, 1982; Whalley and Orford, 1982, 1986; Orford and Whalley, 1983, 1987; Ehrlich *et al.*, 1987; Mazzullo and Anderson, 1987). When Fourier analysis is used, it must be recognized that it tends to be a rather gross shape which is characterized (rather than an estimate of the fine texture) as the number of points used on an outline is rarely more than 4- points, although rather more points may be actually digitized. Hence, peripheral fine detail may be missed as the number of harmonics available is half the number of data points. Conversely, fractal analysis gives most emphasis to small scale information ('textural fractal') and the number of peripheral points is often greater than 200. In either case, it should be emphasized that there is a scale effect and comparisons should ideally be done at the same grain magnification. It is possible to digitize grain outlines directly from the SEM and pass them directly to an image analysis computer. This should be a faster, and cheaper — but not necessarily a more convenient — method than relying on photographic intermediates. As it becomes progressively easier to link digital images from SEMs with image analysis techniques and data bases it is likely that this will become a commonly accepted method of analysis. A digitized view of an SEM screen may occupy 1/4 Mbyte of computer storage, even though much of this will not be needed in analysis. Thus, thought needs to be given to data reduction as well as analysis techniques (Hayward *et al.*, 1989). A video recorder is one means of easy storage although not all SEMs will allow a simple 'hook-up'.

13.2.3.2. Surface textures: replicated environments and operator variance

Traditionally, the first method of analysis of surface textures has been to gain experience using, for example, the Krinsley and Doornkamp (1973) 'Atlas' and then to compare unknowns with these standards. Alternatively, samples from known locations or known transportation pathways etc. can be compared with unknowns (Whalley and Krinsley, 1974; Eyles, 1978; Manker and Ponder, 1978; Whalley, 1978b,

1979). Lists of surface textures can be found in several papers, as mentioned above. The following discussions relate to general environments and detailed analysis of the actual textures and features is kept to a minimum.

It has become more common to identify several individual textures and to count occurrences or evaluate percentage cover for individual grains. Histograms are then drawn up to represent these counts in a semi-quantitative manner (Margolis and Kennett, 1971; Bull, 1978a, b; Bull and Culver, 1979; Higgs, 1979). The last paper is a good example for study as it has a selection of micrographs. Rogerson and Hudson (1983) used a 21-feature table for characterization and then grains were scored as: absent, present or abundant. Principal component analysis was then used to separate environments.

Culver *et al.* (1983) used five independent operators to evaluate this texture element tabulation technique. Very good correspondence was arrived at for each of six samples from several environments. Although 'quantitative' it is time consuming and (usually) micrographs need to be obtained first. Yet it is probable that, once some experience has been achieved, then a purely visual interpretation can suffice surprisingly well. Once experienced, an operator can control individual variance quite well and that the 'mental image processor' is good at remembering distinctive patterns. Automatic feature extraction is, as yet, only in its experimental stage and it is surprisingly time consuming even if the features are distinctive (such as Vs or etch pits). The density of a feature in the viewed surface can be of use; for example, Cremeens *et al.* (1988) have quantified etch pits with image analysis techniques and Williams *et al.* (1987) have provided some indications about the use of digitized areas (rather than outlines) obtained from scanning electron micrographs.

Generally, it will not be possible for individuals to provide detailed analyses of their own subjective assessment and then compare results with those of other workers. Thus, further subjectivity is passed on from one worker to another, to some extent, by means of previously published photographs and interpretations. This may be unavoidable but errors can be controlled or guarded against (e.g. by 'blind' sub-sampling). Interpretation of photographs is then done and, finally, decoding allows comparison with the sample locations. It may even be possible to divorce the roles of interpreter and sample collector (double blind testing).

It is also desirable, but not always possible, to run analyses on samples of known provenance which are related in some way to the unknown. Thus, it would be useful to look at the weathered grus of an input to a glacial system in which sampling and characterization are to be performed. Finally, it should be stressed that SEM surface texture characterization is only one of several interpretational techniques and usually needs to be used in conjunction with other methods for unequivocal interpretation.

13.2.4. Surface Texture Producing Mechanisms

The next sections will outline the major work that has been done in relating erosional mechanisms to transport processes, and the end-product final textures observed. All the main environments for which surface texture analysis is useful are mentioned together with the possible origins or geologic processes involved. Glacial deposits are likely to contain grains from many diverse sources other than strictly glacial in origin. Indeed so called 'glacial textures' can be derived from several sources other than subglacial environments. It is thus important to differentiate subglacial as opposed to supraglacial

origins of material or transport paths. The plates accompanying this chapter (Plates 13.1a–r) are not definitive but rather used to show the range of textures possible in a single sample as well as indicating the types of textures seen in various glacial and proglacial sediments.

13.2.4.1. Original surface textures

The textures displayed on grains released from a rock into the sediment transport–deposition cycle can be thought of as 'original' textures. Igneous, metamorphic or sedimentary rocks may display grain surface textures specific to those origins. It is, therefore, not possible to say what these original textures always look like; authigenic quartz may have euhedral quartz overgrowths, metamorphic quartz grains may be much flattened. Furthermore, surface alteration (e.g. weathering) may take place if grains remain dormant for long periods before being transported. Such uncertainty may be reflected in the textures and how they are interpreted. Ideally, therefore, actual grains from the source should be sampled and compared to modified grains down a transport path (Mazzullo and Magenheimer, 1987). Most of investigations have been done on quartz sand-sized grains but some work on silts has proved useful (Bull, 1978a; Mazzullo *et al.*, 1986; Haines and Mazzullo, 1988) in linking SEM surface texture analysis of silts and fine sands with outline quantification by Fourier techniques.

Plate 13.1. Selected electron micrographs of quartz grains from several environments. They are chosen to illustrate differences as much as typical surface textures and are generally shown as horizontal pairs. The scale is given between a pair of white squares at the bottom of each photograph together with the distance in micrometers. (a) Granite grus; (b) granite grus, (location as (a)); (c) frontal moraine (from supraglacial till) last 1000 years; (d) as (c) but an older moraine in a sequence; (e) subglacial till from present day glacier. Notice slight striations; (f) as (e), same location; (g) subglacial till, note some edge chipping; (h) silt from proglacial stream, collected on filter paper; (i) esker (from mountain glacier system); (j) esker (from a lowland fluvio-glacial complex); (k) river terrace in hill country (from sandstone with euhedral overgrowths on, originally, aeolian grains; (l) same location as (k); (m) possible aeolian/fluvial deposit underlying fluvioglacial sequence; (n) as (m), detail showing etched Vs; (o) loess deposit (bulk); (p) till deposit in bulk; (q) as (p); (r) till individual particles.

At Butterworth-Heinemann we are determined to provide you with a quality service. To help supply you with information on relevant titles as soon as it is available, please fill in the form below and return to us using the FREEPOST facility. Thank you for your help and we look forward to hearing from you.

What title have you purchased? _____

Where was the purchase made ? _____

When was the purchase made? _____

Name (Please Print): _____

Job Title: _____

Street: _____

Town: _____

County: _____   Postcode: _____

Country: _____   Telephone: _____

Company Activity: _____

Signature: _____   Date: _____

(FOR OFFICE USE ONLY)

* Please arrange for me to be kept informed of other books, journals and information services on this and related subjects (* delete if not required).
This information is being collected on behalf of Reed International Books Ltd and may be used to supply information about products produced by companies within the Reed International Books group.

BUTTERWORTH
HEINEMANN
Butterworth-Heinemann Limited ■ Registered Office: Michelin House, 81 Fulham Road, London, SW3 6RB. Registered in England 194776.

Direct Mail Department
Butterworth-Heinemann

FREEPOST

OXFORD

OX2 8BR

UK

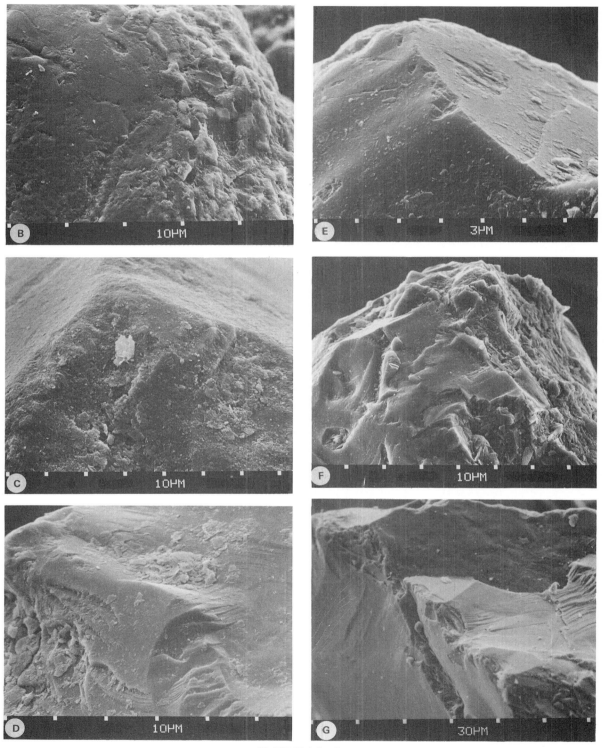

PLATE 13.1 (b–g)

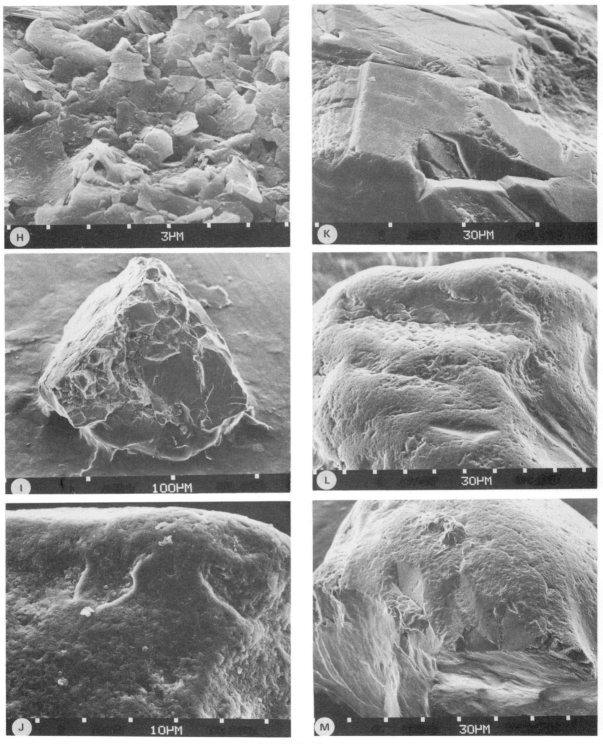

PLATE 13.1 (h–m)

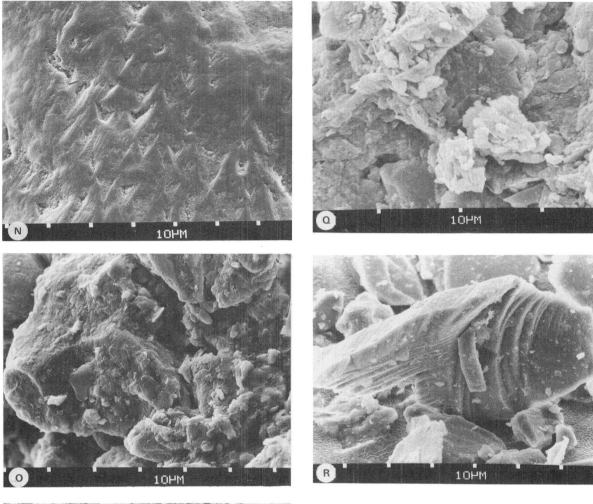

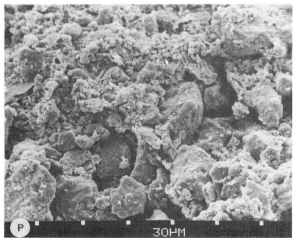

rLATE 13.1 (n–r)

Several papers refer to original textures and show the range of features possible when crushed quartz is used (Krinsley and Doornkamp, 1973; Whalley and Krinsley. 1974; Whalley, 1978b; Gallagher, 1987). Where the original material is grus, rather than artificially produced grains, then additional, complex textures may be formed, for example, surface weathering (Wilson, 1980); crystal growth facets (Wilson, 1978); garnets (Gravenor and Leavitt, 1981); and other heavy minerals (Rahmani, 1973; Lin *et al.*, 1974: Setlow, 1978). Although surprisingly little is directly related to glacial mechanisms, an exception is the formation of chattermarks (Menzies, 1995a, Chapter 7). The study of continental margin deposits from Labrador and western Greenland (Cretaceous and Paleocene) is a good example of quantification methods linked to environment modification (Higgs, 1979).

13.2.4.2. Crushing and grinding mechanisms

Experiments performed to investigate the formation of surface textures by the action of various environments have generally used crushed quartz crystals as starting material. This produces typically angular grains with conchoidal fractures. Other surface textures produced are also typical of brittle fracture mechanisms and are not distinctive markers for other environments (Menzies, 1995a, Chapters 7 and 15).

Sharp and Gomez (1986) reviewed the literature on crushing and grinding mechanisms as applied to SEM analysis of grains and subglacial abrasion and comminution. They suggest several avenues for future investigations by SEM analysis of grains into subglacial mechanisms. However, there does not as yet appear to have been an SEM examination of the complementary bedrock surfaces. Recently, Gomez *et al.* (1988) investigated the nature of quartz particles in relation to textures and the discrimination of transport paths in glaciers (cf. Gallagher, 1987; Goodale and Hampton, 1987). Mahaney *et al.* (1988c) examined both quartz and feldspar grains in a paleosol from Mount Kenya and concluded that both crushing and abrasion mechanism were at work with subsequent aeolian polishing. They also suggest a distinction between valley and continental glaciers (see Section 13.2.4.6). Work on glacial transport by Krinsley and Takahashi (1962b) and later Whalley and Krinsley (1974) suggested that there was no specifically glacial texture. This was largely confirmed by Eyles (1978) and Gomez and Small (1983). The features on original grains resulting from brittle fracture can be produced without any reference to a glacier bed, but have been termed, 'glacial' by some authors (e.g. Gram, 1969; Plafker *et al.*, 1977; D'Orsay and van de Poll, 1985). This inexact use of 'glacial' can yield errors in interpretation. It may be that specific grains have been transported (supraglacially) but there seems to be no differentiation of surface features between these grains and those in low energy fluvial transport or in debris flows. Even where subglacial till has been definitely sampled, distinctive brittle frac-

ture features may not be due to grinding or subglacial crushing but are original rather than superimposed textures (Setlow and Karpovich, 1972; Rehmer and Hepburn, 1974). It is pertinent to be aware that populations from different origins may become mixed or suffer sequential changes in surface features as a consequence of multiple modes of transport; for example, glacial then fluvial transport (Menzies, 1995a, Chapter 8).

13.2.4.3. Edge abrasion and subglacial action

Although the main thrust of surface texture analysis has been the recognition of a variety of individual textures, and to a lesser extent a suite of textures, there are certain features which seem to be particularly significant, for example, the recognition of 'edge abrasion'. The protruberances on individual grains transported in a fluid (uni- or bi- directional) flow are the most likely to be abraded (cf. Goodale and Hampton, 1987). The importance of edge abrasion as an energy transfer indicator is evident from experiments simulating various environments (Mycielska-Dowgiałło, 1978; Robson, 1978; Lindé and Mycielska-Dowgiałło, 1980; Goździk and Mycielska-Dowgiałło, 1982; Lindé, 1983, 1987; Whalley *et al.*, 1982, 1987). Surprisingly, there has been little simulation of 'pure' fluvial action although it has been observed by several workers (Whalley, 1978b; Elzenga *et al.*, 1987).

Any rounding may be initially slight and not visible except under the SEM but becomes progressively more significant and visible in low powered optical microscopes. Edge abrasion can be seen in some (but not all) grains found under a glacier (cf. Whalley, 1978b). Dowdeswell (1982) has also found limited edge rounding in actively transported glacial material. If grinding is in a thin film between ice and bedrock, then edge abrasion is likely to occur on grain edges and corners, and perhaps chattermarks might be expected. If, however, the till layer is thick, then the shear stresses are distributed in a high porewater pressure medium (a mobile bed) and it may be unlikely that such textures form or, if they do, are sporadic. Gemmell *et al.* (1986) reported that grains incorporated into basal ice were believed to be passively transported as there was no evidence of glacial modification – although, surface pitting and weathering was noted. This observation touches upon a general problem of generalizing interpretation when only a few grains examined. It is advisable to use subsets of samples and to view a wide range of particle sizes in a preliminary assessment. Rogerson and Hudson (1983) in their quantitative analysis noted that the glacial environment was no harder to define than most other environments. Finally, there may be potential benefits in linking subglacial erosive processes with bedrock characteristics and abraded grains using the SEM and fractal size analysis (Hooke and Iverson, 1995).

13.2.4.4. Chattermarks

Chattermarks are a special case of abrasion – at least they can be produced by simulated glacial grinding (Whalley, 1978b) and some may well be of 'slip-stick' origin. Grain surface chattermarks may be indicative of glacial action – although they are comparatively rare features (Folk, 1975; Orr and Folk, 1983, 1985; Bull et al., 1980a). The original Folk view was extended by Gravenor (1979, 1980a,b, 1985) but has been disputed by Bull (1977) and Bull et al. (1980a,b). Folk suggested that chattermark patterns on sand-size garnets were due to glacial grinding. Although Gravenor considered that some chattermarks are the result of etching processes of the garnet, the etch patterns tend to be inherited. Gravenor's contention was that the percentage of garnets with chattermark trails is indicative of the distance of glacial transport and can be used for tillites as well as Quaternary sediments (Gravenor and Gostin, 1979; Rocha-Campos et al., 1978; Bull et al., 1980b; Orr and Folk, 1983). Orr and Folk (1983) concluded that chattermark origin is not necessarily environmentally controlled; the visibility and abundance of trails are related to mineralogy and weathering and, 'mechanical grinding may be a reconcilable means of obscure chattermark-trail formation' (p.128). It may be concluded that several modes of formation are possible.

13.2.4.5. Supraglacial, subglacial environments and moraines

Supraglacial deposits are unlikely to show much (if any) evidence of alteration other than weathering;

once material reaches a glacier surface it is likely to be carried passively (Whalley and Krinsley, 1974; Eyles, 1978; Flageollet and Vaskou, 1979; Nakawo, 1979). Gomez and Small (1983) examined grains from supraglacial and englacial sources (the latter originally thought to be subglacial) and concluded that it was possible to distinguish between these transport paths. One criterion used was the presence of the 'microblock' texture of Whalley and Krinsley (1974). This texture can certainly be found but it is probably the result of brittle fracture rather than being a distinctive subglacial texture. Again, it is surprising that more accurate replications of texture formation linked to grain properties are still required.

Examination of englacial material presents some problems of interpretation in that some material may be transported passively, as buried supraglacial grains, but that shear plane development may allow grains to abrade one against another. Yoshida and Watanabe (1983) suggested that 'crushing and or plucking rather than abrasion' took place at the base of Urumqui No. 3 Glacier (TienShan) but this interpretation is equivocal. Whalley (1982) examined shear plane material from the TUTO Tunnel in Greenland but found little evidence of grain interactions that produced edge abrasion.

Deposits from a variety of glacial transport paths can be incorporated into moraines and thus subdivision of pathways is often desirable, though not always possible (e.g. {A} on tills – Whalley and Krinsley, 1974; Kenig, 1980; Corte and Trombotto, 1984; Mellor, 1985; Shakesby, 1989; {B} on tillites – Culver et al., 1978, 1980; Rehmer and Hepburn, 1974; Deynoux and Trompette, 1976).

13.2.4.6. Glacially-related textures — summary

The previous four sections strongly suggest that there is still a need for a full understanding of subglacial crushing and abrasion mechanisms linking both laboratory and field investigations to a variety of conditions, such as till thickness, mineral and bedrock type and the nature of inter-particle contact. This proviso also extends to the nature of chattermarks and their significance in quartz as well as garnet. A recent controversy is the contention by Mahaney et al. (1988b) that it is possible to infer ice thickness and transport distance of Pleistocene ice sheets by the textures on quartz grains as seen under the SEM. Initially, 250 and later 400 till samples were examined by these authors from sites in North America and East Africa. They differentiate between crushing and abrasion features although the former may obliterate the latter. In their response to criticisms by Clark (1989) they state 'we observed a particular pattern of microtextures on quartz grains from several different continental and mountain locales. We attempted to explain these features by using ice thickness and distance of transport as the two variables most likely to cause them' (p. 1204). The use of microtextures to infer ice thickness and activity is probably more difficult than Mahaney et al. (1988b) imply (Menzies, 1995a, Chapter 15).

13.2.4.7. Chemical effects

It was found that in the use of surface textures that certain environments produced chemical etching. This is usually seen as orientated etched pits or Vs on quartz grains. The latter are 'etch Vs' that can be differentiated from mechanically produced features. Crystallographic orientation has a significant role in the formation of chemically-produced textures – they often show distinctive orientations – but there are complexities (Karpovich, 1971; Wilson, 1978, 1979a,b, 1980; Krinsley and Marshall, 1987). Solutional effects ('weathering') as well as overgrowths may obscure mechanically produced textures, but some authors have used the amount of surface alteration as a guide to the length of exposure time and thus as a means of dating deposits. Iron oxides may adhere to surfaces, perhaps associated with clay mineral grains (Bull et al., 1987).

13.2.4.8. Aeolian and fluvial impact features

Despite the fact that moving quartz grains interacting with each other and with a static bed are governed by the same physical principles, there seems to be substantial differences in the surface textures produced by aeolian and fluvial transport paths. This is probably the result of the manner in which particle

interactions take place. The resumé given here is a guide to the types of texture which are mentioned in subsequent sections.

Robson (1978) demonstrated the effects of abrading grains in rotating drums and in a mechanical shaker, the latter being similar to, but more severe than, surf action. Although rounding was achieved there was evidently no production of the impact Vs commonly found on beach grains. As well as chemical etching, certain environments produce distinctive impact V texture. These textures are variable in number, density and size and are produced by impaction of grains under high energy subaqueous conditions. Beaches produce these impactions and also well-rounded edges features (Krinsley and Doornkamp, 1973; Mycielska-Dowgiałło, 1974; Lindé and Mycielska-Dowgiałło, 1980; Krinsley and Marshall, 1987).

Margolis and Kennett (1971) presented evidence that allowed several subaqueous environments to be differentiated in terms of impact V density. A diagram of the number of impact pits per square micrometer plotted against percentage of sand grains with impact pits on more than 10% of the surface allows this differentiation. Because there seems to be a relationship between density of Vs and energy and as they are easily recognized, some quantification has been possible with this texture (cf. Baker, 1976; Darmody, 1985). Cremeens et al.(1988) have used image analysis to quantify etch pits on weathered grain surfaces. This is possible because the V is relatively easily recognized by pattern recognition techniques.

13.2.4.9. Aeolian deposits and loess

In studying these sediments there is a need to look at both sand-sized particles, which is where surface texture identification has largely been performed, and the silt-size range of loess. Various experiments (e.g. Lindé and Mycielska-Dowgiałło, 1980) have confirmed the earlier idea (Krinsley and Takahashi, 1962) that aeolian-derived sand grains rapidly obtain distinctive surface textures. Kolstrup and Jørgensen (1982) used simple surface texture analysis in their lithostratigraphic interpretation of the Older and Younger Coversands in Jutland (Denmark) (cf.

Wilson *et al.*, 1981; Elzenga *et al.*, 1987; Borsy *et al.*, 1987). Once aeolian textures have been recognized they are usually easy to identify, even apparently below the Greenland ice sheet (Whalley and Langway, 1980) and may provide a relatively easy way to identify pockets of loessic material (Vincent and Lee, 1981). Even slight aeolian abrasion produces a characteristic edge rounding (Whalley *et al.*, 1982). As a consequence, first, an aeolian contribution to other deposits is relatively easy to determine if the material is of sand-size; and secondly, abraded debris (from originally angular grains) may provide silt-size material which can be loessic in nature (Whalley *et al.*, 1987). The implications are that a desertic, rather than a glacial, origin for loess is possible. Smalley *et al.* (1979) suggested a glacial origin for loess on the basis of that the prevalence of microfractures in quartz is an important adjunct to subglacial grinding in silt production. Manecki *et al.* (1980) have examined fine-grained material from the Broggi glacier (Peru) in order to test a subglacial theory of loess production. They show micrographs of quartz and heavy mineral grains from several glacier sub-environments but it is not clear which grains come from purely subglacial locations. More work needs to be done on both subglacial formation of loess and the recognition of non-glacial sources of silt in glaciated or periglacial areas (Derbyshire *et al.*, 1988). For instance, does the loess of Mongolia come from mountain glaciers or deserts – or both (cf. Chapter 6)?

There is considerable interest in recognizing the production of aeolian textures, leading to discussion concerning their interpretation and significance. Krinsley and Wellendorf (1980) suggested that features (e.g. 'upturned-plates') seen at high magnification on aeolian-abraded grains are typically aeolian and could be related to wind velocity. Whilst the formative mechanism has been argued about, it does appear that this is a distinctive texture in maturely rounded grains. In a series of experiments, Whalley *et al.* (1982, 1987) show stages in this formation which could be used to mark stages in progressive transport.

Silt-sized particles, as in much loessic materials, have been rather neglected in provenance studies. It appears that they are relatively characterless when compared with sand-sized particles. Nevertheless, experiments have shown that some traces of originally-rounded grains, from which they are chipped, can be found (Whalley *et al.*, 1987). Suggestions about fracture of quartz to give silt-sized particles – and loess by implication – have been advanced (Smalley and Vita-Finzi, 1968; Smalley and Cabrera, 1970; Cegla *et al.*, 1971; Warnke, 1971; Smalley *et al.*, 1973; Whalley, 1974; Bull, 1978a).

A variety of sources of silt production has been proposed and glacial grinding may only be one of these (Smith and Whalley, *in press*). Haines and Mazzullo (1988) show that there are distinguishing aspects of silt that can be used to identify terrigenous deposits in marine deposits. Although first cycle sediments from the coastal plain were an exception, it was possible to determine the sources of silt in marine deposits.

13.2.4.10. Fluvial and glaciofluvial deposits

Although both glacial and glaciofluvial environments are considered to be 'high energy', their impact on sand grains is rather different. This difference may reflect modes of transport but this remains conjectural. Subglacial grinding probably involves high contact pressures at edges and on faces although, if the sediment is more than a few grains thick and especially if silty, high contact stresses may not be easy to achieve. This may be the reason why not all subglacial grains show edge rounding. Rapidly rotating, saltating grains in a subaqueous environment have a high collision frequency but only protuberances come into contact in a manner similar to aeolian transported grains (see Section 13.2.4.9).

There have been few studies of fluvial and glaciofluvial action, either experimentally or by sampling known environments. Recently, Elzenga *et al.* (1987) investigated both periglacial and fluvial sands. Manker and Ponder (1978) show a 'centipede' type solution mark bearing a striking resemblance to some of chattermark trails (see Section 13.2.4.4). Whalley (1979) provides some of the few links between fluvial and glacial environments reflected in surface texture changes (Kuroiwa, 1970; Robson, 1978; Mycielska-Dowgiałło, 1974; Kowalkowski and Mycielska-Dowgiałło, 1983; Richards, 1984; Fenn and Gomez, 1989).

13.2.4.11. Beach and glaciomarine deposits

Beach deposits usually possess a distinctive surface texture typified by etched Vs (Krinsley and Doornkamp, 1973; Middleton and Davis, 1979; Lindé and Mycielska-Dowgiałło, 1980). As grain textures produced by the high energy of the breaker zone are easily recognized, the surface textures technique can be used with some confidence; nevertheless, some mixing of environments (Fig. 13.4) may complicate matters and a statistical examination is desirable. The beach texture may be important in helping to elucidate raised beach sequences but will probably play an ancillary role to other techniques. Several workers have examined the variety of textures seen in coastal and offshore sediments (Strass, 1978).

Distinct differences of texture suites can allow separation of provenance. Thus Geitzenauer et al. (1968) used glacial textures to identify Antarctic glacial activity from the Eocene section of a South Pacific deep sea core. Subsequently, Rex et al. (1970) used grain size analysis together with surface textures, to suggest an interglacial dune sand origin for some Weddell Sea sediments. In both cases, the recognition of unusual textures in a marine environment gives clues to provenance, (e.g. the recognition of textures typical of dune sands). The Margolis and Kennett (1971) graph (see Section 13.2.4.8) proves a way of grading impact energy; thus, deep sea turbidites can be recognized and differentiated from high turbulence beaches. Rogerson and Hudson (1983) used texture feature counting and grain size distribution to identify provenance of coastal sediments in Labrador. An unambiguous distinction of a purely glacial environment is still difficult to uphold unless there is particularly good control of origins – a factor that may distinctly limit its usefulness. Overall, the identification of glacial activity needs to be treated with care. It appears that glacial in this context implies a subglacial origin. The majority of the work done on maritime sediments has been on sand-sized material (Haines and Mazzullo, 1988).

13.2.4.12. Periglacial investigations

Many periglacial studies involve the identification of past environments, often of diverse origins such as aeolian infills, colluvium, solifluctued till and buried beach deposits. Thus, knowledge of a wide variety of environments and their signature on sand (or silt) grains is important. Several of the studies mentioned previously are linked with periglacial conditions. Christensen (1974) provides a few micrographs from original (host) materials as well as sand wedge fills and Elzenga et al. (1987) provide a wide variety of micrographs with a stratigraphic interpretation of periglacial and fluvial deposits from The Netherlands and adjacent Germany (Korotaj and Mycielska-Dowgiałło, 1982). Vincent (1976) examined several sets of grains from periglacial origins but the textures produced do not seem to provide a uniquely distinguishable texture suite from this environment. Thus it may be difficult to separate certain periglacial deposits (e.g. solifluctued material or 'head') from true glacially-transported deposits, especially given the lack of a distinct texture suite even from known tills. This is particularly problematic because, despite Vincent's assertion of features (angular edges, conchoidal fractures) being indicative of glacial abrasion, this is not the case uniquely; grus provides the same 'original' characteristics of brittle fracture. Some of the grains shown by Vincent appear to be aeolian in character. Investigations of many soils outside the tropics may show features related to periglacial activity e.g. Kumai et al. (1978), Legigan and Le Ribault (1974).

13.2.4.13. Volcanic material

Although it may seem unusual to include this section in a chapter concerned with glacial materials it should be noted that tephra can provide useful dating as well as general marker horizons in both glaciers and glacial deposits (Dugmore, 1989). A general review of the wide variety of (quite distinctive) surface textures of volcanic origin can be found in Marshall (1987) (cf. Heiken, 1972; Katsushima and Nishio, 1985; Sheridan and Marshall, 1987). Tephra in ice cores have been studied by several workers (e.g. Kyle and Jezek, 1978; Palais, 1985). As tephra is found quite commonly in a variety of deposits away from near-eruption sources (such as Icelandic tephra) identification with both surface characteristics and chemical properties is valuable.

Generally speaking, glass shards are distinctive and easily recognized.

13.2.4.14. Weathering phenomena and dating applications

Dating of glacial events is an important aspect of many investigations and the SEM can be used, in particular cases, to aid in dating specific materials. Two main areas can be recognised: (A) weathering on mineral grains in soils and sediments; (B) weathering and deposits on exposed rock or boulder surfaces.

(A) Kumai et al. (1978) have used an SEM with an EDS to examine some Antarctic soils of morainic origin which show both weathering effects and salt efflorescences. Wilson (1978, 1979a,b, 1980) illustrated the variety of features visible on regolith unaltered by geomorphic transport processes. Douglas and Platt (1977) describe the relationship between surface morphology and age of soils developed from Wisconsinan, Illinoian and Kansan age parent materials. They also suggest that solution/precipitation effects can be used to indicate the age of these soils. Andrews and Miller (1972) used SEM scanned surface textures to delimit weathering zones associated with glacier retreat in Baffin Island. Darmody (1985) used a single size fraction for examination finding five surface features indicative of 'weathering' and five of 'freshness' on each of 15–30 grains. A binary scoring system was used and a weathering class for the grain and a mean value for the sample was obtained. Similarly, Black and Dudas (1987) investigated soils (chernozem, luvisol and solonetz; A and C horizons) of glacial origin from Alberta identifying a variety of origins by the usual methods of texture interpretation. The degree of surface weathering was assessed by an extension of Darmody's (1985) method.

Vincent (1975), using surface textures to help examine a moraine chronosequence at Nigardsbreen, south Norway, found 'mechanical glacial features' on grains from the younger moraines with solutional features developed on moraines dating from 1840, 1870 and 1750 AD. However, Whalley and Griffey (personal communication), from a similar series of moraines some 500 km farther north failed to find a distinctive weathering sequence on quartz grain surfaces. This ties in with Eyles' (1978) finding of a lack of association between post-depositional diagenesis and time. Mellor (1985) mentions the use of SEM to determine weathering processes on soil chronosequences in south Norway. He suggests that feldspars and mafic minerals are needed to show weathering effects even when the moraines are beyond the Neoglacial limits (cf. Locke, 1979; Mellor, 1986). A potentially important technique, but unexploited so far, is the linking of quantitative measures of grain outline with dissolution and etching of minerals more susceptible than quartz. Dearman and Baynes (1979) found that all stages of etch-pit solution could be present on a single sample so that 'stage' of etch pitting meant very little. This use of grain weathering clearly should be used with care and needs to be used in the same way as other techniques requiring calibration such as lichenometry and weathering rind measurement.

(B) Desert varnish density or colour has been used by Whalley (1983) to delimit glacial stages and related events in an area of the Hindu Kush, but SEM was used as an adjunct to the investigations of varnish origin rather than as an aid to dating. This is a rather specialised application and examination of bedrock weathering with the SEM has not been used to determine weathering or age since deglaciation. Rock varnish has been used (for essentially desertic environments) for dating artifacts and landforms using 'cation-ratio' dating (Dorn et al., 1987a,b) and this has been extended for use with an SEM and EDS (Harrington and Whitney, 1987). For arid–glacial environments this technique might have some potential.

13.2.4.15. Mixed environments

Many workers have used SEM surface texture analysis to help interpret glacial sequences. Frequently, a number of chronological techniques such as lichenometry, boulder weathering and soil development are utilised in the study of glacial sequences (e.g. Wayne and Corte, 1983; Coudé-Gaussen, 1985). Till stratigraphic analysis using SEM texture analysis as an aid has been used by some workers (e.g. Ly, 1978; Perttunen and Hirvas, 1982; Billard and Derbyshire, 1985; Bray et al., 1981).

There are several studies that have attempted to distinguish between glacial sub-environments (e.g. Fillon *et al.*, 1978; Rogerson and Hudson, 1983; Dowdeswell *et al.*, 1985b).

13.3. EXAMINATION OF BULK SAMPLES OF SOILS AND TILLS

In studying tills and other glacially derived materials, there is substantial overlap between glacial geology, pedology and soil mechanics (e.g. Boardman, 1983; Collins and McGown, 1983). The disadvantage of bulk examination is that only small portions can be studied, even if montages covering a few square millimetres are produced (Collins, 1983; Harris, 1985). However, there are some advantages (a) first, there is the enhanced detail and depth of field that can be obtained, similar benefits to that of viewing single grains. (b) Secondly, spatial and linear patterns (or lack of them) can be processed by a variety of techniques. (c) Thirdly, there are the advantages of analytical tools which can be used, such as EDS and BSI. Perhaps not surprisingly, there are difficulties with such methods. The major one is that of preparation, especially if the sample is clayey. As water is usually held in clays, this has to be removed before the sample is placed in a vacuum (Sergeyev *et al.*, 1980a, b; Tovey and Wong, 1973, 1978a; Smart and Tovey, 1982). In general, soils and bulk materials prove to be more difficult to study compared to individual grains (McKee and Brown, 1977).

Once a stable specimen has been acquired the surface to be viewed (either sliced or fractured) can be selected. At one level, visual inspection of this surface may be sufficient. It is possible to use a variety of methods to obtain quantitative or semi-quantitative data about the specimen surface. Recent work linking stratigraphic, sedimentological and geomorphological investigations with the use of the SEM has begun (Fort and Derbyshire, 1988; Owen and Derbyshire, 1988; Owen, 1988b).

13.3.1. Fissures and Pores

Shears and micro-shears in tills have been examined by Derbyshire (1978, 1980) and Derbyshire *et al.* (1985b) the latter emphasising their importance

regarding engineering applications (Gillott, 1969; Barden and Sides, 1971; Smart, 1974; Grabowska-Olszewska, 1975; Tovey, 1978; Derbyshire *et al.*, 1985b; Bruns, 1989). Love and Derbyshire (1985) describe the methods of fabric analysis involving the SEM (cf. Collins and McGown, 1983). Either a density slicer or a scanning lazer device is used to provide the quantification. This work is of particular importance for the investigation of clay and silt structures and pore spaces in tills and loess (Lutenegger, 1981; Derbyshire and Mellors, 1988). Sergeyev *et al.* (1980b) provide a classification for micro-structures in clay soils that may find general applicability.

There are several methods available for quantitative analysis of soil fabrics, whether from a pedological or soil mechanical viewpoint (e.g. Tovey, 1973, 1974, 1986; Smart, 1974; Osipov and Sokolov, 1978; Tovey and Sokolov, 1981; Collins, 1983; Collins and McGown, 1983; Tovey and Wong, 1978b). A sophisticated technique developed by Mirkin *et al.* (1978) and Sokolov *et al.* (1980) uses the images from two adjacent (conjugate) surfaces to provide identification of pores etc. It is also possible to obtain information on the three-dimensional structure of a specimen by serial sections using stereological methods used by cell biologists (e.g. Lin, 1982; Macdonald *et al.*, 1986).

13.4. OTHER INVESTIGATIONS

Much of the foregoing discussion has been related to clastic sediments in glacial and near-glacial environments. However, the facilities offered by the electron microscope (again, in conjunction with various other techniques such as stable isotopes) can be useful in other glacially-orientated investigations.

Subglacial sedimentation of silica and carbonates has been investigated by several workers (Hallet, 1975, 1976). Subglacially produced ferromanganese varnishes have been described (Whalley *et al.*, 1990) and examined with the SEM and electron microprobe. The use of the SEM and associated analytical techniques is important in these investigations which link glacier–bed coupling mechanisms and the nature of the basal water film. This is likely to be an area in which more work will be forthcoming, especially if surface textures of grains within deforming till are examined.

The SEM can be useful for general illustrative purposes as well as specific analysis, pollen and diatoms are obvious material (Burckle *et al.*, 1988) as are analyses of particles of different types from ice cores (Goss *et al.*, 1985).

13.5. FUTURE DEVELOPMENTS

Use of the SEM to characterize grain surface textures and interpret them requires subjective interpretation. Ideally, there needs to be an automatic, rather than human, pattern recognition technique o allow classification of large numbers of grains per sample (and perhaps to quantify size and outline shape). The purpose of these tools would be o speed analysis – and therefore make it cheaper and reproducible. It is likely that this goal is an auto-matic, knowledge-based (expert) system that uses images as training sets. A similar objective can be foreseen for quantification of pores and linear structures in bulk material. In either case, computing power and powerful programs will need to be linked to SEMs, either directly or as post processing adaptions via image analysers.

13.6. CONCLUSIONS

The scanning electron microscope is a useful tool for the glacial sedimentologist and geomorphologist. It provides a suitable means, with appropriate safeguards, for assisting to distinguish sediment paths and possible origins of material. It is important in characterising soils and relating sedimentary to geotechnical properties. Its versatility is matched by its ease of use, although certain precautions need to be taken for some types of investigation.

Chapter 14

GEOCHRONOLOGY OF GLACIAL DEPOSITS

J. Brigham-Grette

14.1. INTRODUCTION

Strong geochronologic control is fundamental to any stratigraphic or paleoenvironmental study. This is especially the case in Quaternary sciences where the concern is with glacial cycles, surficial processes, floral and faunal changes, and the spatial and temporal order in which events occurred. Recent emphasis on the rates and processes of global change highlight the need to accurately quantify the timing and rates of sedimentation and erosion, to understand geomorphic processes, and, ultimately, to link rates and processes to climatically related events. Although the deep-sea marine record provides a relatively continuous, globally-integrated record of glacial and interglacial cycles (Imbrie et al., 1984; Shackleton, 1982, 1987), accurate knowledge of the timing of ice advances and retreats, as well as the history of ice-marginal processes, are germane to understanding ice-sheet dynamics, identifying synchronous and asynchronous events, and recognizing leads and lags between ice sheet fluctuations and changes in the ocean/atmosphere system (cf. Ruddiman, 1987a, b; Broecker and Denton, 1989).

In the last 15 years, our technological understanding of geochronological methods has grown rapidly. For example, uranium-series dating has attained new levels of precision and accuracy and has extended into the range of radiocarbon dating. Cosmogenic-isotope, uranium-trend, and amino acid geochronology are relatively new methods with the potential of providing numerical ages for materials beyond the range of radiocarbon assays. New analytical techniques and knowledge of sedimentological constraints have launched interest in thermoluminescence dating of sediment, whose chronologic clock is zeroed by exposure to sunlight. Accelerator-mass spectrometer technology now enables samples weighing less than a milligram to be dated by radiocarbon assay. In all cases, technological advancements have increased the speed and precision of measurements.

The variety of dating methods available to scientists working on glacial stratigraphy adds to the potential for error and confusion in interpreting age estimates. Each method has its assumptions and limitations that must be understood before it is possible to critically evaluate: (1) whether a particular method should be used in a specific stratigraphic setting, and (2) what a resulting age really means. In all cases, the objective is to select the method(s) that will optimize the information gained from the datable materials available. In most cases, the use of multiple methods is desired as an independent check on the consistency of results.

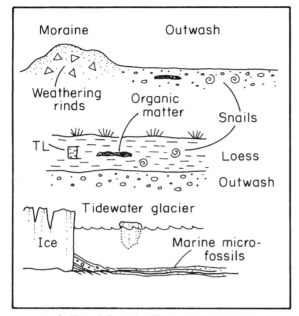

A. *IN SITU* OR CONTEMPORANEOUS

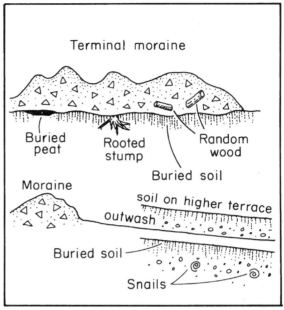

B. MAXIMUM AGE ESTIMATES

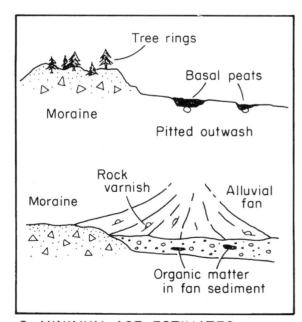

C. MINIMUM AGE ESTIMATES

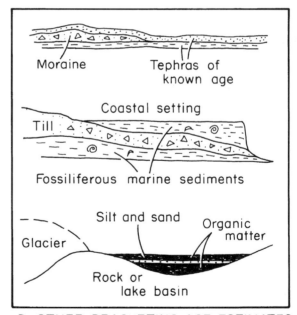

D. OTHER BRACKETING AGE ESTIMATES

FIG. 14.1 Possible stratigraphic relationships between datable materials and glaciogenic sediments. (a) *In situ* materials deposited at the same time as the enclosing sediment providing contemporaneous ages. TL refers to thermoluminescence dating. (b) Buried or reworked materials that provide maximum-limiting age estimates. (c) Deposits and organics that post-date an event providing minimum-limiting age estimates. (d) Datable materials in sediments that overlie and underlie glaciogenic sediments producing bracketing age estimates. (Modified from Porter (1981c) with permission from the author and *Edward Arnold Press, Ltd*)

14.2. RESOLUTION AND STRATIGRAPHIC CONTEXT

The resolution to which one can determine the age of a particular event (e.g. the initiation of loess deposition or the rate of sedimentation in a proglacial lake) depends upon the sampling interval chosen and the accuracy and precision of the method being used. It is important to distinguish between accuracy, which is correctness or how close an estimated age is to the true age, and precision, which is the level of reproducibility to which a measurement can be performed. Most dating methods, after all, simply measure probability, so it is the researcher's task to determine what the 'date' actually means with reference to the stratigraphy and glacial history. An age estimate may be precise (small margin of error) but it may not necessarily be accurate. Conversely an accurate age estimate may be imprecise (large margin of error).

A fundamental aspect of dating glacial deposits is determining the relationship between the material being dated and the geological event of interest (Dean, 1978) (Fig. 14.1). For example, a radiocarbon analysis on wood found in glacial till indicates when the tree lived; the 'event' being dated is the death of the tree. The age estimate does not necessarily indicate when the glacier overran the site, the event of interest. Better constrained stratigraphic context, such as in situ tree stumps overlain by till or buried in outwash, is needed to determine whether the tree was living when the glacier arrived, or whether the tree was reworked from older deposits. In this example, the dated wood provides only a maximum limiting age estimate for the advance of the glacier (Fig. 14.1b); in other words, the glacial advance must be younger than or the same age as the dated wood. Minimum limiting age estimates result from the age of events that post-date the event of interest. For example, an age based on the number of growth rings in a series of small trees on the stabilized surface of a young moraine (Fig. 14.1c) indicates when the trees became established on the surface, but it does not give an estimate of the length of time between deposition of the moraine and the initiation of tree growth. The trees provide a minimum-limiting age for glacial retreat; glacial retreat is older than the age of the trees.

Minimum- and maximum-limiting age estimates on materials that span a depositional event are commonly called bracketing ages (Fig. 14.1d).

Inherent in the study of glacial chronologies is the fact that glacial advances and retreats are regionally time-transgressive. Therefore, the onset of till deposition in one area may lead or lag behind till deposition in another. Although climatic deterioration may have occurred contemporaneously over the Great Lakes region, for example, glacial ice advanced to the U.S/Canadian border thousands of years before it reached central Illinois (Andrews, 1987c; Johnson and Hansel, 1990). Loess deposition can also be time-transgressive (cf. Follmer et al., 1989). The establishment of 'diachronic units' in the North American Stratigraphic Code (NACSN, 1983, Article 91–95, p. 870) provides a means of recognizing the diachronous nature of these and many other types of Quaternary deposits.

Time-parallel or synchronous deposits, i.e. those that are the same age everywhere, are relatively rare in the context of glacial geology. The eruption of tephra across wide regions, for example of western North America, during the Quaternary provide excellent, synchronous stratigraphic markers for dating in glacial and periglacial sequences (Self and Sparks, 1981; Naeser et al., 1981, 1982; Beget 1984; Westgate et al., 1987; Sarna-Wojcicki and Davis, 1991; Sarna-Wojcicki et al., 1991). In addition, evidence for rapid climatic change, such as the Younger Dryas event around the North Atlantic (Broecker et al., 1989; Stea and Mott, 1989; Fairbanks, 1989; Dansgaard et al., 1989; Kennett, 1990), may be recorded by proxy data over large regions (Kallel et al., 1988; Heusser, 1989), and hence provide a nearly contemporaneous datum. Although the precision of this contemporaneity is much lower than that of an ash fall, it is much higher than that of a glacial advance.

Inherent in the stratigraphy of deposits recording repeated glaciation in the same area is the fact that glacial deposition is largely discontinuous and laterally variable. Older deposits are commonly poorly preserved and exposed, even at or beyond the limit of the younger advances. Evidence of some glacial advances may be removed entirely by the obliterative overlap of younger events that are more extensive (Gibbons et al., 1984). The relative position of datable materials and unconformities in a sequence need to be carefully considered to avoid being misled by older materials, such as wood or bone, that have survived reworking (see Chapter 8).

14.3. GEOCHRONOLOGICAL TERMINOLOGY

The advent of new techniques and increased precision in traditional methods has led to confusion over the proper terminology for use in geochronology. Colman *et al.* (1987) recently addressed this issue with recommendations for standardizing many misused terms. Use of the term 'date' has been traditionally misused as a noun to imply some degree of accuracy related to calendar years (Colman *et al.*, 1987). However, any 'date' regardless of its precision is merely an 'age estimate' based upon some decay constant or standard for that particular method. It is not a direct measurement of the passage of calendar years. Dates based on historical records, varves, or tree rings are among the few exceptions where calendar ('sidereal') years are directly implied. The term 'age estimates' or 'ages' should be used rather than 'dates' to describe geochronological results.

Another source of confusion is the word 'absolute', traditionally used to describe the results of radiometric dating methods. Given the assumptions, uncertainties and precision of decay constants associated with all methods, no dating method produces results that should be regarded as 'absolute'. Rather, Colman *et al.* (1987, p. 315) suggest the term 'absolute' be replaced with the term 'numerical' for age estimates 'that provide quantitative estimates of age and uncertainty on a ratio (absolute) time scale'. Table 14.1 outlines the suggested usage of terms.

Not all age estimates can be classified as numerical. Correlated-ages are those that are based on correlation with independently dated deposits at a different site (e.g. paleomagnetic reversals). The age of the deposit of interest is not determined directly. Relative-ages simply indicate the sequence in which sediments were deposited (e.g. soil profile development). Finally, calibrated-ages are those based on process rates that must be calibrated by some independent method. This category overlaps with both numerical and relative-age methods but recognizes the interdependence of many methods on the results of others. For example, a lichenometric growth curve calibrated with radiocarbon age-estimates and historical data from tombstones becomes a technique for producing calibrated-age estimates.

14.4. METHODS AND APPROACHES

Dateable sediments, weathered surfaces, and organic remains are found in a variety of both glacial and proglacial environments. The availability and type of datable materials in any stratigraphic sequence determines which method, or variety of methods, can best be applied. The ultimate value of any age estimate is only as good as the understanding of the stratigraphic context of the sample.

The following discussion is organized according to the classification scheme in Table 14.2. Details concerning the theoretical background, assumptions and inherent variability of these methods can be found in specialized texts (Mahaney, 1984; Aitken, 1985; Bradley, 1985; Rutter, 1985; Thompson and Oldfield,

TABLE 14.1. Suggested use of geochronological terms

Common useage	Suggested useage	Example (see types of results Table 14.2
Date (noun)	Age estimate	^{14}C age estimates
Dates	Ages	^{14}C ages
Absolute age	Numerical-age Calibrated-age Correlated-age	K/Ar numerical age Lichenometric calibrated age Paleomagnetic correlated age
Relative-age	Relative-age	Pedogenic relative-age

TABLE 14.2. Classification of Quaternary geochronological methods (modified from Colman *et al.*, 1987, reprinted with permission of Quaternary Research, University of Washington)

Type of Result					
←———— Numerical Age ————→					
	←———— Calibrated Age ————→				
		←———————— Relative Age ————————→			
					←— Correlated Age —→
Sidereal	Isotopic	Radiogenic	Chemical and Biological	Geomorphic	Correlation
Historical records	^{14}C	Fission-track	Amino acid racemization	Soil profile development	Tephrochronology
Dendro-chronology	K-Ar and ^{39}Ar-^{40}Ar		Obsidian hydration	Rock and mineral weathering	Paleomagnetism
Varve chronology	Uranium series		Lichenometry	Progressive landscape modification	Stable isotopes
	Experimental Methods			Geomorphic position	
	Uranium trend	Thermo-luminescence	Rock varnish chemistry		
	Cosmogenic isotopes	Optically stimulated luminescence			

1986; Taylor, 1987; Easterbrook, 1988a; Rutter *et al.*, 1989). The purpose here is to review the methods available for dating glaciogenic deposits, and to cite examples of specific stratigraphic settings where the advantages of each method are realized. Geochronological methods used to date climatic cycles in non-glaciogenic materials (e.g. uranium-series dating of corals) have been excluded from this discussion.

14.4.1. Sidereal Methods

Geochronological methods that resolve annual events include the use of historical records, dendrochronology, and varve chronology.

14.4.1.1. Historical records

Important data concerning recent fluctuations in existing glaciers and glaciers of the Little Ice Age can be derived from a variety of historical materials

including 16th through 19th century artwork, old photographs (e.g. Kite and Reid, 1977), early maps such as those made by Muir and Reid of the Alaskan tidewater glaciers (Reid, 1892, 1896) and texts that record the positions of glacial margins (cf. Karlén, 1988). Grove's (1988) comprehensive review of The Little Ice Age includes numerous examples of antique maps and sketches that document ice marginal positions throughout portions of Europe, Asia and North America. Instrumental weather records dating back to the turn of the century (Ellsaesser *et al.*, 1986; Jones *et al.*, 1986a, b; Bradley *et al.*, 1987) and climatic proxy data from, for example, diaries, annals and court records as far back as 1000 A.D. in China (Ingram *et al.*, 1978; Wigley *et al.*, 1981; Bradley, 1985; Zhang and Crowley, 1989), also provide useful time series for comparing historical glacier responses to short term, high-frequency changes in climate.

The most serious shortcoming of historical data is that observations are not necessarily continuous or accurate. To use old sketches of glaciers, for example, one must assume that the rendition is accurate and that the date of the work is accurately known. Moreover, the quality of the climate observations or climatic proxy data are often difficult to interpret, compare, and quantify alongside recent instrumental records (Bradley, 1985, 1991; Karl *et al.*, 1989).

14.4.1.2. Dendrochronology

In forested regions, tree rings formed by incremental annual growth provide a useful means of determining the minimum age of glacial deposits (Lawrence, 1950; Parker *et al.*, 1984), particularly moraines (Bray

and Struik, 1963; Carrara and McGimsey, 1981, 1988; Luckman, 1986). By locating and coring the oldest tree(s) on a moraine or landform, and counting the number of annual rings from a core taken above the base of the tree, it is possible to determine the minimum-limiting age of that surface in years (Fig. 14.2). The major drawbacks to this approach are that it is difficult to: (1) estimate the length of time between stabilization of the surface and the establishment of viable tree seedlings (ecesis), (2) determine the number of years required for a tree to grow to the height at which the core was taken (McCarthy *et al.*, 1991), and (3) ensure that samples include the oldest tree. Sigafoos and Hendricks (1969) demonstrated

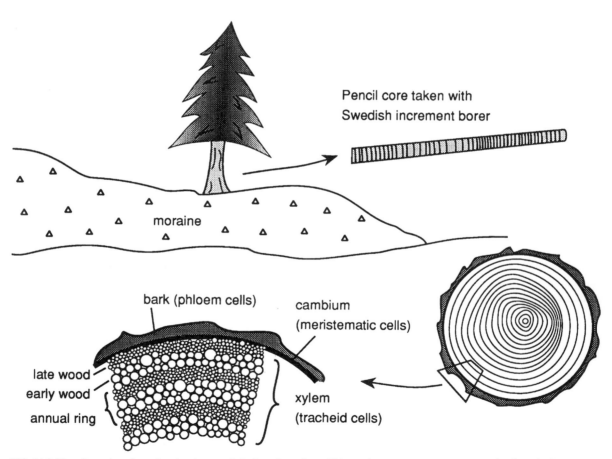

FIG. 14.2. Use of tree rings for estimating the age of glaciogenic surfaces. Either an increment core or a cross-section from the largest trees is collected and the annual rings are counted to determine when trees became established on the deposit. In addition, one must determine the approximate length of time between the emplacement of the deposit and the establishment of trees

that seed germination may require anywhere from 5 to 23 years in parts of the Pacific Northwest, regardless of elevation and aspect. Baden (1952) determined that Engelman spruce (*Picea engelmannii*) 35–40 cm high were about 12 years old in areas of Glacier National Park. Other problems such as missing the pith of the tree with the corer and the identification of missing or false rings also add small errors to most dendrochronological estimates (Luckman, 1986).

Carrara and McGimsey (1981) used an estimate of 15 years to account for both ecesis and the age of the tree at the height of coring in their study of the Neoglacial history of Glacier National Park. Hence, at each sampling site 15 years were added to the number of tree-rings counted to determine the best age estimate for deglaciation from a given limit. In the Canadian Rockies, estimates for ecesis range from 10 to 80 years depending upon microclimatic and pedogenic conditions (Luckman, 1986, 1988). In general, dendrochronology is best applied where knowledge of microclimatic affects and the length of ecesis can be quantified.

More precise dating can be obtained where a tree has been damaged (scarred or tilted) or killed during a glacier advance. These damage events can be dated to the calendar year by counting tree rings since the damage in living trees or by crossdating dead specimens with long tree-ring chronologies (e.g. Luckman, 1988). These techniques may be used for wood found in a variety of stratigraphic situations to date glacier fluctuations (see Fig. 14.3; Ryder and Thomson, 1986; Luckman, 1988; Luckman *et al.*, 1993).

14.4.1.3. Varve chronologies

Lake sediments in glaciated regions often appear as regular, alternating bands of coarse and fine-grained sediment widely known as rhythmites (Ashley, 1975, 1988). In many settings, such as proglacial or ice-marginal lakes, rhythmites consist of alternating layers of sand and silt or silt and clay; summer layers may, in part, be biogenic (ostracodes, diatoms, etc.; see Menzies, 1995a, Chapters 12 and 13). Where it

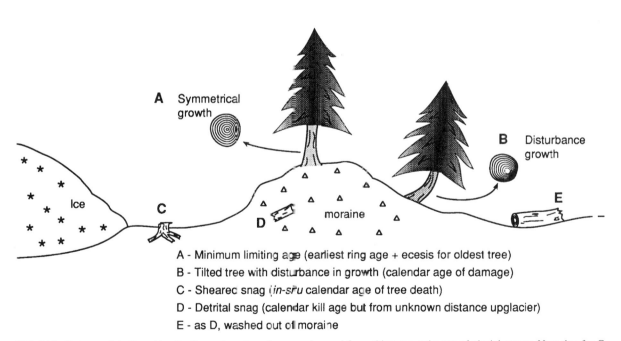

FIG. 14.3. Cartoon of stratigraphic situations where tree rings may be used for making age estimates of glacial events. Note that for C, D, and E, the sample must be cross-dated with a composite tree-ring chronology

PLATE 14.1. Varved sediments from the Canoe Brook site of Glacial Lake Hitchcock, Connecticut Valley, Vermont (Ridge and Larsen, 1990). The darkest bands are annual winter layers. (Photo courtesy of J.C. Ridge)

can be demonstrated that the couplets are created by winter/summer cycles of sedimentation (Plate 14.1), the rhymites are referred to as varves (Swedish *varv*, a periodic repetition) representing true annual layers (O'Sullivan, 1983). Since the mid-19th century, varves have been recognized as a useful means of establishing a geochronology for glacial-lake sedimentation. However, it was not until the work of De Geer (1912, 1921) in Sweden that the technique was widely utilized. By counting each couplet consisting of a light-coloured, coarse-grained (summer) layer and a dark-coloured, fine-grained (winter) layer and correlating the sequence of varve thicknesses from section to section, De Geer established a precise chronology for deglaciation for the last 17,000 years. Although subsequent work has shown that the chronology is accurate only for the last 12,000 years (Lundqvist, 1975, 1980) the findings provided the impetus for work by Antevs (1922, 1925), who established the New England regional varve chronology. Since that time countless studies have been carried out on both modern proglacial lakes and relict deglacial sequences in Europe and North America (e.g. Gustavson, 1975b; Smith, 1978; Gilbert and Shaw, 1981; Renberg, 1981; Perkins and Sims, 1983; Smith and Ashley, 1985; Lotter, 1991).

The most difficult aspect of generating a time scale using lake sediments is establishing that rhythmites are, in fact, true varves. This task is often complicated

by multiple summer inflow events generated by subaqueous density flows or disturbances in sedimentation caused by extremes in rainfall, wind, or thermal stratification of lakes (Ashley, 1975). Before lake sediments can be used for geochronological purposes, it is mandatory that rhythmically layered sediments be identified as annual sediment cycles (cf. Stihler *et al.* 1992).

Recent examples of the use of varve chronologies include studies by Leonard (1986a, c) in lakes of the Canadian Rockies, and by Ridge and Larsen (1990) in the Connecticut Valley, Vermont. Leonard (1986a) examined varves deposited over the last 800 years in cores taken from Hector Lake, Alberta. He then correlated the varves with a tree-ring based moraine chronology, and determined that changes in sedimentation rate closely corresponded with upvalley glacial activity. Ridge and Larsen (1990), in contrast, used exposures of some 500 ice-proximal to ice-distal varves, along with radiocarbon-age estimates, to assess the age of deglaciation in southern Vermont and the timing of basin-wide drainage of Glacial Lake Hitchcock.

14.4.2. Isotopic Methods

Isotopes are species of an element that have the same number of protons and a different number of neutrons, i.e. they have different atomic weights. For example, carbon with 6 protons, exists in three forms – ^{12}C, ^{13}C, ^{14}C – each with an increasing number of neutrons. Although the majority of each element

TABLE 14.3. Half-lives of isotopes commonly used for age estimates of Quaternary glacial deposits

Isotope	3T 1/2 years
^{238}U	4.51×10^9 years
^{40}K	1.31×10^9 years
^{235}U	0.71×10^9 years
^{10}BE	1.60×10^6 years
^{26}Al	7.20×10^5 years
^{36}Cl	3.01×10^5 years
^{14}C	5.73×10^3 years
^{40}Ar	289 years
^{210}Pb	22 years
^{3}He	Stable

occurs as one or more stable forms, some isotopes, such as ^{14}C, are not stable. Through spontaneous radioactive decay by loss of particles from the nucleus (alpha, beta, gamma particles) one isotope of an element may be transformed into another isotope, or into a new element. Naturally occurring, unstable isotopes that decay at known rates (Table 14.3) provide some of the most useful means of measuring geologic time (Faure, 1986). However, because some isotopes are rare and because each isotope has a different half-life, only some stratigraphic settings lend themselves to any one type of isotopic geochronological method.

14.4.2.1. Radiocarbon age estimates

The radioactive decay of ^{14}C is the most widely used isotopic dating method in glacial geology. Produced in the atmosphere by the cosmic ray bombardment of nitrogen, ^{14}C comprises a small percentage of the world's carbon reservoir in the atmosphere, oceans, and most living plant and animal tissues (Taylor, 1987). ^{14}C is used for determining age estimates of organic materials such as wood, peat, seeds, needles, mollusc shells, leaves, bones, etc., found in association with glaciogenic sediments.

While alive, plants and animals assimilate ^{14}C, mostly in equilibrium with the atmosphere, through processes of photosynthesis and respiration. Once an organism dies, assimilation of ^{14}C ceases and the existing ^{14}C decays to stable nitrogen through the loss of a ß particle (an electron).

$$^{14}C \rightarrow {}^{14}N + ß + neutrino \qquad (14.1)$$

The exponential decay rate is expressed as a half-life, the period over which half of the existing isotope transmutates to a new element and half of its radioactivity dissipates (Fig. 14.4). Carbon-14 has a half-life of 5730 ± 40 years BP; however, all radiocarbon age estimates by convention are expressed using the original 'Libby half-life' estimate of 5568 ± 30 years BP (Godwin, 1962; Stuiver and Polach, 1977). By assuming an initial concentration of ^{14}C at death, the concentration of remaining ^{14}C can be used to determine the length of time elapsed since death.

The amount of ^{14}C remaining in a sample can be measured two ways – by 'conventional', indirect methods counting beta decays, or directly by accelerator mass spectrometry (AMS). Details of each method can be found elsewhere (Taylor, 1987; Hedges and Gowlett, 1986; Linick et al., 1989). Conventional radiocarbon age estimates, including those by isotopic enrichment (Stuiver, 1978a; Stuiver et al., 1978) are made by measuring the number of

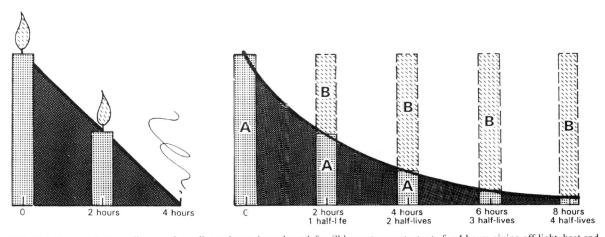

FIG. 14.4. Contrast between linear and non-linear decay. A candle at left will burn at a constant rate for 4 hours giving off light, heat and gases at a linear rate. Radioactive decay (right) occurs at an exponential rate with the parent isotope A decaying to its prodegy product B. After one half life (two hours) one half the parent A remains; after two half lives, only 1/4 remains (from Sawkins et al., 1978, reprinted with permission from the author)

ß-particle emissions given off by a sample (disintegrations/minute/gram). Counting for a number of days is required to achieve a precision of 1%. Using AMS, in contrast, the actual number of ^{14}C atoms themselves are counted, yielding precision of 0.6% in as little as an hour (± 50 year) for a sample 5000 years old (Linick *et al.*, 1989). Most importantly, AMS is capable of making age determinations on samples containing as little as 100 µg of carbon, in contrast to the 0.5–1.0 g of carbon required for conventional counters. This breakthrough is especially important in glacial-stratigraphic studies where organic material is scarce. Moreover, it allows analysis of different carbon fractions of the same sample to test for internal consistency (cf. Nelson *et al.*, 1988). Lowe *et al.* (1988) found that the Cl- and HF/HCL-insoluble fraction of organic-rich late glacial lake sediments yielded consistently older age estimates than other fractions indicating the presence of old mineral carbon derived from ground water. In a similar case, Fowler *et al.* (1986) recommend that a minimum of five AMS age estimates be performed on different chemical extracts of the same sample to test for sources of contamination.

A fundamental assumption inherent in radiocarbon dating is that the atmospheric concentration of ^{14}C has not changed through time. However, this assumption is not strictly valid and ^{14}C production has varied in the past. Changes in the production rate over thousands of years are due largely to changes in the earth's magnetic field; changes on decadal time scales are due largely to changes in solar activity (Bradley, 1985, Stuiver *et al.*, 1991). Changes in the ^{14}C

reservoir in the ocean are due to changes in thermohaline circulation on glacial/interglacial and shorter timescales (Shackleton *et al.*, 1988; Peng, 1989). Calibration of the radiocarbon time scale with calendar years from tree rings is now reliable back to nearly 10,000 years (Becker *et al.*, 1991; Stuiver, 1993). Tree rings and varve records have also been combined to calibrate the radiocarbon timescale back to 28,000 years (Stuiver *et al.*, 1986, 1991); Edwards *et al.* (1986) and Bard *et al.* (1990) have used uranium-thorium (U/Th) age estimates from fossil corals to suggest a cruder calibration back to 30,000 BP (Fig. 14.5). This latter study suggests that by 20,000 BP, radiocarbon age estimates are younger then U/Th ages by as much as 3500 years. This discrepancy is important for comparing ^{14}C -based glacial chronologies with Milankovitch-driven climate parameters (Menzies, 1995a, Chapter 2). Recent work has also demonstrated there are three 450-year plateaus in the tree-ring calibrated Holocene ^{14}C record (Becker et al. 1991) (Fig. 14.6). The ages of organisms living during these ^{14}C plateaus are indistinguishable. The last few centuries are also problematical. Stuiver (1978b) has shown, for example, that a ^{14}C age of 220 ± 50 years actually spans the range of 150–420 calendar years.

Sources of error in radiocarbon age estimates can be subdivided into two broad categories, namely, chemically-derived errors and stratigraphically-

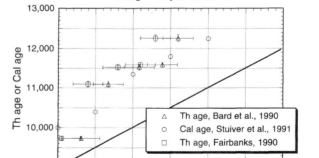

FIG. 14.5. (a) Calibration of radiocarbon years to calendar years back to 30,000 years BP and (b) an enlargement of Late Glacial interval (9–13 Ka). The Bard *et al.* (1990) and Fairbanks (1990) data are based on U-Th data, the data of Stuiver *et al.* (1991) are based on reconstructed geomagnetic dipole fields, varves, and tree-ring records

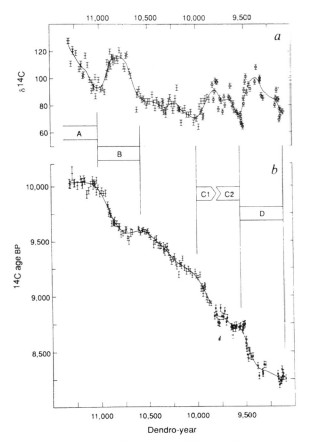

FIG. 14.6. Atmospheric ^{14}C levels and ^{14}C-age calibrated against dendrochronological years. (From Becker *et al.*, 1991, reprinted with permission from Macmillan Magazines Limited. The initial plateau portions of the lower curve at A,B,C,and D are periods when radiocarbon age estimates have low calendar year resolution. These plateaus occur at times of large changes in ^{14}C seen in the upper curve

derived errors. Chemically-derived errors include contamination by either old or young carbon (Olsson, 1974; Taylor, 1987), 'hard-water' effect particularly in terrestrial carbonate fossils caused by the uptake of old carbon from bedrock (Shotton, 1972), 'reservoir' effect in marine organisms caused by sluggish deep-ocean circulation (Mangerud, 1972; Bard, 1988), and fractionation effects caused by the differential metabolic uptake of [14]C (Bradley, 1985). Knowledge of these potential errors is necessary for evaluating all radiocarbon age estimates.

Stratigraphically-derived errors result from mis-interpretations of the context in which the sample was collected. Clayton and Moran (1982) describe a section in North Dakota where wood once growing on an ice-cored moraine later became buried by mass movement of supraglacial debris (re-sedimented debris flow). If the colluvium were to be mis-interpreted as till, then the age would be considered a maximum-limiting age estimate for the moraine, when in fact, it is a minimum-limiting age estimate for deglaciation. Likewise, Lowell and Stuckenrath (1990; Lowell *et al.*, 1990b) describe three strati-graphic units in southern Ohio in which glacially-retransported wood and disseminated organics provide a wide range of age estimates between 19,000 and 27,000 BP. They argue that a single age determination from such units may be misinterpreted and that multiple measurements on a variety of materials are necessary to discriminate reworked material. In their study, stumps in growth position provided the most-closely limiting, maximum age estimate for glacial advance into the Cincinnati region. The detrital wood was older, despite its stratigraphic position above the stumps.

14.4.2.2. K/Ar and Ar/Ar Age Estimates

Glacial sequences in volcanic terrains containing interbedded lava flows, tuffs, or tephras can provide age estimates using a variety of methods. Volcanic deposits are well-known throughout western North America, in particular (Dalrymple, 1964; Birkeland *et al.*, 1976; Pierce *et al.*, 1976; Pierce, 1979; Westgate *et al.*, 1990; Kaufman *et al.*, 1991) where either K/Ar or Ar/Ar dating methods have been used to obtain age estimates.

K/Ar dating is based on the branched decay of [40]K to either [40]Ca (89.5%) by β decay or [40]Ar (10.5%) by electron-capture with a half-life of 1.25×10^9 years (Fig. 14.7). Details of the method can be found elsewhere (Faure, 1986; Bowen, 1988). For K/Ar both the abundance of [40]K and its progeny, [40]Ar, are measured in separate splits of the same sample. Measurements are routinely made on either bulk whole rock samples or high-K mineral separates (e.g. biotite, sanidine and K-feldspar). In general, mineral separates are preferred but may not be available in fine-grained rocks.

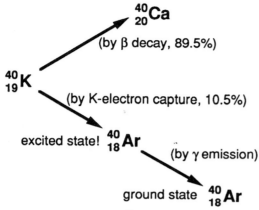

FIG. 14.7. Branched decay of ^{40}K to the ground states of either ^{40}Ca or ^{40}Ar

Nuclide	Half-life	Nuclide	Half-life
Uranium-238	4.51×10^9 years	Uranium-235	7.13×10^8 years
↓		↓	
Uranium-234	2.5×10^5 years	Protactinium-231	3.24×10^4 years
↓		↓	
Thorium-230	7.52×10^4 years	Thorium-227	18.6 years
↓		↓	
Radium-226	1.62×10^3 years	Radium-223	11.1 days
↓		↓	
Radon-222	3.83 days	Lead-207	stable
↓			
Lead-210	22 years		
↓			
Polonium-210	138 days		
↓			
Lead-206	stable		

FIG. 14.8. Decay series of uranium-238 and uranium-235 with half-lives

Two factors limit the usefulness of K/Ar in Quaternary studies. First, the long half-life of ^{40}K and consequent slow buildup of Ar makes determining age estimates on materials less than 100,000 years difficult. Secondly, the uptake of atmospheric Ar or the loss of Ar from the crystal lattice by diffusion or weathering may violate the 'closed system' required for determining the rate of change between parent isotope and progeny. Inherited Ar is another serious problem in young basalts, especially if they erupt through old plutons (Faure, 1986; Kaufman et al., 1991).

Recently a variation of K/Ar dating, called ^{40}Ar/^{39}Ar dating, has been developed to eliminate the need for a separate K analysis (Hall and York 1984, Layer et al., 1987; van den Bogaard et al., 1987, 1989). In this procedure, individual mineral grains are irradiated with neutrons to produce ^{39}Ar from ^{39}K. Because the amount of ^{39}K is proportional to ^{40}K in a sample, age estimates are calculated directly from the ratio ^{40}Ar/^{39}Ar emitted as a gas during sample heating. In addition to greater precision, stepwise heating procedures can also be used to identify Ar loss or gain allowing for more accurate age estimates, especially in young samples (Hall and York, 1984).

14.4.2.3. Uranium-series age estimates

The gradual radioactive breakdown of uranium-235 and uranium-238 (Fig. 14.8) creates a series of decay products that become selectively partitioned in terrestrial and marine carbonates and sediments due to differences in solubility. Uranium is relatively soluble as carbonate or phosphate complexes and easily migrates with ground water, in streams, or sea water; in contrast, the daughter products of uranium (thorium-230 or protactinium-231) are less soluble and readily precipitated or absorbed onto clays. The tendency of short-lived daughter products to be chemically separated from their parents yields a variety of Quaternary dating methods (Schwarcz and Blackwell, 1985; Schwarcz, 1989). Only a few of these methods are applicable specifically to glacial terrains; the dating of Quaternary age travertine, corals, molluscs and deep-sea marine sediments will not be discussed here.

Speleothems have been successfully dated by ^{230}Th/^{234}U and ^{231}Pa/^{235}U for determining the ages of glacial/interglacial cycles from a number of glaciated sites in North America and northwest Europe (Harmon et al., 1978; Lauritzen and Gascoyne, 1980; Gascoyne et al., 1981, 1983; Lauritzen, 1991). Such deposits are only formed in glaciated regions under non-glacial conditions due to the inhibition of ground water flow by permafrost, changes in the partial pressure of ground water CO_2, and the subglacial flooding of cave galleries (Lauritzen, 1991).

Stalactites, stalagmites and laminated flowstone acquire mobile uranium when carbonates precipitate in cave environments. Although detrital ^{230}Th can be incorporated when clay particles are also entombed, the carbonate in deep caves is relatively free of daughter isotopes at the time of formation. The age of the layered, closed-system deposit can then be determined from the *in situ* decay of ^{234}U and the production of ^{230}Th.

$^{230}Th/^{234}U$ dating has also been used successfully to make age estimates of calcareous concretions formed in proglacial lacustrine and marine sediments (Hillaire-Marcel and Causse, 1989a,b). Using this method, they determined numerical-ages for a series of last interglacial and Wisconsinan deposits from the St. Lawrence Lowland, Canada.

Lead-210 (^{210}Pb) is an unstable isotope that may be used to date lake sediments. With a half life of only 22.3 years, the ^{210}Pb method may only be used for determining the age of materials less than 200 years old. This isotope is produced by the decay of radon-222 (^{222}Rn) after the breakdown of radium (^{226}Ra) from thorium-230 (^{230}Th) (Fig. 14.8). Both ^{222}Rn and ^{226}Ra are released into the atmosphere by weathering of rocks where they eventually decay to ^{210}Pb (Fig. 14.9). Lead-210 then returns to Earth with precipita-

tion and is incorporated into lake sediments. Some ^{210}Pb in lake sediments may also be derived from the erosion of ^{226}Ra and ^{210}Pb-bearing sediment from the surrounding catchment (Wise, 1980).

Uranium-trend dating is a relatively new method applicable to a variety of Quaternary sediments, including glacial tills (Rosholt, 1985). This method optimizes the open-system nature of unconsolidated sediments by tracing the time-related accumulation pattern of daughter products of ^{238}U, namely ^{234}U and ^{230}Th, through a deposit. Most ground water contains small amounts of dissolved uranium. As this water migrates through sediment, it produces a 'trail' of radioactive decay products that are readily absorbed onto solid matrix materials (Muhs *et al.*, 1989). The useful age range of uranium-trend dating is 5000 to about 700,000, based on a large number of samples calibrated against independently-dated deposits.

Shroba *et al.* (1983) used uranium-trend dating to cross-check the age of moraines in a portion of the Colorado Rockies. Based on soil development and

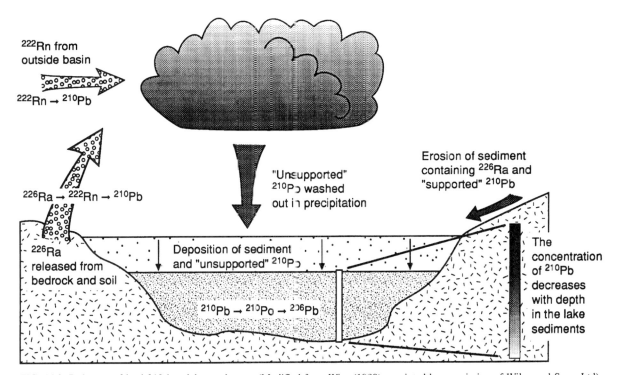

FIG. 14.9. Pathways of lead-210 in a lake catchment. (Modified from Wise (1980), reprinted by permission of Wiley and Sons, Ltd)

other relative-age parameters, the tills were thought to be of Bull Lake age (late Illinoian Glaciation). Uranium-trend dating of splits of the <2 mm fraction from the A2, Bt and Cox- horizons yielded an age of 130,000 ± 40,000 years. This age estimate compares well with an obsidian-hydration age of about 150,000 years (Pierce *et al*., 1976) and uranium-trend age estimates of about 130,000 and 160,000 years on Bull Lake till near West Yellowstone, Montana.

14.4.2.4. Cosmogenic (cosmic-ray produced) isotopes

The development of AMS for precisely measuring various isotopes in rocks exposed to cosmic-rays provides a new means of determining the numerical age of glacial landforms. Unlike radiocarbon dating, which relies on the cosmogenic production of ^{14}C in the atmosphere and its incorporation into organic matter, surface-exposure dating relies on the *in situ* accumulation of cosmogenic isotopes, such as ^{10}Be, ^{36}Cl, ^{3}He and ^{26}Al (cf. Phillips *et al*., 1986; Lal, 1988; Morris, 1991). Suitable rock surfaces are those that have been exposed continuously since the landform was constructed (Phillips *et al*., 1986). Although still in its infancy, surface-exposure dating may become especially useful for dating moraines, outwash surfaces and glacially-eroded surfaces.

Cosmogenic isotopes are also used to date ice cores (Faure, 1986; Stauffer, 1989); conversely, such isotopes in ice cores are also used for evaluating past variability in cosmic ray intensity and solar activity (Lorius *et al*., 1989). Concentrations of ^{10}Be sequestered in loess have recently been used to determine correlated-ages by direct comparisons between the abundance of ^{10}Be and the marine ^{18}O isotope record (Chengde *et al*., 1992).

Beryllium-10, widely produced in the stratosphere by spallation of N and O, is also produced on rock surfaces by nuclear bombardment of ^{16}O in minerals such as quartz and olivine (Lal, 1988; Nishiizumi *et al*., 1989b). With a half-life of 1.6×10^6 years, ^{10}Be should be useful for dating most late Cenozoic rock surfaces. Phillips *et al*. (1986) argue, however, that rock surfaces are susceptible to contamination by atmospherically produced ^{10}Be. They propose that ^{36}Cl, with a half life of 301×10^3 years, is better

suited for dating; ^{36}Cl is more hydrophillic (i.e. soluble in water) than ^{10}Be, allowing the natural discrimination of ^{36}Cl introduced by rain water versus *in situ* produced radionuclides. Chlorine-36 is produced on rock surfaces primarily by thermal neutron activation of ^{35}Cl (Faure, 1986; Leavy *et al*., 1987; Lal, 1988). In all cases, *in situ* cosmogenic nuclide production rates are highly dependent on latitude and altitude (Yokoyama *et al*., 1977; Lal, 1988; Nishiizumi *et al*., 1989b).

Applications of surface-exposure dating to glaciogenic deposits are few at this early stage of development. Phillips *et al*. (1990) recently dated the well-studied middle to late Pleistocene nested moraines at Bloody Canyon, eastern Sierra Nevada (Fig. 14.10a).

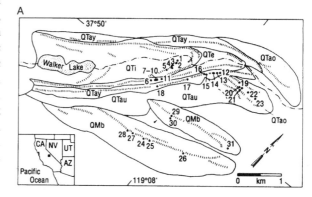

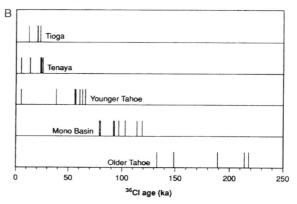

FIG. 14.10. Use of ^{36}Cl age estimates on moraines at Bloody Canyon, eastern Sierra Nevada, California (from Phillips *et al*., 1990, reprinted with permission from the *American Association for the Advancement of Science*). (a) shows samples sites on moraine crests of nested moraines of various age (QMb, Mono Basin; QTa, undifferentiated Tahoe deposits; QTao, older Tahoe; QTay, younger Tahoe; QTe, Tenaya deposits; QTi, Tioga deposits). (b). ^{36}Cl ages by moraine group.

The concentration of ^{36}Cl was measured from the top 5 cm of the largest boulders on moraine crests. Presumably the tallest erratics on the moraine crest are most likely to have been exposed since the deposition of the moraine. Their results showed that, although multiple ages from individual late Pleistocene moraines were tightly grouped, the ages from older moraines were younger than expected based on independent age criteria (Fig. 14.10b). This discrepancy may be attributed to factors such as weathering of the rock surface by spalling, shattering, and grain-by-grain disintegration, gradual exhumation of the erratics, and prolonged snow cover. Surface-exposure ages are best considered minimum age estimates.

Cosmogenic ^{3}He, produced *in situ* in rocks by cosmic ray bombardment, can also be used for determining the exposure age of rock surfaces (Brook and Kurz, 1993). By calibrating the surface-production rate of ^{3}He on independently-dated basalt flows, and other deposits, Cerling (1990) used ^{3}He to date late glacial flood events of Lake Bonneville and the Owens River. Brook *et al.* (1993) used cosmogenic ^{3}He and ^{10}Be to determine exposure ages for boulders on a Pliocene/Pleistocene moraine sequence in the Arena Valley, Antarctica. Like Phillips *et al.* (1990), Brook *et al.* (1993) attributed significant variability in both the ^{3}He and ^{10}Be ages on individual deposits to variable exhumation rates, prior exposure to cosmogenic rays, incorporation of older material in younger moraines, and/or local differences in surface weathering. Although the loss of ^{3}He is also a problem cited by existing studies, the use of a variety of cosmogenic isotopes at a site shows great promise of providing at least minimum-limiting exposure ages on materials lacking any other independent age control.

14.4.3. Radiogenic methods

Radiogenic dating methods are geochronological techniques that are based on the byproducts of radiation in minerals, including crystal lattice damage and the electron-state changes (Faure, 1986; Colman *et al.*, 1987). Although such features are inherently caused by isotopic decay within the mineral, the resulting features and byproducts, rather than the isotopes, are what is actually quantified as a chronological index.

14.4.3.1. Fission track geochronology

Tephras found in association with glacial and proglacial deposits serve as 'golden spikes' in a glacial stratigraphy. They form instantaneous chronological markers, and are often datable by fission track geochronology. A fission track is a microscopic radiation-damaged zone formed in crystalline material by the fission of naturally occurring uranium-238. Fission of ^{238}U creates two lighter energetic nuclei that are propelled through the crystal lattice stripping off electrons from nearby atoms and leaving a tiny track a few microns wide and less than 10–20 m in length (Walter, 1989). Because the rate of fission is proportional to the ^{238}U concentration and the length of time elapsed since crystallization (the volcanic eruption), the number of tracks in a single grain or suite of grains, can be empirically related to geologic time. An electron microscope is used for counting tracks or they are enlarged by chemical etching. Recognizing fission tracks requires an experienced eye. Population and external-detector methods for counting fission tracks are reviewed by Naeser and Naeser (1984, 1988).

Some minerals are unsuitable for fission track analysis, due to variations in track retention and U abundance (Naeser and Naeser, 1984). Volcanic glass, apatite and zircon are abundant in rhyolitic ashes and are most commonly used for dating. Dating glass is hampered by thermal annealing, or healing, of tracks which occurs at ambient temperatures, whereas, in zircon, tracks are stable up to 300°C. As a result, fission track age estimates on glass are almost always younger than those on zircon from the same deposit (Seward, 1979; Naeser *et al.*, 1980) and should be regarded as minimum limiting ages (Naeser and Naeser, 1984).

Partially annealed fission tracks may be detected by their reduced size compared to tracks induced by neutron irradiation in the laboratory. Several methods, including the mean diameter method (Storzer and Wagner, 1969) and the isothermal plateau method (Arias *et al.*, 1981), are routinely used to correct for partial annealing.

The primary advantages of fission track dating are: (1) individual mineral grains are used to detect multiple-age populations and, (2) little material is needed (only 6–12 zircon crystal grains). The primary disadvantages include: (1) annealing, (2) difficulties in counting tracks, (3) zircon is not always available, and (4) low track densities in young or low-U samples. The useful lower age range for dating Quaternary tephras by this method is about 100,000 years (Naeser and Naeser, 1984; Walter, 1989).

The Lake Tapps tephra of the Puget Lowland is a good example of how fission track age estimates have provoked significant revision of the glacial stratigraphy of a well studied region (Easterbrook *et al.*, 1988). The Lake Tapps tephra is overlain and underlain by the Salmon Springs Drift, originally thought to record early Wisconsinan glacial advances of the Cordilleran Ice Sheet based on finite isotopic-enrichment radiocarbon ages from associated peat (Stuiver *et al.*, 1978). Initial fission track ages of about 0.65 Ma on glass and 0.85 Ma on zircon (Easterbrook *et al.*, 1981) have since been superseded by an isothermal-plateau age estimate of about 1.0 Ma on glass from the same deposit (Westgate *et al.*, 1987). The Salmon Springs Drift is now known to be early Pleistocene in age, a fact that required reformation of the regional glacial sequence (Easterbrook *et al.*, 1988).

14.4.3.2. Thermoluminescence dating

In many stratigraphic settings, sediments may be either well beyond the range of radiocarbon methods or lack any other suitable material for dating. In thermoluminescence dating (TL), the time of deposition of the sediment itself is dated. Over the last decade the technique has been used to determine the numerical age of sediments from a variety of depositional environments, in many cases with promising results (Aitken, 1985; Berger, 1988a; Townsend *et al.*, 1988; Forman, 1989; Faïn *et al.*, 1992). Although most glaciogenic sediments do not lend themselves to TL dating (Forman and Ennis, 1992; Berger, 1988b), loess and other aeolian sediments related to glaciation have been widely studied (Wintle, 1990).

Sediments containing the long-lived radioactive isotopes of ^{238}U, ^{235}U, ^{232}Th and/or ^{40}K are continually bombarded by ionizing alpha, beta and gamma particles. As a result, metastable electrons are

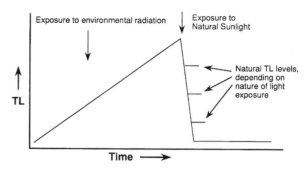

FIG. 14.11. Acquisition and reduction of the natural TL signal in sediments. The natural TL level increases with exposure to ground radiation. TL is reduced with exposure to light during aeolian or fluvial transport (i.e., zeroed). (From Forman (1990) reprinted with permission from *International Association of Sedimentologists*)

displaced and trapped in crystal-lattice defects. Consequently, the longer the mineral is irradiated, the larger the number of metastable electrons trapped in the mineral (Fig. 14.11). These trapped electrons are released either by heating or exposure to sunlight. The eviction of trapped electrons is commonly referred to as zeroing and is required for re-setting the TL 'clock' during deposition.

Thermoluminescence is the light emitted from minerals when heated under controlled conditions; the amount of light is commensurate with the number of trapped electrons, and is proportional to the radiation dose rate and geologic time. Mineral grains, typically quartz and feldspar, are essentially long-term radiation dosimeters, and the TL signal a proxy of accumulated radiation exposure (Forman, 1989). This accumulated radiation is termed the equivalent dose. Age estimates are determined by:

$$\text{TL age estimate} = \frac{\text{Equivalent dose (in Grays)}}{\text{Dose rate (in Grays/year)}} \quad (14.2)$$

In unheated sediments, such as loess or glaciofluvial sediment, the TL signal is composed of two constituents: (1) the easily removed, light-sensitive component, and (2) a residual or light-insensitive component, the baseline TL level at the time of burial. Proper determination of the equivalent dose requires that all of the light-sensitive component be bleached or zeroed. The extent of zeroing is highly dependent upon the depositional environment, which controls

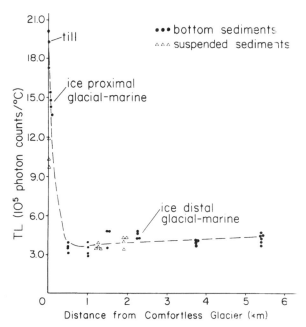

FIG. 14.12. Partial bleaching of glaciogenic sediments in the proglacial environment at Enkelsbukta, Spitsbergen. (From Forman (1990), reprinted with permission from *International Association of Sedimentologists*.) Note that ice proximal sediments are poorly zeroed and bear significant TL; zeroing increases with long transport but is not complete even 5 km from the ice front

(Wintle, 1973). Because the loss may amount to as much as 25% of the TL, an adjustment is necessary to determine an accurate equivalent dose. Laboratories now routinely preheat samples between 75 and 150°C to remove this unstable component (Forman, 1989).

The environment of deposition must be known to be certain that the sediment has been properly rezeroed. Current research with Quaternary sediments is expanding the maximum limit of TL dating beyond 200,000 years. Optically stimulated luminescence (OSL), a new technique similar to TL but employing laser light to evict electrons, may prove to be a more sensitive means of determining accurate age estimates (Huntley et al., 1985; Aitken, 1992), particularly for ice-proximal environments with high sedimentation rates. Electron spin resonance dating (ESR) is also similar to TL in that it measures essentially the same phenomena, however, the accumulation of unpaired electrons on the crystal lattice is measured by determining the number of paramagnetic centres in the mineral (Grün, 1989). This technique has been applied to Quaternary speleothems, travertine, corals, molluscs and tooth enamel, as well as volcanic rock and eolian sediments (cf. Townsend et al., 1988; Faïn et al., 1992).

14.4.4. Chemical and Biological methods

A variety of chemical and biological processes can be used to measure geologic time, especially if their rate is well constrained and stable despite variable conditions. The major drawback to such methods is that past environmental factors such as temperature, moisture and burial history are not always well known, and may have significant effects over periods of one or several glacial/interglacial cycles. Chemically and biologically derived age estimates are most accurate when the rates of these processes are calibrated using other chronological methods.

14.4.4.1. Amino acid geochronology

Amino acid geochronology is used for determining the relative age, and sometimes the calibrated-age, of fossils as old as the late Pliocene. The fundamentals of amino acid geochronology in carbonate fossils have been summarized, for example, by Von Endt

the spectral characteristics, duration of light exposure, and light intensity received by the sediment (Forman, 1989). Numerous studies have shown that while loess and similar aeolian sediments are generally well zeroed in subareal environments (Wintle, 1982, 1990; Berger, 1988a), rapidly deposited water-lain sediments such as subglacial and ice proximal glaciofluvial sediment and glaciomarine mud are only partially bleached (Fig. 14.12) (Forman, 1988, 1990; Gemmell, 1988a; Hütt and Jungner, 1992). Fine-grained sediments in glacial till are saturated with inherited TL and are not zeroed (Berger, 1984; Gemmell, 1988b; Forman, 1990; Forman and Ennis, 1992) although Lamothe (1988) suggests that new electron traps may be created in lodgement tills by communition making it possible to determine age estimates on till with TL.

Anomalous fading, a common problem in TL dating, refers to the minor loss of TL signal in sediments stored in the laboratory prior to analysis

(1979) and Bada (1985) for bones and by Wehmiller (1984, 1986), McCoy (1987a), and Goodfriend (1987) for molluscs (cf. Wehmiller, 1990; Miller and Brigham-Grette, 1989; Bada, 1991; Kaufman and Miller, 1992).

Proteins play a fundamental role in the biomineralization process (Crenshaw, 1980; Lowenstam and Weiner, 1989). Following the death of an organism, proteins preserved in trace amounts in the mineral skeleton begin to break down according to a complex series of decompostional reactions. High-molecular weight proteins are broken down into low-molecular weight polypeptides and eventually into free amino acids (Fig. 14.13). By measuring the extent to which the protein and constituent amino

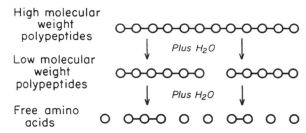

FIG. 14.13. Progressive breakdown by hydrolysis of high-molecular weight proteins into free amino acids. Hydrolysis occurs naturally over time and is routinely induced in the laboratory at high temperatures for preparation of the total hydrolysate mixture for analysis

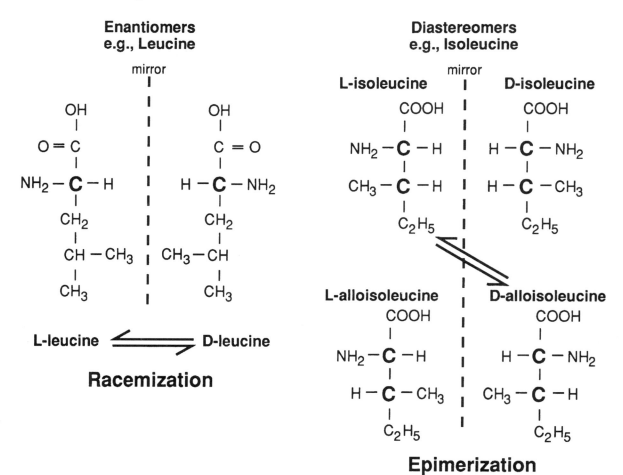

FIG. 14.14. Difference between epimerization (e.g. conversion of L-isoleucine into D-alloisoleucine, racemic at about 1.3) and racemization (e.g. conversion of L-leucine into D-leucine acid, racemic at about 1.0)

acids have degraded, the length of time since death of the organism can be estimated.

Almost all amino acids can exist in two forms, L-(levo) isomers and D-(dextro) isomers. Living organisms exclusively use L-amino acids in the construction of protein. After death L-amino acids undergo a slow interconversion to their D-forms by a reaction called 'racemization', or 'epimerization' in the case of the commonly used amino acid isoleucine (Fig. 14.14). The ratio of D/L increases through time until an equilibrium mixture of the two isomers is achieved.

D/L ratios are routinely determined on the total acid hydrolysate of a sample, that is, a mixture of both free amino acids and those bound as polypeptides. Analysis of the naturally hydrolysed free amino acid fraction provides an alternative measure of relative-age and can be used to calculate the percentage of free amino acids in the sample, an index that also increases with increasing age. Laboratory analyses are routinely done using either liquid or gas chromatography (Miller and Brigham-Grette, 1989).

A number of factors, including taxonomy, affect the rate at which racemization/epimerization occurs. The most critical factor, however, is temperature, or the integrated thermal affects on the sample since deposition (Miller and Brigham-Grette, 1989). For example, between 0 and 20 °C, the rate of isoleucine epimerization doubles for every 4°C increase in temperature. As a result, samples that have experienced large seasonal temperature variations will epimerize at a faster rate than samples that have experienced little seasonal variation at the same mean annual temperature. Depth of burial and burial history are also important, especially for samples enclosed in sediments that were intermittently covered with glacial ice or sea water that effectively insulated the samples from lower ambient air temperatures. This problem is most extreme in Arctic settings and many glaciated continental shelves (Miller and Brigham-Grette. 1989).

The uncertainty involved in estimating past temperature can be circumvented by comparing samples from a limited geographical area, where it can be assumed that samples of the same age have experienced similar post-depositional histories. This application, referred to as aminostratigraphy, allows D/L ratios to be used as an age index correlating disjunct stratigraphic sections in a region (Fig. 14.15). D/L ratios have also been used to correlate late glacial loess (Oches, 1990), and interglacial marine units (Miller and Mangerud, 1985; Wehmiller, et al., 1988; Kaufman and Brigham-Grette, 1993) across north/south latitudinal temperature gradients.

Conversion of amino acid ratios into calibrated-age estimates is done one of two ways, or using a combination of the two methods. Commonly, age estimates are determined by using D/L ratios to extrapolate from or interpolate between the ages of independently dated samples with similar temperature histories. Alternately, the affect of time and temperature on reaction rates are determined empirically by subjecting modern fossils to high-temperature heating experiments. Calibrated-age estimates are made using kinetic models and an assumed temperature history of the sample (McCoy, 1987a,b; Mitterer and Kriausakul, 1989; Wehmiller, 1990; Kaufman and Miller, 1992).

Although a variety of biogenic materials have been used for amino acid geochronology (Hare et al., 1980), molluscs, including marine bivalves and aquatic and terrestrial snails, are most commonly used for the geochronology of glaciogenic deposits. Because the rate of protein breakdown is taxon-dependent, analyses should be restricted to single genera (Miller and Brigham-Grette, 1989; Oches, 1990). Faster racemizing genera are more likely to yield better age resolution for younger deposits than slow racemizers.

Amino acid geochronology has been used on a variety of glaciogenic deposits and marine sediments interbedded with glaciogenic deposits. Applications include those made using marine molluscs from interbedded glaciomarine and marine sequences on Baffin Island (Miller, 1985), Spitsbergen (Miller et al, 1989), Ellesmere Island (Lemmen and England, 1992) and Alaska (Kaufman and Brigham-Grette, 1993). Amino acid data have contributed to arguments concerning the stability of the Laurentide Ice Sheet over Hudson Bay during the Wisconsinan (Andrews et al., 1983; Dyke, 1984; Laymon, 1991). Near the late Wisconsinan ice limit on Cape Cod, amino acids have been used in concert with radiocarbon and uranium-series age estimates to identify marine sediments of last interglacial age and to distinguish

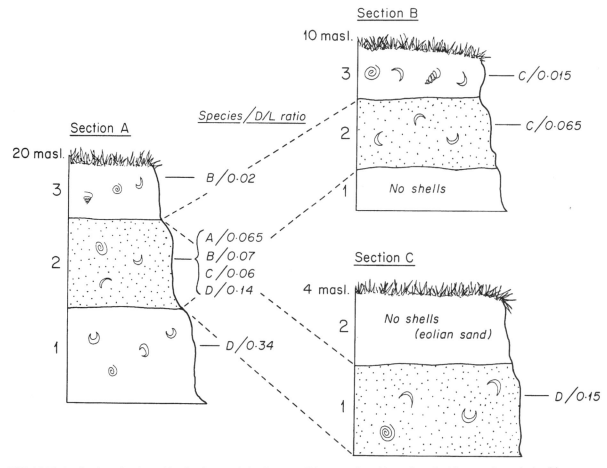

FIG. 14.15. Application of amino acid ratios for correlation between disjunct stratigraphic sections. In this example, analysis of four genera (A,B,C,D) at section A are used to correlate with sections B and C where only one genera is present. (Reprinted from Miller and Brigham-Grette (1989) with kind permission from Pergamon Press Ltd)

Wisconsinan from Illinoian till sheets (Oldale *et al.*, 1982). The subdivision and correlation of European (Miller and Mangerud, 1985) and U.S. Atlantic Coastal Plain (Wehmiller *et al.*, 1992) interglacial deposits has also been refined using amino acid geochronology.

Terrestrial snails collected from loess along the Mississippi River valley, east-central Europe and China are used in conjunction with paleosols to correlate and date disjunct exposures. Miller B.B. *et al.* (1987) first used snails to compare results from Wisconsinan and pre-Wisconsinan sites in central Indiana, demonstrating the application of the method.

Clark *et al.* (1989) later used several genera for comparing amino acid ratios from Late Pleistocene Peoria Loess and Roxanna Silt with ratios from older loess including the Chinatown Silt, Loveland Silt, Sicily Island Loess and the County Line Silt.

14.4.4.2. Obsidian hydration dating

Obsidian is a volcanic rock usually of rhyolitic composition (70% SiO_2) that is ejected and quenched so quickly that no crystalline structure develops; it is essentially volcanic glass. Its hardness and fracture properties made obsidian a popular material among early cultures because it could be easily chipped to form cutting tools and projectile points. In glaciated volcanic regions, obsidian within glacial deposits can be used for relative-dating of glacial events.

A fresh surface of obsidian will absorb water from its surroundings and form a measurable hydration layer, or rind, as it diffuses into the obsidian. Fresh volcanic glass typically contains 0.1 to 0.3 % water by weight. With time, water is absorbed at the surface to form perlite that contains about 3.5% water. This increase in water content causes an increase in density of the outer surface and an increase in the index of refraction that can be easily seen in thin section. If the growth rate of the hydration layer can be established, then the thickness of the layer can be used as an index of calibrated-age. The accuracy of obsidian hydration age estimates depends on the constancy of the hydration rate. The most important factors affecting this rate include the post-depositional temperature history and the composition of the obsidian (Friedman and Long, 1976; Friedman and Obradovich, 1981).

Obsidian-hydration dating of glacial events requires fresh surfaces such as cracks and fractures to have been mechanically created by abrasion and communion so that new hydration layers can form. A classic example of obsidian hydration dating in glaciated terrain is that of Pierce et al. (1976) who used differences in hydration-rind thickness calibrated against K/Ar dated lava flows to subdivide Pinedale and Bull Lake moraines in the Yellowstone region (Fig. 14.16). Obsidian from the 114,000 years West Yellowstone rhyolite flow incorporated in Pinedale-age moraines had hydration rinds 7.2 ± 0.6 µm thick whereas, obsidian eroded from the 179,000 year Cougar Creek rhyolite flow and deposited in Bull Lake-age moraines had rinds measuring 14.2 ± 0.6 µm. This landmark research was critical in reassigning the classic early Wisconsinan Bull Lake moraines to the late Illinoian and the Pinedale moraines to the entire Wisconsinan interval.

14.4.4.3. Lichenometry

Lichens are a symbiotic association of a fungus and green algae; the fungus provides a protective home for the algae who, in turn, supply carbohydrates to the fungus. Although lichens occur in a variety of

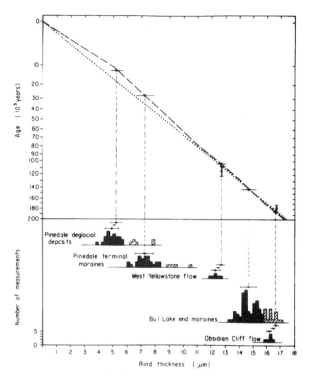

FIG. 14.16. Rate of obsidian-hydration rind growth on West Yellowstone moraines, Montana. (Reprinted from Pierce et al. (1976), with kind permission of the authors). The dotted line gives an average rate curve based on data from in situ flows. The dashed line compensates for differences in hydration rate caused by climatic effects

morphotypes, crustose lichens, most commonly used for geochronology, grow in a radial fashion firmly attached to rocks. The fundamental basis of lichenometry is that the diameter of an individual lichen thalli can be used as an index of the calibrated-age of the landform on which it grows (Plate 14.2).

First proposed by Beschel, (1950, 1961), lichenometry has since been most widely used in alpine and arctic environments for determining the calibrated-age of glaciogenic deposits such as stabilized moraines, outwash surfaces and rock glaciers (e.g. Birkeland, 1973; Luckman and Osborn, 1979; Calkin and Ellis, 1980; Porter, 1981b; Haworth et al., 1986). Other applications include dating glacial trimlines, rockfalls and the extent of former permanent snow banks (Locke and Locke, 1977; Locke et al., 1979; Bradley, 1985; Innes, 1985).

PLATE 14.2. Direct measurement of lichen thalli diameter. *Rhizocarpon geographicum* measured parallel to the longest axis

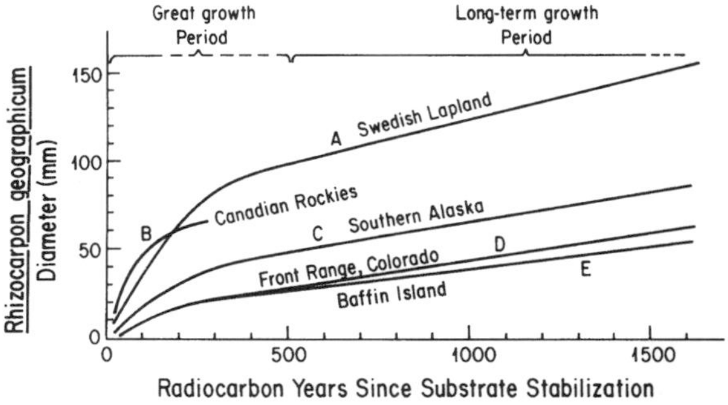

FIG. 14.18. Lichen growth curve from different regions illustrating the effect of climate on the rate of growth of *Rhiocarpon geographicum*. (Curves taken from sources listed in Calkin and Ellis (1980). Reproduced with permission from *Arctic and Alpine Research*, Regents of the University of Colorado

The technique is based on the assumption that the largest lichen is the most optimally growing lichen and provides the closest minimum age estimate of the substrate. Smaller thalli must be later colonizers or growing more slowly. Although there has been some controversy over which diameter of a lichen thalli to measure (cf. Locke *et al.*, 1979), most workers use either the longest axis or the diameter of the largest inscribed circle (Innes, 1986a) (Fig. 14.17). The percentage of total lichen cover on rock surfaces has also been used as a complementary index of relative

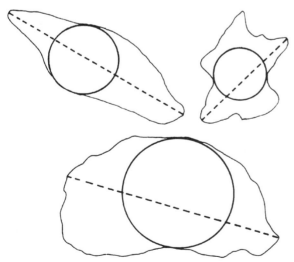

FIG. 14.17. Other workers prefer to measure the largest inscribed circle. (From Innes (1985), reprinted with permission of *Progress in Physical Geography*)

age with debatable success (Innes, 1986b; Werner, 1990).

The use of lichenometry as a geochronologic tool is dependent on establishing the local growth curve for each species (Fig. 14.18). A growth curve is established by: (1) measuring the size of lichens on substrates of known age (buildings, gravestones, monuments), and (2) gathering radiocarbon age estimates on materials associated with the landform. Microclimatic factors can complicate the development of even a simple growth curve (Bradley, 1985).

The useful age range of the method is restricted by the lifespan of the lichen thallus, the stability of the substrate, and competition among neighbouring lichens. In general, lichen grow more slowly in regions of low temperature, short growing season, and low precipitation (Bradley, 1985). In the Colorado Front Range, for example, *Rhizocarpon geographicum* s.l. may only be used to estimate ages of deposits less than about 3000 years old (Benedict, 1967). On Baffin Island, however, this same species has been used to date surfaces over 9000 years old (Miller and Andrews, 1973).

14.4.4.4. Cation-ratio ages of rock varnish

In arid regions, thin manganese-rich coatings commonly accumulate on stable rock surfaces as a result of bacterial action and the accretion of airborne detritus (Elvidge and Collet, 1981; Dorn and Ober-lander, 1981, 1982). The gradual accumulation of 'rock varnish' has long been used as a relative dating method, particularly for estimating the age of desert archaeological materials. Dorn and Oberlander (1981) and Dorn (1983) proposed the use of cation-ratios from rock varnish as a means of estimating the calibrated-age of geomorphic surfaces. More recently Dorn et al., (1989) have suggested that organic matter from varnish could be dated by AMS radiocarbon techniques.

Although of limited use in most glaciated regions, cation-ratio dating has been used to make age estimates of moraines in the Sierra Nevada Mountains (Dorn et al. 1987b) and may be employed on glaciated surfaces in other semi-arid regions. The method is based on the gradual leaching of mino-element cations (i.e. Na, Mg, K and Ca) from the varnish surface relative to less mobile cations (i.e. Ti (Dorn, 1983). The progressive decrease in the ratio of these elements, e.g. (Ca + K)/Ti, is an index of the duration of cation leaching, and hence geologic age. The rate of leaching must be locally calibrated with other geochronological methods to establish a 'cation-leaching curve'.

Cation-ratio dating has been the subject of controversy. Many of the basic assumptions behind the accumulation of varnish and the use of cation ratios have been questioned raising issues concerning both the accuracy and precision of varnish dating (Bierman and Gillespie, 1991; Bierman et al. 1991; Reneau and Raymond, 1991; Reneau et al., 1991). Cation-ratio and AMS dating of desert varnish from geomorphic surfaces clearly has potential and has been shown to provide reproducible results, but remains experimental.

14.4.5. Geomorphic methods

Once a glaciogenic deposit and/or landform has been produced, a number of processes, including chemical and physical weathering, mass movement and fluvial dissection, act on that landform to modify its original shape. Quaternary scientists have long used the progressive weathering and degradation of a landform as an index of relative-age, and in many cases use this index as a qualitative means of correlating deposits over broad regions. This family of methods is based, in part, on the premise that weathering parameters are time dependent and can be used to distinguish episodes of deposition (Black-welder, 1931; Burke and Birkeland, 1979). An inherent problem in most of these semi-quantitative methods is that local differences in microclimate, parent material, vegetation and relief all affect the rate of weathering (cf. Jenny, 1941). Moreover, many rates of weathering are known to be non-linear with modification rates decreasing within creasing age (Colman, 1982). Typically, a variety of relative-weathering methods are combined with classic mor-phostratigraphy, i.e. geomorphic position, to construct a meaningful chronosequence (Burke and Birkeland, 1979).

14.4.5.1. Soil-profile development

Physical and chemical weathering of land surfaces results in a weathering profile of variable depth, depending upon the duration of exposure, micro-climate, parent material, vegetation and relief (Jenny, 1941). Horizonation of weathering profiles as a result of further chemical and mechanical weathering, the translocation of clays, the influx of eolian material and the buildup of carbonates or oxides, creates a 'soil profile' that is superposed on the parent material. The degree of weathering and horizonation is used as a measure of relative-age and for regional correlation, assuming that factors such as parent material, climate, vegetation, etc., are held constant. The fundamental principles of soil stratigraphy for estimating the age of glaciogenic and other geomorphic surfaces are formally recognized as 'pedostratigraphy', an innovative modification of the North American Stratigraphic Code (NACSN, 1983, p. 864).

Details concerning pedogenic processes, the recognition of paleosols and examples of their application are discussed in this volume (Chapter 10) and elsewhere (Birkeland, 1984; Boardman, 1985b; Catt, 1990). As time-parallel or often diachronous stratigraphic marker horizons, the usefulness of soils and paleosols for relative-age estimates and as a tool for regional correlation can not be overstated.

14.4.5.2. Rock and mineral weathering

Rock weathering rates have commonly been used to estimate the relative-age of geomorphic surfaces, including moraines and outwash terraces. Such data have also been used to estimate late Quaternary recurrence intervals of various natural hazards, such as floods and landslides, where other geochronologic control is not available (Colman, 1981).

A variety of relative-weathering parameters are commonly measured on surface boulders of glacial landforms, some more subjective than others. Building upon the early work of Blackwelder (1931) and many others in the mid-1960s, Burke and Birkeland (1979) summarized the utility of parameters including:
(1) the ratio of fresh to weathered boulders,

(2) the ratio of pitted to non-pitted boulders,
(3) weathering pit depth on boulder surfaces,
(4) the maximum height of resistant mafic inclusions on boulder surfaces,
(5) the ratio of fine-grained igneous rocks with and without weathering rinds,
(6) average rind thickness on fine-grained igneous rocks (basalts, andesites, etc.),
(7) hammer-blow weathering ratio based on a ratio of fresh (ringing)/weathered (dull thud)/grusified (disintegrates on impact) boulder counts,
(8) surface boulder frequency in a specified area,
(9) the ratio of granitic boulders to non-granitic boulders (only regional application),
(10) the ratio of split to non-split boulders,
(11) the ratio of oxidized to partially oxidized boulder surfaces.

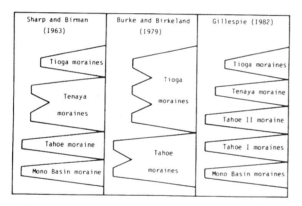

A   End moraines in Bloody Canyon (Walker Creek Valley)

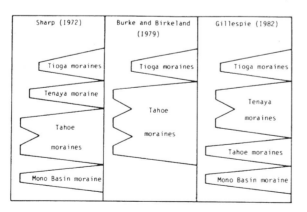

B   End moraines in Green Creek Valley

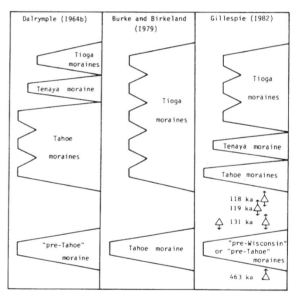

C   End moraines in Sawmill Canyon

FIG. 14.19. Different interpretations of the moraine chronosequence for the eastern Sierra Nevada, California, based on relative-age methods (soils, landforms, rock weathering). (Reprinted from Fullerton (1986a), with kind permission from Pergamon Press Ltd)

Most of these parameters require selecting 50 boulders at each site, then methodical ranking or measuring the relative-weathering characteristics of each. To standardize classifications, the indices are specifically defined. For example, a boulder may be considered 'weathered' if >50% of its exposed surface exhibits single-grain mineral relief (Burke and Birkeland, 1979, p. 24). A surface is 'pitted' if it has >1 concave depression caused by mineral disintegration.

Probably the most common criticism of the relative-weathering methods is that they are subjective and easily biased by the researcher, so that measurements may not be reproducible by other workers. The issue underlies differences of opinion concerning the number and relative-age of glacial advances recorded by moraines preserved in the Sierra Nevada (Fullerton, 1986a) (Fig. 14.19).

Weathering-rind thickness on fine-grained igneous rocks taken from the subsurface is one of the most objective, widely used relative-weathering parameters. Rinds are formed by chemical alteration of a rock surface. Their thickness is measured from the rock surface to the weathering front. Maximum rind thickness, or the mean of two or three measurements on the same rock sample is measured on 25–50 rock samples of uniform lithology collected from a restricted area at or near the ground surface . Variations in rind thickness are attributed to subtle textural and/or compositional variations, microclimatic effects, or reworking of previously weathered clasts (Colman,1981, 1982).

Colman and Pierce (1981, 1984, 1992) compiled rind thickness data from 8 areas of the western United States (Fig. 14.20). In some areas, the rate of rind development was calibrated using independent age control. The non-linear decrease in the rate of rind development (Colman, 1981) indicates that the age resolution of this technique decreases with time and that, in the western US, may be applicable to deposits less than about 250,000 years.

14.4.5.3. Landform modification

Reconstructing the extent of former glaciers relies on proper recognition of penecontemporaneous suites of ice-marginal landforms. While young moraines can be

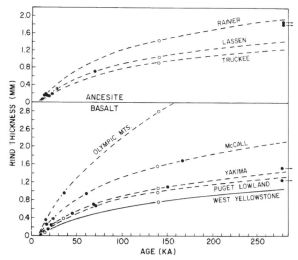

FIG. 14.20 Weathering-rind thickness plotted against time from eight western US study sites. Open circles represent calibration points from West Yellowstone, Montana. (Reproduced from Colman and Pierce (1992), with kind permission of the authors)

readily mapped either on the ground or from aerial photographs, ancient glacial landforms are commonly difficult to identify. With time, erosional processes of slope wash, mass movement, and fluvial dissection conspire to render well-defined landforms into subdued, more amorphous forms (Chamberlin, 1878; Davis, 1899; Blackwelder, 1931). The rate of modification depends upon a variety of factors including, climate, lithology, aspect and relief.

Progressive landform modification of glaciated landforms, especially terminal moraines, has long been used as an index of relative-age (Ruhe, 1950; Sharp, 1969; Pierce, 1979; Colman and Pierce, 1986; Hamilton, 1986b). Traditional field measurements include ice-proximal and ice-distal slope angles, and crest width. The underlying assumption is that all moraines have the same initial form, characterized by 'youthful', sharp-crested, steep-sloped, high-frequency relief. Because moraine morphology is highly variable (Chapters 2 and 3; Menzies, 1995a, Chapter 11) 'youthful appearance' may also be vary from one study area to another.

Recent studies have attempted to quantify moraine morphology using more elaborate criteria and employing statistical tests (Dowdeswell and Morris, 1983; Kaufman and Calkin, 1988; Peck et al., 1990). In their study of a four-fold chronosequence of middle and late Pleistocene moraines in western Alaska, Kaufman and Calkin (1988) found that the mean slope angle of axial-moraine profiles, moraine-crest width, maximum slope along and normal to the moraine crest, and the slope angle of the upper 20 m segment of the distal slope, were the best parameters for distinguishing moraines of different age (Fig. 14.21).

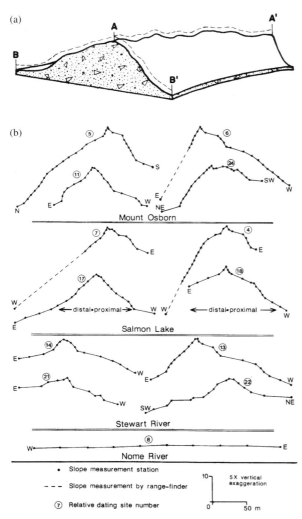

FIG. 14.21. Illustration of cross-sectional moraine profiles used to differentiate moraines on Seward Peninsula, Alaska (from Kaufman and Calkin, 1988). (a) The line from A-A' is the axial crest; B-B' is the cross profile. (b) Comparison of cross profiles for moraines of increasing age. (Reproduced with permission from *Arctic and Alpine Research*, Regents of the University of Colorado)

14.4.5.4. Geomorphic/stratigraphic position

Relative weathering criteria discussed above give information of, not only which deposits are older or younger, but also how much older one is compared to the other, at least in the broadest sense. In addition to relative weathering criteria, the position of moraines in a nested sequence has long been used to determine the relative sequence of events (e.g. Leverett and Taylor, 1915). The oldest moraines are usually the most extensive because they have not been overridden and obliterated by younger glacial advances (Gibbons *et al.*, 1984). This type of 'morphostratigraphy', although not recognized as a formal stratigraphic term, indicates the order of events but is not useful in assigning age estimates. In some cases (e.g. Hamilton, 1986b; Kaufman and Hopkins, 1986; Huston *et al.*, 1990) relative weathering characteristics in conjunction with relative differences in glacial extent, have been used to correlate glacial sequences across wide areas (see Chapter 8).

14.4.6. Correlation methods

Correlation methods are those used to assign age estimates to deposits based upon some measurable physical parameter, or set of parameters, that are compared with the same parameters on deposits dated by independent means. These methods do not provide direct numerical-ages, rather they provide correlation-ages (Table 14.2). Correlation methods include the direct comparison of local time-series data with some form of independently dated regional or global master curve. In this context, 'curve matching' is used to correlate floating sequences (i.e. a sequence with no age control) with more complete, well-dated sequences. This is especially useful for regional tele-correlations and organizing a stratigraphic sequence from a complex series of spatially disjunct field exposures.

14.4.6.1. Stable isotopes

The most widely utilized set of correlation-age techniques in Quaternary studies are stable isotopes, primarily oxygen-18 and carbon-13 (Jansen, 1989); their application has clearly revolutionized the understanding of glacial/interglacial cycles and focused attention on the intimate links between atmospheric and oceanic circulation. Since the early work of Emiliani (1955), Shackleton and Opdyke (1973) and Shackleton (1977), stable isotopes in benthic and planktonic foraminifera are now routinely used for correlating deep-sea sediments all over the world (Imbrie et al., 1984). Likewise, oxygen-18 and deuterium (^{2}H) are routinely used for correlating glacial ice cores (Dansgaard and Oeschger, 1989; Lorius et al., 1989). The down-core isotopic record in sediments or ice is, of course, a climatic proxy without any intrinsic age information. Rather, the chronology of ^{18}O and ^{13}C time-series data are established via independent techniques, such as radiocarbon dating, paleomagnetic stratigraphy, and correlation with Th/U-aged coral terrace sequences in deep sea cores and by counting annual layers and tephrachronology in ice cores. Consequently, changes in the paleomagnetic time scale or decay constants for ^{234}U and ^{14}C have a direct impact on the chronology of stable isotope records. Fine 'tuning' of stable isotope stratigraphies to Milankovitch orbital parameters has also been performed to synchronize global times scales (Imbrie et al., 1984; Shackleton et al., 1990). The widespread use of stable isotopes, particularly oxygen-isotope time series for land/sea tele-correlations, is discussed in Menzies, 1995a, Chapter 2.

14.4.6.2. Tephrachronology

'Tephra' is a collective term for a variety of airborne pyroclastic materials derived from volcanic eruptions (Thorarinsson, 1981). Because highly explosive eruptions can loft fine-grained materials high into the atmosphere, where they are transported long distances by wind, tephra layers form widespread, synchronous stratigraphic markers (Sarna-Wojcicki and Davis, 1991; Sarna-Wojcicki et al., 1991). Sarna-Wojcicki and Davis (1991) characterize tephrachronology as a marriage of two subdisciplines, namely tephrachronometry and tephrastratigraphy. Tephrachronometry refers to numerical-age estimates on tephras based on either Ar/Ar, K/Ar, or fission-track analyses on the tephra itself, or other age estimates (e.g. radiocarbon) on the enclosing sediments. Tephrastratigraphy

implies the use of tephra chemistry, morphology and stratigraphic position for correlation and as an index of relative-age. Because aspects of tephrachronometry are discussed elsewhere in this chapter, tephrastratigraphy will be stressed here. World-wide applications of tephrachronology were recently compiled by Westgate et al. (1992).

The fundamental assumption in tephrastratigraphy is that individual tephras have unique chemical characteristics and shard morphologies that allow them to be 'fingerprinted' and correlated over wide areas. These same characteristics can also be used to trace multiple tephra layers of vastly different age back to common source areas (e.g. Sarna-Wojcicki et al., 1984; Preece et al., 1992). The most common parameters used for characterizing and matching tephras include: (1) the index of refraction of glass shards, (2) phenocryst characteristics (crystal type, abundance, colour), (3) bulk chemical composition, (4) trace-element composition, (5) shard morphology, (6) mineral assemblage, and (7) field characteristics (colour, weathering, thickness, etc.) (Westgate and Gorton, 1981; Sarna-Wojcicki and Davis, 1991). These parameters constitute a chemical and physical signature for an individual tephra that is added to the tephra-stratigraphic data bank (Fig. 14.22). When a new tephra is located, it too can be similarly fingerprinted for comparison with the reference collection. A quantitative measure of similarity is used to determine the best match between the newly discovered tephra and the reference samples. If a satisfactory match is found, the tephra is used for correlating deposits at the two sites. If the reference tephra has been dated by independent means, then the match is used for correlation-age dating of the new tephra site.

The glacial/interglacial stratigraphies over much of western and northwestern North America (downwind of the Pacific rim) (Porter, 1981d, 1991), the North Atlantic, Norwegian/Greenland Sea (Sigurdsson and Loebner, 1981), and portions of northwest Europe (downwind of Iceland) (Mangerud et al., 1984) are reinforced by well-defined tephrastratigraphy. Tephras are also routinely found in ice cores where they provide important numerical-age control and reference horizons for flow model calibration, comparison with detailed visual counts of annual layers,

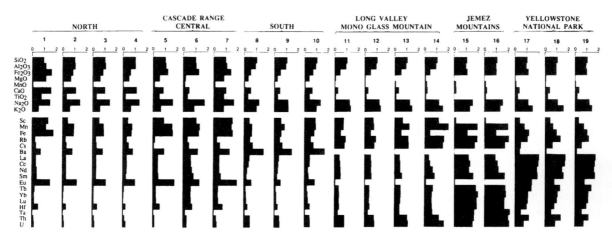

FIG. 14.22. Chemical composition of volcanic glasses from the four major Pliocene and Pleistocene source areas in the conterminous USA as listed below. Values are oxides and elements expressed as a ratio of the concentration of the same oxides and elements in an artificial standard. (Reproduced from Sarna-Wojcicki and Davis (1991) with kind permission of the authors)

Cascade Range (Washington, Oregon, and northeast California)
1. Mount St. Helens, May 18, 1980 layer, Washington.
2. Mount St. Helens Ye layer, Washington, about 3,400 B.P.
3. Mount St. Helens Sg layer, Washington, about 13,000 B.P.
4. Mount St. Helens Cw layer, Washington, about 35,000 B.P.
   5. Rio Dell ash bed of Sarna-Wojcicki *et al.*(1982), specific source unknown; about 1.45 Ma.
6. Loleta ash bed of Sarna-Wojcicki et al.(1982) (=Bend Pumice of Taylor, 1981; Bend Oregon); Humbolt County, California; about 0.3 to 0.4 Ma.
7. Mazama ash bed, Crater lake, Oregon; about 6,850 B.P.
8. Rockland ash bed, Lassen Peak area, California; about 0.4 Ma.
9. Ishi Tuff Member (of the Tuscan Formation), southern Cascade Range, California; about 2.5 Ma.
10. Nomlaki Tuff Member (of the Tehama and Tuscan Formations), southern Cascade Range, California; about 3.4 Ma.

Long Valley-Mono Glass Mountain area (east-central California)
11. Bishop ash bed: about 0.74 Ma.
12. Glass Mountain D ash bed; about 0.9 Ma.
13. Bailey ash bed of Izett (1981); about 1.2 Ma.
14. Tuff of Taylor Canyon; about 2.0 to 2.1 Ma.

Jemez Mountains (New Mexico)
15. Tsankawi pumice bed; about 1.12 Ma.
16. Guaje pumice bed; about 1.45 Ma.

Yellowstone National Park area (Wyoming and Idaho)
17. Huckleberry Ridge ash bed; about 1.98 Ma.
18. Mesa Falls ash bed; about 1.27 Ma.
19. Lava Creek B ash bed; about 0.62 Ma.

and a host of downcore isotopic time-series (Hammer *et al.*, 1980; Hammer, 1989).

A good example of the utility of tephrachronology for refining a glacial chronology is demonstrated by work in the North Cascade Range, Washington (Beget, 1984). In the region of Glacier Peak, a volcano active during Late Pleistocene and Holocene time, Beget used the presence or absence of tephra layers to provide limiting age estimates for four glacial advances. The presence of several late Pleistocene tephras overlying late Wisconsinan moraines suggested that the main valleys were deglaciated by 11,250 BP. Subsequent Holocene advances were bracketed by younger Glacier Peak tephras and distal tephras from Mount St. Helens (3400 BP) and Mount Mazama (~6900 BP) (Fig. 14.23).

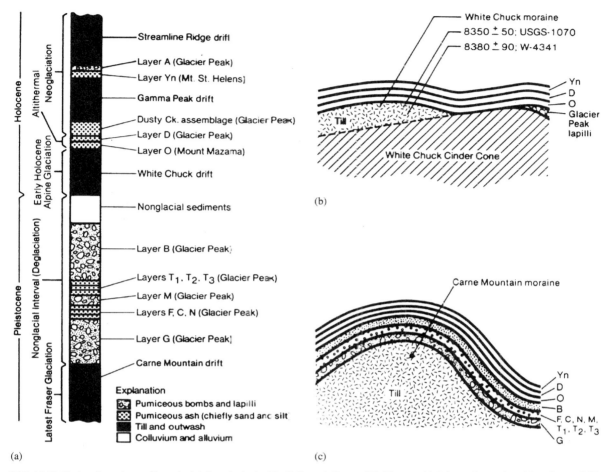

FIG. 14.23. Tephrachronology of late glacial deposits in the North Cascade Range, Washington. (a) Schematic stratigraphic column of tills and interbedded tephra. (b) and (c) Demonstration of the overlap of tephra series over Late Wisconsinan and Holocene tills. (Reproduced from Beget (1984) with kind permission of *Quaternary Research*, University of Washington)

14.4.6.3. Paleomagnetism

Late Cenozoic variations in Earth's magnetic field, as recorded in volcanic rocks and sediments, provide an extremely effective means of determining correlation-ages. Moreover, they can be used as a means of correlating isolated stratigraphic sections over a broad region. Unlike geochronological methods that are based on some time-dependent process and therefore provide numerical or calibrated-age estimates, paleomagnetic measurements indicate only the orientation of the Earth's magnetic field. These data have no inherent temporal information and must be compared

to independently-dated master curves (cf. Stupavsky and Gravenor, 1984; Thompson and Oldfield, 1986; Easterbrook, 1988b; Løvlie, 1989).

Rocks and sediments acquire magnetism in a number of ways. While volcanic rocks acquire magnetization as they cool through the Curie temperature (thermal remanent magnetization, TRM), sediments acquire magnetization by mechanical orientation of magnetic minerals parallel to the Earth's field as they are either: (1) settled from suspension in wind or water (depositional remanent magnetization, DRM), or (2) de-watered and compacted when porewaters allow the reorientation of magnetic minerals parallel to the Earth's field (post-depositional remanent magnetization, p-DRM). Post-depositional chemical, thermal, or viscous magnetization (natural remanent magnetization, NRM) may overprint the primary magnetization. However, these secondary components are routinely removed or minimized in the laboratory with thermal or alternating-field demagnetization (Thompson and Oldfield, 1986).

The Earth's magnetic field varies in intensity and 3-D orientation over several timescales. The parameters traditionally measured include: declination (horizontal departure from true north), inclination (dip

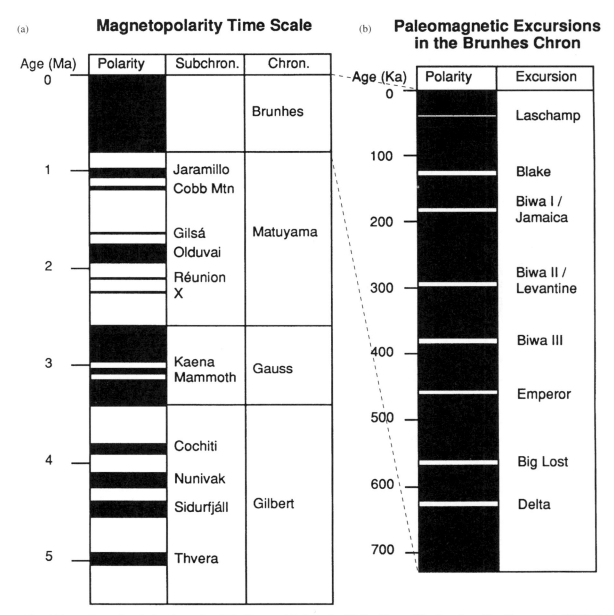

FIG. 14.24. (a) Magnetic polarity timescale of Mankinen and Dalyrmple (1979) with modifications from Shackleton *et al.* (1990). (b) Major excursions within the Bruhnes Chron (adapted from Løvlie, 1989)

from the horizontal), and field intensity. Periods of normal (+ = south or present-day) and reversed (– = north) polarity alternate about every 10^6 years and are known as polarity epochs (or chrons). These complete inversions of the dipole field are recorded globally in volcanic rocks and sediments of a variety of environments. Normal or reversed intervals of shorter duration within epochs are formally recognized as polarity events (or subchron, complete reversals lasting 10^4 to 10^5 years) or polarity excursions (also subchron, but incomplete reversals lasting 10^2 to 10^3 years). The chronology of polarity changes has been continually developed and revised since the early 1960s (Cox, 1969) but is now reasonably well documented (Mankinen and Dalrymple, 1979) (Fig. 14.24). Recent work suggests that many polarity boundaries may be 5–7% older then previously recognized (Champion et al., 1988; Shackleton et al.,

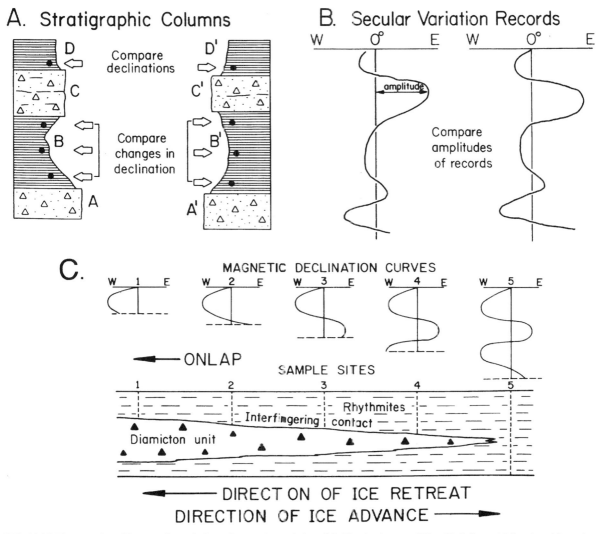

FIG. 14.25. Time-stratigraphic tests of correlation using secular variation of declination in central New York State. (a) Stratigraphic setting and lithostratigraphic correlation between sections. (b) Comparison of records from two different sites. (c) Schematic illustration showing the onlap of lacustrine sediments over glacial diamicton and progressively younger secular variation records to the left. (Reproduced from Ridge et al., 1990 with kind permission of the authors)

1990; Cande and Kent, 1992; Spell and McDougall, 1992).

Polarity changes are best recorded in continuous depositional sequences with a high sedimentation rate, in which short-duration polarity excursions may be recognized. Loess, marine, lacustrine and distal glaciomarine sediments offer the best opportunity for long records in quasi-continuous depositional settings (e.g. Kukla, 1975; Heller *et al.*, 1987; Rolph *et al.*, 1989; Raymo *et al.*, 1989). In practice, however, most analyses are made on discontinuous, complex depositional sequences with nominal data indicating either normal or reversed polarity. Such data require age-estimates by independent means to narrow the possible correlations with the master curve. If a unit is normally magnetized, for example, it says little about its age, unless there is other evidence to indicate that it is older than 780,000 years (the base of the Bruhnes Normal Polarity Chron) but younger than the Pliocene (1.6 Ma). In this case, the sediments were likely deposited during either the Jaramillo subchron (0.99 to 1.07 Ma) or the Olduvai subchron (between 1.77 and 1.95 Ma). Likewise, a reversely magnetized unit is probably older than 780,000 years.

An example of the use of paleomagnetism in a glaciogenic sequence is the study of Easterbrook and Boellstorff (1984) whose work reinforced the revised stratigraphy andof the classical pre-Nebraskan, Nebraskan and Kansan tills (Boellstorff, 1978) (Fig. 14.25). The type-Nebraskan till was found to be magnetically reversed and younger than the 1.27 Ma Pearlette S tephra (now Mesa Falls Ash bed) (Sarna-Wojcicki and Davis, 1991). The type-Kansan till yielded normal polarity and lies stratigraphically above the 0.62 Ma Pearlette O tephra (now Lava Creek B ash bed). Tills from other sections thought to be correlative with the type Nebraskan were found to be normally magnetized, hence, not correlative with the type section. Most importantly, at least two tills (Elk Creek Tills) were found to underlie a 2.2 Ma tephra (Ash at Ashton). This major reinterpretation of the type-sections led to the realization that the oldest tills in the mid-continent region are at least 1 million years older than the original type-Nebraskan section (Boellstorff, 1978; Hallberg, 1986).

Fluctuations in the weaker non-dipole field cause perturbations in both declination and inclination known as secular variations (Banerjee, 1983; Verosub, 1988). These low-amplitude, short-duration (500–3,000 years) variations are recorded across regions of subcontinental scale (Stupavsky and Gravenor, 1984). Secular variation provides greater age resolution than long-term dipole field variations and is a powerful stratigraphic tool for correlating isolated exposures. Ridge *et al.* (1990) used secular variation of magnetic declination in disjunct exposures of Late Wisconsinan glaciolacustrine varves to reconstruct the complex history of deglaciation, readvance, and retreat within the Mohawk Valley of central New York State (Fig. 14.26). Their work demonstrates the high resolution, stratigraphic utility of secular variation and its potential for temporal correlations with other glaciolacustrine sequences.

AGE	STRATIGRAPHY		POLARITY
	Till A1	Type "Kansan" Till	(Normal)
0.62 Ma		Lava Creek B Ash	
	Till A2	Cedar Bluffs Till Clarkson Till	(Normal)
	Till A3	Nickerson Till Santee Till	(Normal)
B/M Boundary			
	Till A4	Type "Nebraskan" Till	(Reversed)
	Till B	Unnamed tills	(Reversed)
1.27 Ma		Mesa Falls ash	
1.98 Ma		Huckleberry Ridge ash	
2.20 Ma		Ash at Ashton	
	Till C1	Elk Creek Till (upper)	(mostly Reversed)
	Till C2	Elk Creek Till (lower)	

FIG. 14.26. Paleomagnetic polarity data and fission-track ages for interbedded tills and tephra in portions of Nebraska, Iowa, and South Dakota where the type-Nebraskan and type-Kansan tills are found. B/M boundary refers to the Brunhes/Matuyama polarity boundary at about 0.78 Ma. Data compiled from Boellstorff (1978), Easterbrook and Boellstorff (1984), and Hallberg (1986)

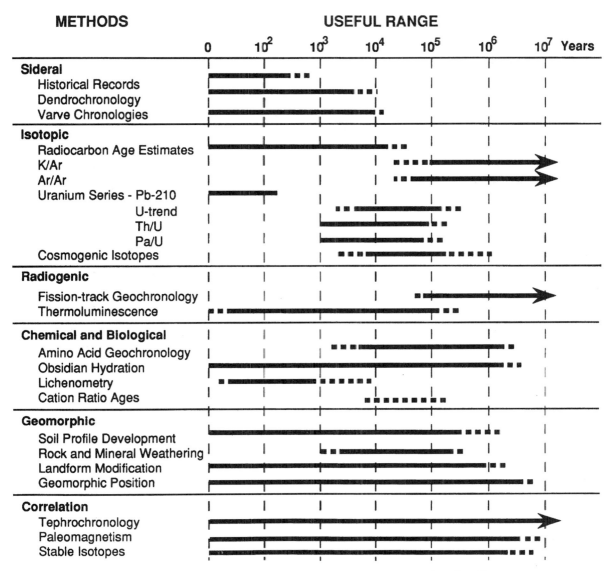

FIG. 14.27. Useful age range of geochronological methods applied to glaciogenic sediments

In recent years, increasing use of magnetic susceptibility as a tool for local and regional correlations has been adopted. Magnetic susceptibility is a measure of the ease with which a material can be magnetized (Thompson and Oldfield, 1986). This parameter is independent of changes in the Earth's magnetic field; rather it is dependent upon the concentration and grain-size of magnetic minerals in the sediment. Bulk magnetic susceptibility is commonly used for determining correlation-ages in loess, glaciomarine and lacustrine sediments. In contrast, the anisotropy of magnetic susceptibility is a measure of the preferred orientation or fabric of larger magnetic minerals in a sample (Easterbrook, 1988b). Preferred orientation in sediments generally results in two types of fabric: (1) a planar fabric caused by random settling through a water column on a bedding plane, and/or (2) a linear fabric caused by effective shear or pressure against the bed (e.g. lodgement tills, Stupavsky *et al.*, 1974; Stupavsky and Gravenor, 1975; Easterbrook, 1988b).

The origin of magnetic susceptibility fluctuations in loess has been a source of controversy. In the long 2.4 Ma old loess record of China, paleosols consistently yield high susceptibilities compared to unweathered loess. The high magnetic susceptibility is attributed to either *in situ* biogenic or pedologic enrichment of magnetite (Maher and Taylor, 1988; Maher and Thompson, 1991), or to the constant flux of magnetite that is otherwise diluted by non-magnetic loess during glacial conditions (Kukla *et al.*, 1990). In contrast, Beget *et al.* (1990) found low susceptibilities in paleosols and high susceptibilities in unweathered loess in central Alaska. They concluded that, in that region, greater wind intensity during glacial episodes increases the concentration of coarse-grained detrital magnetite resulting in higher susceptibility. Together, the studies demonstrate that the interpretation of changes in magnetic susceptibility at any site must be carefully evaluated.

14.5. SUMMARY

The emphasis of this chapter has been geochronological methods that can be directly applied to glaciogenic sediment. A variety of available methods has been discussed that allow numerical-, relative-, calibrated-, or correlation-ages to be determined for many deposits (Fig. 14.27). Clearly, no one method is applicable in all geologic materials and age ranges. An understanding of the potential and limitations of all techniques is needed to identify the best geochronologic approach. Given the assumptions and uncertainties in all geochronological methods, the best strategy is to apply a variety of techniques to assess the age of a deposit. Routinely in Quaternary studies, when multiple methods converge on a single age estimate, some level of confidence in the accuracy of that age estimate can be assumed. This assumption, however, could be erroneous. A cautious interpretation of the data is required before a final age interpretation is made. The uncertainties involved with most Quaternary geochronologic techniques will no doubt provide the impetus for advances in the future.

Chapter 15

DRIFT EXPLORATION

W.W. Shilts

15.1. INTRODUCTION

The composition of glacial sediments has long been of interest in wholly or partially glaciated terrains with bedrock geology favourable for mineralization. In Fennoscandia, tracing glacially displaced mineralized boulders back to their sources has been practised intensely in the twentieth century but is based on observations of boulder transport that predate the general acceptance of glacial theory by a century and a half (Tilas, 1740). In the latter part of the twentieth century, the mapping and interpretation of glacially dispersed mineralized debris increasingly has involved the application of new analytical technologies to measuring the compositions of the finest fractions of glacial sediments and of the proglacial sediments derived from them (Shilts, 1976). As a consequence of the applications of these new technologies, explorationists now rely heavily on geochemical analyses to provide objective data on patterns of dispersal of glacially eroded detritus from bedrock sources of possible economic interest. In addition to or in place of traditional geochemical analyses of sieved fine fractions of glacial sediments, increasingly sophisticated techniques are now used to concentrate clay-sized minerals or coarser minerals with high specific gravities. Separates produced by various concentrating techniques are analyzed geochemically and examined microscopically or by microprobe or scanning electron microscope with backscattering or energy dispersive capabilities (Peuraniemi, 1987; Saarnisto and Tamminen, 1987; DiLabio, 1990).

The purposes of this chapter are: (1) to discuss some of the sedimentological and diagenetic processes that are unique to the glacial and postglacial environment, and (2) to demonstrate present concepts on how to interpret the compositional signals commonly derived from drift. Only after mastering fundamental concepts behind these two aspects of glacial deposition may the explorationist be able to interpret geochemical data with confidence and efficiency.

15.2. SOME CONSTRAINING PRINCIPLES AND CONCEPTS

In most of Canada, northern Europe, north-central and northeastern United States, Antarctica, and in glaciated terrains elsewhere in the world, the chemical and physical signature of bedrock on soils and surficial sediments has been distorted by glacial dispersal. Glacial dispersal is a term that describes the processes through which debris is entrained in and by ice, transported and deposited some distance away.

Glacial origin imparts several important differences to the composition of sediments and soils of glaciated landscapes, when compared to unglaciated landscapes where unconsolidated overburden is formed by pedological and subaerial processes:

(1) In glaciated terrain, enrichment of a mineralogical or chemical component is not confined just to the area where that element is enriched in bedrock; glacial transport may disperse components over wide areas, several times larger than the outcrop area.

(2) Dispersal by glacier ice is usually independent of drainage divides, except in mountainous areas where valley glaciers are often confined to ancestral valley systems. Thus, with the exception of some valley glaciers, detritus from a given source can be distributed throughout several drainage basins.

(3) Because glacial sediments at any given location are composites of detritus eroded from several, often genetically diverse bedrock sources, minerals and, therefore, chemical components of several origins may be found together in one sample. Such seemingly geochemically incompatible groupings as uranium – zinc – chromium may occur, for instance, where a glacier has traversed bedrock including acid, basic and ultrabasic igneous strata in close proximity, a lithologic assemblage that is not uncommon in geologically complex terranes such as the Canadian Shield.

(4) The finest portions of glacial sediments, particularly the chemically reactive clay (<2 µm) sizes, are composed predominantly of easily crushed bedrock detritus, such as phyllosilicate (micaceous) minerals, hematite, serpentine, etc. These components are glacially abraded to clay sizes by virtue of their soft or easily cleavable nature, have relatively low exchange capacities, and generally are not the products of weathering, in contrast to fine fractions of soils in unglaciated areas. That is not to say, however, that some true clay minerals and other products of preglacial, interglacial or interstadial weathering are not eroded and mixed with clay-sized detritus produced by glacial grinding of fresh bedrock or clast surfaces.

(5) In glaciated carbonate terrains, and in areas where the bedrock contains sulphides, olivine-serpentine, or other minerals that are broken down during the first stages of weathering, unweathered glacial

sediments can contain these components in abundance. In contrast to unglaciated areas, mantled by overburden comprising pedologically derived regolith with a relatively simple assemblage of stable primary and secondary minerals, the mineral assemblages of glacial overburden are complex in direct relationship to the complexity of the glacially eroded bedrock terrane. Selective chemical removal of the labile components by weathering processes has occurred only in and just below the thin postglacial solum formed on glacial sediments, and there has been little post-depositional concentration of those secondary weathering products characterized by enhanced exchange capacities, such as Fe–Mn hydroxides or oxides, true clay minerals, etc.

With respect to mineral exploration, the implication of a composition that reflects unaltered, but mixed, mineralogies of eroded bedrock is that even the most labile minerals of economic interest, such as sulphides, are preserved in unweathered glacial deposits. Therefore, if unweathered samples can be obtained, visual examination or geochemical analysis of the sediment can reveal primary mineralogy of source outcrops. Likewise, glacial sediment can be traced directly to its source without having to account for the effects of the low-temperature geochemical alteration of primary minerals that accompanies weathering. In weathered glacial sediments and in regolith derived by chemical degradation of bedrock in unglaciated areas, cations are redistributed by adsorption on or absorption into a variety of secondary mineral phases with high exchange capacity, such as hydrous iron or manganese oxides or clay minerals.

15.3. GEOCHEMICAL PATTERNS ON GLACIATED LANDSCAPES

The ribbon of compositionally distinct debris trailing down-ice from a particular source, once deposited, is called a dispersal train. Because glaciers, in general, and continental ice sheets, in particular, traverse so many different lithologies from their flow centres to points of deposition, each sample of glacial drift can be regarded as having a composition reflecting the overlapping of many dispersal trains (Fig. 15.1). It is a difficult task to filter out the signatures of the various overlapping trains and, in the case of mineral

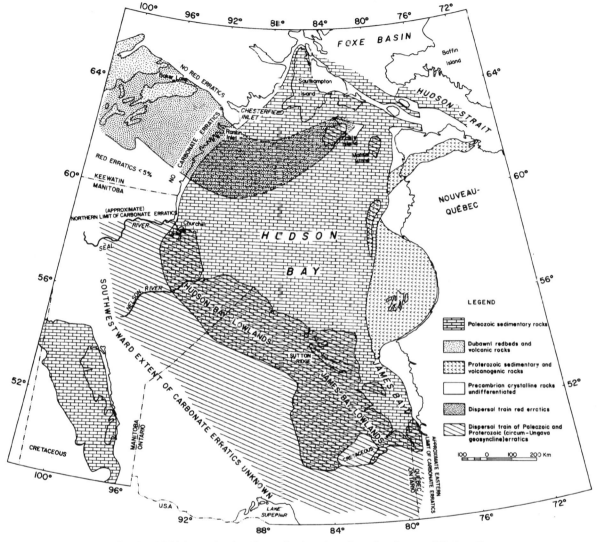

FIG. 15.1. Multiple overlapping dispersal trains of continental scale around Hudson Bay

exploration, to determine a method by which the components of one particular train can be differentiated, mapped, and traced back to their source.

If boulders of mineralized bedrock are found, the task of tracing them to their source, while often not easy, is more straightforward than tracking down the source of a geochemical anomaly. Boulders reveal the 'personality' of their source outcrops – the enclosing lithologies, mineralogy and chemistry of the economic occurrence are often preserved together in a boulder. A geochemical anomaly, on the other hand, has little personality, comprising usually an elemental enrichment above background, which usually can tell the geochemist little about the minerals responsible for the enrichment, much less about the nature of the unmineralized host rocks. By looking at its association with different trace and minor elements, however, (modern analytical techniques make it increasingly inexpensive to do rapid analyses of a wide range of elements) it is often possible to make an educated guess about the significance of an anomaly, and thereby to determine something about the geochemical and petrological personality of the source outcrops.

To use geochemistry of drift for determining patterns of dispersal of cations from their original outcrops and the nature of the outcrops themselves, it is necessary to understand how they and their host mineral phases are distributed in various facies of glacial sediment. Also, even a single till sample may yield a wide range of apparent concentrations of a single chemical component, depending on which size or specific gravity fractions are subjected to analysis. In order to construct a map that truly represents drift provenance, it is necessary to understand the non-provenance-related factors that influence analytical results. In addition to the obvious need to determine the sedimentological nature and stratigraphic position of material sampled for quantitative analysis, there are two major factors, chemical partitioning and post depositional weathering, which must be evaluated before analytical results can be related to provenance.

15.3.1. Chemical Partitioning

Glacial sediments, regardless of the sedimentary processes by which they are ultimately deposited, are generated predominantly by abrasive or crushing processes associated with clast-to-clast or clast-to-bedrock impacts of particles encased within glacier ice or in the ice-bed interface zone. For this reason, each sample of glacial or derived sediment represents a mélange of crushed bedrock from outcrops up-ice from where it is collected. Samples of proglacial or postglacial fluvial or wind-deposited sediments may be further transported in a variety of directions, depending on the topographic gradient before and after isostatic rebound and on prevailing wind direction.

Till represents closely the load entrained by glacial ice. It is composed of rock and mineral fragments with a potential grain size range from the largest fragments broken off source outcrops to the finest 'dust' created by clasts scraping against outcrops or against each other while in transport by glacier ice.

The mineralogical/petrological composition of the

various sizes of particles in till is primarily a function of their physical properties – micas and other soft or cleavable minerals predominate in the finest fractions; quartz, pyrite and other hard, non-cleavable minerals dominate the sand fractions. Thus, there is a marked mineralogical partitioning in the glacially entrained load, in till formed from it, and in fluvial (and aeolian) sediments derived from it either before or after its deposition (Fig. 15.2). This physical, mineralogical partitioning translates geochemically into chemical partitioning for the obvious reason that size fractions dominated by certain mineral phases will reflect the chemical make-up of those phases.

Studies of partitioning in glacial sediments are designed to determine how and where metal is held as well as the optimum grain-size range for providing geochemical analyses that accurately reflect provenance. Although many geochemical partitioning experiments have been carried out on glacial or postglacial sediments, very few have dealt with the reactive, clay-sized ($<2\,\mu m$) fraction. In slightly weathered or unweathered till, this fraction, although commonly referred to as the 'clay' fraction, is actually composed largely of minerals that were easily reduced to clay size by the physical processes of glacial grinding and abrasion. Although quartz, feldspars, carbonates and other minerals can be present in the clay-sized fraction, it is commonly dominated by well crystallized phyllosilicates (mainly chlorite and micas). Locally, it may contain significant amounts of less common 'soft' minerals, such as kaolinite and other clay minerals, serpentine, graphite and hematite.

Nikkarinen et al. (1984) and Shilts (1984) both demonstrated that analyses based on total leaches of a wide range of grain sizes revealed copper and zinc to be significantly enriched in clay-sized fractions, particularly in anomalous samples known to be related to massive sulphide mineralization. Furthermore, selective, weak laboratory leaches applied to the clay-sized fractions of till usually removed only a small percentage of the total metal available, suggesting that the metal is held largely in the structures of primary phyllosilicate minerals and is not adsorbed appreciably or incorporated into secondary mineral phases. Selective leaches caused most oxidized samples to yield significant percentages of

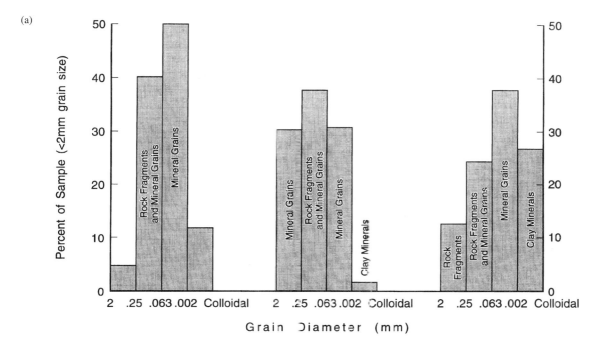

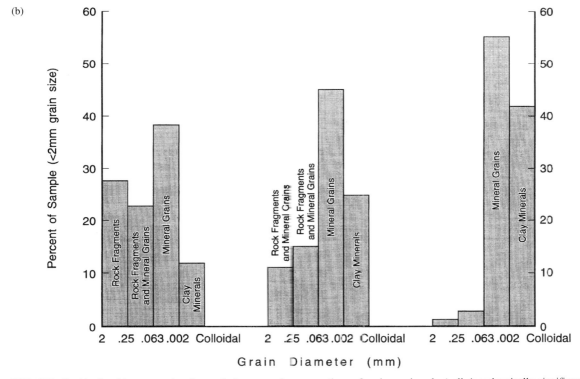

FIG. 15.2. Partitioning histograms showing typical ranges of concentrations of various mineralogically/geochemically significant size fractions in till from (a) Appalachian and (b) Canadian Shield regions of Canada

Fe and Mn from secondary oxide-hydroxide phases, but usually little other metal, suggesting either that the secondary phases generally have not been efficient scavengers of cations or that little metal was released during weathering.

The implications of these results are significant in evaluating geochemical data from till or other glacial or derived sediment (Table 15.1). They suggest that concentrations of many cations in separates containing a wide range of grain sizes are strongly skewed by the amount of <2 µm material, the coarser sand and silt-sized fractions often acting as diluents for some cations of economic interest.

Without adequate attention to the effects of

TABLE 15.1. Examples of Partitioning Tills rich in ultramafic debris: (A) Basal till near an Archaean komatiite, central District of Keewatin; (B) Basal till near Paleozoic ophiolite complex, Quebec Appalachian Mountains; (C) Flow till forming lateral moraine of modern glacier; on Mesozoic ophiolite complex, Swiss Alps, near Zermatt.

Size fraction/element	Cr (ppm)			Ni (ppm)			As (ppm)		
	A	B	C	A	B	C	A	B	C
Bulk Sample	–	284	1744	–	500	1600	–	8	4
2.0–6.0 mm	3320	400	1780	1050	880	1600	160	5	2
0.25–2.0 mm	3220	294	1852	970	556	1500	157	13	2
0.044–0.25 mm	2520	200	1980	745	26	9707	162	4	3
0.004–0.044 mm	1856	236	1424	910	403	1020	245	16	5
0.001–0.004 mm	3560	274	1148	1200	743	2300	553	10	11
<0.001 mm	-	256	1468	–	913	4100	770	13	12

85 SK–22265 Oxidized, sandy, granite-rich till, near Tangier Lake, Nova Scotia

Size fraction/element	W (ppm)	Cu (ppm)	As (ppm)
Bulk Sample	500	106	8
2.0–6.0 mm	60	31	2
0.25–2.0 mm	360	53	3
0.044–0.25 mm	500	90	6
0.004–0.044 mm	550	167	13
0.001–0.004 mm	1800	500	45
<0.001 mm	>2000	609	80

90 KAL–001 Oxidized, slate-rich basal till overlying gold-bearing quartz vein in gabbro, near St Magloire, Quebec Appalachians

Size fraction/element	Au (ppb)	As (ppm)	Sb (ppm)
<0.064 mm	624	207	0.6
0.064–2.0 mm	399	120	0.5
0.002–0.064 mm	838	215	0.7
<0.002 mm	83	341	1.6

80 AR–0772 Sandy, clay-poor pebbly till, Grenville of E. Ontario

Size fraction/element	As (ppm)	Zn (ppm)	Cu (ppm)
Bulk Sample	42	56	15
2.0–6.0 mm	2	40	7
0.25–2.0 mm	21	38	8
0.044–0.25 mm	12	42	8
0.004–0.044 mm	189	65	16
0.001–0.004 mm	330	385	165
<0.001 mm	-	830	440

85 TR–041 Till, composed mostly of weathered, tin-bearing granite, near Rocky Brook, New Brunswick

Size fraction/element	Sn (ppm)	U (ppm)	Mn (ppm)	Fe (%)	As (ppm)
Bulk Sample	20	5.2	175	1.0	<2
2.0–6.0 mm	21	1.9	83	0.3	<2
0.25–2.0 mm	20	3.5	169	1.0	<2
0.044–0.25 mm	41	4.3	400	1.4 2	
0.004–0.044 mm	-	13.3	893	2.8	10
0.001–0.004mm	82	11.8	1255	3.4	25
<0.001 mm	–	11.3	1528	3.2	30

85 NJ–002 Oxidized flow till in ice-contact gravel complex, Franklin Furnace, New Jersey

Size fraction/element	Zn (ppm)	Cd (ppm)
Bulk Sample	1850	0.6
2.0–6.0 mm	1600	0.7
0.25–2.0 mm	1900	0.4
0.044–0.25 mm	1750	0.7
0.004–0.044 mm	925	0.3
0.001–0.004 mm	2778	3.6
<0.001 mm	6000	3.8

80 AR–0283 Gray, blocky basal till, Grenville of E. Ontario

Size fraction/element	U (ppm)	Zn (ppm)
Bulk Sample	11.9	135
2.0–6.0 mm	2.8	93
0.25–2.0 mm	1.9	93
0.044–0.25 mm	3.7	70
0.004–0.044 mm	11.2	138
0.001–0.004 mm	55.0	470
<0.001 mm	104.0	790

80 LAAMK–007 Red clayey basal till, rich in Dubawnt Group erratics, District of Keewatin

Size fraction/element	Cu (ppm)	Pb (ppm)	Cr (ppm)
Bulk Sample	500	1850	132
2.0–6.0 mm	1920	3050	240
0.25–2.0 mm	760	1700	180
0.044–0.25 mm	208	620	66
0.004–0.044 mm	300	540	69
0.001–0.004 mm	2040	3150	198
<0.001 mm	7200	7100	410

80 SMA–192 Near-surface basal till, Central District of Keewatin

Size fraction/element	Pb/ppm	U (ppm)	Mo (ppm)	As (ppm)
Bulk Sample	136	1.7	8	23
2.0–6.0 mm	168	0.9	8	5
0.25–2.0 mm	74	1.1	4	7
0.044–0.25 mm	44	1.2	6	8
0.004–0.044 mm	102	5.5	6	25
0.001–0.004 mm	570	8.0	15	76
<0.001 mm	1300	2.6	26	112

80 AR–0109 Oxidized, ice-contact, gravel, poorly sorted, Grenville of Eastern Ontario

Size fraction/element	Cu (ppm)	Ni ppm	Cr ppm
Bulk Sample	56	326	935
2.0–6.0 mm	36	387	1116
0.25–2.0 mm	27	213	690
0.044–0.25 mm	27	221	1093
0.004–0.044 mm	152	661	1284
0.001–0.004 mm	267	982	1212
<0.001 mm	484	1228	1111

Explanation for Table 15.1
Bulk Sample   total sample finer than 6 mm
2.0–6.0 mm   fraction almost wholly composed of rock fragments
0.25–2.0 mm   fraction composed of mixture of rock fragments and mineral grains
0.044–0.25 mm   sand and coarse silt; fraction composed almost wholly of mineral grains,
                   dominated by quartz and feldspar (usually >90%)
0.004–0.044 mm   Silt; mineral grains dominated by quartz and feldspar
0.001–0.004 mm   Clay; mineral grains dominated by phyllosilicates, and other soft minerals
<0.001   Clay and colloidal particles.
Analysis by Atomic Absorption techniques after HF-HClO4 – HNO3 – HCl leach for Cr, Mn, Fe,
Co, Ni, Cu, Pb, Zn, Mo, Cd. Sn by x-ray fluorescence. U by fluorometric technique. As by
colourimetric technique after HNO3 – HClO4 leach. Au by neutron activation. W by colourimetric
on carbonate sinter. All analyses by Bondar-Clegg and Co. Ltd.

partitioning, a drift geochemical map can become essentially a map of till texture, as graphically illustrated in Figures 15.3 and 15.4, compiled from till sampling around volcanigenic base metal mineralization at Spi Lake, in south central District of Keewatin, Canada. The true picture of glacial dispersal around this potential ore body was provided by direct analyses of the clay fraction of drift (cf. Ridler and Shilts, 1974).

A question that remains unresolved in many areas of volcanogenic or sediment-hosted massive sulphide bodies is why outcropping massive sulphide and other types of mineralization are indicated by geochemical patterns derived from a size fraction that is dominated by silicate minerals and appears to contain few or no sulphides. In addition, the inference that metal detected in the clay fraction is largely structural and not adsorbed and the fact that most dispersal trains from sulphide bodies are far larger than the outcropping ore body could reasonably be expected to produce, pose important questions about the geochemical pathways from ore to host rock to glacial sediments. In the case of syngenetic ores, perhaps the answers to these questions lie in modern geochemical studies of the relationship between actively forming sulphide deposits and the clays of their surrounding sediments near submarine hydrothermal vents in modern oceans (Barrett, T.J. et al., 1988). It is possible that metal in solution near these vents is interstratified with or taken into the lattices of nearby

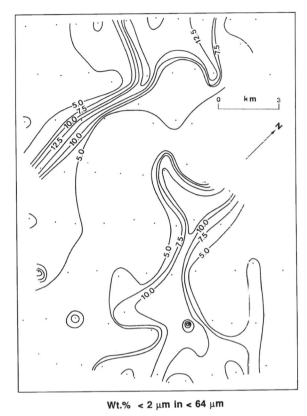

Wt.% < 2 μm in < 64 μm

FIG. 15.3. Pattern of clay concentration, in silt and clay (<64 μm) fractions, Spi Lake sampling grid, District of Keewatin, Canada

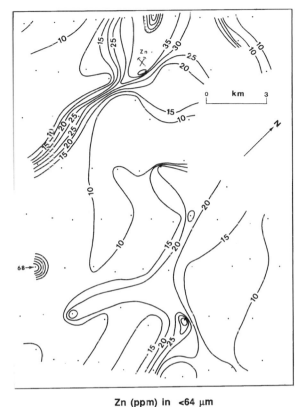

Zn (ppm) in <64 μm

FIG. 15.4. Pattern of Zinc (Zn) concentration in silt and clay fractions from same samples as Fig. 15.3

authigenic clays and is retained in the phyllosilicates derived from them during subsequent diagenesis or metamorphic alteration. Thus, ore bodies formed syngenetically in a sedimentary/volcanic sequence may be surrounded ultimately by a 'halo' of host rocks with metal enriched phyllosilicates.

Once glacial sediments are deposited, post-depositional, low-temperature geochemical processes (mostly weathering) will alter and distort the chemistry of their mineral phases, which were originally sorted on the basis of their resistance to abrasion. However, in general, the pattern of partitioning will persist. The comparison of a sample of highly altered till, collected near an ultramafic source of platinum-group, precious and trace metals, to a sample of 'normal' till 170 m down-ice demonstrates well the persistence of the partitioning pattern (Table 15.2). In this example, the absolute concentrations of metals in

the altered sample range from 2 to 10 times higher than those in the unaltered sample, but the trends through the various grain sizes persist. Pd, Ni, Cu and to a lesser extent Co show the typical distinctive trend of significant enrichment in the <2μm fraction of both samples, as documented in Table 15.1. In this example, Ni actually may be enriched in the <2 μm fraction because of its presence in serpentine, a very soft mineral which readily abrades to very fine sizes.

The general decrease in gold concentrations in finer sizes is probably related to the way this malleable metal abrades during glacial transport; contacts of gold grains with other clasts cause folding or displacement of gold on the grain in a fashion analogous to a knife being drawn across butter, but material is not appreciably flaked off the grain to be redistributed into finer sizes. Furthermore, large

TABLE 15.2. Partitioning in till over ultramafic bedrock, northern Ungava, Quebec*

Raglan 3**	Pd (ppb)	Pt (ppb)	Ni (ppm)	Co (ppm)	Cr (ppm)	Cu (ppm)	Au (ppb)
Bulk Sample (<6.0 mm)	1100	330	2966	91	1305	4460	120
<63 µm (silt + clay)	980	210	3466	84	669	5400	160
2.0–6.0 mm	2100	500	2579	91	2360	4230	620
0.25–2.0 mm	1400	490	3249	117	1760	4430	230
63–250 µm	1100	220	2306	79	904	3380	130
45–63 µm	900	180	2164	62	913	3300	160
2–45 µm	720	300	2418	72	629	3620	210
<2 µm	2300	110	6355	118	882	>10 000	60
Raglan 5*							
Bulk Sample (<6.0 mm)	130	60	679	53	845	477	20
<63 µm (silt + clay)	100	96	592	41	407	433	44
2.0–6.0 mm	130	85	727	60	1440	411	4
0.25–2.0 mm	140	75	674	64	1010	436	6
63–250 µm	110	50	446	40	538	346	8
45–63 µm	68	100	406	34	469	291	34
2–45 µm	78	70	490	37	445	344	26
<2 µm	390	70	1601	104	1045	1405	10

*Samples collected by and donated by Michel Bouchard, Université de Montréal
**Sample of highly altered till collected from a mudboil 8 m down ice from PGE/Sulphide mineralization
***Sample of apparently unaltered till collected from till plain 170 m down ice and downslope from gossan near 'Raglan 3'

flakes of gold, a common morphology in bedrock, are folded and crumpled during glacial transport until they are reduced to smaller, compact balls, again without losing much mass, much in the way that a large piece of paper can be crumpled into a ball (Chapter 16). For chromium, the trend of decreasing metal in finer grain sizes probably reflects the occurrence of Cr predominantly in mineral phases such as magnetite or chromite as opposed to phases such as fuchsite. Because the spinels are hard and non-cleavable, they do not easily reduce to silt and clay sizes by abrasion, whereas easily cleavable, Cr-bearing phyllosilicates, such as fuchsite would tend to be reduced to clay size.

15.3.2. Weathering of Till

Postglacial weathering takes place in the zone of oxidation above the groundwater or permafrost table and can alter drift geochemistry to considerable depths below the ground surface (Rencz and Shilts, 1980; Shilts, 1976, 1984; Peuraniemi, 1984; Shilts and Kettles, 1990). Furthermore, the effects of

weathering on the geochemistry of relatively imper-meable silt and clay-rich till are quite different from those on some of its more permeable silt and clay-poor derivatives, such as esker or other ice-contact gravels (Shilts, 1973a). In an oxidizing environment, labile minerals such as sulphides and carbonates are generally destroyed above the groundwater or perma-frost tables, their chemical constituents being carried away in solution or precipitated or scavenged locally by clay-sized phyllosilicates and by secondary oxides/hydroxides, depending on the element and the local geochemical environment. In poorly-drained sites, where the water table is at or close to the surface and/or the surface is covered with an organic mat, a reducing environment promotes little or no destruc-tion of primary labile minerals(Peuraniemi, 1984).

In porous and permeable glacial sediments, partic-ularly well-sorted sands and gravels, destruction of labile components also takes place, but additional weathering of primary silicate mineral phases that are unstable in the near-surface environment produces a fine debris of secondary mixed-layer clays, hydrox-ides, oxides, etc. that can be physically translocated

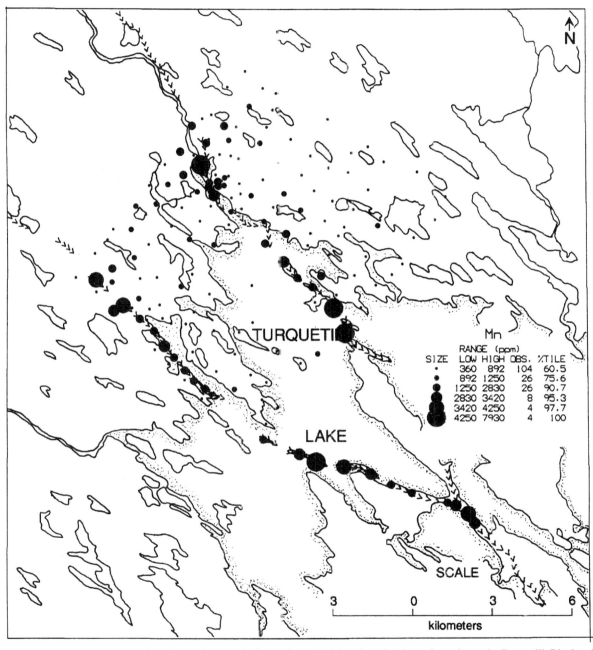

FIG. 15.5. Contrast in concentration of secondary weathering products (Mn) in < 2 μm fractions of an esker and adjacent till, District of Keewatin, Canada

by percolating groundwater from the surface down-ward through the deposit (Shilts, 1973a). This is especially important for eskers or other coarse-grained deposits that have little or no primary fine fraction because of the high-energy fluvial environ-ment in which they were deposited. Furthermore, many such deposits stand as ridges or hummocks, largely above the groundwater table. The enhanced scavenging ability of these secondary minerals gives the fine fractions of glaciofluvial and similar deposits elevated background concentrations of trace elements relative to the same size fractions of nearby till, the fine fractions of which were produced largely by physical crushing of well-crystallized primary miner-als with much lower exchange capacity (Fig. 15.5).

For reasons cited above, weathering restricts the use of heavy mineral or other coarse-grained fractions of near-surface till and derived, coarse-grained, sorted sediments, except in the case of resistate ore minerals (cassiterite, gold, chromite, etc.). The contrast between the mineralogy and chemistry of fine frac-tions of till and sorted sediments derived from till makes accurate sample characterization according to sediment facies absolutely essential to interpreting exploration geochemistry results in glaciated terrain. Mixing oxidized samples of till with those of derived sediments can lead to the classic pattern of non-provenance related geochemical enrichment as depic-ted in Fig. 15.5.

15.3.2.1. A case example from the eastern townships, Quebec

Since till and associated sediments in the Appalachian region of Quebec are rich in sulphide minerals, deep till sections in this region have been studied to determine the effects of post-depositional weathering on mineralogy and chemistry (Shilts, 1975, 1984; Shilts and Kettles, 1990).

Shilts and Kettles (1990) studied weathering pro-cesses in several natural stream banks near Thetford Mines, Quebec. At all sites hard, olive-grey till weathers to a brown to tan colour to a depth of about 2 m below the ground surface. The till contains many cobbles but few boulders and, where unoxidized, has a noticeable component of sand to granule-sized pyrite cubes and fragments. The pyrite is derived from

the underlying and surrounding bedrock, quartz-albite-sericite schists with abundant recrystallized pyrite cubes. A northeast-striking belt of chlorite-epidote schists comprising metabasalts and rhyolites with known base-metal mineralization lies less than 15 km northwest (up ice) from most of the sections (Harron, 1976). The most important glacial dispersal direction, presumably the one that prevailed during deposition of the till, was southeastward.

The effects of weathering on labile minerals are reflected by trace metal concentrations in sand-sized heavy mineral (Sp.G. >3.3gcm-3) separates from till samples collected in vertical profile through typical stream-cut sections (Fig. 15.6a). For all the elements studied, except for chromium, there is a sharp decrease in metal concentration in heavy minerals at and above the oxidized zone, through the upper 2 m of till (Figure 15.7a). Although not apparent in this section, in some nearby sections there is a correspond-ing increase in concentration of some cations in clay fractions of oxidized samples, which indicates that clay-sized phyllosilicates and/or secondary oxides and hydroxides can scavenge some of the metal released by weathering (Shilts and Kettles, 1990). Sulphur and iron also decrease markedly in the zone of oxidation (Fig. 15.7b), supporting the conclusion that pyrite and other sulphide phases were the hosts for metal that has been translocated in the zone of oxidation (Shilts and Kettles, 1990). This and similar studies demonstrate that weathering effects can be important well below the shallow (<1 m thick) postglacial solum.

The effects of most importance in mineral explora-tion are the destruction of primary labile mineral phases and the redistribution of their cations into secondary or other mineral phases or into ground-water. Thus, it is of paramount importance to have some understanding of this source of vertical geo-chemical variation when evaluating geochemical patterns obtained from separates containing the silt and coarser fractions of till, particularly if analyses are performed on heavy mineral concentrates. The postglacial weathering problem is, of course, not nearly so important in exploration for resistate minerals such as chromite, tin, gold, etc., but their apparent concentrations can be augmented in oxidized samples if large concentrations of labile minerals

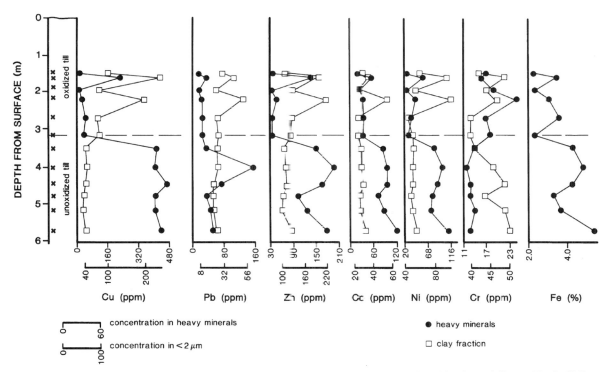

FIG. 15.6. Typical vertical profiles of metal concentrations in < 2 μm and sulphide-rich heavy mineral fractions of till, near Thetford Mines, Québec. Note apparent enrichment of Chromium, derived from the resistate mineral chromite, as a result of the removal of pyrite grains by oxidation of the heavy minerals

present in unaltered samples were removed by weathering processes (Fig. 15.7a).

Assuming that geochemical or other analyses have been carried out in such a way as to avoid non-provenance-related variations, such as those caused by misidentification of sediment facies or stratigraphic position, partitioning, or post depositional weathering, the real variations of geochemical, mineralogical, or lithological parameters may be contoured to provide maps of provenance-related compositional variations. These maps will show, ideally, patterns of glacial dispersal.

15.4. DISPERSAL PATTERNS

Glacial dispersal, where reflected adequately by the available sample spacing, can take several forms, depending on where the source outcrops are located in relation to former centres of outflow and with regard

to certain dynamic features within the former ice sheets (Boulton, 1984; Shilts and Smith, 1989; Bouchard and Salonen, 1990).

As a glacier transports an indicator[1] component away from a particular source, the concentration of the component is attenuated, both by addition of debris eroded from the dispersal area and by deposition of indicator components along the way. Generally, the decline in concentration of indicator components with distance can be plotted as a negative exponential curve, high frequencies declining rapidly to low frequencies, which are then maintained at levels above background for distances several times greater than the width of the source outcrop. The zone of rapid decrease in indicator frequencies may be defined as the 'head' of dispersal and the extended

'Indicator' is a glacially transported rock, mineral, or chemical component that is derived from and can be traced to a specific source area.

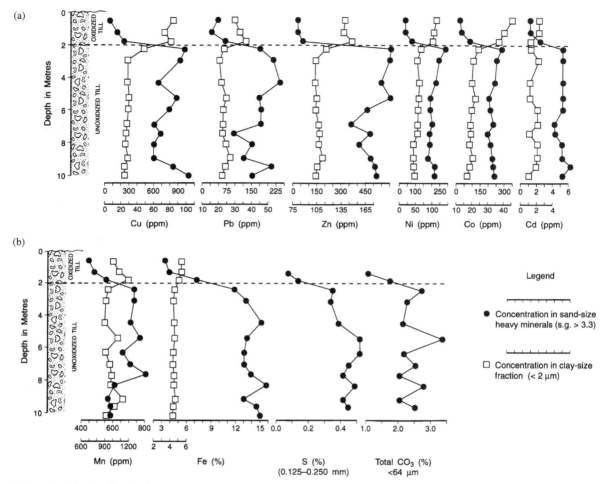

FIG. 15.7. Typical profile showing sharp decrease in sulphur and iron in pyrite-rich heavy mineral fractions of till, near Thetford Mines. Note that carbonate leaching depths correspond to depth of pyrite oxidation

zone of lower frequencies as the 'tail' of dispersal. The area of dispersal is called the 'dispersal train' (Fig. 15.8), and the curve itself the 'dispersal curve' (Shilts, 1976) (Fig. 15.9).

15.4.1. Factors influencing Dispersal Patterns

The apparent shape and dimensions of the dispersal curve and dispersal train are influenced in turn by a number of factors:

(1) The lithologic nature and topography of the source area influences how much of a component will be eroded and available for transport. If the source

area is a topographically positive feature and is composed of 'soft' rock (limestone, serpentinized peridotite, etc.) or rock that is highly fractured as a result of structural (jointing) or periglacial (frost heaving) disruption, it is likely to provide a source of debris through repeated glacial events. Hard, massive outcrops, such as rhyolites or basalts, provide comparatively less debris, even if they stand as positive features. If the source outcrops are in narrow depressions aligned with the general direction of glacial flow, the increased velocity of ice flowing through the constriction may generate much more debris than flow on adjacent, flatter terrain. In such

NICKEL

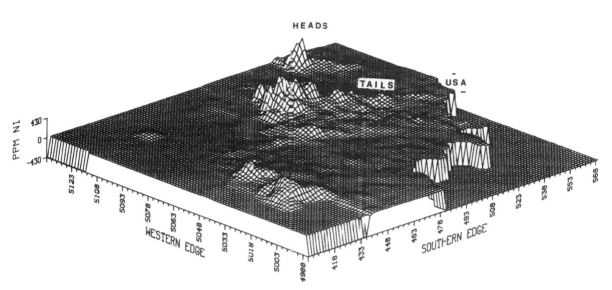

FIG. 15.8. 3-dimensional plot of nickel dispersal showing head and tail of dispersal from ultramafic bedrock, Thetford Mines area, Québec

cases dispersal trains may be particularly well deve-oped (Shilts and Smith, 1989, p. 51).

(2) The topography of the dispersal area has an important effect on the shape and continuity of both the dispersal curve and dispersal train. In an ideal case, the dispersal area may have relatively little relief; the shapes of the curve and the train are, therefore, controlled principally by (a) the rate of dilution by debris eroded from the dispersal area and mixed with the indicator component and (b) the rate of deposition of indicator components in the dispersal area. In a topographically irregular dispersal area, ridges, escarpments, valleys, and other features may block or divert debris carried along by the ice, destroying, displacing, or truncating dispersal curves or trains. Blocking or diversion is particularly common in geologically (and topographically) complex areas such as parts of the Canadian and Fenno-scandian Shield, the Appalachians, and the Cordillera (Fig. 15.10).

(3) Since there are several sedimentary environments associated with deposition of till, it is important to recognize the circumstances under which debris that

forms the till is transported and deposited by glacial ice. In general, glaciers entrain three types of sediment load: (1) a tractive debris layer between the ice and its immobile bed; (2) a dense concentration of debris near their base; and (3) a less dense load of debris scattered throughout or on the rest of the ice mass. Basal debris forms relatively dense, compact till which is 'smeared on' beneath a glacier or is melted out of the sole of the glacier during the waning stages of glaciation (Chapter 2). Debris carried higher in the ice (englacial debris) or on the ice surface (supraglacial debris) is melted out with or without accompanying deformation, or it slumps off the glacier ice by various mass-wasting processes during retreat of an ice sheet or valley glacier. These till facies, basal and supraglacial, are important to recognize during dispersal studies, as they can vary radically in composition (Fig. 15.10) (Shilts, 1973b).

In general, supraglacial deposits tend to be dominated by the lithologies of the topographically higher (often local) or more distant elements of the dispersal area, while basal deposits tend to be dominated by topographically lower elements of the dispersal area.

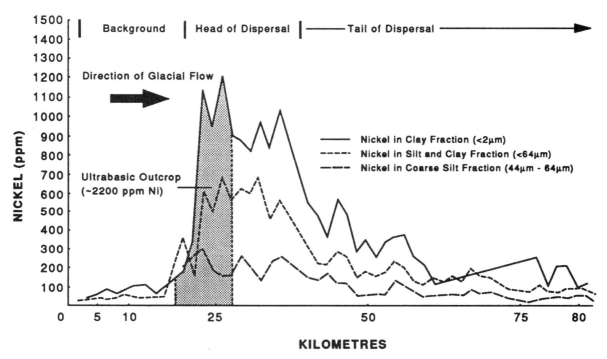

FIG. 15.9. Dispersal curves of various fractions of till from main dispersal train in Thetford Mines area

However, where glaciers were advancing up river valleys cut deep into highlands, they may have carried their basal load far up the valleys, in which case the resulting basal deposits may contain significant amounts of components of distant origin (Holmes, 1952; Shilts, 1976; Aber, 1980).

(4) Where studied in the vicinity of ice divides in the southern District of Keewatin and southern Quebec (Canada) regional trends of dispersal trains are not necessarily parallel to other indicators of latest ice movements – striae in particular (Shilts, 1984; Shilts and Smith, 1989). Similarly, in some regions where directions of ice flow are known to have shifted through considerable angles during multiple or single glacial events, individual or multiple tills can show considerable vertical variations, especially in geologically complex areas (Fig. 15.10) (Shilts, 1976, 1978). The reasons for vertical and spatial variations, or lack thereof, during a single glacial event are presently poorly known, but probably are related to the changing dynamics of the base of the glacier in a given region with time. In other words, erosion may

be enhanced at some stage(s) of the glacier's occupation of a particular piece of terrain. The length of time that a glacier was flowing at a particular azimuth also may have affected the distance that debris could be transported, a consideration that is particularly important in the vicinity of ice divides, many of which came into existence or moved to their last position late in a glacial event. It is known that basal ice flow velocities increase exponentially away from centres of outflow, with the result that much of the debris in transit near the centre takes a great deal of time to move any appreciable distance (Boulton, 1984, p. 219).

(5) The magnitude of dispersal and the proportion of 'far-travelled' to 'local' components in till (and its various varieties and derivatives) are concepts about which it is difficult to generalize (cf. Clark P.U., 1987; Bouchard and Salonen, 1990). The complexity of factors governing actual and apparent dispersal in any particular region precludes the development of universal rules. Many studies have drawn conflicting conclusions with respect to these concepts, the conflicts being created largely by the local nature of

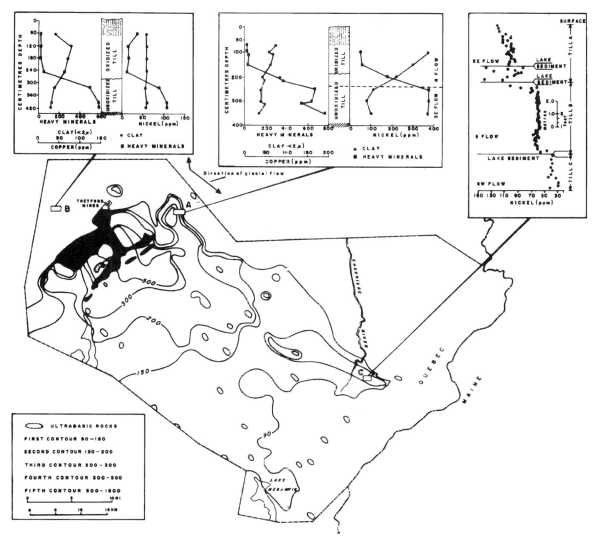

FIG. 15.10. (a) Examples of various types of vertical variation in till sections in geologically complex terrain: (A) shows shift of ice flow (clay) and weathering (heavy minerals); (B) shows effects of weathering; (C) shows effects of different directions of ice flow affecting tills of different glaciations. Contour map shows concentrations of Nickel (Ni) in < 64 μm fraction of surface till

the studies with attendant influence of local factors on the compositions of the till studied.

Because of the negative exponential character of the ideal dispersal curve, the common observation that most material is dispersed a short distance from its source, and that the bulk of any till sample consists mostly of 'local' debris can be considered to be generally true. However, the presence of abundant local pebbles and cobbles should not be taken to mean

that the finer fractions of the till are of similar local origin, nor should the opposite be taken on faith. For instance, some modern glaciers on Bylot Island, Arctic Canada, have been observed to carry almost 100% coarse Precambrian debris at their snouts, which are at least 10 km down ice from the nearest Precambrian outcrop. The local bedrock comprises unconsolidated or poorly consolidated Cretaceous sediments which are so easily disaggregated that they

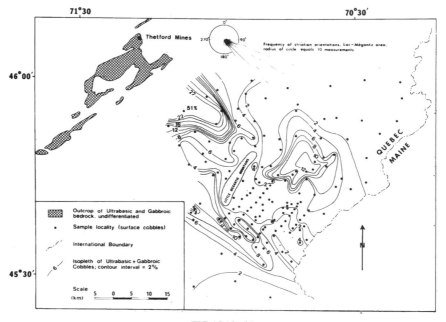

FIG 15.10. (b)

rarely form clasts coarser than granule size. Analysis of the sand and finer sizes of the glacier load reveals that it is made up predominantly of detritus derived from the underlying bedrock (DiLabio and Shilts, 1979), underscoring the caveat against proposing universal rules to quantify dispersal patterns (Fig. 15.11).

(6) Concentration of a component declines gradually in the down ice direction until it merges with and becomes indistinguishable from natural or analytical variations in background. Generally, the more distinctive a component is with respect to the chemistry or lithology of rocks of the dispersal area, the farther it can be traced. Chromium, for instance, is depleted in most crustal rocks; therefore, chromium anomalies generally can be traced for long distances from ophiolitic or other ultramafic complexes that have high chromium concentrations. Likewise, distinctive red volcaniclastic pebbles derived from outcrops of late Proterozoic Dubawnt Group bedrock west of Hudson Bay have been traced for long distances, even where present in very small amounts, across terrain underlain by Archean-age gneissic bedrock and by light coloured Paleozoic limestones of northern Hudson Bay (Shilts, 1982) (Fig. 15.12).

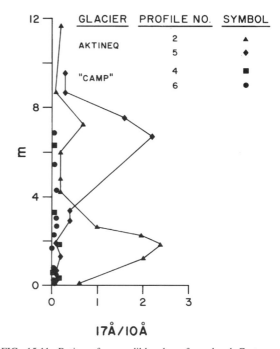

FIG. 15.11. Ratios of expandible clays from local Cretaceous-Tertiary bedrock to illite largely derived from Precambrian terrain, 10 km up-ice, in basal debris zones of glaciers, Bylot Island, Arctic Canada. Granule size and larger debris at these sites is nearly 100% Precambrian

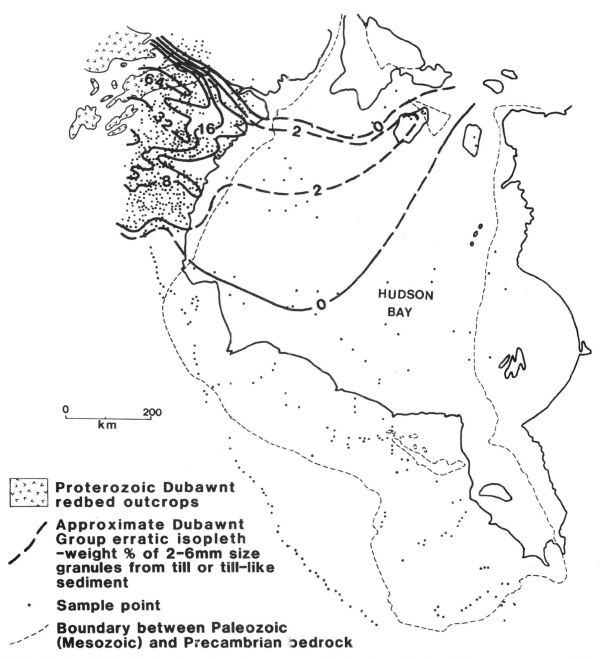

FIG. 15.12. Dispersal of visually distinctive, red, granule size and larger volcanic and volcaniclastic erratics from outcrops of Precambrian Dubawnt Group, Arctic Canada

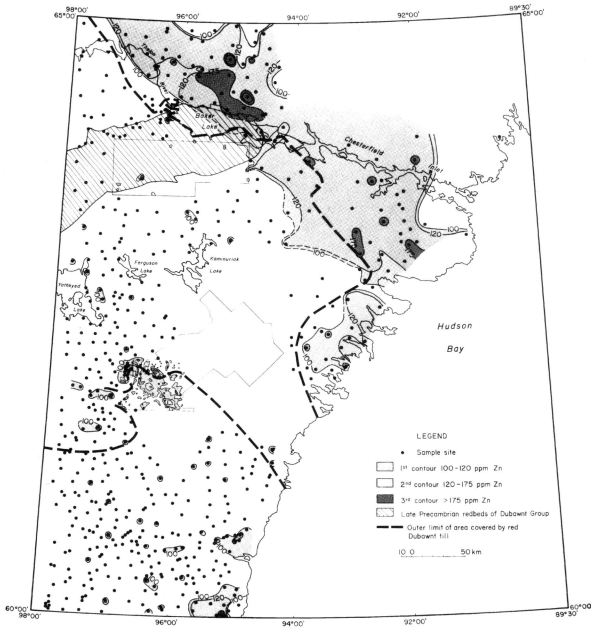

FIG. 15.13. Regional patterns of Zinc (Zn) dispersal in the clay fractions of till, Arctic Canada. Quasi-rectangular outlined areas were sampled at a spacing of 1 per 1.6 km and represent geochemistry of ~2500 till samples each

15.5. SCALES OF DISPERSAL

To understand and interpret the significance of the sometimes subtle geochemical tails of dispersal, it is necessary to understand, along with those factors discussed in previous sections, the scales of glacial dispersal. In glacial terrain, the composition of a sample is a composite of many overlapping dispersal trains. The trains occur at a variety of scales and emanate from sources up-ice from where the sample was collected. For convenience of discussion, four geologically meaningful scales of dispersal have been defined (Shilts 1984): (1) Continental, (2) Regional, (3) Local and (4) Small-Scale.

15.5.1. Continental scale

Dispersal on a continental scale can be measured in hundreds to more than a thousand kilometres (Prest and Nielsen, 1987) (Figs.15.1 and 15.12). Although not usually important in mineral exploration, very far-travelled coarse clasts and fine-grained components, if detected and not properly related to a distant source, may be thought to come from local sources, creating severe exploration problems. The occurrence of diamonds in drift of the American midwest (Stewart et al., 1988) is perhaps the best example of this problem. Although very rare, diamonds are easily identifiable and may come from as far away as northern Ontario (Brummer, 1978) and the Hudson Bay region, from which an immense train of Paleozoic debris has been dispersed southward and westward during repeated glaciations (Fig. 15.1).

15.5.2. Regional Scale

Geochemical dispersal on a regional scale can be measured in tens or hundreds of kilometres. This scale of dispersal has been mapped in the District of Keewatin, Canada for till that overlies Archean and younger Precambrian bedrock (Fig. 15.13). The reasons for the consistently elevated metal levels in some parts of this region are not clearly understood. At this low sample density some areas of metal enrichment may have been caused by the areal homogenization, with distance, of several relatively small bodies of geochemically distinctive rocks. For instance, elevated nickel, cobalt and chromium concentrations could result from dispersal from a cluster

of komatiitic or other ultramafic bodies. In other cases, trace element levels in till are known to be suppressed as a result of far-travelled, metal-poor debris being mixed with more metal-rich debris from local rocks (Shilts and Wyatt, 1989). The large dispersal train of clay-sized hematite and kaolin (Fig. 15.12) from the easily eroded clastic sedimentary formations of the Dubawnt group (Donaldson, 1965) has depressed background and anomalous levels of metal from local rocks over a large area, making the higher levels of metal in till outside the train appear anomalous by contrast (Fig. 15.13). Also, within the area of the Dubawnt dispersal train, geochemical expression of mineralized outcrops is depressed because of the diluting effects of Dubawnt detritus.

15.5.3. Local Scale

Local glacial dispersal can be detected by reconnaissance sampling at the scale of one sample per one to four square kilometres. Sampling of over 10,000 km² of terrain in southern Keewatin, Canada at this scale has been accomplished. Reconnaissance sample spacing in Finland and on the northern Fennoscandian Nordkalott project (Kautsky, 1986) is similar. Geochemical anomalies detected at the local scale are much more easily related to mineralization than are those of the larger scales of dispersal. At this scale of sampling, the tails of dispersal trains from potential ore bodies are likely to be detected, but sample and analytical control (particularly sediment facies, partitioning and weathering effects) must be precise enough to allow them to be differentiated from background or from the tails of trains of regional or continental scale. Figures 15.14 and 15.15 illustrate some examples of local dispersal. In Fig. 15.14, zinc anomalies from known zones of zinc mineralization are superimposed on a regional train of zinc from an unknown source (see Fig. 15.13 for regional context). Figure 15.15 shows a 'negative' dispersal train in which metal-poor kaolin that cements the poorly consolidated Precambrian (Dubawnt Group) sediments that underlie Pitz Lake has been dispersed southeastward, depressing regional metal levels so much that a distinct train of metal-poor debris has been formed. This is a local expression of the dilution phenomenon illustrated by Fig. 15.13.

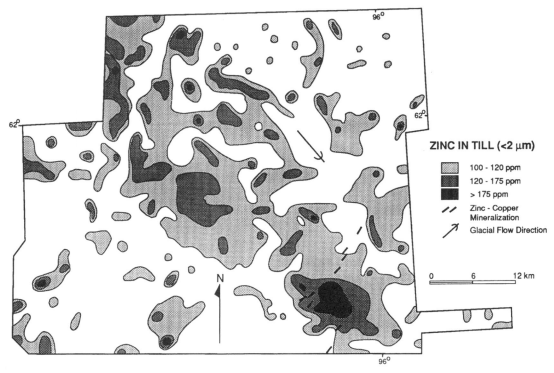

FIG. 15.14. Dispersal of Zinc (Zn) in clay from till samples, Henninga Lake area, District of Keewatin. Note more intense, superimposed train down-ice from Zinc mineralization. Refer to southwest end of lower detailed grid, Fig. 15.13 for location

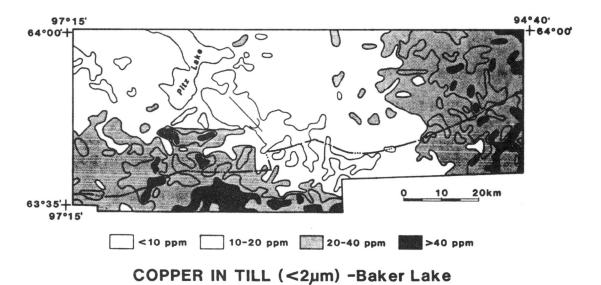

FIG. 15.15. Negative dispersal train of depressed Zinc values in clay fractins dominated by kaolin eroded from kaolin-cemented sandstones of the Pitz Lake Basin. Area is the upper rectangular area on Fig. 15.13 (note change of contour intervals)

15.5.4. Small-Scale

Small-scale dispersal is usually encountered in the last stages of mineral exploration. The boulder train tracing routinely carried out in Finland (Hirvas, 1989) is designed to map this scale of dispersal. There are many published case histories showing examples of small-scale dispersal (cf. Kujansuu and Saarnisto, 1990). An example of a small nickel dispersal train extending down-ice from outcrops of presently uneconomic nickel-copper mineralization in District of Keewatin (Fig. 15.16) is used to illustrate the

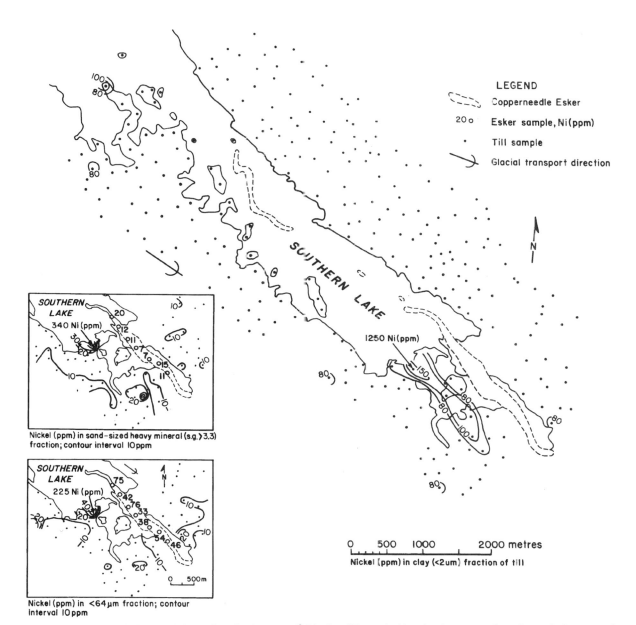

FIG. 15.16. Dispersal of Nickel (Ni) from mineralized outcrops in District of Keewatin. Note the clear pattern from clay analysis compared to the non-indicative pattern from silt/clay and heavy minerals caused by partitioning and weathering effects, respectively

importance of correct identification of sediment facies and of choosing proper sample processing and analytical techniques. The train of nickel that is so clearly defined by analysis of the <2 μm fraction of till (AAS after hot HNO_3–HCL leach) is very poorly expressed by analyses of sand-sized heavy minerals (Sp.G. >3.3g cm-3) and of -250-mesh (<64μm) fractions, the former because weathering has destroyed the labile, Ni-bearing sulphides, and the latter because the metal-poor, quartz-feldspar-rich silt fraction dilutes the sample in an unpredictable pattern. Note also that the Ni concentrations of the <64 μm fractions of samples from the esker are 4 to 7 times higher than these for similar fractions of nearby till, as predicted in the discussion of weathering.

15.6. SOME IMPORTANT EXCEPTIONS TO CLASSICAL DISPERSAL PATTERNS

The classic 'head and tail' form of dispersal is a useful model for interpreting or indicating the composition of a glaciated landscape, but recent research has shown that there are other important types of dispersal trains, particularly in regions formerly or presently covered by the great continental ice sheets. It has long been known that shifting ice flow directions throughout a major glaciation may form fan-shaped dispersal patterns although, within a fan, there may be one or more distinctive, ribbon-shaped dispersal trains. The edges of a fan are really the absolute limits of dispersal of a component and reflect the maximum range of ice flow directions across an area during one or more glaciations (Flint, 1971; Salonen, 1987).

One variation of the fan pattern is the amoeboid pattern (author's terminology) described by Klassen and Thompson (1989) for dispersal trains near or on the major ice dispersal centre in Labrador— Nouveau Québec. Similar examples have been described by Blais (1989), Lowell et al. (1990a), and Stea et al. (1989) in areas of late glacial ice divides in Quebec, Maine and Nova Scotia, respectively. As these ice divides migrated in a complex way around and across source outcrops or suddenly developed in areas of former unidirectional ice flow as a result of drawdown of parts of the glacier margin into rising marine waters, debris was dispersed first one way, then

another. The ultimate results of this constant shifting of positions of ice flow centres were irregular or 'amoeboid' areal patterns of dispersal (Fig. 15.17).

Even far from centres of ice flow, where lobes of ice from different centres coalesced, the zone of coalescence probably shifted with time, depending on the relative competition between ice masses (Veillette, 1989). Irregular or amoeboid dispersal can result from such shifting where the 'suture' marking the point of coalescence swept back and forth across source outcrops. Kaszycki (pers. commun; Kaszycki et al., 1988) has shown the potential for this in northern Manitoba, where the Keewatin and Labradorean Ice Sheets merged. The suture marking the zone of confluence shifted continually according to the climatic ascendency of one or the other ice sheets, producing, in the case of the Wheatcroft Lake arsenic dispersal train, westward striae formed by Labradorean ice in an area where southward dispersal was effected by Keewatin ice.

Recently, a third major type of dispersal train, formed by rapidly flowing streams of ice within the decaying North American continental ice sheet, has been recognized (Dyke and Dredge, 1989, pp. 198–199; Hicock, 1988; Aylsworth and Shilts, 1989; Hicock et al., 1989; Thorleifson and Kristjansson, in preparation). The areal compositional pattern of this type of dispersal train is similar to that of the classical dispersal ribbon, and it is commonly expressed geomorphologically by trains of drumlins and dense development of eskers. The compositional profile, however, is radically different from that of the ideal dispersal curve, and the width of the train often bears little relationship to the width of the source outcrops.

In profile, the composition of a train formed by 'streaming' is flat, maintaining high concentrations of the dispersed component from the source outcrops to the end of the train (Fig. 15.18). Little bedrock in the area of dispersal is incorporated into the train. The end of this type of train is marked by a sharp drop to background of the source component and probably marks the position of the ice front of the retreating glacier at the time the ice stream was activated. The width of the dispersal train usually reflects the width of the ice stream and not that of the source outcrops. Compositions change abruptly at the sides of the

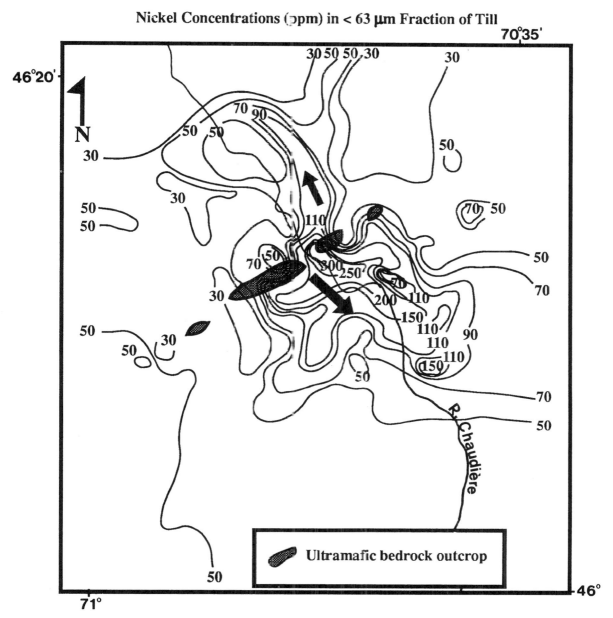

FIG. 15.17. Example of 'amoeboid', multidirectional dispersal train around ultramafic outcrops, Québec Appalachians

trains also. For instance, on Boothia Peninsula in northern District of Keewatin, Canada, a sharp-sided train of carbonate detritus several tens of kilometres wide, emerges from a small area within a large Paleozoic carbonate basin (Dyke and Dredge, 1989). Likewise, several narrow trains of Paleozoic carbon- ate debris extend southwestward across the Canadian Shield like fingers from the large Paleozoic platform that underlies Hudson Bay and the Hudson Bay Lowlands. The till in these trains, even 100 km from the source outcrops, is almost totally composed of clasts eroded from Paleozoic and Proterozoic out-

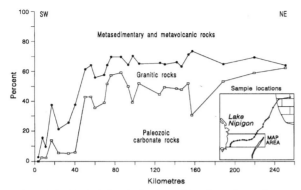

FIG. 15.18. Typical longitudinal profile of a dispersal curve associated with one of the many ice streams that formed in the last stages of breakup of the Laurentide Ice Sheet

crops within and adjacent to Hudson Bay (Hicock, 1988; Hicock *et al.*, 1989).

The ice stream dispersal trains in Ontario have important implications for the application of drift compositional studies to mineral exploration. The fact that exotic ice stream drift is relatively undiluted by local bedrock makes it a mask that conceals the compositional signal of bedrock beneath areas that it covers. In many places, the exotic till overlies till composed of more local components, presumably with dispersal characteristics reflecting the normal head-and-tail type of dispersal train (Fig. 15.19). The local till is assumed to have been deposited earlier in the same glaciation for which the retreat phase was marked by ice sheet instability resulting in ice streaming. Till with significant admixtures of local components also forms surface deposits adjacent to former ice streams, but may be encountered only in boreholes in areas covered by exotic, ice stream drift.

The late glacial ice sheet failures which resulted in narrow, rapidly flowing ice streams, probably represent instabilities created by excessive water trapped over strata of low permeability at the base of the decaying ice sheet (Hicock *et al.*, 1989), or deformable beds under active fast ice streams (Boulton and Hindmarsh, 1987; Menzies, 1989a; Alley, 1992) and may be much more common in continental glaciers than formerly recognized. They are presently recognized only where drift compositions are strongly anomalous or where they are marked by distinct trains

of drumlins (Fig. 15.20) and esker fields (Aylsworth and Shilts, 1989).

Finally, one more peculiarity of glacial dispersal with presently unevaluated implications for mineral exploration and drift composition should be mentioned – the 'conveyor belt effect'. Taken as a whole, a glacier works in a way somewhat analogous to a conveyor belt, continuously entraining material from many source outcrops and transporting it to the end of the glacier. During transport, much of the debris from one type of outcrop is deposited or is diluted by incorporation of debris from other outcrops. This results in the head and tail type of dispersal curve, with the tail eventually being obscured within background compositions. However, there is the potential for debris from every outcrop along the path a glacier follows, from its source or centre to its end, to reach the glacier's margin or snout, albeit in small amounts. If the glacier front stands in one place for some time, either at its terminal position or during halts during the active phase of its retreat, there is the opportunity for small amounts of debris from distant sources to be dumped and concentrated at its terminus (Stewart and Broster, 1990). With time, small amounts of material will increase in concentration in fluvial and ice-lain sediments associated with either terminal or end moraines, particularly if the debris has high specific gravity or is too large to be transported far in the proglacial fluvial system. High concentrations of gold in outwash and in alluvium derived from outwash at the Wisconsinan terminal moraine in the midwestern United States as well as diamond occurrences in moraines around the driftless area in Wisconsin may be products of this type of conveyor-belt concentration (Smith, 1992).

Because of the lack of lateral mixing of debris transported in ice, sampling of terminal moraines in modern or ancient valley glaciers – in the Cordillera for instance – can provide an efficient way to find clues to mineralization in their catchment basins (DiLabio and Shilts, 1979; Evenson and Clinch, 1987). The concentrations of far-travelled components in water-sorted sediments of the terminal moraines are likely to be much higher than those in the drift deposited over most of the glacier's bed. If carried out properly, a single sampling traverse along an end moraine should yield detailed information

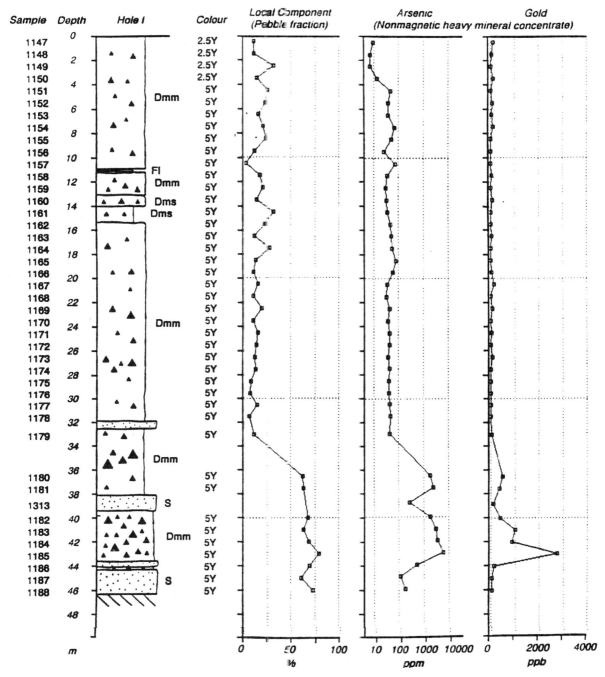

FIG. 15.19. Geochemistry and lithology of samples from a borehole drilled through thick till in the area represented by Fig. 15.18. Note the contrast in lithological and geochemical signals between the upper ice stream till and lower, local drift complex

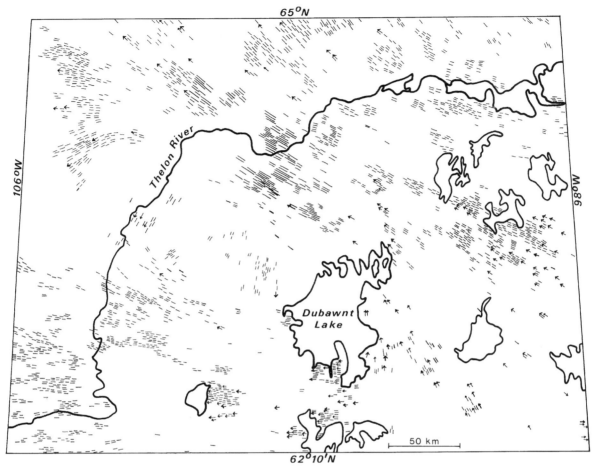

FIG. 15.20. Trains of drumlins formed by ice stream activity west of the Keewatin Ice Divide during the last stages of the Laurentide Ice Sheet

about mineralization and composition of bedrock in the glacier's catchment – whether the glacier is of continental scale, in which case the interpretation would be complex, or of the local scale typical of Alpine terrain, in which case relatively precise geochemical information about its catchment may be gleaned from limited, judicious sampling.

15.7. CONCLUSION

This overview of the relationship of drift geochemistry to the principles of dispersal draws heavily on Canadian examples, and particularly on work carried out at the Geological Survey of Canada. Similar glacial sedimentological and geochemical work, which has led to similar conclusions, has been carried out in many glaciated regions, that of geoscientists in Finland being probably the most varied and most prolifically published. Developments in the field of drift prospecting, the geochemical aspects of which have only blossomed since the late 1960s (Kauranne, 1975; Coker and DiLabio, 1989), are summarized in many Fennoscandian publications and in dedicated publications such as the semi-annual Prospecting in Areas of Glaciated Terrain (PAGT) series, sponsored by the Institution of Mining and Metallurgy, London, U.K. Recent texts dedicated to mineral prospecting in glaciated areas have been published or sponsored by the Geological Surveys of Canada (DiLabio and Coker, 1989), Ontario (Garland, 1989) and Finland (Kujansuu and Saarnisto, 1990) or in special issues of journals (Bjorklund, 1984).

Finally, it should be recognized that in this era of environmental concerns, the very geochemical data that can lead to discovery of mineralization in a glaciated area, may serve as an invaluable source of baseline data for planning environmentally friendly mining and for evaluating and mitigating the ultimate effects of mining activities. Drift geochemical surveys, in general, can and should be used to provide a chemical baseline from which to evaluate the plethora of environmental geochemical problems that will continue to beset industrialized areas in the decade and century to come.

Chapter 16

GEOLOGY OF PLACER DEPOSITS IN GLACIATED ENVIRONMENTS

V. M. Levson and S. R. Morison

16.1. INTRODUCTION

This chapter examines the sedimentary and geomorphic characteristics of placer deposits in glaciated environments. A discussion of placer deposits is included in this book since they are an important economic resource and, as they are formed at the surface, their physical characteristics are strongly affected by glacial and other surficial sedimentary processes. Many of these physical properties, such as depth of burial, grain size, ore grade and spatial distribution, directly affect mining and exploration practices. In addition, since productive placers are restricted mainly to sorted sand and gravel lithofacies, many of the sedimentary concepts discussed here in reference to placers can also be applied to aggregate resources.

Placer deposits are accumulations of heavy minerals that have been eroded from bedrock lode sources and concentrated through sedimentation processes involving gravity, water, wind or ice (Morison, 1989). Sedimentary accumulations of placer gold or other important heavy minerals such as diamonds, tin and platinum group elements are of economic importance in many areas of the world. The emphasis here is on gold placers as they dominate other types of heavy mineral deposits and, although some references to the latter are made, most descriptions of these deposits in the literature, particularly of tin and diamond placers, come from unglaciated areas.

Placer deposits, particularly gold, are of economic interest as they can be utilized to help locate lode sources that may supply a larger and more certain supply of gold or other minerals than the placer deposit itself. Although some gold placers may have a chemically-derived component formed by solution and re-precipitation (Boyle, 1979), the search for lode sources of all types of heavy mineral deposits is based on the premise that placer deposits are primarily detrital in origin. Thus, the tracing of source areas is dependent mainly on the dynamics and physical characteristics of the concentrating environment.

Although most placer-producing areas in the world occur in unglaciated areas, placer deposits in glaciated terrains support mining industries in northern Asia, British Columbia, Yukon Territory (Canada), and Alaska and numerous other regions (Boyle, 1979). This chapter discusses placer deposits in glaciated areas with emphasis on those occurring in Canada, particularly British Columbia, since they are among the best known and most productive in the world (Morison,1983a, b, 1985a, 1989; Levson et al., 1990; Levson, 1991a, 1992a,b; Levson and Giles, 1991, 1993).

16.2. SEDIMENTARY ENVIRONMENTS

Placer deposits in glaciated areas occur in a wide variety of sedimentary environments including alluvial, colluvial, glacial, glaciofluvial, coastal and marine settings. In the following sections the genesis of auriferous lithofacies typical of each of these environments will be discussed. The lithofacies' characteristics and origin of alluvial placers are considered in the context of buried placers of which they are a dominant component.

16.2.1. Alluvial Placer Deposits

Alluvial placers occur in current and abandoned water courses. Their formation is controlled by hydraulic conditions of the stream and the supply of gold and other sediment (Boyle, 1979). Fluvial settings, such as creeks, gulches, meandering and braided streams, and alluvial fans and fan-deltas, all have potential for heavy mineral concentration and preservation. Gold is most concentrated in coarse-grained pebble and boulder gravel and conglomerate facies (Fayzullin, 1969). On the basis of geomorphic and sedimentary characteristics, alluvial placers can be grouped into: (1) creek or gulch placers formed in relatively high-gradient, tributary valleys; (2) river placers formed in relatively low-gradient, main valleys; and (3) alluvial fan and fan-delta placers. Placer deposits in the second category are the most common.

PLATE 16.1. Modern channel of Beggs Gulch (Location 9, Fig. 16.5) largely mined out in the 1800s

16.2.1.1. Creek and gulch placers

Creek and gulch placer deposits are associated with deeply dissected, high-relief terrains with weathered and colluviated slopes; valleys are narrow, restricted, deeply incised and commonly occur as tributaries to larger main valleys (Plate 16.1). The longitudinal profiles are relatively steep compared to river placers (see below). Discharge is seasonal in nature and gradients are moderate to high. They typically carry large volumes of water and sediment during episodic flood events generated by snow melt and storm induced freshets. Large increases in the amount and size of material moved by the stream occur as a result of increased discharge and velocity. High stream capacity and competence are ideal for concentrating

coarse placer gold and, consequently, gulch placers have proven to be most productive.

The resulting alluvial deposits are generally thin and confined to narrow channels. The gravelly sediments typically are crudely stratified, poorly sorted and locally derived. Clasts are angular and commonly are up to large boulder size. Pay zones occur on or close to the bedrock surface in the thalweg of the gulch stream bed. The gold is generally coarse grained and the accompanying heavy mineral suite is closely allied to the local bedrock mineralogy (Boyle, 1979).

In glaciated areas, paleogulch placers are typically buried by till, glaciofluvial and glaciolacustrine sediments (Levson and Giles, 1993; Levson, 1992b). Holocene streams, however, may re-occupy preglacial

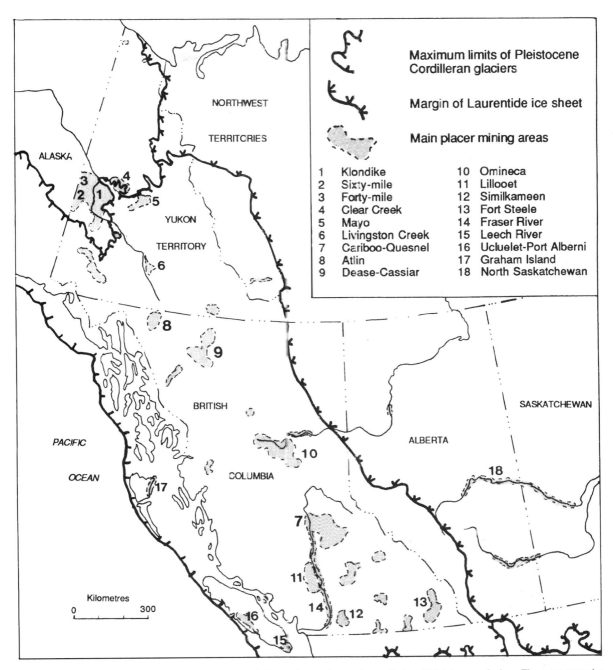

FIG. 16.1. Major placer mining areas in Western Canada in relation to the maximum limits of Pleistocene glaciers. The most extensive placer mining region occurs on the northwest periphery of Pleistocene glaciations in west-central Yukon, but numerous placer areas also occur within the region covered by the Cordilleran ice sheet, particularly in British Columbia. The general northwest trend of placer occurrences reflects the regional strike of mineralized bedrock terranes in the Canadian Cordillera. Numbered areas are discussed in the text. (Placer areas modified from Boyle, 1979; ice limits from Clague, 1989c and, for central Yukon, from Hughes, 1987)

and interglacial channels, eroding the overburden and, in some cases, re-concentrating older placers. These Holocene gulch placers were depleted in most regions by early miners due to ease of exploration and mining. However, the poor sorting and large clast size typical of these deposits locally prevented deeper mining and, in some areas such as in central and northern British Columbia (Fig. 16.1, Locations 7 and 8), Holocene gulch placers of this type remain unexploited (Levson, 1991a, 1992a).

16.2.1.2. River placers

River placers are associated with fluvial environments such as meandering and braided streams. Valley bottoms are commonly wide with relatively gentle longitudinal-profiles. In meandering river environments, heavy minerals may accumulate in main channel bedforms such as dunes or as lag deposits, or in point bars and chute cutoffs on the inside curve of meanders, especially during flood conditions. The distribution of detrital gold deposited in a braided river environment is sporadic and discontinuous (Smith and Minter, 1980). In this environment, favourable areas for heavy mineral concentrations include transverse bars, channel bends, channel junctions and segments where stream flow is convergent. In addition, entrapment of heavy minerals during aggradation of proximal channel gravels, as either diffuse gravel sheets (Hein and Walker, 1977) or as unit bar forms, is possible. In summary, the pay zone in river placers ranges from linear paystreaks that are parallel to the channel system, to discontinuous concentrations, which are not necessarily limited to the base of the alluvial fill (Morison, 1989). In general, coarse lag placer concentrations occur where stream flow is concentrated, such as in the thalweg and at channel junctions, and finer gold is deposited at sites where stream competence initially decreases, such as in the lee of obstacles or on point bar margins.

River placer gravels typically are well-sorted, horizontally stratified, imbricated, well-rounded pebble to cobble gravels (Plate 16.2). Planar and trough cross-stratified gravel beds occur locally and sandy interbeds and lenses are common. Terrace gravel

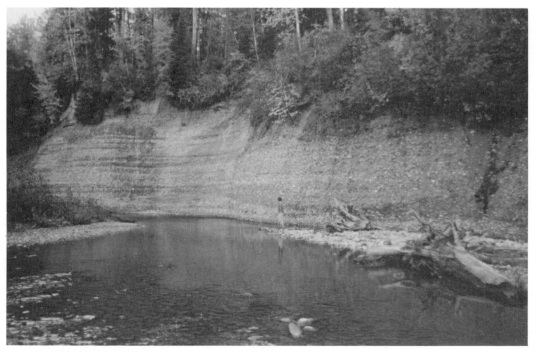

PLATE 16.2. Typical river channel gravels exposed along the Cottonwood River (location 3 on Fig. 16.5)

sequences are typically capped by overbank fines exhibiting weak horizontal laminae and containing abundant organic material. Scoured lower contacts in gravel beds are often overlain by concentrations of coarse clasts. Channel lags formed during periods of relative channel stability are primary placer targets. Overlying bedded gravel sequences contain less gold and reflect bar sedimentation during aggradational phases (Levson and Giles, 1993).

River placers are the most common type of placer deposit in both glaciated and unglaciated areas. In recent river alluvium and on low terraces near the placer source they are easily located and evaluated. In addition, relatively coarse gold in the upper reaches of gold-bearing streams is easily recovered. In the Cariboo placer mining region of British Columbia (Fig. 16.1, Location 7), for example, placer deposits of these types were quickly depleted by mining activity within several years of discovery (Johnston and Uglow, 1926). More recent mining of fluvial deposits has focused on more distal terrace and active river placers where progressively finer gold has been

carried. Mining activities in downstream placers of this type are limited by the efficiency of fine gold recovery and the ability of the miner to locate sedimentary environments and specific facies where fine gold is concentrated (Levson, 1991a). In active river channels fine gold is commonly mined from sands on the surface of gravel bedforms and point bars exposed during low water and from areas such as the upstream end of ephemeral flood channels where flow depth and velocity decrease suddenly (Plate 16.3).

16.2.1.3. Alluvial fan placers

Gold in alluvial fan placers is found in debris-flow sediments and in interstratified gravel, sand and silt of subaerial and subaqueous origin (Krapez, 1985). The deposits typically are dominated by diamicton with some interbedded, poorly-sorted gravels and sands (Plate 16.4). Resedimentation of unconsolidated deposits along valley slopes in the drainage basin occurs by gravity-dominated processes. Debris-flow

PLATE 16.3. Small placer mining operation (photo centre) on the Fraser River (location 47 on Fig. 16.5) exploiting modern point bar sediments

PLATE 16.4. Gold-bearing alluvial fan sediments on Spanish Mountain (location 21 on Fig. 16.5). Note the crude stratification and poor sorting of debris-flow deposits and interbedded gravels and sands

sedimentation occurs mainly in periods immediately following deglaciation when lack of vegetation and climatic conditions enhance slope instability. The debris flow deposits are commonly reworked by intermittent fluvial activity. Gold values are highest in regions where the resedimented deposits incorporated older auriferous alluvial gravels or other gold-bearing sediments.

Gold values typically are relatively uniform throughout alluvial fan sequences but overall low to moderate grades contrast sharply with higher grade fluvial deposits. For example, gold content is generally consistent throughout mined sequences of alluvial fan sediments in central British Columbia (Fig. 16.1, Location 7) and averages about 1 gm^{-3}, not including gold finer than 100 mesh (0.149 mm) (Levson and Giles, 1991). The fan deposits consist of poorly sorted and crudely stratified coarse gravels interpreted as debris-flow deposits and interbedded lenses of better sorted gravel, sand and silt interpreted as fluvial channel deposits (Plate 16.4). Gold within the fan sequences is believed to be mainly local in

origin as indicated by its typically high angularity and by the abundance of locally derived clasts in associated sediments (Levson and Giles, 1993).

Levson *et al.* (1990) provided a detailed description of alluvial fan and fan-delta placer deposits (Fig. 16.2) exposed at an operating mine in the Cariboo region of central British Columbia (Fig. 16.1, Location 7). The auriferous deposits consist of a complex unit of sands and gravels (Fig. 16.2, Unit 4) with gold concentrations of approximately 2 gm^{-3}. The uppermost deposits in the gold-bearing sequence are clast-supported, poorly sorted diamictons (Fig. 16.2, Unit 5) that contain about 0.6 gm^{-3} of gold. They exhibit sedimentary characteristics typical of modern debris-flow deposits such as crude horizontal-bedding, weak imbrication, poor sorting, disorganized fabric, gradational bed contacts and folded and boudinaged beds (Bull, 1972; Burgisser, 1984; Kochel and Johnson, 1984). Interbedded stratified-gravel beds indicate some fluvial activity, probably between debris-flow events. The temporal increase in debris-flow activity, possibly due to the reduction in vegetative cover

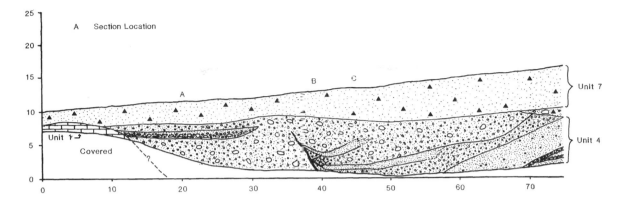

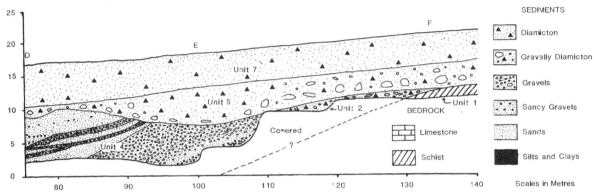

FIG. 16.2. Geologic cross-section showing stratigraphic units exposed along the upper part of the main highwall at the Ballarat Mine (location 29, FIG. 16.5). All units except the upper till (unit 7) are mined for gold (from Levson *et al.*, 1990, reproduced with permission of *British Columbia Ministry of Energy, Mines and Petroleum Resources*)

associated with glacial conditions, resulted in progradation of alluvial-fan sediments over the area. The uppermost diamicton at the mine (Fig. 16.2, Unit 7) is interpreted as a till deposited during the Late Wisconsinan glaciation.

16.2.2. Glacial and Glaciofluvial Placers

The subglacial erosional characteristics of an advancing wet-based ice sheet are such that pre-existing auriferous gravel deposits, occurring in valleys oriented parallel to the direction of ice advance, are commonly eroded (Morison, 1989). Gold is thus incorporated into the resulting glacial debris and dilution of gold concentrations occurs due to mixing of distally and locally derived sediments (Chapter 15).

Although glaciers may generally destroy or dilute pre-existing placer concentrations, some subglacial processes can locally deposit gold from bedrock or paleoplacer sources in mineable quantities. Tills, deposited by glaciers overriding older stream placers for example, can be productive where the gold concentration in the original placer was exceptionally high (Maurice, 1988; Levson and Giles, 1991, 1993). Till and glacigenic debris-flow deposits (Plate 16.5) contain mineable quantities of gold at a few sites in the Cariboo area of British Columbia (Fig. 16.1, Area 7). Glaciers probably incorporated gold from rich paleochannel placers up-valley of the mines. Gold concentrations in till and incorporated auriferous sediments in most areas, however, are too low to mine (e.g. $\leq 0.1–0.2\,\mathrm{g\,m^{-3}}$) but may locally be worth processing with improved recovery systems (Levson,

PLATE 16.5. Auriferous sequence of till and debris-flow deposits of marginal productivity at an abandoned mine in the Cariboo (location 32 on Fig. 16.5)

Zealand, for example, esker-like deposits have been mined for their gold content (Williams, 1974).

Due to typically low gold concentrations, economic glaciofluvial placers are usually restricted to large volume outwash deposits with minimal overburden thicknesses. For example, the economic production from glaciofluvial placers in New Zealand has been mainly a result of exploitation of near surface outwash gravels by large-scale dredging operations (Williams, 1974). Similarly, in the Cariboo region of British Columbia, near surface outwash gravels (Plate 16.6), containing gold concentrations of approximately $0.15–0.35 \, \text{gm}^{-3}$, have recently been mined on a large scale (Levson and Giles, 1991) (Plate 16.6).

Glaciofluvial placer gravels commonly occur as terraces deposited either by ice-marginal glaciofluvial streams or in proglacial braided streams shortly after local deglaciation. Sedimentation in these proximal streams is largely aggradational and characterized by multiple shifting channels. Consequently, the resultant gravels form large volume deposits with low gold concentrations. Gold content is generally highest at the base of channel fill sequences but otherwise may be distributed throughout the deposits. In the Cariboo, terrace gravels occurring well above modern river levels (Plate 16.7) contain significant quantities of gold only where they overlie older gulch placer deposits. Elsewhere, the gravels are almost devoid of gold.

16.2.3. Coastal and Marine Placers

Glaciation can increase the potential and, in some cases, is a prerequisite for the formation of coastal and marine placer deposits. High energy glaciofluvial and glacial processes are capable of carrying large volumes of heavy minerals into marine environments where they can be concentrated by nearshore processes into lag and strandline deposits (Barker et al., 1989). Similarly, fluctuations in sea level, common in glaciated areas, allow for stream transport of auriferous alluvium into formerly submerged areas. For example, Nelson and Hopkins (1972) described detrital gold accumulations in the sediments of the Bering Sea off Nome, Alaska. The concentrations are related to marine regressions and transgressions with associated reworking of auriferous debris deposited

1991a). Generally, detrital gold is widely dispersed in tills and reconcentration by postglacial streams is required before an economic placer deposit is created.

Economic gold concentrations are also rare in glaciofluvial deposits and are usually restricted to areas where meltwaters have reworked older placers. Glaciofluvial placers are known from the Yukon, British Columbia, Quebec, Siberia and elsewhere (Boyle, 1979). Ice contact deposits such as kames, eskers and proglacial outwash sediments may locally contain low quantities of gold (Morison, 1989). In rare cases, subglacial meltwaters may form mineable concentrations of gold from previously deposited sediments, but these placers are usually small. In New

PLATE 16.6. Cobble to boulder, glaciofluvial gravels exposed in a former large open-pit mine at Tregillus Lake (location 37 on Fig. 16.5) (pick is 65 cm long)

PLATE 16.7. Typically well-stratified and well-sorted high-terrace gravels near Burns Creek (location 41 on Fig. 16.5). Gravels in this pit are barren but a few hundred metres west they support a productive placer mine. Paleocurrent data indicate that gold occurs only downstream of the intersection of the glaciofluvial channel and an older, auriferous paleochannel

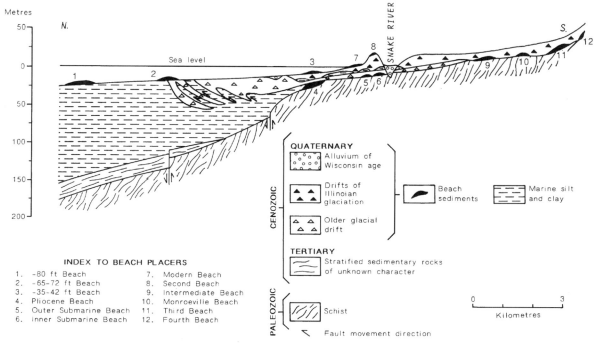

FIG. 16.3. Geologic cross-section showing coastal and offshore paleoplacer deposits in the vicinity of Nome, Alaska, based on study of mine exposures, drill records and seismic-reflection profiles (from Nelson and Hopkins, 1972 and Boyle, 1979). See Fig. 16.11 for location

on the shelf during glacial intervals. Khitrov and Khershberg (1982) described Russian marine placers that formed under conditions of non-uniform marine transgression during the Pleistocene and Holocene.

Coastal and marine placers include beach or strandline deposits, buried marine scarps, drowned fluvial deposits and submerged residual or lag deposits (Samson, 1984; Barker *et al.*, 1989). Beach placers are formed on wave-dominated shorelines through the winnowing action associated with long-shore currents and wave action. Paystreaks follow strandlines and gold grains are generally small and of high fineness (Boyle, 1979). Beach placers can be divided into those resulting from erosion of gold-bearing bedrock by wave action and those that formed by wave and tidal reworking of detrital sediments, principally auriferous tills and glaciofluvial deposits (Samson, 1984). Beach placers have been mined extensively near Nome, Alaska, since the turn of the century

(Cobb, 1973). Three submerged beaches occur on the sea floor offshore of Nome, and two, including the modern beach, occur onshore (Fig. 16.3).

On northeast Graham Island on the west coast of British Columbia, black beach sands containing very fine gold and platinum formed by wind and wave action reworking Pleistocene, stratified sands and gravels exposed in coastal bluffs (Mandy, 1934). Crudely-stratified, auriferous gravel lenses within the sands may represent interglacial re-concentrations of gold from morainic deposits. The auriferous tills may be derived from gold- and platinum-bearing source rocks in Alaska (Carmichael and Moore, 1930). The black sands occur as lenses up to about 120 m long, 15 m wide and 15 cm thick. They do not lie on impervious strata such as clays but are distinctly interbedded with white sands and their extent and distribution is subject to wave action during storms (Carmichael and Moore, 1930; Mandy, 1934). The beach placers are best developed in areas subjected to violent southeasterly gales, but one placer on the north side of the Island at Masset Inlet has yielded unusually coarse gold for the area, with strata containing as much as 20 g (0.6 oz) of gold and 70 g (2 oz) of platinum per tonne (Carmichael and Moore, 1930).

Submarine residual placers are most common in high energy marine settings such as along the Scotian Shelf or in the Hecate Strait off the Queen Charlotte Islands in British Columbia (Samson, 1984). Heavy mineral concentrations on the continental shelf sea bed off northern British Columbia include thin lag deposits formed by hydrodynamic reworking during extreme yearly storms (Barrie et al., 1988). Gold in marine placers in the Country Harbour area of southeastern Nova Scotia is coarsest and most concentrated in lag gravels on bathymetric highs exposed to the winnowing action of storm waves (Hopkins, 1985).

Offshore placers can accumulate on the sea bed as relict deposits formed during glacially-induced low sea level stands or under present oceanographic conditions (Barrie et al., 1988). Titaniferous rich sands up to 20 m thick in Queen Charlotte Sound formed 10,500 BP at 100 m below present sea level (Barrie, 1990). Surface relict gravels offshore of Nome (Fig. 16.3) contain gold reworked mainly from auriferous glacial drift and locally from bedrock by nearshore processes. The gravels have an average gold content of 920 ppb with most of the gold ≥1 mm in diameter (Nelson and Hopkins, 1972). During periods of eustatic low sea level, induced by glaciation in the Nome area, streams dissected exposed beaches and reconcentrated gold in alluvial channels. During high sea level stands in interglacial and interstadial times, nearshore bottom currents buried the former channels and other topographic lows with non-auriferous fine-grained sediments but simultaneously prevented burial of more elevated auriferous relict beach gravels (Nelson and Hopkins, 1972). Offshore paleochannel placers, formed during isostatic low sea levels after ice retreat, have also been recorded in Nova Scotia (Samson, 1984; Hopkins, 1985). Some of the rivers may have inherited their courses from Tertiary channels and re-concentrated gold from the older placers (Samson, 1984).

16.2.4. Colluvial Placer Deposits

Colluvial placer deposits typically overlie bedrock and are thickest near the base of slopes. They generally consist of unconsolidated, matrix-supported diamicton with thin intrabeds and lenses of poorly sorted sand and gravel. They typically contain a high proportion of locally derived angular clasts. Clast-supported, sandy gravel units directly over bedrock are the most productive.

16.3. REGIONAL DISTRIBUTION OF PLACER DEPOSITS IN GLACIATED AREAS

It is important to note that although placer deposits are mined in many glaciated areas of the world, the most productive placers occur in unglaciated terrains. Bundtzen (1991), for example, noted that 80% of the commercial placer gold and other heavy mineral deposits in Alaska and Yukon Territory and 90% of those in the Magadan region of northeast Asia occur in unglaciated areas where the placers evolved over several millions of years. It is generally assumed that, even in glaciated areas, most placer minerals were released from bedrock by weathering and erosion during the Tertiary and, to a lesser extent, during non-glacial and/or interglacial intervals of the Quaternary.

Areas glaciated during the Quaternary generally have diminished placer productivity for several reasons. First, there was less time for colluvial and fluvial processes to concentrate placers during interglacial and postglacial periods than in preglacial times. Second, rich placer deposits that accumulated over long preglacial periods may be eroded and re-deposited during glaciation by ice and glacial melt-waters. The resultant glaciogenic sediments typically have relatively low gold contents due to dilution of the original placer deposits with glacially derived debris. Since exposed areas of auriferous bedrock are also susceptible to glacial erosion, successive glaciations progressively may decrease the availability of easily erodable gold point-source concentrations (Smith and Minter, 1980). Placers that do escape erosion by glaciers generally are buried by glaciogenic sediments making them difficult to locate and mine. Finally, glaciogenic and associated processes spatially (vertically and horizontally) disperse and sedimentologically alter preglacial placers. The resultant stratigraphically and geomorphologically complex distribution and sedimentologic variability of placer deposits in glaciated terrains creates numerous exploration, evaluation and mining problems. Yet, in spite of these difficulties, placer deposits in glaciated areas continue to be important commercial sources of gold and other heavy minerals.

16.3.1. Canadian Placer Deposits

Placer deposits in Canada range in age from the Precambrian uranium paleoplacers of the Blind River–Elliot Lake area in northern Ontario to modern gold deposits on Fraser River bars in British Columbia. Although this discussion is concerned primarily with gold and with placer deposits of Quaternary age, some consideration is given to other types of placers, particularly Tertiary placers that, in some cases, served as sources for Quaternary deposits. Quaternary placers include Pleistocene preglacial, interglacial and glacial deposits and Holocene post-glacial sediments.

16.3.1.1. Cordillera

A large number of placer deposits occur in the Canadian Cordillera within, and along the margins of, regions glaciated in the Pleistocene (Fig. 16.1). Most placer mining areas occur east of the Coast Mountains and west of the Rocky Mountains, in a region of high, bedrock mineral potential defined by the northwest trending Intermontane and Omineca tectonic belts in central British Columbia and Yukon Territory.

Yukon Territory: Yukon placers have contributed up to 5% of total Canadian gold production, and are the most economically significant placer deposits in Canada (Morison, 1989). According to royalty records, Yukon placer deposits produced nearly 12 million fine oz (385,200 kg) of gold between 1885 and 1982 (Debicki, 1983). The Yukon Territory is unique in Canada in that the west-central and northern regions escaped glaciation (Hughes *et al.*, 1969, 1972; Hughes, 1987) and as a result there is a high degree of placer deposit preservation in these regions, notably in the Klondike, Sixty-mile and Forty-mile areas (Fig.

16.1, Locations 1 to 3). However, in central and southern regions, Quaternary glaciation has dispersed preglacial and interglacial placer deposits in valleys oriented parallel to ice flow and buried those in valleys transverse or oblique to ice direction.

Placer mining throughout the glaciated areas of central and southern Yukon occurs in a variety of placer deposit settings, each with a unique depositional history. For example, in Clear Creek drainage basin (Fig. 16.1, Location 4) placer gold can be found in valley bottom creek and gulch gravel, glacial gravel and buried preglacial fluvial gravel (Morison, 1983a,b, 1985a). In the Dublin Gulch area near Mayo (Fig. 16.1, Location 5) small colluvial placers have developed from local bedrock sources. Although the Dublin Gulch region is beyond the limits of the last glaciation, remnants of weathered tills and hematite erratics provide evidence of an earlier glacial incursion (Boyle, 1965, 1979; Boyle and Gleeson, 1972). In the Mayo region, gold also occurs in other valleys that contain till and glaciofluvial sediments, such as Duncan, Haggart and Lynx Creek valleys. Gold in these areas is believed to be derived from pre-glacial auriferous gravels that escaped severe glacial erosion (Boyle and Gleeson, 1972). Interglacial placer deposits buried by thick glaciolacustrine, glaciofluvial and glacial deposits have been mined for many years along several streams in the Livingstone Creek area (Fig. 16.1, Location 6), about 125 km within the limits of Pleistocene glacial advances (Levson, 1992b).

Auriferous gravel in the Klondike (Fig. 16.1, Location 1) was first described by McConnell (1905, 1907). He divided these deposits into stream gravel, terrace gravel, 'White Channel Gravel', and high level river (glacial) gravel. The White Channel deposit (Pliocene?) forms terraces 50 to 100 m above present-day stream levels and is divided into white and yellow gravel units (McConnell, 1907). White Channel clastic sediment is characterized by a complex assemblage of lithofacies types ranging from laminated silt and clay to massive and disorganized boulder gravel (Morison and Hein, 1987), and gold concentration generally increases with depth (Gleeson, 1970). White Channel Gravel is thought to represent Pliocene to early Pleistocene gravelly sedimentation in a braided river environment (Morison, 1985b; Morison and Hein, 1987).

The Clear Creek drainage basin (Fig. 16.1, Location 4) is a well studied example of placer deposit settings in glaciated parts of the Yukon. Proven placer settings include stream and gulch gravel and buried preglacial fluvial gravel (Fig. 16.4). Favourable placer deposit settings that have not been explored extensively include alluvial fan gravel, glacial gravel deposits and buried Pliocene (?) gravelly sediments (Morison, 1983a, 1985a).

The upper reaches of the Clear Creek drainage contains remnant occurrences of glacial drift deposited during the pre-Reid glacial advance (Morison, 1983a). The pre-Reid glaciation is the oldest Cordilleran ice advance in the Yukon (pre-Illinoian) and includes both the Nansen and Klaza glacial periods. Preglacial fluvial gravel or 'buried channels' may be preserved beneath these pre-Reid deposits and the following description is typical of the stratig-

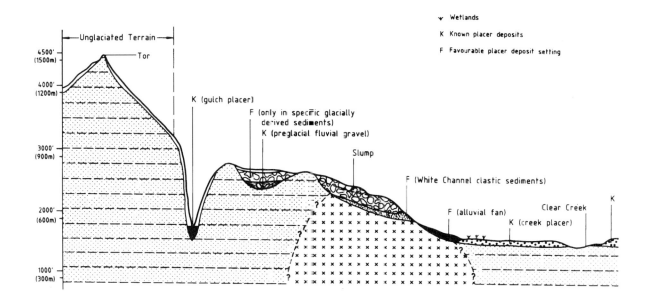

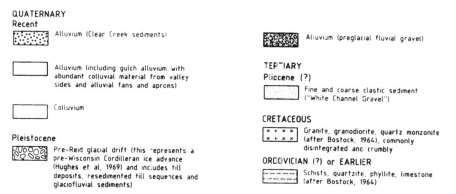

FIG. 16.4. Schematic profile of Clear Creek drainage basin (area 4 on Fig. 16.1) showing types of placer deposits and surficial materials (from Morison, 1985, reproduced with permission from *Yukon Exploration and Geological Services*)

raphy associated with this type of buried placer deposit:

The lowermost deposits, underlying pre-Reid glacial drift, are gold-bearing, clast-supported, matrix-filled gravels with abundant trough cross-stratification. Occasional disorganized, openwork gravelly beds and trough cross-stratified sandy beds are also present. The placer gravels were deposited in a multi-channelled stream system before the onset of pre-Reid glaciation. The gravels occur at relatively high levels in relation to the modern stream system indicating a significant change in relative base level since their deposition. The auriferous gravels are commonly overlain by a few metres of diamicton with lithofacies characteristics that resemble basal melt-out till at the margin of Matanuska Glacier in Alaska (Lawson, 1981b). This lower diamicton in some areas is gradationally overlain by several metres of inter-bedded massive diamicton and sorted, fine clastic sediments. Diamicton beds are up to 1 m thick, have a matrix texture of sandy silt to fine sand and contain subangular coarse clasts (distinctly lacking in the underlying melt-out till unit). The sorted interbeds are up to 10 cm thick and range from fine sand to clay. These general lithofacies characteristics suggest that these deposits originated as a series of debris flows (massive diamicton), possibly on ice marginal surfaces, with associated ponding (clay beds) and/or meltwater reworking (fine sand and silt beds) between sediment flow events (Lawson, 1982). Surface diamicton deposits that contain local bedrock fragments and drape the underlying slope morphology are interpreted as recent colluvial deposits.

British Columbia: Between 1856 and 1981, placer mining districts in British Columbia produced approximately 164,000 kg (4.3 million fine oz) of gold (Debicki, 1984). Most of this gold has come from the Cariboo Mining District (Fig. 16.1, Location 7) which has produced 75,000 to 100,000 kg (about 2.5 to 3 million ounces) of gold since its discovery in 1860 and continues to be the major placer producing area in the glaciated regions of Canada. Over 350 mining operations and exploration projects are active in the Cariboo. A second important placer mining region is the Atlin area in northwest British Columbia (Fig. 16.1, Location 8). Other well known placer streams in the province are shown on Fig. 16.1.

Although British Columbia was glaciated repeatedly during the Pleistocene, sediments predating the Late Wisconsinan Fraser Glaciation occur at many sites and include fluvial gold-bearing gravels of Late Tertiary to Mid Wisconsinan (Late Pleistocene) age (Levson, 1992a; Levson and Giles, 1991, 1993). Studies of pollen collected from interstadial sediments near the centre of the Cordilleran ice sheet indicate ice-free conditions in a radiocarbon dated interval from 40,000 to >51,000 BP (Clague et al., 1990). Paleoplacers in a few areas are also buried by till and ice stagnation deposits of the penultimate glaciation (Clague, 1991).

Although most of the productive placers in British Columbia formed in preglacial or interglacial times, economic post-glacial placers also occur where Holocene fluvial processes have reconcentrated gold from pre-existing placers. The period of time since deglaciation and the amount of erosion of auriferous bedrock seems to have been inadequate in most areas to allow for the formation of economically viable placers primarily from bedrock sources.

Recent studies of the trace element geochemistry of gold in British Columbia have demonstrated a spatial and genetic relationship between lode and placer gold at a number of sites (Knight and McTaggart, 1990). Although oxidation of host minerals (e.g. pyrite) may release trace gold to solutions that can, under rare circumstances, reprecipitate gold to form nuggets, this process typically results in the formation of nearly pure gold (Boyle, 1979; Mann, 1984; Vasconcelos and Kyle, 1989). However, very few of the several hundred analyses of gold fineness presented by Knight and McTaggart (1990) indicated gold of this purity, and for this reason as well as several others, such as the geochemical association of known lodes and nearby placers, the typically abraded shape of the grains and the presence of angular quartz or pyrite rather than fluvially rounded inclusions, they concluded that the gold was of detrital origin. Spatial relationships between placer deposits and lode sources, such as paleoflow direction, distance of transport and proximity to source, can be determined largely from the sedimentologic characteristics and inferred depositional environments of the placers (Levson, 1991c).

Cariboo Mining District: The Cariboo region is one of the largest and best known placer gold producing

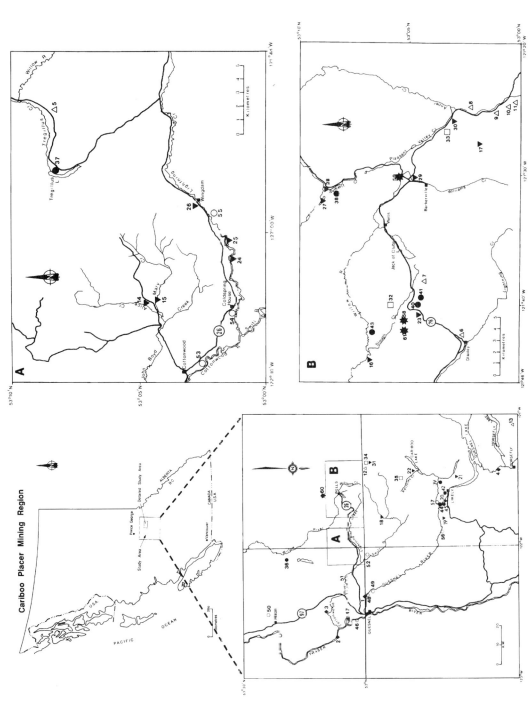

FIG. 16.5. Location of the Cariboo placer mining area and study sites, with detailed maps of the Wells-Barkerville region (A) and the lower Lightning Creek area (B). Symbols indicate the type of placer: diamonds – Tertiary placers; open triangles – buried gulch placers; solid triangles – buried river-valley placers; open squares – glacial placers; solid circles – glacial placers; open circles – low terrace placers. The most productive placers are buried alluvial deposits (triangles). (Modified from Levson and Giles, 1991, 1992)

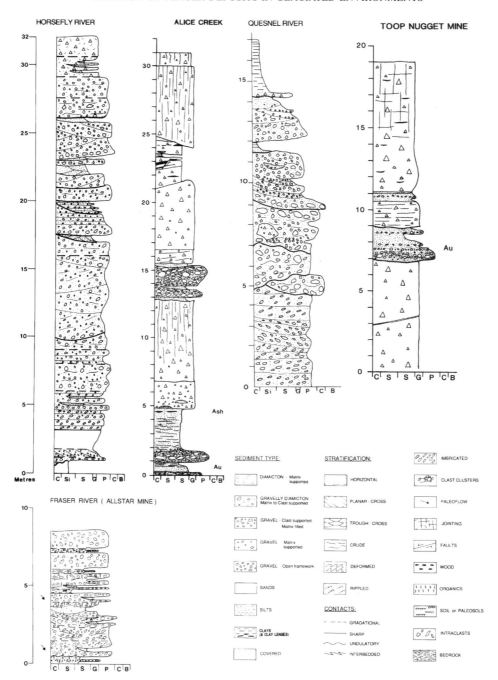

FIG. 16.6. Stratigraphic sections of Tertiary and Quaternary deposits at placer mines in the Cariboo region (see Fig. 16.5 for locations). Gold-bearing, cemented, gravels and sands at the Allstar Mine (location 2) and at the base of exposures along Horsefly and Quesnel rivers are Tertiary in age. They are stratigraphically overlain by glaciofluvial and glacial deposits, as shown in the Horsefly River section (location 4), and by multiple Quaternary gravel units and glaciolacustrine deposits as at the Quesnel River section (location 1). Preglacial gravels at Alice Creek (location 14) and the Toop Nugget Mine (location 15) are overlain by up to 30 metres of till and intertill sands and gravels. Vertical scale in metres on all sections. Horizontal scale indicates mean grain size: C-clay S-sand Si-silt G-granule P-pebble C-cobble B-boulder. (Modified from Levson et al., 1990 and Levson and Giles, 1992)

areas in glaciated terrain in the world. This area is discussed, in detail, to illustrate the stratigraphic, sedimentologic and geomorphic characteristics of placers in glaciated areas. Many comprehensive descriptions have been made of placer deposits in the Cariboo (e.g. Johnston and Uglow, 1926; Clague, 1987a,b, 1989a,b, 1991; Eyles and Kocsis, 1988, 1989a,b; Clague et al., 1990; Levson et al., 1990; Levson and Giles, 1991, 1993; Levson, 1992a). The locations of 61 sites investigated by Levson and Giles (1991, 1993), are shown in Fig. 16.5. Each site is classified by its stratigraphic, sedimentologic and geomorphic characteristics; and they are organized on Fig. 16.5, into placers of presumed or known Tertiary age, pre-Late Wisconsinan buried gulch, alluvial fan and river placers, Late Wisconsinan glacial placers, glaciofluvial and high-terrace placers, Holocene (postglacial) low-terrace placers and colluvial and alluvial fan placers. The most productive of these deposits, Tertiary and Pleistocene preglacial and interglacial placers, are generally buried below till

and other non-auriferous glaciogenic sediments of one or more glaciations.

At the Alice Creek and Toop Nugget mines, for example, gold-bearing paleochannel gravels are overlain by up to 30 m of Pleistocene overburden including two diamicton (till) units (Fig. 16.6). Although the lowermost gravels are of probable Tertiary age, inter-till sands and gravels and early postglacial gravels also contain some gold, probably reconcentrated from underlying units (Levson et al., 1990). Similarly, in the Horsefly and Fraser River areas (Fig. 16.5), buried auriferous gravels are overlain by thick sequences of Tertiary and Quaternary deposits (Plate 16.8; Fig. 16.6). The lowermost pay zones may contain up to ~25 g of gold per tonne. They typically are comprised of partially cemented, massive to crudely stratified, sands and gravels with scoured lower contacts (Plate 16.9) interpreted as fluvial channel deposits with interbedded sandy bar and overbank sediments (Fig. 16.6).

PLATE 16.8. Incised valley of the Fraser River north of Quesnel near location 2 on Fig. 16.5). Exposures in the background are Tertiary and Quaternary sediments

PLATE 16.9. Exposure along the Quesnel River at the Allstar Mine (location 2 on Fig. 16.5) of partially-cemented Tertiary gravels and interbedded sands. Note the scour structure on the lower right, (rock hammer for scale)

The stratigraphic complexities of preglacial and interglacial gravel sequences are illustrated by a succession of texturally mature river gravels, believed to be preglacial in age, and at least two other distinct younger gravel units (Fig. 16.6), exposed along the Quesnel River (Fig. 16.5, Location 1). Postglacial terrace gravels, unconformably overlying the older gravel units, are currently being mined, but preglacial gravels are not mined at this location owing to their induration. Much of the placer gold in the region is believed to have been reconcentrated from preglacial gravels by younger river systems (Levson and Giles, 1991, 1993).

Atlin Mining District: In the Atlin area of northwest B.C. (Fig. 16.7), gold placers occur mainly in fluvial gravel deposits that are almost everywhere overlain by Late Wisconsinan glacial drift (Levson, 1992a; Levson and Kerr, 1992). The Atlin placers are exceptionally coarse and have yielded some of the largest nuggets in the province (Plate 16.10). The principal placer deposits are in the valleys of Spruce

and Pine Creeks, the two highest gold producing streams in British Columbia.

In the Spruce Creek valley (Fig. 16.7), several distinct stratigraphic units occur, including possible preglacial, interglacial and postglacial deposits. Up to 4 metres of the lowermost auriferous gravel units overlying bedrock have been mined (Fig. 16.8). The gold-bearing gravels are generally poorly sorted, clast supported and crudely stratified. They are interpreted as high-energy fluvial channel deposits (Levson, 1992a). The paleochannel orientation is oblique to the trend of the modern valley as indicated by paleo-current measurements (Fig. 16.8). Stratigraphically higher gravel units are largely barren and are believed to be mainly glaciofluvial in origin. Strata consisting of diamictons in the uppermost part of the sequence are interpreted as till and ice-proximal debris-flow deposits.

In the Pine Creek valley (Fig. 16.7), gold-bearing gravels, interpreted as fluvial channel and debris-flow deposits overlie bedrock. They are overlain succes-

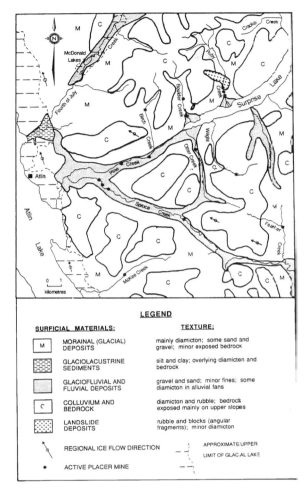

FIG. 16.7. Generalized surficial geology of the Atlin area (location 8 on Fig. 16.1) showing the locations of active placer mines (from Levson, 1992a, reproduced with permission of *British Columbia Ministry of Energy Mines and Petroleum Resources*). The deposits are mainly paleochannel gravels confined to modern valleys. They are buried by till and glaciofluvial sediments at most mines and by Pleistocene basalts in the Ruby Creek valley

sively by till, glacially derived debris-flow and proximal glaciofluvial deposits (Levson, 1992a). Other large placer producing streams in the Atlin area (Fig. 16.7) include Birch, Boulder, McKee, Otter, Ruby and Wright creeks (Levson and Kerr, 1992). The placers at Ruby Creek are notable in that they underlie Pleistocene basalts (Plate 16.11).

In the Princeton district in southern B.C. (Fig. 16.1, Location 12), the Similkameen and Tulameen drainage basins are unique in that they have produced commercial quantities of platinum in addition to gold (Raicevic and Cabri, 1976). Rice (1947) noted that the amount of bedrock erosion during the Holocene in the Tulameen area was insufficient to produce the placers of the area. He concluded that glacial and post-glacial streams reworked remnants of rich preglacial placers that either escaped the dissipating effect of glaciation or were only slightly disturbed by ice.

There are 27 known coastal placer occurrences in British Columbia, ten of which are beach placers located on Vancouver Island and the Queen Charlotte Islands (Samson, 1984). The remainder are mainly coastal stream placers such as the Leech River and China Creek placers on Vancouver Island (Fig. 16.1, Location 15). The most notable beach placers include those at Florencia Bay near Ucluelet on western Vancouver Island (Fig. 16.1, Location 16) and on Graham Island (Fig. 16.1, Location 17) in the Queen Charlottes (Mandy, 1934). Beach placers derived from reworked Pleistocene sediments at Florencia Bay produced in excess of 45,000 g (1500 oz) of gold (Holland, 1950; Samson, 1984).

16.3.1.2. Plains

The most significant alluvial placer drainage basin on the western plains is the North Saskatchewan River (Tyrrell, 1915). In Alberta, additional drainage basins with known placer gold deposits include the McLeod, Athabasca, Peace, Red Deer, Milk and Redwater rivers (Halferdahl, 1965; Giusti, 1983). Recorded gold production from Alberta placers from 1887 to 1981 was nearly 100,000 g (30,788 ounces) (Giusti, 1983). Gold on the North Saskatchewan River (Fig. 16.1, Location 18) occurs as extremely small flakes and particles in bar-top gravels and is most concentrated in areas underlain by Cretaceous pyritic coals (Tyrrell, 1915; Boyle, 1979). In Saskatchewan, alluvial placer occurrences are found on the North Saskatchewan, Waterhen and Poplar rivers, and in the Leaf Lake area (Coombe, 1984). Placer gold on the western plains is found mainly in Holocene stream gravels, and specific source areas are not well understood.

PLATE 16.10. Large gold nuggets recently recovered from McKee Creek (see Fig. 16.7 for location). (Photo courtesy of John Harvey)

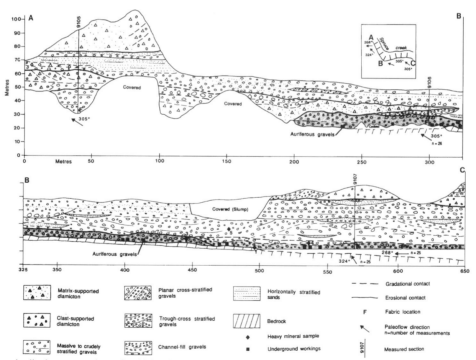

FIG. 16.8. Longitudinal cross-section of gold-bearing placer gravels (heavy shading) and the overlying stratigraphic succession of Quaternary sediments exposed on lower Spruce Creek (see FIG. 16.7 for location). Although numerous underground mine adits are shown, the lower gravels have been recently open-pit mined. (From Levson, 1992a, reproduced with permission of *British Columbia Ministry of Energy, Mines and Petroleum Resources*)

PLATE 16.11. Pleistocene basalts overlying gold-bearing gravel on Ruby Creek (see FIG. 16. 7 for location)

16.3.1.3. Central Canada

Ontario: In Ontario some gold has been obtained from river gravels and glacial sediments in drainage basins such as Vermilion River, Wanapietei River, Meteor Lake, Lake Manitou, Savante Lake and Grassy River (Boyle, 1979). Glaciofluvial sediments in the Timmins-Kirkland Lake mining area also contain placer gold (Ferguson and Freeman, 1978). The source of the gold in these various occurrences is unknown but because it is fine-grained, it was possibly derived through glacial erosion of gold-bearing lodes.

Ferguson and Freeman (1978) described 27 gold placer occurrences in Ontario. Much of the gold was eroded from till and glaciofluvial deposits by Holocene streams (Boyle, 1979). Few of these occurrences have been extensively mined, with the exception of the Afton gold and silver mine in the Sudbury District that produced 1412 kg (45,414 oz) of gold and 258 kg (8299 oz) of silver between 1935 and 1941. Prest (1949) recorded drilling results from Fraser Lake that showed 4.6 m of high grade placer gravels. Ferguson and Freeman (1978) recorded a silver placer in beach gravels from the former shore of Cobalt Lake in the Timiskaming District.

Quebec: In Quebec a number of small placers are known in the Eastern Townships of Gaspé, with most mining activity occurring in the Chaudière and Sherbrooke districts (Boyle, 1979; LaSalle, 1980 Shilts and Smith, 1988). The Beauceville placer area in the Rivière Chaudière drainage basin (Fig. 16.9)

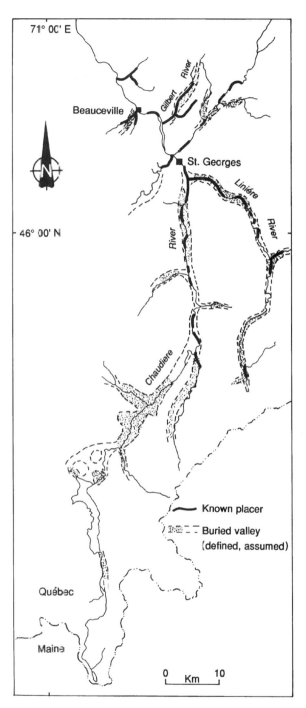

FIG. 16.9. Placer gold deposits in the Beauceville region, Quebec (after Boyle, 1979 and Shilts and Smith, 1988). Buried valley locations were determined by refraction seismic methods and drilling (Shilts and Smith, 1986)

has produced more than 3100 kg (100,000 oz) of gold making it the richest placer area in eastern Canada (MacKay, 1921; Boyle, 1979). Deeply buried preglacial channels in Gilbert River, Meule Creek and Rivière des Plantes have produced most of the placer gold in the Chaudière district (LaSalle, 1980). Numerous buried placers still exist in the region as indicated by seismic and drilling results (Shilts and Smith, 1986).

Placer gold in the Appalachian region was probably released from host rocks and initially concentrated during Tertiary weathering and dissection. The auriferous sediments are up to 10 m thick, iron-oxide stained and commonly limonite-cemented. Gravel beds are bouldery and do not contain erratics, suggesting that they are preglacial in age (MacKay, 1921; Boyle, 1979). In places, twenty or more metres of till and glaciofluvial sands overlie auriferous alluvial gravels resting on bedrock or clay (MacKay, 1921). Shilts and Smith (1988) found that the main pay gravels in the Gilbert River deposits are overlain by glacial and interstadial (or interglacial) deposits representing three or four distinct glacial events. Many of the paleoplacer deposits were destroyed during glaciation, but the nature of the terrain in relation to ice movement permitted preservation of placers in valleys oriented transverse to glacier flow.

Postglacial placers in the region occur in modern river bars, stream beds, floodplain sediments and terraces. They contain mainly fine-grained gold in non-economic quantities. Interglacial and post-glacial fluvial systems in the area are enriched in gold where they intersect older placers or where gold, dispersed throughout glacial deposits, has been reconcentrated (Boyle, 1979; Maurice, 1986, 1988).

16.3.1.4. Canadian Maritime Provinces

A comprehensive summary of coastal and marine placer deposits in Nova Scotia was provided by Samson (1984). Gold in continental shelf deposits on the eastern Nova Scotian coast, interpreted to be abandoned stream channel and relict strandline deposits, have been the focus of some recent exploration activity (Libby, 1969; Boyle, 1979; Hopkins, 1985). Beach placers have been derived from auriferous tills

at Red Head near Isaacs Harbour. Marine gold placers in the Country Harbour area occur in detrital lag deposits, relict drowned beaches, fluvial and glacio-fluvial concentrations and in glaciomarine sediments (Hopkins, 1985). Tills on the continental shelf, with gold contents up to $1.8 \, \text{gm}^{-3}$, were reworked by fluvial and marine transgressive processes to form placers. Offshore and beach placer deposits in Nova Scotia occur mainly in areas underlain by gold-bearing rocks of the Meguma Group (Boyle, 1979; Hopkins, 1985).

Other placers in the Maritimes include a lower Carboniferous placer deposit in the Gays River district, Nova Scotia (Malcolm and Faribault, 1929) and a beach sand placer in New Brunswick at Saints Rest on Taylor Peninsula (Samson, 1984). Recent studies in Newfoundland have centred on modern beach and marine chromium placers in the Port au Port Bay area, especially near the mouth of Fox Island River (Emory-Moore et al., 1988; Solomon et al., 1990) and on the placer gold potential of the Newfoundland shelf (Emory-Moore and Solomon, 1989; Emory-Moore, 1991).

16.3.2. Placer Deposits in North-central Asia

16.3.2.1. Siberia

Although Siberia continues to be a major placer producing area in the world, little information on the sedimentology and origin of the placers has been published. Kartashov (1971), however, discussed different types of alluvial placers, categorizing them into (1) coarse, concentrated, autochthonous or proximal placers and (2) finer, lower grade allochthonous placers formed in surficial alluvium far downstream from their source. Khitrov and Khershberg (1982) surveyed the paleogeographic setting of Neogene to Quaternary marine placers. The lithofacies and gold characteristics of residual, colluvial and alluvial placer deposits of Jurassic to Holocene age were briefly described by Fayzullin (1969) and Fayzullin and Tolmacheva (1972). Kazakevich (1972) considered the origin of gold placers and classified them into residual (eluvial), colluvial (slope, solifluction and landslide), alluvial (stream, deltaic and lacustrine), glacial and interglacial

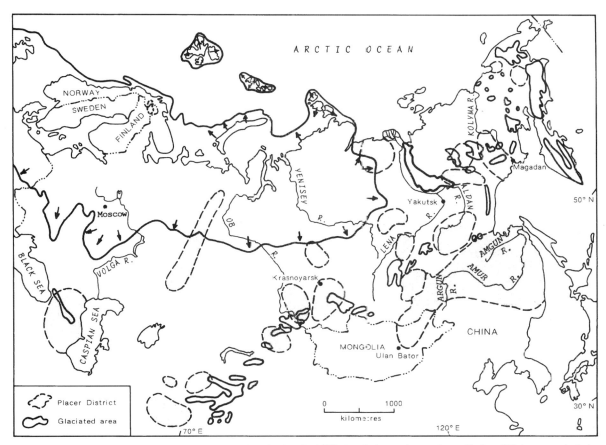

FIG. 16.10. Placer mining districts in north-central Asia (after Boyle, 1979) in relation to the maximum limits of Pleistocene glaciation as compiled by Flint (1971). Note the abundance of placer mining areas in unglaciated regions, particularly in Siberia. The isolated placer area in northern Finland, discussed in the text, was covered by the Scandinavian ice-sheet

(morainal and glaciofluvial), eolian and marine (beach, lagoonal and submerged) placers.

Bundtzen (1991) recently compared placer gold production in the Magadan region in the far northeast part of Asia (Fig. 16.10), which was not extensively glaciated during the Pleistocene, with south-central Alaska where geologically similar terranes were buried by ice during the Late Wisconsinan and, locally, are being modified by active glacial processes. Although mining in the two regions is not directly comparable (due, for example, to differences in economic factors such as mining subsidies), he found a wide disparity in placer production between the two areas that can be attributed in part to geomorphic processes associated with glaciation that destroy, dilute and bury placer deposits. The same conclusion was reached by comparing gold production in areas with igneous source rocks in unglaciated placer districts of central Alaska with geologically similar, but placer poor, glaciated areas of the Magadan region. The Kolyma River region in the Magadan placer district (Fig. 16.10) produced in excess of 2×10^6 kg (65 million oz) between 1929 and 1989, with about 90% of the commercial deposits occurring in unglaciated areas (Bundtzen, 1991).

The largest river placers in the world, in the Lena, Aldun and Amur drainage basins (Fig. 16.10), have produced more than 1.24×10^6 kg (40 million oz) of gold since their discovery in 1829 (Boyle, 1979). Fayzullin and Turchinova (1973) described five types of Cenozoic placer deposits from this region: those that occur in feather-like, parallel and radial geometries, eroded from primary bedrock sources, and those in radial and lattice-like patterns, developed from conglomerate and gravel formations that acted as intermediate collectors of gold. Areas where the drainage pattern has changed little throughout the Cenozoic, contain large, productive, terrace, buried and recent-stream placers. Pleistocene and Holocene placers have been reconcentrated from Pliocene pebble conglomerates in Cenozoic basins within Mesozoic depressions. The richest placers in these areas occur where channels of younger streams coincide with the thalwegs of Tertiary streams and their tributaries (Fayzullin and Turchinova, 1973).

Although most of the placer districts in north-central Asia are in unglaciated areas, small ice caps affected many placer regions (Fig. 16.10). For example, placers of the Bodaibo River, a tributary of the Vitim River in the Lena district, occur in preglacial alluvial channels and benches buried by up to 150 m of interglacial deposits and till deposited during the most extensive glaciation in the area (Early Illinoian?). Placer deposits in this area also occur in interglacial channel gravels and Holocene alluvium. The preglacial channel placers are the most productive with pay streaks 1–5 m thick and 10–300 m wide, generally much wider than modern channels (Boyle, 1979).

16.3.2.2. Finland

Saarnisto et al. (1991) recently investigated a rich placer area in the Ivalojoki area in Finnish Lapland. The placers are believed to be glaciofluvial gravels containing gold derived from granulite basement rocks that extend from Norway in the west to the eastern border of Lapland (Fig. 16.10). In addition to meltwater channel gravels, some gold has been recovered from fluvial terrace and stream channel deposits and, in lesser amounts, from till. The region has produced approximately 1×10^6 g of gold with average grades of 1.8 gm^{-3}. The gold is derived mainly from epigenetic quartz-carbonate veins and also from quartz-hematite veins and arsenopyrite-bearing shear zones (Saarnisto et al., 1991).

16.3.3. Placer Deposits in the United States

16.3.3.1. Contiguous states

Almost all the placer gold produced in the United States has come from California (2.12×10^6 kg), Alaska (650,000 kg), Montana (226,000 kg) and Idaho (132,000 kg) with relatively minor contributions from Oregon, Nevada, Colorado, Arizona and South Dakota (Yeend and Shawe, 1989). Glaciation in virtually all of these areas was confined to high Alpine ranges. Major placer areas (Fig. 16.11) generally occur outside or on the peripheries of large ice complexes such as those that formed in the Cascade Range of Oregon and Washington, the Aleutian, Alaska and Brooks ranges in Alaska and in the Sierra Nevada (Porter et al., 1983).

In California, placer districts in the western foothills of the Sierra Nevada have produced gold from both Quaternary and Tertiary gravels (Yeend and Shawe, 1989). Additional gold has been produced from Tertiary gravels in the western Sierra Nevada, from Quaternary and Tertiary gravels in the Trinity River basin, and from Quaternary gravels along the Salmon River (Fig. 16.11) (Yeend and Shawe, 1989). Glaciers in the Sierra Nevada were confined mainly to cirques and the upper parts of valleys and had only a small effect on the distribution of Tertiary gold (Boyle, 1979).

16.3.3.2. Alaska

Although Alaska has been extensively glaciated in the past, large portions of the state have always remained ice free (Fig. 16.11). The main placer mining areas within the state occur outside the maximum limits of glaciation, but a large number of mines have exploited placers in glaciated areas (Cobb, 1973; Yeend and Shawe, 1989). The following discussion emphasizes two of the most important glaciated regions that have had significant placer mining activity in recent years: the Valdez Creek area in south-central Alaska and offshore placers in the Nome area.

The Valdez Creek mine currently produces more gold than any other placer or lode gold mine in Alaska (T. Bundtzen, pers. commun., 1993) (Reger and

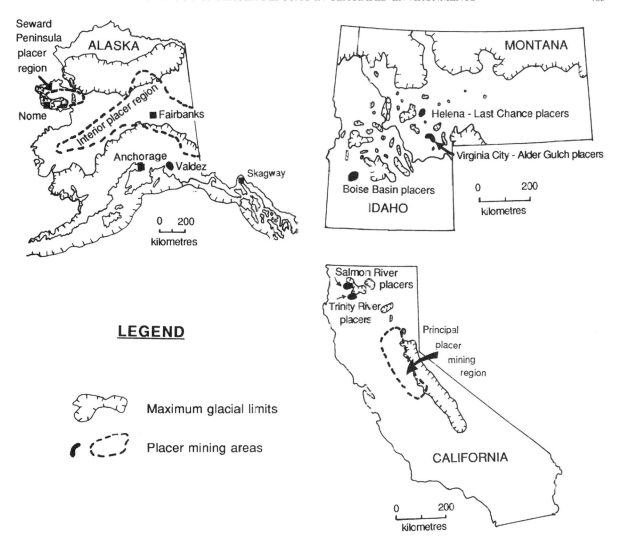

FIG. 16.11. Placer mining areas in Alaska, Idaho, Montana and California (from Yeend and Shawe, 1989) in relation to the maximum limits of glaciation as compiled by Porter *et al.* (1983). The most significant placer deposits within glaciated areas occur in Alaska, most notably in the Valdez and Nome areas

Bundtzen, 1990). The region has a complex Pleistocene history characterized by local base level changes resulting from interactions between main valley and tributary valley glaciers (Cobb, 1973). Reger and Bundtzen (1990) found evidence for four glaciations in the area which they interpreted as pre-Illinoian (pre-glacial?), Illinoian, Wisconsinan and Holocene (Fig. 16.12). Gold, believed to have been scoured from lode sources by glaciers during the Illinoian glaciation was deposited in tills in the Valdez Creek valley and later reworked to form the rich paleo-channel placer deposits of the area. Due to differences in the synchroneity and extent of glaciers in main and tributary valleys, drainage in the Valdez Creek valley was commonly blocked and ice-marginal lakes formed. During deglaciation, significant lowering of

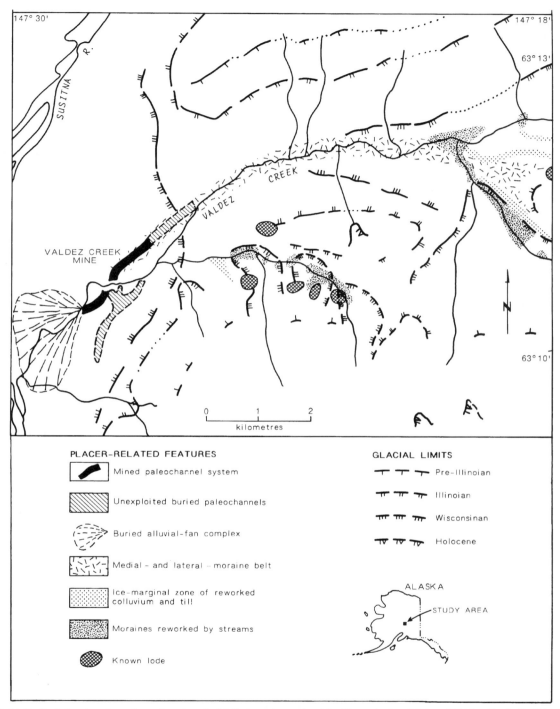

FIG. 16.12. Buried and surface placer deposits, glacial limits and associated landforms in the vicinity of Valdez Creek. Gold-bearing deposits include buried channel and alluvial-fan sediments, medial and lateral moraines down-valley from lode sources and glaciofluvially reworked colluvium and morainal deposits. The Valdez Creek mine is exploiting buried paleochannel deposits and is the largest gold producer in Alaska. (Modified from Reger and Bundtzen, 1990)

base level occurred and resulted in channel incision and formation of the placers. During the Late Wisconsinan, glaciers in the Valdez Creek valley did not coalesce with ice in the Susitna valley (Fig. 16.12) and a sequence of glaciolacustrine sediments was deposited in a lake impounded between the glaciers. These sediments were overlain by till deposited as glaciers expanded in the area (Reger and Bundtzen, 1990).

Gold in the Valdez Creek area occurs in at least three different paleochannels cut by the modern river canyon. The paleochannels, from youngest to oldest, are believed to date from a Mid-Wisconsinan interstadial, the Sangamonian interglacial and a Late-Illinoian interstadial. The gold-bearing gravels are up to 10 m thick with an average thickness of 3–4 m. They exhibit cut-and-fill structures near the base and horizontal bedding in the upper part. They are well imbricated, contain discontinuous cobble beds and have a basal lag-gravel comprised of boulders up to 3 m in diameter. Gold is concentrated in the lower few metres of the gravels and in the underlying fractured bedrock. Gold also overlies thin silt and clay layers that occur locally within the gravels. Reger and Bundtzen (1990) recognized six targets for new placer discoveries in the glaciated terrain surrounding the Valdez Creek mine (Fig. 16.12). Those with moderate to high placer potential include: (1) extensions of known buried paleochannels; (2) new unexploited paleochannels; (3) buried alluvial fans downstream of known paleochannels; and (4) medial moraine deposits down-valley of major known lode gold sources. Placers of low to moderate potential include: (1) gold-bearing valley-side colluvium and till reworked by ice-marginal meltwater streams, and (2) gravels in zones where late-glacial and postglacial streams breached and reworked gold-bearing moraines (cf. Smith T.E., 1970, 1981b; Bressler et al., 1985; Hughes, 1989).

A second major placer mining area in glaciated parts of Alaska is in the Nome district (Fig. 16.11). The region has produced more than 100,000 kg (3.45 million oz) of refined gold from strandline deposits and another 35,000 kg (1.11 million oz) from stream and glacial deposits (Barker et al., 1989). Placer deposits of nearly every type have been mined in the region including residual gold and scheelite placers,

placers formed by mass-wasting processes, wave-cut platform placers, and gulch, alluvial terrace, ice-marginal channel, beach and modern stream placers (Cobb, 1973). Glaciers overrode the coastal plain near Nome during early Pleistocene and Illinoian glaciations but did not reach the area in the Wisconsinan (Nelson and Hopkins, 1972). During low sea level stands in glacial periods, glaciers from Seward Peninsula (and Siberia's Chukotka Peninsula) eroded gold from bedrock, alluvium and beach placers and deposited auriferous drift many kilometres beyond the present shorelines. Subsequent transgressions and regressions of the sea resulted in reworking of the drift into rich placers, some of which are now buried and some are submerged. Nelson and Hopkins (1972) recognized 12 beach deposits in the region including buried, offshore, onshore and active beaches (Fig. 16.3). Offshore and onshore beach placers near Nome have recently been mined by large scale dredging operations (Bundtzen et al., 1990). The world's largest bucket-line offshore-dredge has been operating in the area and produced approximately 1120 kg (36,000 oz) of gold in each of 1987 and 1988 (Barker et al., 1989).

Other glaciated areas of Alaska (Fig. 16.11) with major placer mining activity in recent years include the Yentna-Cache Creek district northwest of Anchorage and the Christochina district northeast of Valdez (Yeend and Shawe, 1989). Gold has been recovered from Tertiary quartz-rich conglomerates, Pleistocene glaciofluvial deposits and Recent stream and terrace gravels in the Yentna district and from reconcentrated Tertiary conglomerate and glacial deposits in the Christochina district (Cobb, 1973). Bundtzen (1986) described placer concentrations in modern stream and terrace gravels, alluvial fan deposits and till in a heavily glaciated area of Alaska about 40 km northwest of Skagway (Fig.16.11). About 2500 kg (80,000 oz) of gold were produced from the area, with about 95% of the gold coming from Porcupine Creek and its tributaries. Potential nearshore placers in glaciated regions of Alaska include platinum group mineral placers at Chagvan, Goodnews and Kuskokwim Bays, gold placers at Cook Inlet and gold and garnet beach placers in the Yakutat–Yakataga area (Barker et al., 1989).

16.3.4. New Zealand Placer Deposits

Placer gold mining has been a relatively important part of the New Zealand economy, producing approximately 465,000 kg (15 million oz) of gold, compared to lode gold production, totalling about 350,000 kg (Williams, 1974). Nearly two-thirds of the placer gold has come from unglaciated areas including Quaternary schistose gravels and Tertiary conglomerates in Otago, deeply buried Quaternary deposits in Marlborough and near surface Quaternary gravels and Tertiary conglomerates in Nelson (Fig. 16.13). Most of the gold in Otago came from unglaciated areas since placer deposits in ice-covered regions were diluted by glacial debris eroded from barren rocks in the Southern Alps. In contrast, glaciation had an important role in the formation of placers in Westland and southwestern Nelson. Gold, originally derived from lode sources in Precambrian greywackes, Alpine

schists and early Pleistocene, Tertiary and Cretaceous conglomerates, was incorporated into glaciofluvial and morainal deposits by Pleistocene glaciers. The gold was subsequently reworked into auriferous gravels and sands by interglacial and Holocene streams and beaches.

During the Pleistocene, glaciers from the Southern Alps reached tidewater on the southwest coast, but farther north they terminated as inland piedmont glaciers (Fig. 16.13). The most important placer deposits in Westland occur along the northwest periphery of Pleistocene glaciation. Most placer gold has been derived from Pleistocene glaciofluvial deposits, Holocene stream gravels and modern beach sands. Alluvial placer deposits are uncommon in southern Westland where Pleistocene glaciers terminated seaward of the present coastline. However, raised and submerged beach sand placers occur along the entire length of the west coast. Gold at the

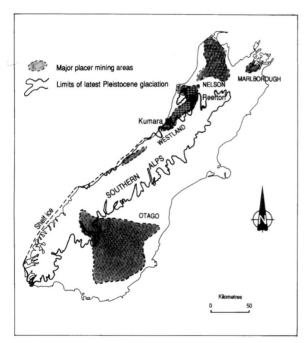

FIG. 16.13. Placer mining areas and late Pleistocene glacial limits in New Zealand (modified from Williams, 1974 and Flint, 1971). The most important placers in glaciated parts of the island occur on the northwest periphery of the ice sheet where glaciers transported gold from bedrock sources in the Southern Alps. Placers occur mainly in glaciofluvial, fluvial and beach deposits. (See Fig. 16.14 for details of the Reefton area)

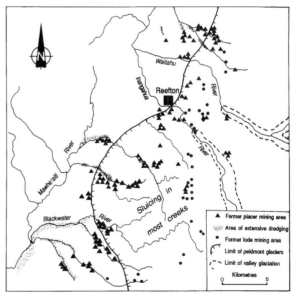

FIG. 16.14. Placer mining areas near Reefton, New Zealand (see Fig. 16.13 for location), showing the approximate western limit of Pleistocene piedmont glaciers flowing out from the Southern Alps. Placers are confined to areas east of the glacial limit and to creeks draining those areas, reflecting glacial transport of gold into the region. (From Williams, 1974)

southwestern most end of South Island has been recovered from raised beach and river placers formed by reworking of glacial deposits (Williams, 1974).

Glacial and glaciofluvial gold placers have been extensively mined at a number of sites in Westland and Nelson (Fig. 16.13). In the Kumara area, in northern Westland, gold occurs mainly in outwash gravels deposited outside the limits of Late Pleistocene moraines and to a lesser extent in kame and esker deposits. The outwash gravels are richest where they overlie Miocene marine clays (Williams, 1974). In contrast, in the Reefton area (Fig. 16.14), gold has been recovered mainly within the limits of piedmont glaciation. This relationship between the distribution of gold and morainal deposits presumably reflects glacial erosion of gold from bedrock sources to the east and redeposition in till. The gold was then reworked from the till by glaciofluvial streams close to the ice margin. Outwash from younger valley glaciers and postglacial streams further reconcentrated the gold from the older glaciofluvial and morainal deposits. The gold is fine grained and progressively diminishes in size with distance from the ice margins.

16.3.5. South American Placer Deposits

Pleistocene glaciers in South America were restricted to the Andes Cordillera extending both easterly across the Patagonian Plains and westerly to sea level along the southern Chilean Coast. The ice sheet became progressively narrower to the north breaking up into small ice caps and valley and cirque glaciers north of about 30°. Evidence for multiple glacial advances has been recognized in a number of areas including southern Argentina, on the Chilean flank of the Andes, near La Paz, Bolivia and in Peru and Ecuador (Flint, 1971; Rabassa and Clapperton, 1990; Schubert and Clapperton, 1990).

Placer deposits in glaciated areas of Peru and Bolivia have recently been described (Delaune et al., 1988; Hérail et al., 1989a,b; Hirvas, 1991). In Brazil, Quaternary alluvial placer deposits were formed during two stages of semi-arid depositional cycles which may be correlated with glaciations at higher latitudes (Tadeu et al., 1991). Geomorphologically, there are alluvial sequences that occupy present

drainage basins and are correlated in time with the Late Wisconsinan glacial interval. Older buried paleodeposits are characteristic of the western Amazon and are found in terraces buried by colluvium. Paleoplacers in Brazil are found in association with greenstone belts and are typically quartz-pebble pyritiferous and auriferous conglomerates (Scarpelli, 1991). It is noteworthy that pervasive Late Cenozoic nonglacial gravels, termed 'Tehuelche-type gravels' (Flint, 1971), occur as thin veneers on pediment surfaces over large areas east of the Andes in Argentina. The gravels typically underlie glacial drift and are presumed to be Pleistocene in age.

The following discussion of placers in South America is not comprehensive but rather focuses on some recent studies describing geologic relationships between placer deposits and glaciation (cf. Hérail, 1991).

16.3.5.1. Peru

Delaune et al. (1988) and Hérail et al. (1989a) described the gold distribution, gold content, associated heavy mineral suites, and changes in morphological characteristics of detrital gold in till and glaciofluvial sediments in the Ancocala-Ananea basin in the southeastern Andes of Peru (cf. Hirvas, 1991). They concluded that: (1) economic quantities of placer gold in glacial drift are only possible where primary mineralized zones have been eroded directly by glacial ice; (2) gold content in glacial drift decreases as the distance from primary bedrock sources increases and the transition between till and glaciofluvial sediments is characterized by an increase in gold content; (3) gold distribution in glacial sediments has no relationship with sedimentary structures and there is no discernable increase in gold content at the bedrock contact; and (4) glacially transported placer gold acquires morphologic characteristics that may be considered diagnostic of glacial erosion such as rolled edges and crush marks (Chapter 13; Menzies, 1995a, Chapter 16).

16.3.5.2. Bolivia

Hérail et al. (1990) studied the morphology and chemistry of placer gold from fluvial gravel in the Andes of Bolivia as a tool for predicting the distance travelled from primary lode sources. They found that characteristics such as the disappearance of primary crystalline outlines and surface modifications such as striations and folds correspond to distance travelled from original lode sources and that there appears to be no relationship between changes in chemical composition of placer gold and distance travelled.

Hérail *et al.* (1989b) documented the occurrence of placer gold in both Tertiary basins and Quaternary terraces in the Tipuani-Mapiri Basin in Bolivia. Tertiary conglomerates of the Cangalli Formation contain placer gold with grades in proximal 'fluvio-torrential' deposits of 21.3 to 112 gm^{-3} (cf. Biste *et al.*, 1991; Delaune *et al.*, 1991; Morteani, 1991).

16.4. BURIED PLACER DEPOSITS

In this section of the chapter the relationship of placer deposits to the glacial overburden will be investigated. Numerous recent studies illustrate the complex geologic origin of placers in glaciated terrains (e.g. Morison, 1985a,b; Shilts and Smith, 1988; Eyles and Kocsis 1989a,b; Levson *et al.*, 1990; Reger and Bundtzen, 1990; Levson and Giles, 1991, 1993; Levson, 1992a,b). Glacially-induced environmental

and base-level changes control the dominant erosional and depositional processes. Preservation and burial of preglacial and interglacial placers is largely controlled by these processes and is recorded mainly in the glacigenic sediments that overlie the auriferous deposits.

Consequently, an understanding of the glacial history, as revealed by overburden studies, is needed to develop mining strategies and to identify new sites where gold-bearing placers have been deposited and preserved (Chapter 15). The current trend toward an increased exploitation of preglacial and interglacial placers, deeply buried by till deposits, further emphasizes this need for detailed geological analyses. The following discussion focuses mainly on buried placers in the Cariboo region of central British Columbia where stratigraphic and sedimentologic analyses of this type recently have been conducted.

16.4.1. Buried Valley Settings

Four types of buried-valley placer deposits (Figs 16.15 and 16.16), distinguished by differences in size, topographic position and paleostream gradient, have been recognized (Levson and Giles, 1991, 1993) in

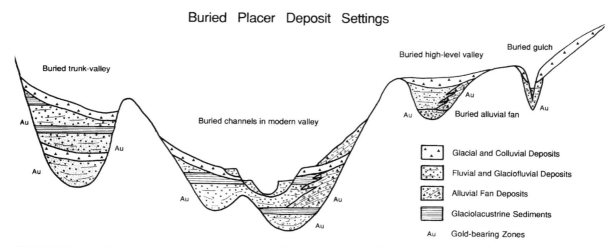

FIG. 16.15. Common buried placer deposit settings in the Cariboo. Gold occurs mainly at the base of alluvial channel sequences in the buried valleys and, to a lesser extent, in buried alluvial-fan deposits. (From Levson and Giles, 1991, reproduced with permission of *British Columbia Ministry of Energy, Mines and Petroleum Resources*)

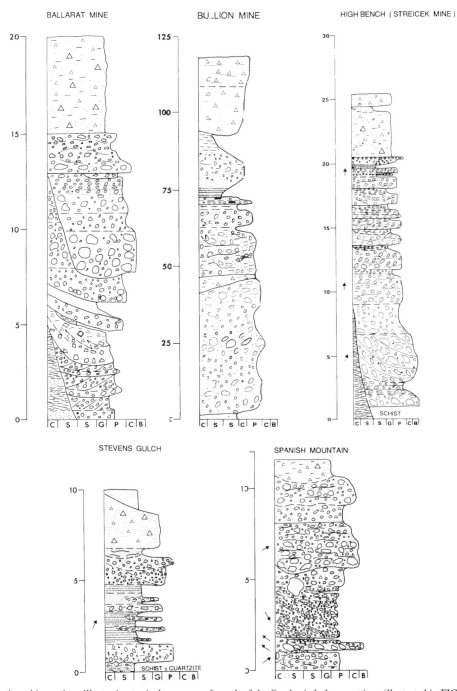

FIG. 16.16. Stratigraphic sections illustrating typical sequences in each of the five buried placer settings illustrated in FIG. 16.15 (see Fig. 16.5 for locations). Buried channel deposits in a modern valley were exploited at the Ballarat Mine (location 29); the Bullion (location 20) and Streicek (location 23) mines provide examples of buried trunk and high-level (tributary?) valley deposits, respectively; a buried gulch placer is exposed at Stevens Gulch (location 11), and; buried alluvial-fan deposits are mined at Spanish Mountain (location 21). (See Fig. 16.6 for legend; vertical scale on all sections is in metres.) (From Levson and Giles, 1992, reproduced with permission of *British Columbia Ministry of Energy, Mines and Petroleum Resources*)

central British Columbia. These include broad, low-gradient, abandoned trunk-valleys, abandoned high-level valleys, buried channels in modern valleys and relatively steep, narrow, buried-gulch channels. Paleoalluvial fan sediments also partially infill some buried valleys (Fig. 16.15).

16.4.1.1. Buried trunk-valleys

Abandoned trunk valleys are usually bedrock walled and infilled with stratigraphically complex sequences of gravel, sand, silt, clay and diamicton. For example, the valley excavated by hydraulic operations at the Bullion mine (Fig. 16.5, Location 20) is an excellent illustration of a large buried Pleistocene valley (Levson, 1991b). The mine occupies a bedrock-walled valley about 1 kilometre long and over 100 m deep (Plate 16.12) that is truncated at the north end by the modern Quesnel River. Sediments from two glacial periods (tills and glaciofluvial sands and gravels) and intervening nonglacial deposits are locally preserved (Clague, 1987b). Gold production

that has exceeded 3860 kg (120,000 oz) was recovered mainly from the lowermost gravels on bedrock (Sharpe, 1939), but some gold also occurs in stratigraphically higher gravel units, probably overlying till or lacustrine deposits that are resistant to erosion (Fig. 16.16).

16.4.1.2. Buried high-level valleys

Abandoned valleys at elevations substantially higher than modern streams may represent channel remnants from periods of relatively high base-level prior to more recent valley incision. A fine example of a high-level buried-valley placer is provided by the Streicek mine (Figs. 16.5b and 16.16, Location 23, Plates 16.13 and 16.14).

16.4.1.3. Buried-channel deposits in modern valleys

Many paleochannel placer deposits occur directly underneath, or adjacent to, modern alluvial channels (Figs 16.15 and 16.16, Plate 16.15). Buried channels

PLATE 16.12. The Bullion mine near Likely (location 20 on Fig. 16.5). This Pleistocene buried-valley was hydraulically excavated for gold contained mainly within gravels at the base of the sequence and, to a lesser extent, at stratigraphically higher levels. The hill on the far side of the mine is mainly bedrock but stratified buried-channel sediments are still exposed on its flanks

PLATE 16.13. Mining high-level, buried, paleochannel gravels on a bench above Devils Lake Creek at the Streicek Mine (location 23 on FIG. 16.5)

PLATE 16.14. Typical buried-channel gravels exposed near the base of the mine shown in Plate 16.13

PLATE 16.15. Poorly sorted gold-bearing gravels at the base of a highwall exposure at the Gallery Gold Mine along Lightning Creek (location 24 on FIG. 16. 5). The base of the pit is on bedrock and the main auriferous zone occurs in the overlying two to four metres of gravel (measuring rod is 4 m high)

adjacent to modern streams are frequently segmented by recent erosion and their lateral extent is therefore often poorly defined. Paleochannels below modern streams commonly parallel the modern stream course but may be deeply buried by alluvium and glacial deposits. Both these paleoplacer types invariably occur below the water table.

16.4.1.4. Buried gulch deposits

Buried, high-gradient, stream channel placers are commonly mined in the Cariboo region (Levson and Giles, 1991, 1993). A typical example is Stevens Gulch (Fig. 16.5b, Location 10, Fig. 16.16) where poorly sorted pebble gravels occur on bedrock and contain coarse nuggets (commonly 8 to 16 g). They are inferred to be debris-flow deposits because of their massive structure, chaotic fabric, dominance of angular local clasts and poor sorting. Buried gulch placers may also locally contain fluvial sediments, typically moderately-well sorted sands and gravels.

They also typically are overlain by colluvial diamicton, till and, in some areas, glaciofluvial sands and gravels and glaciolacustrine sediments.

16.4.2. Exploring for Buried Placers

The depletion of shallow, easily mined deposits and the trend towards increased exploitation of placers buried by glacial sediments requires an increased reliance on geologic information to locate new reserves and to evaluate exploration targets in glaciated areas. Some paleochannel systems are entirely buried by glacial sediments, having little or no topographic expression or relation to the modern drainage pattern. Exploration activities should focus on regional and detailed airphoto interpretation, geophysical (seismic, ground penetrating radar and magnetometer) surveys and follow-up drilling to locate and test these buried placer deposits. For example, numerous fluvial channels, buried by as much as 100 m of glaciolacustrine sediments and till, have been detected by seismic refraction methods and drilling along several rivers in the Beauceville placer area of southern Quebec (Fig. 16.9) (Shilts and Smith, 1986).

Paleogeographic reconstructions are required to understand the drainage patterns of preglacial and interglacial rivers and therefore identify probable gold-bearing buried channels. In areas of active exploration, detailed stratigraphic and sedimentologic data will also aid in locating and identifying the extent, volume and overburden thickness of deeply buried, gold-bearing strata.

Gold concentrations in buried placers are most easily located in gulches and narrow tributary channels where they tend to be concentrated in a relatively narrow zone at the channel base (Fig. 16.15). In broad valley river placers, pay streaks do not necessarily occur in the lowest depression or gutter and may have little relation to the modern channel (Boyle, 1979). Gold in floodplain, alluvial fan and deltaic deposits tends to be dispersed throughout the sediments.

Exploration for deeply buried auriferous gravels may be targeted initially on meltwater channels and other drainage routes where the overburden has been removed by natural processes. Concentrating exploration and mining activities, on sites where these valleys coincide with ancient gold-bearing drainage courses, may be a cost-effective means of locating and mining buried placers. For example, a mine discovered in 1972 in the Cariboo in an area that has been frequently reworked since the 1860s, is located in a meltwater channel (Levson et al., 1990; Levson and Giles 1993). Glacial meltwater at the site incised a channel and removed much (at least 20 m) of the overburden. The channel contains a small misfit stream but was never explored even though thousands of prospectors followed the old Cariboo gold rush trail through the property.

16.4.3. Preservation of Buried Placers

Due to the erosive nature of glaciers and subglacial meltwaters, the distribution of placers in glaciated areas is generally highly discontinuous. Exploration programs can benefit by focusing on geologic settings with good placer preservation potential (Levson, 1990). These include: (1) areas where proglacial lakes

formed depositional sinks and reduced ice erosion; (2) sites in the lee of bedrock highs where glacial erosion was minimal; (3) tributary valleys not subject to valley glaciation or oriented transversely to the regional direction of ice flow; and (4) regions near glacial limits where reduced subglacial erosion may result in placer deposit preservation. Narrow channels oriented oblique to former glacier flow, supporting thin overburden, are the best targets.

Preservation of a large placer deposit buried by till on Spanish Mountain, near Likely, British Columbia (Fig. 16.5), was attributed to the oblique orientation of the deposit relative to the regional direction of glacial ice flow (Levson and Giles, 1993). In addition, based on determinations of glacial flow direction in the area, the placer is located in the lee of a large bedrock high. Placer deposits in the Livingstone Creek area in central Yukon Territory (Fig. 16.1) are preserved along several tributary streams where glaciolacustrine depositional sinks formed as a result of damming by main valley glaciers (Levson, 1992b). The tributaries are oriented obliquely to the former ice flow direction of the trunk valley glacier and they were not inundated by local valley glaciers prior to the damming event. Consequently, erosion of the placer deposits by ice or meltwater did not occur.

Regions proximal to the limits of glaciation may also contain promising areas to explore for buried placer deposits. For example, preserved preglacial fluvial placers in the Clear Creek area in central Yukon (Fig. 16.1, Location 4) commonly occur at elevations near the upper limits of glaciation where subglacial erosion is decreased (Morison, 1983a,b, 1985a). The association of melt-out till, resedimented deposits, ponded fine sediments and meltwater deposits in the area is typical of a stagnant glacier environment. In addition, glacial drift in the region is not auriferous further indicating minimal subglacial erosion of the underlying stream gravel.

16.4.4. Mining Buried Placer Deposits

Placer mining continues to be active in many glaciated areas, even in heavily exploited, traditional gold mining regions, indicating that the potential for locating new, hidden, gold-bearing placer deposits is good (Plate 16.16). Higher exploration and mining

PLATE 16.16. Thick sequence of glacial and glaciofluvial overburden overlying gold-bearing gravels (dark unit at the base of the section) in the high-wall of an active mine at Otter Creek (see Fig. 16.7 for location). The measuring rod beside the person in the lower centre of the photo is 4 m high. Overburden is removed in benches and hauled to the dumping area in the background

costs can be offset by the typically high gold contents in these deeply buried gravels. The sedimentologic, geomorphic and stratigraphic characteristics of different types of placers directly affect mining practices (Levson, 1991a). Table 16.1 summarizes the sedimentary characteristics of different types of placers that influence exploration, evaluation and production, and outlines some common mining and processing methods used in each setting.

16.5. CONCLUSIONS

The largest and best known placer deposits in glaciated terrains occur in British Columbia, Alaska, the Yukon Territory, Siberia and New Zealand. Placer deposits in glaciated regions occur in a wide variety of geologic settings including fluvial channel, alluvial fan, deltaic, beach, colluvial, glacial and glaciofluvial environments, ranging in age from Tertiary to Recent. The most productive are Tertiary and Pleistocene preglacial and interglacial fluvial channel deposits.

They are typically buried by thick sequences of Quaternary sediments and are most commonly mined in areas where they have been exposed by meltwater or postglacial fluvial erosion. Buried paleochannels may have little relation to modern drainage patterns and commonly occur as dissected channel remnants, sometimes at elevations well above modern stream levels. They may also be buried at depth in valley bottoms below Holocene alluvium and Pleistocene sediments. Buried alluvial fan deposits are large in volume but generally lower grade than fluvially deposited placers of similar age. Although smaller in size than buried-channel and fan deposits, paleogulch gravels commonly contain significantly higher gold concentrations. Historically, they have been rich gold producers, but they are often difficult to locate. Consequently, there is good potential for locating new high-grade paleogulch channels, especially in high-relief, till covered regions.

Geophysical (seismic, ground-penetrating radar and magnetometer) studies and drilling programs are

TABLE 16.1. Geologic setting, sedimentary characteristics relevant to exploration and mining, and common production methods for placers in glaciated terrains (modified from Levson, 1991a)

SEDIMENTARY CHARACTERISTICS   PRODUCTION METHODS

TERTIARY PLACERS

- cementation
- thick Tertiary and Quaternary overburden

- underground mining

BURIED INTERGLACIAL PLACERS

- unconsolidated overburden
- over-consolidation of pay gravels
- thin or discontinuous pay zones
- clay-rich or compact deposits interbedded with pay

- open pit mines commonly in meltwater channels and leeside settings
- disaggregation systems

a) Buried River Channel Settings

- excessive overburden in abandoned trunk valleys
- water shortages in topographically high valleys
- excess ground-water and surface-water problems in buried channels within modern valleys
- pay zones segmented by recent erosion

- hydraulic mining (formerly) and large open pit operations
- recycled water supplies
- water pumps

b) Buried Gulch and Stream Settings

- dilution of pay gravels by colluvium
- poor sorting
- abundant cobble and boulder sized clasts

- specialized excavating equipment, explosives, and screening equipment

c) Buried Alluvial Fan Settings

- textural diversity
- interbedded fine-grained & poorly-sorted deposits
- lower grade than fluvial placers

- large volume open pit mines
- jig-trommel-sluice processing systems

GLACIAL PLACERS

- relatively low grade
- over-consolidation
- high clay content
- large boulders

- open pit mines (usually down valley of rich fluvial placers)
- specialized disaggregation and pre-screening equipment

GLACIOFLUVIAL AND HIGH TERRACE PLACERS

- often low grade
- large boulders
- poorly sorted strata
- water shortages

- large volume open pit mines
- dredging operations
- pre-screening equipment

HOLOCENE LOW TERRACE PLACERS

- mainly fine gold (< 1 mm) in downstream locations
- silty overburden and abundant organics

- floating dredge or shovel systems

HOLOCENE COLLUVIAL PLACERS

- thin (< 2 m) and localized pay zones

- shallow open pit mines
- portable processing plants

needed to locate and evaluate placer deposits buried by thick glacigenic overburden. Detailed stratigraphic and sedimentologic data, such as thickness, depth and geometry of strata, paleochannel orientation and paleoflow direction, are also required to trace gold-bearing units.

Placers deposited during glacial periods are relatively rare and of lower grade than nonglacial placers. Economic quantities of gold contained within till and glaciofluvial deposits occur mainly where ice or glacial meltwaters eroded pre-existing, relatively rich fluvial placers.

Gold-bearing, postglacial alluvial-fan deposits and aggradational braided-stream gravels deposited shortly after deglaciation typically support large volume, low-grade placer mines. Gold is commonly distributed widely throughout the deposits. Placer deposits in active alluvial channels and on low terraces have been largely depleted by mining except on downstream reaches where mainly fine-grained gold is recovered. Productive Holocene placers often occur in areas with underlying Tertiary or interglacial auriferous gravels.

Chapter 17

PROBLEMS AND PERSPECTIVES

J. Menzies

17.1. INTRODUCTION

The intent of this closing chapter is not to sprinkle the text with references or to state dogmatically where the discipline as a whole should go to seek new answers but rather give a perspective on the field by highlighting some of the persisting problems that dog the search for even greater understanding. It is apparent that although much information is shared and broadly disseminated too often members studying one sub-environment take little heed of anything less than dazzling discoveries in another. Yet much of this science, like all others, is founded upon the slow and, in itself, uneventful accumulation of a myriad of minor details that only when synthesised move our understanding a minute step forward. However, too often these minor steps are not shared or their significance is not conveyed from one environment to another. A superb instance is the different approaches adopted by those interested in processes of diamicton and diamictite deposition. For example, the sum of the efforts of both Precambrian and Pleistocene researchers, both groups working on a similar problem, will undoubtably add up to more than their separate endeavours; however such collaboration is limited at present.

Perhaps only twenty years ago it was possible for a single researcher to have a moderate grasp of all the subdisciplines concerned with glacial environments, while today only an elementary understanding of many of these subunits is possible. As with any form of specialisation breadth of knowledge is lost in pursuit of depth, thus an ever-increasing need arises for general perspectives to be expressed to avoid either needless repetition or pointless explorations. Likewise, the derivation of countless untested hypotheses has strewn the paths of glacial geo(morpho)ology in almost careless abandon. The need to develop 'pet' hypotheses as *de rigueur* must be replaced by testing and re-testing of the few sound ideas that survive preliminary examination. Only by this means can a sound body of theories on processes, mechanics and correlationships within glacial environments be developed from which new ideas can be checked against and the science advanced. Linked to this sound footing is the requirement of reliability testing and validation of techniques, discriminant tools and characteristic signatures by as objective means as possible.

There are several areas within glacial geo(morphology) that are less well studied and therefore leave gaps in our general understanding and may hold keys to existing problems that remain to be disclosed. An example has been the limited emphasis, beyond general description, that has, in the past, characterised glacial erosion. Likewise, the contribution of proglacial lakes to ice mass marginal instability, to proximal modes of deposition, to isostatic crustal movement and the link with ice mass balance gradient are profitable areas requiring greater research effort.

New and exciting fields of research into glacial environments are being developed often by cross-fertilisation from existing sciences. For example, the use of micromorphology in glacial sediment microstructure research shows enormous potential. Likewise, optically stimulated luminescence (OSL) as an added tool in dating methods should permit details of chronology and stratigraphy hitherto only inferred. By explaining aspects of the rheology of slurries much can be gleaned that is immediately pertinent to deformable bed conditions, porewater contents and stress field states. Similarly, from rock mechanics and knowledge of brittle fracture processes within tribology, a different perspective on the processes of glacial erosion can be acquired.

Finally, there is a need to 'step-back' from the minutiae of research areas and examine past glacial environments on regional and continental scales. With advances in GIS and increasingly discriminant remote sensing imagery combined with mega-geomorphology, an understanding of macroscale relationships generated by ice masses both in front of and at their beds may be gained.

17.2. PARADIGM SHIFTS

Over the past two decades interest in glacial environments has grown immensely as has the sophistication of techniques and methods used in their study. During this same time period a fundamental shift in the methodology of studying these environments has emerged. Where, in the past, a geographical/morphological approach was employed; today a movement to a sedimentological/glaciological method has gained ascendancy. This paradigm shift has been discussed elsewhere (see Chapter 9) and need not be repeated other than to note that in shifting from the former to the latter approach a geographical perspective has often been lost in the minutiae of detail utilised in process-mechanistic studies. Where the geographical/morphological methodology emphasised the landform as its basic unit of study, the sedimentological/glaciological method begins with the individual particle or sub-facies unit as a basic building unit.

The methodological stages that have occurred in studying glacial environments are typical of any developing science – beginning with a descriptive stage and moving to a more detailed 'deconstructionist' approach where a process/mechanistic methodology is employed to examine in ever-increasing detail the fundamental components of study. This development from a descriptive to process-oriented stage has been triggered by the development of increasingly sophisticated and reliable tools (e.g. dating methods, SEM and sedimentological facies models). A third stage can now be anticipated in which the individual elements or facets of the glacial environment can be re-assembled within a paradigm that will emphasise both a geographical and mechanistic perspective. Such a paradigm will take into account the geographical conditions within which glacial and non-glacial sedimentary processes operate yet are constrained by glaciological, geotechnical and rheological parameters.

17.3. QUESTIONS ILLUSTRATIVE OF PROBLEMS

As our understanding of glacial environments has increased, so novel and intriguing questions have arisen. The emergence of these questions is testimony to the vigour and intensity of future research in glacial environments. Some of the many questions that need to be answered have been categorised within individual sub-environments; however, many of questions transgress these artificial boundaries and are pertinent to several or all glacial environments. To define these questions as problems is more than semantics since without adequate answers further scientific progress in understanding glacial environments will be limited and restricted.

17.3.1. General

There are several questions that are relevant to all glacial environments and are perhaps of broader significance than those solely applicable to specific environments.

(a) In general, recognition and differentiation of facies types persists as a continuing problem. The ability to discriminate between the varied environments within which diamictons can be deposited is perhaps the best example at present. It remains problematic to distinguish between diamictons from subglacial and subaqueous environments in many critical instances.

(b) Where significant environmental transformations caused by climatic change such as periglacial activity or diagenesis have influenced sediment types, recognition of these changes as signatures within the sediments remains illusive.

(c) In the last few years considerable interest has been engendered in trying to understand the many complex structures both at the micro and macro-level found within all glacial diamictons. At present only limited explanations exist for these intriguing structures. Structures within glacial sediments may be developed as a result of intrinsic or extrinsic stress conditions or a combination of both; however, discrimination between these forms of structure generation are equivocal.

(d) A general understanding of the impact of porewater and porewater pressures on sedimentation mechanisms, rates of sedimentation and the structures imprinted by porewater in sediments in terms of physical and chemical effects remains limited. Porewater in directly influencing effective stress levels must have a major part to play in glaciotectonism and sediment deformation at all scales. In translocating clays and soluble minerals, porewater alters the chemical signature and, in some instances, the geotechnical properties of some sediments. The extent of this impact remains only inferred. Related to porewater are the many dewatering and syneresis structures that are attributable to the effects of high rates of saturated sediment deposition, overloading, glacial tectonism and neo-tectonism.

(e) It is typically stated that particles that have passed through the glacial system are distinctive in shape, surface morphology and size distribution, yet little research has closely investigated particle shape evolution, fracture mechanisms and surface morphological imprinting other than in a limited sense. The field of tribology and rock mechanics is a potential rich source of new insights into these aspects of particle development that needs to be considered.

(f) There is a determined need to relate glaciological parameters to those linked with glacial depositional mechanics. For example, it is as yet unknown how ice streaming and surging grossly influences subglacial and proximal glacial sediments

on land and subaqueously both in terms of previously deposited sediments and pencontemporaneous sedimentation processes.

(g) In terms of stratigraphic correlation, there remains a need to better understand paleosol preservation and pedogenic formative processes under different interglacial and interstadial climatic conditions. The value of dating techniques in stratigraphic correlation and surface age dating continues to require improvements in reliability and validity.

(h) As noted above under item (d), the impact and varying likely response of glacial sediments to neoseismic activity is relatively little studied. Where, in the past, intense deformation or surface turbations have been regarded as pedogenic or glaciotectonic in origin, new evidence has linked many of these structures and intercalations of glacial sediments with recent seismic activity. From an applied aspect it is crucial that a clearer knowledge of this response by glacial sediments to seismic events be obtained.

(i) As a final general concern, it remains relatively unknown how glacial sediments alter or change over time as a result of subaerial diagenesis. Some work in front of present Icelandic glaciers point to very rapid changes in clay and silt content with marked changes in geotechnical properties occurring but how penetrative these changes are over time has been little studied. Today, engineers deal with past glacial sediments that have undergone the effects of diagenesis over several thousands of years – thus such questions are imperative.

17.3.2 Subglacial Environments

The following questions have not been arranged in any order of significance or pertinence but reflect many of the remaining issues that need to be tackled in understanding subglacial environments. Several questions or parts of questions are repeated in the subsections deliberately to re-emphasise their pertinence. Typically, many of these questions generate other related questions that need answers.

(a) In recognising subglacial sediment facies types a major problem remains in differentiating between sub-facies types and also sub-environments not only within subglacial environments but between the subglacial and other environments. For example, the differentiation between subglacial debris flows and those generated in supraglacial or proglacial regimens remains conjectural. The key to recognition and discrimination lies in determining if diagnostic signatures can be found that would permit, for example, the distinction between sediments characteristic of soft bed deformation and those deposited under non-deforming subglacial bed conditions.

(b) As yet relatively few studies have concentrated upon fully exploring and understanding the development of subglacial hydraulic systems both in the spatial context beneath ice masses of varying thermal and stress field regimens and in the temporal in terms of hydraulic system maturation, disruption and re-stabilisation as must occur under surging conditions and other changing bed states. Similarly, the mechanisms of subglacial meltwater erosion and the influence of meltwater on overall ice sheet stability require considerable research.

(c) As the importance and ubiquity of surging glacier conditions becomes more apparent it is critical that the signature, if any, be explored in detail, both within the subglacial sediments and on landscape in terms of possible bedforms.

(d) A related issue to both subglacial hydraulics and surging behaviour is the question of recognising jökulhlaup events within the subglacial sedimentary record. At present evidence is beginning to accumulate that should permit recognition but still more work needs to be done both in modern glacial environments and on past glacial sediments.

(e) Only limited work has been accomplished in recent years on the mechanics of subglacial glacio-fluvial deposition whether in cavities, channels or sheets. Problems concerning when and how deposition occurs especially under full-pipe flow conditions remains enigmatic. Equally critical are the relationships between glaciofluvial sediments and diamicton deposition and the general variations in subglacial bed conditions over time and space beneath an ice mass.

(f) A typical characteristic of most diamictons is a clast fabric of varying degrees of strength. In the past accepting that all subglacial diamictons were of a lodgement, melt-out or flow till type, clast fabric was used as a discriminant tool but, since depositional conditions beneath an ice mass are more complex than hitherto conceived, clast fabric and its sig-

nificance must be re-examined, as must the mechanism(s) of fabric development given the complex rheological and thermal conditions found beneath ice masses.

(g) Within the subglacial environment, in particular, the enormous variation and transience in bed conditions must impact upon deposition mechanisms, structures and stratigraphy – yet this reflection of changing conditions remains undetected within existing subglacial sediments.

(h) Little is known concerning post-depositional diagenesis of subglacial sediments. The alteration of geotechnical properties is especially significant; however, the timing and extent of diagenetic change remains generally unknown.

(i) It is well known that on approaching an ice front exposed bedrock may exhibit evidence of iron oxide deposition in the lee side of boulders and bedrock protrusions. Likewise translocation of carbonates can be seen as calcite deposits and in many glaciofluvial gravels cementation by calcium carbonate is often reported. Both instances reveal the effectiveness of low temperature geochemistry in dissolution and re-precipitation; however, the details of the geochemical processes, the rates of chemical reaction and the influences of temperature and pressure are insufficiently understood.

(j) Questions such as 'how are drumlins formed' have been examined for over a century yet their origin remains enigmatic. The subglacial forms and possibly associated forms such as Rogen and fluted moraines encapsulate at the larger scale the problems of understanding subglacial depositional processes. It has been suggested that these forms are part of a bedform continuum and as such are intrinsically related in origin and development, however, this idea remains only one of several hypotheses on their genesis. If these forms are bedform assemblages what are the bedforming mechanisms, why do sets of these forms seem to co-exist, and if bedform transition occurs how is it manifest, and what are the controls on bedformation and maturation?

(k) As a final example, the details of glaciotectonism and the relationship of this complex process to imprinting structures and other diagnostic signatures in subglacial sediments need further investigation. Likewise discrimination between neotectonic seismic events within subglacial sediments and contemporaneous glaciotectonism requires further detailed study.

17.3.3. Englacial Environments

Englacial, like subglacial, environments are normally obscured from view. When such an environment is penetrated invariably the environment is altered however slightly. Knowledge of englacial environments are therefore largely derived from indirect evidence.

(a) The development of englacial hydraulic systems remains largely obscure. The stages and style of development are largely unknown as are the impact of changes in ice velocity especially the effect of surging and subsequent englacial hydraulic system re-establishment.

(b) The discrimination of englacial sediments from either supra- or subglacial remains in many cases an insurmountable problem. Only when the facies context can be ascertained is distinction possible.

(c) The mechanics of sedimentation in englacial tunnels and conduit remains obscure. In many instances sedimentation would appear to occur under full-pipe flow meltwater regimes presumably under high hydraulic pressure gradients. Likewise knowledge of sedimentation processes in partially filled englacial drainage systems is insufficient. As a consequence both the mechanics and timing of the formation of eskers and kames and associated glaciofluvial englacial deposits is inadequately modelled and understood.

17.3.4. Supraglacial Environments

As interest in valley glaciation as an analog for past glaciations waned over the last 20 years, in many instances close attention to supraglacial environments declined. It was generally accepted, for example, that the impact and contribution from supraglacial sediment sources along the southern margins of the Laurentide Ice Sheet in southern Canada was limited; yet evidence from the mid-western states in the USA contradicted this assumption. As a result, in the past few years, an increasing realisation has dawned that a considerable volume of subglacial sediments were elevated over the last few kilometres towards the margins of the Laurentide Ice Sheet and appeared as supraglacial sediments sliding and flowing off the ice margins into morainal and other proximal proglacial environments both on land and into lakes. Thus the need to better comprehend these environments must again been viewed as essential.

(a) Perhaps the greatest challenge is in distinguishing supraglacial sediments from all other subfacies. Where sediment originally was subglacially derived, as was the case virtually all along southern margins of the Laurentide Ice Sheet, facies differentiation is especially difficult.

(b) It is likely that in the event of a surge, supraglacial sediments should reflect the impact of such a dynamic process but as yet recognition of such events has not been possible.

(c) As our understanding of sediment delivery to the margins increases, it is apparent that there is a considerable need to be able to link changes in ice mass balance and thus balance gradient with changes in sediment delivery especially in the supraglacial environmental system. If the variations in magnitude and frequency of changes in ablation/accumulation are sufficiently great, then thresholds that will influence sediment delivery to the margins will occur and be reflected in facies type, volume, possibly depositional mechanics and size and spatial/temporal distribution of moraines.

17.3.5. Proglacial Environments

It is perhaps fair to say that interest in proglacial environments has been the realm of only a few dedicated researchers and that all to commonly the environment has been virtually ignored to the detriment of our understanding of the general glacial system.

(a) The proglacial environment is invariably dominated by meltwater streams, thus the mechanics of stream development in relation to glacial hydrologic regimes is critical to an understanding of the dominating processes in the environment.

(b) The impact of jökulhlaups is of such magnitude both distally downstream and in the immediate proximal zone in any proglacial environment, that recognition of such infrequent but catastrophic events is an integral part in understanding ice mass behaviour and sedimentation within a glaciated basin. At present discrete characteristics that can be used to distinguish jökulhlaup events are still being sought both in proximal and distal locations.

(c) Attempts to understand in-channel sedimentation processes such as braid development has become an increasingly important area of study with proglacial environments. Due to the unique hydraulic regimes of glacial meltwater streams, the ability to understand the evolution of in-channel forms has the potential to permit recognition and differentiation of specific proglacial sedimentological processes and hydraulic regimens.

(d) Within proglacial environments there are many forms and structures such as boulder rings that are indicative of both meltwater flow regimes and the impact of melting buried ice. At present our knowledge of many of these structures is limited thus what these forms are symptomatic of remains hidden.

(e) As is often the case ice masses advance over the proglacial zone resulting in sediment re-incorporation into the glacial system. The effect of ice advancement is multi-fold, causing ice shoving, pushing, squeezing and general deformation of proglacial sediments. In some cases sediments may be completely reworked back into the subglacial system losing all attributes of any previous depositional history, while in other cases sediments may be strongly tectonised, folded and faulted and/or proglacial wedges may be constructed.

17.3.6 Glaciomarine/Glaciolacustrine Environments

In the past decade advances in our knowledge of glaciomarine environments and past glaciomarine sediments has increased dramatically. As exploration of oceanic basins has progressed, data pertinent to glaciomarine sediments have proliferated at a immense rate. However, details of sedimentation processes and rates remain fragmentary at best. The extent and thickness of glaciomarine sediments both from Pleistocene and pre-Pleistocene glaciations has meant that major reassessment of the importance of subaqueous glacial sedimentation.

Much of what has been learnt from glaciomarine investigations can be applied to the lacustrine environment but other aspects of lacustrine sedimentation, water column thermal stratification, and debris release mechanisms differ.

Of immense importance is the realization that the interaction between ice masses and lacustrine and marine environments is crucial in understanding many of the processes and styles of sedimentation that occur at the boundary between these environments. For example, in the Great Lakes region of North America and around the North Sea Basin in Europe the interaction of land ice and subaqueous environments is critical in understanding not only the forms of sedimentation that occurred where these different environments interacted but also in explaining the subsequent dynamics of the ice masses themselves.

(a) Although much is known concerning subaqueous sedimentation under glacial conditions, there are still many details of sedimentation rates and the impact of ice stresses along grounding lines that remain to be fully explained. For example, the impact varying forms of basal ice movement and thermal conditions have on sediment facies types within a lake or ocean basin is poorly understood.

(b) Where ice masses have surged into bodies of water sedimentation patterns and rates are not well known. The tectonic effects of surging on glaciomarine/lacustrine sediments is only known for a few locations. Detection of surging events as portrayed in these sediment is not possible at present.

17.4. EPILOGUE

Science has made enormous progress in our understanding of glacial environments and it is confirmation of the achievements and ongoing vitality of the discipline that so many new and intriguing questions remain to be solved. Today, the relevance of understanding glaciers and glacial sediments has never been greater where, for example, the utilisation of ice masses for irrigation purposes, or sediments for agriculture, toxic waste disposal, building materials, foundation uses have become singularly applicable.

Research in glacial environments has reached new and exciting phases on many fronts while the relevance of these environments has become heightened as our awareness of global environmental change and potential long-term implications to the earth have increased. As threats to the environment from numerous sources and activities increase, it is only too evident in those areas of the world underlain by past glacial sediments that a vastly improved understanding of these materials and their origins is indispensable.

REFERENCES

Aario, R. (1972). Associations of bed forms and pale-ocurrent patterns in an esker delta, Haapajärvi, Finland. *Ann. Acad. Scientiarum Fennicae*, A, pt. III, **111**, 55 pp.

Aario, R. (1977a). Classification and terminology of morainic landforms in Finland. *Boreas*, **6**, 87–100.

Aario, R. (1977b). Associations of flutings, drumlins, hummocks and transverse ridges. *Geojournal*, **1**, 65–72.

Aario, R. (1977c). Flutings, drumlins and Rogen-landforms. *Nordia*, **2**, 5–14.

Aario, R. (1987). Drumlins of Kuusamo and Rogen-ridges of Ranua, northeast Finland. In *Drumlin Symposium*: (J. Menzies and J. Rose Eds), pp. 87–102, A.A. Balkema Publishers, Rotterdam.

Aario, R. (1990). Morainic landforms in Northern Finland. In (*Glacial Heritage of Northern Finland, Excursions Guide III: Drumlin Symposium* (R. Aario ed.) pp. 13–28. Nordia Tiedonantoja Sarja A No. 1. Oulu: Oulun Yliopisto Maantieteen Laitos.

Aario, R., Forsstrom, L. and Lahermo, P. (1974). Glacial landforms with special reference to drumlins and flutings in Koillismaa, Finland. *Bulletin Geological Survey of Finland*, **273**, 1–30.

Aarseth, I. and Mangerud, J. (1974). Younger Dryas and moraines between Harangerfjorden and Sognefjorden, Western Norway. *Boreas*, **3**, 3–22.

Aartolahti, T. (1972). On deglaciation in southern and western Finland. *Fennia*, **114**, 1–84.

Aartolahti, T. (1974). Ring ridge hummocky moraines in northern Finland. *Fennia*, **134**, 22 pp.

Aber, J. S. (1980). Nature and origin of exotic-rich drift in the Appalachian Plateau. *American Journal of Science*, **280**, 363–384.

Aber, J. S. (1982). Model for glaciotectonism. *Geological Society of Denmark Bulletin*, **30**, 79–90.

Aber, J. S. (1985). Definition and model for Kansan glaciation. *Institute for Tertiary-Quaternary Studies – TER-QUA Symposium Series*, **1**, 53–60.

Aber, J. S. (1988). Bibliography of glaciotectonic references. In *Glaciotectonics: Forms and Processes* (D. G. Croot ed.) pp. 195–210, Rotterdam: A.A. Balkema Publishers.

Aber, J. S (1993). Expanded bibliography of glaciotectonic references. In *Glaciotectonics and Mapping Glacial Deposits. Proceedings of the INQUA Commission on the Formation and Properties of Glacial Deposits.* (J. S. Aber ed.) 99–137, University of Regina, Regina: Canadian Plains Research Centre.

Aber, J. S., Croot, D. G. and Fenton, M. M. (1989). *Glaciotectonic Landforms and Structures*, Boston, Dordrecht: Kluwer Academic Publisher, 200 pp.

Acton, D. F. and Fehrenbacher, J. B. (1976). Mineralogy and topography of glacial tills and their effect on soil formation in Saskatchewan. In (*Glacial Till: An Interdisciplinary Study* (R. F. Legget ed.) pp. 170–185, Ottawa: The Royal Society of Canada, Special Publication No. 12.

Addison, K. (1981). The contribution of discontinuous rock-mass failure to glacier erosion. *Annals of Glaciology*, **2**, 3–10.

Ahlbrandt, T. S. and Andrews, S. (1978). Distinctive sedimentary features of cold-climate eolian deposits, North Park, Colorado. *Palaeogeography, Palaeoclimatology, Palaeoecology*, **25**, 327–351.

Aitchison, J. C., Bradshaw, M. A. and Newman, J. (1988). Lithofacies and origin of the Buckeye Formation: Late Paleozoic glacial and glaciomarine sediments, Ohio Range, Transantarctic Mountains, Antarctica. *Palaeo-*

geography, Palaeoclimatology, Palaeoecology, **64**, 93–104.

Aitken, M. J. (1985). *Thermoluminescence Dating*. London: Academic Press, 359 pp.

Aitken, M. J. (1992). Optical dating. *Quaternary Science Reviews*, **11**, 127–131.

Akerman, J. (1980). Studies on periglacial geomorphology in West Spitsbergen. *Meddelanden Fran Lunds Universitets Geografiska Institution Avhanglingrar*, **89**, 1–297.

Aksu A. E. and Mudie, P. J. (1985). Magnetostratigraphy and palynology demonstrate at least 4 million years of Arctic Ocean sedimentation. *Nature*, **318**, 280–283.

Albino, K. and Dreimanis, A. (1988). A time-transgressive kinetostratigraphic sequence spanning 180o in a single section at Bradtville, Ontario, Canada. In *Glaciotectonics* (D. G. Croot, ed.) pp. 11–20, Rotterdam: A.A. Balkema Publishers.

Alden, W. C. (1911). Radiation in glacial flow as a factor in drumlin formation. *Geological Society of America Bulletin*, **22**, 733–734.

Alden, W. C. (1918). The Quaternary geology of south-east Wisconsin. United States Geological Survey Professional Paper 106.

Alean, P., Braun, S., Iken, A., Schram, K. and Zwosta, G. (1986). Hydraulic effects at the glacier bed and related phenomena. International Workshop 16–19 September, 1985, Interlaken, Switzerland. *Mitteilungen der Versuchsanstalt für Wasserbau, Hydrologie und Glaziologie*, 90, 148 pp.

Alexander, H. S. (1932). Pothole erosion. *Journal of Geology*, **40**, 305–337.

Allen, J. R. L. (1971). Transverse erosional marks of mud and rock: their physical basis and geological significance. *Sedimentary Geology*, **5**, 167–385.

Allen, J. R. L. (1982). *Sedimentary Structures: Their Character and Physical Basis*. 2 vol, Amsterdam: Elsevier Science Publishing, 611 and 679 pp.

Allen, P. (1991). Deformation structures in British Pleistocene sediments. In *Glacial Deposits in Great Britain and Ireland* (J. Ehlers, P. L. Gibbard and J. Rose, eds) pp. 455–469, Rotterdam: A.A. Balkema.

Allen, P., Cheshire, D. A., and Whiteman, C. A. (1991). The tills of southern East Anglia. In *Glacial Deposits in Great Britain and Ireland* (J. Ehlers, P. L. Gibbard and J. Rose, eds) pp. 255–278, Rotterdam: A.A. Balkema.

Allen, P. A. and Collinson, J. D. (1986). Lakes. In *Sedimentary Environments and Facies*, 2nd edn, (H. G. Reading ed.) pp. 63–94, Oxford: Blackwell Scientific Publishers.

Alley, R. B. (1989a). Water-pressure coupling of sliding and bed deformation: I. Water system. *Journal of Glaciology*, **35**, 108–118.

Alley, R. B. (1989b). Water-pressure coupling of sliding and bed deformation: II. Velocity-depth profiles. *Journal of Glaciology*, **35**, 119–129.

Alley, R. B. (1991). Deforming-bed origin for southern Laurentide till sheets? *Journal of Glaciology*, **37**, 67–76.

Alley, R. B. (1992). Flow-law hypothesis for ice-sheet modelling. *Journal of Glaciology*, **38**, 245–256.

Alley, R. B., Blankenship, D. D., Bentley, C. R. and Rooney, S. T. (1986). Deformation of till beneath ice stream B, West Antarctica. *Nature*, **322**, 57–59.

Alley, R. B., Blankenship, D. D., Rooney, S. T. and Bentley, C. R. (1987a). Continuous till deformation beneath ice sheets. *International Association of Hydrological Sciences*, **170**, 81–91.

Alley, R. B., Blankenship, D. D., Bentley, C. R. and Rooney, S. T. (1987b). Till beneath ice stream B. 3. Till deformation: evidence and implications. *Journal of Geophysical Research*, **92B**, 8921–8929.

Alley, R. B. and Whillans, I. M. (1984). Response of the East Antarctica ice sheet to sea-level rise. *Journal of Geophysical Research*, **89**, 6487–6493.

Allison, C. W., Young, G. M., Yeo, G. M. and Delaney, G. D., (1981). Glacigenic rocks of the Upper Tindir Group, east-central Alaska. In *Earth's pre-Pleistocene Glacial Record*, (M. J. Hambrey and W. B. Harland, eds) pp. 720–723, Cambridge: Cambridge University Press.

Åmark, M. (1980). Glacial flutings at Isfallsglaciären, Tarfala, Swedish Lapland. *Geologiska Föreningen i Stockholm Förhandlingar*, **102**, 251–259.

Andersland, O. B. (1989). General report on mechanical properties of frozen soil. In *Ground Freezing 88* Proceedings of the 5th International Symposium on Ground Freezing, Nottingham, England, 26–28 July 1988, vol. 2, (R. H. Jones and J. T. Holden, eds) pp. 433–441.

Andersland, O. B. and Al-Moussawi, H. M. (1988). Cyclic thermal strain and crack formation in frozen soils. In *Ground Freezing 88*, Proceedings of the 5th International Symposium on Ground Freezing, Nottingham, England, 26–28 July 1988, Volume 1, (R. H. Jones and J. T. Holden, eds) pp. 167–172.

Anderson, J. B. (1983). Ancient glacial-marine deposits: their spatial and temporal distribution. In *Glacial-Marine Sedimentation*. (B. F. Molnia, ed.) pp. 3–92, New York: Plenum Press.

Anderson, J. B. and Ashley, G. M. (1991). Glacial marine sedimentation: Paleoclimatic significance. *Geological Society of America Special Paper*, 261, 232 pp.

Anderson, J. B., Brake, C., Domack, E., Meyers, N. and

Wright, R. (1983). Development of a polar glacial-marine sedimentation model from Antarctic Quaternary deposit and glaciological information. In *Glacial-Marine Sedimentation* (B. F. Molnia, ed.) pp. 223–264, New York Plenum Press.

Anderson, J. B., C. F. Brake, and N. C. Myers. (1984) Sedimentation in the Ross Sea continental shelf, Antarctica. *Marine Geology*, 57, 295–333.

Anderson, J. B., Domack, E. W. and Kurtz, D. D. (1980a) Observations of sediment-laden icebergs in Antarctic waters: implications to glacial erosion and transport. *Journal of Glaciology*, 25, 387–396.

Anderson, J. B., Kurtz, D. D., Domack E. W. and Balshaw, K. M. (1980b). Glacial and glacial marine sediments of the Antarctic continental shelf. *Journal of Geology*, 88, 399–414.

Anderson, J. L. and Sollid, J. L. (1971). Glacial chronology and glacial geomorphology in the marginal zones of the glaciers, Mitdalsbreen and Nigardsbreen, south Norway. *Norsk Geografisk Tidsskrift*, 25, 1–38.

Anderson, L. W., (1978). Cirque glacier erosion rates and characteristics of Neoglacial tills, Pangnirtung Fjord area, Baffin Island, North West Territories, Canada. *Arctic and Alpine Research*, 10: 749–760.

Anderson, M. G. and K. S. Richards. (1987) *Slope Stability Geotechnical Engineering and Geomorphology*. Chichester: John Wiley & Sons, 648 pp.

Anderson, W. F. (1972). The geotechnical properties of the till of the Glasgow region and the development of a constant rate of expansion pressure-meter suitable for measuring the undrained strength and deformation characteristics of this till. Unpublished Ph.D thesis, University of Strathclyde, Glasgow.

Anderson, W. F. (1983). Foundation engineering in glaciated terrain. In *Glacial Geology* (N. Eyles ed.) Chapter 12, 275–301, Oxford: Pergamon Press.

Andrews, D. E. (1980). Glacially thrust bed rock – an indication of late Wisconsin climate in western New York. *Geology*, 8, 97–101.

Andrews, J. T. (1963) The cross-valley moraines of north-central Baffin Island, NWT: a descriptive analysis. *Geographical Bulletin*, 19, 49–77.

Andrews, J. T. (1970). A geomorphological study of post-glacial uplift with particular reference to Arctic Canada. Institute of British Geographers, Special Publication, 2, 156pp.

Andrews, J. T. (1985). Quaternary environments of Eastern Canada Arctic, Baffin Bay and west Greenland. Boston: Allen and Unwin publishers, 774 pp.

Andrews, J. T. (1987a). Postface: The Laurentide Ice Sheet: research problems. *Géographie Physique et Quaternaire*, 41, 315–318.

Andrews, J. T. (1987b). Glaciation and sea-level: a case study. In *Sea Surface Studies: A Global View* (R. J. N. Devoy, ed.) pp. 95–126, London, New York, Sydney: Croom Helm.

Andrews, J. T. (1987c). The late Wisconsin glaciation and deglaciation of the Laurentide ice sheet. In *North America and Adjacent Oceans During the Last Deglaciation* (W. F. Ruddiman and H. E. Wright, Jr., eds) pp. 13–37, Boulder, Colorado: Geological Society of America, The Geology of North America, K–3.

Andrews, J. T. and King, C. A. M. (1968). Comparative till fabrics and till fabric variability in a till sheet and a drumlin: a small scale study. *Proceedings of the Yorkshire Geological Society*, 36, 435–461.

Andrews, J. T. and Miller, G. H. (1972). Chemical weathering of tills and surficial deposits in east Baffin Island, N.W.T. Canada. In *International Geography 1972*, vol. 1, Paper 0102, (W. P. Adams and F. M. Helleiner Eds) pp. 5–7, Toronto: University of Toronto Press.

Andrews, J. T. and Smithson, B. B. (1966). Till fabrics of the cross-valley moraines north-central Baffin Island, North West Territories, Canada. *Geological Society of America Bulletin*, 77, 271–290.

Andrews, J. T., Shilts, W. W. and Miller, G. H. (1983). Multiple deglaciations of the Hudson Bay Lowlands, Canada, since deposition of the Missinaibi (last-interglacial?) Formation. *Quaternary Research*, 19, 18–37.

Aniya, M. and Naruse, R. (1985). Structure and morphology of the Solar Glacier. In *Glaciological Studies in Patagonia L Northern Icefield, 1983–1984* (C. Nakajima, ed.), Data Center for Glacier Research, Japanese Society of Snow and Ice, 70–79.

Aniya, M. and Welch, R. (1981). Morphological analyses of glacial valleys and estimates of sediment thickness on the valley floor: Victoria Valley system, Antarctica. *The Antarctic Record*, 71, 76–95.

Anonymous, (1985). Glaciers, ice sheets and sea level: effect of CO_2-induced climatic change. Report of a workshop held in Seattle, Washington, September 1984. United States Department of Energy, Washington, D.C., pp. 241–247.

Antevs, E. (1922). The recession of the last ice sheet in New England. *American Geographical Society Research Series*, No. 11, 120 pp.

Antevs, E. (1925). Retreat of the last ice sheet in eastern Canada. *Canadian Geological Survey Memoir* 146, 142 pp.

Anundsen, K. (1989). Late Weichselian relative sea levels in southwest Norway: observed strandline tilts and neotectonic activity. *Geloogiska Foreningens i Stockholm*

Forhandlingar, **111**, 288–292.

Arias, C., Bigazzi, G. and Bonadonna, F. P. (1981). Size corrections and plateau age in glass shards. *Nuclear Tracks and Radiation Measurements*, **283**, 368–372.

Ashley, G. M. (1975). Rhythmic sedimentation in glacial Lake Hitchcock, Massachusetts-Connecticut. In *Glacio-fluvial and Glaciolacustrine Sedimentation* (A. V. Jopling and B. C. McDonald, eds), Society of Economic Paleontologists and Mineralogists, Special Publication No. 23, pp. 304–320.

Ashley, G. M. (1988). Classification of glaciolacustrine sediments. In *Genetic classification of Glacigenic Deposits* (R. P. Goldthwait and C. L. Matsch, eds) pp. 243–260, Rotterdam: A.A. Balkema Publishers.

Ashley, G. M., Boothroyd, J. C., and Borns, H. W. Jr. (1991). Sedimentology of late Pleistocene (Laurentide) deglacial-phase deposits, eastern Maine; an example of a temperate marine grounded ice-sheet margin. In *Glacial marine sedimentation; Paleoclimatic Significance* (J. B. Anderson and G. M. Ashley, eds), Geological Society of America Special Paper 261, pp. 107–125.

Ashton, N. M., Cook, J. Lewis, S. G. and Rose, J. (eds) (1992). *High Lodge: Excavations by G. de G. Sieveking 1962–68 and J. Cook 1988*. London: British Museum Press, 192 pp.

Attewell, P. B. and Farmer, I. W. 1975. *Principles of Engineering Geology*. New York: Chapman & Hall Publishers, 1045 pp.

Attig, J. W. (1985). Pleistocene Geology of vilas County, Wisconsin. Wisconsin Geological and Natural History Survey. *Information Circular 50*. 32 pp.

Attig, J. W. and Clayton, L. (1993). Stratigraphy and origin of an area of hummocky glacial topography, northern Wisconsin, U.S.A. *Quaternary International*, **18**, 61–68.

Attig, J. W., Mickelson, D. M. and Clayton, L. (1989). Late Wisconsin landform distribution and glacier-bed conditions in Wisconsin. *Sedimentary Geology*, **62**, 399–405.

Augustinus, P. C. (1992). The influence of rock mass strength on glacial valley cross-profile morphometry: a case study from the Southern Alps, New Zealand. *Earth Surface Processes and Landforms*, **17**, 39–51.

Augustinus, P. C. and Selby, M. J. (1990). Rock slope development in McMurdo Oasis, Antarctica, and implications for interpretations of glacial history. *Geografiska Annaler*, **72A**, 55–62.

Aylsworth, J. M. and Shilts, W. W. (1985). Glacial features of the west-central Canadian Shield. Geological Survey of Canada, Current Research, part B, Paper 85–1B, 375–381.

Aylsworth, J. M. and Shilts, W. W. (1989). Bedforms of the Keewatin Ice Sheet, Canada. *Sedimentary Geology*, **62**, 407–428.

Babcock, E. A. (1977). A comparison of joints in bedrock and fractures in overlying Pleistocene lacustrine deposits, central Alberta. *Canadian Geotechnical Journal*, **14**, 357–366.

Babcock, E. A., Fenton, M. M. and Andriashek, L. D. (1978). Shear phenomena in ice-thrust gravels, central Alberta. Canadian *Journal of Earth Sciences*, **15**, 277–283.

Bada, J. L. 1985. Amino acid racemization dating of fossil bones. *Annual Review of Earth and Planetary Sciences*, **13**, 241–268.

Bada, J. L. (1991). Amino acid cosmogeochemistry. *Philosophical Transactions of the Royal Society of London*, **333(B)**, 349–358.

Baden, G. M. (1952). Plant succession following glacial recession in Glacier National Park, Montana. Unpublished paper, US National Park Service, Glacier National Park, 36 pp.

Baker, H. W. Jr. (1976). Environmental sensitivity of submicroscopic surface textures on quartz sand grains – a statistical evaluation. *Journal of Sedimentary Petrology*, **46**, 871–880.

Bakker, J. P. (1965) A forgotten factor in the interpretation of glacial stairways. *Zeitschrift für Geomorphologie*, **NF 9**, 18–34.

Ballantyne, C. K. (1989). The Loch Lomond readvance on the Isle of Skye, Scotland: glacier reconstruction and palaeoclimatic implications. *Journal of Quaternary Science*, **4**, 95–108.

Ballantyne, C. K. and Gray, J. M. (1984). The Quaternary geomorphology of Scotland: the research contribution of J. B. Sissons. *Quaternary Science Reviews*, **3**, 259–289.

Ballantyne, C. K. and Whittington, G. (1987). Niveo-aeolian sand deposits on An Teallach, Wester Ross, Scotland. *Transactions of the Royal Society of Edinburgh, Earth Sciences*, **78**, 51–63.

Banerjee, I. and MacDonald, B. C. (1975). Nature of esker sedimentation. In *Glaciofluvial and Glaciolacustrine Sedimentation* (A. V. Jopling and B. C. MacDonald, eds). SEPM Special Publication 23, 132–154.

Banerjee, S. K. (1983). The Holocene paleomagnetic record in the United States. In *Late Quaternary environments of the United States, Volume 2, The Holocene* (H. E. Wright Jr., ed.) pp. 78–85, Minneapolis: University of Minnesota Press.

Banham, P. H. (1975). Glacitectonic structures: a general discussion with particular reference to the contorted drift of Norfolk. In *Ice Ages: Ancient and Modern*. Geological Journal Special Issue 6., (A. E. Wright and F. Moseley, eds) pp. 69–94, Liverpool: Seel House Press.

Banham, P. H. (1977). Glacitectonites in till stratigraphy. *Boreas*, **6**, 101–105.

Banham, P. H. (1988a). Thin-skinned glaciotectonic structures. In *Glaciotectonics: Forms and Processes* (D. C. Croot ed.) pp. 21–25, Rotterdam: A. A. Balkema Publishers.

Banham, P. H. (1988b). Polyphase glaciotectonic deformation in the Contorted Drift of Norfolk. In *Glaciotectonics: Forms and Processes* (D. G. Croot ed.), pp. 27–32, Rotterdam: A.A. Balkema Publishers.

Baranowski, S. (1969). Some remarks on the origin of drumlins. *Geografia Polonica*, **17**, 197–208.

Baranowski, S. (1970). The origin of fluted moraine at the fronts of contemporary glaciers. *Geografiska Annaler*, **52A**, 68–75.

Baranowski, S. (1979). The origin of drumlins as an ice-rock interface problem. *Journal of Glaciology*, **23**, 435–436.

Bard, E. (1988). Correction of accelerator mass spectrometry 14C ages measured in planktonic foraminifera: paleoceanographic implications. *Paleoceanography*, **3**, 635–645.

Bard, E. and Broecker, W. S. (1992). *The Last Deglaciation: Absolute and Radiocarbon Chronologies*. Berlin: Springer-Verlag, 344 pp.

Bard, E., Hamelin, B., Fairbanks, R. G. and Zindler, A. (1990). Calibration of the 14C timescale over the past 30,000 years using mass spectrometric U–Th ages from Barbados corals. *Nature*, **345**, 405–410.

Barden, L. and Sides, G. (1971). Sample disturbance in the investigation of clay structure, *Géotechnique*, **21**, 211–222.

Barker J. C., Robinson, M. S. and Bundtzen, T. K. (1989). Marine placer development and opportunities in Alaska. *Mining Engineering*, **42**, 551–558.

Barkla, H. (1935). The drumlins of Türi (Estonia). Geological Institute, University of Türi Publication 48.

Barnes, H. L. (1956). Cavitation as a geological agent. *American Journal of Science*, **254**, 493–505.

Barnes, P. W., Asbury, J. L., Rearic, D. M. and Ross, C. R. (1987). Ice erosion of a sea-floor knickpoint at the inner edge of the stamukhi zone, Beaufort Sea, Alaska. *Marine Geology*, **76**, 207–222.

Barnes, P. W. and Lien, R. (1988). Icebergs rework shelf sediments to 500 m off Antarctica. *Geology*, **15**, 1130–1133.

Barnes, P. W. and Reimnitz, E. (1979). Ice gouge obliteration and sediment redistribution event: 1977–1978,

Beaufort Sea, Alaska. United States Geological Survey Open File Report 79–848.

Barnes P. W., Reimnitz, E. and Fox, D. (1982a). Ice rafting of fine-grained sediment, a sorting and transport mechanism, Beaufort Sea, Alaska. *Journal of Sedimentary Petrology*, **52**, 493–502.

Barnes, P. W., Reiss, T. and Reimnitz, E. (1982b). Sediment content of nearshore fast ice: fall 1980, Beaufort Sea, Alaska. United States Geological Survey, Open-File Report, 82–716, 19 pp.

Barnes, P. W., Rearic, D. M. and Reimnitz, E. (1984). Ice gouging characteristics and processes. In *The Alaskan Beaufort Sea – Ecosystems and Environments* (P. W. Barnes, D. M. Schell and E. Reimnitz eds) pp. 185–212, Orlando: Academic Press.

Barnett, H. F. and Finke, P. G. (1971). Morphometry of landforms: drumlins. United States Army Natick Lab., Earth Science Lab. Technical Report. No. Es–63.

Barrett, P. J. (1980). The shape of rock particles, a critical review. *Sedimentology*, **27**, 291–303.

Barnett, P. J. (1990). Tunnel valleys: evidence of catastrophic release of subglacial meltwater, central-southern Ontario, Canada. *Geological Society of America, Abstracts with Programs*, **22**, 3.

Barrett, P. J., Hambrey, M. J., Harwood, D. M. Pyane, A. R. and Webb, P. N. (1989). General Synthesis. In *Antarctic Cenozoic History of the CIROS-I Drillhole* (P. J. Barrett ed.) pp 241–251, Bulletin in the Miscellaneous Series of the New Zealand Department of Scientific and Industrial Research, No. 245.

Barrett, P. J., Pyne, A. R. and Ward, B. L. 1983. Modern sedimentation in McMurdo Sound, Antarctica. In *Antarctic Earth Science* (R. L. Oliver, P. R. James and J. B. Jago eds) pp. 550–554, Canberra: Australian Academy of Science, Cambridge: Cambridge University Press.

Barrett, T. J., Sherlock, R. L., Juras, S. J., Wilson, G. C. and Allen, R. (1994). Geological investigations of the H-W Deposit, Buttle Lake Camp, central Vancouver Island. Geological fieldwork 1993; a summary of field activities and current research B. C. Ministry of Energy, Mines and Petroleum Resources, pp. 339–344.

Barrie, J. V. (1990). Contemporary and relict titaniferous sand facies on the western Canadian continental shelf. *Continental Shelf Research*, **11**, 67–80.

Barrie, J. V., Emory-Moore, M., Luternauer, J. L. and Bornhold, B. D. (1988). Origin of modern heavy mineral deposits, northern British Columbia continental shelf. *Marine Geology*, **84**, 43–51.

Barron, J. A., Larsen, B. and Baldauf, R. B. (1991). Evidence for late Eocene to early Oligocene Antarctic glaciation and observations on late Neogene glacial history of Antarctica: results from Leg 119. In *Proceedings of the Ocean Drilling Program, Kerguelen Plateau-Prydz Basin* (J. A. Barron, B. Larsen, J. G. Baldauf, C. Alibert, S. Berkowitz, J. -P. Caulet, S. R. Chambers, A. K. Cooper, R. E. Cranston, W. U. Dorn, W. U. Ehrmann, R. D. Fox, G. A. Fryxell, M. J. Hambrey, B. T. Huber, C. J. Jenkins, S. -H. Kang, B. H. Keating, K. W. Mehl, I.Noh, G. Ollier, A. Pittenger, H. Sakai, C. J. Schroder, A. Solheim, D. A. Stockwell, H. R. Thierstein, B. Tocher, B. R. Turner, W. Wei, E. K. Mazzullo and N. J. Stewart eds),. *Proceedings of the Ocean Drilling Program, Scientific Results*, **119**, 869–891.

Barry, R. G. (1985). Snow cover, sea ice, and permafrost. In Glaciers, Ice Sheets and Sea Level: Effect of a CO2-induced Climatic Change. Report of a workshop held in Seattle, Washington, September 1984. Washington, D.C: United States Department of Energy, 330 pp.

Becker, B., Kromer, B. and Trimborn, P. (1991). A stable-isotope tree-ring timescale of the Late Glacial/Holocene boundary. *Nature*, **353**, 647–649.

Beddow, J. K. (1980). Particle morphological analysis. In *Advanced Particulate Morphology* (J. K. Beddow and T. P. Meloy eds) pp. 31–50, Boca Raton, Florida: CRC Press.

Beget, J. E. (1984). Tephrochronology of late Wisconsin deglaciation and Holocene glacier fluctuations near Glacier Peak, North Cascade Range, Washington. *Quaternary Research*, **21**, 304–316.

Beget, J. E. (1986). Modelling the influence of till rheology on the flow and profile of the Lake Michigan lobe, southern Laurentide ice sheet, U.S.A. *Journal of Glaciology*, **32**, 235–41.

Beget, J. E. (1987). Low profile of the northwest Laurentide ice sheet. *Arctic and Alpine Research*, **19**, 81–88.

Beget, J. E., Stone, D. B. and Hawkins, D. B. (1990). Paleoclimatic forcing of magnetic susceptibility variations in Alaskan loess during the late Quaternary. *Geology*, **18**, 40–43.

Bélanger, S. and Filion, L. (1991). Niveo-aeolian sand deposition in subarctic dunes, eastern coast of Hudson Bay, Québec, Canada. *Journal of Quaternary Science*, **6**, 27–37.

Belknap, D. F., Kelley, J. T. and Shipp, R. C. (1986). Quaternary stratigraphy of representative Maine estuaries: initial examination by high resolution seismic reflection profiling. In *A Treatise on Glaciated Coasts* (D. Fitzgerald and P. Rosen eds) pp. 178–207, New York: Academic Press.

Benedict, J. B. (1967). Recent glacial history of an alpine area in the Colorado Front Range, USA. I: Establishing a lichen-growth curve. *Journal of Glaciology*, **6**, 817–832.

Benn, D. I. (1989). Controls on sedimentation in a late Devensian ice-dammed lake, Achnasheen, Scotland. *Boreas*, **18**, 31–42.

Bennekom, van A. J., Berger, J. W., van der Gaast, S. J. and deVries, R. T. P. (1988). Primary productivity and the silica cycle in the southern ocean (Atlantic Sector). *Paleogeography, Paleoclimatology, Paleoecology*, **67**, 19–30.

Bennett, M. R. and Boulton, G. S. (1993). A reinterpretation of Scottish 'hummocky moraine' and its signficance for the deglaciation of the Scottish Highlands during the Younger Dryas or Loch Lomond Stadial. *Geological Magazine*, **130**, 301–318.

Berg, M. W. van den and Beets, D. J. (1987). Saalian glacial deposits and morphology in The Netherlands. In *Tills and Glaciotectonics* (J. J. M. van der Meer ed.) pp. 235–251, Rotterdam: A.A. Balkema Publishers.

Berger, A. and Loutre, M. F. (1988). New insolation values for the climate of the last 10 million years. Scientific Report 1988/13, Institut d'Astronomie et de Geophysique Georges Lemaitre. Universite Catholique de Louvain-la-Neuve, Belgium.

Berger, G. W. (1984). Thermoluminescence dating studies of glacial silts from Ontario. *Canadian Journal of Earth Sciences*, **21**, 1393–1399.

Berger, G. W. (1988a). Dating Quaternary events by luminescence. In *Dating Quaternary Sediments, Geological Society of America Special Paper*, 227 (D. J. Easterbrook ed.) pp. 13–50.

Berger, G. W. (1988b). TL dating studies of tephra, loess and lacustrine sediments. *Quaternary Science Reviews*, **7**, 295–303.

Berger W. (1973). Deep-sea carbonates: Pleistocene dissolution cycles. *Journal of Foraminiferal Research*, **3**, 187–195.

Berger, W. H. (1990). The Younger Dryas cold spell – a quest for causes. *Palaeogeography, Palaeoclimatology, Palaeoecology* (Global and Planetary Change Section), **89**, 219–237.

Berkson, J. M and Clay, C. S. (1973). Microphysiography and possible iceberg grooves on the floor of western Lake Superior. *Geological Society of America Bulletin*, **84**, 1315–1328.

Bernard, C. (1971a). Les marques sous-glaciaires d'aspect plastique sur la roche enplace (p-forms): Observation sur la bordure du Bouclier canadien et examen de la question (I). *Revue de Géographie de Montréal*, **25**, 111–127.

Bernard, C. (1971b). Les marques sous-glaciaires d'aspect plastique sur la roche enplace (p-forms): Observation sur la bordure du Bouclier canadien et examen de la question

(II). *Revue de Géographie de Montréal*, **25**, 265–279.

Berner, R. A. (1981). A new geochemical classification of sedimentary environments. *Journal of Sedimentary Petrology*, **51**, 359–365.

Berthelsen, A. (1973). Weichselian ice advance and drift succession in Denmark. *Bulletin of the Geological Institute University of Uppsala*, New Series **5**, 21–29.

Berthelsen, A. (1978). The methodology of kineto-stratigraphy as applied to glacial geology. *Bulletin of the Geological Society of Denmark*, Special Issue, **27**, 25–38.

Berthelsen, A. (1979). Recumbent -folds and boudinage structures formed by subglacial shear: an example of gravity tectonics. *Geologie en Mijnbouw*, **58**, 253–60.

Beschel, R. E. (1950). Flechten als Altersmasstab rezenter Moränen. *Zeitschrift für Gletscherkunde und Glazialgeologie*, **1**, 152–161.

Beschel, R. E. (1961). Dating rock surfaces by lichen growth and its application to glaciology and physiography (lichenometry). In *Geology of the Arctic 2*, (G. C. Raasch ed.) pp. 1044–1062, Toronto: University of Toronto Press.

Beuf, S., Biju-Duval, B. , de Charpal, O., Rognon, P, Gariel, O. and Bennacef, A. (1971). Les grès du paléozoique inférieur au Sahara: sedimentation et discontinuites evolution structurale d'un craton. *Publications de l'Institut Français du Pétrole, collection 'Science et technique du pétrole, No. 18'*, Paris: Éditions Technip.

Bezinge, A. (1987). Glacial meltwater streams, hydrology and sediment transport: the case of the Grande Dixence hydroelectricity scheme. In *Glacio-Fluvial Sediment Transfer – an Alpine Perspective* (A. M. Gurnell and M. J. Clark eds) pp. 473–498, Chichester: John Wiley & Sons.

Bierman, P. R. and Gillespie, A. R. (1991). Accuracy of rock-varnish chemical analyses: Implications for cation-ratio dating. *Geology*, **19**, 196–199.

Bierman, P. R., Gillespie, A. R. and Kuehner, S. 1991. Precision of rock-varnish chemical analyses and cation-ratio ages. *Geology*, **19**, 135–138.

Billard, A. and Derbyshire, E. (1985). Pleistocene stratigraphy and morphogenesis of La Dombes: an alternative hypothesis. *Bulletin de l'Association française pour l'étude du Quaternaire*, **22**, 85–96.

Bilzi, A. F. and Ciolkosz, E. J. (1977). A field morphology rating scale for evaluating pedological development. *Soil Science*, **124**, 45–48.

Birch, F. S. and Trask, R. (1978). Geophysical study of bedrock cores under some drumlins in the seacoast region of New Hampshire. *Geological Society of America,*

Abstracts with Programs, **10**, 2, 33.

Bird, J. B. (1967). *The Physiography of Arctic Canada*. Baltimore: Johns Hopkins Press, 336 pp.

Birkeland, P. W. (1973). Use of relative-dating methods in a stratigraphic study of rock glacier deposits, Mt. Sopris, Colorado. *Arctic and Alpine Research*, **5**, 401–416.

Birkeland, P. W. (1978). Using soils to date quaternary deposits in Eastern Baffin Island, N.W.T., Canada. *Arctic and Alpine Research*, **10**. 733–747.

Birkeland, P. W. (1982). Subdivision of Holocene glacial deposits, Ben Ohau Range, New Zealand, using relative-dating methods. *Geological Society America Bulletin*, **93**, 433–449.

Birkeland, P. W. (1984). *Soils and Geomorphology*. New York: Oxford University Press, 372 pp.

Birkeland, P. W. and Burke, R. M. (1988). Soil catena chronosequences on Eastern Sierra Nevada moraines, California, U.S.A. *Arctic and Alpine Research*, **20**, 473–484.

Birkeland, P. W. and Janda, R. J. (1971). Clay mineralogy of soils developed from Quaternary deposits of the eastern Sierra Nevada, California. *Geological Society America Bulletin*, **82**, 2495–2514.

Birkeland, P. W., Burke, R. M. and Yount, J. C. (1976). Preliminary comments on late Cenozoic glaciations in the Sierra Nevada. In *Quaternary Stratigraphy of North America* (W. C. Mahaney ed.) pp. 283–295, New York: Halsted Press.

Birkeland, P. W., Burke, R. M. and Walker, A. L. (1980). Soils and subsurface rock-weathering features of Sherwin and pre-Sherwin glacial deposits, eastern Sierra Nevada, California. *Geological Society America Bulletin*, **91**, 238–244.

Birkeland, P. W., Burke, R. M. and Benedict, J. B. (1989). Pedogenic gradients for iron and aluminium accumulation and phosphorus depletion in Arctic and Alpine soils as a function of time and climate. *Quaternary Research*, **32**, 193–204.

Birkeland, P. W. and Shroba, R. R. (1974). The status of the concept of quaternary soil forming intervals in the western United States. In *Quaternary Environments: Proceedings of a symposium* (W. C. Mahaney ed.), York University, Atkinson College, Toronto, Ontario, pp. 241–276.

Birkenmajer, K. (1982). Pliocene tillite-bearing successions on King George Island (South Shetland Islands), *Antarctica Studia Geologica Polonica*, **74**, 7–72.

Bischof, J., Koch, J., Kubisch, M., Spielhagen, R. F. and Thiede, J. (1990). Nordic Seas surface ice drift reconstructions: evidence from ice-rafted coal fragments during oxygen isotope stage 6. In *Glaciomarine Environ-*

ments: Processes and Sediments. (J. A. Dowdeswell and J. D. Scourse eds), Geological Society of London, Special Publication No. 53. pp. 235–251.

Bisdom, E. B. A. (1981). *Submicroscopy of soils and weathered rocks*. Wageningen, Netherlands: Centre for Agricultural Publishing and Documentation, 320 pp.

Biste, M. H., Bufler, R. and Friedrich, G. (1991). Geology and exploration of gold placer deposits of the Precambrian shield of eastern Bolivia. In *International Symposium on Alluvial Gold Placers*; (G. Hérail, ed.) pp. 17, Abstracts, La Paz, Bolivia.

Bjelm, L. (1976). Deglaciation of the Smaland Highland, with special reference to deglaciation dynamics, ice thickness and chronology. Unpublished thesis, University of Lund, Sweden.

Björklund, A. J. (1984). Geochemical exploration – 1983. *Journal of Geochemical Exploration*, pp. 21, 501.

Bjørlykke K. (1985). Glaciations, preservation of their sedimentary record and sea level changes – A discussion based on Late Precambrian and Lower Palaeozoic sequences in Norway. *Palaeogeography, Palaeoclimatology, Palaeoecolology*, **51**, 197–207.

Bjørlykke K., Bue, B. and Elverhøi, A. (1978). Quaternary sediments in the northwestern part of the Barents Sea and their relation to the underlying Mesozoic bedrock. *Sedimentology*, **25**, 227–246.

Björnsson, S. (1953). Drumlins in the region of Sommen and Åsunden, south-east Sweden. *Sveriges Geografisk Årsbok*, **29**, 104–126.

Black, J. M. W. and Dudas, M. J. (1987). The scanning electron microscopic morphology of quartz in selected soils from Alberta. *Canadian Journal of Soil Science*, **67**, 965–971.

Black, R. F. (1970). Glacial geology of Two Creeks Forest Bed, Valderan Tyoe Locality, and Northern Kettle Moraine State Forest. *Wisconsin Geological and Natural History Survey Information Circular*, pp. 44.

Blackwelder, E. (1931). Pleistocene glaciation in the Sierra Nevada and Basin Ranges. *Geological Society of America Bulletin*, **42**, 865–922.

Blais, A. (1989). Lennoxville glaciation of the middle Chaudière and Etchemin valleys, Beauce region, Québec. Unpublished MSc thesis, Carleton University, Ottawa, pp. 137.

Blake, W., Jr. (1970). Studies of glacial history in Arctic Canada. I. Pumice, radiocarbon dates, and differential postglacial uplift in the eastern Queen Elizabeth Islands. *Canadian Journal of Earth Sciences*, **7**, 634–664.

Blake, W., Jr. (1978). Aspects of glacial history, southeastern Ellesmere Island, District of Franklin. *Geological Survey of Canada*, Paper 78–1A, 175–182.

Blankenship, D. D., Bentley, C. R., Rooney, S. T. and Alley, R. B. (1986). Seismic measurements reveal a saturated, porous layer beneath an active Antarctic Ice Stream. *Nature*, **322**, 54–57.

Blankenship, D. D., Bentley, C. R., Rooney, S. T. and Alley, R. B. (1987). Till beneath ice stream B, 1. properties derived from seismic travel times. *Journal of Geophysical Research*, **92**, 8903–8911.

Blankenship, D. D., Rooney, S. T., Alley, R. B. and Bentley, C. R. (1989). Seismic evidence for a thin basal layer at a second location on ice stream B, Antarctica. (Abstract), *Annals of Glaciology*, **12**, 200.

Blatt, H. (1987). Oxygen isotopes and the origin of quartz. *Journal of Sedimentary Petrology*, **57**, 373–377.

Blatt, H., Middleton, G. V. and Murray, R. C. (1980). *Origin of Sedimentary Rocks*. Englewood Cliffs: Prentice-Hall Publishing, pp. 782.

Blewett, W. L., and Rieck, R. L. (1987). Reinterpretation of a portion of the Munising moraine in northern Michigan. *Bulletin of the Geological Society of America*, **98**, 169–175.

Bloom, A. L. (1967). Pleistocene shorelines: a new test of isostasy. *Geological Society of America Bulletin*, **78**, 1477–1494.

Bloom, A. L., Broecker, W. S., Chappell, J. M. A., Matthews, R. K. and Mesolella, K. J. 1974. Quaternary sea level fluctuations on a tectonic coast: New ^{230}Th/^{234}U dates from the Huon Peninsula, New Guinea. *Quaternary Research*, **4**, 185–205.

Bluemle, J. P. and Clayton, L. (1984). Large-scale glacial thrusting and related processes in North Dakota. *Boreas*, **13**, 279–299.

Bluemle, J. P., Lord, M. L. and Hunke, N. T. (1993). Exceptionally long narrow drumlins formed in subglacial cavities. *Boreas*, **22**, 15–24.

Boardman, J. (1983). The role of micromorphological analysis in an investigation of the Troutbeck Paleosol, Cumbria, England. In *Soil Microscopy* (P. Bullock and C. P. Murphy eds), vol. 1, pp. 281–288, Berkhamstead: A B Academic Press.

Boardman, J. (1985a). Comparison of soils in midwestern United States and Western Europe with the interglacial record. *Quaternary Research*, **23**, 62–75.

Boardman, J. (ed.) (1985b). *Soils and Quaternary Landscape Evolution*. Chichester: John Wiley & Sons, pp. 391.

Boardman, J. (ed.) (1987). *Periglacial Processes and Landforms in Britain and Ireland*. Cambridge: Cambridge University Press, pp. 296.

Boellstorff, J. (1978). North American Pleistocene stages reconsidered in light of probable Pliocene-Pleistocene continental glaciation. *Science*, **202**, 305–307.

Bogacki, M. (1970). Eolian processes on the forefield of the Skeidarárjökull (Iceland). *Académie Polonaise des Sciences Bulletin, Série des Sciences Géologiques et Géographiques*, **18**, 279–287.

Bond, G., Broecker, W., Johnsen, S., McManus, J., Labeyrie, L., Jouzel, J. and Bonani, G. (1993). Correlations between climate records from North Atlantic sediments and Greenland ice. *Nature*, **365**, 143–147.

Bond, G. and Fernandes, T. R. C. (1974). Scanning electron microscopy applied to quartz grains from Kalahari type sands. *Transactions of the Geological Society of South Africa*, **77**, 191–199.

Booth, D. B. (1994) Glaciofluvial infilling and scour of the Puget Lowland, Washington, during ice-sheet glaciation. *Geology*, **22**, 695–698.

Booth, D. B. and Hallet, B. (1993). Channel networks carved by subglacial water: Observations and reconstructions in the eastern Puget Lowland of Washington. *Geological Society of America Bulletin*, **105**, 671–683.

Bordonau, J. (1993). The Upper Pleistocene ice-lateral till complex of Cerler (Esera Valley, central southern Pyrenees: Spain). *Quaternary International*, **18**, 5–14.

Bordonau, J. and Meer, J. J. M. van der (1994). An example of a kinking microfabric in Upper-Pleistocene glaciolacustrine deposits from Llavorsi (Central Southern Pyrenees, Spain). *Geologie en Mijnbouw*, **73**, 23–30.

Borns, H. W. (1973). Late Wisconsin fluctuations of the Laurentide Ice Sheet in southern and eastern New England. In *The Wisconsinan Stage. Geological Society of America Memoir*, (R. F. Black, R. P. Goldthwait, and H. B. Willman eds), **136**, 37–45.

Borns, H. W., Jr., and Matsch, C. L. (1988). A provisional genetic classification of glaciomarine environments, processes, and sediments. In *Genetic Classification of Glacigenic Deposits* (R. P. Goldthwait and C. L. Matsch eds) pp. 261–266, Rotterdam: A.A. Balkema Publishers.

Borsy, Z., Felszerfalvi, J., Loki, J. and Franyo, F. (1987). Electron microscopic investigations of sand material in the core drillings in the Great Hungarian Plain. *Geo-Journal*, **15**, 185–195.

Bouchard, M. A. (1980). Late Quaternary geology of the Témiskimie area, central Québec, Canada. Unpublished PhD Thesis, McGill University, Montréal. pp. 284.

Bouchard, M. A. (1986) Géologie des dépôts meubles de la région de Témiscamie, Territoire du Nouveau-Québec. Ministere de L'Energie et des Ressources du Québec, Mémoire MM 83–03. 90 pp.

Bouchard, M. A. (1989). Subglacial landforms and deposits in central and northern Québec, Canada, with emphasis on Rögen moraines. *Sedimentary Geology*, **62**, 293–308.

Bouchard, M. A. and Marcotte, C. (1986). Regional glacial dispersal patterns in Ungava, Nouveau-Québec. *Geological Survey of Canada*, Paper 86–1B, 295–304.

Bouchard, M. A. and Salonen, V. P. (1989). Glacial dispersal of boulders in the James Bay Lowlands of Québec, Canada. *Boreas*, **18**, 189–200.

Bouchard, M. A. and Salonen, V. -P. (1990). Boulder transport in Shield areas. In *Glacial Indicator Tracing* (R. Kujansuu and M. Saarnisto eds) pp. 87–107, Rotterdam: A.A. Balkema Publishers.

Boulton, G. S. (1968). Flow tills and some related deposits on some Vestspitsbergen glaciers. *Journal of Glaciology*, **7**, 391–412.

Boulton, G. S. (1970). On the deposition of subglacial and melt-out tills at the margins of certain Svalbard glaciers. *Journal of Glaciology*, **9**, 231–245.

Boulton, G. S. (1971). Till genesis and fabric in Svalbard, Spitsbergen. In *Till: A Symposium* (R. P. Goldthwait ed.) pp. 41–72, Columbus: Ohio State University Press.

Boulton, G. S. (1972a). The role of thermal regime in glacial sedimentation. In *Polar Geomorphology* (R. J. Price and D. E. Sugden eds), Institute of British Geographers Special Publication, **4**, 1–19.

Boulton, G. S. (1972b). Modern Arctic glaciers as depositional models for former ice sheets. *Quarterly Journal of the Geological Society of London*, **128**, 361–393.

Boulton, G. S. (1974). Processes and patterns of glacial erosion. In *Glacial Geomorphology* (D. R. Coates ed.) pp. 41–87, New York: SUNY at Binghampton.

Boulton, G. S. (1975). Processes and patterns of subglacial sedimentation: a theoretical approach. In *Ice Ages: Ancient and Modern. Geological Journal, Special Issue*, 6, (A. E. Wright and F. Moseley, eds) pp. 7–42, Liverpool: Seel House Press.

Boulton, G. S. (1976a). A genetic classification of tills and criteria for distinguishing tills of different origin. In *Till: its Genesis and Diagenesis* (W. Stankowski ed.), Uniwersytet im. Adama Mickiewicza, Poznan. *Seria Geografia*, **12**, 65–80.

Boulton, G. S. (1976b). The development of geotechnical properties in glacial tills. In *Glacial Till; An Interdisciplinary Study* (R. F. Legget ed.), The Royal Society of Canada Special Publication No. 12, Ottawa, pp. 292–303.

Boulton, G. S. (1976c). The origin of glacially fluted surfaces observations and theory. *Journal of Glaciology*, **17**, 287–309.

Boulton, G. S. (1978). Boulder shapes and grain-size distributions of debris as indicators of transport paths through a glacier and till genesis. *Sedimentology*, **25**, 773–799.

Boulton, G. S. 1979. Processes of glacier erosion on

different substrata. *Journal of Glaciology*, **23**, 15–37.

Boulton, G. S. (1980). Genesis and classification of glacial sediments. In *Tills and Glacigene Deposits* (W. Stankowski, ed.),. Uniwersytet im. Adama Mickiewicza, Poznan. *Seria Geografia*, **20**, 15–18.

Boulton, G. S. (1982). Subglacial processes and the development of glacial bedforms. In *Research in Glacial, Glacio-fluvial and Glacio-lacustrine Systems* (R. Davidson-Arnott, W. Nickling and B. D. Fahey eds) pp. 1–31, 6th Guelph Symposium on Geomorphology. Norwich: Geo Books Publishing.

Boulton, G. S. (1984). Development of a theoretical model of sediment dispersal by ice sheets. In *Prospecting in Areas of Glaciated Terrain, 1984*. London: Institution of Mining and Metallurgy, pp. 213–223.

Boulton, G. S. (1986). Push-moraines and glacier-contact fans in marine and terrestrial environments. *Sedimentology*, **33**, 677–698.

Boulton, G. S. (1987a). A theory of drumlin formation by subglacial sediment deformation. In *Drumlin Symposium* (J. Menzies and J. Rose eds),pp. 25–80, Rotterdam: A.A. Balkema Publishers.

Boulton, G. S. (1987b). Progress in glacial geology during the last fifty years. *Journal of Glaciology*, **Special Issue**, 25–32.

Boulton, G. S. (1990). Sedimentary and sea level changes during glacial cycles and their control on glacimarine facies architecture. In *Glaciomarine Environments: Processes and Sediments* (J. A. Dowdeswell and J. D. Scourse eds), Geological Society of London, Special Publication No. 53, pp. 15–52.

Boulton, G. S. and Clark, C. D. (1990a). A highly mobile Laurentide ice sheet revealed by satellite images of glacial lineations. *Nature*, **346**, 813–817.

Boulton, G. S. and Clark, C. D. (1990b). The Laurentide Ice Sheet through the last glacial cycle: drift lineations as a key to the dynamic behaviour of former ice sheets. *Transactions of the Royal Society of Edinburgh, Earth Sciences*, **81**, 327–347.

Boulton, G. S. and Dent, D. L. (1974). The nature and rates of post- depositional changes in recently deposited till from south-east Iceland. *Geografiska Annaler*, **56A**, 121–134.

Boulton, G. S. and Deynoux, M. (1981). Sedimentation in glacial environments and the identification of tills and tillites in ancient sedimentary sequences. *Precambrian Research*, **15**, 397–422.

Boulton, G. S. and Eyles, N. (1979). Sedimentation by valley glaciers; A model and genetic classification. In *Moraines and Varves: Origin/Genesis/Classification*, (Ch. Schlüchter ed.) pp. 11–23, Rotterdam: A.A. Balkema Publishers.

Boulton, G. S. and Hindmarsh, R. C. A. (1987). Sediment deformation beneath glaciers: rheology and geological consequences. *Journal of Geophysical Research*, **92B**, 9059–9082.

Boulton, G. S. and Jones, A. S. (1979). Stability of temperate ice caps and ice sheets resting on beds of deformable sediment. *Journal of Glaciology*, **24**, 29–43.

Boulton, G. S. and Paul, M. A. (1976). The influence of genetic processes on some geotechnical properties of glacial tills. *Quarterly Journal of Engineering Geology*, **9**, 159–194.

Boulton, G. S. and Spring, U. (1986). Isotopic fractionation at the base of polar and sub-polar glaciers. *Journal of Glaciology*, **32**, 475–485.

Boulton, G. S., Dent, D. L. and Morris, E. M. (1974). Subglacial shearing and crushing, and the role of water pressures in tills from south-east Iceland. *Geografiska Annaler*, **56A**, 134–145.

Boulton, G. S., Cox, F. C., Hart, J. K. and Thornton, M. (1984). The glacial geology of Norfolk. *Bulletin of the Geological Society of Norfolk*, **34**, 103–122.

Boulton, G. S., Jones, A. S, Clayton, K. M. and Kenning, M. J. (1977). A British ice-sheet model and patterns of glacial erosion and deposition in Britain. In *British Quaternary Studies: Recent Advances* (F. W. Shotton ed.) pp. 231–246, Oxford: Clarendon Press.

Boulton, G. S., Baldwin, C. T., Peacock, J. D., McCabe, A. M., Miller, G., Jarvis, J., Horsefield, B., Worsley, P., Eyles, N., Chroston, P. M., Day, T. E., Gibbard, P. L., von Brunn, P. E. and von Brunn, V. (1982). A glacio-isostatic facies model and amino acid stratigraphy for late Quaternary events in Spitsbergen and the Arctic. , **298**, 437–441.

Boulton, G. S., Smith, G. S., Jones, A. S. and Newsome, J. (1985). Glacial geology and glaciology of the last mid-latitude ice sheets. *Journal of Geological Society of London*, **142**, 447–474.

Bowen, D. Q. (1978). *Quaternary Geology: A Stratigraphic Framework for Multidisciplinary Work*. Oxford: Pergamon Press, 221 pp.

Bowen, R. (1988). *Isotopes in the Earth Sciences*. New York: Elsevier, 647 pp.

Bowen, D. Q. (1991). Time and space in the glacial sediment systems of the British Isles. In *Glacial Deposits in Great Britain and Ireland* (J. Ehlers, P. L. Gibbard and J. Rose eds) pp. 3–12, Rotterdam: A.A. Balkema Publishers.

Bowen, D. Q. (1994). The Pleistocene of north west Europe. *Science Progress*, **76**, 209–223.

Bowen, D. Q., Rose, J., McCabe, A. M and Sutherland, D. G. (1986). Correlation of Quaternary glaciations in

England, Ireland, Scotland and Wales. *Quaternary Science Reviews*, **5**, 299–340.

Boyce, J. I. and Eyles, N. (1991). Drumlins carved by deforming till streams below the Laurentide ice sheet. *Geology*, **19**, 787–790.

Boyd, R., Scott, D. B. and Douma, M. (1988) Glacial tunnel valleys and Quaternary history of the outer Scotian shelf. *Nature*, **333**, 61–64.

Boyle, E. A. (1988). Cadmium: chemical tracer of deep-water paleoceanography. *Paleoceanography*, **3**, 471–489.

Boyle, R. W. (1965). Geology, geochemistry, and origin of the lead-zinc-silver deposits of the Keno Hill – Galena Hill area, Yukon Territory. *Geological Survey of Canada*, Bulletin **111**, 302 pp.

Boyle, R. W. (1979). The geochemistry of gold and its deposits. *Geological Survey of Canada*, Bulletin **280**, 584 pp.

Boyle, R. W. and Gleeson, C. F. (1972). Gold in the heavy mineral concentrates of stream sediments, Keno Hill area, Yukon Territory. *Geological Survey of Canada*, Ottawa, Paper 71–51, 8 pp.

Boyle, R. W., Alexander, W. M. and Aslin, G. E. M. (1975). Some observations on the solubility of gold. *Geological Survey of Canada*, Paper 75–24, 6pp.

Bradley, R. S. (1985). *Quaternary Paleoclimatology, Methods of Paleoclimatic Reconstruction*. London: Allen and Unwin, 472 pp.

Bradley, R. S. (1991). Instumental records of past global change; lessons for the analysis of noninstrumental data. In *Global Changes of the Past* (R. S. Bradley, ed), UCAR/Office for Interdisciplinary Earth Studies, Global Change Institute Vol. 2, Boulder, Colorado, pp. 105–116.

Bradley, R. S., Diaz, H. F., Eischeid, J. K., Jones, P. D., Kelly, P. M. and Goodess, C. M. (1987). Precipitation fluctuations over northern hemisphere land areas since the mid-19th century. *Science*, **237**, 171–175.

Bradley, W. C., Fahnestock, R. K. and Rowekamp, E. T. (1972). Coarse sediment transport by flood flows on Knik river, Alaska. *Geological Society of America Bulletin*, **83**, 1261–1284.

Bradley, W. C., Hutton, J. T. and Twidale, C. R. (1978). Role of salts in development of granitic tafoni, South Australia. *Journal of Geology*, **86**, 647–654.

Brand, G., Pohjola, V. and Hooke, LeB. R. (1987). Evidence for a till layer beneath Storglaciären, Sweden, based on electrical resistivity measurements. *Journal of Glaciology*, **33**, 311–314.

Bray, D. A., Freeman, S. M. C., Read, D., Walsh, P. T. and Wilson, P. (1981). Quaternary sediments at Westfield

Farm, Fimber, North Humberside. *Proceedings of the Yorkshire Geological Society*, **43**, 377–393

Bray, J. R. and Struik, G. J. (1963). Forest growth and glacial chronology in eastern British Columbia, and their relation to recent climatic trends. *Canadian Journal of Botany*, **41**, 1245–1271.

Breitkopf, J. H. (1988). Iron formations related to mafic volcanism and ensialic rifting in the southern margin zone of the Damara orogen, Namibia. *Precambrian Research*, **38**, 111–130.

Brennand. T. C. and Sharpe, D. R. (1993). Ice-sheet dynamics and subglacial meltwater regime inferred from form and sedimentology of glaciofluvial systems: Victoria Island, District of Franklin, Northwest Territories. *Canadian Journal of Earth Sciences*, **30**, 928–944.

Bressler, J. R., Jones, W. C. and Cleveland, G. (1985). Geology of a buried channel system at the Denali placer gold mine. *The Alaska Miner*, **13**, 9.

Brewer, R. (1976). Fabric and mineral analysis of soils. Huntington: Krieger, 482 pp.

Brigham, J. K. (1983). Stratigraphy, amino acid geochronology, and correlation of Quaternary sea-level and glacial events. Broughton Island, arctic Canada. *Canadian Journal of Earth Sciences*, **20**, 577–598.

Brodzikowski, K. and Van Loon, A. J. (1985). Penecontemporaneous non-tectonic brecciation of unconsolidated silts and muds. *Sedimentary Geology*, **41**, 269–282.

Brodzikowski, K. and Van Loon, A. J. (1987). A systematic classification of glacial and periglacial environments, facies and deposits. *Earth Science Reviews*, **24**, 297–381.

Brodzikowski, K. and Van Loon, A. J. (1991). *Glacigenic Sediments. Developments in Sedimentology, 49*. Amsterdam: Elsevier Science Publications, 674 pp.

Brodzikowski, K., Gotowa_a, R., Kasza, L. and Van Loon, A. J. (1987a). The Kleszczów Graben (central Poland): reconstruction of the deformational history and inventory of the resulting soft-sediment deformational structures. In *Deformation of Sediments and Sedimentary Rocks* (M. E. Jones and R. M. F. Preston, eds) pp. 241–254, Geological Society Special Publication 29.

Brodzikowski, K., Gotowa_a, R., Ha_uszczak, A., Krzyszkowski. D. and Van Loon, A. J. (1987b). Soft-sediment deformations from glaciodeltaic, glaciolacustrine and fluviolacustrine sediments in the Kleszczów Graben (central Poland). In *Deformation of Sediments and Sedimentary Rocks*. (M. E. Jones and R. M. F. Preston, eds) pp 255–267, Geological Society Special Publication 29.

Broecker, W. S. and Denton, G. H. (1989). The role of

ocean-atmosphere reorganizations in glacial cycles. *Geochimica et Cosmochimica Acta*, **53**, 2465–2501.

Broecker, W. S. and Denton, G. H. (1990a). What drives glacial cycles? *Scientific American*, **262**, 1, 49–56.

Broecker, W. S. and G. H. Denton. (1990b). The role of ocean-atmosphere reorganizations in glacial cycles. *Quaternary Science Reviews*, **9**, 305–341.

Broecker, W. S., and Peng, T. -H. (1982). *Tracers in the Sea*. Palisades: Eldigio Press, 690 pp.

Broecker, W. S., Andree, M., Wölfli, W., Oeschger, H., Bonani, G. Kennett, J. P. and Peteet, D. (1988). The chronology of the last deglaciation: Implications to the cause of the Younger Dryas event. *Paleoceanography*, **3**, 1–19.

Broecker, W. S., Kennett, J. P., Flower, B. P., Teller, J. T., Trumbore, S., Bonani, G. and Wölfli, W. (1989). Routing of meltwater from the Laurentide Ice Sheet during the Younger Dryas Cold Episode. *Nature*, **341**, 318–321.

Brook, E. J. and Kurz, M. D. (1993). Surface-exposure chronology using in-situ cosmogenic ^{3}He in Antarctic quartz sandstone boulders. *Quaternary Research*, **39**, 1–10.

Brook, E. J., Kurz, M. D., Ackert Jr., R. P., Denton, G. H., Brown, E. T., Raisbeck, G. M. and Yiou, F. (1993). Chronology of Taylor Glacier advances in Arena Valley, Antarctica, using in situ cosmogenic ^{3}He and ^{10}Be. *Quaternary Research*, **39**, 11–23.

Broster, B. E. and Hicock, S. R. (1985). Multiple flow and support mechanisms and the development of inverse grading in a subaquatic glacigenic debris flow. *Sedimentology*, **32**, 645–657.

Brown, N. E., Hallet, B. and Booth, D. B. (1987). Rapid soft bed sliding of the Puget Glacial Lobe. *Journal of Geophysical Research*, **92**, 8985–8997.

Brugman, M. M. (1985). The effects of a volcanic eruption on Shoestring Glacier, Mt. St. Helens. Unpublished PhD thesis (part I), California Institute of Technology.

Bruins, H. J. and Yaalon, D. H. (1979). Stratigraphy of the Netivot section in the desert loess of the Negev (Israel). *Acta Geologica, Academia Scientiarum Hungaricae*, **22**, 161–69.

Brummer, J. J. (1978). Diamonds in Canada. *Canadian Institute of Mining and Metallurgy Bulletin*, **71**, 64–79.

Bruns, J. (1989). Stress indicators adjacent to buried channels of Elsterian age in North Germany. *Journal of Quaternary Science*, **4**, 267–272.

Bryant, I. D. (1991). Sedimentology of glaciofluvial deposits. In *Glacial Deposits of Britain and Ireland* (J. Ehlers, P. L. Gibbard, and J. Rose, eds) pp. 437–442, Rotterdam: A.A. Balkema Publishers.

Budd, W. F., Jensen, D. and Radok, U. (1971). Derived physical characteristics of the Antarctic Ice Sheet. Australian National Antarctic Research Expedition, Interim Report, Ser. A., *Glaciology*, Publication No. 120, 178 pp.

Bugge, T., Knarud, R. and Mørk, A. (1984). Bedrock geology on the mid-Norwegian continental shelf. In *Petroleum Geology of the North European Margin* (A. M. Spencer, S. O. Johsen, A. Mørk, E. Nysaether, P. Songstad and A. Spinnangr, eds) pp. 271–283, London: Graham & Trotman.

Bugge, T., Belderson, R. H. and Kenyon, N. H. (1988). The Storegga slide. *Philosophical Transactions of the Royal Society of London, Series A*, **325**, 357–388.

Bull, P. A. (1977). Glacial deposits identified by chattermark trails in detrital garnets – Comment. *Geology*, **5**, 248–249.

Bull, P. A. (1978a). A quantitative approach to scanning electron microscope analysis of cave sediments. In *Scanning Electron Microscopy in the Study of Sediments* (W. B. Whalley, ed.) pp. 201–226, Norwich: Geo Abstracts.

Bull, P. A. (1978b). Observations on small sedimentary quartz particles analysed by SEM. *Scanning Electron Microscopy*, **1978/1**, 821–828.

Bull, P. A. (1981), Environmental reconstruction by electron microscopy. *Progress In Physical Geography*, **5**, 368–397.

Bull, P. A. and Culver, S. J. (1979). An application of scanning electron microscopy to the study of ancient sedimentary rocks from the Saionia Scarp, Sierra Leone. *Palaeogeography, Palaeoclimatology, Palaeoecology*, **26**, 159–172.

Bull, P. A., Culver, S. J. and Gardner, R. (1980a). Chattermark trails as palaeoenvironmental indicators. *Geology*, **8**, 318–322.

Bull, P. A., Culver, S. J. and Gardner, R. (1980b). The nature of the Late Paleozoic glaciation in Gondwana as determined from an analysis of garnets and other heavy minerals: Discussion. *Canadian Journal of Earth Sciences*, **17**, 282–284.

Bull, P. A., Whalley, W. B. and Magee, A. W., (1986). An annotated bibliography of environmental reconstruction by SEM, 1962–1985. *British Geomorphological Research Group, Technical Bulletin*, No. 35, Norwich, UK: Geo Books, 94 pp.

Bull, P. A., Goudie, A. S., Price Williams, D. and Watson, A. (1987). Colluvium: a scanning electron microscope analysis of a neglected sediment type. In *Clastic Particles* (J. R. Marshall, ed.) pp. 16–35, New York: Van Nostrand Reinhold.

Bull, W. B. (1972). Recognition of alluvial-fan deposits in

the stratigraphic record. In *Recognition of Ancient Sedimentary Environments* (J. K. Rigby and W. K. Hamblin, eds), Society of Economic Paleontologists and Mineralogists, Special Publication 16, pp. 63–83.

Bullock, P. (1985). The role of micromorphology in the study of Quaternary soil processes. In *Soils and Quaternary Landscape Evolution* (J. Boardman, ed.) pp. 45–68, Chichester: John Wiley & Sons.

Bullock, P. and Murphy, C. P. (1979). Evolution of a paleo-argillic brown earth (paleudalf) from Oxfordshire, England. *Geoderma*, **22**, 225–252.

Bullock, P., Federoff, N., Jongerius, A., Stoops, G., and Tursina T. (1985). *Handbook for Soil Thin Section Description*. Albrighton: Waine Research Publications, 152 pp.

Bundtzen T. K. (1986). Placer geology of the Porcupine mining district, Skagway B–4 quadrangle, Alaska. *Alaska Division of Geological and Geophysical Surveys*, Public-data File 86–27, 26 pp.

Bundtzen T. K. (1991). Comparison of placer and lode deposits of central and western Alaska with northeast USSR. *Alaska Division of Geological and Geophysical Surveys*, unpublished abstract, presented at the Canadian Institute of Mining, 1 p.

Bundtzen, T. K., Swainbank, R. C., Deagen, J. R. and Moore, J. L. (1990). Alaska's mineral industry, 1989. *Alaska Division of Geological and Geophysical Surveys*, Special Report 44, 100 pp.

Burbank, D. W. and Kang, Jian Cheng. (1991). Relative dating of Quaternary moraines, Rongbuk valley, Mount Everest, Tibet: implications for an ice sheet on the Tibetan Plateau. *Quaternary Research*, **36**, 1–18.

Burckle, L. H., Gayley, R. I., Ram, M. and Petit, J.-R. (1988). Diatoms in Antarctic ice cores: some implications for the glacial history of Antarctica. *Geology*, **16**, 326–329.

Burgisser, H. M. (1984). A unique mass flow marker bed in a Miocene streamflow molasse sequence, Switzerland. In *Sedimentology of Gravels and Conglomerates* (E. H. Koster and R. J. Steel, eds), Canadian Society of Petroleum Geologists, Memoir 10, pp. 147–163.

Burke, M. J. (1969). The Forth valley: an ice-moulded lowland. *Transactions of the Institute of British Geographers*, **48**, 51–59.

Burke, R. M. and Birkeland, P. W. (1979). Reevaluation of multiparameter relative dating techniques and their application to the glacial sequence along the eastern escarpment of the Sierra Nevada, California. *Quaternary Research*, **11**, 21–51.

Burkholder, P. R. and Mandelli, E. F. (1965). Carbon assimilation of marine phytoplankton in Antarctica.

Proceedings of the National Academy of Sciences, **54**, 437–444.

Burnham, C. P. (1964). The occurrence of eluviated clay in soils of Central England. Proceedings 8th International Congress Soil Science, Bucharest, Romania, **3**, 1303–1310.

Butler, D. R. (1987). A Pinedale/Bull Lake interglacial paleosol and its implications, central Lemhi Mountains, Idaho. *Physical Geography*, **8**, 57–71.

Cai, J., Powell, R. D. and Loubere, P. (1992). Source components of glacimarine sediments defined by Fourier shapes of quartz silt particles. *Geological Society of America, Abstracts with Programs*, **24**, 84.

Cailleux, A. (1942). Les actions éoliennes periglaciaires en Europe. *Société Géologique de France, Mémoires*, **46**, 1–176.

Cailleux, A. (1968). Periglacial of McMurdo Strait (Antarctica). *Biuletyn Peryglacjalny*, **17**, 57–90.

Cailleux, A. (1972). Les formes et dépôts nivéo-éoliens actuels en Antarctique et au Nouveau-Québec. *Cahiers de Géographie de Québec*, **16**, 377–409.

Cailleux, A. (1974). Formes précoces et albédos du nivéo-éolien. *Zeitschrift für Geomorphologie*, **18**, 437–459.

Cailleux, A. and Calkin, P. E. (1963). Orientation of hollows in cavernously weathered boulders in Antarctica. *Biuletyn Peryglacialny*, **12**, 147–150.

Cailleux, A. and Schneider, H. (1968). L'usure des sables vue au microscope électronique é balayage. *Science Progrés, La Nature*, **3395**, 92–94.

Calkin, P. E. and Ellis, J. M. (1980). A lichenometric dating curve and its application to Holocene glacier studies in the central Brooks Range, Alaska. *Arctic and Alpine Research*, **12**, 245–264.

Calkin, P. E. and Rutford, R. H. (1974). The sand dunes of Victoria Valley, Antarctica. Geographical Review, **64**, 189–216.

Calvert S. E. (1983). Sedimentary geochemistry of silicon. In *Silicon Geochemistry and Biogeochemistry* (S. R. Aston, ed.) pp. 143–186, New York: Academic Press.

Calvert, S. E. and Veevers, J. J. (1962). Minor structures of unconsolidated marine sediments revealed by X-radiography. *Sedimentology*, **1**, 287–295.

Cameron, T. D. J., Stoker, M. S. and Long, D. (1987). The history of Quaternary sedimentation in the UK sector of the North Sea Basin. *Journal of the Geological Society of London*, **144**, 43–58.

Campbell, I. B. and Claridge, G. G. C. (1987). *Antarctica: Soils, Weathering Processes and Environment. Developments in Soil Science 16*, Amsterdam: Elsevier, 368 pp.

Cande, S. C. and Kent, D. V. (1992). A new geomagnetic

polarity time scale for the Late Cretaceous and Cenozoic. *Journal of Geophysical Research*, **97B**, 13917–13951.

Canfield, H. E., Hallberg, G. R. and Kemmis, T. J. (1984). A unique exposure of Quaternary deposits in Johnson County, Iowa. *Proceedings of the Iowa Academy of Science*, **91**, 98–111.

Carey S. W. and Ahmad, N. (1961). Glacial marine sedimentation. In *Geology of the Arctic*, vol. II. (G. O. Raasch, ed.) pp. 865–894, Toronto: University of Toronto Press.

Carl, J. D. (1978). Ribbed moraine-drumlin transition belt, St Lawrence Valley. *Geology*, **6**, 562–566.

Carmichael, H. and Moore, C. W. (1930). Placer mining in British Columbia, with special reports on Atlin, Queen Charlotte, Cariboo, Quesnel and Omineca Mining Divisions. British Columbia Ministry of Energy, Mines and Petroleum Resources, Victoria, *Bulletin 2*, 66 pp.

Carol, H. (1947). The formation of roches moutonées. *Journal of Glaciology*, **1**, 57–59.

Carrara, P. E. and McGimsey, R. G. (1981). The Late-Neoglacial histories of the Agassiz and Jackson Glaciers, Glacier National Park, Montana. *Arctic and Alpine Research*, **13**, 183–196.

Carrara, P. E. and McGimsey, R. G. (1988). Map showing distribution of moraines and extent of glaciers from the mid-19th century to 1979 in the Mount Jackson area, Glacier National Park, Montana. U.S. *Geological Survey, Miscellaneous Investigations Series*, Report I–1508–C, map scale 1:24,000.

Carruthers, R. G. (1953). Glacial drifts and the undermelt theory. Newcastle-upon-Tyne, England: Harold Hill Publishers, 42 pp.

Carter, R. W. G. (1992). Sea-level: past, present and future. In *Applications of Quaternary Research, Quaternary Proceedings 2*. (J. M. Gray, ed.) pp. 111–132, Cambridge: Quaternary Research Association.

Catt, J. A. (1979). Soils and Quaternary geology in Britain. *Journal Soil Science*, **30**, 607–642.

Catt, J. A. (1986). *Soils and Quaternary Geology: A Handbook for Field Scientists*. Monographs on Soil and Resource Surveys, 11, Oxford: Clarendon, 267 pp.

Catt, J. A. (1987). Palaeosols. *Progress in Physical Geography*, **11**, 487–510.

Catt, J. A. (1988a). Soils of the Plio-Pleistocene: do they distinguish types of interglacial? *Philosophical Transactions of the Royal Society London B*, **318**, 539–557.

Catt, J. A. (1988b). *Quaternary Geology for Scientists and Engineers*. New York: John Wiley, Halsted Press, 340 pp.

Catt, J. A. (ed.) (1990). Paleopedology manual. *Quaternary International*, **6**, 94 pp.

Catt, J. A. and Penny, L. F. (1966). The Pleistocene deposits of Holderness, east Yorkshire. *Proceedings of the Yorkshire Geological Society*, **35**, 375–420.

Catt, J. A., Corbett, W. M., Hodge, C. A. H., Madgett, P. A., Tatler, W. and Weir, A. H. (1971). Loess in the soils of north Norfolk. *Journal of Soil Science*, **22**, 444–452.

Cegla, J., Buckley, T. and Smalley, I. J. (1971). Microtextures of particles from some European loess deposits. *Sedimentology*, **17**, 129–134.

Cegla, J., Dzulynski, S. and Rzechowski, J. (1976). Experiments on gradational contacts between varves and till. In *Till: Its Genesis and Diagenesis* (M.Stankowski, ed.), Uniwersytet im Adama Mickiewicza, Poznan, *Seria Geografia*, **12**, 161–165.

Cepek, A. G. (1986). Quaternary stratigraphy of the German Democratic Republic. In *Quaternary Glaciations in the Northern Hemisphere* (V. Sibrava, D. Q. Bowen and G. M. Richmond, eds), *Quaternary Science Reviews*, **5**, 359–364.

Cerling, T. E. (1990). Dating geomorphologic surfaces using cosmogenic ^{3}He. *Quaternary Research*, **33**, 148–156.

Chamberlin, T. C. (1878). The extent and significance of the Wisconsin kettle moraines. *Wisconsin Academy of Science Transactions*, **4**, 201–234.

Chambers, R. (1853). On glacial phenomena in Scotland and parts of England. *Edinburgh New Philosophical Journal*, **54**, 229–281.

Champion, D. E., Lanphere, M. A. and Kuntz, M. A. (1988). Evidence for a new geomagnetic reversal from lava flows in Idaho: discussion of short polarity reversals in the Brunhes and Late Matuyama polarity chrons. *Journal of Geophysical Research*, **93**: 11667–11680.

Chandler, R. J. (1973). A study of structural discontinuities in stiff clays using a polarising microscope. *Proceedings Roc. International Symposium on Soil Structure*, Gothenburg, Sweden: 78–85.

Chappell, J. (1983). A revised sea-level record for the last 300,000 years from Papua, New Guinea. *Search*, **14**, 99–101.

Chappell, J. and Shackleton, N. J. (1986). Oxygen isotopes and sea level. *Nature*, **324**, 137–140.

Chappell, J. and Thom, B. G. (1978). Termination of last interglacial episode and the Wilson Antarctic surge hypothesis, *Nature*, **272**, 809–810.

Charlesworth, J. K. (1957). *The Quaternary Era, with Special References to its Glaciation*. London: Edward Arnold, 1700 pp.

Cheel, R. J. and Rust, B., (1982). Coarse grained facies of glaciomarine deposits near Ottawa, Canada. In *Research in Glacial, Glacio-fluvial and Glacio-lacustrine Systems*.

6th Guelph Symposium on Geomorphology (R. David-son-Arnott, W. Nickling and B. D. Fahey, eds) pp. 279–295, Norwich: Geo Books Publishing.

Cheel, R. J. and Rust. B. R. (1986). A sequence of soft-sediment deformation (dewatering) structures in late Quaternary subaqueous outwash near Ottawa, Canada. *Sedimentary Geology*, **47**, 77–93.

Cheng, D. C. -H. and Richmond, R. A. (1978). Some observations on the rheological behaviour of dense suspensions. *Rheologica Acta*, **17**, 446–453.

Chengde, S., Beer, J., Tungsheng, L., Oeschger, H., Bonari, G., Suter, M. and Wölfli, W. (1992). ^{10}Be in Chinese loess. *Earth and Planetary Science Letters*, **109**, 169–177.

Chinn, T. J. H. and Dillon, A. (1987). Observations on a debris-covered polar glacier 'Whisky Glacier', James Ross Island, Antarctic Peninsula, Antarctica. *Journal of Glaciology*, **33**, 300–310.

Chorley, R. J. (1959). The shape of drumlins. *Journal of Glaciology*, **3**, 339–344.

Christensen, L. (1974). Crop marks revealing large-scale patterned ground structures in cultivated areas, south-western Jutland, Denmark. *Boreas*, **3**, 153–180.

Christiansen, E. A. (1971). Tills in southern Saskatchewan. In *Till, a Symposium* (R. P. Goldthwait, ed.) pp. 167–183, Columbus: Ohio State University Press.

Christiansen, E. A. and Whitaker, S. H. (1976). Glacial thrusting of drift and bedrock. In *Glacial Till, An Interdisciplinary Study* (R. F. Legget, ed.) pp. 121–130, The Royal Society of Canada, Ottawa, Special Publication No. 12.

Christie-Blick, N. (1982). Upper Precambrian (Eocambrian) mineral fork tillite of Utah: a continental glacial and glaciomarine sequence: discussion. *Geological Society of America Bulletin*, **93**, 184–186.

Chumakov, N. M. and Elston, D. P. (1989). The paradox of Late Proterozoic glaciations at low latitudes. *Episodes*, **12**, 115–120.

Church, M. (1972). Baffin Island Sandurs: a study of Arctic fluvial processes. *Geological Survey of Canada, Bulletin*, **216**, 280 pp.

Church, M. and Gilbert, R. (1975). Proglacial fluvial and lacustrine environments. In *Glaciofluvial and Glaciolacustrine Sedimentation* (A. V. Jopling and B. C. McDonald, eds), Society of Economical Paleontologists and Minerologists, Special Publication No.23, Tulsa, pp. 22–100.

Clague, J. J. (1987a). Quaternary stratigraphy and history, Williams Lake, British Columbia. *Canadian Journal of Earth Sciences*, **24**, 147–158.

Clague, J. J. (1987b). A placer exploration target in the Cariboo district, British Columbia. *Geological Survey of Canada Paper*, **87–1A**, 177–180.

Clague, J. J. (1989a). Gold in the Cariboo district, British Columbia. *Mining Review*, September/October, 26–32.

Clague, J. J. (1989b). Placer gold in the Cariboo district, British Columbia. *Geological Survey of Canada Paper*, **89–1E**, 243–250.

Clague, J. J. (1989c). Cordilleran ice sheet. In *Quaternary Geology of Canada and Greenland* (R. J. Fulton, ed.), Geology of Canada, No. 1, *Geological Survey of Canada*, Ottawa, pp. 40–42.

Clague, J. J. (1991). Quaternary stratigraphy and history of Quesnel and Cariboo River valleys, British Columbia: Implications for placer gold production. *Geological Survey of Canada*, Paper **91–1A**, 1–5.

Clague, J. J. and Bornhold, B. C. (1980). Morphology and littoral processes of the Pacific coast of Canada. In *Coastlines of Canada* (S. B. McCann, ed.), *Geological Survey of Canada*, Paper **80–10**, 339–380.

Clague, J. J., Hebda, R. J. and Mathewes, R. W. (1990). Stratigraphy and paleoecology of Pleistocene interstadial sediments, central British Columbia. *Quaternary Research*, **34**, 208–226.

Clapperton, C. M. (1968). Channels formed by the superimposition of glacial meltwater streams, with special reference to the east Cheviot Hils, Northeast England. *Geografiska Annaler*, **50A**, 207–220.

Clapperton, C. M. (1971). The pattern of deglaciation in part of Northumberland. *Transactions of the Institute of British Geographers*, **53**, 67–78.

Clapperton, C. M. (1975). The debris content of surging glaciers in Svålbard and Iceland. *Journal of Glaciology*, **14**, 395–406.

Clapperton, C. M. (1989). Asymmetrical drumlins in Patagonia, Chile. *Sedimentary Geology*, **62**, 387–398.

Clapperton, C. M. (1993). *Quaternary Geology and Geomorphology of South America*. Amsterdan: Elsevier Science Publishers, 779pp.

Clark, C. D. (1993). Mega-scale glacial lineations and cross-cutting ice-flow landforms. *Earth Surface Processes and Landforms*, **18**, 1–29.

Clark, C. D. (1994). Large-scale ice moulding: a discussion of the genesis and glaciological significance. *Sedimentary Geology*, **91**, 253–268.

Clark, G. M. and Ciolkosz, E. J. (1988). Periglacial geomorphology of the Appalachian Highlands and Interior Highlands south of the glacial border – a review. *Geomorphology*, **1**: 191–220.

Clark, J. A., Farrell, W. E. and Peltier, W. R. (1978). Global changes in postglacial sea level: a numerical calculation. *Quaternary Research*, **9**, 265–287.

Clark, J. I., Konuk, I., Poorooshasb, F., Whittick, J. and Woodworth-Lynas, C. (1990). Ice scouring and the design of offshore pipelines. Proceedings of an invited workshop, April 18–19th, Calgary, Alberta. Canada Oil and Gas Lands Administration and Centre for Cold Ocean Resources Engineering, 499 pp.

Clark, P. U. (1987). Subglacial sediment dispersal and till composition. *Journal of Geology*, **95**, 527–541.

Clark, P. U. (1989). Relative differences between glacially crushed quartz transported by mountain and continental ice – some examples from North America and East Africa: discussion. *American Journal of Science*, **289**, 1195–1198. [Reply by Mahaney et al., 1198–1205].

Clark, P. U. (1991). Striated clast pavements: products of deforming subglacial sediment? *Geology*, **19**, 530–533.

Clark, P. U. (1992). Surface form of the southern Laurentide ice sheet and its implications to ice-sheet dynamics. *Geological Society of America Bulletin*, **104**, 595–605

Clark, P. U. and Hansel, A. K. (1989). Clast ploughing, lodgement and glacier sliding over a soft glacier bed. *Boreas*, **18**, 201–207.

Clark, P. U. and Lea, P. D. (1992). The last interglacial-glacial transition in North America. Geological Society of America, Special Paper 270, 317 pp.

Clark, P. U. and Walder, J. S. (1994). Sublacial drainage, eskers, and deforming beds beneath the Laurentide and Eurasian ice sheets. *Bulletin of the Geological Society of America*, **106**, 304–316.

Clark, P. U., Nelson, A. R., McCoy, W. D., Miller, B. B. and Barnes, D. K. (1989). Quaternary aminostratigraphy of Mississippi Valley loess. *Geological Society of America Bulletin*, **101**, 918–926.

Clarke, G. K. C. (1987a). A short history of scientific investigations on glaciers. *Journal of Glaciology*, Special Issue, 4–24.

Clarke, G. K. C. (1987b). Subglacial till: a physical framework for its properties and processes. *Journal of Geophysical Research*, **92B**, 9023–9036.

Clarke, G. K. C., Collins, S. G. and Thompson, D. E. (1984). Flow, thermal structure, and subglacial conditions of a surge-type glacier. *Canadian Journal of Earth Sciences*, **21**, 232–240.

Clayton, L. (1964). Karst topography on stagnant glaciers. *Journal of Glaciology*, **5**, 107–112.

Clayton, L. (1967). Stagnant-glacier features of the Missouri Coteau in North Dakota. *North Dakota Geological Survey*, Miscellaneous Series 30, 25–46.

Clayton, L. and Cherry, J. A. (1967). Pleistocene superglacial and ice-walled lakes of west-central North America. *North Dakota Geological Survey*, Miscellaneous Series 30, 47–52.

Clayton, L. and Moran S. R. (1974). A glacial process-form model. In *Glacial Geomorphology* (D. R. Coates, ed.), SUNY at Binghamton, pp. 89–119.

Clayton, L. and Moran, S. R. (1982). Chronology of late Wisconsinan glaciation in middle North America. *Quaternary Science Reviews*, **1**, 55–82.

Clayton, L., Attig, J. W. and Mickelson, D. M. (1987). Glacier-bed conditions in Wisconsin during late Wisconsin time. International Union for Quaternary Research, 12th International Congress, July 1987. Programme with Abstracts, pp. 145.

Clayton, L., Laird, W. M., Klassen, R. W. and Kupsch, W. O. (1965). Intersecting minor lineations on Lake Agassiz plain. *Journal of Geology*, **73**, 652–656.

Clayton, L., Mickelson, D. M. and Attig, J. W. (1989). Evidence against pervasively deformed bed material beneath rapidly moving lobes of the southern Laurentide Ice Sheet. *Sedimentary Geology*, **62**, 203–208.

Clayton, L., Moran, S. R., and Bluemle, J. P. (1980). Explanatory text to accompany the geologic map of North Dakota. *North Dakota Geological Survey R. I.*, **69**, 93 pp.

Coates, D. R. (1991). Glacial deposits. In The heritage of engineering geology: The first hundred years, G. A. Kiersch, (ed.), *Geological Society of America*, centennial Special Volume 3, 299–322.

Cobb, E. H. (1973). Placer deposits of Alaska. *United States Geological Survey Bulletin*, **1374**, 213 pp.

Coker, W. B. and DiLabio, R. N. W. (1989). Geochemical exploration in glaciated terrain: geochemical responses. In *Proceedings of Exploration '87*, (G. D. Garland, ed.), Ontario Geological Survey, Special Volume 3, pp. 336–383.

Colbeck, S. C. and Gow, A. J. (1979). The margin of the Greenland ice sheet at Isua. *Journal of Glaciology*, **24**, 155–165.

Collins, K. (1983). Scanning electron microscopy of engineering soils. *Geoderma*, **30**, 243–252.

Collins, K. and McGown, A. (1983). Micromorphological studies in soil engineering. In *Soil Micromorphology*, Vol. 1. (P. Bullock and C. P. Murphy, eds), AB Academic Publishers, pp. 195–217.

Collinson, J. D. (1979). Deserts. In *Sedimentary Environments and Facies* (H. G. Reading, ed.) pp. 80–96, New York: Elsevier.

Collinson, J. D. and Thompson, D. B. (1982). *Sedimentary Structures*. Boston: George Allen & Unwin Publishing, 194 pp.

Colman, S. M. (1981). Rock-weathering rates as functions of time. *Quaternary Research*, **15**, 250–264.

Colman, S. M. (1982). Chemical weathering of basalts and

andesites: evidence from weathering rinds. *United States Geological Survey Professional Paper* **1246**, 51 pp.

Colman, S. M. and Pierce, K. L. (1981). Weathering rinds on andesitic and basaltic stones as a Quaternary age indicator, western United States. *United States Geological Survey Professional Paper* **1210**, 56 pp.

Colman, S. M. and Pierce, K. L. (1984). Correlation of Quaternary glacial sequences in the western United States based on weathering rinds and related studies. In *Correlation of Quaternary Chronologies* (W. C. Mahaney, ed.) pp. 437–454, Norwich: Geobooks Ltd.

Colman, S. M. and Pierce, K. L. (1986). Glacial sequence near McCall, Idaho: weathering rinds, soil development, morphology, and other relative-age criteria. *Quaternary Research*, **25**, 25–42.

Colman, S. M. and Pierce, K. L. (1992). Varied records of early Wisconsinan alpine glaciation in the western United States derived from weathering-rind thicknesses. In *The Last Interglacial-Glacial Transition in North America*. Geological Society of America. (P. U. Clark and P. D. Lea, eds), Special Paper, **270**, pp. 269–278.

Colman, S. M., Pierce, K. L. and Birkeland, P. W. (1987). Suggested terminology for Quaternary dating methods. *Quaternary Research*, **28**, 314–319.

Cooke, R. U. (1981). Salt weathering in deserts. *Proceedings of the Geologists' Association*, **92**, 1–16.

Coombe, W. (1984). Gold in Saskatchewan. *Saskatchewan Geological Survey*, Open File Report 84–1, 134 pp.

Cooper, A. J., Funk, G. H. and Anderson, E. G. (1989). Using Quaternary stratigraphy to help locate a hazardous waste treatment site. In *Applied Quaternary Research* (E. F. J. De Mulder and B. P. Hageman, (Eds) A.A. Balkema Publishers, Rotterdam: pp. 1–13.

Corbato, C. E. (1965). Theoretical gravity anomalies of glaciers having parabolic cross sections. *Journal of Glaciology*, **5**, 255–258.

Cornwell, J. D. and Carruthers, R. M. (1986). Geophysical studies of a buried valley system near Ixworth, Suffolk. *Proceedings of the Geologists' Association*, **97**, 357–364.

Corte, A. E. and Trombotto, D. (1984). Quartz grain surface textures in laboratory experiments and in field conditions of rock glaciers. *Microscopia Electronica y Biologia Celular*, **8**, 71–79.

Cotton, C. A. (1941). The shoulders of glacial troughs. *Geological Magazine*, **78**, 113–128.

Coudé-Gaussen, G. (1985). Observations au microscope electronique a balayage de grains de quartz issus de depots glaciaires Pleistocenes de basse altitude dans les montagnes du nord-ouest du Portugal. *Actas (Grupo de Trabalho Portugues para o Estudio do Quaternario;*

Grupo Espanol de Trabajo del Cuaternario) Lisbon, **1**, 185–189.

Coutard, J. P. and Mücher, H. J. (1985). Deformations of laminated silt loam due to repeated freezing and thawing cycles. *Earth Surface Processes and Landforms*, **10**, 309–319.

Cowan, D. S. (1982). Deformation of partly dewatered and consolidated Franciscan sediments near Piedras Blancas Point, California. In *Trench Forearc Geology*. (J. K. Leggett, ed.), Geological Society of London Special Publication, No. 10, pp. 439–457.

Cowan, D. S. (1985). Structural styles in Mesozoic and Cenozoic mélanges in the western Cordillera of North America. *Geological Society of America Bulletin*, **96**, 451–462.

Cowan, E. A. and Powell, R. D. (1990). Suspended sediment transport and deposition of cyclically interlaminated sediment in a temperate glacial fjord, Alaska, U.S.A. In *Glaciomarine Environments: Processes and Sediments*. (J. A. Dowdeswell and J. D. Scourse, eds), Geological Society of London, Special Publication No. 53, 75–89.

Cowan,W. R. (1968). Ribbed moraine: till-fabric analysis and origin. *Canadian Journal of Earth Sciences*, **15**, 1145–1159.

Cowan, W. R., Sharpe, D. R., Feenstra, B. H. and Gwyn, Q. H. J. (1978). Glacial geology of the Toronto-Owen Sound area. In *Toronto '78 Field Trips Guidebook* (A. L. Currie and W. O. Mackasey eds), Geological Association of Canada, 1–16.

Cowie, J W., Zeigler, W., Boucot, A. J., Bassett, M. and Remane, E. (1986). Guidelines and statutes of the International Commission on Stratigraphy (ICS). *Courier Forschungsinstitut Senckenberg*, **83**, 1–14.

Cox, A. (1969). Geomagnetic reversals. *Science*, **163**, 237–247.

Cox, F. C. (1985). The tunnel valleys of Norfolk, East Anglia. *Proceedings of the Geologists' Association*, **96**, 357–369.

Cremeens, D. L., Darmody, R. G. and Jansen, I. J. (1987). SEM analysis of weathered grains: pretreatment effects. *Geology*, **15**, 401–404.

Cremeens, D. L., Norton, L. D., Darmody, R. G. and Jansen, I. J. (1988). Etch-pit measurements on scanning electron micrographs of weathered grain surfaces. *Soil Science Society of America Journal*, **52**, 883–885.

Crenshaw, M. A. (1980). Mechanisms of shell formation and dissolution. In *Skeletal Growth of Aquatic Organisms*. (D. C. Rhoads and R. A. Lutz, eds) pp. 115–132, New York: Plenum Press.

Croot, D. G. (1987). Glacio-tectonic structures: a mesoscale model of thin-skinned thrust sheets? *Journal of Structural Geology*, **9**, 797–808.

Croot, D. G. (ed.) (1988). *Glaciotectonics, Forms and Processes*. Rotterdam: A.A. Balkema Publishers, 212 pp.

Crowell, J. C. (1982). Continental glaciation through geologic times. In *Climate in Earth History*. (National Research Council) pp. 77–82, Washington: National Academy Press.

Crozier, M. J. (1975). On the origin of the Peterborough drumlin field testing the dilatancy theory. *Canadian Geographer*, **19**, 181–195.

Culver, S. J., Williams, H. R. and Bull, P. A., (1978). Infracambrian glaciogenic sediments from Sierra Leone. *Nature*, **274**, 49–51.

Culver, S. J., Williams, H. R. and Bull, P. A., (1980). Late Precambrian glacial deposits from the Rokelide fold belt, Sierra Leone. *Palaeogeography, Palaeoclimatology, Palaeoecology*, **30**, 65–81.

Culver, S. J., Bull, P. A., Campbell, S., Shakesby, R. A. and Whalley, W. B., (1983). Environmental discrimination based on quartz grain surface textures: a statistical investigation. *Sedimentology*, **30**, 129–136.

Cunningham, C. M. (1980). The glacial geology of the southeastern area of the District of Keewatin, Northwest Territories, Canada. Unpublished PhD thesis, Univeristy of Massachusetts. 286 pp.

Curry, B. B., Troost, K. C. and Berg, R. C. (1994). Quaternary geology of the Martinsville Alternative Site, Clark County, Illinois. *Illinois State Geological Survey*, Circular 556, 85 pp.

Czarnecka, E. and Gillott, J. E., (1980). Roughness of limestone and quartzite pebbles by the modified Fourier method. *Journal of Sedimentary Petrology*, **50**, 857–868.

Czerwinski, J. (1973). Certain elements of the microrelief in the forefield of Brei_amerkurjökull Glacier (Iceland) and the problem of the so-called fluted moraines. *Czasopismo Geograficzne*, **44**, 305–314.

Dahl, R. (1965). Plastically sculptured detail forms on rock surfaces in northern Nordland, Norway. *Geografiska Annaler*, **47**, 83–140.

Dalrymple, G. B. (1964). Potassium-argon dates of three Pleistocene interglacial basalt flows from the Sierra Nevada, California. *Geological Society of America Bulletin*, **75**, 753–758.

Daniel, E. (1975). Glacialgeologi inom kartbladet Moskosel i mellersta Lappland. *Sveriges Geologiska Undersökning*, **Ba 25**, 121 pp.

Dansgaard, W. and Oeschger, H. (1989). Past environmental long-term records from the Arctic. In *The Environmental Record in Glaciers and Ice Sheets*. (H. Oeschger and C. C. Langway, Jr., eds) pp. 287–318, New York: J. Wiley & Sons.

Dansgaard, W., White, J. W. C. and Johnsen, S. J. (1989). The abrupt termination of the Younger Dryas Climate event. *Nature*, **339**, 532–534.

Dansgaard, W., Johnsen, S. J., Clausen, H. B., Dahl-Jensen, D., Gundestrup, N. S., Hammer, C. U., Hvidberg, C. S., Steffensen, J. P., Sveinbjornsdottir, A. E., Jouzel, J. and Bond, G. (1993). Evidence for general instability of past climate from a 250-kyr ice-core record. *Nature*, **364**, 218–220.

Dardis, G. F. (1985). Till facies associations in drumlins and some implications for their mode of formation. *Geografiska Annaler*, **67A**, 13–22.

Dardis, G. F. (1987). Sedimentology of late-Pleistocene drumlins in south-central Ulster, Northern Ireland. In *Drumlin Symposium*. (J. Menzies and J. Rose, eds) pp. 215–224, Rotterdam: A.A. Balkema Publishers.

Dardis, G. F. and McCabe, A. M. (1983). Facies of subglacial channel sedimentation in late-Pleistocene drumlins, Northern Ireland. *Boreas*, **12**, 263–278.

Dardis, G. F. and McCabe, A. M. (1987). Subglacial sheetwash and debris flow deposits in Late-Pleistocene drumlins, Northern Ireland. In *Drumlin Symposium*. (J. Menzies and J. Rose, eds) pp. 225–240, Rotterdam: A.A. Balkema Publishers.

Dardis, G. F., McCabe, A. M. and Mitchell, W. I. (1984). Characteristics and origins of lee-side stratification sequences in late-Pleistocene drumlins, Northern Ireland. *Earth Surface Processes and Landforms*, **9**, 409–424.

Dare-Edwards, A. T. (1984). Loessic clays of south-east Australia. *Loess Letters Supplement*, **2**, 3–16.

Darmody, R. G. (1985). Weathering assessment of quartz grains: a semiquantitative approach. *Soil Science Society of America Journal*, **49**, 1322–1324.

Davenport, C. A. and Ringrose, P. S. (1985). Fault activity and palaeoseismicity during Quaternary time in Scotland – preliminary results. In *Earthquake Engineering in Britain*, Thomas Telford Ltd, London, 81–93.

Davenport, C. A. and Ringrose, P. S. (1987). Deformation of Scottish Quaternary sediment sequences by strong earthquake motions. In *Deformation of Sediments and Sedimentary Rocks*. (M. E.Jones and R. M. F. Preston, eds), Geological Society Special Publication No. 29, pp.299–314.

Davidson-Arnott, R., Nickling, W. and Fahey, B. D. (eds) (1982). *Research in Glacial, Glacio-fluvial and Glaciolacustrine Systems. 6th Guelph Symposium on Geomorphology*. Norwich: Geo Books Publishing, 318 pp.

Davis, W. M. (1899). Geographical cycle. *Geographical Journal*, **14**, 481–504.

Davis, W. M. (1916). The Mission Range, Montana. *Geographical Review*, **2**, 267–288.

De Geer, G. (1889). Ändmoräner i trakten mellan Spånga och Sundbyberg. *Geologiska Föreninger: i Stockholm Förhandlingar*, **11**, 395–397.

De Geer, G. (1912). A geochronology of the last 12,000 years. 11th International Geological Congress, Stockholm, 1910 *Compte Rendu*, **1**, 241–258.

De Geer, G. (1921). Correlation of late glacial clay varves in North America with the Swedish time scale. *Geologiska Föreningen i Stockholm Förhandlingar*, **43**, 70–73.

De Jong, M. G. G., Rappol, M. and Raupke, J. (1982). Sedimentology and geomorphology of drumlins in western Allgau, southern Germany. *Boreas*, **11**, 37–45

De Mulder, E. F. J. and Hageman, B. P. (eds) (1989). *Applied Quaternary Research*. Rotterdam: A.A. Balkema Publishers, 185 pp.

Dean, J. S. (1978). Independent dating in archaeological analysis. In *Advances in archaeological method and theory*. (M. B. Schiffer, ed.) pp. 223–255, New York: Academic Press.

Dearman, W. R. and Baynes, F. J. (1979). Etch-pit weathering of feldspars. *Proceedings of the Ussher Society*, **4**, 390–401.

Debicki, R. L. (1983). An overview of the placer mining industry in Yukon, 1978 to 1982. In *Yukon Placer Mining Industry 1978–1982*. (R. L. Debicki, ed.), Exploration and Geological Services Northern Affairs Program, Department of Indian and Northern Affairs Canada, Whitehorse, pp. 7–14.

Debicki, R. L. (1984). An overview of the placer mining industry in Atlin mining division, 1978 to 1982. British Columbia Ministry of Energy, Mines and Petroleum Resources, Geological Branch, Paper 1984–2, 19 pp.

Deere, D. U. and Patton, F. D. (1971). Slope stability in residual soils. *Proceedings of 4th Pan American Conference on Soil Mechanics and Foundation Engineering, Puerto Rico*, **1**, 87–170.

Delaune, M., Fornari, M., Hérail, G., Laubacher, G. and Rouhier, M. (1988). Correlation between heavy mineral distribution and geomorphological features in the Plio-Pleistocene gold-bearing sediments of the Peruvian eastern Cordillera through principal component analysis. *Bulletin of the Geological Society of France*, **4**, 133–144.

Delaune, M., Fornari, M., Hérail, G., Viscarra, G. and Miranda, G. (1991). Heavy mineral suites in the gold placers from the Real Cordillera and the Tipuani Mapiri basin. In *International Symposium on Alluvial Gold Placers: Abstracts*. (G. Hérail, ed.), La Paz, Bolivia, pp. 36

Demek, J. and Kukla, J. (1969). *Periglazialzone, Loess und Paleolithikum der Tschekoslowakei*. Tschechoslowakische Akademie der Wissenschaften, Geographisches Institut, Brno, 157 pp.

DeNee, P. B. (1978). Collecting, handling and mounting of particles for the S.E.M. *Scanning Electron Microscopy*, **1978/I**, 479–486.

Denton, G. H. and Hughes, T. J. (eds) (1981a). *The Last Great Ice Sheets*. New York: Wiley-Interscience, 484 pp.

Denton G. and Hughes, T. J. (1981b). The Arctic Ice sheet: An outrageous hypothesis. In *The Last Great Ice Sheets*. (G. H. Denton and T. J. Hughes, eds) pp. 440–467, New York: Wiley-Interscience.

Derbyshire, E. (1958). The identification and classification of glacial drainage channels from aerial photographs. *Geografiska Annaler*, **40**, 188–195.

Derbyshire, E. (1961). Subglacial col gullies and the deglaciation of the North-East Cheviots. *Transactions of the Institute of British Geographers*, **29**, 31–46.

Derbyshire, E. (1978). A pilot study of till microfabrics using the scanning electron microscope. In *Scanning Electron Microscopy in the Study of Sediments* (W. B. Whalley, ed.) pp. 41–59, Norwich: Geo Abstracts.

Derbyshire, E. (1980). The relationship between depositional mode and fabric strength in tills: schema and test from two temperate glaciers. In *Tills and glacigene deposits*. (W. Stankowski, ed.), Uniwersytet im Adama Mickiewicza, Poznan. *Seria Geografia*, **20**, 41–48.

Derbyshire, E. (1983). The Lushan dilemma: Pleistocene glaciation south of the Chang Jiang (Yangtze River). *Zeitschrift für Geomorphologie*, NF **27**, 445–471.

Derbyshire, E. (1987). A history of glacial stratigraphy in China. *Quaternary Science Reviews*, **6**, 301–314.

Derbyshire, E. (1992). Origine et remise en cause de la théorie des glaciations pléistocècenes de la Chine du sud-est: le cas des massifs de Lushan et Huangshan. *Annales de Géographie*, **566**, 472–490.

Derbyshire, E. and Mellors, T. W. (1988). Geological and geotechnical characteristics of some loess and loessic soils from China and Britain: a comparison. *Engineering Geology*, **25**, 135–175.

Derbyshire, E., McGown, A. and Radwan, A. (1976). 'Total' fabric of some till landforms. *Earth Surface Processes*, **1**, 17–26.

Derbyshire, E., Li, J. J., Perrott, F. A., Xu, S. Y. and Waters, R. S. (1984). Quaternary glacial history of the Hunza valley, Karakoram Mountains, Pakistan. In *The International Karakoram Project 1980*, Vol. 2, (K. J. Miller, ed.) pp. 456–495, Cambridge: Cambridge University Press.

Derbyshire, E., Love, M. A and Edge, M. J. (1985a). Fabrics of probable segregated ground-ice origin in some sediment cores from the North Sea Basin. In *Soils and Quaternary Landscape Evolution*. (J. Boardman, ed.) pp. 261–280, Chichester: John Wiley & Sons.

Derbyshire, E., Edge, M. J. and Love, M. (1985b). Soil fabric variability in some glacial diamicts. In *Glacial Tills 85. Proceedings of the International Conference on Construction in Glacial Tills and Boulder Clays* (M. C. Forde, ed.) pp. 169–175, Edinburgh, Scotland: Engineering Technics Press.

Derbyshire, E., Billard, A., Van Vliet-Lanoë, B., Lautridou, J. -P. and Cremaschi, M. (1988). Loess and palaeoenvironment: some results of a European joint programme of research. *Journal of Quaternary Science*, 3, 147–169.

Derbyshire, E., Shi Yafeng, Li Jijun, Zheng Benxing, Li Shijie and Wang Jingtai, (1991). Quaternary glaciation of Tibet: the geological evidence. *Quaternary Science Reviews*, 10, 485–510.

Deynoux, M. (1985). Terrestrial or waterlain glacial diamictites? Three case studies from the Late Precambrian and Late Ordovician glacial drifts in West Africa. *Palaeogeography, Palaeoclimatology, Palaeoecology*, 51, 97–141.

Deynoux, M. and Trompette, R. (1976). Late Precambrian mixites: glacial and/or nonglacial? Dealing especially with the mixtites of west Africa. *American Journal of Science*, 276, 1302–1315.

Deynoux, M. and Trompette, R. (1981). Late Ordovician tillites of the Taoudeni Basin, West Africa. In *Earth's pre-Pleistocene Glacial Record*. (M. J. Hambrey and W. B. Harland, eds) pp. 89–96, Cambridge: Cambridge University Press.

Deynoux, M,. Miller, J. M. G., Domack, E. W., Eyles, N., Fairchild, I. J. and Young, G. M. (1994). *Earth's Glacial Record*. Cambridge: Cambridge University Press, 266 pp.

DiLabio, R. N. W. (1990). Classification and interpretation of the shapes and surface textures of gold grains from till on the Canadian Shield. *Geological Survey of Canada*, Paper 90–1C, pp. 323–329.

DiLabio, R. N. W. and Coker, W. B. (1989). Drift prospecting. *Geological Survey of Canada*, Paper 89–20, 169 pp.

DiLabio, R. N. W. and Shilts, W. W. (1979). Composition and dispersal of debris by modern glaciers, Bylot Island, Canada. In *Moraines and Varves: Origin/Genesis/Classification*. (Ch. Schlüchter, ed.) pp. 145–155, Rotterdam: A.A. Balkema Publishers.

Dilks, A. and Graham, S. C. (1985). Quantitative mineralogical characterization of sandstones by back-scattered electron image analysis. *Journal of Sedimentary Petrology*, 55, 347–355.

Dingle, R. V. (1971). Buried tunnel valleys off the Northumberland coast, western North Sea. *Geologie en Mijnbouw*, 50, 679–686.

Dionne, J. -C. (1977). Relict iceberg furrows on the floor of glacial Lake Ojibwa, Quebec and Ontario. *Maritime Sediments*, 13, 79–81.

Domack, E. W. (1983). Facies of Late Pleistocene glacial-marine sediments on Whidbey Island, Washington: an isostatic glacial-marine sequence. In *Glacial-Marine Sedimentation* (B. F Molnia, ed.) pp. 535–570, New York: Plenum Press.

Domack, E. W. (1988). Biogenic facies in the Antarctic glacimarine environment: basis for a polar glacimarine summary. *Palaeogeography, Palaeoclimatology, Palaeoecology*, 63, 357–372.

Domack, E. W. and Domack, C. R. (1991). *Cenozoic Glaciation: The Marine Record Established by Ocean Drilling*. Washington D.C.: Joint Oceanographic Institutions, 49 pp.

Domack, E. W. and Lawson, D. E. (1985). Pebble fabric in an ice-rafted diamicton. *Journal of Geology*, 93, 577–591.

Domack E. W., Jull, A. J. T. , Anderson, J. B., Linick, T. W. and Williams, C. R.(1989). Application of tandem-accelerator mass spectrometer dating to Late Pleistocene-Holocene sediments of the East-Antarctic continental shelf. *Quaternary Research*, 31, 277–287.

Donaldson, J. A. (1965). The Dubawnt Group, District of Keewatin and Mackenzie. *Geological Survey of Canada*, Paper 64–20, 11 pp.

Donovan, R. N. and Fosterm, R. J. (1972). Subaqueous shrinkage cracks from the Caithness flagstone series (Middle Devonian) of Northeast Scotland. *Journal of Sedimentary Petrology*, 42, 309–317.

Doornkamp, J. C. and King, C. A. M. (1971). *Numerical Analysis in Geomorphology*. London: Edward Arnold, 372 pp.

Dorn, R. I. (1983). Cation-ratio dating: a new rock varnish age-determination technique. *Quaternary Research*, 20, 49–73.

Dorn, R. I. and Oberlander, T. M. (1981). Microbial origin of desert varnish. *Science*, 213, 1245–1247.

Dorn, R. I. and Oberlander, T. M. (1982). Rock Varnish. *Progress in Physical Geography*, 6, 317–367.

Dorn, R. I., Tanner, D., Turin, B. D. and Dohrenwend, J. C. (1987a). Cation-ratio dating of Quaternary materials in the east-central Mojave desert, California. *Physical*

Geography, **8**, 72–81.

Dorn, R. I., Turrin, B. D., Jull, A. J. T., Linick, T. W. and Donahue, D. J. (1987b). Radiocarbon and cation-ratio ages for rock varnish on Tioga and Tahoe morainal boulders of Pine Creek, Eastern Sierra Nevada, California, and their paleoclimatic implications. *Quaternary Research*, **28**, 38–49.

Dorn, R. I., Jull, A. J. T., Donahue, D. J., Linick, T. W. and Toolin, L. J. (1989). Accelerator mass spectrometry radiocarbon dating of rock varnish. *Geological Society of America Bulletin*, **101**, 1363–1372.

Dorr, J. Van N., 2nd, (1945). Manganese and iron deposits of Morro do Urucum, Mato Grosso, Brazil. *United States Geological Survey Bulletin*, **946A**, 47 pp.

D'Orsay, A. M. and van de Poll, H. W., (1985). Quartz-grain surface textures: Evidence for middle Carboniferous glacial sediment input to the Parrsboro Formation of Nova Scotia. *Geology*, **13**, 285–287.

Dort, W. Jr. (1967). Internal structure of Sandy Glacier, Southern Victoria Land, Antarctica. *Journal of Glaciology*, **6**, 529–540.

Dort, W. Jr., Roots, E. F. and Derbyshire, E. (1969). Firn–ice relationships, Sandy Glacier, Southern Victoria Land, Antarctica. *Geografiska Annaler*, **51(A)**, 104–111.

Dott, R. H. Jr. (1992). Eustasy: the historical ups and downs of a major geological concept. *Geological Society of America Memoir 180*, 120 pp.

Douglas, L. A. and Platt, D. W. (1977). Surface morphology of quartz and age of soils. *Soil Science Society of America Journal*, **41**, 641–645.

Dowdeswell, J. A. (1982). Scanning electron micrographs of quartz sand grains from cold environments examined using Fourier shape analysis. *Journal of Sedimentary Petrology*, **52**, 1315–1323.

Dowdeswell, J. A. and Morris, S. E. (1983). Multivariate statistical approaches to the analysis of late Quaternary relative age data. *Progress in Physical Geography*, **7**, 157–176.

Dowdeswell, J. A. and Scourse, J. D. (Editors) (1990). *Glaciomarine Environments: Processes and Sediments*. Geological Society Special Publication No. 53. 423 pp.

Dowdeswell, J. A. and Sharp, M. J. (1986). Characterization of pebble fabrics in modern terrestrial glacigenic sediments. *Sedimentology*, **33**, 699–710.

Dowdeswell, J. A., Hambrey, M. J and Ruitang Wu. (1985a). A comparison of clast fabric and shape in Late Precambrian and modern glacigenic sediments. *Journal of Sedimentary Petrology*, **55**, 691–704.

Dowdeswell, J. A., Osterman, L. E. and Andrews, J. T., (1985b). Quartz sand grain shape and other criteria used to distinguish glacial and non-glacial events in a marine

core from Frobisher Bay, Baffin Island, N.W.T., Canada. *Sedimentology*, **32**, 119–132.

Dowdeswell, J. A., Villinger, H., Whittington, R. J. and Marienfeld, P. (1993). Iceberg scouring in Scoresby Sund and on the east Greenland continental shelf. *Marine Geology*, **3**, 37–53.

Drake, L. D. (1972). Mechanisms of clast attrition in basal till. *Geological Society of America Bulletin*, **83**, 2159–2166.

Dredge, L. A. (1982). Relict ice-scour marks and late phases of Lake Agassiz in northernmost Manitoba. *Canadian Journal of Earth Sciences*, **19**, 1079–1087.

Dredge, L. A. and Grant, D. R. (1987). Glacial deformation of bedrock and sediment, Magdalen Islands and Nova Scotia, Canada: Evidence for a regional grounded ice sheet. In *Tills and Glaciotectonics* (J. J. M. van der Meer, ed.) pp. 183–195, Rotterdam: A.A. Balkema Publishers.

Dreimanis, A. (1953). Studies of friction cracks along the shores of Cirrus Lake and Kasakokwog Lake, Ontario. *American Journal of Science*, **251**, 769–783.

Dreimanis, A. (1969). Selection of genetically significant parameters for investigation of tills. Zeszyty Naukowe Uniwersytetu Adama Mickiewicza w Poznaniu. *Seria Geografia*, **8**, 15–29.

Dreimanis, A. (1976). Tills: their origin and properties. In *Glacial Till; An Inter-disciplinary Study*. (R. F. Legget, ed.) pp. 11–49, Ottawa: The Royal Society of Canada, Special Publication No. 12.

Dreimanis, A. (1977). Correlation of Wisconsin glacial events between the eastern Great Lakes and the St. Lawrence Lowlands. *Géographie Physique et Quaternaire*, **31**, 37–51.

Dreimanis, A. (1979). The problem of waterlain tills. In *Moraines and Varves: Origin/Genesis/Classification*. (Ch. Schlüchter, ed.) pp. 167–177, Rotterdam: A.A. Balkema Publishing.

Dreimanis, A. (1982). Two origins of the stratified Catfish Creek Till at Plum Point, Ontario, Canada. *Boreas*, **11**, 173–180.

Dreimanis, A. (1983). Penecontemporaneous partial disaggregation and/or resedimentation during the formation and deposition of subglacial till. *Acta Geologica Hispanica*, **18**, 153–160.

Dreimanis, A. (1984a). Comments on 'Sedimentation in a large lake: an interpretation of the Late Pleistocene stratigraphy at Scarborough Bluffs, Ontario, Canada'. *Geology*, **12**, 185–186.

Dreimanis, A. (1984b). Lithofacies types and vertical profile models; an alternative approach to the description and environmental interpretation of glacial diamict and diamictite sequences – discussion. *Sedimentology*, **31**,

885–886.

Dreimanis, A. (1987). Genetic complexity of a subaquatic till tongue at Port Talbot, Ontario, Canada. In INQUA Report on Activities 1982–1987, (R. Kujansuu and M. Saarinisto, eds), Zurich: ETH, pp. 68–78.

Dreimanis, A. (1988). Tills: their genetic terminology and classification. In *Genetic classification of glacigenic deposits*. (R. P. Goldthwait and C. L. Matsch, eds) pp. 17–83, Rotterdam: A.A. Balkema Publishers.

Dreimanis, A. (1993). Small to medium-sized glacitectonic structures in till and in its substratum and their comparison with mass movement structures. *Quaternary International*, **18**, 69–80.

Dreimanis, A. and Karrow, P. F. (1972). Glacial history of the Great Lakes – St. Lawrence Region, the classification of the Wisconsin(an) Stage, and its correlatives. *Proceedings of the 24th International Geological Congress*, Section 12, 5–15.

Dreimanis, A. and Lundqvist, J. (1984). What should be called till? *Striae*, **20**, 5–10.

Dreimanis, A. and Schlüchter, Ch. (1985). Field criteria for the recognition of till or tillite. *Palaeogeography, Palaeoclimatology, Palaeoecology*, **51**, 7–14

Dreimanis, A., and Vagners, U. J. (1971). Bimodal distribution of rock and mineral fragments in basal tills. In *Till: A Symposium*. (R. P. Goldthwait, ed.) pp. 237–250, Columbus, Ohio: Ohio State University Press.

Dreimanis, A. and Vagners, U. J. (1972). The effect of lithology upon texture of till. In *Research Methods in Pleistocene Geomorphology. Proceedings of the 2nd Guelph Symposium on Geomorphology, 1971*. (E. Yatsu and A. Falconer, eds) pp. 66–82, Norwich: GeoAbstracts.

Dreimanis, A., Hamilton, J. P, and Kelly, P. E. (1987). Complex subglacial sedimentation of Catfish Creek till at Bradtville, Ontario, Canada. In *Tills and Glaciotectonics*. (J. J. M. van der Meer, ed.) pp. 73–87, Rotterdam: A.A. Balkema Publishers.

Drewry, D. J. (1986). *Glacial Geologic Processes*. London: Edward Arnold, 276 pp.

Drewry, D. J. and Cooper, A. P. R. (1981). Processes and models of Antarctic glaciomarine sedimentation. *Annals of Glaciology*, **2**, 117–122.

Drozdowski, E. (1979). The patterns of deglaciation and associated depositional environments of till. In *Moraines and Varves*, (Ch. Schlüchter ed.), 237–248, Rotterdam: A.A. Balkema Publishers.

Dugmore, A. (1989). Icelandic volcanic ash in Scotland. *Scottish Geographical Magazine*, **105**, 168–172.

Dunbar R. B., Leventer, A. R. and Stochton, W. L. (1989). Biogenic sedimentation in McMurdo Sound, Antarctica.

Marine Geology, **85**, 155–180.

Duplessy, J. -C. (1978). Isotope studies. In *Climatic Changes*. (J. R. Gribbin, ed.) pp. 46–67, Cambridge: Cambridge University Press.

Dyke, A. S. (1984). Multiple deglaciations of the Hudson Bay Lowlands, Canada, since deposition of the Missinaibi (last-interglacial?) Formation: Discussion. *Quaternary Research*, **22**, 247–252.

Dyke, A. S. and Dredge, L. A. (1989). Quaternary geology of the northwestern Canadian Shield. In *Quaternary Geology of Canada and Greenland*. (R. J. Fulton, ed.), Ch. 3, pp. 189–235, *Geological Survey of Canada, Geology of Canada*, no. 1.

Dyke, A. S. and Morris, T. F. (1988). Drumlin fields, dispersal trains, and ice streams in Arctic Canada. *Canadian Geographer*, **32**, 86–90.

Dylik, J. (1969). L'action du vent pendant le dernier âge froid sur le territoire de la Pologne Centrale. *Biuletyn Peryglacjalny*, **20**, 29–44.

Dylikowa, A. (1969). Le problème des dunes intérieures en Pologne à la lumière des études de structure. *Biuletyn Peryglacjalny*, **20**, 45–80.

Dyson, J. L. (1952). Ice-ridged moraines and their relation to glaciers. *American Journal of Science*, **250**, 204–211.

Easterbrook, D. J. (1964). Void ratios and bulk densities as means of identifying Pleistocene till. *Geological Society of America Bulletin*, **75**, 745–750.

Easterbrook, D. J. (1988a). *Dating Quaternary Sediments*. Geological Society of America Special Paper 227, 165 pp.

Easterbrook, D. J. (1988b). Paleomagnetism of Quaternary deposits. In *Dating Quaternary Sediments*. (D. J. Easterbrook, ed.), Geological Society of America Special Paper 227, pp. 111–122.

Easterbrook, D. J. and Boellstorff, J. (1984). Paleomagnetism and chronology of chronology of early Pleistocene tills in the central United States. In *Correlation of Quaternary Chronologies*. (W. C. Mahaney, ed.) pp. 73–90, Norwich: Geobooks.

Easterbrook, D. J., Briggs, N. D., Westgate, J. A. and Gorton, M. P. (1981). Age of the Salmon Springs glaciation in Washington. *Geology*, **9**, 87–93.

Easterbrook, D. J., Roland, J. L., Carson, R. J. and Naeser, N. D. (1988). Application of paleomagnetism, fission-track dating, and tephra correlation to Lower Pleistocene sediments in the Puget Lowland, Washington. In *Dating Quaternary Sediments* (D. J. Easterbrook, ed.), Geological Society of America Special Paper 227, pp. 139–165.

Echelmeyer, K. and Wang Zhongxiang. (1987). Direct observation of basal sliding and deformation of basal drift at sub-freezing temperatures. *Journal of Glaciology*, 33, 83–98.

Edwards, M. B. (1976). *Sedimentology of Late Precambrian Sveanor and Kapp Sparre Formations at Aldousbreen Wahlenbergfjorden, Nordaustlandet*. Norsk Polarinstitut Arbok, 1974, pp. 51–61.

Edwards, M. B. (1979). Late Precambrian glacial loessites from North Norway and Svalbard. *Journal of Sedimentary Petrology*, 49, 85–92.

Edwards, M. B. (1986). *Glacial Environments*. In *Sedimentary Environments and Facies*, 2nd ed. (H. G. Reading, ed.) pp. 445–470, Oxford: Blackwell Scientific Publishers.

Edwards, R. L., Chen J. H. and Wasserburg, G. J. (1986). ^{238}U–^{234}U–^{230}Th–^{232}Th systematics and the precise measurement of time over the past 500,000 years. *Earth and Planetary Science Letters*, 81, 175–192.

Ehlers, J. (1981). Some aspects of glacial erosion and deposition in north Germany. *Annals of Glaciology*, 2, 143–146.

Ehlers, J. (ed.) (1983a). The glacial history of north-west Germany. In *Glacial Deposits in North-West Europe*. (J. Ehlers, ed.) pp. 229–238, Rotterdam: A.A. Balkema Publishers.

Ehlers, J. (ed.) (1983b). *Glacial Deposits in North-West Europe*. Rotterdam: A.A. Balkema Publishers. 470 pp.

Ehlers, J. and Grube, F. (1983). Meltwater deposits in north-west Germany. In *Glacial Deposits in North-West Europe*. (J. Ehlers, ed.) pp. 249–256, Rotterdam: A.A. Balkema Publishers.

Ehlers, J., Gibbard, P. L. and Rose, J. (eds) (1991). *Glacial Deposits in Great Britain and Ireland*, Rotterdam: A.A. Balkema Publishers, 580 pp.

Ehlers, J., Meyer, K. -D. and Stephan, H. -J. (1984). The pre-Weichselian glaciations of North-West Europe. *Quaternary Science Reviews*, 3, 1–40.

Ehrlich, R., Kennedy, S. K. and Brotherhood, C. D. (1987). Respective roles of Fourier and SEM techniques in analyzing sedimentary quartz. In *Clastic Particles* (J. R. Marshall, ed.) pp. 292–301, New York: Van Nostrand Reinhold.

Eidvin, T. and Riis, F. (1989). Nye dateringer av de re vestligste borehullene i Barentshavet. Resultater og konsekvenser for den tertiære hevningen. Oljedirektoratet Stavanger, Norway, NPD–contribution 27, 43 pp.

Einarsson, T. and Albertsson, K. J. (1988). The glacial history of Iceland during the past three million years. In *The Past Three Million Years: Evolution of Climatic Variability in the North Atlantic Region* (N. J. Shackleton, R. G West and D. Q. Bowen, eds) pp. 227–234, Cambridge: Cambridge University Press.

Eisbacher, G. H. (1985). Late Proterozoic rifting, glacial sedimentation and sedimentary cycles in the light of Windermere deposition, western Canada. *Palaeogeography, Palaeoclimatology, Palaeoecology*, 51, 231–254.

Eisenhauer, A., Mangini, A., Botz, R., Walter, P., Beer, J.,Bonani, G., Suter, M. , Hofmann, H. J. and Wölfli, W. (1990). High resolution 10Be 230Th stratigraphy of late Quaternary sediments from the Fram Strait (Core 23235). In *Geological history of the polar oceans: Arctic versus Antarctic* (U. Bleil and J. Thiede, eds), NATO ASI Series, C 308. Dordrecht: Kluwer Academic Publishers, pp. 475–487.

Eldholm, O., Thiede, J. and Taylor et al. (1987). *Proceedings of the Ocean Drilling Program*. Initial Reports, (PtA), ODP, 104.

Elfström, A. (1987). Large boulder deposits and catastrophic floods: A case study of the Bàldakatj area, Swedish Lapland. *Geografiska Annaler*, 69A, 101–121.

Ellsaesser, H. W., MacCracken, M. C., Walton, J. J. and Grotch, S. L. (1986). Global climatic trends as revealed by the recorded data. *Reviews in Geophysics*, 24, 745–792.

Elson, J. A. (1957). Origin of washboard moraines. *Geological Society of America Bulletin*, 70, 1721.

Elson, J A. (1961). *Geology of Tills. Proceedings of the Fourteenth Soil Mechanics Conference 13 and 14 October, 1960*. National Research Council of Canada, Associate Committee on Soil and Snow mechanics, Technical Memorandum No. 69: 5–17.

Elson, J. A. (1968). Washboard moraines and other minor moraine types. In *Encyclopedia of Geomorphology* (R. W. Fairbridge, ed.), Reinhold, 1213–1219.

Elson, J A. (1988). Comment on glacitectonite, deformation till, and comminution till. In *Genetic Classification of Glacigenic Deposits* (R. P. Goldthwait and C. L. Matsch, eds) pp. 85–91, Rotterdam: A.A. Balkema Publishers.

Elverhøi, A. (1984). Glacigenic and associated marine sediments in the Weddell Sea, fjords of Spitsbergen and the Barents Sea: a review. *Marine Geology*, 57, 53–88.

Elverhøi, A., Liestøl, O. and Nagy, T. (1980). Glacial erosion, sedimentation and microfauna in the inner part of Kongsfjorden, Spitsbergen. *Norsk Polarinstitut Skrifter*, 172, 33–61.

Elverhøi, A., Pfirman, S., Solheim, A. and Larsen, B. B. (1989). Glaciomarine sedimentation on epicontinental seas – exemplified by the northern Barents Sea. *Marine Geology*, 85: 225–250.

Elverhøi, A., Nyland-Berg, M., Russwurm, L. and Solheim,

A. (1990). Late Weichselian ice recession in the central Barents Sea. In *Geological History of the Polar Oceans; Arctic versus Antarctic* (U. Bleil and J. Thiede, eds), NATO ASI Series, C 308. Dordrecht: Kluwer Academic Publishers, pp. 289–307.

Elvidge, C. D. and Collet, C. J. (1981). Desert varnish in Arizona: Distribution and spectral characteristics. In *Technical Papers of the American Society of Photogrammetry*, ASP–ACSM Fall Technical Meeting , San Francisco, September 9–11, pp. 215–222.

Elzenga, W., Schwan, J., Baumfalk, Y. A., Vandenberghe, J. and Krook, L. (1987). Grain surface characteristics of periglacial aeolian and fluvial sands. *Geologie en Mijnbouw*, **65**, 273–286.

Embleton, B. J. J. and Williams, G. E. (1986). Low Palaeolatitude of deposition for late Precambrian periglacial varvites in South Australia: implications for palaeoclimatology. *Earth and Planetary Science Letters*, **79**, 419–430.

Embleton, C. and King, C. A. M. (1968). *Glacial and Periglacial Geomorphology*, London: Edward Arnold, 608 pp.

Embleton, C. and King, C. A. M. (1975). *Periglacial Geomorphology*, 2nd edn, London: Edward Arnold, 203 pp.

Emiliani, C. (1955). Pleistocene temperatures. *Journal of Geology*, **63**, 538–578.

Emiliani, C. (1966). Palaeotemperature analysis of Caribbean cores P6304–8 and P6304–9 and a generalized temperature curve for the past 425,000 years. *Journal of Geology*, **74**, 109–126.

Emiliani, C. and Shackleton, N. J. (1974). The Brunhes epoch: Isotopic paleotemperatures and geochronology. *Science*, **183**, 511–514.

Emory-Moore, M. (1991). Placer gold potential of the northern Newfoundland shelf. Geological Survey of Canada, Open File 2417, 113 pp.

Emory-Moore, M. and Solomon, S. (1989). Placer gold potential offshore Newfoundland: a preliminary assessment. Newfoundland Department of Mines, Mineral Development Division, Report of Activities **89–1**, 229–236.

Emory-Moore, M., Solomon, S. and Dunsmore, D. (1988). Placer potential of Fox Island River and east-central Port au Port Bay: a preliminary assessment. Newfoundland Department of Mines, Mineral Development Division, Current Research, **88–1**: 343–355.

Engelhardt, H. F., Harrison, W. D. and Kamb, B. (1978). Basal sliding and conditions at the glacier bed as revealed by bore-hole photography. *Journal of Glaciology*, **20**, 469–508.

England, J. (1986). Glacial erosion of a High Arctic valley. *Journal of Glaciology*, **32**, 60–64.

Eriksson, K. A. (1983). Siliciclastic-hosted iron-formations in the early Archaean Barberton and Pilbara sequences. *Journal of the Geological Society of Australia*, **30**, 473–482.

Eronen, M. (1983). Late Weichselian and Holocene shore displacement in Finland. In *Shorelines and Isostasy* (D. E. Smith and A. G. Dawson, eds), Institute of British Geographers, Special Publication, 16. London: Academic Press, pp. 183–207.

Evans, I. S. (1987). A new approach to drumlin morphometry. In *Drumlin Symposium* (J. Menzies and J. Rose, eds) pp. 119–130, Rotterdam: A.A. Balkema Publishers.

Evans, L. J. (1982). Dating methods of Pleistocene deposits and their problems: VII Paleosols. *Geoscience Canada*, **9**, 155–160.

Evans, L. J. and Cameron, B. H. (1979). A chronosequence of soils developed from granitic morainal material, Baffin Island, N.W.T. *Canadian Journal Soil Science*, **59**, 203–210.

Evenson, E. B. (1970). A method for 3-dimensional microfabric analysis of tills obtained from exposures or cores. *Journal of Sedimentary Petrology*, **40**, 762–764.

Evenson, E. B. (1971). The relationship of macro- and microfabrics of till and the genesis of glacial landforms in Jefferson County, Wisconsin. In *Till, a Symposium* (R. P. Goldthwait, ed.) pp. 345–364, Columbus: Ohio State University Press.

Evenson, E. B. and Clinch, J. M. (1987). Debris transport mechanisms at active alpine glacier margins: Alaskan case studies. In *INQUA Till Symposium, Finland 1985* (R. Kujansuu and M. Saarnisto, eds) pp. 111–136, Geological Survey of Finland Special Paper 3.

Evenson, E. B., Dreimanis, A. and Newsome, J. W. (1977). Subaquatic flow tills: a new interpretation for the genesis of some laminated till deposits. *Boreas*, **6**, 115–133.

Evenson, E. B., Pasquini, T. A., Stewart, R. A., and Stephens, G. (1979). Systematic provenance investigations in areas of alpine glaciation: applications to glacial geology and mineral exploration. In *Moraines and Varves: Origin/Genesis/Classification*, (Ch. Schlüchter, ed.) pp. 25–42, Rotterdam: A.A. Balkema Publishers.

Evenson, E. B., Schlüchter, Ch. and Rabassa, J. (eds) (1983). *Tills and Related Deposits: Genesis, Petrology, Application, Stratigraphy*. Rotterdam: A.A. Balkema Publishers, 454 pp.

Eybergen, F. A. (1987). Glacier snout dynamics and contemporary push moraine formation at the Turtmannglacier, Wallis, Switzerland. In *Tills and Glaciotectonics* (J. J. M. van der Meer,, ed.) pp. 217–231, Rotterdam:

A.A. Balkema Publishers.

Eyles, C. H. (1987). Glacially influenced submarine-channel sedimentation in the Yakataga Formation, Middleton Island, Alaska. *Journal of Sedimentary Petrology*, **57**, 1004–1017

Eyles, C. H. (1988a). Glacially- and tidally-influenced shallow marine sedimentation of the Late Precambrian Port Askaig Formation, Scotland. *Palaeogeography, Palaeoclimatology, Palaeoecology*, **68**, 1–25.

Eyles, C. H. (1988b). A model for striated boulder pavement formation on glaciated, shallow-marine shelves: an example from the Yakataga Formation, Alaska. *Journal of Sedimentary Petrology*, **58**, 62–71.

Eyles, C. H. and Eyles, N. (1983a). Sedimentation in a large lake: a reinterpretation of the late Pleistocene stratigraphy at Scarborough Bluffs, Ontario, Canada. *Geology*, **11**, 146–152.

Eyles, C. H. and Eyles, N. (1983b). Glaciomarine model for upper Precambrian diamictites of the Port Askaig Formation, Scotland. *Geology*, **11**, 692–696.

Eyles, C. H. and Lagoe, M. B. (1990). Sedimentation patterns and facies geometries on a temperate glacially-influenced continental shelf: the Yakataga Formation, Middleton Island, Alaska. In *Glaciomarine Environments: Processes and Sediments* (J. A. Dowdeswell and J. D. Scourse, eds), Geological Society of London Special Publication. No. 53, pp. 363–386.

Eyles, C. H., Eyles, N. and Miall, A. D. (1985). Models of glaciomarine sedimentation and their application to the interpretation of ancient glacialsequences. In Glacial Record, Proceedings of the Till Mauretania '83 Symposium. (M. Deynoux, ed.), *Palaeogeography, Palaeoclimatology, Palaeoecology*, **51**, 15–84.

Eyles, N. (1977). Late Wisconsinan glacitectonic structures and evidence of postglacial permafrost in north-central Newfoundland. *Canadian Journal of Earth Sciences*, **14**, 2797–2806.

Eyles, N. (1978). Scanning electron microscopy and particle size analysis of debris from a British Columbian glacier; a comparative report. In *Scanning Electron Microscopy in the Study of Sediments* (W. B. Whalley, ed.) pp. 227–242, Norwich: Geo Abstracts.

Eyles, N. (1979) Facies of supraglacial sedimentation on Icelandic and Alpine temperate glaciers. *Canadian Journal of Earth Sciences*, **16**, 134–1361.

Eyles, N. (ed.) (1983a). *Glacial Geology: An Introduction for Engineers and Earth Scientists*. Oxford: Pergamon Press, 409 pp.

Eyles, N. (1983b). Modern Icelandic glaciers as depositional models for 'hummocky moraine' in the Scottish Highlands. *In Tills and Related Deposits* (E. B. Evenson,

Ch. Schlüchter, and J. Rabassa, eds) pp. 47–59, Rotterdam: A.A. Balkema Publishers.

Eyles, N. (1987). Late Pleistocene debris-flow deposits in large glacial lakes in British Columbia and Alaska. *Sedimentology*, **53**, 33–71.

Eyles, N. and Clark, B. M. (1985). Gravity-induced soft-sediment deformation in glaciomarine sequences of the Upper Proterozoic Port Askaig Formation, Scotland. *Sedimentology*, **32**, 789–814.

Eyles, N. and Clark, B. M. (1988). Storm-influenced deltas and icescouring in a late Pleistocene glacial lake. *Geological Society of America Bulletin*, **100**, 793–809.

Eyles, N. and Kocsis, S. P. (1988). Gold placers in Pleistocene glacial deposits, Barkerville, British Columbia. *Canadian Institute of Mining and Metallurgy Bulletin*. **81**, 71–79.

Eyles, N and Kocsis, S. P. (1989a). Sedimentological controls on gold distribution in Pleistocene placer deposits of the Cariboo Mining District, British Columbia, Canada. British Columbia Ministry of Energy, Mines and Petroleum Resources Paper, 1989–1, 377–385.

Eyles, N and Kocsis, S. P. (1989b). Sedimentological controls on gold in a late Pleistocene glacial placer deposit, Cariboo Mining District, British Columbia, Canada. *Sedimentary Geology*, **65**, 45–68.

Eyles, N. and McCabe, A. M. (1989). The Late Devensian (<22,000BP) Irish Basin: the sedimentary record of a collapsed ice sheet margin. *Quaternary Science Reviews*, **8**, 307–351.

Eyles, N. and Menzies, J. (1983). The subglacial landsystem In *Glacial Geology* (N. Eyles, ed.) pp. 19–67, Oxford: Pergamon Press.

Eyles, N. and Miall, A. D. (1984). Glacial facies. In *Facies models Geoscience Canada Reprint Series 1*, (R. G. Walker, ed.), 2nd ed., pp. 15–38.

Eyles, N. Clark, B. M., and Clague, J. J. (1987a). Coarse-grained sediment gravity flow facies in a large supraglacial lake. *Sedimentology*, **34**, 93–216.

Eyles, N., Day, T. E. and Gavican, A. (1987b). Depositional controls on the magnetic characteristics of lodgement tills and other glacial diamict facies. *Canadian Journal of Earth Sciences*, **24**, 2436–2458.

Eyles, N., Eyles, C. H. and McCabe, A. M. (1989). Sedimentation in an ice-contact subaqueous setting: The Mid-Pleistocene 'North Sea Drifts' of Norfolk, U.K. *Quaternary Science Reviews*, **8**, 57–74.

Eyles, N., Eyles, C. H. and Miall, A. D. (1983). Lithofacies types and vertical profile models; an alternative approach to the description and environmental interpretation of glacial diamict and diamictite sequences. *Sedimentology*, **30**, 393–410.

Eyles, N., Mullins, H. T. and Hine, A. C. (1990). Thick and fast: sedimentation in a Pleistocene fiord lake of British Columbia, Canada. *Geology*, **18**, 1153–1157.

Eyles, N., Sladen, J. A. and Gilroy, S. (1982). A depositional model for stratigraphic complexes and facies superimposition in lodgement tills. *Boreas*, **11**, 317–333.

Faegri, K. (1952). On the origin of pot-holes. *Journal of Glaciology*, **2**, 24–25.

Fahnestock, R. K. and Bradley, W. C. (1973). Knik and Matanuska rivers, Alaska: A contrast in braiding. In *Fluvial Geomorphology* (M. Morisawa, ed.) pp. 220–250, London: Allen & Unwin.

Faïn, J., Miallier, D., Aitken, M. J., Bailiff, I. K., Grün, R., Mangini, A., Mejdahl, V., Rendell, H. M., Townsend, P. D., Valladas, G., Visocekas, R. and Wintle, A. G. (eds) (1992). Proceedings of the 6th international specialist seminar on thermoluminescence and electron spin resonance dating. *Quaternary Science Reviews*, **11(1/2)**, 274 pp.

Fairbanks, R. G. (1989). A 17,000-year glacio-eustatic sea level record: influence of glacial melting rates on the Younger Dryas event and deep-ocean circulation. *Nature*, **342**, 637–642.

Fairbanks, R. G. (1990). The age and origin of the 'Younger Dryas climatic event' in Greenland ice cores. *Paleoceanography*, **5**, 937–948.

Fairbridge, R. W. (1961). Eustatic changes in sea level. *Physics and Chemistry of the Earth*, **4**, 99–185.

Fairbridge, R. W. (1983). Isostasy and eustasy. In *Shorelines and Isostasy* (D. E. Smith and A. G. Dawson, eds), Institute of British Geographers, Special Publication, 16. London: Academic Press, pp. 3–25.

Fairchild, H. L. (1911). Radiation of glacial flow as a factor in drumlin formation (discussion). *Geological Society of America Bulletin*, **2**, 734.

Fairchild, I. J. (1983). Effects of glacial transport and neomorphism on Precambrian dolomite crystal sizes. *Nature*, **304**, 714–716.

Fairchild, I. J. (1991). Origins of carbonate in Neoproterozoic stromatolites and the identification of modern analogues. *Precambrian Research*, **53**, 281–299.

Fairchild, I. J. and Hambrey, M. J. (1984). The Vendian succession of northeastern Spitsbergen: petrogenesis of a dolomite-tillite association. *Precambrian Research*, **26**, 111–167.

Fairchild, I. J. and Spiro, B. (1990). Carbonate minerals in glacial sediments: geochemical clues to paleoenvironment. In *Glaciomarine Environments: Processes and Sediments* (J. A. Dowdeswell and J. D. Scourse, eds), Geological Society of London Special Publication. No.

53, pp. 201–216.

Farrington, A. and Synge, F. M. (1970). The eskers of Tullamore district. In *Irish Geographical Studies* (N. Stephens and R. E. Glasscock eds) 49–52.

Faure, G. (1986). *Principles of Isotope Geology*. New York: John Wiley & Sons, 589 pp.

Fayzullin, R. M. (1969). Lithological-facies characteristics and the gold contents of a series of conglomerates of the Kamensk deposit. *Chemical Abstracts*, **70**, 108202b.

Fayzullin, R. M. and Tolmacheva, T. Y. (1972). Nature of native gold from the placers of the Chikoi-Menzinsk interfluve (western Transbaikalia). *Chemical Abstracts*, **77**, 154966W.

Fayzullin, R. M. and Turchinova, D. M. (1973). On relationships between gold placers and primary sources or intermediate collectors. *Doklady Earth Science Sections*, **212**, 242–243.

Fedoroff, N. and Goldberg, P. (1982). Comparative micromorphology of two Late Pleistocene paleosols (in the Paris Basin). *Catena*, **9**, 227–251.

Feeser, V. (1988). On the mechanics of glaciotectonic contortion of clays. In *Glaciotectonics: Forms and Processes*, (D. G. Croot, ed.) pp. 63–76, Rotterdam: A.A. Balkema Publishers.

Felix-Henningsen, P. and Urban, B. (1982). Paleoclimatic interpretation of a thick intra-Saalian paleosol the 'Bleached Loam' on the Drenthe moraines of northern Germany. *Catena*, **9**, 1–8.

Fenn, C. R. and Gomez, B. (1989). Particle size analysis of the sediment suspended in a proglacial stream: Glacier de Tsidjiore Nouve, Switzerland. *Hydrological Processes*, **3**, 123–135.

Ferguson, S. A. and Freeman, E. B. (1978). Ontario occurrences of float, placer gold and other heavy minerals. *Ontario Geological Survey, Mineral Deposits Circular 17*, 214 pp.

Fernlund, J. M. R. (1988). The Halland coastal moraines: Are they end moraines or glaciotectonic ridges? In *Glaciotectonics: Forms and Processes* (D. G. Croot, ed.) pp. 77–90, Rotterdam: A.A. Balkema Publishers.

Fernlund, J. M. R. (1994). Tills and non-till diamictons in glacial sequences. In *Formation and Deformation of Glacial Deposits* (W. P. Warren and D. G.Croot, eds) pp. 29–38, Rotterdam: A.A. Balkema Publishers.

Fillon, R. H., Ferguson, C. and Thomas, F. (1978). Cenozoic provenance and sediment cycling: Hamilton Bank, Labrador Shelf. *Journal of Sedimentary Petrology*, **48**, 253–268.

Finch, T. F. (1977). Drumlins between Kilfenora and Doolin, Co. Clare. *Proceedings of the Royal Dublin Society, Science Series A*, **6**, 1–7.

Firth, C. R. and Haggart, B. A. (1989). Loch Lomond Stadial and Flandrian shorelines in the inner Moray Firth area, Scotland. *Journal of Quaternary Science*, **4**, 37–50.

Fischbein, S. A. (1987). Analysis and interpretation of ice-deformed sediments from Harrison Bay, Alaska. *United States Geological Survey*, Open File: 87–0262, 73 pp.

Fischer, A. G., (1982). Long-term climatic oscillations recorded in stratigraphy. In *Climate in Earth History. Studies in Geophysics*. Washington, DC: National Academy Press, pp. 97–104.

Fisher, D. A., Reeh, N. and Langley, K. (1985). Objective reconstructions of the Late Wisconsinan Laurentide Ice Sheet and the significance of deformable beds. *Géographie Physique et Quaternaire*, **39**, 229–238.

Fisher, T. G. and Shaw, J. (1992). A depositional model for rogen moraine, with examples from the Avalon Peninsula, Newfoundland. *Canadian Journal of Earth Sciences*, **29**, 669–686.

FitzPatrick, E. A. (1984). Micromorphology of soils. London: Chapman and Hall, 433 pp.

Fitzsimons, S. J. (1990). Ice-marginal depositional processes in a polar maritime environment, Vestfold Hills, Antarctica. *Journal of Glaciology*, **36**, 279–286.

Flageollet, J. C. and Vaskou, P. (1979). Aspects exoscopiques de quartz de moraines des Vosges moyennes au microscope electronique a balayage. *Revue de Geographie Physique et de Géologie Dynamique*, **21**, 307–313.

Flemal, R. C., Hinkley, K. C. and Hesler, J. L. (1973). DeKalb Mounds: a possible Pleistocene (Woodfordian) pingo field in north-central Illinois. In *The Wisconsian Stage* (R. F. Black, R. P. Goldthwait and H. B. Willman, eds), Geological Society of America, Memoir **136**, 229–250.

Flint, R. F. (1947). *Glacial Geology and the Pleistocene Epoch*, New York: John Wiley & Sons, 589 pp.

Flint, R. F. (1957). *Glacial and Pleistocene Geology*, New York: John Wiley & Sons, 553 pp.

Flint R. F. (1971). *Glacial and Quaternary Geology*. New York and London: John Wiley & Sons, 892 pp.

Foldvik, A. and Gammelsrød, T. (1988). Notes on Southern Ocean hydrography, sea-ice and bottom water formation. *Palaeogeography, Palaeoclimatology, Palaeoecology*, **67**, 3–17.

Folk, R. L. (1975). Glacial deposits identified by chattermark trails in detrital garnets, *Geology*, **8**, 473–475

Follmer, L. R. (1978). The Sangamon Soil in its type area – a review. In *Quaternary Soils* (W. C. Mahaney, ed.), 125–165, Norwich, England: Geoabstracts.

Follmer, L. R. (1982). The geomorphology of the Sangamon surface: its spatial and temporal attributes. In *Space and Time in Geomorphology* (C. E. Thorn, ed.) pp. 117–146, London: George Allen and Unwin.

Follmer, L. R. (1983). Sangamon and Wisconsinan pedogenesis in the midwestern United States. In *Late-Quaternary Environments of the United States, Volume 1 – The Late Pleistocene* (S. C. Porter, ed.) pp. 138–144, London: Longman.

Follmer, L. R., McKenna, D. P. and King, J. E. (1989). Quaternary records of central and northern Illinois. *Illinois State Geological Survey Guidebook 20*, 86 pp.

Fookes, P. G. (1965). Orientation of fissures in stiff overconsolidated clay of the Siwalik system. *Géotechnique*, **15**, 195–206.

Fookes, P. G. (1978). Terminology of discontinuities. *The Engineering Behaviour of Glacial Materials. Proceedings of the Symposium at the University of Birmingham, April 1975*, pp. 225–226.

Fookes, P. G. and Denness, B. (1969). Observational studies on fissure patterns in Cretaceous sediments of south-east England. *Géotechnique*, **19**, 453–477.

Fookes, P. G. and Parrish, D. G. (1969). Observations on small-scale structural discontinuities in the London Clay and their relationship to regional geology. *Quarterly Journal of Engineering Geology*, **1**, 217–240.

Forde, M. C. (1985). Glacial Tills 85. *Proceedings of the International Conference on Construction in Glacial Tills and Boulder Clays, March 1985*, Edinburgh: Engineering Technics Press, 256 pp.

Forman, S. L. (1988). The solar resetting of thermoluminescence of sediments in a glacier-dominated fiord environment in Spitsbergen: geochronologic implications. *Arctic and Alpine Research*, **20**, 243–253.

Forman, S. L. (1989). Applications and limitations of thermoluminescence to date Quaternary sediments. *Quaternary International*, **1**, 47–59.

Forman, S. L. (1990). Thermoluminescence properties of fiord sediments from Engelskbukta, western Spitsbergen, Svalbard; a new tool for deciphering depositional environment? *Sedimentology*, **37**, 377–384.

Forman, S. L. and Ennis, G. (1992). Limitations of thermoluminescence to date waterlain sediments from glaciated fiord environments of western Spitsbergen, Svalbard. *Quaternary Science Reviews*, **11**, 61–70.

Forsström, L., Aalto, M., Eronen, M. and Grönlund, T. (1988). Stratigraphic evidence for Eemian crustal movements and relative sea-level changes in eastern Fennoscandia. *Palaeogeography, Palaeoclimatology, Palaeoecology*, **68**, 317–335.

Fort, M. and Derbyshire, E. (1988). Some characteristics of tills in the Annapurna Range, Nepal. In *The Palaeoenvironment of East Asia from the Mid-Tertiary, vol. 1*. (P.

Whyte, ed.), Centre of Asian Studies – Occasional Papers and Studies, Hong Kong: University of Hong Kong, pp. 195–214.

Fort, M., Burbank, D. W. and Freytet, P. (1989). Lacustrine sedimentation in a semiarid alpine setting: an example from Ladakh, northwestern Himalaya. *Quaternary Research*, **31**, 332–350.

Foscolos, A. E., Rutter, N. W. and Hughes, O. L. (1977). The use of pedological studies in interpreting the Quaternary history of central Yukon Territory. *Geological Survey of Canada Bulletin*, **271**.

Fowler, A. J., Gillespie, R. and Hedges, R. E. M. (1986). Radiocarbon dating of sediments by accelerator mass spectrometry. *Physics of the Earth and Planetary Interiors*, **44**, 15–20.

Frakes, L. A. (1975). Geochemistry of Ross Sea diamicts. In *Initial Reports of Deep Sea Drilling Project*, **28**. Washington, DC: United States Government Printing Office, pp. 789–792.

Frakes, L. A. (1985). A preliminary model for subaqueous-glacial and post-glacial sedimentation in intra-continental basins. *Palaeogeography, Palaeoclimatology, Palaeoecology*, **51**, 347–356.

Frakes, L. A. (1986). Mesozoic-Cenozoic climatic history and causes of the glaciation. In *Mesozoic and Cenozoic Oceans* (K. J. Hsü, ed.) pp. 33–48 Geodynamic Series, 15. Washington: American Geophysical Union.

Francis, E. A. (1975). Glacial sediments: a selective review. In *Ice Ages: Ancient and Modern. Geological Journal, Special Issue 6* (A. E. Wright and F. Moseley, eds) pp. 43–68, Liverpool: Seel House Press.

Francis, E. A. (1981). On the classification of glacial sediments. In *The Quaternary in Britain* (J. Neale and J. Flenley eds) Oxford: Pergamon Press, pp. 237–247.

Freeze, R. A. and J. A. Cheery. (1979). *Groundwater*. Englewood Cliffs: Prentice Hall, 604 pp.

Freiwald, A., Henrich R., Schaefer, P. and Willkomm, H. (1991). The significance of high-boreal to subarctic maerl deposits in northern Norway to reconstruct Holocene climatic changes and sea level oscilliations. *Facies*, **25**, 315–340.

French, H. M. (1976). *The Periglacial Environment*. London: Longmans Publishing, 309 pp.

French, H. M. and Gozdzik, J. S. (1988). Pleistocene epigenetic and syngenetic frost fissures, Be_chatów, Poland. *Canadian Journal of Earth Sciences*, **25**, 2017–2027.

Friedman, I. and Long, W. (1976). Hydration rate of obsidian. *Science*, **191**, 347–352.

Friedman, I. and Obradovich, J. (1981). Obsidian hydration dating of volcanic events. *Quaternary Research*, **16**, 37–47.

Fristrup, B. (1952). Wind erosion within the Arctic deserts. *Geografisk Tidsskrift* 51–56.

Frye, J. C. and Willman, H. B. (1962). Morphostratigraphic units in Pleistocene stratigraphy. *American Association of Petroleum Geologists*, **46**, 112–113.

Frye, J. C. and Willman, H. B. (1973) Wisconsinan climatic history interpreted from Lake Michigan Lobe deposits and soils. In *The Wisconsian Stage*. (R. F. Black, R. P. Goldthwait and H. B. Willman, eds), *Geological Society of America, Memoir* **136**, 135–152.

Fullerton, D. S. (1986a). Stratigraphy and correlation of glacial deposits from Indiana to New York and New Jersey. In *Quaternary Glaciations in the Northern Hemisphere*. (V. Sibrava, D. Q. Bowen and G. M. Richmond, eds), *Quaternary Science Reviews* **5**, 23–37.

Fullerton, D. S. (1986b). Chronology and correlation of glacial deposits in the Sierra Nevada, California. In *Quaternary Glaciations in the Northern Hemisphere*. (V. Sibrava, D. Q. Bowen and G. M. Richmond, eds), *Quaternary Science Reviews*, **5**, 161–169.

Fulton, R. J. (ed.). (1984). *Quaternary Stratigraphy of Canada – A Canadian Contribution to IGCP Project 24*. Geological Survey of Canada Paper, 84–10, 210 pp.

Fulton, R. J. (1986). Quaternary stratigraphy of Canada. In *Quaternary Glaciation in the Northern Hemisphere*. (V. Sibrava, D. Q. Bowen and G. M. Richmond, eds), *Quaternary Science Reviews*, **5**, 207–209.

Fulton, R. J. (1989). *Quaternary Geology of Canada and Greenland*. Geological Survey of Canada, Geology of Canada No. 1, 839 pp.

Fulton, R. J. and Andrews, J. T. (1987). *The Laurentide Ice Sheet. Géographie Physique et Quaternaire*, **41**, 179–318.

Fyfe, G. (1990). The effect of water depth on ice-proximal glaciolacustrine sedimentation: Salpausselkä I, southern Finland. *Boreas*, **19**, 147–164.

Gale, S. J. and Hoare, P. G. (1986). Blakeney ridge sands and gravels. In *The Nar Valley and North Norfolk* (R. G. West and C. A. Whiteman eds), Quaternary Association Field Guide, Coventry, 94–95.

Gallagher, J. J. Jr. (1987). Fractography of sand grains broken by uniaxial compression. In *Clastic Particles* (J. R. Marshall, ed.) pp. 189–228, New York: Van Nostrand Reinhold.

Galloway, R. W. (1956). The structure of moraines in Lyngsdalen, North Norway. *Journal of Glaciology*, **2**, 730–733.

Galon, R. (1959). New investigations of inland dunes in Poland. *Przeglad Geogr*, **31**, 93–110.

Garcia Ambrosiani, K. and Robertsson, A. M. (1992) Early Weichselian interstadial sediments at Härnösand, Sweden. *Boreas*, **21**, 305–317.

Gard, G. (1988). Late Quaternary calcareous nannofossil biochronology and paleo-oceanography of Arctic and Subarctic seas. *Meddelanden från Stockholms Universitets Geologiska Institution*, **275**, 8–45.

Gard, G. and J. Backman. (1990). Synthesis of Arctic and Sub-Arctic Coccolith biochronology and history of North Atlantic drift water influx during the last 500 000 years. In *Geological history of the polar oceans: Arctic verses Antarctic*. (U. Bleil and J. Thiede, eds) pp. 417–435, NATO ASI Series C, Dordrecht: Kluwer Academic Publishing.

Garland, G. D. (1989). *Proceedings of Exploration '87: Third Decennial International Conference on Geophysical and Geochemical Exploration for Minerals and Groundwater*. Ontario Geological Survey, Special Volume 3, 960 pp.

Gascoyne, M., Ford, D. C. and Schwarcz, H. P. (1981). Late Pleistocene chronology and paleoclimate of Vancouver Island determined from cave deposits. *Canadian Journal of Earth Sciences*, **18**, 1643–1652.

Gascoyne, M., Schwarcz, H. P. and Ford, D. C. (1983). Uranium-series ages of speleothem from northwest England: correlation with Quaternary climate. *Philosophical Transactions of the Royal Society of London*, **301(B)**, 143–164.

Gaunt, G. D. (1981). Quaternary history of the southern part of the Vale of York. In *The Quaternary in Britain* J. Neale and J. Flenley, eds) pp. 82–97, Oxford: Pergamon Press.

Geikie, A. (1863). On the phenomena of the glacial drift of Scotland. *Transactions of the Geological Society of Glasgow*, **1**, 1–190.

Geikie, J. (1894). *The Great Ice Age*. 3rd edn., London: Edward Stanford, 850 pp.

Geitzenauer, K. R., Margolis, S. V. and Edwards. D. S. (1968). Evidence consistent with Eocene glaciation in a South Pacific deep sea sedimentary core. *Earth and Planetary Science Letters*, **4**, 173–177.

Gelinas, P. J. (1974). Contributions to the study of erosion along the North Shore of Lake Erie. Unpublished PhD thesis, University of Western Ontario. London, Ontario, 264 pp.

Gellatly, A. F. (1985). Phosphate retention: relative dating of Holocene soil development. *Catena*, **12**, 227–240.

Gemmell, A. M. D. (1988a). Thermoluminescence dating of glacially transported sediments: some considerations. *Quaternary Science Reviews*, **7**, 277–285.

Gemmell, A. M. D. (1988b). Zeroing of the TL signal in sediment undergoing fluvioglacial transport. An example from Austerdalen, western Norway. *Quaternary Science Reviews*, **7**, 339–345.

Gemmell. A. M. D., Sharp, M. J. and Sugden, D. E. (1986). Debris from the basal ice of the Agassiz Ice Cap, Ellesmere Island, Arctic Canada. *Earth Surface Processes and Landforms*, **11**, 123–130.

Geonautics Limited (1985). Design of an iceberg scour repetitive mapping network for the Canadian east coast. Environmental Studies Revolving Funds Report No. 043, Ottawa, 50 pp.

Geonautics Limited (1989). Regional ice scour data base update studies. Environmental Studies Research Funds Report No. 105, Ottawa, 168 pp.

Gerrard, A. J. (1990). *Mountain Environments*, London: Belhaven Press, 317 pp.

Gibbard, P. L. (1980). The origin of stratified Catfish Creek Till by basal melting. *Boreas*, **9**, 71–85.

Gibbard, P. L. and Turner, C. (1988). In defence of the Wolstonian stage. *Quaternary Newsletter*, **54**, 9–14.

Gibbons, A. B., Megeath, J. D. and Pierce, K. L. (1984). Probability of moraine survival in a succession of glacial advances. *Geology*, **12**, 327–330.

Gilbert, G. K. (1906). Crescentic gouges on glaciated surfaces. *Geological Society of America Bulletin*, **17**, 303–316.

Gilbert, G. and Pedersen, K. (1986). Ice scour data base for the Beaufort Sea. Environmental Studies Revolving Funds, Report No. 055, Ottawa, 99 pp.

Gilbert. R. (1982). Contemporary sedimentary environments on Baffin Island, N.W.T., Canada: glaciomarine processes in fjords of eastern Cumberland Peninsula. *Arctic and Alpine Research*, **14**, 1–12.

Gilbert, R. (1990). Rafting in glaciomarine environments. In *Glaciomarine Environments: Processes and Sediments* (J. A. Dowdeswell and J. D. Scourse, eds), Geological Society of London, Special Publication No. 53, pp. 105–120.

Gilbert, R. and Shaw, J. (1981). Sedimentation in proglacial Sunwapta Lake, Alberta. *Canadian Journal of Earth Sciences*, **18**, 81–93.

Gilbert, R., Handford, K. J. and Shaw, J. (1992). Ice scours in the sediments of glacial Lake Iroquois, Prince Edward County, eastern Ontario. *Géographie Physique et Quaternaire*, **46**, 189–194.

Gillette, D. A., Adams, J., Endo, A., Smith, D. and Kihl, R. (1980). Threshold velocities for input of soil particles into the air by desert soils. *Journal of Geophysical Research*, **85**, 5621–5630.

Gillott, J. E. (1969). Study of the fabric of fine-grained sediments with the Scanning Electron Microscope. *Journal of Sedimentary Petrology*, **39**, 90–105.

Gipp, M. R. (1993). The orientation of buried iceberg scours and other linear phenomena. *Marine Geology*, **114**, 263–272.

Giusti, L. (1983). The distribution, grades and mineralogical composition of gold-bearing placers in Alberta. Unpublished MA thesis, University of Alberta, Edmonton, Alberta, 397 pp.

Gjessing, J. 1965. On _plastic scouring' and _subglacial erosion'. *Norsk Geografisk Tidsskrift*, **20**, 1–37.

Gjessing, J. (1966). Some effects of ice erosion on the development of Norwegian valleys and fjords. *Norsk Geografisk Tidsskrift*, **20**, 273–299.

Gjessing, J. (1967). Potholes in connection with plastic scouring forms. *Geografisk Annaler*, **49A**, 178–187.

Gleeson, C. F. (1970). Heavy mineral studies in the Klondike area, Yukon Territory. Geological Survey of Canada, Bulletin 173, 63 pp.

Glückert, G. (1973). Two large drumlin fields in central Finland. *Fennia*, 120.

Godwin, H. (1962). Half-life of radiocarbon. *Nature*, **195**, 984.

Goldstein, B. (1989). Lithology, sedimentology, and genesis of the Wadena drumlin field, Minnesota, U.S.A. *Sedimentary Geology*, **62**, 241–280.

Goldstein, J. I., Newbury, D. E., Echlin, P., Joy, D. C., Fiori, C. and Lifshin, E. (1981). *Scanning Electron Microscopy and X-ray Microanalysis: A Text for Biologists, Material Scientists, and Geologists*. New York: Plenum Press, 673 pp.

Goldthwait, R. P. (ed.) (1971). *Till: A Symposium*. Columbus: Ohio State University Press, 402 pp.

Goldthwait, R. P. and Matsch, C. L. (eds) (1988). *Genetic Classification of Glacigenic Deposits*. Rotterdam: A.A. Balkema Publishers, 294 pp.

Gomez, B. and Small, R. J. (1983). Genesis of englacial debris within the Lower Glacier de Tsidjiore Nouve, Valais, Switzerland, as revealed by scanning electron microscopy. *Geografiska Annaler*, **65A**, 45–51.

Gomez, B., Dowdeswell, J. A. and Sharp, M. J. (1988). Microstructural control of quartz sand grain shape and texture: implications for the discrimination of debris transport pathways through glaciers. *Sedimentary Geology*, **57**, 119–129.

Goodale, J. L. and Hampton, M. A. (1987). An experimental and theoretical study of some quartz grain surface features. In *Clastic Particles* (J. R. Marshall, ed.) pp. 229–241, New York: Van Nostrand Reinhold.

Goodfriend, G. A. (1987). Evaluation of amino-acid racemization/epimerization dating using radiocarbon-dated fossil land snails. *Geology*, **15**, 698–700.

Goodhew, P. J. and Humphreys, F. J. (1988). Electron microscopy and analysis. 2nd edn, London: Taylor and Francis, 243 pp.

Goodman, D. J., King, G. C. P., Millar, D. H. M. and Robin, G. De Q. (1979). Pressure-melting effects in basal ice of temperate glaciers: laboratory studies and field observations under Glacier d'Argéntiere. *Journal of Glaciology*, **23**, 259–271.

Goodwin, C. R., Finley, J. C. and Howard, L. M. (1985). Ice scour bibliography. Environmental Studies Revolving Funds Report No. 010, Ottawa, 99 pp.

Gordon, J. E., Darling, W. G., Whalley, W. B and Gellatly, A. F. (1988). δD–$\delta^{18}O$ relationships and the thermal history of basal ice near the margins of two glaciers in Lyngen, north Norway. *Journal of Glaciology*, **34**, 265–68.

Goss, E., Mayewski, P. A. and Lyons, W. B. (1985). Examination of selected microparticles from the Sentik Glacier core, Ladakh Himalaya, India. *Journal of Glaciology*, **31**, 196–197.

Goudie, A. S. (1974). Further experimental investigation of rock weathering by salt and other mechanical processes. *Zeitschrift für Geomorphologie*, Supplementband **21**, 1–12.

Goudie, A. S. 1983. Dust storms in space and time. *Progress in Physical Geography*, **7**, 502–530.

Goudie, A. S. and Day, M. J. (1981). Disintegration of fan sediments in Death Valley, California, by salt weathering. *Physical Geography*, **1**, 126–137.

Goudie, A. S., Cooke, R. U. and Evans, I. (1970). Experimental investigation of rock weathering by salts. *Area*, **2**, 42–48.

Goudie, A. S., Cooke, R. U. and Doornkamp, J. C. (1979). The formation of silt from quartz dune sand by salt-weathering processes in deserts. *Journal of Arid Environments*, **2**, 105–112.

Gow, A. J., Epstein, S. and Sheehy, W. (1979). On the origin of stratified debris in ice cores from the bottom of the Antarctic ice sheet. *Journal of Glaciology*, **23**, 185–192.

Gozdźik, J. and Mycielska-Dowgiallo, E. (1982). Badanie wplywu niektórych procesów geologicznych na przeksztalcenie powierzchni ziarn kwarcowych. *Przeglad Geograficzny*, **54**, 219–241.

Grabowska-Olszewska, B. (1975). SEM analysis of microstructures of loess deposits. *Bulletin of the International Association of Engineering Geologists*, **11**, 45–48

Graf, W. L. (1971). Geomorphology of the glacial valley cross-section. *Arctic and Alpine Research*, **2**, 303–312.

Gram, R. (1969). Grain surface features in Eltanin cores and Antarctic glaciation. *Antarctic Journal of the United States*, **4**, 174–175.

Grant, P. (1978). The role of the scanning electron microscope in cathodoluminescence petrology. In *Scanning Electron Microscopy in the Study of Sediments* (W. B. Whalley, ed.) pp. 1–11 Norwich: Geo Abstracts.

Grass, J. D. (1984). Ice scour and ice ridging studies in Lake Erie. Proceedings of the International Association of Hydraulic Research, 7th International Symposium on Ice, Hamburg, **2**, 33–43.

Grass, J. D. (1985). Lake Erie cable crossing – ice scour study. In *Workshop on Ice Scouring*, 15 – 19 February, 1982, National Research Council of Canada Associate committee on geotechnical research, Technical memorandum No. 136, 1–10.

Gravenor, C. P. (1955). The origin and significance of prairie mounds. *American Journal of Science*, **253**, 475–481.

Gravenor, C. P. (1979). The nature of the Late Paleozoic glaciation in Gondwana as determined from an analysis of garnets and other heavy minerals. *Canadian Journal of Earth Sciences*, **16**, 1137–1153.

Gravenor, C. P. (1980a). Chattermarked garnets and heavy minerals from the Late Paleozoic glacial deposits of southeastern Brazil. *Canadian Journal of Earth Sciences*, **17**, 156–159

Gravenor, C. P. (1980b). Heavy minerals and sedimentological studies on the glaciogenic Late Precambrian Gaskiers formation of Newfoundland. *Canadian Journal of Earth Sciences*, **17**, 1331–1341.

Gravenor, C. P. (1985). Chattermarked garnets found in soil profiles and beach environments. *Sedimentology*, **32**, 295–306.

Gravenor, C. P. and Gostin, V. A. (1979). Mechanisms to explain the loss of heavy minerals from the upper Palaeozoic tillites of South Africa and Australia and the late Precambrian tillites of Australia. *Sedimentology*, **26**, 707–717.

Gravenor, C. P. and Kupsch, W. O. (1959). Ice-disintegration features in Western Canada. *Journal of Geology*, **67**, 48–67.

Gravenor, C. P. and Leavitt, R. K. (1981). Experimental formation and significance of etch patterns on detrital garnets. *Canadian Journal of Earth Sciences*, **18**, 765–775.

Gravenor, C. P. and Meneley, W. A. (1958). Glacial flutings in central and northern Alberta. *American Journal of Science*, **256**, 715–728.

Gravenor, G. P., von Brunn, V. and Dreimanis, A. (1984). Nature and classification of waterlain glaciogenic sediments, exemplified by Pleistocene, Late Paleozoic and Late Precambrian deposits. *Earth Science Reviews*, **20**, 105–166.

Gray, J. M. (1974). The Main Rock Platform of the Firth of Lorn, western Scotland. *Transactions of the Institute of British Geographers*, **61**, 81–99.

Gray, J. M. (1981). P-forms from the Isle of Mull. *Scottish Journal of Geology*, **17**, 39–47.

Gray, J. M. (1982). The last glaciers (Loch Lomond Advance) in Snowdonia, N. Wales. *Geological Journal*, **17**, 111–133.

Gray, J. M. (1983). The measurement of relict shoreline altitudes in areas affected by glacio-isostasy, with particular reference to Scotland. In *Shorelines and Isostasy* (D. E. Smith and A. G. Dawson, eds) pp. 97–127, Institute of British Geographers, Special Publication, 16. London: Academic Press.

Gray, J. M. (1985). Glacio-isostatic shoreline development in Scotland: an overview. Department of Geography, Queen Mary & Westfield College, University of London, Occasional Paper, 24, 61 pp.

Gray, J. M. (1988). Glaciofluvial channels below the Blakeney esker. *Quaternary Newsletter*, **55**, 8–12.

Gray, J. M. (1991). Glaciofluvial landforms. In *Glacial Deposits of Britain and Ireland* (J. Ehlers, P. L. Gibbard, and J. Rose, eds) pp. 443–453, Rotterdam: A.A. Balkema Publishers.

Gray, J. M. and Lowe, J. J. (eds) (1977). *Studies in the Scottish Late Glacial Environments*. Oxford: Pergamon Press, 197 pp.

Gregory, J. W. (1913). *The Nature and Origin of Fjords*. London: J. Murray publ., 542 pp.

Grisak, G. E., and Cherry, J. A. (1975). Hydrologic characteristics and response of fractured till and clay confining a shallow aquifer. *Canadian Geotechnical Journal*, **12**, 23–43.

Grisak, G. E., Cherry, J. A., Vonhof, J. A. and Bluemle, J. P. (1976). Hydrogeologic and hydrochemical properties of fractured till in the interior plains region. In *Glacial Till: An Inter-disciplinary Study* (R. F. Legget, ed.), The Royal Society of Canada, Special Publication No. 12, Ottawa, pp. 304–335.

Grobe, H. (1986). Sedimentation processes on the Antarctic continental margin at Cape Norvegia during the Late Pliocene. *Geologische Rundschau*, **75**, 97–104.

Grobe, H. (1987). Facies classification of glacio-marine sediments in the Antarctic. *Facies*, **17**, 99–109.

Grobe, H., Mackensen, A., Hubberten, H. -W., Spiess, V. and Fütterer, D. K.(1990). Stable isotope record and late Quaternary sedimentation rates at the Antarctic continental margin. In *Geological History of the Polar Oceans: Arctic versus Antarctic* (U. Bleil and J. Thiede, eds), NATO ASI Series, C 308, Dordrecht: Kluwer Academic Publishers, pp. 539–572.

Grove, J. M. (1988). *The Little Ice Age*. London: Methuen, 498 pp.

Grube, F. (1983). Tunnel valleys. In *Glacial Deposits in North-West Europe* (J. Ehlers, ed.) pp. 257–258, Rotterdam: A.A. Balkema Publishers.

Grün, R. (1989). Electron spin resonance (ESR) dating. *Quaternary International*, **1**, 65–109.

Gurnell, A. M. and Clark, M. J. (eds) (1987). *Glacio-Fluvial Sediment Transfer – an Alpine Perspective*. Chichester: John Wiley & Sons, 524 pp.

Gustavson, T. C. (1975a). Bathymetry and sediment distribution in proglacial Malaspina Lake, Alaska. *Journal of Sedimentary Petrology*, **45**, 450–461.

Gustavson, T. C. (1975b). Sedimentation and physical limnology in proglacial Malispina Lake, southeastern Alaska. In (A. V. Jopling and B. C. McDonald, eds), Glaciofluvial and glaciolacustrine sedimentation. Society of Economic Paleontologists and Mineralogists Special Publication No. 23, pp. 249–263.

Gustavson, T. C. and Boothroyd, J. C. (1981). Subglacial fluvial erosion: a major source of stratified drift, Malaspina glacier, Alaska. In *Research in Glacial, Glaciofluvial, and Glaciolacustrine Systems, Proceedings of the 6th Guelph Symposium in Geomorphology*. (R. Davidson-Arnott, W. Nickling and B. D. Fahey, eds) pp. 83–116 Norwich: Geo Books.

Gustavson, T. C. and Boothroyd, J. C. (1987). A depositional model for outwash, sediment sources, and hydrologic characteristics, Malaspina Glacier, Alaska; A modern analog of the southeastern margin of the Laurentide Ice Sheet. *Bulletin of the Geological Society of America*, **99**, 187–200.

Haake, F. -W. and Pflaumann, U. (1989). Late Pleistocene foraminiferal stratigraphy on the Vøring Plateau, Norwegian Sea. *Boreas*, **18**, 343–356.

Habbe, K. -A. (1989). On the origin of the drumlins of the South German Alpine Foreland. *Sedimentary Geology*, **62**, 357–369.

Hafsten, U. 1983. Biostratigraphical evidence for Late Weichselian and Holocene sea-level changes in southern Norway. In *Shorelines and Isostasy*. (D. E. Smith and A. G. Dawson, eds) pp. 161–181, Institute of British Geographers, Special Publication, 16. London: Academic Press.

Haines, J. and Mazzullo, J. (1988). The original shapes of quartz silt grains: a test of the validity of the use of quartz grain shape analysis to determine the sources of terrigenous silt in marine sedimentary deposits. *Marine Geology*, **78**, 227–240.

Hald, M. and Vorren, T. O. (1987). Stable isotope stratig-

raphy and paleoceanography during the last deglaciation on the continental shelf off Troms, Northern Norway. *Paleoceanography*, **2**, 583–599.

Hald, M., Sættem, J. and Nesse, E. (1990). Middle and Late Weichselian stratigraphy in shallow drillings from the southwestern Barents Sea: foraminiferal, amino acid and radiocarbon evidence. *Norsk Geologisk Tidsskrift*, **70**, 241–257.

Haldorsen, S. (1981). Grain-size distribution of subglacial till and its relation to glacial crushing and abrasion. *Boreas*, **10**, 91–105.

Haldorsen, S. (1983). Mineralogy and geochemistry of basal till and their relationship to till-forming processes. *Norsk Geologisk Tidsskrift*, **63**, 15–25.

Haldorsen, S. and Shaw, J. (1982). The problem of recognizing melt-out till. *Boreas*, **11**, 261–277.

Haldorsen, S., Deinboll Jenssen, P., Koler, J. Chr. and Myhr, E. (1983). Some hydraulic properties of Sandy-silty Norwegian tills. *Acta Geologica Hispanica*, **18**, 191–198.

Haldorsen, S., Jørgensen, P., Rappol, M. and Riezebos, P. A. (1989). Composition and source of the clay-sized fraction of Saalian till in The Netherlands. *Boreas*, **18**, 89–97.

Halferdahl, L. B. (1965). The occurrence of gold in Alberta rivers. Alberta Research Council, Economic Mineral Files, Open File Report 65–11, 22 pp.

Hall, C. M. and York, D. (1984). The applicability of ^{40}Ar/^{39}Ar dating to young volcanics. In *Quaternary Dating Methods: Developments in Paleontology and Stratigraphy*. Vol. 7. (W. C. Mahaney, ed.) pp. 67–74 New York: Elsevier.

Hall, M. G. and Lloyd, G. E. (1981). The SEM examination of geological samples with a semiconductor backscattered electron detector. *American Mineralogist*, **66**, 362–368.

Hallberg, G. R. (1980). Pleistocene stratigraphy in east-central Iowa. *Iowa Geological Survey Technical Information Series*, **10**, 168 pp.

Hallberg, G. R. (1986). Pre-Wisconsin glacial stratigraphy of the central plains region in Iowa, Nebraska, Kansas, and Missouri. In *Quaternary Glaciation in the Northern Hemisphere*. (V. Sibrava, D. Q. Bowen and G. M. Richmond, eds), *Quaternary Science Reviews*, **5**, 11–15.

Hallet, B. (1975). Subglacial silica deposits. *Nature*, **254**, 682–683.

Hallet, B. (1976). Deposits formed by subglacial precipitation of $CaCO_3$. *Geological Society of America Bulletin*, **87**, 1003–1015.

Hallet, B. (1981). Glacial abrasion and sliding: their dependence on the debris concentration in basal ice.

Annals of Glaciology, **2**, 23–28.

Hambrey, M. J. and Harland, W. B. (1979). Analysis of pre-Pleistocene glacigenic rocks: aims and problems. In *Moraines and Varves: Origin/Genesis/Classification*, (Ch. Schlüchter, ed.) pp. 271–275, Rotterdam: A.A. Balkema Publishers.

Hambrey, M. J. and Harland, W. B. (eds) (1981a). *Earth's Pre-Pleistocene Glacial Record*. Cambridge: Cambridge University Press, 1004 pp.

Hambrey, M. J. and Harland, W. B. (1981b). Summary of Earth's pre-Pleistocene glacial record. In Earth's Pre-Pleistocene Glacial Record (M. J. Hambrey and W. E. Harland, eds) pp. 943–954, Cambridge: Cambridge University Press.

Hambrey, M. J. and Spencer, A. M. (1987). Late Precambrian glaciation of central East Greenland. *Medaelelser om Groenland*, Greenland Geoscience 19, 50 pp.

Hambrey, M. J., Barrett, P. J. and Robinson, P. H. (1989). Stratigraphy. In Antarctic Cenozoic history from the CIROS-1 drillhole, McMurdo Sound. (P. J. Barrett, ed.), *DSIR Bulletin*, **245**, 23–48.

Hamilton, T. D. (1986a). Correlation of Quaternary glacial deposits in Alaska. In *Quaternary Glaciations in the Northern Hemisphere*. (V. Sibrava, D. Q. Bowen and C. M. Richmond, eds), *Quaternary Science Reviews*, **5**, 171–180.

Hamilton, T. D. (1986b). Late Cenozoic glaciation of the central Brooks Range. In *Glaciation in Alaska – The Geologic Record* (T. D. Hamilton, K. M. Reed, and R. M. Thorson, eds) pp. 9–50, Anchorage: Alaska Geological Society.

Hammer, C. U. (1989). Dating by physical and chemical seasonal variations and reference horizons. In *The Environmental Record in Glaciers and Ice Sheets* (H. Oeschger and C. C. Langway Jr., eds) pp. 99–121, New York: John Wiley & Sons.

Hammer, C. U., Clausen, H. B. and Dansgaard, W. (1980). Greenland ice sheet evidence of post-glacial volcanism and its climatic impact. *Nature*, **288**, 230–235.

Hansel, A. K., and Johnson, W. H. (1987). Ice marginal sedimentation in a late Wisconsinan end moraine complex, Northeastern Illinois, USA. In *Tills and Glaciotectonics* (J. J. M. van der Meer, ed.) pp. 97–104, Rotterdam: A.A. Balkema Publishers.

Hanvey, P. M. (1987). Sedimentology of lee-side stratification sequences in late-Pleistocene drumlins, north-west Ireland. In *Drumlin Symposium*. (J. Menzies and J. Rose, eds) pp. 241–253, Rotterdam: A.A. Balkema Publishers.

Hanvey, P. M. (1989). Stratified flow deposits in a late Pleistocene drumlin in northwest Ireland. *Sedimentary Geology*, **62**, 211–221.

Harbor, J. M. (1992). Numerical modeling of the development of U-shaped valleys by glacial erosion. *Geological Society of America Bulletin*, **104**, 1364–1375.

Harbor, J. M. and Wheeler, D. A. (1992). On the mathematical description of glaciated valley cross sections. *Earth Surface Processes and Landforms*, **17**, 477–485.

Harbor, J. M., Hallet, B. and Raymond, C. F. (1988). A numerical model of landform development by glacial erosion. *Nature*, **333**, 347–349.

Harden, J. W. (1982). A quantitative index of soil development from field descriptions: examples from a chronosequence in central California. *Geoderma*, **28**, 1–28.

Hare, P. E., Hoering, T. C. and King, K. Jr. (1980). *Biogeochemistry of Amino Acids*. New York: John Wiley & Sons, 558 pp.

Hargraves, R. B. (1976). Precambrian geologic history. *Science*, **193**, 363–371.

Harland, W. B. (1992). Stratigraphic regulation and guidance: A critique of current tendencies in stratigraphic codes and guides. *Geological Society of America Bulletin*, **104**, 1231–1235.

Harland, W. B. and Herod, K. N. (1975). Glaciations through time. In *Ice Ages: Ancient and Modern*. (A. E. Wright and F. Moseley, eds) pp. 189–216, *Geological Journal, Special Issue 6*, Liverpool: Seel House Press.

Harland, W. B., Herod, K. N. and Krinsley, D. (1966). The definition and identification of tills and tillites. *Earth Science Review*, **2**, 225–256.

Harmon, R. S., Thompson, P., Schwarcz, H. P. and Ford, D. C. (1978). Late Pleistocene paleoclimates of North America as inferred from stable isotope studies of speleothems. *Quaternary Research*, **9**, 54–70.

Harrington, C. D. and Whitney, J. W. (1987). Scanning electron microscope method for rock-varnish dating. *Geology*, **15**, 967–970.

Harris, C. (1985). Geomorphological applications of soil micromorphology with particular reference to periglacial sediments and processes. In *Geomorphology and Soils*. (K. S. Richards, R. R. Arnett and S. Ellis, eds) pp. 219–232, London: George Allen and Unwin.

Harris, C. (1987). Solifluction and related periglacial deposits in England and Wales. In *Periglacial Processes and Landforms in Britain and Ireland*. (J. Boardman, ed.) pp. 209–223, Cambridge: Cambridge University Press.

Harris, C. and Bothamley, K. (1984). Englacial deltaic sediments as evidence for basal freezing and marginal shearing, Leirbreen, southern Norway. *Journal of Glaciology*. **30**, 30–34.

Harris, C. and Donnelly, R. (1991). The glacial deposits of South Wales. In *Glacial Deposits in Great Britain and*

Ireland (J. Ehlers, P. L. Gibbard, and J. Rose, eds) pp. 279–290, Rotterdam: A.A. Balkema Publishers.

Harris, S. A. (1967). Origin of part of the Guelph drumlin field and the Galt-Paris moraines, Ontario – a reinterpretation. *Canadian Geographer*, **11**, 16–34.

Harris, S. E. (1943). Friction cracks and the direction of glacial movement. *Journal of Geology*, **51**, 244–258.

Harrison, P. W. (1957). A clay-till fabric: its character and origin. *Journal of Geology*, **65**, 275–308.

Harron, G. A. (1976). Metallogénèse des gîtes de sulphures des Cantons de l'Est. *Ministère des Richesses Naturelles du Québec*, **ES-27**, 42 pp.

Hart, J. K. (1990). Proglacial glaciotectonic deformation and the origin of the Cromer Ridge push moraine complex, North Norfolk, England. *Boreas*, **19**, 165–180.

Hart, J. K. (1992). Sedimentary environments associated with Glacial Lake Trimingham, Norfolk, UK. *Boreas*, **21**, 119–136.

Hart, J. K. (1994). Till fabric associated with deformable beds. *Earth Surface Processes and Landforms*, **19**, 15–32.

Hart, J. K. and Boulton, G. S. (1991). The interrelation of glaciotectonic and glaciodepositional processes within the glacial environment. *Quaternary Science Reviews*, **10**, 335–350.

Hart, J. K., Hindmarsh, R. C. A. and Boulton, G. S. (1990). Styles of subglacial glaciotectonic deformation within the context of the Anglian ice-sheet. *Earth Surface Processes and Landforms*, **15**, 227–241.

Hartford, D. 1985. Properties and behaviour of Irish glacial till. In *Glacial Tills 85. Proceedings of the International Conference on Construction in Glacial Tills and Boulder Clays, March 1985* (M. C. Forde, ed.) pp. 93–98, Edinburgh, Scotland: Engineering Technics Press.

Harry, D. G. and Gozdzik, J. S. (1988). Ice wedges: growth, thaw transformation, and palaeoenvironmental significance. *Journal of Quaternary Sciences*, **3**, 39–55.

Hartshorn, J. H. (1958). Flowtill in southeast Massachusetts. *Geological Society of America Bulletin*, **69**, 477–482.

Hattin, D. E. (1958). New evidence of high-level glacial drainage in the White Mountains, N.H. *Journal of Glaciology*, **3**, 315–319.

Haworth, L. A., Calkin, P. E. and Ellis, J. M. (1986). Direct measurement of lichen growth in the central Brooks Range, Alaska, U.S.A., and its application to lichenometric dating. *Arctic and Alpine Research*, **18**, 289–296.

Haynes, V. M. (1972). The relationship between drainage areas and sizes of outlet troughs of the Sukertoppen ice cap, Greenland. *Geografiska Annaler*, **54A**, 66–75.

Hayward, J., Orford, J. D. and Whalley, W. B. (1989). Three implementations of fractal analysis of particle outlines. *Computers and Geosciences*, **15**, 199–207.

Hedges, E. M. and Gowlett, J. A. J. (1986). Radiocarbon dating by accelerator mass spectrometry. *Scientific American*, **254**, 100–107.

Heezen, B. C. and Ewing, M. (1952). Turbidity currents and submarine slumps, and the 1929 Grand Banks earthquake. *American Journal of Science*, **250**, 849–873.

Heidenreich, C. (1964). Some observations on the shape of drumlins. *Canadian Geographer*, **8**, 101–107.

Heiken, G. (1972). Morphology and petrography of volcanic ashes. *Geological Society of America Bulletin*, **83**, 1961–1988.

Heikkinen, O. and Tikkanen, M. (1979). Glacial flutings in northern Finnish Lapland. *Fennia*, **15**, 1–12.

Heikkinen, O. and Tikkanen, M. (1989). Drumlins and flutings in Finland: their relationship to ice movement and to each other. *Sedimentary Geology*, **62**, 349–356.

Hein, F. J. and Walker, R. G. (1977). Bar evolution and development of stratification in the gravelly, braided, Kicking Horse River, British Columbia. *Canadian Journal of Earth Sciences*, **14**, 562–570. Heinrich, H. (1988). Origin and consequences of cyclic ice rafting in the Northeast Atlantic ocean during the past 130,000 years. *Quaternary Research*, **29**, 142–152.

Heller, F., Beat, M., Wang, J., Li, H. and Liu Tungsheng, (1987). Magnetization and sedimentation history of loess in the Central Loess Plateau of China. In *Aspects of Loess Research* (T. S. Liu, ed.) pp. 147–163, Beijing: China Ocean Press.

Hencher, S. R. (1987). The implications of joints and structures for slope stability. In *Slope Stability* (M. G. Anderson and K. S. Richards, eds) pp. 145–186, Chichester: John Wiley & Sons.

Henderson, P. J. (1988). Sedimentation in an esker system influenced by bedrock topography near Kingston, Ontario. *Canadian Journal of Earth Sciences*, **25**, 987–999.

Henderson-Sellers, A. and Henderson-Sellers, B. (1988). Equable climate in the early Archean. *Nature*, **336**, 117–118.

Hendry, H. E. and Stauffer, M. R. (1975). Penecontemporaneous recumbent folds in through cross-bedding of Pleistocene sands in Saskatchewan, Canada. *Journal of Sedimentary Petrology*, **45**, 932–943.

Hendry, J. M. (1982). Hydraulic conductivity of a glacial till in Alberta. *Ground Water*, **20**, 162–169.

Henrich, R. (1986). A calcite dissolution pulse in the

Norwegian-Greenland Sea during the last deglaciation. *Geologische Rundschau*, **75**, 805–827.

Henrich, R. (1989). Glacial/interglacial cycles in the Norwegian Sea: sedimentology, paleoceanography and evolution of Late Pliocene to Quaternary northern hemisphere climate. In *Proc. Final Reps. Ocean Drilling Program 104*, (O. Eldholm, J. Thiede and E. Taylor, eds) pp. 189–232.

Henrich, R. (1990). Cycles, rhythms and events in quaternary Arctic and Antarctic glaciomarine deposits. In *Geological history of the polar oceans: Arctic versus Antarctic*. (U. Bleil and J. Thiede, eds), NATO ASI Series, C 308, pp. 213–244, Dordrecht: Kluwer Academic Publishers.

Henrich, R. (1991). Cycles, rhythms, and events on high input and low input glaciated continental margins. In *Cycles and Events in Stratigraphy* (G. Einsele, W. Ricken and A. Seilacher, eds) pp. 751–772, Berlin: Springer-Verlag.

Henrich, R., Kassens, H., Vogelsang, E., and Thiede, J. (1989a). Sedimentary facies of glacial/interglacial cycles in the Norwegian Sea during the last 350 ka. *Marine Geology*, **86**, 283–319.

Henrich, R., Wolf, T., Bohrmann, G. and Thiede, J. (1989b). Cenozoic paleoclimatic and plaeoceanographic changes in the Northern Hemisphere revealed by variability of coarse-fraction composition in sediments from the Vøring Plateau – ODP Leg 104 drill sites. In *Scientific Results, Ocean Drilling Program 104* (O. Eldholm, J. Thiede and E. Taylor et. al., eds) pp. 75–188.

Hérail, G. (ed.). (1991). *International Symposium on Alluvial Gold Placers; Abstracts*. La Paz, Bolivia.

Hérail, G., Fornari, M. and Rouhier, M. (1989a). Geomorphological control of gold distribution and gold particle evolution in glacial and fluvioglacial placers of the Ancocala-Ananea basin – southeastern Andes of Peru. *Geomorphology*, **2**, 369–383.

Hérail, G., Fornari, M., Viscarra, G., Laubacher, G., Argollo, J. and Miranda, V. (1989b). Geodynamic and gold distribution in the Tipuani-Mapiri basin, Bolivia. In *International Symposium on Intermontane Basins: Geology and Resources*. Thailand: Chaing Mai, pp. 342–352.

Hérail, G., Fornari, M., Viscarra, G. and Miranda, V. (1990). Morphological and chemical evolution of gold grains during the formation of a polygenic fluviatile placer: the Mio-Pleistocene Tipuani placer example, Andes, Bolivia. *Chronique de la Recherche Miniere*, **500**, 41–9.

Heusser, C. J. (1989). Climate and chronology of Antarctica and adjacent South America over the past 30,000 yrs. *Palaeogeography, Palaeoclimatology, Palaeoecology*, **76**, 31–37.

Hicock, S. R. (1988). Calcareous till facies north of Lake Superior, Ontario: implications for Laurentide ice streaming. *Géographie Physique et Quaternaire*, **42**, 120–135.

Hicock, S. R. (1990). Genetic till prism. *Geology*, **18**, 517–519.

Hicock, S. R. (1991). On subglacial stone pavements in till. *Journal of Geology*, **99**, 607–619.

Hicock, S. R. (1992). Lobal interactions and rheologic superposition in subglacial till near Bradtville, Ontario, Canada. *Boreas*, **21**, 73–88.

Hicock, S. R. and Dreimanis, A. (1985). Glaciotectonic structures as useful ice-movement indicators in glacial deposits: four Canadian case studies. *Canadian Journal of Earth Sciences*, **22**, 339–346.

Hicock, S. R. and Dreimanis, A. (1989). Sunnybrook drift indicates a grounded early Wisconsin glacier in the Lake Ontario basin. *Geology*, **17**, 169–172.

Hicock, S. R. and Dreimanis, A. (1992a). Sunnybrook drift in the Toronto area, Canada: re-investigation and re-interpretation. In *The Last Interglacial/Glacial in North America* (P. U. Clark and P. D. Lea, eds) Geological Society of America, Special Paper 270, pp. 139–162.

Hicock, S. R. and Dreimanis, A. (1992b). Deformation till in the Great Lakes region: implications for rapid flow along the south-central margin of the Laurentide Ice Sheet. *Canadian Journal of Earth Sciences*, **29**, 1565–1579.

Hicock, S. R., Dreimanis, A. and Broster, B. E. (1981). Submarine flow tills at Victoria, British Columbia. *Canadian Journal of Earth Sciences*, **18**, 71–80.

Hicock, S. R., Kristjansson, F. J. and Sharpe, D. R. (1989). Carbonate till as a soft bed for Pleistocene ice streams on the Canadian Shield north of Lake Superior. *Canadian Journal of Earth Sciences*, **26**, 2249–2254.

Higgins, C. G. (1957). Origin of potholes in glaciated regions. *Journal of Glaciology*, **3**, 11–12.

Higgins, C. G. and Coates, D. R. (eds) (1990). Groundwater geomorphology: the role of subsurface water in earth surface processes and landforms. Geological Society of America Special Publication 252, 368 pp.

Higgs, R. (1979). Quartz-grain surface features of Meso-zoic-Cenozoic sands from the Labrador and Western Green and continental margins. *Journal of Sedimentary Petrology*, **49**, 599–610.

Hill, A. R. (1968). An analysis of the spatial distribution and origin of drumlins in north Down and south Antrim, Northern Ireland. Unpublished PhD thesis, Queen's University, Belfast.

Hill, A. R. (1971). The internal composition and structure of drumlins in North Down and South Antrim, Northern Ireland. *Geografiska Annaler*, **53A**, 14–31.

Hill, A. R. (1973). The distribution of drumlins in County Down, Ireland. *Annals of the Association of American Geographers*, **63** (2), 226–240.

Hillaire-Marcel, C. and Causse, C. (1989a). Chronologie Th/U des concrétions calcaires des varves du lac glaciaire de Deschaillons (Wisconsinien inférieur). *Canadian Journal of Earth Sciences*, **26**, 1041–1052.

Hillaire-Marcel, C. and Causse, C. (1989b). The late Pleistocene Laurentide Glacier: Th/U dating of its major fluctuations and $\delta^{18}O$ range of the ice. *Quaternary Research*, **32**, 125–138.

Hillefors, Å. (1983). The Dösebacka and Ellesbo drumlins – Morphology and stratigraphy. In *Glacial Deposits in North-West Europe* (J. Ehlers, ed.) pp. 141–150, Rotterdam: A.A. Balkema Publishers.

Hindmarsh, R. C. A., Boulton, G. S and Hutter, K. (1989). Modes of operation of thermo-mechanically coupled ice sheets. *Annals of Glaciology*, **12**, 57–69.

Hinsch, W. 1979. Rinnen an der Basis des glaziären Pleistozäns in Schleswig-Holstein. *Eiszeitalter und Gegenwart*, **29**, 173–178.

Hirano, M. (1981). An explantion of the U-shaped profile of the glacier valley based on the variation principle. *Bulletin of the Faculty of Literature (Jinbun-Kenkyu), Osaka City University*, **33**, 1–14.

Hirano, M., and Aniya, M. (1988). A rational explanation of cross- profile morphology for glacial valleys and of glacial valley development. *Earth Surface Processes and Landforms*, **13**, 707–716.

Hirano, M. and Aniya, M. (1989). A rational explanation of cross-profile morphology for glacial valleys and of glacial valley development: A further note. *Earth Surface Processes and Landforms*, **14**, 173–174.

Hirvas, H. (1989). Application of glacial geological studies in prospecting in Finland. In *Drift Prospecting* (R. N. W. DiLabio and W. B. Coker, eds), Geological Survey of Canada, Paper 89–20, pp. 1–6.

Hirvas, H. (1991). Glacial transport of gold in overburden in Ananea, Peru. In *International Symposium on Alluvial Gold Placers; Abstracts* (G. Hérail, ed.), La Paz, Bolivia, pp. 77.

Hirvas, H., Kujansuu, R. and Tynni, R. (1976). Till stratigraphy in Northern Finland. In Report nr. 3 IGCP Project 73/1/24 (Quaternary Glaciations in the Northern Hemisphere) (D. J. Easterbrook and V. Sibrava (eds) pp. 256–273.

Hirvas, H., Lagerbäck, R., Mäkinen, K., Nenonen, K., Olsen, L., Rodhe, L. and Thoresen, M. (1988). The Nordkalott Project: studies of Quaternary geology in northern Fennoscandia. *Boreas*, **17**, 431–437.

Hobbs, W. H. 1933. The origin of the loess associated with continental glaciation based upon studies in Greenland. International Geographical Congress, Paris 1931. *Compte Rendu*, **2**, 408–411.

Hobbs, W. H. (1942). Wind – the dominant transportation agent within extra-marginal zones to continental glaciers. *Journal of Geology*, **50**, 556–559.

Hobbs, W. H. (1943a). The glacial anticyclones and the continental glaciers of North America. *American Philosophical Society, Proceedings*, **86**, 368–402.

Hobbs, W. H. (1943b). The glacial anticyclone and the European continental glacier. *American Journal of Science*, **241**, 333–36.

Hodge, S. M. (1979). Direct measurement of basal water pressures: progress and problems. *Journal of Glaciology*, **23**, 309–319.

Hodgson, D. M. (1982). Hummocky and fluted moraine in part of North-West Scotland. Unpublished PhD Thesis, University of Edinburgh.

Hodgson, D. M. (1986). A study of fluted moraines in the Torridon area, NW Scotland. *Journal of Quaternary Sciences*, **1**, 109–118.

Hodgson, G. J., Lever, J. H., Woodworth-Lynas, C. M. T and Lewis, C. F. M. (1988). The dynamics of iceberg grounding and scouring (DIGS) experiment and repetitive mapping of the eastern Canadian continental shelf. Environmental Studies Research Funds Report No. 094, Ottawa, pp. 316.

Hoffman, P. F. (1988). United plates of America, the birth of a craton: Early Proterozoic assembly and growth of Laurentia. *Annual Review of Earth and Planetary Sciences*, **16**, 543–603.

Hoffman, P. F. (1989). Speculations on Laurentia's first gigayear (2.0 to 1.0 Ga). *Geology*, **17**, 135–138.

Högbom, I. (1923). Ancient inland dunes of northern and middle Europe. *Geografiska Annaler*, **5**: 113–243.

Holdsworth, G. (1973). Ice deformation and moraine formation at the margin of an ice cap adjacent to a proglacial lake. In *Research in Polar and Alpine Geomorphology*, pp. 187–199 (B. D. Fahey and R. D. Thompson, eds), Third Guelph Symposium, 1973. Norwich: Geo-Abstracts.

Holland, S. S. (1950). Placer gold production of British Columbia. British Columbia Ministry of Energy, Mines and Petroleum Resources, Bulletin 28, 89 pp.

Hollingworth, S. E. (1931). The glaciation of western Edenside and adjoining areas and the drumlins of Edenside and the Solway Basin. *Quarterly Journal of the Geological Society of London*, **87**, 281–357.

Holmes, A. (1965). *Principles of Physical Geology*. Edinburgh: Nelson, 532 pp.

Holmes, C. D. (1941). Till fabric. *Geological Society of*

America Bulletin, **52**, 1299–1354.

Holmes, C. D. (1952). Drift dispersion in west-central New York. *Geological Society of America Bulletin*, **63**, 993–1010.

Holmes, C. D. (1960). Evolution of till-stone shapes, central New York. *Geological Society of America Bulletin*, **71**, 1645–1660.

Holtedahl, H. (1967). Notes on the formation of fjords and fjord-valleys. *Geografiska Annaler*, **49A**, 183–203.

Holtedahl, O. (1960). Geology of Norway. *Norges Geologiske Undersøkelse*, **208**, 540 pp.

Homci, H. (1974). Jungpleistozäne Tunneltälter im Nordosten von Hamburg (Rahlstedt-Meiendorf). *Mittleungen Geologie-Paläontologie*. Institüt Universität Hamburg, **43**, 99–126.

Honjo, S., Manganini, J. and Wefer, G. (1988). Annual particle flux and a winter outburst of sedimentation in the northern Norwegian Sea. *Deep Sea Research*, **35**, 1224–1234.

Hooke, R. LeB. (1968). Comments on 'The formation of shear moraines; an example from South Victoria Land, Antarctica'. *Journal of Glaciology*, **7**, 351–352.

Hooke, R. LeB. (1984) On the role of mechanical energy in maintaining subglacial water conduits at atmospheric pressure. *Journal of Glaciology*, **30**, 180–187

Hooke, R. LeB. and Iverson, N. R. (1995). Grain size distribution in deforming subglacial tills: role of grain fracture. *Geology*, **23**, 57–60.

Hopkins, R. (1985). Placer gold potential offshore of Country Harbour, Nova Scotia. Energy, Mines and Resources Canada, Ocean Mining Division, Canada Oil and Gas Lands Administration, Internal Publication, 89 pp.

Hoppe, G. (1951). Drumlins i Nordöstra Norbotten. *Geografiska Annaler*, **33**, 1299–1354.

Hoppe, G. (1952). Hummocky moraine regions with special reference to the interior of Norbotten. *Geografiska Annaler*, **34**, 1–72.

Hoppe, G. (1957). Problems of glacial morphology and the ice age. *Geografiska Annaler*, **39A**, 1–18

Hoppe, G. (1959). Glacial morphology and inland ice recession in northern Sweden. *Geografiska Annaler*, **41**, 193–212.

Hoppe, G. (1968). Tärnasjö-områdets geomorfologi. En översiktlig orientering med särskild hänsyn till de glaciala och postglaciala formelementen. Stockholm University Naturgeografisk Institut Forskninsrapp, **2**, 24 pp.

Hoppe, G. (1974). The glacial history of the Shetland Islands. Institute of British Geographers, Special Publication, **7**, 197–210.

Hoppe, G. and Schytt, V. (1953). Some observations on fluted moraine surfaces. *Geografiska Annaler*, **35**, 105–115.

Horberg, L. (1951). Intersecting minor ridges and periglacial features in the Lake Agassiz basin, North Dakota. *Journal of Geology*, **59**, 1–18.

Hotzel, I. S. and Miller. J. D. (1983). Icebergs: their physical dimensions and the presentation and application of measured data. *Annals of Glaciology*, **4**, 116–123.

Houmark-Nielson, M. (1981). Glacialstratigrafi i Danmark øst for Hovedopholdslinien. Dansk Geologiska Föreningen Årsskrift for 1980, pp. 61–76.

Houmark-Nielson, M. (1983). Glacial stratigraphy and morphology of the northern Bælthav region. In *Glacial Deposits in North-West Europe*. (Ehlers, J., ed.) pp. 211–218, Rotterdam: A.A. Balkema Publishers.

Houmark-Nielson, M. (1987). Pleistocene stratigraphy and glacial history of the central part of Denmark. *Bulletin of the Geological Society of Denmark*, **36**, 389 pp.

Hubbert, M. K. and Willis, D. G. (1957). Mechanics of hydraulic fracturing. *Transactions American Institute of Mechanical Engineers*, **210**, 153–68.

Huddart, D. (1983). Flow tills and ice-walled lacustrine sediments, the Petteril Valley, Cumbria, England. In *Tills and Related Deposits* (E. B. Evenson, Ch. Schlüchter, and J. Rabassa, eds) pp. 81–94, Rotterdam: A.A. Balkema Publishers.

Huggett, Q. J. and Kidd, R. B. (1984). Identification of ice-rafted and other exotic material in deepsea dredge hauls. *Geo-Marine Letters*, **3**, 23–29.

Hughes, O. L. (1964). Surficial geology, Nichicun-Kaniapiskau map-area, Québec. Geological Survey of Canada, Bulletin 106.

Hughes, O. L., Campbell, R. B., Muller, J. E. and Wheeler, J. O. (1969). Glacial limits and flow patterns, Yukon Territory, south of 65° North latitude. Geological Survey of Canada, Paper 68–34, 9 pp.

Hughes, O. L., Rampton, V. N. and Rutter, N. W. (1972). Quaternary geology and geomorphology, southern and central Yukon (Northern Canada). *24th International Geological Congress, Montreal, Excursion Guidebook*, A–11. 30–34.

Hughes, R. A. (1987). The Denali placer gold mine. *The Alaska Miner*, **17**, 14–15, 19.

Hughes, T. J. (1981). Numerical reconstruction of paleo-ice sheets. In *The Last Great Ice Sheets* (G. H. Denton and T. J. Hughes, eds) pp. 221–261, New York: Wiley-Interscience.

Huntley, D. J., Godfrey-Smith, D. I. and Thewalt, M. L. W. (1985). Optical dating of sediments. *Nature*, **313**, 105–107.

Huston, M. M., Brigham-Grette, J. and Hopkins, D. M. (1990). Paleogeographic significance of middle Pleistocene glaciomarine deposits on Baldwin Peninsula, Northwest Alaska. *Annals of Glaciology*, **14**, 111–114.

Hütt, G. and Jungner, H. (1992). Optical and TL dating on glaciofluvial sediments. *Quaternary Science Reviews*, **11**, 161–163.

Hutter, K. (1983). *Theoretical Glaciology: Material Science of Ice and the Mechanics of Glacier and Ice Sheets.* Dordrecht: D. Reidel Publishing, 510 pp.

Ignatius, H. C. (1958). On the late-Wisconsin deglaciation in Eastern Canada, 1. Glacial geological observations from north-central Québec. *Acta Geographia (Helsinki)*, **16**, 34 pp.

Imbrie, J. and Imbrie, K. P. (1979). *Ice Ages: Solving the Mystery.* Hillside, NJ: Enslow, 224 pp.

Imbrie, J., and Imbrie, J. Z. (1980). Modelling the climatic response to orbital variations. *Science*, **207**, 943–953.

Imbrie, J., Hays, J. D., Martinson, D. G., McIntyre, A., Mix, A. C., Morley, J. J., Pisias, N. G., Prell, W. and Shackleton, N. J. (1984). The orbital theory of Pleistocene climate: Support from a revised chronology of the marine ∂18O record. In *Milankovitch and Climate, Part I.* (A. Berger, J. Imbrie, J. Hays, G. Kukla, and B. Saltzman, eds) pp.269–305, Hingham, MA: D. Reidel.

Ingólfsson, O. (1988). Large-scale glaciotectonic deformation of soft sediments: a case study of a late Weichselian sequence in Western Iceland. In *Glaciotectonics: Forms and Processes* (D. G. Croot, ed.) pp.101–107, Rotterdam: A.A. Balkema Publishers.

Ingram, M. J., Underhill, D. J. and Wigley, T. M. L. (1978). Historical climatology. *Nature*, **276**, 329–334.

Innes, J. L. (1985). Lichenometry. *Progress in Physical Geography*, **9**, 187–254.

Innes, J. L. (1986a). Dating exposed rock surfaces in the Arctic by lichenometry: the problem of thallus circularity and its effect on measurement errors. *Arctic*, **39**, 253–259.

Innes, J. L. (1986b). The use of percentage cover measurements in lichenometric dating. *Arctic and Alpine Research*, **18**, 209–16.

Iverson, N. R. (1993). Regelation of ice through subglacial debris: implications for sediment transport. *Geology*, **21** (6), 559–567.

Jackson, T. A. (1965). Power-spectrum analysis of two 'varved' argillites in the Huronian Cobalt Series (Precambrian) of Canada. *Journal of Sedimentary Petrology*, **35**, 877–886.

Jaeger, J. C. and Cook, N. G. W. (1976). *Fundamentals of Rock Mechanics*, (2nd edn). London: Chapman and Hall Publishing, Halsted Press Book, 585 pp.; 1979 (3rd edn), 593 pp.

Jahns, R. H. (1941). Outwash chronology in northeastern Massachusetts (abstract). *Geological Society of America Bulletin*, **52**, 1910.

Jamieson, T. F. (1865). On the history of the last geological changes in Scotland. *Quarterly Journal of the Geological Society of London*, **21**, 161–203.

Jansen, E. (1989). The use of stable oxygen and carbon isotope stratigraphy as a dating tool. *Quaternary International*, **1**, 151–166.

Jansen, E. and Bjørklund, K. (1985). Surface ocean circulation in the Norwegian Sea 15.000 B.P. to present. *Boreas*, **14**, 243–257.

Jansen, E. and Erlenkeuser, H. (1985). Ocean circulation in the Norwegian-Greenland sea during the last deglaciation: isotopic evidence. *Palaeogeography, Palaeoclimatology, Palaeoecology*, **49**, 189–206.

Jansen, E. and Sjøholm, J. (1991). Reconstruction of glaciation over the past 6 Myr from ice-borne deposits in the Norwegian Sea. *Nature*, **349**, 600–603.

Jansen, E., Befring, S., Bugge, T., Eidvin, T., Holtedahl, H. and Sejrup, H. P. (1987). Large submarine slides on the Norwegian continental margin: sediments, transport and timing. *Marine Geology*, **78**, 77–107.

Jansen, E., Bleil, U., Henrich, R., Kringstad, L. and Slettemark, B. (1988). Paleoenvironmental changes in the Norwegian Sea and the northeastern Atlantic during the last 2.8 m.u.: ODP/DSDP Sites 610, 642, 643 and 644. *Paleoceanography*, **3**, 563–581.

Jansen, E., Sjøholm, J., Bleil, U. and Erichsen, J. A. (1990). Neogene and Pleistocene glaciations in the Northern Hemisphere and Late Miocene-Pliocene global ice volume fluctuations: Evidence from the Norwegian Sea. In *Geological History of the Polar Oceans: Arctic versus Antarctic* (U. Bleil and J. Thiede, eds) pp.677–705, NATO ASI Series, C 308, Dordrecht: Kluwer Academic Publishing.

Jauhiainen, E. (1975). Morphometric analysis of drumlin fields in northern central Europe. *Boreas*, **4**, 219–230.

Jenkins, D. A. 1985. Chemical and mineralogical composition in the identification of palaeosols. In *Soils and Quaternary Landscape Evolution* (J. Boardman, ed.) pp.23–43, Chichester: John Wiley & Sons.

Jenny, H. (1941). *Factors of Soil Formation. A System of Quantitative Pedology.* New York: McGraw-Hill Publishers, 281 pp.

Jim, C. Y. (1990). Stress, shear deformation and micromorphological clay orientation: synthesis of various concepts. *Catena*, **17**, 431–447.

Johansson, H. G. (1972). Moraine ridges and till stratigraphy in Västerbotten, northern Sweden. *Sveriges Geologiska Undersökning*, **C673**, 50 pp.

Johansson, H. G. (1983). Tills and moraines in northern Sweden. In *Glacial Deposits in North-West Europe* (J. Ehlers, ed.) pp. 123–130, Rotterdam: A.A. Balkema Publishers.

Johansson, P. (1988). Deglaciation pattern and ice-dammed lakes along the Saariselkä mountain range in northeastern Finland. *Boreas*, **17**, 541–552.

Johnsen, S. J., Clausen, H. B., Dansgaard, W., Fuhrer, K., Gundestrup, N., Hammer, C. U., Iversen, P., Jouzel, J., Stauffer, B. and Stefensen, J. P. (1992). Irregular glacial interstadials in a new Greenland ice core. *Nature*, **359**, 311–313.

Johnson, A. M. (1970). *Physical Processes in Geology*. San Francisco: Freeman & Cooper, 577 pp.

Johnson, M. D. (1980). Origin of the Lake Superior Red Clay and glacial history of Wisconsin's Lake Superior shoreline west of the Bayfield peninsula. Unpublished MSc thesis, University of Wisconsin, Madison, 108 pp.

Johnson, M. D. (1983). The origin and microfabric of Lake Superior Red Clay. *Journal of Sedimentary Petrology*. **53**, 859–873.

Johnson, M. D. (1990). Fabric and origin of diamictons in end moraines, Animas River valley, Colorado, U.S.A. *Arctic and Alpine Research*, **22**, 14–25.

Johnson, W. H. (1986). Stratigraphy and correlation of the glacial deposits of the Lake Michigan lobe prior to 14 ka BP. In *Quaternary Glaciations in the Northern Hemisphere* (V. Sibrava, D. Q. Bowen and G. M. Richmond, eds), *Quaternary Science Reviews*, **5**, 17–22.

Johnson, W. H. (1990). Ice-wedge casts and relict patterned ground in Central Illinois and their environmental significance. *Quaternary Research*, **33**, 51–72.

Johnson, W. H. and Hansel, A. K. (1990). Upper Wisconsinan glacigenic sequences at Wedron, Illinois. *Journal of Sedimentary Petrology*, **60**, 26–41.

Johnsson, G. (1956). Glacialmorfologiska studier i södra Sverige med särkild hänsyn till glaciala riktningselement och periglaciala frostenonmen. Meddelanden från Lunds Geografiska Institiionen Avhandlingar.

Johnston, W. A. and Uglow, W. L. (1926). Placer and vein gold deposits of Barkerville, Cariboo District, British Columbia. Geological Survey of Canada, Memoir 149, 246 pp.

Jones, G. A. and Keigwin, L. D. (1988). Evidence from Fram Strait (78Ñ) for early deglaciation. *Nature*. **336**, 56–59.

Jones, M. E. and Preston, R. M. F. (eds) (1987). *Deformation of Sediments and Sedimentary Rocks*. Geological Society Special Publication No. 29, 350 pp.

Jones, N. (1982). The formation of glacial flutings in east-central Alberta. In *Research in Glacial, Glaciofluvial and Glaciolacustrine Systems* (R. Davidson-Arnott, W. Nickling and B. D. Fahey, eds) pp. 49–70, Norwich: GeoBooks.

Jones, P. D., Raper S. C. B., Bradley, R. S., Diaz, H. F., Kelly. P. M. and Wigley, T. M. L. (1986a). Northern Hemisphere surface air temperature variations: 1851–1984. *Journal of Climate and Applied Meteorology*, **25**, 161–179.

Jones, P. D., Raper, S. C. B. and Wigley, T. M. L. (1986b). Southern hemisphere surface air temperature variations: 1851–1984. *Journal of Climate and Applied Meteorology*, **25**, 1213–1230.

Josenhans, H. W., Zevenhuizen, J. and Klassen, R. A. (1986). The Quaternary geology of the Labrador Shelf. *Canadian Journal of Earth Sciences*, **23**, 1190–1213.

Jurgaitis, A. and Juozapavicius, G. (1988). Genetic classification of glaciofluvial deposits and criteria for their recognition. In *Genetic Classification of Glacigenic Deposits* (R. P. Goldthwait and C. L. Matsch, eds) pp. 227–242, Rotterdam: A.A. Balkema Publishers.

Jurgaitis, A., Mikalauskas, G. and Juozapavicius, G. (1982). *Bedded Structures of Glaciofluvial Deposits in the Baltic Area* (in Russian). Vilnius: Mokslas publ., 52 pp.

Kallel. N., Labeyrie, L. D., Arnold, M., Okada, H., Dudley, W. C and Duplessey, J. -C. (1988). Evidence of cooling during the Younger Dryas in the western north Pacific. *Oceanologica Acta*, **11**, 369–375.

Kamb, B. (1991). Rheological nonlinearity and flow instability in the deforming bed mechanism of ice stream motion. *Journal of Geophysical Research*, **96B**, 16,585–16,595.

Kamb, B. and LaChapelle, E. (1964). Direct observations of the mechansim of glacier sliding over bedrock. *Journal of Glaciology*, **5**, 159–172.

Kamb, B., Raymond, C. F., Harrison, W. D., Engelhardt, H., Echelmeyer, K. A., Humphrey, N., Brugman, M. M. and Pfeffer, T. (1985). Glacier surge mechanism: 1982–1983 surge of Variegated Glacier, Alaska. *Science*, **227**, 469–479.

Kanasewich, E. R. (1963). Gravity measurements on the Athabasca Glacier, Alberta, Canada. *Journal of Glaciology*, **4**, 617–631.

Karl, T. R., Tarpley, J. D., Quayle, R. G., Diaz, H. F., Robinson, D. A. and Bradley, R. S. (1989). The recent climate record: what it can and cannot tell us. *Reviews of Geophysics*, **27**, 405–430.

Karlén, W. (1988). Scandinavian glacial and climatic fluctuations during the Holocene. *Quaternary Science*

Reviews, **7**, 199–209.

Karlstrom, E. T. (1988). Multiple paleosols in pre-Wisconinan drift, northwestern Montana and southwestern Alberta. *Catena*, **15**, 147–178.

Karpovich, R. P. (1971). Surface features of quartz sand grains, northeast coast, Gulf of Mexico. *Gulf Coast Association of Geological Societies Transactions*, **21**, 451–461.

Karrow, P. F. (1967). Pleistocene geology of the Scarborough area. Ontario Dept. of Mines, Geological Report 46.

Karrow, P. F. (1974). Till straigraphy in parts of southwestern Ontario. *Geological Society of America Bulletin*, **85**, 761–768.

Karrow, P. F. (1984a). Comments on 'Sedimentation in a large lake: an interpretation of the Late Pleistocene stratigraphy at Scarborough Bluffs, Ontario, Canada'. *Geology*, **12**, 185.

Karrow, P. F. (1984b). Lithofacies types and vertical profile models: an alternative approach to the description and environmental interpretation of glacial diamict and diamictite sequences – discussion. *Sedimentology*, **31**, 883–884.

Karrow, P. F. (1989). Quaternary geology of the Great Lakes subregion. In *Quaternary Geology of Canada and Greenland* (R. J. Fulton, ed.), Geological Survey of Canada No. 1 *also* Geological Society of America, *The Geology of North America v. K-1*, Ch. 4, pp. 326–350.

Karrow, P. F. (1992). Carbonates, granulometry, and color of tills on the south-central Canadian Shield and their implications for straigraphy and radio-carbon dating. *Boreas*, **21**, 379–391.

Karrow, P. F. and Calkin, P. E. (1985). *Quaternary Evolution of the Great Lakes*. Geological Association of Canada Special Publication No. 30, 258 pp.

Kartashov, I. P. (1971). Geological features of alluvial placers. *Economic Geology*, **66**, 879–885.

Kaszycki, C. A. (1987). A model for glacial and proglacial sedimentation in the shield terrane of southern Ontario. *Canadian Journal of Earth Sciences*, **24**, 2373–2391.

Kaszycki, C. A. (1989). Quaternary geology and glacial history of the Haliburton Region, South Central Ontario, Canada: a model for glacial and proglacial sedimentation. Unpublished PhD dissertation, University of Illinois at Urbana-Champaign, 219 pp.

Kaszycki, C. A., Suttner, W. and DiLabio, R. N. W. (1988). Gold and arsenic in till, Wheatcroft Lake dispersal train, Manitoba. Geological Survey of Canada, Paper 88–1C, pp. 341–351.

Katsushima, T. and Nishio, F. (1985). Volcanic ash in dirt layers from the Allan Hills Bare Ice Area in Victoria Land, Antarctica. *Proceedings of the Seventh Symposium on Polar Meterology and Glaciology, Memoirs of National Institute of Polar Research*, Special Issue, **39**, 193–208.

Kaufman, D. S. (1992). Aminostratigraphy of Pliocene–Pleistocene high-sea-level deposits, Nome coastal plain and adjacent nearshore area, Alaska. *Geological Society of America Bulletin*, **104**, 40–52.

Kaufman, D. S. and Brigham-Grette, J. (1993). Aminostratigraphic correlations and paleotemperature implications, Pliocene-Pleistocene high-sea-level deposits, northwestern Alaska. *Quaternary Science Reviews*, **12**, 21–33.

Kaufman, D. S. and Calkin, P. E. (1988). Morphometric analysis of Pleistocene glacial deposits in the Kigluaik Mountains, Northwestern Alaska, U.S.A. *Arctic and Alpine Research*, **20**, 273–284.

Kaufman, D. S. and Hopkins, D. M. (1986). Glacial history of the Seward Peninsula. In *Glaciation in Alaska – the Geologic Record* (T. D. Hamilton, K. M. Reed and R. M. Thorson, eds) pp. 51–78, Anchorage: Alaska Geological Society.

Kaufman D. S. and Miller, G. H. (1992). Overview of amino acid geochronology. *Comparative Biochemistry and Physiology*, **102B**, 199–204.

Kaufman, D. S., Walter, R. C., Brigham-Grette, J. and Hopkins, D. M. 1991. Middle Pleistocene age of the Nome River Glaciation, Northwestern Alaska. *Quaternary Research*, **36**, 277–293.

Kauranne, K. L. (1975). Regional geochemical mapping in Finland. In *Prospecting in Areas of Glaciated Terrain – 1975* (M. J. Jones, ed.) pp. 71–81, London: Institution of Mining and Metallurgy.

Kautsky, G. (1986). Geochemical atlas of northern Fennoscandia: Nordkalott Project. Geological Survey of Sweden, Uppsala, 19 pp., 155 maps.

Kaye, B. H. (1981). Direct characterization of fine particles. New York: John Wiley & Sons, 398 pp.

Kaye, B. H. and Naylor, A. G. (1972). An optical information procedure for characterising the shape of fine particle images. *Pattern Recognition*, **4**, 195–199.

Kazakevich, Y. P. (1972). Conditions of formation and preservation of buried gold placers. Izdatelstvo 'Nedra', Moscow, 216 pp.

Kazi, A. and Knill, J. L. (1969). The sedimentation and geotechnical properties of the Cromer Till between Happisburgh and Cromer, Norfolk. *Quarterly Journal of Engineering Geology*, **2**, 64–86.

Kazi, A. and Knill, J. L. (1973). Fissuring in glacial lake clays and tills on the Norfolk coast, United Kingdom. *Engineering Geology*, **7**, 35–48.

Kearsley, A. and Wright, P. (1988). Geological applications

of scanning cathodoluminescence imagery. *Microscopy and Analysis*, September 1988, 49–51.

Kemmis, T. J. (1991). Glacial landforms, sedimentology, and depositional environments of the Des Moines Lobe, Northern Iowa. Unpublished PhD dissertation. University of Iowa, 384 pp.

Kemmis, T. J., Hallberg, G. R. and Lutenegger, A. J. (1981). Depositional environments of glacial sediments and landforms on the Des Moines Lobe, Iowa. *Iowa Geological Survey Guidebook Series, 6*, 132 pp.

Kemp, R. A. (1985). Soil Micromorphology and the Quaternary. *Quaternary Research Association, Technical Guide No. 2*, 80 pp.

Kemp, R. A. (1987). Genesis and environmental significance of a buried Middle Pleistocene soil in eastern England. *Geoderma*, **41**, 49–77.

Kempe, S. and Degens, E. T. (1985). An early soda ocean? *Chemical Geology*, **53**, 95–108.

Kempe, S., Kazmierczak, J. and Degens, E. T. (1989). The soda ocean concept and its bearing on biotic evolution. In *Origin, Evolution and Modern Aspects of Biomineralization in Plants and Animals. Proceedings of the 5th International Symposium on Biomineralization* (R. E. Crick, ed.), New York: Plenum Press, 517 pp.

Kenig, K. (1980). Mineral composition of deposits of Gaas Glacier frontal moraine (Spitsbergen) with reference to SEM analysis of quartz grain surface (in Polish). *Kwartalnik Geologiczny*, **24**, 711–740.

Kenn, M. J. and Minton, P. (1968). Cavitation induced by vorticity at a smooth flat wall. *Nature*, **217**, 633–634.

Kennett, J. P. 1990. The Younger Dryas event: an introduction. *Paleoceanography*, **5**, 891–895.

Khitrov, V. V. and Khershberg, L. B. (1982). Paleogeographic setting of Neogene-Quaternary sedimentation and distribution pattern of native gold. *Chemical Abstracts*, **98**, 6526W.

King, D. T. Jr. and Banholzer, G. S. Jr. (1978). Technique for mounting and SEM study of individual particles less than 10 micrometers in diameter. *Journal of Sedimentary Petrology*, **48**, 625–627.

King, L. H. and Fader, G. B. J. (1986). Wisconsinan glaciation of the Atlantic continental shelf of southeast Canada. *Geological Survey of Canada, Bulletin, 363*, 72 pp.

King, L. H., Rokoengen, K., Fader, G. B. J. and Gunleksrud, T. (1991). Till-tongue stratigraphy. *Geological Society of America Bulletin*, **103**, 637–659.

Kite, G. W. and Reid, I. A. (1977). Volumetric change of the Athabasca Glacier over the last 100 years. *Journal of Hydrology*, **32**, 279–294.

Klassen, R. A. (1982). Glaciotectonic thrust plates, Bylot Island, District of Franklin. Geological Survey of Canada. Current Research, part A, Paper 82–1A, pp. 369–373.

Klassen, R. W. (1989). Quaternary geology of the southern Canadian Interior Plains. In *Quaternary Geology of Canada and Greenland* (R. J. Fulton, ed.) *Geology of Canada No. 1*, Geological Survey of Canada, DNAG Volume K-1, Geological Society of America, pp. 138–156.

Klassen, R. A., and Fisher, D. A. (1988). Basal-flow conditions at the northeastern margin of the Laurentide ice sheet, Lancaster Sound. *Canadian Journal of Earth Sciences*, **25**, 1740–1750.

Klassen, R. A. and Thompson, F. J. (1989). Ice flow history and glacial dispersal patterns, Labrador. In *Drift Prospecting* (R. N. W. DiLabio and W. B. Coker, eds), Geological Survey of Canada. Paper 89–20, pp. 21–29.

Kleman, J. (1994). Preservation of landforms under ice sheets and ice caps. *Geomorphology*, **9**, 19–32.

Kluiving, S. J. (1989). Glaciotektoniek in de stuwwal van Itterbeck-Uelsen. Unpublished MSc thesis, University of Amsterdam, Amsterdam, The Netherlands, 117 pp.

Knight J. and McTaggart, K. C. (1990). Lode and placer gold of the Coquihalla and Wells areas, British Columbia (92H, 93H). In *Exploration in British Columbia 1989* (B. Grant and J. Newell, eds) British Columbia Ministry of Energy, Mines and Petroleum Resources, pp. 105–118.

Knutsson, G. (1971). Studies of groundwater flow in till soils. *Geologiska Föreningen i Stockholm Förhandlingar*, **93**, 1–22.

Kochel, R. C. and Johnson, R. A. (1984). Geomorphology and sedimentology of humid-temperate alluvial fans, central Virginia. In *Sedimentology of Gravels and Conglomerates* (E. H. Koster and R. J. Steel, eds), Canadian Society of Petroleum Geologists, Memoir 10, pp. 109–122.

Koc Karpuz, N. and Jansen, E. (1992). A high resolution diatom record of the last deglaciation from the SE Norwegian Sea: documentation of rapid climatic changes. *Paleoceanography*, **7**, 499–520.

Kocurek, G. and Fielder, G. (1982). Adhesion structures. *Journal of Sedimentary Petrology*, **52**, 1229–1241.

Koerner, R. M. and Fisher, D. A. (1979). Discontinuous flow, ice texture, and dirt content in the basal layers of the Devon Island ice cap. *Journal of Glaciology*, **23**, 209–221.

Kolstrup, E. and Jørgensen, J. B. (1982). Older and Younger coversand in southern Jutland (Denmark). *Bulletin of the Geological Society of Denmark*, **30**, 71–77.

Königsson, L. -K. (ed.) (1984). Ten years of Nordic Till research. *Striae*, **20**, 108 pp.

Kor, P. S. G., Shaw, J. and Sharpe, D. R. (1991). Erosion of bedrock by subglacial meltwater, Georgian Bay, Ontario: a regional view. *Canadian Journal of Earth Sciences*, **28**, 623–642.

Korina, N. A. and Faustova, M. A. (1964). Microfabrics of modern and old moraines. In *Soil Micromorphology* (A. Jongerius, ed.) pp. 333–338, Amsterdam: Elsevier.

Korotaj, M. and Mycielska-Dowgia_o, E. (1982). Wurmian periglacial processes on the Kolno Plateau in the light of sedimentological investigations with the use of the scanning electron microscope. *Biuletyn Periglacjalny*, **29**: 53–57.

Koshechkin, B. I. (1958). Traces of ice floe action on the seabed in shallow waters of the northern Caspian Sea. *Akademiia Nauk SSR Laboratoria Astrometodov, Trudy*, **6**, 227–234.

Koster, E. A. (1988). Ancient and modem cold-climate aeolian sand deposition: a review. *Journal of Quaternary Sciences*, **3**, 69–83.

Koster, E. A. and Dijkmans, J. W. A. (1988). Niveo-aeolian deposits and denivation forms, with special reference to the Great Kobuk Sand Dunes, Northwestern Alaska. *Earth Surface Processes and Landforms*, **13**, 153–170.

Koteff, C. (1974). The morphologic sequence concept and deglaciation of southern New England. In *Glacial Geomorphology* (D. R. Coates, ed.) pp. 121–144, SUNY at Binghamton.

Koteff, C. and Pessl, Jr., F. (1981). Systematic ice retreat in New England. United States Geological Survey Professional Paper 1179, 20 pp.

Kowalkowski, A. and Mycielska-Dowgia_o, E. (1983). Stratigraphy of fluvial and eolic deposits in valley of the Kopanica River on the basis of sedimentologic and pedologic investigation. *Geologisches Jahrbuch*, **71A**, 119–148.

Krajewski, K. (1977). Póznoplejstocenskie i Holocenskie procesy wydmotwórcze w Pradolinie Warszawsko-Berlinskiej w wid_ach Warty i Neru. *Acta Geographica Lodziensia*, **39**, 87 pp.

Krall, D. B. (1977). Late Wisconsinan ice recession in East-central New York. *Geological Society of America Bulletin*, **88**, 1697–1710.

Krapez, B. 1985. The Ventersdorp contact placer: a gold-pyrite placer of stream and debris flow origins from the Archean Witwatersrand basin of South Africa. *Sedimentology*, **32**, 223–234

Krinsley, D. H., and Donahue, J. (1968). Environmental Interpretation of sand grain surface textures by electron microscopy. *Geological Society of America Bulletin*, **79**, 743–748.

Krinsley, D. H. and Doornkamp, J. C. (1973). *Atlas of Quartz Sand Surface Textures*. Cambridge: Cambridge Univ. Press, 91 pp.

Krinsley, D. H. and Marshall, J. R. (1987). Sand grain textural analysis: an assessment. In *Clastic Particles* (J. R. Marshall, ed.) pp. 2–15, New York: Van Nostrand Reinhold.

Krinsley, D. and Takahashi, T. (1962a). An application of electron microscopy to geology. N.Y. Academy of Science Transcripts, **25**, 3–22.

Krinsley, D. and Takahashi, T. (1962b). Surface textures of sand grains – an application of electron microscopy: glaciation. *Science*, **138**, 1262–1264.

Krinsley, D. and Tovey, N. K. (1978). Cathodoluminescence in quartz sand grains. *Scanning Electron Microscopy*, **1978/I**, 887–892.

Krinsley, D. and Wellendorf, W. (1980). Wind velocities determined from the surface textures of sand grains, *Nature*, **283**, 372–373.

Kröner, A. and Rankama, K. (1972). Late Precambrian glaciogenic sedimentary rocks in southern Africa: a compilation with definitions and correlations. *Geological Society of Finland Bulletin*, **45** (1), 79–102.

Krüger, J. (1969). Landskabsformer i sydlige Sjaelland. *Geografisk Tidsskrift*, **68**, 105–212.

Krüger, J. (1979). Structures and textures in till indicating subglacial deposition. *Boreas*, **8**, 323–340.

Krüger, J. 1983. Glacial morphology and deposits in Denmark. In *Glacial Deposits in North-West Europe* (J. Ehlers, ed.) pp. 181–191, Rotterdam: A.A. Balkema Publishers.

Krüger, J. (1984). Clasts with stoss-lee form in lodgement tills: a discussion. *Journal of Glaciology*, **30**, 241–243.

Krüger, J. (1987). Relationship of drumlin shape and distribution to drumlin stratigraphy and glacial history, Myrdalsjökull, Iceland. In *Drumlin Symposium* (J. Menzies and J. Rose, eds) pp. 257–266, Rotterdam: A.A. Balkema Publishers.

Krüger, J. (1994). Glacial processes, sediments, landforms and stratigraphy in the terminus region of Myrdalsjökull, Iceland. *Folia Geographica Danica*, **21**, pp. 233.

Krüger, J. and Thomsen, H. H. (1981). Till fabric i et recent bundmoraenelandskab, Island. *Dansk Geologiska Föreningen Arssbok for 1980*, 19–28.

Krüger, J. and Thomsen, H. H. (1984). Morphology, stratigraphy, and genesis of small drumlins in front of the glacier Myrdalsjökull, south Iceland. *Journal of Glaciology*, **30**, 94–105.

Krygowski, B., Rzechowski, J. and Stankowski, W. (1969). Project of classifying glacial tills. *Bulletinn des Amis des Sciences Lettres Poznan*, Ser. V, **31**, 141–154.

Kuenen, P. H. (1960). Experimental abrasion 4: Eolian

action. *Journal of Geology*, **68**, 427–449.

Kuhle, M. (1987). Subtropical mountain- and high-land-glaciation as ice age triggers and the waning of the glacial periods in the Pleistocene. *GeoJournal*, **14**, 393–421

Kujansuu, R. (1967). On the deglaciation of western Finnish Lapland. *Bulletin de la Commission Géologique de Finlande*, **232**, 98 pp.

Kujansuu, R. and Saarnisto, M. (eds) (1987). INQUA Till Symposium, Finland 1985. *Geological Survey of Finland Special Paper*, 3, 194 pp.

Kujansuu, R. and Saarnisto, M. (eds) (1990). *Glacial Indicator Tracing*. Rotterdam: A.A. Balkema Publishers, 252 pp.

Kukla, G. J. (1975). Loess stratigraphy of central Europe. In *After the Australopithecines; Stratigraphy, Ecology, and Culture in the Middle Pleistocene* (K. W. Butzer and G. L. Isaac, eds) pp. 99–188, The Hague: Mouton Publishers.

Kukla, G., An, Z. S., Melice, J. L., Gavin, J. and Xiao, J L. (1990). Magnetic susceptibility record of Chinese loess. *Transactions of the Royal Society of Edinburgh, Earth Sciences*, **81**, 263–288.

Kulig, J. J. (1985). A sedimentation model for the deposition of glacigenic deposits in central Alberta. Unpublished MSc thesis, University of Alberta, Edmonton, 245 pp.

Kumai, M. (1977). Electron microscope analysis of aerosols in snow and deep ice cores from Greenland. *International Association of Hydrological Sciences*, **118**, 341–350

Kumai, M., Anderson, D. M. and Ugolini, F. C. (1978). Antarctic soil studies using a scanning electron microscope. *Proceedings of the Third International Conference on Permafrost*, Vol. 1, Edmonton, Canada, National Research Council of Canada, pp. 107–112.

Kupsch. W. O. (1955). Drumlins with jointed boulders near Dollard, Saskatchewan. *Geological Society of America Bulletin*, **66**, 327–338.

Kupsch, W. O. (1962). Ice-thrust ridges in Western Canada. *Journal of Geology*, **70**, 582–594.

Kurimo, H. (1974). Linear features as indicators of ice movement in the Posio-Kuusamo area, Finland. *Terra*, **86**, 52–61.

Kurimo, H. (1977). Pattern of dead-ice deglaciation forms in western Kemijärvi, northern Finland. *Fennia*, **153**, 43–56.

Kuroiwa, D. (1970). Surface topography and mineral composition of silt and sand discharged from Nigardsbreen in Norway [in Japanese]. *Low Temperature Science*, **28(A)**, 97–104.

Kurtz, D. D. and Anderson, J. B. (1979). Recognition and sedimentologic description of recent debris flow deposits from the Ross and Weddell seas, Antarctica. *Journal of Sedimentary Petrology*, **49**, 1159–1170.

Kuster, H. and Meyer, K. -D. (1979). Glaziäre Rinnen im mittleren und nodöslichen Niedersachsen. Eiszeitalter und Gegenwart, **29**, 125–156.

Kyle, P. R. and Jezek, P. A. (1978). Compositions of three tephra layers from the Byrd Station ice core, Antarctica. *Journal of Volcanology and Geothermal Research*, **4**, 225–232.

Lachenbruch, A. H. (1961). Depth and spacing of tension cracks. *Journal of Geophysical Research*, **66**, 4273–4292.

Lachenbruch, A. H. 1962. Mechanics of thermal contraction cracks and ice-wedge polygons in permafrost. *Geological Society of America, Special Publication 70*, 69 pp.

Lachenbruch, A. H. (1966). Contraction theory of ice wedge polygons: a qualitative discussion. Permafrost International Conference, Lafayette, Indiana, U.S.A., pp. 63–71.

Lafeber, D. (1964). Soil fabric and soil mechanics. In *Soil Micromorphology* (A. Jongerius, ed.) pp. 351–360, Amsterdam: Elsevier.

Lagerbäck, R. (1988). The Veiki moraines in northern Sweden – widespread evidence of an Early Weichselian deglaciation. *Boreas*, **17**, 469–486.

Lagerlund, E. 1987. An alternative Weichselian glaciation model, with special reference to the glacial history of Skåne, South Sweden. *Boreas*, **16**, 433–459.

Lagerlund, E. and Meer, J. J. M. van der. (1990). Micromorphological observations on the Lund Diamicton. Extended Abstract in Methods and problems of till stratigraphy – INQUA-88 Proceedings (E. Lagerlund, ed.) Lundqua Report 32, pp. 37–38.

Lal. D. (1988). In situ-produced cosmogenic isotopes in terrestrial rocks. *Annual Review of Earth and Planetary Sciences*, **16**, 355–388.

Lambeck, K. (1990). Glacial rebound, sea-level change and mantle viscosity. *Quarterly Journal of the Royal Astronomical Society*, **31**, 1–30.

Lambeck, K. (1991a). A model for Devensian and Flandrian glacial rebound and sea level change in Scotland. In *Glacial Isostasy, Sea-level and Mantle Rheology* (R. Sabatini, K. Lambeck and E. Boschi, eds) pp. 33–61, Dordrecht: Kluwer Academic Publishers.

Lambeck, K. (1991b). Glacial rebound and sea-level change in the British Isles. *Terra Nova*, **3**, 379–389.

Lambert, I. B., Donnelly, T. H. and Rowlands, N. J. (1980). Genesis of upper Proterozoic stratabound copper mineralization, Kapunda, South Australia. Mineralium Deposita, **15**, 1–18.

Lamothe, M. (1988). Dating till using thermoluminescence. *Quaternary Science Reviews*, **7**, 273–276.

Langohr, R. and Van Vliet, B. (1981). Properties and distribution of Vistulan permafrost traces in today's surface soils of Belgium with special reference to the data provided by soil survey. *Biuletyn Peryglacjalny*, **28**, 137–148.

LaSalle, P. (1980). L'or dans les sédiments meubles: formation des placers, extraction et occurrences dans le sud-est du Québec. Ministère de l'Énergie et des Resources du Québec, DPV–745, 26 pp.

LaSalle, P. and Shilts, W. W. (1993). Younger Dryas-age readvance of Laurentide ice into the Champlain Sea. *Boreas*, **22**, 25–37.

Lauritzen, S. E. (1991). Uranium series dating of speleothems: A glacial chronology for Nordland, Norway, for the last 600 Ka. In Late Quaternary Stratigraphy in the Nordic Countries 150,000 – 15,000 B.P. (B. G. Anderson and L. -K. Kõngsson, eds), *Striae*, **34**, 127–133.

Lauritzen, S. E. and Gascoyne, M. (1980). The first radiometric dating of Norwegian stalagmites: evidence of pre-Weichselian karst caves. *Norsk Geografisk Tidsskrift*, **34**, 77–82.

Laverdière, C. and Dionne, J. -C. (1969). Les Roches dissymetriques de l'ést du lac Saint-Jean. *Revue des Géographie de Montréal*, **23**, 358–365.

Laverdière, C., Guimont, P. and Dionne, J. -C. (1985). Les formes et les marques de l'érosion glaciaire du plancher rocheaux: signification, terminologie, illustration. *Palaeogeography, Palaeoclimatology, Palaeoecology*, **51**, 365–387.

Lavrushin, Yu. A. (1968). Features of deposition and structure of the glacial-marine deposits under conditions of a fjord coast. *Lit. Pol. Isk.*, **3**, 63–79.

Lavrushin, Yu. A. (1971). Dynamische Fazies und Subfazies der Grundmoräne. *Zeitschrift für Angewandte Geologie*, **17**, 337–343.

Lavrushin, Yu. A. (1976). Structure and development of ground moraines of continental glaciation. Academy of Sciences of the USSR Transactions, **288**, 237 pp.

Lawrence, D. B. (1950). Estimating dates of recent glacier advances and recession rates by studying tree growth layers. *American Geophysical Union Transactions*, **31**, 243–248.

Lawson, D. E. (1976). Observations on flutings at Spencer Glacier, Alaska. *Arctic and Alpine Research*, **8**, 289–296.

Lawson, D. E. (1979a). Sedimentological analysis of the western terminus region of Matanuska Glacier, Alaska. Cold Regions Research and Engineering Laboratory Report, Hanover, New Hampshire **79–9**, 122 pp.

Lawson, D. E. (1979b). A comparison of the pebble orientations in ice and deposits of the Matanuska glacier, Alaska. *Journal of Geology*, **87**, 629–645.

Lawson, D. E. (1981a). Sedimentological characteristics and classification of depositional processes and deposits in the glacial environment. Cold Regions Research and Engineering Laboratory Report, Hanover, New Hampshire 81–27.

Lawson, D. E. (1981b). Distinguishing characteristics of diamictons at the margin of the Matanuska glacier, Alaska. *Annals of Glaciology* **2**, 78–84.

Lawson, D. E. (1982). Mobilization, movement and deposition of active subaerial sediment flows, Matanuska Glacier, Alaska. *Journal of Geology*, **90**, 279–300.

Lawson, D. E. (1988). Glacigenic resedimentation: Classification concepts and application to mass-movement processes and deposits. In *Genetic Classification of Glacigenic Deposits* (R. P. Goldthwait and C. L. Matsch, eds) pp. 147–169, Rotterdam: A.A. Balkema Publishers.

Layer, P. W., Hall, C. M. and York, D. (1987). The derivation of ^{40}Ar/^{39}Ar age spectra of single grains of hornblende and biotite by laser step-heating. *Geophysical Research Letters*, **14**, 757–760.

Laymon, C. A. (1991). Marine episodes in Hudson Strait and Hudson Bay, Canada, during the Wisconsin Glaciation. *Quaternary Research*, **35**, 53–62.

Le Ribault, L. (1975a). L'exoscopie, methode et application. Notes et Memoires, 12, (Compagnie Française des Pétroles), Paris, 230 pp.

Le Ribault, L. (1975b). Application de l'exoscopie des quartz à quelques échantillons prélevés en Manche orientale. *Philosophical Transactions of Royal Society of London*, **279A**, 279–288.

Le Ribault, L. (1977). *L'Exoscopie des Quartz*. Paris: Masson, 150 pp.

Le Ribault, L. (1978). The exoscopy of quartz sand grains. In *Scanning Electron Microscopy in the Study of Sediments* (W. B. Whalley, ed.) pp. 319–328, Norwich: Geo Abstracts.

Leavy, B. D., Phillips, F. M., Elmore, D., Kubik, P. W. and Gladney, E. (1987). Measurement of cosmogenic ^{36}Cl/Cl in young volcanic rocks: an application of accelerator mass spectrometry in geochronology. *Nuclear Instruments and Methods in Physics Research*, **B29**, 246–250.

Lee, H. A. (1959). Surficial geology of southern District of Keewatin and the Keewatin Ice divide, North West Territories. *Geological Survey of Canada Bulletin*, **51**, 42 pp.

Lee, H. A., Craig, B. G. and Fyles, J. G. (1957). Keewatin Ice Divide, (abstract). *Geological Society of America*

Bulletin, **68**, 1760–1761.

Leeder, M. (1987). Sediment deformation structures and the palaeotectonic analysis of sedimentary basins, with a case-study from the Carboniferous of northern England. In *Deformation of Sediments and Sedimentary Rocks* M. E. Jones and R. M. F. Preston, eds), Geological Society Special Publication No. 29, pp. 137–146.

Legget, R. F. (1961). *Soils in Canada*. Special Publication, 3, Ottawa: Royal Society of Canada, 240 pp.

Legget, R. F. (ed.) (1965). *Soils in Canada*. Toronto: University of Toronto Press, 240 pp.

Legget, R. F. (ed.) (1976). *Glacial Till. An Inter-disciplinary Study*. The Royal Society of Canada, Special Publication No. 12, Ottawa, 412 pp.

Legigan, P. and Le Ribault, L. (1974). Evolution des quartz dans un podzol humo-ferrugineux développé sur le Sable des Landes. *Comptes Rendu de l'Academie de Sciences, Paris*, **279**, 799–802.

Lehr, J. D. and Hobbs, H. C. (1992). Glacial geology of the Laurentide Divide area, St. Louis and Lake Counties, Minnesota. *Minnesota Geological Survey Field Trip Guidebook 18*, 73 pp.

Lehr, J. D. and Matsch, C. L. (1987). The late Wisconsinan Vermilion moraine in northeastern Minnesota: An ice-marginal complex of multiple origin. *Geological Society of America Abstracts with Program, 19*, 231.

Leinen, M., Cwienk, D., Heath, G. R., Biscaye, P. E., Kolla, V., Thiede, J. and Dauphin, J. P. (1986). Distribution of biogenic silica and quartz in recent deep-sea sediments. *Geology*, **14**, 199–203.

Lemke, R. W. (1958). Narrow linear drumlins near Velva, North Dakota. *American Journal of Science*, **256**, 270–283.

Lemmen, D. S. and England, J. (1992). Multiple glaciations and sea level changes, northern Ellesmere Island, high arctic Canada. *Boreas*, **21**, 137–152.

Leonard, E. M. (1986a). Varve studies at Hector Lake, Alberta, Canada and the relationship between glacial activity and sedimentation. *Quaternary Research*. **25**, 199–214.

Leonard, E. M. (1986b). Use of lacustrine sedimentary sequences as indicators of Holocene glacial history, Banff National Park, Alberta, Canada. *Quaternary Research*, **26**, 218–231.

Leonard, E. M. (1986c). Glaciological and climatic controls on lake sedimentation, Canadian Rocky Mountains. *Zeitschrift für Gletscherkunde und Glazialgeologie*, **21**, 35–42

Lessig, H. D. (1961). The soils developed on Wisconsin and Illinoian age glacial outwash terraces along Little Beaver Creek and the adjoining upper Ohio Valley, Columbiana

County, Ohio. *Ohio Journal of Science*, **61**. 286–294.

Leverett, F. (1899). The Illinois glacial lobe. United States Geological Survey, Monograph **38**, 817 pp.

Leverett, F. and Taylor, F. B. (1915). The Pleistocene of Indiana and Michigan and the history of the Great Lakes. United States Geological Survey Monograph, **53**, 529 pp.

Levine, E. R. and Ciolkosz, E. J. (1983). Soil development in tills of various ages in northeastern Pennsylvania. *Quaternary Research*, **19**, 85–99.

Levson, V. M. (1990). Geologic settings of buried placer gold deposits in glaciated regions of the Canadian Cordillera. Geological Association of Canada – Mineralogical Association of Canada, Program with Abstracts, **15**, 76.

Levson, V. M. (1991a). Influence of geology on alluvial gold mining in central British Columbia, Canada. In *Alluvial Mining*. Published for the Institute of Mining and Metallurgy and Elsevier Applied Science, London, pp. 245–267.

Levson, V. M. (1991b). The Bullion mine; a large buried-valley gold placer. *The Cariboo Miner*, **2**, 8–9.

Levson, V. M. (1991c). Geology of Cariboo gold placers and implications for lode gold exploration. British Columbia Ministry of Energy, Mines and Petroleum Resources, Release Notification RN90–7, 5.

Levson, V. M. (1992a). Quaternary geology of the Atlin area (104N/11W, 104N/12E). In *Geological Fieldwork 1991* (B. Grant and J. M. Newell, eds), British Columbia Ministry of Energy, Mines and Petroleum Resources, Paper 1992–1, pp. 375–390.

Levson, V. M. (1992b). The sedimentology of Pleistocene deposits associated with placer gold bearing gravels in the Livingstone Creek area, Yukon Territory. In *Yukon Geology* (T. Bell, ed.) pp. 99–132, Whitehorse: Exploration and Geological Services Division.

Levson, V. M. and Giles, T. R. (1991). Stratigraphy and geologic settings of gold placers in the Cariboo mining district (93A, B, G, H). British Columbia Ministry of Energy, Mines and Petroleum Resources, Paper 1991–1, pp. 331–344.

Levson, V. M. and Giles, T. R. (1993). Geology of Tertiary and Quaternary gold-bearing placers in the Cariboo Region, British Columbia. British Columbia Ministry of Energy, Mines and Petroleum Resources Paper, 155 pp.

Levson, V. M. and Kerr, D. K. (1992). Surficial Geology and Placer Gold Settings of the Atlin – Surprise Lake area, NTS 104 N/11, 12; British Columbia Ministry of Energy, Mines and Petroleum Resources, Open File 1992–7, 1:50 000 map.

Levson, V. and Rutter, N. W. (1986). A facies approach to

the stratigraphic analysis of Late Wisconsinan sediments in the Portal Creek area, Jasper National Park, Alberta. *Géographie Physique et Quaternaire*, **40**, 129–144.

Levson, V. M. and Rutter, N. W. (1988). A lithofacies analysis and interpretation of depositional environments of montane glacial diamictons, Jasper, Alberta, Canada. In *Genetic Classification of Glacigenic Deposits* (R. P. Goldthwait and C. L. Matsch, eds) pp. 117–142.

Levson, V. and Rutter, N. W. (1989). Late Quaternary stratigraphy, sedimentology, and history of the Jasper townsite area, Alberta, Canada. *Canadian Journal of Earth Sciences*, **26**, 1325–1342.

Levson, V. M., Giles, T. R., Bobrowsky, P. T. and Matysek, P. F. (1990). Geology of placer deposits in the Cariboo mining district, British Columbia; implications for exploration (93A, B, G, H). Geological Fieldwork 1989, British Columbia Ministry of Energy, Mines and Petroleum Resources, Report No. 1990–1, pp. 519–529.

Lewis, C. F. M and Blasco, S. M. (1990). Character and distribution of sea-ice and iceberg scours. In *Ice Scouring and the Design of Offshore Pipelines* (J. I. Clark, I. Konuk, F. Poorooshasb, J. Whittick and C. Woodworth-Lynas, eds), Proceedings of an invited workshop, April 18–19th, Calgary, Alberta. Canada Oil and Gas Lands Administration and Centre for Cold Ocean Resources Engineering, pp. 57–101.

Lewis, C. F. M and Woodworth-Lynas, C. M. T. (1990). Ice scour. In Geology of the continental margin of eastern Canada (M. J. Keen and G. L. Williams, eds), Geology of Canada No. 2. *Geological Survey of Canada*, pp. 785–793.

Lewis, C. F. M., Josenhans, H. W, Simms, A., Sonnichsen, G. V. and Woodworth-Lynas, C. M. T. (1989). The role of seabed disturbance by icebergs in mixing and dispersing sediment on the Labrador shelf: a high latitude continental margin. In program with abstracts, *Canadian Continental Shelf Symposium (C2S3)*, Bedford Institute of Oceanography, Dartmouth, October pp. 2–7.

Lewis, S. G. (1992). High Lodge; stratigraphy and depositional environments. In *High Lodge: Excavations by G. de G. Sieveking 1962–68 and J. Cook 1988* (Ashton, N. M., Cook, J. Lewis, S. G. and Rose, J. eds) pp. 51–85, London: British Museum Press.

Lewis, W. V. (1954). Pressure release and glacial erosion. *Journal of Glaciology*, **2**, 417–422.

Li, J. J., Derbyshire, E. and Xu, S. Y. (1984). Glacial and paraglacial sediments of the Hunza valley, northwest Karakoram, Pakistan: a preliminary analysis. In *The International Karakoram Project 1980, 2,* (K. Miller ed.) pp. 496–535, Cambridge: Cambridge University Press.

Libby, F. (1969). Gold in the sea. *Sea Frontiers*, **15**, 232–241.

Libby, W. F. (1955). *Radiocarbon Dating*. 2nd edn, Chicago: University of Chicago Press, 175 pp.

Lill, G. O. and Smalley, I. J. (1978). Distribution of loess in Britain. *Proceedings of the Geologists' Association*, **88**, 57–65.

Lin, C. 1982. Microgeometry I. Autocorrelation and rock microstructure. *Journal of the International Association for Mathematical Geology*, **14**, 343–360.

Lin, I. J., Rohrlich, V. and Slatkine, A. (1974). Surface microtextures of heavy minerals from the Mediterranean coast of Israel. *Journal of Sedimentary Petrology*, **44**, 1281–1295.

Lindé, K. (1983). Some surface textures of experimental and natural sands of Icelandic origin. *Geografiska Annaler*, **65A**, 193–200.

Lindé, K. (1987). Experimental aeolian abrasion of different sand-size materials: some preliminary results. In *Clastic Particles* (J. R. Marshall, ed.) pp. 242–247, Van Nostrand Reinhold.

Lindé, K. and Mycielska-Dowgia__o, E. (1980). Some experimentally produced microtextures on grain surfaces of quartz sand. *Geografiska Annaler*, **62A**, 171–184.

Lindsey, D. A. (1969). Glacial sedimentology of the Precambrian Gowganda Formation, Ontario, Canada. *Geological Society of America Bulletin*, **80**, 1685–1702.

Lindström, E. (1985). The Uppsala Esker: The Åsby-Drälinge Exposures. *Striae*, **22**, 27–32.

Lingle, C. S. and Brown, T. J. (1987). A subglacial aquifer bed model and water pressure dependent basal sliding relationship for a West Antarctic ice stream. In *Dynamics of the West Antarctic Ice Sheet* (C. J. Van der Veen and J. Oerlemans, eds) pp. 249–285, Dordrecht: D. Reidel Publishing.

Linick, T. W., Damon, P. E., Donahue, D. J. and Jull, A. J. T. (1989). Accelerator Mass spectrometry; the new revolution in radiocarbon dating. *Quaternary International*, **1**, 1–6.

Link, P. K. and Gostin, V. A. (1981). Facies and paleogeography of Sturtian glacial strata (Late Precambrian), South Australia. *American Journal of Science*, **281**, 353–374.

Linton, D. L. (1963). The forms of glacial erosion. *Transactions of the Institute of British Geographers*, **33**, 1–28.

Ljunger, E. (1930). Spaltektonik und Morphologie der Schwedischen Skagerrak-Küsle, 3: Die Erosionsformen. *Bulletin of the Geological Institution of the University of Uppsala*, **21**, 255–478.

Lliboutry, L. (1968). General theory of subglacial caviation and sliding of temperate glaciers. *Journal of Glaciology*,

7, 21–58.

Lliboutry, L. (1983). Modifications to the theory of intraglacial waterways for the case of subglacial ones. *Journal of Glaciology*, **29**, 216–226.

Locke, C. W. and Locke III, W. W. (1977). Little Ice Age snow-cover extent and paleoglaciation thresholds: North-Central Baffin Island, N.W.T., Canada. *Arctic and Alpine Research*, **9**, 291–300.

Locke III, W. W. (1979). Etching of hornblende grains in Arctic soils: an indicator of relative age and paleoclimate. *Quaternary Research*, **11**, 197–212.

Locke III, W. W. (1986). Fine particle translocation in soils developed on glacial deposits, southern Baffin Island, N.W.T., Canada. *Arctic and Alpine Research*, **18**, 33–43.

Locke III, W. W., Andrews, J. T. and Webber, P. J. (1979). A manual for lichenometry. *British Geomorphological Research Group, Technical Bulletin 26*, 47 pp.

Longva, O. and Bakkejord, K. J. (1990). Iceberg deformation and erosion in soft sediments, Southeast Norway. *Marine Geology*, **92**, 87–104.

Longva, O. and Thoresen, M. K. (1991). Iceberg scours, iceberg gravity craters and current erosion marks from a gigantic Preboreal flood in southeastern Norway. *Boreas*, **20**, 47–62.

Lorius, C., Jouzel, J., Ritz, C., Merlivat, L., Barkov, N. I., Korotkevich, Y. S. and Kotlyakov, V. M. (1985). A 150,000 year climatic record from Antarctic ice. *Nature*, **316**, 591–596.

Lorius, C., Raisbeck, G., Jouzel, J. and Raynaud, D. (1989). Long-term environmental records from Antarctic ice cores. In *The Environmental Record in Glaciers and Ice Sheets* (H. Oeschger and C. C. Langway, Jr., eds) pp. 343–361, New York: John Wiley & Sons.

Lotter, A. F. (1991). Absolute dating of the Late-Glacial period in Switzerland using annually laminated sediments. *Quaternary Research*, **35**, 321–330.

Love, M. A. and Derbyshire, E. (1985). Microfabric of glacial soils and its quantitative measurement. In *Glacial Tills 85, Proceedings of the International Conference on Construction in Glacial Tills and Boulder Clays, March 1985* (M. C. Forde, ed.) pp. 129–135, Edinburgh: Engineering Technics Press.

Løvlie, R. (1989). Paleomagnetic stratigraphy: a correlation method. *Quaternary International*, **1**, 129–149.

Lowe, D. R. (1975). Water escape structures in coarse-grained sediments. *Sedimentology*, **22**, 157–204.

Lowe, D. R. and LoPiccolo, R. D. (1974). The characteristics and origins of dish and pillar structures. *Journal of Sedimentary Petrology*, **44**, 484–501.

Lowe, J. J. and Gray, J. M. (1980). The stratigraphic subdivision of the lateglacial in northwest Europe: a discussion. In *Studies in the Lateglacial of Northwest Europe* (J. J. Lowe, J. M. Gray and J. E. Robinson eds) pp 157–175, Oxford: Pergamon Press.

Lowe, J. J. and Walker, M. J. C. (1984). *Reconstructing Quaternary Environments*. London: Longman publ, 389 pp.

Lowe, J. J., Gray, J. M. and Robinson, J. E. (eds) (1980). *Studies in the Lateglacial of Northwest Europe*. Oxford: Pergamon Press.

Lowe, J. J., Lowe, S., Fowler, A. J., Hedges, R. E. M. and Austin, T. J. F. (1988). Comparison of accelerator and radiometric radiocarbon measurements obtained from late Devensian Lateglacial lake sediments from Llyn Gwernan, North Wales, UK. *Boreas*, **17**, 355–369.

Lowe, J. J., Ammann, B., Birks, H. H., Bjork, S., Coope, G. R., Cwynar, L., De Beaulieu, J. L., Mott, R. J., Peteet, D. M. and Walker, M. J. C. (1994). Climatic changes in areas adjacent to the North Atlantic during the last glacial-interglacial transition (14–9 ka BP): a Contribution to IGCP–253. *Journal of Quaternary Science*, **9**, 185–198.

Lowell, T. V. and Stuckenrath, R. (1990). Late Wisconsin advance and retreat pattern in the Miami Sublobe, Laurentide Ice Sheet. *Annals of Glaciology*, **14**, 172–175.

Lowell, T. V., Kite, J. S., Calkin, P. E. and Halter, E. F. (1990a). Analysis of small-scale erosional data and a sequence of late Pleistocene flow reversal, northern New England. *Geological Society of America Bulletin*, **102**, 74–85.

Lowell, T. V., Savage, K. M., Brockman, C. S. and Stuckenrath, R. (1990b). Radiocarbon analyses from Cincinnati, Ohio, and their implications for glacial stratigraphic Interpretations. *Quaternary Research*, **34**, 1–11.

Lowenstam, H. A. and Weiner, S. (1989). On Biomineralization. New York: Oxford University Press, 324 pp.

Luckman, B. H. (1986). Reconstruction of Little Ice Age events in the Canadian Rocky Mountains. *Géographie Physique et Quaternaire*, **40**, 17–28.

Luckman, B. H. (1988). Dating the moraines and recession of Athabasca and Dome glaciers, Alberta, Canada. *Arctic and Alpine Research*, **20**, 40–4.

Luckman, B. H. and Osborn, G. D. (1979). Holocene glacier fluctuations in the middle Canadian Rocky Mountains. *Quaternary Research*, **11**, 52–77.

Luckman, B. H., Holdsworth, G. and Osborn, G. D. (1993). Neoglacial fluctuations in the Canadian Rockies. *Quaternary Research*, **39**, 144–153.

Lundqvist, G. (1943). Norrlands jordarter. *Sveriges Geologiska Undersökning*, **C457**, 165 pp.

Lundqvist, J. (1958). Beskrivning till jordartskarta över Värmlands län. *Sveriges Geologiska Undersökning*, **Ca38**, 229 pp.

Lundqvist, J. (1969). Problems of the so-called Rogen Moraine. *Sveriges Geologiska Undersökning*, **C648**, 32 pp.

Lundqvist, J. (1970). Studies of drumlin tracks in central Sweden. *Acta Geographia Lodziensia*, **24**, 317–326.

Lundqvist, J. (1975). Ice recession in central Sweden, and the Swedish Time Scale. *Boreas*, **4**, 47–54.

Lundqvist, J. (1977). Till in Sweden. *Boreas*, **6**, 73–85.

Lundqvist, J. (1980). The deglaciation of Sweden after 10,000 B.P. *Boreas*, **9**, 229–238.

Lundqvist, J. (1981). Moraine morphology – terminological remarks and regional aspects. *Geografiska Annaler*, **63a**, 127–138.

Lundqvist, J. (1987). Till and glacial landforms in a dry, polar region. *Zeitschrift für Geomorphologie*, NF, **33**, 27–41.

Lundqvist, J. (1989). Røgen (ribbed) moraine – identification and possible origin. *Sedimentary Geology*, **62**, 281–292.

Lundqvist, J. (1990). The Younger Dryas event in Scandinavia. In Termination of the Pleistocene, IGCP Project 253, Field Conference Guide 31, (J. Lundqvist and M. Saarnisto, eds), *Geological Survey of Finland*, pp. 5–24.

Lundqvist, J., Clayton, L. and Mickelson, D. M. (1993). Deposition of the late Wisconsin Johnstown moraine, south-central Wisconsin. *Quaternary International*, **18**, 53–59.

Lunkka, J. P. (1988). Sedimentation and deformation of the North Sea Drift Formation in the Happisburgh area, North Norfolk. In *Glaciotectonics: Forms and Processes* (D. G. Croot, ed.) pp. 109–122, Rotterdam: A.A. Balkema Publishers.

Lutenegger, A. J. (1981). Stability of loess in light of inactive particle theory. *Nature*, **291**, 360.

Luttig, G., Paepe, R., West, R. G. and Zagwijn, W. H. (1969). Key to the interpretation and nomenclature of Quarternary stratigraphy. Hanover.

Ly, C. K. (1978). Grain surface features in environmental determination of late Quaternary deposits in New South Wales. *Journal of Sedimentary Petrology*, **48**, 1219–1226.

Maarleveld, G. C. (1960). Wind directions and cover sands in The Netherlands. *Biuletyn Peryglacjalny*, **8**, 49–58.

Maarleveld, G. C. (1964). Periglacial phenomena in The Netherlands during different parts of the Würm time. *Biuletyn Peryglacjalny*, **14**, 251–256.

Maarleveld, G. C. (1983). Ice-pushed ridges in the Central Netherlands. In *Glacial Deposits in North-West Europe* (J. Ehlers, ed.) pp. 393–397, Rotterdam: A.A. Balkema Publishers.

Mabbutt, J. A. (1977). *Desert Landforms*. Cambridge: MIT Press, 340 pp.

MacAyeal, D. R. (1987). Ice-shelf backpressure: form drag versus dynamic drag. In *Dynamics of the West Antarctic Ice Sheet* (C. J. Van der Veen and J. Oerlemans, eds) pp. 141–160, Dordrecht: D. Reidel Publishing.

MacAyeal, D. R. (1989a). Ice-shelf response to ice-stream discharge fluctuations: III. The effects of ice-stream imbalance on the Ross Ice Shelf, Antarctic. *Journal of Glaciology*, **35**, 38–42.

MacAyeal, D. R. (1989b). Large-scale ice flow over a viscous basal sediment: theory and application to ice stream B, Antarctica. *Journal of Geophysical Research*, **94**, 4071–4087.

McCabe, A. M. (1989). The distribution and stratigraphy of drumlins in Ireland. In *Glacial Deposits in Great Britain and Ireland* (J. Ehlers, P. L. Gibbard, and J. Rose, eds) pp. 421–436, Rotterdam: A.A. Balkema Publisher.

McCabe, A. M., and Dardis, G. F. (1989). A geological view of drumlins in Ireland. *Quaternary Science Reviews*, **8**, 169–177.

McCabe, A. M. and Eyles, N. (1988). Sedimentology of an ice-contact glaciomarine delta, Carey Valley, Northern Ireland. *Sedimentary Geology*, **59**, 1–14.

McCarroll, D. (1989). Potential and limitations of the Schmidt hammer for relative-age dating: field tests on Neoglacial moraines, Jotunheimen, southern Norway. *Arctic and Alpine Research*, **21**, 268–275.

McCarthy, D. P., Luckman, B. H. and Kelly, P. E. (1991). Sampling height-age error correction for spruce seedlings in glacial forelands, Canadian Cordillera. *Arctic and Alpine Research*, **23**: 451–455.

MacClintock, P. and Dreimanis, A. (1964). Reorientation of till fabric by overriding glacier in the St. Lawrence Valley. *American Journal of Science*, **262**, 133–142.

McConnell, R. G. (1905). Report on the Klondike gold fields. *Geological Survey of Canada, Annual Report*, **14(B)**, 1–71.

McConnell, R. G. (1907). Report on gold values in the Klondike high level gravels. *Geological Survey of Canada, Publication 979*, 34 pp.

McCoy, W. D. (1987a). The precision of amino acid geochronology and paleothermometry. *Quaternary Science Reviews*, **6**, 43–54.

McCoy, W. D. (1987b). Quaternary aminostratigraphy of

the Bonneville Basin, western United States. *Geological Society of America Bulletin,* **98**, 99–112.

Macdonald, I. F., Kaufmann, P. and Dullien, F. A. L. (1986). Quantitative image analysis of finite porous media. I. Development of genus and pore map software. *Journal of Microscopy,* **144**, 277–296.

McGee, W. J. (1883). Glacial cañons. *Science,* **2**, 315–316.

McGee, W. J. (1894). Glacial cañons. *Journal of Geology,* **2**, 350–364.

McGown, A. (1985). Construction problems associated with the tills of Strathclyde. In *Glacial Tills 85. Proceedings of the International Conference on Construction in Glacial Tills and Boulder Clays, March 1985* (M. C. Forde, ed.) pp. 177–186, Edinburgh Scotland: Engineering Technics Press.

McGown, A. and Derbyshire, E. (1977). Genetic influences on the properties of till. *Quarterly Journal of Engineering Geology,* **10**, 389–410.

McGown, A. and Radwan, A. M. (1975). The presence and influence of fissures in the boulder clays of west central Scotland. *Canadian Geotechnical Journal,* **12**, 84–97.

McGown, A., Radwan, A. M. and Gabr, A. W. A. (1977). Laboratory testing of fissures and laminated soils. *Proceedings of the 9th International Conference on Soil Mechanics and Foundation Engineering,* **1**, 205–210.

McGown, A., Salvidar-Sali, A. and Radwan, A. M. (1974). Fissure patterns and slope failures in till at Hurlford, Ayrshire. *Quarterly Journal of Engineering Geology,* **7**, 1–26.

MacKay, B. R. (1921). Beauceville Map-area, Quebec. *Geological Survey of Canada, Memoir 127,* 105 pp.

Mackay, J. R. and Mathews, W. H. (1964). The role of permafrost in ice-thrusting. *Journal of Geology,* **72**, 378–380.

McKee, T. R. and Brown, J. L. (1977). Preparation of specimens for electron microscopic examination. In *Minerals in Soil Environments* (J. B. Dixon ed.) pp. 809–846, Madison: Soil Science Society of America.

McKenna, D. P. and Follmer, L. R. (1990). Farmdale and Sangamon soils at the Wempleton southeast section. In *Quaternary Records of Central and Northern Illinois* (L. R. Follmer, D. P. McKenna and J. E. King, eds) *Guidebook 20, Illinois State Geological Survey,* 33–41.

McKenna-Neuman, C. and Gilbert, R. (1986). Aeolian processes and landforms in glaciofluvial environments of southeastern Baffin Island, N.W.T. Canada. In *Aeolian Geomorphology* (W. G. Nickling, ed.) pp. 213–235, Boston: Allen & Unwin.

Mackiewicz, N. E., Powell, R. D., Carlson, P. R. and Molnia, B. F. (1984). Interlaminated ice-proximal glaci-

marine sediments in Muir Inlet, Alaska. *Marine Geology,* **57**, 113–147.

McKinlay D. G., McGown, A., Radwan, A. and Hossain, D. (1978). Representative sampling and testing in fissured lodgement tills. *Symposium on Engineering Behaviour of Glacial Materials, Birmingham,* pp. 129–140.

McPherson, H. J. and Gardner, J. S. (1969). The development of glacial landforms in the vicinity of the Saskatchewan Glacier. *Canadian Alpine Journal,* **52**, 90–96.

Madgett, P. A. 1975. Re-interpretation of Devensian till stratigraphy in eastern England. *Nature,* **253**, 105–107.

Madgett, P. A. and Catt, J. A. (1978). Petrography, stratigraphy and weathering of Late Pleistocene tills in East Yorkshire, Lincolnshire and north Norfolk. *Proceedings of the Yorkshire Geological Society,* **42**, 55–108.

Magny, M. (1993). Solar influences on Holocene climatic changes illustrated by correlations between past lake-level fluctuations and the atmospheric ^{14}C record. *Quaternary Research,* **40**, 1–9.

Mahaney. W. C. (ed.) (1984). *Quaternary Dating Methods. Developments in Palaeontology and Stratigraphy, 7,* Amsterdam: Elsevier Science Publishers, 431 pp.

Mahaney. W. C., Vortisch, W. B. and Julig, P. (1988a). Relative differences between glacially-crushed quartz transported by mountain and continental ice: some examples from North America and East Africa. *American Journal of Science,* **288**, 810–826.

Mahaney. W. C., Vortisch, W. B. and Spence, J. R. (1988b). Pollen study and scanning electron microscopy of aeolian grains in a compound paleosol in the Mutonga drainage, Mount Kenya. *Journal of African Earth Sciences,* **7**, 895–902.

Maher, B. A. and Taylor, R. M. (1988). Formation of ultrafine-grained magnetite in soils. *Nature,* **336**, 368–370.

Maher, B. A. and Thompson, R. (1991). Mineral magnetic record of the Chinese loess and paleosols. *Geology,* **19**, 3–5.

Maher, B. A. and Thompson, R. (1992). Paleoclimatic significance of the mineral magnetic record of the Chinese loess and paleosols. *Quaternary Research,* **37**, 155–170.

Malcolm, W. and Faribault, E. R. (1929). Gold fields of Nova Scotia. *Geological Survey of Canada, Memoir 156,* 253 pp.

Maltman, A. (1987). Shear zones in argillaceous sediments – an experimental study. In *Deformation of Sediments and Sedimentary Rocks* (M. E. Jones and R. M. F. Preston, eds) pp. 77–87, Geological Society Special Publication No. 29.

Maltman, A. J. (1988). The importance of shear zones in

naturally deformed wet sediments. *Tectonophysics*, **145**, 163–175.

Mandy, J. T. (1934). Gold bearing black-sand deposits of Graham Island, Queen Charlotte Islands. *Transactions of the Canadian Institute of Mining and Metallurgy*, **37**, 563–572.

Manecki, A., Muszynski, M. and Wrzak, J. (1980). Fine-grained deposits from the bottom of Broggi Glacier and its foreland. *Prace Mineralogiczne*, **64**, 27–45.

Mangerud, J. (1972). Radiocarbon dating of marine shells, including a discussion of apparent age of Recent shells from Norway. *Boreas*, **1**, 143–172.

Mangerud, J., Andersen, S. T., Berglund, B. E. and Donner, J. J. (1974). Quaternary stratigraphy in Norden, a proposal for terminology and classification. *Boreas*, **3**, 109–127.

Mangerud, J., Lie, S. E., Furnes, H., Kristiansen, I. L. and Lømo, L. (1984). A Younger Dryas ash bed in western Norway, and its possible corelations with tephra in cores from the Norwegian Sea and the North Atlantic. *Quaternary Research*, **21**, 85–104.

Manker, J. P. and Ponder, R. D. (1978). Quartz grain surface features from fluvial environments of northeastern Georgia. *Journal of Sedimentary Petrology*, **48**, 1227–1232.

Mankinen, E. A. and Dalrymple, G. B. (1979). Revised geomagnetic polarity time scale for the interval 0–5 m.y. B.P. *Journal of Geophysical Research*, **84**, 615–626.

Mann, A. W. (1984). Mobility of gold and silver in lateritic weathering profiles: some observations from Western Australia. *Economic Geology*, **79**, 38–49.

Mannerfelt, C. M. (1945). Några glacialmorfologiska Formelement. *Geografiska Annaler*, **27**, 1–239.

Mannerfelt, C. M. (1949). Marginal drainage channels as indicators of the gradients of Quaternary ice caps. *Geografiska Annaler*, **31**, 194–199.

Marcussen, I. (1973). Studies on flow till in Denmark. *Boreas*, **2**, 213–231.

Marcussen, I. (1975). Distinguishing between lodgement till and flow till in Weichselian deposits. *Boreas*, **4**, 113–123.

Margolis, S. V. and Kennett, J. P. (1971). Cenozoic paleoglacial history of Antarctica recorded in subantarctic deep-sea cores. *American Journal of Science*, **271**, 1–36.

Margolis, S. V. and Krinsley, D. H. (1974). Processes of formation and environmental occurrence of microfeatures on detrital quartz grains. *American Journal of Science*, **274**, 449–464.

Markgren, M. and Lassila, M. (1980). Problems of moraine morphology: Rogen moraine and Blattnick moraine. *Boreas*, **9**, 271–274.

Marmo, J. S. and Ojakangas, R. W. (1984). Lower Proterozoic glaciogenic deposits, eastern Finland. *Geological Society of America Bulletin*, **95**, 1055–1062.

Marshall, J. R. (ed.) (1987). *Clastic Particles*. New York: Van Nostrand Reinhold, 346 pp.

Martin, H. (1965). *The Precambrian geology of South West Africa and Namaqualand*. Precambrian Research Unit, University of Cape Town, 159 pp.

Martin, J. H. (1990). Glacial-interglacial CO_2 change: the iron hypothesis. *Paleoceanography*, **5**: 1–13.

Martinson, D. G., Pisias, N. G., Hays, J. D., Imbrie, J., Moore, T. C. and Shackleton, N. J. (1987). Age dating and the orbital theory of the ice ages: development of a high-resolution 0 to 300, 000-year chronostratigraphy. *Quaternary Research*, **27**, 1–29.

Mathews, W. H. (1974). Surface profiles of the Laurentide ice sheet in its marginal areas. *Journal of Glaciology*, **13**, 37–43.

Matthes, F. E. (1930). Geological history of Yosemite Valley. *United States Geological Survey Professional Paper, 160*, 137 pp.

Matthews, J. A. (1985). Radiocarbon dating of surface and buried soils: principles, problems and prospects. In *Geomorphology and Soils* (K. S. Richards, R. R. Arnett and S. Ellis, eds) pp. 269–288, London: George Allen and Unwin.

Matthews, J. A. (1992). *The Ecology of Recently-Deglaciated Terrain*. Cambridge: Cambridge University Press, 386 pp.

Maurice, Y. T. (1986). Distribution and origin of alluvial gold in southwest Gaspésie, Québec. *Geological Survey of Canada, Paper 86–1B*, pp. 785–795.

Maurice, Y. T. (1988). Regional alluvial heavy mineral geochemistry as a prospecting method in glaciated Appalachian terrain: a case history from the southern Quebec placer-gold belt. In *Prospecting in Areas of Glaciated Terrain – 1988* (D. R. MacDonald, ed.) pp. 185–203, Halifax: Canadian Institute of Mining and Metallurgy.

May, R. W. (1980). The formation and significance of irregularly shaped quartz grains in till. *Sedimentology*, **27**, 325–331.

Mazzullo, J. and Anderson, J. B. (1987). Grain shape and surface texture analysis of till and glacial-marine sand grains from the Weddell and Ross Seas, Antarctica. In *Clastic Particles* (J. R. Marshall, ed.) pp. 314–327, New York: Van Nostrand Reinhold.

Mazzullo, J. and Magenheimer, S. (1987). The original shapes of quartz sand grains. *Journal of Sedimentary Petrology*, **57**, 479–487.

Mazzullo, J., Sims, D. and Cunningham, D. (1986). The

effects of eolian sorting and abrasion upon the shapes of fine quartz sand grains. *Journal of Sedimentary Petrology*, **56**, 45–56.

Meer, J. J. M. van der (1982). The Fribourg area, Switzerland: a study in Quaternary geology and soil development. PhD thesis, University of Amsterdam. Amsterdam: Publicaties van het Fysisch Geografisch en Bodemkundig Laboratorium 32, 203 pp.

Meer, J. J. M. van der (1983). A recent drumlin with fluted surface in the Swiss Alps. In *Tills and Related Deposits: Genesis, Petrology, Application, Stratigraphy* (E. B. Evenson, Ch. Schlüchter and J. Rabassa eds) pp. 105–110, Rotterdam: A.A. Balkema Publishers.

Meer, J. J. M. van der (1987a). Micromorphology of glacial sediments as a tool in genetic varieties of till. In INQUA Till Symposium, Finland 1985 (R. Kujansuu and M. Saarnisto, eds), *Geological Survey of Finland Special Paper, 3*, pp. 77–89.

Meer, J. J. M. van der (ed.) (1987b). *Tills and Glaciotectonics*. Rotterdam: A.A. Balkema Publishers, 270 pp.

Meer, J. J. M. van der (1987c). Field trip 'Tills and end moraines in The Netherlands and NW Germany'. In *Tills and Glaciotectonics* (J. J. M. van der Meer, ed.) pp. 261–268, Rotterdam: A.A. Balkema Publishers.

Meer, J. J. M. van der (1992). Micromorphology of Pleistocene sediments from the southern North Sea; Boreholes BH 89.2 and BH 89.3. *Fysisch Geografisch Bodemkundig Laboratorium*. University of Amsterdam, 82 pp.

Meer, J. J. M. van der (1993). Microscopic evidence of subglacial deformation. *Quaternary Science Reviews* **12**, 553–587.

Meer, J. J. M. van der and Laban, C. (1990). Micromorphology of some North Sea till samples, a pilot study. *Journal of Quaternary Science*, **5**, 95–101.

Meer, J. J. M. van der, Muecher, H. J. and Hoefle H. Ch. (1994). Observations in some thin sections of Antarctic glacial deposits. *Polarforschung, 62*, 57–65.

Meer, J. J. M. van der Rappol, M. and Semeijn, J. N (1983). Micromorphological and preliminary X-ray observations on a basal till from Lunteren, The Netherlands. *Acta Geologica Hispanica*, **18**, 199–205.

Meer, J. J. M. van der Rappol, M. and Semeijn, J. N (1985). Sedimentology and genesis of glacial deposits in the Goudsberg, Central Netherlands. *Mededelingen van de Rijks Geologische Dienst*, **39**, 1–29.

Meer, J. J. M. van der Rabassa, J. O. and Evenson, E. B. (1992). Micromorphological aspects of glaciolacustrine sediments in northern Patagonia, Argentina. *Journal of Quaternary Sciences*, **7**, 31–44.

Meer, J. J. M. van der Verbers, A. L. L. M. and Warren, W.

P. (1994). The micromorphological character of the Ballycroneen Formation (Irish Sea Till): A first assessment. In *Formation and Deformation of Glacial Deposits* (W. P. Warren and D. G. Croot, eds) pp. 39–49, Rotterdam: A.A. Balkema Publishers.

Melhorn, W. N. and Kempton, J. P. (1991). Geology and hydrogeology of the Teays-Mahomet Bedrock Valley system. *Geological Society of America, Special Paper 258*, 128 pp.

Mellor, A. (1985). Soil chronosequences on Neoglacial moraine ridges, Jostedalsbreen and Jotunheimen, southern Norway: a quantitative pedogenicapproach. In *Geomorphology and Soils* (K. S. Richards, R. R. Arnett and S. Ellis, eds) pp. 288–308, London: George Allen & Unwin.

Mellor, A. (1986). Textural and scanning electron microscope observations of some arctic-alpine soils developed in Weichselian and Neoglacial till deposits in southern Norway. *Arctic and Alpine Research*, **18**, 327–336.

Menzies, J. (1979a). The mechanics of drumlin formation with particular reference to the change in pore-water content of the till. *Journal of Glaciology*, **22**, 373–384.

Menzies, J. (1979b). A review of the literature on the formation and location of drumlins. *Earth Science Reviews*, **14**, 315–359.

Menzies, J. (1981). Investigations into the Quaternary deposits and bedrock topography of central Glasgow. *Scottish Journal of Geology*, **17**, 155–168.

Menzies, J. (1984). *Drumlins: A Bibliography*. Norwich: Geobooks, 117 pp.

Menzies, J. (1986). Inverse-graded units within till in drumlins near Caledonia, southern Ontario. *Canadian Journal of Science*, **23**, 774–86.

Menzies, J. (1987). Towards a general hypothesis on the formation of drumlins. In *Drumlin Symposium* (J. Menzies and J. Rose, eds) pp. 9–24, Rotterdam: A.A. Balkema Publishers.

Menzies, J. (1989a). Subglacial hydraulic conditions and their possible impact upon subglacial bed formation. *Sedimentary Geology*, **62**, 125–150.

Menzies, J. (1989b). Drumlins – Products of Controlled or Uncontrolled Glaciodynamic Response? *Quaternary Science Reviews*, **8**, 151–158.

Menzies, J. (1990a). Brecciated diamictons from Mohawk Bay. S. Ontario, Canada. *Sedimentology*, **37**, 481–493.

Menzies, J. (1990b). Evidence of cryostatic dessication processes associated with sand, intraclasts, within diamictons, southern Ontario, Canada. *Canadian Journal of Earth Sciences*, **27**, 684–693.

Menzies, J. (1990c). Sand Intraclasts within a diamicton mélánge, southern Niagara Peninsula, Ontario, Canada. *Journal of Quaternary Science*, **5**, 189–206.

Menzies, J. (ed.) (1995). *Modern Glacial Environments – Processes, Dynamics and Sediments*. Oxford: Butterworth-Heineman.

Menzies, J. and Habbe, K. A.. (1992). A cryogenic wedge within gravels, north of Kempten, Bavaria, F.R.G. *Zeitschrift für Geomorphologie*, **36**, 365–374.

Menzies, J. and Maltman, A. J. (1992). Microstructures in diamictons – evidence of subglacial bed conditions. *Geomorphology*, **6**, 27–40.

Menzies, J. and Rose, J. (eds) (1987). *Drumlin Symposium*. Rotterdam: A.A. Balkema Publishers, 360 pp.

Menzies, J. and Rose, J. (eds) (1989). Subglacial bedforms – drumlins, Rogen moraine and associated subglacial bedforms. *Sedimentary Geology*, **62**, 117–407.

Menzies, J. and Woodward, J. (1993). Preliminary study of subglacial diamicton microstructures as reflected in drumlin sediments at Chimney Bluffs, New York. In *Glaciotectonics and Mapping Glacial Deposits. Proceedings of the INQUA Commission on Formation and Properties of Glacial Deposits, vol. 1* (J. S. Aber, ed.) pp. 36–45, Regina: Canadian Plains Research Centre, University of Regina.

Menzies, J. and Suppiah, A. (in press). *Structural Discontinuities in Glacial Tills: A Review*.

Meyer, K. -D. (1987). Ground and end moraines in Lower Saxony. In *Tills and Glaciotectonics* (J. J. M. van der Meer, ed.) pp. 197–204, Rotterdam: A.A. Balkema Publishers.

Miall, A. D. (1983). Glaciomarine sedimentation in the Gowganda Formation (Huronian), northern Ontario. *Journal of Sedimentary Petrology*, **53**, 477–491.

Miall, A. D. (1984). *Principles of Sedimentary Basin Analysis*. Berlin: Springer-Verlag, 490 pp.

Mickelson, D. M. (1971). Glacial geology of the Burroughs Glacier area, southeastern Alaska. Ohio State University Institute of Polar Studies, Report No. 40.

Mickelson, D. M. (1973). Nature and rate of basal till deposition in a stagnating ice mass, Burroughs Glacier, Alaska. *Arctic and Alpine Research*, **5**, 7–27.

Mickelson, D. M. and Clayton, L. (1981). Subglacial conditions and processes during Middle Woodfordian time in Wisconsin. Geological Society of America. North Central Section. Abstracts with Programs, p. 30.

Mickelson, D. M., Clayton, L. Fullerton, D. S. and Borns, H. W. Jr. (1983). The late Wisconsin glacial record of the Laurentide ice sheet in the United States. In *The Late Pleistocene. Late Quaternary Environments of the United States, vol. I* (S. C. Porter, ed.) pp. 3–37, Minneapolis: University of Minnesota.

Middleton, G. V. (1973). Johannes Walther's law of correlation of facies. *Geological Society of America*

Bulletin, **84**, 979–988.

Middleton, G. V. and Davis, P. M. (1979). Surface textures and rounding of quartz sand grains in intertidal sandbars, Bay of Fundy, Nova Scotia. *Canadian Journal of Earth Sciences*, **16**, 2071–2085.

Midttun, L. (1985). Formation of dense bottom water in the Barents Sea. *Deep Sea Research*, **32**, 1233–1241.

Miller, B. B., McCoy, W. D. and Bleuer, N. K. (1987). Stratigraphic potential of amino acid ratios in Pleistocene terrestrial gastropods: an example from West-Central Indiana, USA. *Boreas*, **16**, 133–138.

Miller, G. H. (1985). Aminostratigraphy of Baffin Island shell-bearing deposits. In *Quaternary Environments: Eastern Canadian Arctic, Baffin Bay, and Western Greenland* (J. T. Andrews, ed.) pp. 394–427, Boston: Allen and Unwin.

Miller, G. H and Andrews, J. T. (1973). Quaternary history of northern Cumberland Peninsula, east Baffin Island, Northwest Territories, Canada. Part VI: Preliminary lichen growth curve for Rhizocarpon Geographicum. *Geological Society of America Bulletin*, **83**, 1133–1138.

Miller, G. H. and Brigham-Grette, J. (1989). Amino acid geochronology: resolution and precision in carbonate fossils. *Quaternary International*, **1**, 111–128.

Miller, G. H. and Mangerud, J. (1985). Aminostratigraphy of European marine interglacial deposits. *Quaternary Science Reviews*, **4**, 215–278.

Miller, G. H., Sejrup, H. P., Lehman, S. J. and Forman, S. L. (1989). Glacial history and marine environmental change during the last interglacial-glacial cycle, western Spitsbergen, Svalbard. *Boreas*, **18**, 273–296

Miller, J. M. G. (1989). Glacial advance and retreat sequences in a Permo-Carboniferous section, central Transantarctic Mountains. *Sedimentology*, **36**, 419–430.

Miller, J. W. Jr. (1972). Variations in New York drumlins. *Annals of the Association of American Geographers*, **62**, 418–423.

Miller, K. G., Fairbanks, R. G. and Mountain, G. S. (1987). Tertiary oxygen isotope synthesis, sea level history, and continental margin erosion. *Paleoceanography*, **2**, 1–19.

Milligan, V. (1976). Geotechnical aspects of glacial tills. In *Glacial Till: An Inter-disciplinary Study* (R. F. Legget, ed.) pp. 269–291, The Royal Society of Canada, Special Publication No. 12, Ottawa.

Mills, H. H. (1980). An analysis of drumlin form in the northeastern and north-central United States. *Geological Society of America Bulletin*, **91**, 2214–2289.

Mills, H. H. (1987). Morphometry of drumlins in the northeastern and north-central USA. In *Drumlin Symposium* (J. Menzies and J. Rose, eds) pp. 131–148, Rotterdam: A.A. Balkema Publishers.

Mills, P. C. (1983). Genesis and diagnostic value of soft-sediment deformation structures – a review. *Sedimentary Geology*, **35**, 83–104.

Minervin, A. V. (1984). Cryogenic processes in loess formation in central Asia. In *Late Quaternary Environments of the Soviet Union* (A. A. Velichko, ed.) pp. 133–140, London: Longman.

Minell, H. (1979). The genesis of tills in different moraine types and the deglaciation in a part of central Lapland. *Sveriges Geologiska Undersökning*, **C754**, 83 pp.

Mirkin, C. R., Osipov, V. I., Romm, E. S., Sokolov, V. N. and Tolkachev, M. D. (1978). The use of the scanning electron microscope for the investigation of the properties of porous bodies. In *Scanning Electron Microscopy in the Study of Sediments* (W. B. Whalley, ed.) pp. 13–16, Norwich: Geo Abstracts.

Mitchell, G. F., Penny, L. F., Shotton, F. W. and West, R. G. 1973. A correlation of Quaternary deposits in the British Isles. *Geological Society of London Special Report*, **4**, 99 pp.

Mitterer, R. M. and Kriausakul, N. (1989). Calculation of amino acid racemization ages based on apparent parabolic kinetics. *Quaternary Science Reviews*, **8**, 353–357.

Mix, A. C. and Ruddiman, W. R. (1984). Oxygen isotope analysis and Pleistocene ice volumes. *Quaternary Research*, **21**, 1–20.

Möbus, G. and Peterss, K. (1983). Kluftstatische Spannungsanalyse im Geschiebemergel, Analysis of glacitectonic structures. 4th Glacitectonic Symposium. Wyzsza Szko_a Inzynierska, pp. 139–153.

Mollard, J. D. (1983). The origin of reticulate and orbicular patterns on the floor of the Lake Agassiz basin. *Geological Association of Canada Special Paper*, 26, pp. 355–374.

Moller, P. (1987). Moraine morphology, till genesis, and deglaciation pattern in the Asnen Area, South-central Småland, Sweden. LUNDQUA Thesis 20, 146 pp.

Molnia, B. F. (1983). *Glacial-Marine Sedimentation*. New York: Plenum Press, 844 pp.

Moncrieff, A. C. M. and Hambrey, M. J. (1988). Late Precambrian glacially-related grooved and striated surfaces in the tillite group of Central East Greenland. *Palaeogeography, Palaeoclimatology, Palaeoecology*, **65**, 183–200.

Moncrieff, A. C. M. and Hambrey, M. J. (1990). Marginal-marine glacial sedimentation in the late Precambrian succession of East Greenland. In *Glaciomarine Environments: Processes and Sediments* (J. A. Dowdeswell and J. D. Scourse, eds) pp. 387–410, Geological Society of London Special Publication No. 53.

Montes, A. S. L., Gravenor, C. P. and Montes, M. L. (1985). Glacial sedimentation in the Late Precambrian Bebedourc Formation, Bahia, Brazil. *Sedimentary Geology*, **44**, 349–358.

Mooers. H. D. (1989a). On the formation of tunnel valleys of the Superior lobe, central Minnesota. *Quaternary Research*, **32**, 24–35.

Mooers. H. D. (1989b). Drumlin formation: a time transgressive model. *Boreas*, **18**, 99–107.

Moran, S. R. (1971). Glacitectonic structures in drift. In *Till, a Symposium* (R. P. Goldthwait, ed.) pp. 127–148, Ohio State University Press.

Moran, S. R., Clayton, L., Hooke, R. LeB., Fenton, M. M. and Andriashek, L. D. (1980). Glacier-bed landforms of the prairie region of North America. *Journal of Glaciology*, **25**, 457–476.

Morawski, W. (1984). Osady wodnomorenowe (Waterlaid sediments). *Prace Inst. Geol. Warszawa*, **108**, 74 pp.

Morawski, W. (1985). Pleistocene glaciogenic sediments of the watermorainic facies. *Quaternary Studies in Poland*, **6**, 99–115.

Morawski, W. (1988). Watermorainic sediments: origin and classification. In *Genetic Classification of Glacigenic Deposits* (R. P. Goldthwait and C. L. Matsch, eds) pp. 143–144, Rotterdam: A.A. Balkema Publishers.

Morel, F. and Irving, E. (1978). Tentative paleocontinental maps for the early Phanerozoic and Proterozoic. *Journal of Geology*, **86**, 535–561.

Morgan, A. V. (1973). The Pleistocene geology of the area north and west of Wolverhampton, Staffordshire, England. *Philosophical Transactions of the Royal Society of London*, **265B**, 233–297.

Morison, S. R. (1983a). Surficial geology of Clear Creek drainage basin, Yukon Territory (NTS sheets 115 P, 11, 12, 13, 14). Exploration and Geological Services Division, Northern Affairs Program, Whitehorse, Yukon, Open File Release 1983–2, map scale 1:50,000.

Morison, S. R. 1983b. A sedimentologic description of Clear Creek fluviatile sediments 115P, central Yukon. In *Yukon Exploration and Geology 1982*. Exploration and Geological Services Northern Affairs Program, Indian and Northern Affairs Canada, Whitehorse, pp. 50–54.

Morison, S. R. (1985a). Placer deposits of Clear Creek drainage basin 115P, central Yukon. In *Yukon Exploration and Geology 1983*. Exploration and Geological Services Northern Affairs Program, Indian and Northern Affairs Canada, Whitehorse, pp. 88–93.

Morison, S. R. (1985b). Sedimentology of White Channel Placer Deposits, Klondike Area, West-central Yukon. Unpublished MSc. thesis, University of Alberta, Edmonton. 149 pp.

Morison, S. R. (1989). Placer deposits in Canada. In *Quaternary Geology of Canada and Greenland* (R. J. Fulton, ed.) pp. 687–697. Geology of Canada No. 1. Geological Survey of Canada.

Morison, S. R. and Hein, F. J. (1987). Sedimentology of the White Channel gravels, Klondike area, Yukon Territory: fluvial deposits of a confined valley. In *Recent Developments in Fluvial Sedimentology* (F. G. Ethridge, R. M. Flores and M. D. Harveys, eds) pp. 205–216, Society of Economic Paleontologists and Mineralogists, Special Publication 39.

Morley, J. J. (1983). Identification of density-stratified waters in the late Pleistocene North Atlantic: A faunal derivation. *Quaternary Research*, **20**, 374–386.

Mörner, N. -A. (1976). Eustasy and geoid changes. *Journal of Geology*, **84**, 123–151.

Morris, E. M. and Morland, L. W. (1976). A theoretical analysis of the formation of glacial flutes. *Journal of Glaciology*, **17**, 311–323.

Morris, J. D. (1991). Applications of cosmogenic 10Be to problems in the earth sciences. *Annual Review of Earth and Planetary Sciences*, **19**, 313–350.

Morris, T. H., Clark, D. and Blasco, S. M. (1985). Sediments of the Lomonosov Ridge and Makarov Basin: A Pliocene stratigraphy for the North Pole. *Geological Society of America Bulletin*, **96**, 901–910.

Morrison, R. B. (1965). Means of Time-Stratigraphic division and long-distance correlation of Quaternary successions. In *Means of Correlation of Quaternary Successions* (R. B. Morrison and H. E. Wright Jr., eds) pp. 1–113, University of Utah Press.

Morteani, G. (1991). Grain size distribution of placer gold in the Rio Kaka area, Bolivia – its bearing on gravimetric concentration processes. In *International Symposium on Alluvial Gold Placers; Abstracts*. (G. Hérail, ed.) pp. 129–130, La Paz, Bolivia.

Mortlock, R. A., Charles, C. D., Froelich, P. N., Zibello, M. A.,Saltzman, J., Hays, J. D. and Burckle, L. H. (1991). Evidence for lower productivity in the Antarctic Ocean during the last glaciation. *Nature*, **351**, 220–223.

Moss, A. J. (1966). Origin, shaping and significance of quartz sand grains. *Journal of the Geological Society of Australia*, **13**, 97–136.

Moss, A. J. and Green, P. (1975). Sand and silt grains: predetermination of their formation and properties by microfractures in quartz. *Journal of the Geological Society of Australia*, **22**, 485–495.

Moss, A. J., Walker, P. H. and Hutka, J. (1973). Fragmentation of granitic quartz in water. *Sedimentology*, **20**, 489–511.

Moss, V. A. (1988). Image processing and image analysis. *Proceedings of the Royal Microscopical Society*, **23**, 83–88.

Mücher, H. J. and Vreeken, W. J. (1981). (Re)deposition of loess in southern Limbourg, The Netherlands: 2. Micromorphology of the Lower Silt Loam complex and comparison with deposits produced under laboratory conditions. *Earth Surface Processes and Landforms*, **6**, 355–363.

Muhs, D. R., Rosholt, J. N. and Bush, C. A. (1989). The uranium-trend dating method: principles and application for southern California marine terrace deposits. *Quaternary International*, **1**, 19–34.

Muller, E. H. (1963). Geology of Chautauqua County, New York – Part 2, Pleistocene geology. *New York State Museum and Scientific Service Bulletin 392*.

Muller, E. H. (1977). Late glacial and early postglacial environments in western New York. *Annals of the New York Academy of Sciences*, **288**, 223–233.

Muller, E. H. (1983a). Dewatering during lodgement of till. In *Tills and Related Deposits: Genesis, Petrology, Application, Stratigraphy* (E. B. Evenson, Ch. Schlüchter and J. Rabassa, eds) pp. 13–18, Rotterdam: A.A. Balkema Publishers.

Muller, E. H. (1983b). Till genesis and the glacier sole. In *Tills and Related Deposits: Genesis, Petrology, Application, Stratigraphy.* (E. B. Evenson, Ch. Schlüchter and J. Rabassa, eds) pp. 19–22, Rotterdam: A.A. Balkema Publishers.

Murphy, C. P. (1986). *Thin Section Preparation of Soils and Sediments*. Berkhamsted: AB Academic, 149 pp.

Murphy, J. A. (1982). Considerations, materials, and procedures for specimen mounting prior to scanning electron microscopic examination. *Scanning Electron Microscopy*, 1982, **II**, 657–696.

Murray, T. and Dowdeswell, J. A. (1992). Water throughflow and the physical effects of deformation on sedimentary glacier beds. *Journal of Geophysical Research*, **97B**, 8993–9002.

Mustard, P. S. and Donaldson, J. A. (1987). Early Proterozoic ice-proximal deposition: The lower Gowganda Formation at Cobalt, Ontario, Canada. *Geological Society of America Bulletin*, **98**, 373–387.

Mycielska-Dowgia__o, E. (1974). The micro relief of quartz grain surfaces from fluvial and dune sands. *Annales de la Societé Géologique de Polande*, **44**, 2–3.

Mycielska-Dowgia__o, E. (1978). A scanning electron microscope study of quartz grain surface textures from boulder clays of North and Central Poland. In *Scanning Electron Microscopy in the Study of Sediments* (W. B. Whalley, ed.) pp. 243–248, Norwich: Geo Abstracts.

Myhre, A. M. and Eldholm, O. (1988). The western

Svalbard margin ($74°-80\mathring{N}$). *Marine and Petroleum Geology*, **5**, 134–156.

Naeser, C. W. and Naeser, N. D. (1988). Fission-track dating of Quaternary events. In *Dating Quaternary Sediments* (D. J. Easterbrook, ed.) pp. 1–11, Geological Society of America Special Paper, 227.

Naeser, C. W., Izett, G. A. and Obradovich, J. D. (1980). Fission-track and K–Ar ages of natural glasses. *United States Geological Survey Bulletin*, **1489**, 39 pp.

Naeser, C. W., Briggs, N. D., Obradovich, J. D. and Izett, G. A. (1981). Geochronology of Quaternary tephra deposits. In *Tephra Studies* (S. Self and R. S. J. Sparks, eds.) pp. 13–47, NATO Advanced Studies Institute Series. C. Dordrecht: D. Reidel Publishing Company.

Naeser, N. D. and Naeser, C. W. (1984). Fission-track dating. In *Quaternary Dating Methods: Developments in Palaeontology and Stratigraphy, 7* (W. C. Mahaney, ed.) pp. 87–100, New York: Elsevier.

Naeser, N. D., Westgate, J. A., Hughes, O. L. and Péwé, T. L. (1982). Fission-track ages of late Cenozoic distal tephra beds in the Yukon Territory and Alaska. *Canadian Journal of Earth Sciences*, **19**, 2167–2178.

Nahon, D. and Trompette, R. (1982). Origin of siltstones: glacial grinding versus weathering. *Sedimentology*, **29**, 25–35.

Nakada, M. and Lambeck, K. (1988). The melting history of the late Pleistocene Antarctic ice sheet. *Nature*, **333**, 36–40.

Nakawo, M. (1979). Supraglacial debris of G2 glacier in Hidden Valley, Mukut Himal, Nepal. *Journal of Glaciology*, **22**, 273–283.

Nance, R. D., Worsley, T. R. and Moody, J. B. (1986). Post-Archean biogeochemical cycles and long-term episodicity in tectonic processes. *Geology*, **14**, 514–518.

Nance, R. D., Worsley, T. R. and Moody, J. B. (1988). The Supercontinent cycle. *Scientific American*, **259**, 72–79.

Nelson, C. H. and Hopkins, D. M. (1972). Sedimentary processes and distribution of particulate gold in the northern Bering Sea. *United States Geological Survey*, Professional Paper 689, 27 pp.

Nelson, R. E., Carter, L. D. and Robinson, S. W. (1988). Anomalous radiocarbon ages from a Holocene detrital organic lens in Alaska and their implications for radiocarbon dating and paleoenvironmental reconstructions in the Arctic. *Quaternary Research*, **29**, 66–71.

Nesbitt, H. W. and Young, G. M. (1982). Early Proterozoic climates and plate motions inferred from major element chemistry of lutites. *Nature*, **299**, 715–717.

Nesbitt, H. W. and Young, G. M. (1984). Prediction of some weathering trends of plutonic and volcanic rocks based on thermodynamic and kinetic considerations. *Geochi-*

mica et Cosmochimica Acta, **48**, 1523–1534.

Nesje, A. and Whillans, I. M. (1994). Erosion of Sognefjord, Norway. *Geomorphology*, **9**, 33–45.

Newbury, D. E., Joy, D. C., Echlin, P., Fiori, C. E. and Goldstein, J. I. (1986). Advanced Scanning Electron Microscopy and X-Ray Microanalysis. New York: Plenum, 466 pp.

Nichols, R. L. (1966). Geomorphology of Antarctica. In *Antarctic Soils and Soil Forming Processes* (J. C. F. Tedrow, ed.) pp. 1–46, American Geophysical Union, Antarctic Research Series, 8.

Nicholson, R. (1963). A note on the relation of rock fracture and fjord direction. *Geografiska Annaler*, **45**, 303–304.

Nickling, W. G. (1978). Eolian sediment transport during dust storms: Slims River Valley, Yukon Territory. *Canadian Journal of Earth Sciences*, **15**, 1069–1084.

Nielsen, P. E. (1983). The lithology and genesis of the Danish tills. In *Glacial Deposits in North-West Europe* (J. Ehlers, ed.) pp. 193–196, Rotterdam: A.A. Balkema Publishers.

Niessen, F., Lister, F. and Giovanoli, F. (1992). Dust transport and palaeoclimate during the Oldest Dryas in Central Europe – implications from varves (Lake Constance). *Climate Dynamics* (in press).

Nikkarinen, M., Kallio, E., Lestinen, P. and Äyräs, M. (1984). Mode of occurrence of Cu and Zn in till over three mineralized areas in Finland. In *Geochemical Exploration – 1983* (A. J. Björklund, ed.) *Journal of Geochemical Exploration*, **21**, pp. 239–247.

Nilsen, T. H. (1973). The relation of joint patterns to the formation of fjords in Western Norway. *Norsk Geologisk Tidsskrift*, **53**, 183–194.

Nilsson, T. (1983). *The Pleistocene; Geology and Life in the Quaternary Ice Age*. Dordrecht: D. Reidel, 651 pp.

Nishiizumi, K., Elmore, D. and Kubik, P. W. (1989a). Update on terrestrial ages of Antarctic Meteorites. *Earth and Planetary Science Letters*, **93**, 299–313.

Nishiizumi, K., Winterer, E. L., Kohl, C. P., Klein, J., Middleton, R., Lal, D. and Arnold, J. R. (1989b). Cosmic ray production rates of 10Be and 26Al in quartz from glacially polished rocks. *Journal of Geophysical Research*, **94**, 17907–17915.

Nobles, L. H. and Weertman, J. (1971). Influence of irregularities of the bed of an ice sheet on deposition rate of till. In *Till: A Symposium* (R. P. Goldthwait, ed.) pp. 117–126, Columbus: Ohio State University Press.

Nordstrom, C. E. and Margolis, S. V. (1972). Sedimentary history of Central California Shelf sands as revealed by scanning electron microscopy. *Journal of Sedimentary Petrology*, **42**, 527–536.

North American Commission on Stratigraphic Nomenclature (NACSN), (1983). North American Stratigraphic Code. *American Association of Petroleum Geologists Bulletin*, **67**, 841–875.

Nye, J. (1965). The flow of a glacier in a channel of rectangular, elliptic or parabolic cross-section. *Journal of Glaciology*, **5**, 661–690.

Nystuen J. P. (1985). Facies and preservation of glacigenic sequences from the Varanger Ice Age in Scandinavia and other parts of the North Atlantic Region. *Palaeogeography, Palaeoclimatology, Palaeoecology*, **51**, 209–229.

Oberbeck, V. R., Marshall, J. R. and Aggarawal, H. (1993). Impacts, tillites and the breakup of Gondwanaland. *Journal of Geology*, **101**, 1–19.

Oches, E. A. (1990). Aminostratigraphy of the Peoria Loess, Mississippi Valley, U.S.A., with Late Quaternary paleotemperature estimates. Unpublished MSc. thesis, University of Massachusetts, Amherst, 193 pp.

Ojakangas, R. W. and Matsch, C. L. (1980). Upper Precambrian (Eocambrian) Mineral Fork Tillite of Utah: a continental glacial and glaciomarine sequence. *Geological Society of America Bulletin*, **91**, 495–501.

Ojakangas, R. W. and Matsch, C. L. (1982). Upper Precambrian (Eocambrian) Mineral Fork Tillite of Utah: a continental glacial and glaciomarine sequence: reply. *Geological Society of America Bulletin*, **93**, 186–187.

Okko, V. (1955). Glacial drift in Iceland, its origin and morphology. *Bulletin Commission de Géologique de Finlande*, **170**, 1–133.

Oldale, R. N., Valentine, P. C., Cronin, T. M., Spiker, E. C., Blackwelder, B. W., Belknap, D. F., Wehmiller, J. F. and Szabo, B. J. (1982). Stratigraphy, structure, absolute age, and paleontology of the upper Pleistocene deposits at Sankaty Head, Nantucket Island, Massachusetts. *Geology*, **10**, 246–252.

Olsson, G. (1965). Subakvatisk åsbildning, med exempel från sydöstra Dalarna. Lic. avh. Uppsala University Naturgeografisk Institute, 120 pp.

Olsson, I. U. (1974). Some problems in connection with the evaluation of 14C dates. *Geologiska Föreningen i Stockholm Förhandlingar*, **96**, 311–320.

Olsson, T. (1977). Ground water in till soils. *Striae*, **4**, 13–16.

Orford, J. D. and Whalley, W. B. (1983). The use of the fractal dimension to quantify the morphology of irregular-shaped particles. *Sedimentology*, **30**, 655–668.

Orford, J. D. and Whalley, W. B. (1987). The quantitative description of highly irregular sedimentary particles: the use of the fractal dimension. In *Clastic Particles* (J. R. Marshall, ed.) pp. 267–280, New York: Van Nostrand Reinhold.

Orford, J. D. and Whalley, W. B. (1991). Quantitative grain form analysis. In Principles, *Methods and Application of Particle Size Analysis* (J. P. M. Syvitski, ed.) pp. 88–108, Cambridge: Cambridge University Press.

Orheim, O. and Elverhøi, A. (1981). Model for submarine glacial deposition. *Annals of Glaciology*, **2**, 123–128.

Orr, E. D. and Folk, R. L. (1983). New scents on the chattermark trail: weathering enhances obscure microfractures. *Journal of Sedimentary Petrology*, **53**, 121–129.

Orr, E. D. and Folk, R. L. (1985). Discussion: Chattermarked garnets found in soil profiles and beach environments. *Sedimentology*, **32**, 307–308.

Osborn, G. and Karlstrom, E. T. (1989). Holocene moraine and paleosol stratigraphy, Bugaboo Glacier, British Columbia. *Boreas*, **18**, 311–322.

Osipov, V. I. and Sokolov, V. N. (1978). Microstructure of recent clay sediments examined by scanning electron microscopy. In *Scanning Electron Microscopy in the Study of Sediments* (W. B. Whalley, ed.) pp. 29–40, Norwich: GeoAbstracts.

Osterman, L. E. and Andrews, J. T. (1983). Changes in glacial-marine sedimentation in core HU 77–159, Frobisher Bay, Baffin Island, NWT: a record of proximal, distal and ice-rafting glacial-marine environments. In *Glacial-Marine Sedimentation* (B. F. Molnia, ed.) pp. 451–493, New York: Plenum Press.

Ostry, R. C. and Deane, R. E. (1963). Microfabric analyses of till. *Geological Society of America Bulletin*, **74**, 165–168.

O'Sullivan, P. E. (1983). Annually-laminated lake sediments and the study of Quaternary environmental changes – a review. *Quaternary Science Reviews*, **1**, 245–313.

Ovenshine, A. T. (1970). Observations of iceberg rafting in Glacier Bay, Alaska, and the identification of ancient ice-rafted deposits. *Geological Society of America Bulletin*, **81**, 891–894.

Owen, G. (1987). Deformation processes in unconsolidated sands. In *Deformation of Sediments and Sedimentary Rocks* (M. E. Jones and R. M. F. Preston, eds), Geological Society of London Special Publication No. 29, pp. 11–24.

Owen, L. A. (1988a). Terraces, Uplift and Climate, Karakoram Mountains, Northern Pakistan. Unpublished PhD thesis, University of Leicester, Leicester, UK, 399 pp.

Owen, L. A. (1988b). Wet-sediment deformation of Quaternary and Recent sediments in the Skardu Basin, Karakoram Mountains, Pakistan. In *Glaciotectonics: Forms and Processes* (D. G. Croot, ed.) pp. 123–147, Rotterdam: A.A. Balkema Publishers.

Owen, L. A. and Derbyshire, E. (1988). Glacially deformed diamictons in the Karakoram Mountains, northern Pakistan. In *Glaciotectonics: Forms and Processes*. (D. G. Croot, ed.) pp. 149–176, Rotterdam: A.A. Balkema Publishers.

Owen, T., Cess, R. D. and Ramanathan, V. (1979). Enhanced CO2 greenhouse to compensate for reduced solar luminosity on early Earth. *Nature*, **277**, 640–642.

Palais, J. M. (1985). Particle morphology, composition and associated ice chemistry of tephra layers in the Byrd Ice Core: evidence for hydrovolcanic eruptions. *Annals of Glaciology*, **7**, 42–48.

Palmer, A. S. (1982). Stratigraphy and selected properties of loess in Wairarapa, New Zealand. Unpublished PhD thesis, Victoria University, Wellington, New Zealand.

Parizek, R. R. (1964). Geology of the Willow Bunch area, Saskatchewan. Saskatchewn Research Council Geology Division, Report SRC 4.

Parizek, R. R. (1969). Glacial ice-contact rings and ridges. *Geological Society of America*, Special Paper 123, 49–102.

Parker, M. L., Jozsa, L. A., Johnson, S. G. and Bramhall, P. A. (1984). Tree-ring dating in Canada and the northwestern U.S. In *Quaternary Dating Methods. Developments in Palaeontology and Stratigraphy, 7* (W. C. Mahaney, ed.) pp. 211–226, Amsterdam: Elsevier.

Parkin, G. W. and Hicock, S. R. (1988). Sedimentology of a Pleistocene glacigenic diamicton sequence near Campbell River, Vancouver Island, British Columbia. In *Genetic Classification of Glacigenic Deposits* (R. P. Goldthwait and C. L. Matsch, eds) pp. 97–116, Rotterdam: A.A. Balkema Publishers.

Parsons, R. B., Weisel, C. J. Logan, G. H. and Nettleton, W. D. (1981). The soil sequence of Late Pleistocene glacial outwash terraces from Spokane floods in the Idaho Panhandle. *Journal Soil Science Society America*, **45**, 925–930.

Pasierbski, M. (1979). Remarks on the genesis of subglacial channels in northern Poland. *Eiszeitalter und Gegenwart*, **29**, 189–200.

Paterson, W. S. B. (1981). *The Physics of Glaciers*, 2nd ed, Oxford: Pergamon Press, 380 pp.

Patterson, C. J. (1994). Tunnel-valley fans of the St. Croix moraine, east-central Minnesota, USA. In *Formation and Deformation of Glacial Deposits* (W. P. Warren and D. C. Croot, eds) pp. 69–88, Rotterdam: A.A. Balkema Publishers.

Paul, M. A. (1981). *Soil Mechanics in Quaternary Science*. Cambridge: Quaternary Research Association.

Paul, M. A. and Evans, H. (1974). Observations on the internal structure and origin of some flutes in glaciofluvial sediments, Blomstrandbreen, north-west Spitsbergen. *Journal of Glaciology*, **13**, 393–400.

Paul, M. A. and Eyles, N. (1990). Constraints on the preservation of diamict facies (melt-out tills) at the margins of stagnant glaciers. *Quaternary Science Reviews*, **9**, 51–69.

Peach, B. N. and Horne, J. (1880). The glaciation of the Orkney Islands. *Quarterly Journal of the Geological Society*, **36**, 648–663.

Peach, B. N. and Horne, J. (1930). *Chapters in the Geology of Scotland*. Oxford University Press, London. 21 pp.

Peck, B. J., Kaufman, D. S. and Calkin, P. E. (1990). Relative dating of moraines using moraine morphometric and boulder weathering criteria, Kigluaik Mountains, Alaska. *Boreas*, **19**, 227–239.

Pedersen, S. A. S. (1993). The glaciodynamic event and glaciodynamic sequence. In *Glaciotectonics and Mapping Glacial Deposits* (J. S. Aber, ed.), Proc. INQUA Comm. Formation and Properties of Glacial Deposits. Canadian Plains Research Centre, University of Regina, Regina, pp. 67–85.

Pedersen, S. A. S. and Petersen, K. S. (1988). Sand-filled frost wedges in glaciotectonically deformed mo-clay on the island of Fur, Denmark. In *Glaciotectonics: Forms and Processes* (D. G. Croot, ed.) pp. 185–190, Rotterdam: A.A. Balkema Publishers.

Peltier, W. R. (1982). Dynamics of the Ice Age earth. In *Advances in Geophysics* (B. Saltzman, ed.), New York: Academic Press, 24, pp. 1–146.

Peltier, W. R. (1987a). Glacial isostasy, mantle viscosity, and Pleistocene climatic change. In *North America and Adjacent Oceans During the Last Deglaciation* (W. F. Ruddiman and H. E. Wright Jr., eds) pp. 155–182, The Geology of North America, Vol. K–3. Boulder, Colorado: Geological Society of America.

Peltier, W. R. (1987b). Mechanisms of relative sea-level change and the geophysical responses to ice-water loading. In *Sea Surface Studies: A Global View* (R. J. N. Devoy, ed.) pp. 57–94 London, New York and Sydney: Croom Helm.

Peltier, W. R. (1991). The ICE–3G model of late Pleistocene deglaciation: construction, verification and applications. In *Glacial Isostasy, Sea-Level and Mantle Rheology* (R. Sabatini, K. Lambeck and E. Boschi, eds) pp. 95–119, Dordrecht: Kluwer Academic Publishers.

Peltier, W. R. and Andrews, J. T. (1976). Glacial-isostatic adjustment. I. The forward problem. *Geophysical Journal of the Royal Astronomical Society*, **46**, 605–646.

Peltier, W. R. and Andrews, J. T. (1983). Glacial geology

and glacial isostasy of the Hudson Bay region. In *Shorelines and Isostasy* (D. E. Smith and A. G. Dawson, eds) pp. 285–319, Institute of British Geographers, Special Publication, 16. London: Academic Press.

Peng, T. H. (1989). Changes in ocean ventilation rates over the last 7,000 years based on 14C variations in the atmosphere and oceans. *Radiocarbon*, **31**, 481–492.

Penck, A. and Brückner, E. (1909). *Die Alpen im Eiszeitalter.* Leipzig: Tauchnitz, Bd. 1–3, 1199 pp.

Penny, L. F. and Catt, J. A. (1967). Stone orientation and other structural features of tills in East Yorkshire. *Geological Magazine*, **104**, 344–360.

Penny, L. F., Coope, G. R. and Catt, J. A. (1969). Age and insect fauna of the Dimlington Silts, East Yorkshire. *Nature*, **224**, 65–67.

Perkins, J. A. and Sims, J. D. (1983). Correlation of Alaskan varve thickness with climatic parameters, and use in paleoclimatic reconstruction. *Quaternary Research*, **20**, 308–321.

Perrin, R. M. S., Rose, J. and Davies, H. (1979). The distribution, variation and origins of Pre-Devensian tills in eastern England. *Philosophical Transactions of the Royal Society of London*, **287B**, 535–570.

Perttunen, M. and Hirvas, H. (1982). An attempt to use the roundness of quartz grains for till stratigraphy. *Bulletin of the Geological Society of Finland*, **54**, 25–33.

Peteet, D. (ed.) (1993). Global Younger Dryas? *Quaternary Science Reviews*, **12**, 277–355.

Pettijohn, F. J. (1975). *Sedimentary Rocks.* 3rd edn, New York: Harper and Row Publishing, 628 pp.

Peuraniemi, V. (1984). Weathering of sulphide minerals in till in some mineralized areas in Finland. *Prospecting in Areas of Glaciated Terrain: 1984*; London; Institution of Mining and Metallurgy, pp. 127–135.

Peuraniemi, V. (1987). Interpretation of heavy mineral geochemical results from till. In *INQUA Till Symposium, Finland, 1985* (R. Kujansuu and M. Saarnisto, eds), Geological Survey of Finland Special Paper, 3, pp. 169–179.

Péwé, T. L. (1955). Origin of the upland silt near Fairbanks, Alaska. *Geological Society of America Bulletin*, **66**, 699–724.

Pfirman, S. and Solheim, A. (1989). Subglacial meltwater discharge in the open marine tidewater glacier environment: observations from Nordaustlandet, Svalbard Archipelago. *Marine Geology*, **86**, 265–281.

Pfirman, S., Lange, M. A., Wollenburg, I. and Schlosser, P. (1990). Sea ice characteristics and the role of sediment inclusions in deep-sea deposition: Arctic-Antarctic comparisons. In *Geological History of the Polar Oceans: Arctic versus Antarctic* (U. Bleil and J. Thiede, eds),

NATO ASI Series, C 308, Dordrecht: Kluwer Academic Publishers, pp. 187–211.

Phillips, F. M., Leavy, B. D., Jannik, N. O., Elmore, D. and Kubik, P. W. (1986). The accumulation of cosmogenic chlorine-36 in rocks: a method for surface exposure dating. *Science*, **231**, 41–43.

Phillips, F. M., Zreda, M. G., Smith S. S., Elmore, D., Kubik, P. W. and Sharma, P. (1990). Cosmogenic Chlorine-36 chronology for glacial deposits at Bloody Canyon, eastern Sierra Nevada. *Science*, **248**, 1529–1532.

Pierce, K. L. (1979). History and dynamics of glaciation in the northern Yellowstone Park area. United States Geological Survey Professional Paper 729–F, 90 pp.

Pierce, K. L., Obradovich J. D. and Friedman, I. (1976). Obsidian hydration dating and correlation of Bull Lake and Pinedale glaciations near West Yellowstone, Montana. *Geological Society of America Bulletin*, **87**, 703–710.

Piotrowski, J. A. (1991). Quartär- und hydrogeologische Untersuchungen im Bereich der Born höveder Seenkette, Schleswig-Holstein. Reports Geologisch-Paläontologisches Institut und Museum, No. 43, 194 pp.

Piotrowski, J. A. (1994). Tunnel-valley formation in northwest Germany – geology, mechanisms of formation and subglacial bed conditions for the Bornnhöved tunnel valley. *Sedimentary Geology*, **89**, 107–141.

Piotrowski, J. A. and Smalley, I. J. (1987). The Woodstock drumlin field, southern Ontario, Canada. In *Drumlin Symposium* (J. Menzies and J. Rose ed.) pp. 309–322, Rotterdam: A.A. Balkema Publishers.

Piper, D. J. W. (1991). Seabed geology of the Canadian eastern continental shelf. *Continental Shelf Research*, **11**, 1013–1035.

Piper, D. J. W., Mudie, P. J., Fader, G. B. Josenhans, H. W.MacLean, B. and Vilks, G. (1990). Quaternary geology (Chapter 11). In *Geology of the Continental Margin of Eastern Canada* (M. J. Keen and G. L. Williams, eds), Geology of Canada No. 2. *Geological Survey of Canada*, pp. 473–607.

Pissart, A. (1966). Le rôle géomorphologique du vent dans la région de Mould Bay (Ile Prince Patrick – N.W.T. – Canada). *Zeitschrift für Geomorphologie*, **10**, 226–236.

Piteau, D. R. (1970). Geological factors significant to the stability of slopes cut in rock. Symposium on Planning of Open Pit Mines, Johannesburg, pp. 33–53.

Piteau, D. R. (1973). Characterising and extrapolating rock joint properties in engineering practice. Rock Mechanics, Supplement No. 2, pp. 5–31.

Plafker, G. and Addicott, W. O. (1976). Glaciomarine deposits of Miocene through Holocene age in the

Yakataga Formation along the Gulf of Alaska margin Alaska. In *Recent and Ancient Sedimentary Environments in Alaska, Symposium Proceedings* (T. P. Miller. ed.) pp. 1–22, Anchorage: Alaska Geological Society.

Plafker, G., Bartsch-Winkler, S. and Ovenshine, A. T (1977). Paleoglacial implications of coarse detritus in DSDP Leg 36 Cores. *Initial Reports of the Deep Sea Drilling Project*, **36**, 857–864.

Pons, A., Guiot, J., De Beaulieu, J. L. and Reille, M. (1992) Recent contributions to the climatology of the last Glacial–Interglacial Cycle based on French poller sequences. *Quaternary Science Reviews*, **11**, 439–448.

Poorooshasb, F., Clark, J. I. and Woodworth-Lynas, C. M. T (1989). Small scale modelling of iceberg scouring of the seabed. In *Tenth International Conference on Port and Ocean Engineering under Arctic Conditions (POAC '89), Lulea, Sweden*. Vol. 1, pp. 133–145.

Porter, S. C. (1975). Weathering rinds as a relative-age criterion: applications to subdivision of glacial deposits in the Cascade Range. *Geology*, **3**, 101–104.

Porter, S. C. (1976). Pleistocene glaciation in the southern part of the North Cascade Range, Washington. *Geological Society America Bulletin*, **87**, 61–75

Porter, S. C. (1981a). Recent glacier variations and volcanic eruptions. *Nature*, **291**, 139–142.

Porter, S. C. (1981b). Lichenometric studies in the Cascade Range of Washington: establishment of Rhizocarpon Geographicum growth curves at Mt. Rainier. *Arctic and Alpine Research*, **13**, 11–23.

Porter, S. C. (1981c). Glaciological evidence of Holocene climatic change. In *Climate and History: Studies in Past Climates and their Impact on Man* (T. M. L. Wigley, M. J. Ingram and G. Farmer, eds) pp. 82–110, Cambridge: Cambridge University Press.

Porter, S. C. (1981d). Use of tephrochronology in the Quaternary geology of the United States. In *Tephra Studies* (S. Self and R. S. J. Sparks, eds) pp. 135–160, Dordrecht: D. Reidel.

Porter, S. C. (ed.) (1983). *Late Quaternary Environments of the United States, Volume 1. – The Late Pleistocene*. Minneapolis: University of Minnesota, 407 pp.

Porter, S. C. (1991). Volcanic records and loess deposits. In *Global Changes of the Past* (R. S. Bradley, ed.) pp. 295–320, Boulder, Colorado: University Corporation for Atmospheric Research/Office for Interdisciplinary Earth Studies.

Porter, S. C., Pierce, K. L. and Hamilton, T. D. (1983). Late Wisconsinan mountain glaciation in the western United States. In *Late Quaternary Environments of the United States, Volume 1* (S. C. Porter, ed.) pp. 71–111, Minneapolis: University of Minnesota.

Poser, H. (1932). Einige Untersuchungen zur Morphologie Ostgrönlands. *Meddelelser om Grønland*, **94**, 55 pp.

Poser, H. (1948). Aolische Ablagerungen und Klima des Spätglazials in Mittel-und Westeuropa. Die Naturwissenschaften, 35, 269–276 and 307–312.

Poser, H. (1950). Zur Rekonstruktion der Spätglazialen Luftdruckverhaltnisse in Mittel und Westeuropa auf Grund der Vorzeitlichen Binrendünen. *Erdkunde*, **4**, 81–88.

Powell, R. D. (1981a). A model for sedimentation by tidewater glaciers. *Annals of Glaciology*, **2**, 129–134.

Powell, R. D. (1981b). Sedimentation conditions in Taylor Valley, Antarctica, inferred from textural analysis of DVDP cores. *American Geophysical Union Antarctic Research Series* **33**, 331–349.

Powell R. D. (1983). Submarine flow tills at Victoria, British Columbia: discussion. *Canadian Journal of Earth Sciences*, **20**, 509–510.

Powell, R. D. (1984). Glacimarine processes and inductive lithofacies modelling of ice shelf and tidewater glacier sediments based on Quaternary examples. *Marine Geology*, **57**, 1–52.

Powell, R. D. (1988). Processes and facies of temperate and subpolar glaciers with tidewater fronts. Short Course Notes, Geological Society of America, Denver, 114 pp.

Powell, R. D. (1990). Glacimarine processes at grounding-line fans and their growth to ice-contact deltas. In *Glaciomarine Environments: Processes and Sediments*. (J. A. Dowdeswell and J. D. Scourse, eds) Geological Society of London, Special Publication No. 53, pp. 53–73.

Powell, R. D. (1991). Grounding-line systems as second-order controls on fluctuations of tidewater termini of temperate glaciers. In *Glacial Marine Sedimentation; Palaeoclimatic Significance* (J. B. Anderson and G. M. Ashley, eds), Geological Society of America Special Paper, 251, pp. 75–93.

Powell, R. D. and Gostin, V. A. (1990). A glacially-influenced, storm-dominated continental shelf system on the Permian Australian-Gondwana margin. Abstract of Papers, 13th International Sedimentological Congress, Nottingham, UK, August 26–31, pp. 435–436.

Powell, R. D. and Molnia, B. F. (1989). Glacimarine sedimentary processes, facies and morphology of the south-southeast Alaska shelf and fjords. *Marine Geology*, **85**, 359–390.

Prebble, M. M. (1967). Cavernous weathering in the Taylor dry valley, Victoria Land, Antarctica. *Nature*, **216**, 1194–1195.

Preece, S. J., Westgate, J. A. and Gorton, M. P. (1992). Compositional variation and provenance of Late Cenozoic distal tephra beds, Fairbanks area, Alaska. In *Tephrachronology: Stratigraphic Applications of Tephra* (J. A. Westgate, R. C. Walter and N. Naeser, eds) Quaternary International, 13/14, 97–102.

Preiss, W. V. (1987). The Adelaide geosyncline – Late Proterozoic stratigraphy, sedimentation, palaeontology and tectonics. *Bulletin of the Geological Survey of South Australia*, **53**, 438 pp.

Prest, V. K. (1949). The Pleistocene geology of the Vermillion River system near Capreol, district of Sudbury, Ontario. Ontario Department of Mines, Preliminary Report 1949–2, 8 pp.

Prest, V. K. (1984). The late Wisconsinan glacier complex. In *Quaternary Stratigraphy – a Canadian Contribution to IGCP Project 24*. (R. J. Fulton, ed.) Geological Survey of Canada, Paper 84–80, pp. 21–36.

Prest, V. K. and Nielsen, E. (1987). The Laurentide Ice Sheet and long-distance transport. In *INQUA Till Symposium, Finland 1985* (R. Kujansuu and M. Saarnisto, eds), Geological Survey of Finland Special Paper, 3, pp. 91–101.

Prest, V. K., Grant, D. R. and Rampton, V. N. (1969). Glacial map of Canada. Geological Survey of Canada Map 1253–A

Price, R. J. (1960). Glacial meltwater channels in the upper Tweed drainage basin. *Geographical Journal*, **126**, 483–489.

Price, R. J. (1963). A glacial meltwater drainage system in Peebleshire, Scotland. *Scottish Geographical Magazine*, **79**, 133–141.

Price, R. J. (1969). Moraines, sandar, kames and eskers near Brei_amerkurjokull, Iceland. *Transactions of the Institute of British Geographers*, **46**, 17–43.

Price, R. J. (1973). Glacial and Fluvioglacial Landforms. London: Longman, 242pp.

Pronk, A. G., Bobrowsky, P. T. and Parkhill, M. A. (1989). An interpretation of late Quaternary glacial flow indicators in the Baie des Chaleurs region, northern New Brunswick. *Géographie Physique et Quaternaire*, **43**, 179–190.

Prudic, D. E. (1982). Hydraulic conductivity of a fine-grained till, Cattaraugus County, New York. *Ground Water*, **20**, 194–204.

Punkari, M. (1980). The ice lobes of the Scandinavian Ice Sheet during the deglaciation in Finland. *Boreas*, **9**, 307–310.

Punkari, M. (1982). Glacial geomorphology and dynamics in the eastern parts of the Baltic Shield interpreted using Landsat imagery. *Photogrammetric Journal of Finland*, **9**, 77–93.

Punkari, M. (1984). The relations between glacial dynamics and tills in the eastern parts of the Baltic Shield. In Ten years of Nordic Till research (L. -K. Königsson, ed.), *Striae*, **20**, 49–54.

Pye, K. (1984). Loess. *Progress in Physical Geography*, **8**, 176–217.

Pye, K. and Krinsley, D. (1983). Mudrocks examined by backscattered electron microscopy. *Nature*, **301**, 412–413.

Rabassa, J. (1987). Drumlins and drumlinoid forms in northern James Ross island, Antarctic Peninsula. In *Drumlin Symposium* (J. Menzies and J. Rose eds) pp. 267–290, Rotterdam: A.A. Balkema Publishers.

Rabassa, J. and Clapperton, C. M. (1990). Quaternary glaciations in the southern Andes. In *Quaternary Glaciations in the Southern Hemisphere* (C. M. Clapperton, ed.), *Quaternary Science Reviews*, **9**, 153–174.

Rahmani, R. A. (1973). Grain surface etching features of some heavy minerals. *Journal of Sedimentary Petrology*, **43**, 882–888.

Raicevic, D. and Cabri, L. J. (1976). Mineralogy and concentration of Au- and Pt-bearing placers from the Tulameen River area in British Columbia. *Canadian Institute of Mining and Metallurgy Bulletin*, **69**, 111–119.

Rainbird, R. H., Heaman, L. M. and Young, G. (1992). Sampling Laurentia: detrital zircon geochronology offers evidence for an extensive Neoproterozoic river system originating from Grenville orogen. *Geology*, **20**, 351–354.

Rains, B., Shaw, J., Skoye, R., Sjogren, D. and Kvill, D. (1993). Late Wisconsin subglacial megaflood paths in Alberta. *Geology*, **21**, 323–326.

Ramm, M. (1989). Late Quaternary carbonate sedimentation and paleoceanography in the eastern Norwegian Sea. *Boreas*, **18**, 255–272.

Randall, B. A. O. (1961). On the relationship of valley and fjord direction to the fracture pattern of Lyngen, Troms, N. Norway. *Geografiska Annaler*, **43**, 336–338.

Ranson, C. E. (1967). An assessment of the glacial deposits of north-east Norfolk. *Bulletin of the Geological Society of Norfolk*, **16**, 1–16.

Rao, C. P. (1981). Criteria for recognition of cold-water carbonate sedimentation: Berriedale Limestone (Lower Permian), Tasmania, Australia. *Journal of Sedimentary Petrology*, **51**, 491–506.

Rappol, M. (1983). Glacigenic properties of till. Studies in glacial sedimentology from the Allgäu Alps and The Netherlands. PhD thesis, University of Amsterdam, Publicaties van het Fysisch Geografisch en Bodemkundig Laboratorium 34, Amsterdam, 225 pp.

Rappol, M. (1985). Clast-fabric strength in tills and debris flows compared for different environments. *Geologie en Mijnbouw*, **64**, 327–332.

Rappol, M. (1986). Aspects of ice flow patterns, glacial sediments, and statigraphy in northwest New Brunswick. In *Current Research, Part B, Geological Survey of Canada*, Paper 86–1B, 223–237.

Rappol, M. (1987). Saalian till in The Netherlands: a review. In *Tills and Glaciotectonics* (J. J. M. van de Meer, ed.) pp. 3–21, Rotterdam: A.A. Balkema Publishers.

Rappol, M. (1989). Glacial history and stratigraphy of northwestern New Brunswick. *Géographie Physique & Quaternaire*, **43**, 191–206.

Rappol, M. (1993). Ice flow and glacial transport in lower St. Lawrence, Quebec. *Geological Survey of Canada Paper 90–19*, 28 pp.

Rappol, M. and Stoltenberg, H. M. P. (1985). Compositional variability of Saalian till in The Netherlands and its origin. *Boreas*, **14**, 33–50.

Rappol, M., Haldorsen, S., Jørgensen, P., van der Meer, J. J M. and Stoltenberg, H. M. P. (1989). Composition and origin of petrographically stratified thick till in the northern Netherlands and a Saalian glaciation model for the North Sea Basin. *Medelingen van de Werkgroep Tertiair en Kwartair Geologie*, **26**, 31–64.

Rasmussen, L. Å. (1975). Kineto-stratigraphic glacial drift units on Hindsholm, Denmark. *Boreas*, **4**, 209–217.

Rastas, J. and Seppälä, M. (1981). Rock jointing and abrasion forms on rouches moutonnées, SW Finland. *Annals of Glaciology*, **2**, 159–163.

Raukas, A. (1969). Composition and genesis of Estonian tills. *Zeszyty Naukowe Uniwersytetu A. Mickiewicza* Seria Geografia **8**, 167–176.

Raukas, A. (1977). Ice-marginal formations and the main regularities of the deglaciation on Estonia. *Zeitschrift für Geomorphologie, Supplementbans*, **27**, 68–78.

Raukas, A., Haldorsen, S. and Mickelson, D. M. (1988). On the comparison and standardization of investigation methods for the identification of genetic varieties of glacigenic deposits. In *Genetic Classification of Glacigenic Deposits* (R. P. Goldthwait and C. L. Matsch, eds) pp. 211–216, Rotterdam: A.A. Balkema Publishers.

Raymo, M. E., Ruddiman, W. F., Backman, J., Clement, B M. and Martinson, D. G. (1989). Late Pliocene variation in Northern Hemisphere ice sheets and North Atlantic deep water circulation. *Paleoceanography*, **4**, 413–446.

Reading, H. G. (1978). *Sedimentary Environments and Facies*. Oxford: Blackwell, 557 pp.

Reading, H. G. and Walker, R. G. (1966). Sedimentation of Eocambrian tillites and associated sediments in Finn-

mark, northern Norway. *Palaeogeography, Palaeoclimatology. Palaeoecology*, **2**, 177–212.

Reger, R. D. and Bundtzen, T. K. (1990). Multiple glaciation and gold-placer formation, Valdez Creek Valley, western Clearwater Mountains, Alaska. Alaska Department of Natural Resources, Division of Geological and Geophysical Surveys, Professional Report 107, 29 pp.

Rehmer, J. A. and Hepburn, J. C. (1974). Quartz sand surface textural evidence for a glacial origin of the Squantum 'Tillite', Boston Basin, Massachusetts. *Geology*, **2**, 413–415.

Reid, C. (1882). The Geology of the country around Cromer. *Memoirs of the Geological Survey of England and Wales*.

Reid H. F. (1892). Studies of the Muir Glacier, Alaska. *National Geographic Magazine*, **4**, 19–84.

Reid, H. F. (1896). Glacier Bay and its glaciers. United States Geological Survey, 16th Annual Report, Part **1**, 415–461.

Reid, I. R (1970). Geomorphology and glacial geology of the Martin River Glacier, Alaska. *Arctic*, **23**, 245–267.

Reimer, G. E. (1984). The sedimentology and stratigraphy of the southern basin of glacial Lake Passaic, New Jersey. Unpublished MS thesis, Rutgers University, New Brunswick, NJ.

Reimnitz, E., Barnes, P. W. and Phillips, R. L. 1984. Geological evidence for 60 meter deep pressure ridge keels in the Arctic Ocean. *Proceedings of the International Association of Hydraulic Research, 7th International Symposium on Ice, Hamburg*, **2**. 189–206.

Reiter, E. R. (1961). *Meterologie der Strahlströme (Jet Streams)*. Wien: Springer Verlag, 473 pp.

Renberg, I. (1981). Formation, structure, and visual appearance of iron-rich varved lake sediments. *Verhandlungen internationen Vereinigung für Limnologies*, **21**, 94–101.

Rencz, A. N. and Shilts, W. W. (1980). Nickel in soils and vegetation of glaciated terrains. In *Nickel in the Environment* (J. O. Nriagu, ed.) pp. 151–188, New York: John Wiley and Sons.

Reneau, S L. and Raymond, R. Jr. (1991). Cation-ratio dating of rock varnish: Why does it work? *Geology*, **19**, 937–940.

Reneau, S. L., Oberlander, T. M. and Harrington, C. D. (1991). Accelerator mass spectrometry radiocarbon dating of rock varnish: discussion and reply. *Geological Society of America Bulletin*, **103**, 310–314.

Repo, R. (1954). Om forhallandet mellan rafflor och asar. *Geologi*, **6**, 45 pp.

Rex, R. W., Margolis, S. V. and Murray, B. (1970). Possible interglacial dune sands from 300 meters water depth in the Weddell Sea, Antarctica. *Geological Society of*

America Bulletin, **81**, 3465–3472.

Rice, H. M. A. (1947). Geology and mineral deposits of the Princeton map-area, British Columbia. *Geological Survey of Canada, Memoir 243*, 136 pp.

Richards, K. (1984). Some observations on suspended sediment dynamics in Storbregrova, Jotunheimen. *Earth Surface Processes and Landforms*, **9**, 101–112.

Ridge, J. C. and Larsen, F. D. (1990). Re-evaluation of Antevs' New England varve chronology and new radiocarbon dates of sediments from glacial Lake Hitchcock. *Geological Society of America Bulletin*, **102**, 889–899.

Ridge, J. C., Brennan, W. J. and Muller, E. H. (1990). The use of paleomagnetic declination to test correlations of late Wisconsinan glaciolacustrine sediments in central New York. *Geological Society of America Bulletin*, **102**, 26–44.

Ridky, R. W. and Bindschadler, R. A. (1990). Reconstruction and dynamics of the Late Wisconsin 'Ontario' ice dome in the Finger Lakes region, New York. *Geological Society of America Bulletin*, **102**, 1055–1064.

Ridler, R. H. and Shilts, W. W. (1974). Mineral potential of the Rankin Inlet. *Canadian Mining Journal*, **95** (July), 32–42.

Rieck, R. L. (1979). Ice stagnation and paleodrainage in and near an interlobate area. *Michigan Academician*, **11**, 219–235.

Riezebos, P. A. and van der Waals, L. (1974). Silt-sized quartz particles: a proposed source. *Sedimentary Geology*, **12**, 279–285.

Riezebos, P. A., Boulton, G. S., van der Meer, J. J. M, Ruegg, G. H. J. Beets, D. J., Castel, I. I. Y., Hart, J., Quinn, I., Thornton, M. and van der Wateren, F. M. (1986). Products and effects of modern aeolian activity on a nineteenth-century glacier-pushed ridge in West Spitsbergen, Svalbard. *Arctic and Alpine Research*, **18**, 389–396.

Riley, N. W. (1982). Rock wear by sliding ice. Unpublished PhD thesis, University of Newcastle, Newcastle upon Tyne, U.K.

Ringberg, B. (1983). Till stratigraphy and glacial rafts of chalk at Kvarnby, southern Sweden. In *Glacial Deposits in North-West Europe* (J. Ehlers, ed.) pp. 151–54, Rotterdam: A.A. Balkema Publishers.

Roberts, D. (1974). A discussion. the relation of joint patterns to the formation of fjords in western Norway. *Norsk Geologisk Tidsskrift*, **54**, 213–215.

Roberts, M. C. and Mark, D. M. (1970). The use of trend surfaces in till fabric analysis. *Canadian Journal of Earth Sciences*, **7**, 1179–1184.

Roberts, M. C. and Rood, K. M. (1984). The role of the ice contributing area, the morphology of transverse fjords,

British Columbia. *Geografiska Annaler*, **66A**, 381–393.

Robertson-Rintoul, M. S. E. (1986). A quantitative soil-stratigraphic approach to the correlation and dating of Post-Glacial river terraces in Glen Feshie, western Cairngorms. *Earth Surface Processes and Landforms*, **11**, 605–617.

Robinson, B. W. and Nickel, E. H. (1979). A useful new technique for mineralogy: the back-scattered electron/low vacuum mode of SEM operation. *American Mineralogist*, **64**, 1322–1328.

Robson, D. A. (1978). Laboratory experiments on the abrasion of detrital quartz grains. *Proceedings of the Yorkshire Geological Society*, **42**, 217–227.

Rocha-Campos, A. C., Farjallat, J. E. S. and Yoshida, R. (1968). New glacial features of the upper Paleozoic Itararé Subgroup in the State of São Paulo, Brazil. *Geological Society of Brazil Bulletin*, **17**, 47–57.

Rocha-Campos, A. C., Krauspenhar, E. B. and Folk, R. L. (1978). Glacial deposits identified by chattermark trails in detrital garnets: Comment. *Geology*, **6**, 8–10.

Rocha-Campos, A. C., Santos, P. R. and Canuto, J. R. (1990). Ice-scouring structures in Late Paleozoic glacial lake sediments, Parana Basin, Brazil. *Geological Society of America*, Abstracts with program, **22**, 66.

Rochette, J. -C. and Cailleux, A. (1971). Dépôts nivéo-éoliens annuels à Poste-de-la-Baleine, Nouveau-Québec. *Revue de Géographie de Montréal*, **25**, 35–41.

Rodhe, L. (1988). Glaciofluvial channels formed prior to the last deglaciation: examples from Swedish Lapland. *Boreas*, **17**, 511–516.

Rogerson, R. J. and Hudson, H. M. (1983). Quartz surface microtextures and grain-size characteristics of Quaternary sediments in the Porcupine Strand area of coastal Labrador, Newfoundland, Canada. *Canadian Journal of Earth Sciences*, **20**, 377–387.

Rolph, T. C., Shaw, J., Derbyshire, E. and Jingtai, W. (1989). A detailed geomagnetic record from Chinese loess. *Physics of the Earth and Planetary Interiors*, **56**, 151–164.

Romans, J. C. C., Robertson, L. and Dent, D. L. (1980). The micromorphology of young soils from south-east Iceland. *Geografiska Annaler*, **62A**, 93–103.

Rosberg, J. E. (1925). Jättegrytor i södra Finland. *Fennia*, **46**, 1–103.

Rose, J. (1974). Small scale spatial variabilty of some sedimentary properties of lodgement and slumped till. *Proceedings of the Geologists' Association*, **85**, 223–237.

Rose, J. (1981). Field Guide to the Quaternary Geology of the Loch Lomond basin. *Proceedings of the Geological Society of Glasgow*, **1980/81**, 3–19.

Rose, J. (1985). The Dimlington Stadial/Dimlington Chronozone: a proposal for naming the main glacial episode of the Late Devensian in Britain. *Boreas*, **14**, 225–230.

Rose, J. (1987a). Drumlins as part of a glacier bedform continuum. In *Drumlin Symposium* (J. Menzies and J. Rose, eds) pp. 103–116, Rotterdam: A.A. Balkema Publishers.

Rose, J. (1987b). Status of the Wolstonian Glaciation in the British Quaternary. *Quaternary Newsletter*, **53**, 1–9.

Rose, J. (1988a). Stratigraphic nomenclature for the British Middle Pleistocene – procedural dogma or stratigraphic common sense? *Quaternary Newsletter*, **54**, 15–20.

Rose, J. (1988b). Discussion in: The evolution of oxygen isotope variability in the North Atlantic over the past three million years (N. J. Shackleton, J. Imbrie and N. G. Pisias) *Philosophical Transactions of the Royal Society of London*, **318B**, 687–688.

Rose, J. (1989a). Stadial type sections in the British Quaternary. In *Quaternary Type Sections: Imagination or Reality?* (J. Rose and Ch. Schlüchter, eds) pp. 45–67, Rotterdam: A.A. Balkema Publishers.

Rose, J. (1989b). Glacier stress patterns and sediment transfer associated with the formation of superimposed flutes. *Sedimentary Geology*, **62**, 151–176.

Rose, J. (1991). Stratigraphic basis of the 'Wolstonian Glaciation', and retention of the term 'Wolstonian' as a chronostratigraphic Stage name – a discussion. In *Central East Anglian and the Fen Basin, Field Guide* (S. G. Lewis, C. A. Whiteman and D. R. Bridgland, eds) pp. 15–20, London: Quaternary Research Association.

Rose, J. and Allen, P. (1977). Middle Pleistocene stratigraphy in south-east Suffolk. *Journal Geological Society London*, **133**, 83–102.

Rose, J. and Letzer, J. M. (1977). Superimposed drumlins. *Journal of Glaciology*, **18**, 471–480.

Rose, J. and Schlüchter, Ch. (eds) (1989). *Quaternary Type Sections: Imagination or Reality?* Rotterdam: A.A. Balkema Publishers, 208 pp.

Rose. J., Lowe J. J. and Switzer, R. (1988). A radiocarbon date on plant detritus beneath till from the type area of the Loch Lomond Readvance. *Scottish Journal of Geology*, **24**, 113–124.

Rose, J., Boardman, J., Kemp, R. A. and Whiteman, C. A. (1985a). Palaeosols and the interpretation of the British Quaternary stratigraphy. In *Geomorphology and Soils* (K. S. Richards, R. R. Arnett and S. Ellis, eds), pp. 348–375, London: George Allen and Unwin.

Rose, J., Allen, P., Kemp, R. A., Whiteman, C. A. and Owen, N. (1985b). The early Anglian Barham soil of Eastern England. In *Soils and Quaternary Landscape Evolution* (J. Boardman, ed.) pp. 197–229, Chichester:

John Wiley & Sons.

Rosholt, J. N. (1985). Uranium-trend systematics for dating Quaternary sediments. *United States Geological Survey, Open-File Report*, 85–298, 34 pp.

Röthlisberger, H. (1972). Water pressure in intra- and subglacial channels. *Journal of Glaciology*, **11**, 177–203.

Röthlisberger, H. and Iken, A. (1981). Plucking as an effect of water-pressure variations at the glacier bed. *Annals of Glaciology*, **2**, 57–62.

Röthlisberger, H. and Lang, H. (1987). Glacial hydrology. In *Glacio-Fluvial Sediment Transfer – an Alpine Perspective*. (A. M. Gurnell and M. J. Clark, eds) pp. 207–284. John Wiley & Sons, Chichester.

Rotnicki, K. (1976). The theoretical basis for and a model of the origin of glaciotectonic deformations. *Quaestiones Geographicae*, **3**, 103–139.

Rõuk, A. -M. (1974). On the till structures in the drumlins of central Estonia. *Izvestia Akad. Nauk Eesti S.S.R. Chim. Geol. (Tallinn)*, **23**, 149–160.

Rõuk, A. -M. and Raukas, A. (1989). Drumlins of Estonia. *Sedimentary Geology*, **62**, 371–384.

Rudberg, S. (1954). Västerbottens berggrundsmorfologi. Ett försök till rekonstruktion av preglaciala erosionsgenerationer i Sverige. *Geographica*, **25**, 457 pp.

Rudberg, S. (1973). Glacial erosion forms of medium size – a discussion based on four Swedish case studies. *Zeitschrift für Geomorphologie*, **17**, 33–48.

Ruddiman, W. F. (1987a). Northern oceans. In *North America and Adjacent Oceans During the Last Deglaciation. The Geology of North America, Volume K–3* (W. F. Ruddiman and H. E. Wright Jr., eds) pp. 137–154, Boulder, Colorado: Geological Society of America.

Ruddiman, W. F. (1987b). Synthesis: The Ocean ice-sheet record. In *North America and Adjacent Oceans During the Last Deglaciation. The Geology of North America, Volume K–3* (W. F. Ruddiman and H. E. Wright Jr., eds) pp. 463–478, Geological Society of America, Boulder, Colorado.

Ruddiman, W. F. and Reymo, M. E. (1988). Northern hemisphere climate regimes during the past 3 Ma: possible tectonic connections. *Philosophical Transactions of the Royal Society of London*, **318B**, 411–430.

Ruegg, G H. J. (1981a). Sedimentary features and grain size of glaciofluvial and periglacial Pleistocene deposits in The Netherlands and adjacent parts of Western Germany. *Verh. Naturwiss. Verein Hamburg*, **24**, 133–154.

Ruegg, G H. J. (1981b). Ice-pushed Lower and Middle Pleistocene deposits near Rhenen (Kwintelooijen): sedimentary-structural and lithological/granulometrical investigations. In *Geology and Archaeology of Pleistocene Deposits in the Ice-pushed Ridge Near Rhenen and Veenendaal* (G. H. J. Ruegg and J. G. Zandstra, eds), *Mededelingen Rijks Geologische Dienst*, **35**, 165–177.

Ruegg, G. H. J. (1983). Periglacial aeolian evenly laminated sandy deposits in the Late Pleistocene of NW Europe, a facies unrecorded in modem sedimentological handbooks. In *Aeolian Sediments and Processes* (M. S.Brookfield and T. S. Ahlbrandt, eds) pp. 455–482, Amsterdam: Elsevier Scientific Publications.

Ruhe, R. V. (1950). Graphic analysis of drift topographies. *American Journal of Science*, **248**, 435–443.

Ruhe, R. V. (1969). *Quaternary Landscapes in Iowa*. Ames, Iowa: Iowa State University Press, 255 pp.

Ruland, W. W., Cherry, J. A. and Feenstra, S. (1991). The depth of fractures and active ground-water flow in a clayey till plain in southwestern Ontario. *Groundwater*, **29**, 405–417.

Rust, B. R. (1978). Depositional models for braided alluvium. In *Fluvial Sedimentology* (A. D. Miall, ed.) Canadian Society of Petroleum Geologists Memoir 5, 605–625.

Rust, B. R. and Romanelli, R. (1975). Late Quaternary subaqueous outwash deposits near Ottawa, Canada. In *Glaciofluvial and Glaciolacustrine Sedimentation* (A. V. Jopling and B. C. McDonald, eds), Society of Economic Paleontologists and Mineralogists, Special Publication No. 23, pp. 177–192.

Ruszczynska-Szenajch, H. (1976). Glacitectonic depressions and glacial rafts in mid-eastern Poland. *Studia Geologica Polonica*, **50**, 1–106.

Ruszczynska-Szenajch, H. (1983). Lodgement tills and syndepositional glacitectonic processes related to subglacial thermal and hydrologic conditions. In *Tills and Related Deposits: Genesis, Petrology, Application, Stratigraphy* (E. B. Evenson, Ch. Schlüchter and J. Rabassa, ed.) pp. 113–117, Rotterdam: A.A. Balkema Publishers.

Ruszczynska-Szenajch, H. (1987). The origin of glacial rafts: detachment, transport, deposition. *Boreas*, **16**, 101–112.

Ruszczynska-Szenajch, H. (1988). Glaciotectonics and its relationship to other glaciogenic processes. In *Glaciotectonics: Forms and Processes* (D. G. Croot, ed.) pp. 191–193, Rotterdam: A.A. Balkema Publishers.

Ruszczynska-Szenajch, H. (1993). Relationship of large-scale glaciotectonic features to substratum and bedrock conditions in Central and eastern Poland. In *Glaciotectonics and Mapping Glacial Deposits* (J. S. Aber, ed.), Proc. INQUA Comm. Formation and Properties of Glacial Deposits. Canadian Plains Research Centre, Univeristy of Regina, Regina, pp. 51–66.

Rutter, N. W. (1985). Dating methods of Pleistocene deposits and their problems. *Geoscience Canada Reprint Series, 2*, 87 pp.

Rutter, N. W., Brigham-Grette, J. and Catto, N. (eds) (1989). Applied Quaternary dating methods. *Quaternary International*, **1**, 166 pp.

Ryder, J. M. (1981). Geomorphology of the southern part of the Coast Mountains of British Columbia. *Zeitschrift für Geomorphologie*, **37**, 120–147.

Ryder, J. M. and Thomson, B. (1986). Neoglaciation in the Southern Coast Mountains of British Columbia: Chronology prior to the last Neoglacial Maximum. *Canadian Journal of Earth Sciences*, **23**, 273–287.

Rzechowski, J. (1986). Pleistocene till stratigraphy in Poland. In *Quaternary Glaciations in the Northern Hemisphere* (V. Sibrava, D. Q. Bowen and G. M. Richmond, eds), *Quaternary Science Reviews*, **5**, 365–372.

Saarnisto, M. and Tamminen, E. (1987). Placer gold in Finnish Lapland. In INQUA Till Symposium, Finland 1985 (R. Kujansuu and M. Saarnisto, eds) pp. 181–194, *Geological Survey of Finland Special Paper, 3*.

Saarnisto, M., Tamminen, E. and Vaasjoki, M. (1991). Gold in bedrock and glacial deposits in the Ivalojoki area, Finnish Lapland. *Journal of Geochemical Exploration*, **39**, 303–322.

Sættem, J. (1990). Glaciotectonic forms and structures on the Norwegian continental shelf: Observations, processes and implications. *Norsk Geologisk Tidsskrift*, **70**, 81–94.

Sagan, C. and Mullen, G. (1972). Earth and Mars: evolution of atmospheres and surface temperatures. *Science*, **177**, 52–56.

Saint-Onge, D. A. (1965). La Géomorphologie de l'Île Ellef Ringnes, Territoires du NordOuest, Canada. Canada Department of Mines and Technical Surveys, Geographical Branch Geographical Paper, Ottawa, No. 38, 46 pp.

Saint-Onge, D. A. (1984). Surficial deposits of the Redrock Lake area. *Geological Survey of Canada, Paper 84–1A*, 271–277.

Sakshaug, E. and Holm-Hansen, O. (1984). Factors governing pelagic production in polar oceans. In *Marine Phytoplankton and Productivity. Lecture Notes on Coastal and Estuarine Studies, 8* (O. Holm-Hansen, L. Bolis and R. Gilles, eds) pp. 1–17, New York: Springer.

Salonen, V. P. (1987). Observations on Boulder Transport in Finland. In INQUA Till Symposium, Finland 1985 (R. Kujansuu and M. Saarnisto, eds) pp. 179–181, *Geological Survey of Finland Special Paper, 3*.

Salvador, A. (1994). *International Stratigraphic Guide*, 2nd edn, Boulder: Geological Society of America, 214 pp.

Samson, J. (1984). An overview of coastal and marine gold placer occurrences in Nova Scotia and British Columbia. Energy, Mines and Resources Canada, Ocean Mining Division, Canada Oil and Gas Lands Administration Internal Publication, 162 pp.

Samtleben, C. and Bickert, T. (1990). Coccoliths in sediment traps from the Norwegian Sea. *Marine Micropaleontology*, **16**, 39–64.

Samuelson, C. (1926). Studien über die Wirkungen des Windes in den Kalten und gemässigten Erdtielen. *Bulletin of the Geological Institute, University of Uppsala*, **20**, 57–230.

Sarna-Wojcicki, A. M. and Davis, J. O. (1991). Quaternary tephrochronology. In *Quaternary Nonglacial Geology: Conterminous United States Geological Society of America, Boulder* (R. B. Morrison, ed.), The Geology of North America, K–2, pp. 93–116.

Sarna-Wojcicki, A. M., Bowman, H. R., Meyer, C. E., Russell, P. C., Woodward, M. J., McCoy, G., Rowe Jr., J. J., Baedecker, P. A., Asaro, F. and Michael, H. (1984). Chemical analyses, correlations, and ages of upper Pliocene and Pleistocene ash layers of east-central and southern California. United States Geological Survey Professional Paper, 1293, 40 pp.

Sarna-Wojcicki, A. M., Lajoie, K. R., Meyer C. E., Adam, D. P. and Rieck, H. J. (1991). Tephrochronologic correlation of upper Neogene sediments along the Pacific margin, conterminous United States. In *Quaternary Nonglacial Geology: Conterminous U.S. Geological Society of America, Boulder* (R. B. Morrison, ed.), The Geology of North America, K–2, pp. 117–140.

Sarntheim, M. (1978). Sand deserts during glacial maximum and climatic optimum. *Nature*, **272**, 43–46

Sarnthein, M. and Tiedemann, R. (1989). Toward a high-resolution stable isotope stratigraphy of the last 3.4 million years: Sites 658 and 659 off northwest Africa. *Proceedings ODP, Scientific Results*, **108**, 167–185.

Sarnthein, M., Erlenkeuser, H, von Grafstein, R. and Schroeder, C. (1984). Stable-isotope stratigraphy for the last 750,000 years: 'Meteor' core 13519 from eastern equatorial Atlantic. *Meteor Forschung-Ergebnisse C*, **38**, 9–24.

Sarnthein, M., Jansen, E., Arnold, M., Duplessy, J. C., Erlenkruser, H., Flatoy, A., Veum, T., Vogelsang, E. and Weinelt, M. S. (1992). $\partial18O$ time-slice reconstruction of meltwater anomalies at termination I in the North Atlantic between 50 and 80N. In The last deglaciation: absolute and radiocarbon chronologies. (F. Bard and W. S. Broecker, eds), NATO Series, Volume, 12. Springer,

Berlin. pp. 184–200.

Sauer, E. K. (1978). The engineering significance of glacier ice-thrusting. *Canadian Geotechnical Journal*, **15**, 457–472.

Saunderson, H. C. (1977). The sliding bed facies in esker sands and gravels: a criterion for full-pipe (tunnel) flow? *Sedimentology*, **24**, 623–638.

Saunderson, H. C. and Jopling, A. V. (1980). Palaeohydraulics of a tabular, cross-stratified sand in the Brampton esker, Ontario. *Sedimentary Geology*, **25**, 169–188.

Sauter, D. (1967). Till fabric analysis of a drumlin. Unpublished MS thesis, Syracuse University, NY.

Savage, N. M. (1972). Soft-sediment glacial grooving of Dwyka age in South Africa. *Journal of Sedimentary Petrology*, **42**, 307–308

Savage, W. Z. (1968). Application of plastic flow analysis to drumlin formation (Pleistocene). Unpublished MS thesis, Syracuse University, NY.

Sawkins, F. J., Chase, C. G., Darby, D. G., and Rapp, G., Jr. (1978). *The Evolving Earth* (2nd edn). New York: Macmillan, 558 pp.

Scarpelli, W. (1991). Auriferous quartz-pebble conglomerates in Brazil. In *International Symposium on Alluvial Gold Placers; Abstracts* (G. Hérail, ed.) pp. 151–152, La Paz, Bolivia.

Schack Pedersen, S. A. (1988). Glacitectonite: brecciated sediments and cataclastic sedimentary rocks formed subglacially. In *Genetic Classification of Glacigenic Deposits* (R. P. Goldthwait and C. L. Matsch, eds) pp. 89–91, Rotterdam: A.A. Balkema Publishers.

Schermerhorn, L. J. G. (1974). Late Precambrian mixtites: glacial and/or nonglacial? *American Journal of Science*, **274**, 673–824.

Schermerhorn, L. J. G. (1983). Late Proterozoic glaciation in the light of CO2 depletion in the atmosphere. In Proterozoic geology: selected papers from an international symposium (L. G. Medaris Jr., C. W. Byers, D. M. Mickelson and W. C. Shanks, eds), *Geological Society of America, Memoir 161*, pp. 309–315.

Schlüchter, Ch. (ed.) (1979). *Moraines and Varves: Origin/Genesis/Classification*. Rotterdam: A.A. Balkema Publishers, 441 pp.

Schlüchter, Ch. (1992). Terrestrial Quaternary stratigraphy. *Quaternary Science Reviews*, **11**, 603–607.

Schou, A (1949). The landscapes. In *Atlas of Denmark* (N. Nielsen ed.) Copenhagen.

Schubert, C. and Clapperton, C. M. (1990). Quaternary glaciations in the northern Andes (Venezuela, Columbia and Equador). In *Quaternary Glaciations in the Southern Hemisphere* (C. M. Clapperton, ed.), *Quaternary Science Reviews*, **9**, 123–135.

Schultz, C. B. and Frye, J. C. (1968). *Loess and Related Eolian Deposits of the World*. Lincoln: University of Nebraska Press, 369 pp.

Schwan, J. (1986). The origin of horizontal alternating bedding in Weichselian aeolian sands in Northwestern Europe. *Sedimentary Geology*, **49**, 73–108.

Schwan, J. (1987). Sedimentologic characteristics of a fluvial to aeolian succession in Weichselian talsand in the Emsland (FRG). *Sedimentary Geology*, **52**, 273–298.

Schwarcz, H. P. (1989). Uranium series dating of Quaternary deposits. *Quaternary International*, **1**, 7–17.

Schwarcz, H. P. and Blackwell, B. (1985). Dating methods of Pleistocene deposits and their problems: II Uranium-series disequilibrium dating. In *Dating Methods of Pleistocene Deposits and their Problems* (N. W. Rutter, ed.) pp. 9–18, Geoscience Canada, Reprint Series 2, Toronto: Geological Association of Canada.

Schwertmann, U., Murad, E. and Schulze, D. G. (1982). Is there Holocene reddening (hemitite formation) in soils of axeric temperate areas? *Geoderma*, **27**, 209–223.

Schytt, V. (1963). Fluted moraine surfaces. *Journal of Glaciology*, **4**, 825–827.

Scott, D. (ed.) (1979). *Wear, Treatise on Materials Science and Technology*, vol. 13. New York: Academic Press, 498 pp.

Scott, J. S. (1976). Geology of Canadian tills. In *Glacial Till* (R. F. Legget, ed.), The Royal Society of Canada Special Publication No. 12, Ottawa, pp. 50–66.

Seed, H. B. (1968). Landslides during earthquakes due to soil liquefaction. *Journal of Soil Mechanics Foundation Division, Proceedings of American Society of Civil Engineers*, **94**, 1053–1122.

Seed, H. B. (1979). Soil liquefaction and cyclic mobility evaluation for level ground during earthquakes. *Journal of Geotechnical Engineering Division Proceedings, American Society of Civil Engineers*, **105**, 201–255.

Seilacher, A. (1984). Sedimentary structures tentatively attributed to seismic events. *Marine Geology*, **55**, 1–12.

Sejrup, H. P., Miller, G. H., Brigham-Grette, J., Løvlie, R. and Hopkins, D. M. (1984). Amino acid epimerization implies rapid sedimentation rates in Arctic Ocean cores. *Nature*, **310**, 772–775.

Sekyra, J. (1969). Periglacial phenomena in the oases and the mountains of the Enderby Land and the Dronning Maud Land (East Antarctica). *Biuletyn Peryglacjalny*, **19**, 277–289.

Selby, M. J., Rains, R. B and Palmer, R. W. P. (1974). Eolian deposits of the ice-free Victoria Valley, southern Victoria Land, Antarctica. *New Zealand Journal of Geology and Geophysics*, **17**, 543–562.

Self, S. and Sparks, R. S. J. (1981). *Tephra Studies. Proceedings of the NATO Advance Study Institute Series*. Boston: D. Reidel, 495 pp.

Selley, R. C. (1982). *An Introduction to Sedimentology*, 2nd edn. London: Academic Press Publishing, 417 pp.

Serebryanny, L. R. and Orlov, A. V. (1982). Genesis of marginal moraines in the Caucasus. *Boreas*, **11**, 279–289.

Seret, G. (1979). The genesis of drumlins. In *Moraines and Varves* (Ch. Schlüchter, ed.) pp. 189–196, Rotterdam: A.A. Balkema Publishers.

Seret, G. (1983). Microstructures in thin sections of several kinds of tills. *Quaternary International*, **18**, 97–101.

Seret, G., Dricot, E. and Wansard, G. (1990). Evidence for an early glacial maximum in the French Vosges during the last glacial cycle. *Nature*, **346**, 453–456.

Sergeyev, Y. M., Grabowska-Olszewska, B., Osipov, V. I., Sokolov, V. N. and Kolomenski, Y. N. (1980a). The classification of microstructures of clay soils. *Journal of Microscopy*, **120**, 237–260.

Sergeyev, Y. M., Spivak, G. V., Osipov, V. I., Rau, E. I., Sokolov, V. N. and Filippov, M. N. (1980b). The application of the SEM in the study of soils. *Scanning*, **3**, 262–272.

Setlow, L. W. (1978). Age determination of reddened coastal dunes in northwest Florida, USA, by use of scanning electron microscopy. In *Scanning Electron Microscopy in the Study of Sediments* (W. B. Whalley, ed.) pp. 283–305, Norwich: Geo Abstracts.

Setlow, L. W. and Karpovich, R. P. (1972). 'Glacial' micro-textures on quartz and heavy mineral sand grains from the littoral environment. *Journal of Sedimentary Petrology*, **42**, 864–875.

Seward, D. (1979). Comparison of zircon and glass fission-track ages from tephra horizons. *Geology*, **7**, 479–482.

Seyfried, W. and Bischoff, J. L. (1977). Hydrothermal transport of heavy metals by seawater: the role of seawater/basalt ratio. *Earth and Planetary Science Letters*, **34**, 71–77.

Shackleton, N. J. (1977). Carbon–13 in Uvigerina: tropical rainforest history and the Equatorial Pacific carbonate dissolution cycles. In *The Fate of Fossil CO_2 in the Oceans* (N. R. Anderson and A. Malahoff, eds) pp. 401–427, New York: Plenum Press.

Shackleton, N. J. (1982). The deep-sea sediment record of climate variability. *Progress in Oceanography*, **11**, 199–218.

Shackleton, N. J. (1987). Oxygen isotopes, ice volume and sea level. *Quaternary Science Reviews*, **6**, 183–190.

Shackleton, N. J. and Opdyke, N. D. (1973). Oxygen isotope and palaeomagnetic stratigraphy of equatorial Pacific core V28–238: oxygen isotope temperatures and ice volumes on a 105 year and a 106 year scale. *Quaternary Research*, **3**, 39–55.

Shackleton, N. J., Duplessy, J. -C., Arnold, M , Maurice, P., Hall, M. A. and Cartlidge, J. (1988). Radiocarbon age of last glacial Pacific deep water. *Nature*, **335**, 708–711.

Shackleton, N. J., Berger, A. and Peltier, W. A. (1990). An alternative astronomical calibration of the lower Pleistocene timescale based on ODP Site 667. *Transactions of the Royal Society of Edinburgh, Earth Sciences*, **81**, 251–261.

Shackleton, N. J., Imbrie, J. and Hall, M. A. (1983). Oxygen and carbon isotope record of East Pacific core V19–30: implications for the formation of deep water in the North Atlantic. *Earth and Planetary Science Letters*, **65**, 233–244.

Shakesby, R. A. (1989). Variability in Neoglacial moraine morphology and composition, Storbreen, Jotunheimen, Norway: within-moraine patterns and their implications. *Geografiska Annaler*, **71A**, 17–29.

Sharp, M. J. (1982). A comparison of the landforms and sedimentary sequences produced by surging and non-surging glaciers in Iceland. Unpublished PhD thesis, University of Aberdeen, Aberdeen, Scotland.

Sharp, M. J. (1985a). Sedimentation and stratigraphy at Eyjabakkajökull – an Icelandic surging glacier. *Quaternary Research*, **24**, 268–284.

Sharp, M. J. (1985b). 'Crevasse-fill' ridges – a landform type characteristic of surging glaciers? *Geografiska Annaler*, **67A**, 213–220.

Sharp, M. J. and Gomez, B. (1986). Processes of debris comminution in the glacial environment and implications for quartz sand-grain micromorphology. *Sedimentary Geology*, **46**, 33–47.

Sharp, R. P. (1969). Semiquantitative differentiation of glacial moraines near Convict Lake, Sierra Nevada, California. *Journal of Geology*, **77**, 68–91.

Sharpe, D. R. (1985). The stratified nature of deposits n streamlined glacial landforms on southern Victoria Island, N.W.T. *Geological Survey of Canada*, Current research Part A, **85–1A**, 365–371.

Sharpe, D. R. (1987). The stratified nature of drumlins from Victoria Island and southern Ontario, Canada. In *Drumlin Symposium* (J. Menzies and J. Rose, eds) pp. 185–214, Rotterdam: A.A. Balkema Publishers.

Sharpe, D. R. (1988a). The internal struture of glacial landforms: an example from the Halton till plain, Scarborough Bluffs, Ontario. *Boreas*, **17**, 15–26.

Sharpe, D. R. (1988b). Late glacial landforms of Wollaston Peninsula, Victoria Island, Northwest Territories: product of ice-marginal retreat, surge, and mass stagnation. *Canadian Journal of Earth Sciences*, **25**, 262–279.

Sharpe, D. R. and Barnett, P. J. (1985). Significance of sedimentological studies on the Wisconsinan stratigraphy of southern Ontario. *Géographie Physique et Quaternaire*, **39**, 255–273.

Sharpe, D R. and Cowan, W. R. (1990). Moraine formation in northwestern Ontario: product of subglacial fluvial and glaciolacustrine sedimentation. *Canadian Journal of Earth Sciences*, **27**, 1478–1486.

Sharpe, D. R. and Shaw, J. (1989). Erosion of bedrock by subglacial meltwater, Cantley, Quebec. *Geological Society of America Bulletin*, **101**, 1011–1020.

Sharpe, R. F. (1939). The Bullion hydraulic mine. *The Miner*, pp. 37–40.

Shaw, J. (1975). The formation of glacial flutings. In *Quaternary Studies* (R. P. Suggate and M. M. Cresswell, eds) pp. 253–258, The Royal Society of New Zealand, Wellington.

Shaw, J. (1977a). Tills deposited in arid polar environments. *Canadian Journal of Earth Sciences*, **14**, 1239–1245.

Shaw, J. (1977b). Till body morphology and structure related to glacier flow. *Boreas*, **6**, 189–201.

Shaw, J. (1977c). Sedimentation in an alpine lake during deglaciation. *Geografiska Annaler*, **59A**, 221–240.

Shaw, J. (1979). Genesis of the Sveg tills and Rögen moraines of central Sweden: a model of basal melt out. *Boreas*, **8**, 409–426.

Shaw, J. (1980). Drumlins and large-scale flutings related to glacier folds. *Arctic and Alpine Research*, **12**, 287–298.

Shaw, J. (1982). Melt-out till in the Edmonton area, Alberta, Canada. *Canadian Journal of Earth Sciences*, **19**, 1548–1569.

Shaw, J. (1983a). Forms associated with boulders in melt-out till In *Tills and Related Deposits: Genesis, Petrology, Application, Stratigraphy* (E. B. Evenson, Ch. Schlüchter and J. Rabassa, eds) pp. 3–12, Rotterdam: A.A. Balkema Publishers.

Shaw, J. (1983b). Drumlin formation related to inverted meltwater erosional marks. *Journal of Glaciology*, **29**, 461–479.

Shaw, J. (1985). Subglacial and ice marginal environments. In *Glacial Sedimentary Environments* (G. M. Ashley and N. D. Smith, eds), SEPM Short Course 16, Tulsa, OK, pp. 7–84.

Shaw, J. (1987). Glacial sedimentary processes and environmental reconstruction based on lithofacies. *Sedimentology*, **34**, 103–116.

Shaw, J. (1988a). Coarse-grained sediment flow facies in a large supraglacial lake, discussion. *Sedimentology*, **35**, 527–530.

Shaw, J. (1988b). Models of terrestrial glacial sedimentation. In *Glacial Facies Models, Continental Terrestrial*

Environments (J. Shaw and G. Ashley, eds), Geological Society of America Short Course, 1–78.

Shaw, J. (1988c). Subglacial erosion marks, Wilton Creek, Ontario. *Canadian Journal of Earth Sciences*, **25**, 1256–1267.

Shaw, J. and Freschauf, R. C. (1973). A kinematic discussion of the formation of glacial flutings. *Canadian Geographer*, **17**, 19–35.

Shaw, J. and Kvill, D. (1984). A glaciofluvial origin for drumlins of the Livingstone Lake area, Saskatchewan. *Canadian Journal of Earth Sciences*, **12**, 1442–1459.

Shaw, J. and Sharpe, D. R. (1987a). Drumlins and erosion marks in southern Ontario. *XIIth INQUA Congress Field Excursion C–25*. Ottawa: National Research Council of Canada, 17 pp.

Shaw, J. and Sharpe, D. R. (1987b). Drumlin formation by subglacial meltwater erosion. *Canadian Journal of Earth Sciences*, **24**, 2316–2322.

Shaw, J., Gilbert, R. and Archer, J. J. J. (1978). Proglacial lacustrine sedimentation during winter. *Arctic and Alpine Research*, **10**, 689–699.

Shaw, J., Kvill, D. and Rains, B. (1989). Drumlins and catastrophic subglacial floods. *Sedimentary Geology*, **62**, 177–202.

Shawe, D. R. and Wier, K. L. (1989). Gold deposits in the Virginia City-Alder Gulch District, Montana. *United States Geological Survey Bulletin*, **1857(G)**, 14–19.

Shennan, I. (1987). Holocene sea-level changes in the North Sea Region. In *Sea Level Changes* (M. J. Tooley and I. Shennan, eds) pp. 109–151, Institute of British Geographers, Special Publication, 20. Oxford: Blackwell.

Shennan, I. (1989). Holocene crustal movements and sea-level changes in Great Britain. Journal of *Quaternary Science*, **4**, 77–89.

Sheridan, M. F. and Marshall, J. R. (1987). Comparative charts for quantitative analysis of grain-textural elements on pyroclasts. In *Clastic Particles* (J. R. Marshall, ed.) pp. 98–121, New York: Van Nostrand Reinhold.

Shetsen, I. (1984). Application of pebble lithology to the differentiation of glacial lobes in southern Alberta. *Canadian Journal of Earth Sciences*, **21**, 920–933.

Shi, Yafeng, Ren, Binghui, Wang, Jingtai and Derbyshire, E. (1986). Quaternary glaciation in China. In *Quaternary Glaciations in the Northern Hemisphere* (V. Sibrava, D. Q. Bowen and G. M. Richmond, eds), *Quaternary Science Reviews*, **5**, 503–507.

Shilts, W. W. (1971). Till studies and their application to regional drift prospecting. *Canadian Mining Journal*, **92** (April), 45–50.

Shilts, W. W. (1973a). Drift prospecting; geochemistry of eskers and till in permanently frozen terrain: District of

Keewatin, Northwest Territories. *Geological Survey of Canada, Paper 72–45*, 34 pp.

Shilts, W. W. (1973b). Glacial dispersal of rocks, minerals, and trace elements in Wisconsinan till, southeastern Québec, Canada. *Geological Society of America, Memoir 136*, 189–219.

Shilts, W. W. (1975). Principles of geochemical exploration for sulphide deposits using shallow samples of glacial drift. *Canadian Institute of Mining and Metallurgy Bulletin*, **68** (May), 73–80.

Shilts, W. W. (1976). Glacial till and mineral exploration. In *Glacial Till; An Inter-disciplinary Study* (R. F. Legget, ed.), The Royal Society of Canada, Special Publication No. 12, Ottawa, pp. 205–224.

Shilts, W. W. (1978). Detailed sedimentological study of till sheets in a stratigraphic section, Samson River, Quebec. *Geological Survey of Canada Bulletin*, **285**, 30 pp.

Shilts, W. W. (1980). Flow patterns in the central North American ice sheet. *Nature*, **286**, 213–218.

Shilts, W. W. (1982). Quaternary evolution of the Hudson/James Bay region. *Le Naturaliste Canadien*, **109**, 309–332.

Shilts, W. W. (1984). Till geochemistry in Finland and Canada. *Journal of Geochemical Exploration*, **21**, 95–117.

Shilts, W. W. and Kettles, I. M. (1990). Geochemical/mineralogical profiles through fresh and weathered till. In *Glacial Indicator Tracing* (R. Kujansuu and M. Saarnisto, eds) pp. 187–216, Rotterdam: A.A. Balkema Publishers.

Shilts, W. W. and Smith, S. L. (1986). Stratigraphy of placer gold deposits; overburden drilling in Chaudière Valley, Quebec. *Geological Survey of Canada, Paper 86–1A*, pp. 703–712.

Shilts, W. W. and Smith, S. L. (1988). Glacial geology and overburden drilling in prospecting for buried gold placer deposits, southeastern Quebec. In *Prospecting in Areas of Glaciated Terrain – 1988* (D. R. Macdonald, ed.) pp. 141–169, Halifax: Canadian Institute of Mining and Metallurgy.

Shilts, W. W. and Smith, S. L. (1989). Drift prospecting in the Appalachians of Estrie-Beauce, Quebec. In *Drift Prospecting* (R. N. W. DiLabio and W. B. Coker, eds) pp. 41–59, *Geological Survey of Canada, Paper 89–20*.

Shilts, W. W. and Wyatt, P. H. (1989). Gold and base metal exploration using drift as a sample medium, Kaminak Lake-Turquetil Lake area, District of Keewatin. *Geological Survey of Canada*, Open File 2132, 149 pp.

Shilts, W. W., Aylsworth, J. M., Kaszycki, C. A. and Klassen, R. A. (1987). Canadian shield. In *Geomorphic Systems of North America* (W. L. Graf ed.), Geological

Society of America Centennial Special Volume, 2, 119–161.

Shoemaker, E. M. (1986a). The formation of fjord thresholds. *Journal of Glaciology*, 32, 65–71.

Shoemaker, E. M. (1986b). Subglacial hydrology for an ice sheet resting on a deformable aquifer. *Journal of Glaciology*, 32, 20–30.

Shoemaker, E. M. (1991). On the formation of large subglacial lakes. *Canadian Journal of Earth Sciences*, 28, 1975–1981.

Shoemaker, E. M. (1992a). Subglacial floods and the origin of low-relief ice-sheet lobes. *Journal of Glaciology*, 38, 105–112.

Shoemaker, E. M. (1992b). Water sheet outburst floods from the Laurentide ice sheet. *Canadian Journal of Earth Sciences*, 29, 1250–1264.

Shotton, F. W. (1972). An example of hard-water error in radiocarbon dating of vegetable matter. *Nature*, 240, 460–461.

Shreve, R. L. (1972). Movement of water in glaciers. *Journals of Glaciology*, 11, 205–214.

Shreve, R. L. (1984). Glacier sliding at subfreezing temperatures. *Journal of Glaciology*, 30, 341–347.

Shreve, R. L. (1985). Esker characteristics in terms of glacier physics, Katahdin Esker System, Maine. *Geological Society of America Bulletin*, 96, 639–646.

Shroba, R. R. and Birkeland, P. W. (1983a). Trends in Late-Quaternary soil development in the Rocky Mountains and Sierra Nevada of the western United States. In *Late Quaternary Environments of the United States. Volume 1 TheLate Pleistocene* (S. C. Porter, ed.) pp. 145–156, Minneapolis: University of Minnesota Press.

Shroba, R. R., Rosholt, J. N. and Madole, R. F. (1983b). Uranium-trend dating and soil B horizon properties of till of Bull Lake age, North St. Vrain drainage basin, Front Range, Colorado. *Geological Society of America Abstracts with Programs*, 15, 431.

Sibrava, V. (1986). Correlation of European glaciations and their relation to the deep-sea record. In *Quaternary Glaciations in the Northern Hemisphere* (V. Sibrava, D. Q. Bowen and G. M. Richmond, eds). *Quaternary Science Reviews*, 5, 433–441.

Sibrava, V., Bowen, D. Q. and Richmond, G. M. (eds) (1986). *Quaternary Glaciations in the Northern Hemisphere. Quaternary Science Reviews*, 5, 510 pp.

Sigafoos, R. S. and Hendricks, E. L. (1969). The time interval between stabilization of alpine glacial deposits and establishment of tree seedlings. *United States Geological Survey Professional Paper*, 650–B, 89–93.

Sigmundsson, F. (1991). Post-glacial rebound and asthenosphere viscosity in Iceland. *Geophysical Research Letters*, 18, 1131–1134.

Sigurdsson, H. and Loebner, B. (1981). Deep-sea record of Cenozoic explosive volcanism in the North Atlantic. In *Tephra Studies* (S. Self and R. S. J. Sparks, eds) pp. 289–316, Dordrecht: D. Reidel.

Simpson, J. B. (1933). The late-glacial readvance moraines of the Highland border west of the river Tay. *Transactions of the Royal Society of Edinburgh*, 57, 633–645.

Sims, J. (1975). Determining earthquake recurrence intervals from deformational structures in young lacustrine sediments. *Tectonophysics*, 29, 141–152.

Sinitsin, Y. M. (1958). *Central Asia*. Moscow: Central Publishing House.

Sissons, J. B. (1958a). Subglacial stream erosion in southern Northumberland. *Scottish Geographical Magazine*, 74, 163–174.

Sissons, J. B. (1958b). Supposed ice-dammed lakes in Britain with particular reference to the Eddleston valley, southern Scotland. *Geografiska Annaler*, 40, 159–187.

Sissons, J. B. (1960a). Some aspects of glacial drainage channels in Britain, part I. *Scottish Geographical Magazine*, 76, 131–146.

Sissons, J. B. (1960b). Subglacial, marginal and other glacial drainage in the Syracuse-Oneida area, New York. *Geological Society of America Bulletin*, 71, 1575–1588.

Sissons, J B. (1961a). A subglacial drainage system by the Tinto Hills, Lanarkshire. *Transactions of the Edinburgh Geological Society*, 18, 113–123.

Sissons, J. B. (1961b). Some aspects of glacial drainage channels in Britain, part II. *Scottish Geographical Magazine*, 77, 15–36.

Sissons, J. B. (1963). The glacial drainage system around Carlops, Peebleshire. *Transactions of the Institute of British Geographers*, 32, 95–111.

Sissons, J. B. (1967). *The Evolution of Scotland's Scenery*. Edinburgh: Oliver & Boyd, 259 pp.

Sissons, J B. (1972). Dislocation and non-uniform uplift of raised shorelines in the western part of the Forth valley. *Transactions of the Institute of British Geographers*, 55, 145–159.

Sissons, J. B. (1976). *Scotland*. London: Methuen & Co., 150 pp.

Sissons, J. B. (1977). The Loch Lomond readvance in the northern mainland of Scotland. In *Studies in the Scottish Lateglacial Environment* (J. M. Gray, and J. J. Lowe, eds) pp. 45–59, Oxford: Pergamon.

Sissons, J. B. (1979). The Loch Lomond Stadial in the British Isles. *Nature*, 280, 199–203.

Sissons, J. B. (1982). A former ice-dammed lake and associated glacier limits in the Achnasheen area, central

Ross-shire. *Transactions of the Institute of British Geographers*, New Series **7**, 205–216.

Sissons, J. B. (1983). Shorelines and isostasy in Scotland. In *Shorelines and Isostasy* (D. E.Smith and A. G. Dawson, eds) pp. 209–225, Institute of British Geographers, Special Publication, 16. London: Academic Press.

Sissons, J. B. and Cornish, R. (1982). Differential glacio-isostatic uplift of crustal blocks at Glen Roy, Scotland. *Quaternary Research*, **18**, 268–288.

Sitler, R. F. (1968). Glacial till in oriented thin section. *XXIII International Geological Congress*, **8**, 283–295.

Sitler, R. F. and Chapman, C. A. (1955). Microfabrics of till from Ohio and Pennsylvania. *Journal of Sedimentary Petrology*, **25**, 262–269.

Skempton, A. W., Schuster, R. L. and Petley, D. J. (1969). Joints and fissures in the London Clay at Wraysbury and Edgware. *Géotechnique*, **19**, 205–217.

Sladen, J. A. and Wrigley, W. (1983). Geotechnical properties of lodgement till. In *Glacial Geology* (N. Eyles, ed.), Oxford: Pergamon Press, Ch. 8, 184–212.

Smalley, I. J. (1966a). The properties of glacial loess and the formation of loess deposits. *Journal of Sedimentary Petrology*, **36**, 669–676.

Smalley, I. J. (1966b). Drumlin formation; a rheological model. *Science*, **151**, 1379–1380.

Smalley, I. J. (1968). The loess deposits and Neolithic culture of Northern China. *Man*, **3**, 224–241.

Smalley, I. J. (1971). 'In-situ' theories of loess formation and the significance of the calcium carbonate content of loess. *Earth Science Reviews*, **7**, 67–85.

Smalley, I. J. (1972). The interaction of great rivers and large deposits of primary loess. *New York Academy of Science, Transactions*, **34**, 534–542.

Smalley, I. J. (1990). Possible formation mechanisms for the modal coarse-silt quartz particles in loess deposits, *Quaternary International*, **7/8**, 23–27.

Smalley, I. J. and Cabrera, J. G. (1970). The shape and surface texture of loess particles. *Geological Society of America Bulletin*, **81**, 1591–1596.

Smalley, I. J. and Krinsley, D. H. (1978). Loess deposits associated with deserts. *Catena*, **5**, 53–66.

Smalley, I. J. and Piotrowski, J. A. (1987). Critical strength/stress ratios at the ice-bed interface in the drumlin forming process: from 'dilatancy' to 'cross-over'. In *Drumlin Symposium* (J. Menzies and J. Rose, eds) pp. 81–86, Rotterdam: A.A. Balkema Publishers.

Smalley, I. J. and Smalley, V. (1983). Loess material and loess deposits: formation, distribution and consequences. In *Eolian Sediments and Processes* (M. E. Brookfield and T. S. Ahlbrandt, eds) pp. 51–68, Amsterdam: Elsevier.

Smalley, I. J. and Unwin, D. J. (1968). Formation and shape of drumlins and their orientation and distribution. *Journal of Glaciology*, **7**, 377–480.

Smalley, I. J. and Vita-Finzi, C. (1968). The formation of fine particles in sandy deserts and the nature of 'desert' loess. *Journal of Sedimentary Petrology*, **38**, 766–774.

Smalley, I. J., Krinsley, D. H. and Vita-Finzi, C. (1973). Observations on the Kaiserstühl loess. *Geological Magazine*, **110**, 29–36.

Smalley, I. J., Krinsley, D. H., Moon, C. F. and Bentley, S. P. (1979). Processes of quartz fracture in nature and the formation of clastic sediments. In*Mechanisms of Deformation and Fracture* (K. E. Easterling, ed.) pp. 119–127, Oxford: Pergamon Press.

Smart, P. (1974). Electron microscope methods in soil micromorphology. In *Soil Microscopy* (G. K. Rutherford, ed.) pp. 190–206, Kingston, Ontario: The Limestone Press.

Smart, P. and Tovey, N. K. (1981). *Electron Microscopy of Soils and Sediments: Examples*. Oxford, UK: Oxford University Press, 177 pp.

Smart, P. and Tovey, N. K. (1982). *Electron Microscopy of Soils and Sediments: Techniques*. Oxford, UK: Clarendon Press, 264 pp.

Smith, B. J. and Whalley, W. B. (in press). Non-glacial, loess-sized quartz silt: Implications for the origins of 'desert loess': review and further data. In *Loess and China, Essays for Liu Tung-sheng* (I. J. Smalley and E. Derbyshire, eds), Leicester: Leicester University Press.

Smith, C. A. S., Tarnocai, C. and Hughes, O. L. (1986). Pedological investigations of Pleistocene glacial drift surfaces in the central Yukon. *Géographie Physique et Quaternaire*, **15**, 29–37.

Smith, G. W. (1982). End moraines and the pattern of last ice retreat from central and south coastal Maine. In *Late Wisconsinan Glaciation of New England* (G. J. Larson and B. D. Stone, eds) pp. 195–209, Dubuque, Iowa: Kendall-Hunt.

Smith, H. T. U. (1965). Dune morphology and chronology in central and western Nebraska. *Journal of Geology*, **73**, 557–578.

Smith, K. C. (1992). Source and occurrence of placer gold in central Ross County, Ohio. Unpublished MSc thesis, Ohio State University, Columbus, Ohio. 169 pp.

Smith, N. D. (1978). Sedimentation processes and patterns in a glacier-fed lake with low sediment input. *Canadian Journal of Earth Sciences*, **15**, 741–756.

Smith, N. D. and Ashley, G. M. (1985). Proglacial Lacustrine Environment. In *Glacial Sedimentary Environments* (G. M. Ashley and N. D. Smith, eds), SEPM Short Course 16, Tulsa, OK, pp. 135–216.

Smith, N. D. and Minter, W. E. L. (1980). Sedimentologic

controls of gold and uranium in two Witwatersrand palaeoplacers. *Economic Geology*, **75**, 1–14.

Smith N. D., Vendl, M. A. and Kennedy, S. K. (1982). Comparison of sedimentation regimes in four glacier-fed lakes of western Alberta. In *Research in Glacial, Glacio-fluvial, and Glacio-lacustrine Systems* (R. Davidson-Arnott, W. Nickling, and B. D. Fahey, eds) pp. 203–233, 6th Guelph Symposium on Geomorphology. Norwich: Geo Books Publishing.

Smith, R. F. and Boardman, J. (1989). The use of soil information in the assessment of the incidence and magnitude of historical flood events in upland Britain. In *Floods: Hydrological, Sedimentological and Geomorphological Implications* (K. Beven and P. Carling, eds) pp. 185–197, Chichester: John Wiley & Sons.

Smith, T. E. (1970). Gold resource potential of the Denali bench gravels, Valdez Creek mining district, Alaska. *United States Geological Survey, Professional Paper 700-D*, pp. 146–152.

Smith, T. E. (1981b). Geology of the Clearwater Mountains, south-central Alaska. *Alaska Division of Geological and Geophysical Surveys, Geologic Report 60*, 72 pp.

Soil Survey Staff (1951). *Soil Survey Manual.* Washington: United States Department of Agriculture, 503 pp.

Soil Survey Staff (1975). *Soil Taxonomy. United States Department of Agriculture Handbook 436.* Washington, DC: United States Government Printing Office. 754 pp.

Sokolov, V. N. Osipov, V. I. and Tolkachev, M. D. (1980). The electron microscopic studies of pore space of solids by a method of conjugate surfaces. *Journal of Microscopy*, **120**, 363–366.

Solheim, A. (1991). The depositional environment of surging sub-polar tidewater glaciers: a case study of the morphology, sedimentation and sediment properties in a surge affected marine basin outside Nordaustlandet, the Northern Barents Sea. *Norsk Polarinstitutt, Skrifter*, **194**, 97 pp.

Solheim, A. and Kristoffersen, Y. (1984). The Physical Environment Western Barents Sea, 1:1,500 000, Sheet 3. Sediments above the upper regional unconformity: thickness, seismic stratigraphy and outline of the glacial history. *Norsk Polarinstitutt Skrifter*, **179B**, 26 pp.

Solheim, A., Russwurm, L., Elverhøi, A. and Nyland Berg, M. (1990). Glacial geomorphic features in the northern Barents Sea: direct evidence for grounded ice and implications for the pattern of deglaciation and late glacial sedimentation. In *Glaciomarine Environments: Processes and Sediments* (J. A. Dowdeswell and J. D. Scourse, eds), Geological Society of London, Special Publication No. 53, pp. 253–268.

Sollid, J. L. and Carlsson, A. B. (1984). De Geer moraines and eskers in Pasvik, North Norway. In Ten years of Nordic till research (L. -K. Königsson, ed.), *Striae*, **20**, 55–61.

Sollid, J. L. and Sørbel, L. (1984). Distribution and genesis of moraines in Central Norway. In Ten years of Nordic till research (L. -K. Königsson, ed.), *Striae*, **20**, 63–67.

Sollid, J. L. and Sørbel, L. (1988). Influence of temperature conditions in formation of end moraines in Fennoscandia and Svalbard. *Boreas*, **17**, 553–558.

Sollid, J. L., Andersen, S., Hamre, N. Kjeldsen, O., Salvigsen, O., Stuerød, S., Tveitå, T. and Wilhelmsen, A. (1973). Deglaciation of Finnmark, North Norway. *Norsk Geografisk Tidsskrift*, **27**, 233–325.

Solomon, S. M., Swinden, S., Emory-Moore, M. and Proudfoot, D. M. (1990). Spatial analysis in marine placer exploration: preliminary results from St. George's and Port au Port Bays, Newfoundland. Newfoundland Department of Mines and Energy, Mineral Development Division, Current Research, Report No. 90-1, pp. 77–83.

Sonett, C. P. and Finney, S. A. (1990). The spectrum of radiocarbon. *Philosophical Transactions of the Royal Society of London*, **330(A)**, 413–426.

Sønstegaard, E. (1979). Glaciotectonic deformation structures in unconsolidated sediments at Os, south of Bergen. *Norsk Geologisk Tidsskrift*, **59**, 223–228.

Souchez, R. A. and Lorrain, R. D. (1991). *Ice Composition and Glacier Dynamics*. Berlin: Springer-Verlag, 207 pp.

Sparks, B. W. and West, R. G. (1964). The drift landforms around Holt, Norfolk. *Transactions of the Institute of British Geographers*, **35**, 27–35.

Spell, T. L. and McDougall, J. (1992). Revisions to the age of the Brunhes-Matuyama boundary and the Pleistocene geomagnetic polarity timescale. *Geophysical Research Letters*, **19**, 1181–1184.

Spencer, A. M. (1971). Late Precambrian glaciation in Scotland. *Memoirs of the Geological Society of London*, **6**, 1–48.

Spencer, A. M. (1975). Late Precambrian glaciation in the North Atlantic region. In *Ice Ages: Ancient and Modern* (A. E. Wright and F. Moseley, eds) pp. 217–240, Geological Journal, Special Issue 6, Liverpool: Seel House Press.

Spencer, A. M. (1981). The Late Precambrian Port Askaig tillite in Scotland. In *Earth's Pre-Pleistocene Glacial Record* (M. J. Hambrey and W. B. Harland, eds) pp. 632–636, Cambridge: Cambridge University Press.

Sperling, C. H. B. and Cooke, R. U. (1985). Laboratory simulation of rock weathering by salt crystallization and hydration processes in hot, arid environments. *Earth Surface Processes and Landforms*, **10**, 541–555.

Spiegler, D. (1989). Ice-rafted Cretaceous and Tertiary fossils in Pleistocene/Pliocene Sediments, OCD Leg 104 sites. In *Scientific Results of ODP 104, Ocean Drilling Program* (O. Eldholm, J. Thiede and Taylor, E. et. al., eds), College Station, pp. 739–744.

Spindler, M. (1990). A comparison of Arctic and Antarctic sea ice and the effects of different properties on sea ice biota. In Geological history of the polar oceans; Arctic versus Antarctic. NATO ASI Series, C 308 (U. Bleil and J. Thiede, eds), Dordrecht: Kluwer Academic Publishers, pp. 173–186.

Stabell, B. (1986). A diatom maximum horizon in upper Quaternary deposits. *Geologische Rundschau*, **75**, 175–184.

Stahman, D. (1992). Composition and shape of fluted and equant drumlins in thee north-central New York drumlin field. Unpublished MSc thesis, Lehigh University, Bethlehem, Pennsylvania.

Stalker, A. MacS. (1960). Ice-pressed drift forms and associated deposits in Alberta. *Geological Survey of Canada Bulletin 57*, 38 pp.

Stalker, A. MacS. (1973). The large interdrift bedrock blocks of the Canadian Prairies. *Geological Survey of Canada, Paper 75–1A*, pp. 421–422.

Stalker, A. S. (1976). Megablocks, or the enormous erratics of the Albertan Prairies. *Geological Survey of Canada, Paper 76–1C*, pp. 185–188.

Stanford, S. D and Mickelson, D. M. (1985). Till fabric and deformational structures in drumlins near Waukesha, Wisconsin, U.S.A. *Journal of Glaciology*, **31**, 220–228.

Stankowski, W. (1976). Till – its genesis and diagenesis. Zeszyty Naukowe Uniwersytetu im Adama Mickiewicza, Poznan, Poland, *Geograficzna Seria*, **12**, 266 pp.

Stauffer, B. R. (1989). Dating of ice by radioactive isotopes. In *The Environmental Record in Glaciers and Ice Sheets* (H. Oeschger, and C. C. Langway Jr., eds) pp. 123–139, New York: John Wiley & Sons.

Stea, R. R. and Brown, Y. (1989). Variation in drumlin orientation, form and stratigraphy relating to successive ice flows in southern and central Nova Scotia. *Sedimentary Geology*, **62**, 223–240.

Stea, R. R. and Mott, R. J. (1989). Deglaciation environments and evidence for glaciers of Younger Dryas age in Nova Scotia, Canada. *Boreas*, **18**, 169–187.

Stea, R. R., Turner, R. G., Finck, P. W. and Graves, R. M. (1989). Glacial dispersal in Nova Scotia: a zonal concept. In *Drift Prospecting* (R. N. W. DiLabio and W. B. Coker, eds), *Geological Survey of Canada, Paper 89–20*, pp. 155–169.

Steidtmann, J. R. (1973). Ice and snow in eolian sand dunes of southwestern Wyoming. *Science*, **179**, 796–798.

Steiner, J. and E. Grillmair. (1973). Possible galactic causes for periodic and episodic glaciations. *Geological Society of America Bulletin*, **84**, 1003–1018.

Stephan, H. -J. (1985). Deformations striking parallel to glacier movement as a problem in reconstructing its direction. *Geological Society of Denmark Bulletin*, **34**, 47–53.

Stephan, H. -J. (1987). Form, composition and origin of drumlins in Schleswig-Holstein. In *Drumlin Symposium* (J. Menzies and J. Rose, eds) pp. 335–345, Rotterdam: A.A. Balkema Publishers.

Stephan, H. -J. (1988). Origin of a till-like diamicton by shearing. In *Genetic Classification of Glacigenic Deposits* (R. P. Goldthwait and C. L. Matsch, eds) pp. 93–96, Rotterdam: A.A. Balkema Publishers.

Stephan, H. -J. and Ehlers, J. (1983). North German till types. In *Glacial Deposits in North-West Europe* (J. Ehlers, ed.) pp. 239–247, Rotterdam: A.A. Balkema Publishers.

Stephens, G. C., Evenson, E. B., Tripp, R. B. and Detra, D. (1983). Active alpine glaciers as a tool for bedrock mapping and mineral exploration: a case study from Trident Glacier, Alaska. In *Tills and Related Deposits: Genesis, Petrology, Application, Stratigraphy* (E. B. Evenson, Ch. Schlüchter and J. Rabassa, eds) pp. 195–204, Rotterdam: A.A. Balkema Publishers.

Stevens, R. L. (1990). Proximal and distal glacimarine deposits in southwestern Sweden: contrasts in sedimentation. In *Glaciomarine Environments: Processes and Sediments* (J. A. Dowdeswell and J. D. Scourse, eds), *Geological Society of London, Special Publication No. 53*, pp. 307–316.

Stewart, R. A. and Broster, B. E. (1990). Compositional variability of till in marginal areas of continental glaciers. In *Glacial Indicator Tracing* (R. Kujansuu and M. Saarnisto, eds) pp. 123–149, Rotterdam: A.A. Balkema Publishers.

Stewart, R. A., Mayberry, S. W. and Pickerill, M. J. (1988). Composition of till in the vicinity of the Lake Ellen Kimberlite and implications for the source of diamonds in glacial sediments of eastern Wisconsin. In *Prospecting in Areas of Glaciated Terrain – 1988*, Institution of Mining and Metallurgy, London and Canadian Institute of Mining and Metallurgy, Nova Scotia Department of Mines and Energy, Halifax, pp. 103–120.

Stieglitz, R. D. (1969). Surface textures of quartz and heavy-mineral grains from fresh-water environments: An application of Scanning Electron Microscopy. *Geological Society of America Bulletin*, **80**, 2091–2094.

Stihler, S. D., Stone, D. B. and Beget, J. E. (1992). 'Varve' counting vs. tephrochronology and 137Cs and 210Pb dating: a comparative test at Skilak Lake, Alaska. *Geology*, **20**, 1019–1022.

Stone, J. C. (1959). A description of glacial retreat features in mid-Nithsdale. *Scottish Geographical Magazine*, **75**, 164–168.

Storzer, D. and Wagner, G. A. (1969). Correction of thermally lowered fission track ages of tektites. *Earth and Planetary Science Letters*, **5**, 463–468.

Strass, I. F. (1978). Microtextures of quartz sand grains in coastal and shelf sediments, Møre, Western Norway. *Marine Geology*, **28**, 107–134.

Straw, A. (1960). The limit of the 'Last' Glaciation in north Norfolk. *Proceedings of the Geologists' Association*, **71**, 379–390.

Streeter, S. S., Belanger, P. E., Kellogg, T. B. and Duplessy, J. C. (1982). Late Pleistocene paleo-oceanography of the Norwegian-Greenland Sea: Benthic foraminiferal evidence. *Quaternary Research*, **18**, 72–90.

Streiff-Becker, R. (1951). Pot-holes and glacier mills. *Journal of Glaciology*, **1**, 488–490.

Strömberg, B. (1971). Isrecessionen i området kring Ålands hav. *Naturgeogrisk Institutionen Stockholm University, Forskningsrapp*, **10**, 1–156.

Stuiver, M. (1978a). Carbon–14 dating: a comparison of beta and ion counting. *Science*, **202**, 881–883.

Stuiver, M. (1978b). Radiocarbon timescale tested against magnetic and other dating methods. *Nature*, **273**, 271–274.

Stuiver, M. (ed.) (1993). Calibration 1993. *Radiocarbon*, **35**, 244 pp.

Stuiver, M. and Polach, H. A. (1977). Discussion: Reporting of 14C data. *Radiocarbon*, **19**, 355–363.

Stuiver, M., Heusser C. J. and Yang, I. C. (1978). North American glacial history extended to 75,000 years ago. *Science*, **200**, 16–21.

Stuiver, M., Kromer, B., Becker, B. and Ferguson, C. W. (1986). Radiocarbon age calibration back to 13,300 years BP and the 14C age matching of the German Oak and U.S. Bristlecone Pine chronologies. *Radiocarbon*, **28**, 969–979.

Stuiver, M., Braziunas, T. F., Becker, B. and Kromer, B. (1991). Climatic, solar, oceanic and geomagnetic influences on late-glacial and Holocene atmospheric 14C/12C change. *Quaternary Research*, **35**, 1–24.

Stupavsky, M. and Gravenor, C. P. (1975). Magnetic fabric around boulders in till. *Geological Society of America Bulletin*, **86**, 1534–1536.

Stupavsky, M. and Gravenor, C. P. (1984). Paleomagnetic dating of Quaternary sediments: a review. In *Quaternary Dating Methods. Developments in Palaeontology and Stratigraphy*, 7 (W. C. Mahaney, ed.) pp. 123–140,

Amsterdam: Elsevier.

Stupavsky, M., Gravenor, C. P. and Symons, D. T. A. (1974). Paleomagnetism and magnetic fabric of the Leaside and Sunnybrook tills near Toronto, Ontario. *Geological Society of America Bulletin*, **85**, 1233–1236.

Sturdy, R. G., Allen, R. H., Bullock, P., Catt, J. A. and Greenfield, S. (1979). Paleosols developed on Chalky Boulder Clay in Essex. *Journal of Soil Science*, **30**, 117–137.

Sturm, M. (1979). Origin and composition of clastic varves. In *Moraines and Varves: Origin/Genesis/Classification* (Ch. Schlüchter, ed.) pp. 281–285, Rotterdam: A.A. Balkema Publishers.

Sugden, D. E. (1970). Landforms of deglaciation in the Cairngorm mountains, Scotland. *Transactions of the Institute of British Geographers*, **51**, 201–219.

Sugden, D. E. (1977). Reconstruction of the morphology, dynamics, and thermal characteristics of the Laurentide Ice Sheet at its maximum. *Arctic and Alpine Research*, **9**, 21–47.

Sugden, D. E. (1978). Glacial erosion by the Laurentide Ice Sheet. *Journal of Glaciology*, **20**, 367–391.

Sugden, D. E. and John, B. S. (1976). *Glaciers and Landscape – A Geomorphological Approach*. London: Edward Arnold. 320 pp.

Sumartojo, J. and Gostin, V. A. (1976). Geochemistry of the Late Precambrian Sturt Tillite, Flinders Ranges, South Australia. *Precambrian Research*, **3**, 243–252.

Sun, T. C. and Yang, H. J. (1961). The great Ice Age glaciation in China. *Acta Geologica Sinica*, **41**, 234–244 (in Chinese).

Sutherland, D. G. (1980). Problems of radiocarbon dating deposits from newly deglaciated terrain: examples from the Scottish Lateglacial. In *Studies in the Lateglacial of Northwest Europe* (J. J. Lowe, J. M. Gray and J. E. Robinson, eds) pp. 139–149, Oxford: Pergamon Press.

Sutherland, D. G. (1981). The high-level marine shell beds of Scotland and the build-up of the last Scottish ice-sheet. *Boreas*, **10**, 247–254.

Sutherland, D. G. (1984). Modern glacier characteristics as a basis for inferring former climates with particular reference to the Loch Lomond Stadial. *Quaternary Science Reviews*, **3**, 291–309.

Sutherland, D. G. (1986). A review of Scottish marine shell radiocarbon dates, their standardization and interpretation. *Scottish Journal of Geology*, **22**, 145–164.

Sutinen, R. (1985). On the subglacial sedimentation of hummocky moraines and eskers in Northern Finland. *Striae*, **22**, 21–26.

Svensson, H. (1958). Morphometrisher Beitrag zur Charakterisierung von Glazialtalern. *Zeitschrift für Gletscherskunde und Glazialgeologie*, **4**, 99–105.

Svensson, H. (1959). Is the cross-section of a glacial valley a parabola? *Journal of Glaciology*, **3**, 362–363.

Svensson, H. (1988). Ice-wedge casts and relict polygonal patterns in Scandinavia. *Journal of Quaternary Science*, **3**, 57–67.

Svensson, H. and Frisen, R. (1964). Hallmorfolgi och isrorelser inom ett alvaromraade vid Degerhamn. *Seriges Geografisk Årsbok*, **40**, 19–30.

Swanson, D. K. (1985). Soil catenas on Pinedale and Bull Lake moraines, Willow Lake, Wind River Mountains, Wyoming. *Catena*, **12**, 329–342.

Swineford, A. and Frye, J. C. (1955). Petrographic comparison of some loess samples from western Europe with Kansas loess. *Journal of Sedimentary Petrology*, **25**, 3–23.

Syvitski, J. P. M. (1991a). Towards an understanding of sediment deposition on glaciated continental shelves. *Continental Shelf Research*, **11**, 897–937.

Syvitski, J. P. M. (1991b). Principles, methods, and application of particle size analysis. Cambridge: Cambridge University Press, 368 pp.

Syvitski, J. P. M. and Murray, J. W. (1981). Particle interaction in fjord suspended sediment. *Marine Geology*, **39**, 215–242.

Syvitski, J. P. M., Burrell, D. C. and Skei, J. M. (1987). *Fjords: Processes and Products*. New York: Springer-Verlag, 379 pp.

Tadeu, A., Veiga, C., Jorge, G. and Barros, C. (1991). A genetic-exploration model for the alluvial gold placers of the Brazilian Amazon. In *International Symposium on Alluvial Gold Placers; Abstracts* (G. Hérail, ed.), La Paz, Bolivia, pp. 163–164.

Talbot, C. J. and Von Brunn, V. (1987). Intrusive and extrusive (micro)melange couplets as distal effects of tidal pumping by a marine ice sheet. *Geological Magazine*, **124**, 513–525.

Tang, J, Fu, H. and Yu, Z. (1987). Stratigraphy, type and formation condition of the Late Precambrian banded iron ores in South China. *Chinese Journal of Geochemistry*, **6**, 331–341.

Tang Yongyi (1987). A preliminary research on the genesis and environment of the sediments in the Lu Shan Mountains from some of their microtextures and microstructures (in Chinese). *Journal of Glaciology and Geocryology*, **9**, 165–170.

Taylor, R. E. (1987). *Radiocarbon Dating: An Archaeological Perspective*. New York: Academic Press, 212 pp.

Taylor, S. R. and McLennan, S. M. (1985). The continental crust: its composition and evolution. Oxford: Blackwell Scientific, 312 pp. Teller, J. T. (1987). Proglacial lakes and the southern margin of the Laurentide Ice Sheet. In *NorthAmerica and Adjacent Ocean During the Last Deglaciation* (W. F. Ruddiman and H. E. Wright, Jr., eds), *The Geology of North America, K–3*, Geological Society of America, pp. 39–69.

Teller, J. and Clayton, L. (1983). *Glacial Lake Agassiz. Geological Association of Canada Special Paper, 26*, 451 pp.

Terzaghi, K. and Peck, R. B. (1967). *Soil mechanics in Engineering Practice*. New York: John Wiley & Sons, 729 pp.

Theakstone, W. H. (1967). Basal sliding and movement near the margin of the glacier Østerdalsisen, Norway. *Journal of Glaciology*, **6**, 805–816.

Thomas, G. S. P. (1984a). A late Devensian glaciolacustrine fan-delta at Rhosesmor, Clwyd, North Wales. *Geological Journal*, **19**, 125–141.

Thomas, G. S. P. (1984b). Sedimentation of a subaqueous esker-delta at Strabathie, Aberdeenshire. *Scottish Journal of Geology*, **20**, 9–20.

Thomas, G. S. P. and Connell, R. J. (1985). Iceberg drop, dump and grounding structures from Pleistocene glaciolacustrine sediments, Scotland. *Journal of Sedimentary Petrology*, **55**, 243–249.

Thomas, R. H. (1979). The dynamics of marine ice sheets. *Journal of Glaciology*, **24**, 167–177.

Thompson, K. S. R. (1972). The last glaciers in western Perthshire. Unpublished PhD thesis, University of Edinburgh, Edinburgh, Scotland.

Thompson, R. and Oldfield, F. (1986). *Environmental Magnetism*. London: Allen and Unwin, 227 pp.

Thomsen E. and Vorren, T. O. (1986). Macrofaunal palaeoecology and stratigraphy in late Quaternary shelf sediments off northern Norway. *Palaeogeography, Palaeoclimatology, Palaeoecology*, **56**, 103–150.

Thorarinsson, S. (1981). The application of tephrochronology in Iceland. In *Tephra Studies* (S. Self and R. S. J. Sparks, eds) pp. 109–134, Dordrecht: D. Reidel.

Thorleifson, L. H. and Kristjansson, R. J. (In prep). Surficial geology of the Beardmore–Geraldton area, District of Thunder Bay, Northern Ontario. Geological Survey of Canada in cooperation with the Ontario Geological Survey.

Thorp, P. W. (1986). A mountain icefield of Loch Lomond Staial age, western Grampians, Scotland. *Boreas*, **15**, 83–97.

Thorp, P. W. (1991). The glaciation and glacial deposits of the western Grampians. In *Glacial deposits in Great*

Britain and Ireland (J. Ehlers, P. L. Gibbard and J. Rose, eds) pp. 137–149, Rotterdam: A.A. Balkema.

Tilas, D. (1740). Tanckar om Malmletande, i anledning af löse gråstenar. Kongl. Svenska Vetenskaps – *Academiens Handlingar 1739–1740*, I, 190–193.

Tilmann, S. E. (1973). The effect of grain orientation on Fourier shape analysis. *Journal of Sedimentary Petrology*, **43**, 867–869.

Titus, J. G. (1987). The greenhouse effect, rising sea level and society's response. In *Sea Surface Studies: A Global View* (R. J. N. Devoy, ed.) pp. 499–528, London: Croom Helm.

Totten, S. M. (1969). Overridden recessional moraines n north-central Ohio. *Bulletin of the Geological Society of America*, **80**, 1931–1946.

Tovey, N. K. (1973). Quantitative analysis of electron micrographs of soil structure. *Proceedings of the International Symposium on Soil Structure*. Stockholm: Swedish Geotechnical Society, pp. 50–57.

Tovey, N. K. (1974). Some applications of electron microscopy to soil engineering. In *Soil Microscopy* (G. K. Rutherford, ed.) pp. 119–142, Kingston, Ontario: The Limestone Press.

Tovey, N. K. (1978). Potential developments in stereoscopic scanning electron microscope studies of sediments. In *Scanning Electron Microscopy in the Study of Sediments* (W. B. Whalley, ed.) pp. 105–117, Norwich: Geo Abstracts.

Tovey, N. K. (1986). Microfabric, chemical and mineralogical studies of soils: techniques. *Geotechnical Engineering*, **17**, 131–166.

Tovey, N. K. and Smart, P. (1986). Intensity gradient techniques for orientation analysis of electron micrographs. *Scanning*, **8**, 75–90.

Tovey, N. K. and Sokolov, V. N. (1981). Quantitative SEM methods for soil fabric analysis. *Scanning Electron Microscopy*, 1981 **I**, 537–554.

Tovey, N. K. and Wong, K. Y. (1973). The preparation of soils and other geological materials for the electron microscope, *Proceedings of International Symposium on Soil Structure*, Stockholm: Swedish Geotechnical Society, pp. 58–66.

Tovey, N. K. and Wong, K. Y. (1978a). Preparation, selection and interpretation problems in scanning electron microscope studies of sediments. In *Scanning Electron Microscopy in the Study of Sediments* (W. B. Whalley, ed.) pp. 181–199, Norwich: Geo Abstracts.

Tovey, N. K. and Wong, K. Y. (1978b). Optical techniques for analysing scanning electron micrographs. *Scanning Electron Microscopy*, 1978, **I**, 381–392.

Tovey, N. K., Eyles, N. and Turner, R. (1973). Sand grain

selection procedures for observations in the SEM. *Scanning Electron Microscopy*, 1978, **I**, 393–399.

Townsend, P. D., Rendell, H. M., Aitken, M. J., Bailiff, I. K., Durrani, S. A., Faïn, J., Grün, R., Mangini, A, Mejdahl, V. and Smith, B. W. (1988). Thermoluminescence and Electron-Spin-Resonance Dating. *Quaternary Science Reviews*, **7 (3/4)**, 291 pp.

Trenhaile, A. S. (1971). Drumlins; their distribution, orientation and morphology. *Canadian Geographer*, **15**, 113–126.

Trenhaile, A. S. (1975). The morphology of a drumlin field. *Annals of the Association of American Geographers*, **65**, 297–312.

Tricart, J. (1970). Geomorphology of Cold Environments. London: Macmillan, 320 pp.

Tricart, J. and Cailleux, A. (1962). Le Modele Glaciare et Nival. Paris: Sedes, 508 pp.

Trinkler, E. (1930). The ice-age on the Tibetan Plateau and in the adjacent regions. *Geographical Journal*, **75**, 225–232.

Tsui, P. C., Cruden, D. M. and Thomson, S. (1988). Mesofabric, microfabric and submicrofabric of ice-thrust bedrock, Highvale mine, Wabamun Lake area, Alberta. *Canadian Journal of Earth Sciences*, **25**, 1420–1431.

Tsui, P. C., Cruden, D. M. and Thomson, S. (1989). Fabric studies of ice-thrust shear zones as applied to problems in geotechnical engineering. In *Applied Quaternary Research* (E. F. J. De Mulder and B. P. Hageman, eds) pp. 147–164, Rotterdam: A.A. Balkema Publishers.

Tucker, M. E. (1986). *The Field Description of Sedimentary Rocks*. Geological Society of London Handbook Series, Open University Press, 112 pp.

Tufnell, L. (1984). *Glacier Hazards*. London: Longman, 97 pp.

Turner, C. (1989). Type sections and Quaternary deposits. In *Quaternary Type Sections: Imagination or Reality?* (J. Rose and Ch. Schlüchter, eds) pp. 41–44, Rotterdam: A.A. Balkema.

Turner, C. and West, R. G. (1968). The subdivision and zonation of interglacial periods. *Eiszeitalter und Gegenwart*, **19**, 93–101.

Tyrrell, J. B. (1915). Gold on the North Saskatchewan River. *Canadian Mining Institute Transactions*, **18**, 160–173.

Ussing, N. V. (1903). Om Jyllands Hedesletter og Teorierne om deres Dannelse. *Copenhagen, Overs k. danske Vid. Selsk. Förh.*, **2**, 99–165.

Valentine, K. W. G. and Dalrymple, J. B. (1975). The identification, lateral variation, and chronology of two buried paleocatenas at Woodhall Spa and West Runton, England. *Quaternary Research*, **5**, 551–590.

van den Bogaard, P., Hall, C. M., Schmincke, H. -U. and York, D. (1987). 40Ar/39Ar laser dating of single grains; ages of Quaternary tephra from the east Eifel volcanic field, FRG. *Geophysical Research Letters*, **14**, 1211–1214.

van den Bogaard, P., Hall, C. M., Schmincke, H. -U. and York, D. (1989). Precise single-grain 40Ar/39Ar dating of a cold to warm climate transition in Central Europe. *Nature*, **342**, 523–525.

van Gijssel, K. (1987). A lithostratigraphic and glaciotectonic reconstruction of the Lamstedt Moraine, Lower Saxony (FRG). In *Tills and Glaciotectonics* (J. J. M. van der Meer, ed.) pp. 145–155, Rotterdam: A.A. Balkema Publishers.

van Ginkel, M. (1991). Vertikale variabiliteit in de micromorfologie van dikke till afzettingen in Nederland. Unpublished MSc thesis, University of Amsterdam, Amsterdam, The Netherlands.

Van Vliet, B. and Langohr, R. (1981). Correlation between fragipans and permafrost with special reference to silty Weichselian deposits in Belgium and Northern France. *Catena*, **8**, 137–154.

Van Vliet-Lanöe, B. (1985). Frost effects in soils. In *Soils and Quaternary Landscape Evolution* (J. Boardman, ed.) pp. 117–158, Chichester: John Wiley & Sons.

Van Vliet-Lanöe, B. (1986). Le pedocomplexe du dernier interglaciare (de 125 000 à 75 000 B.P.). Variations de faciès et signification paléoclimatique du sud de la Pologne à l'ouest de la Bretagne. *Bulletin de l'Association Francaise Quaternaire*, **1/2**, 139–150.

Van Vliet-Lanöe, B. (1990). The genesis and age of the argillic horizon in Weichselian loess of northwestern Europe. *Quaternary International*, **5**, 49–56.

Van Vliet-Lanöe, B. and Hequette, A. (1987). Activité éolienne et sables limoneaux sur les versants exposés au nord-est de la Peninsule du Brogger, Spitzberg du Nord-Ouest (Svalbard). In *Loess and Periglacial Phenomena* (M. Pecsi and H. M. French, eds) pp. 103–123, Budapest: Akademia Kiado.

Vasconcelos, P. and Kyle, R. J. (1989). Supergene geochemistry and crystal morphology of gold in a semi-arid weathering environment, application to gold prospecting. 13th International Geochemical Exploration Symposium, Rio de Janeiro 1989, Association of Exploration Geochemists, pp. 22–24.

Veillette, J. J. (1989). Ice movements, till sheets and glacial transport in Abitibi-Timiskaming, Quebec and Ontario. In *Drift prospecting* (R. N. W. DiLabio and W. B. Coker, eds), *Geological Survey of Canada, Paper 89–20*, pp. 139–154.

Velichko, A. A. and Faustova, M. A. (1986). Glaciations in the East European region of the USSR. In *Quaternary Glaciations in the Northern Hemisphere* (V. Sibrava, D. Q. Bowen and G. M. Richmond, eds) *Quaternary Science Reviews*, **5**, 447–461.

Velichko, A. A. and Morozova, T. D. (1969). Structure of the loess mass of the Russian plain. *Akademiya Nauk SSSR, Izvestiya, Seriya Geograficheskaya*, **4**, 18–29 (in Russian).

Verosub, K. L. (1988). Geomagnetic secular variation and the dating of Quaternary sediments. In *Dating Quaternary Sediments* (D. J. Easterbrook, ed.), *Geological Society of America Special Paper, 227*, pp. 123–138.

Vincent, P. J. (1975). Urzezbienie powierzchni ziarn kwarcowych z osadów morenowych róznego wieku oraz wynikajace z tego ogólne wnioski geomorfologiczne (Surface textures of quartz grains from a morainic chronosequence and some general geomorphological implications). *Przeglad Geograficzny*, **47**, 577–582.

Vincent, P. J. (1976). Some periglacial deposits near Aberystwyth, Wales, as seen with a scanning electron microscope. *Biuletyn Peryglacjalny*, **25**, 59–64.

Vincent, P. J. and Lee, M. P. (1981). Some observations on the loess around Morecambe Bay, Northwest England. *Proceedingas of the Yorkshire Geological Society*, **43**, 281–294.

Vinje, T. E. (1980). Some satellite-tracked iceberg drifts in the Antarctic. *Annals of Glaciology*, **1**, 83–87.

Virkkala, K. (1952). On the bed structure of till in eastern Finland. *Commission Géologique de Finlande Bulletin*, **157**, 97–109.

Virkkala, K. (1960). On the striations and glacier movements in the Tampere region, southern Finland. *Societé Géologique de Finlande, Comptes Rendus*, **32**, 159–176.

Visser, J. N. J. (1983). Submarine debris flow deposits from the Upper Carboniferous Dwyka Tillite Formation in the Kalahari Basin, South Africa. *Sedimentology*, **30**, 511–523.

Visser, J. N. J. (1985). The Dwyka Formation along the north-western margin of the Karoo Basin in the Cape Province, *South Africa. Transactions of the Geological Society of South Africa*, **88**, 37–48.

Visser, J. N. J. (1989). The Permo-Carboniferous Dwyka Formation of Southern Africa: Deposition by a predominantly subpolar marine ice sheet. *Palaeography, Palaeoclimatology, Palaeoecology*, **70**, 377–391.

Visser, J. N. J. (1990). Glacial bedforms at the base of the Permo-Carboniferous Dwyka Formation along the western margin of the Karoo Basin, South Africa. *Sedimentol-*

ogy, **37**, 231–245.

Visser, J. N. J. and Hall, K. J. (1984). A model for the deposition of the Permo-Carboniferous Kruitfontein boulder pavement and associated beds, Elandsvlei, South Africa. *Transactions of the Geological Society of South Africa*, **87**, 161–168.

Visser, J. N. J., Colliston, W. P. and Terblanche, J. C. (1984). The origin of soft-sediment deformation structures in Permo-Carboniferous glacial and proglacial beds, South Africa. *Journal of Sedimentary Petrology*, **54**, 1183–1196.

Vivian, R. and Bocquet, G. (1973). Subglacial cavitation phenomena under the Glacier d'Argentière, Mont Blanc, France. *Journal of Glaciology*, **12**, 439–451.

Vogt, P. R., Crane, K. and Sundvor, E. (1994). Deep Pleistocene iceberg plowmarks on the Yermak Plateau: Sidescan and 3.5 kHz evidence for thick calving ice fronts and a possible marine ice sheet in the Arctic Ocean. *Geology*, **22**, 403–406.

Von Endt, D. W. (1979). Techniques of amino acid dating. In Pre-Llano Cultures of the Americas: Paradoxes and Possibilities (R. L. Humphrey and D. Stanford, eds.) pp. 71–100, Washington D.C.: The Anthropological Society of Washington.

Von Huene, R., Larson, E. and Crough, J. (1973). Preliminary study of ice-rafted erratics as indicators of glacial advances in the Gulf of Alaska. In *Initial Reports of the Deep Sea Drilling Project, 18*. Washington, D.C: United States Government Printing Office, pp. 835–842.

Vorren, T. O. (1973). Glacial geology of the area between Jostedalsbreen and Jotunheimen, South Norway. *Norges Geologiske Undersoekelse*, **291**, 46 pp.

Vorren, T. O. (1977). Grain-size distribution and grain-size parameters of different till types on Hardangervidda, south Norway. *Boreas*, **6**, 219–227.

Vorren, T. O., Hald, M., Edvardsen, M. and Lind-Hansen, O. -W. (1983). Glacigenic sediments and sedimentary environments on continental shelves: general principles with a case study from the Norwegian shelf. In *Glacial Deposits in North-West Europe* (J. Ehlers, ed.) pp. 61–73, Rotterdam: A.A. Balkema Publishers.

Vorren, T. O., Hald, M. and Thomsen, E. T. (1984). Quaternary sediments and environments on the continental shelf off Northern Norway. *Marine Geology*, **57**, 229–257.

Vorren, T. O., Hald, M. and Lebesbye, E (1988). Late Cenozoic environments in the Barents Sea. *Paleoceanography*, **3**, 601–612.

Vorren, T. O., Lebesbye, E., Andereasson, K. and Larsen, K. B. (1989). Glacigenic sediments on a passive continental margin as exemplified by the Barents Sea. *Marine*

Geology, **85**, 251–272.

Vorren, T. O., Lebesbye, E. and Larsen, K. B. (1990). Geometry and genesis of the glacigenic sediments in the southern Barents Sea. In *Glaciomarine Environments: Processes and Sediments* (J. A. Dowdeswell and J. D. Scourse, eds.), Geological Society of London, Special Publication No. 53, pp. 269–288.

Vorren, T. O., Richardsen, G., Knutsen, S. M. and Henriksen, E. (1991). Cenozoic erosion and sedimentation in the western Barents Sea. *Marine and Petroleum Geology*, **8**, 317–340.

Vortisch, W. B., Mahaney, W. C. and Fecher, K. (1987). Lithology and weathering in a paleosol sequence on Mount Kenya: East Africa. *Geologica et Paleontologica*, **21**, 245–255.

Vreeken. W. J. and Mücher, H. J. (1981). (Re)deposition of loess in southern Limbourg, The Netherlands. 1. Field evidence for conditions of deposition of the lower silt loam complex. *Earth Surface Processes and Landforms*, **6**, 337–354.

Walcott, R. I. (1970). Isostatic response to loading of the crust in Canada. *Canadian Journal of Earth Sciences*, **7**, 716–727.

Walder, J. S. (1982). Stability of sheet flow of water beneath temperate glaciers and implications for glacier surging. *Journal of Glaciology*, **28**, 273–293.

Walder, J. S. (1986). Hydraulics of subglacial cavities. *Journal of Glaciology*, **32**, 439–445.

Walker, D. A. (1978). Preparation of geological samples for Scanning Electron Microscopy. *Scanning Electron Microscopy*, 1978, **I**, 185–190.

Walker, M. J. C. (1973). The nature and origin of a series of elongated ridges in the Morley Flats area of the Bow Valley, Alberta. *Canadian Journal of Earth Sciences*, **10**, 1340–1346.

Walker, M. J. C., Coope, G. R. and Lowe, J. J. (1993). The Devensian (Weichselian) Lateglacial palaeoenvironmental record from Gransmoor, East Yorkshire, England. *Quaternary Science Reviews*, **12**, 659–680.

Walker, R. G. (1984). General introduction: Facies, facies sequences and facies models. In *Facies Models, Second ean* (R. G. Walker, ed.) pp. 1–9, Geoscience Canada, Reprint Series 1, Toronto, Geological Association of Canada.

Walter, M. R. and Bauld, J. (1983). The association of sulphate evaporites, stromatolitic carbonates and glacial sediments: examples from the Proterozoic of Australia and the Cainozoic of Antarctica. *Precambrian Research*, **21**, 129–148.

Walter, R. C. (1989). Application and limitation of fission-track geochronology to Quaternary tephras. *Quaternary International*, **1**, 35–46.

Warnke, D. A. (1971). The shape and surface texture of loess particles: discussion. *Geological Society of America Bulletin*, **82**, 2357–2360.

Warnke, D. A. and Gram, R. (1969). The study of mineral-grain surfaces by interference microscopy. *Journal of Sedimentary Petrology*, **39**, 1599–1604.

Warren, W. P. (1991). Fenitian (Midlandian) glacial deposits and glaciation in Ireland and the adjacent offshore regions. In *Glacial Deposits in Great Britain and Ireland* (J. Ehlers, P. L. Gibbard and J. Rose, eds) pp. 79–88, Rotterdam: A.A. Balkema Publishers.

Warren, W. P. and Ashley, G. M. (1994). Origin of the ice-contact stratified ridges (eskers) of Ireland. *Journal of Sedimentary Research*, **64**, 433–449.

Warren, W. P. and Croot, D. G. (eds) (1994). *Formation and Deformation of Glacial Deposits*, Rotterdam: A.A. Balkema Publishers, 223 pp.

Washburn, A. L. (1969). Weathering, frost action and patterned ground in the Mesters Vig District, Northeast Greenland. *Meddelelser om Grønland*, **176**, 303 pp.

Washburn, A. L. (1980). Geocryology: a Survey of Periglacial Processes and Environments. London: Edward Arnold Publishing, 406 pp.

Wateren, D. F. M. van der (1987). Structural geology and sedimentology of the Dammer Berge push moraine, FRG. In *Tills and Glaciotectonics* (J. J. M. van der Meer, ed.) pp. 157–182, Rotterdam: A.A. Balkema Publishers.

Wateren, D. F. M. van der (1992). Structural geology and sedimentology of push moraines – processes of soft sediment deformation in a glacial environment and the distribution of glaciotectonic styles. PhD thesis, University of Amsterdam, Amsterdam, The Netherlands, 238 pp.

Watson, R. A. and Wright, H. E., Jr. (1980). The end of the Pleistocene: a general critique of chronostratigraphic classification. *Boreas*, **9**, 153–163.

Wayne, W. J. and Corte, A. E. (1983). Multiple glaciations of the Cordon del Plata, Mendoza, Argentina. *Palaeogeography, Palaeoclimatology, Palaeoecology*, **42**, 185–209.

Webb, P. N. and McKelvey, P. C. (1959). Geological investigations in South Victoria Land, Antarctica. *New Zealand Journal of Geology and Geophysics*, **2**, 120–136.

Weber, J. N. (1958). Recent grooving in lake bottom sediments at Great Slave Lake, Northwest Territories. *Journal of Sedimentary Petrology*, **28**, 333–341.

Weddle, T. K. (1992). Late Wisconsinan stratigraphy in the lower Sandy valley, New Sharon, Maine. *Geological Society of America Bulletin*, **104**, 1350–1363.

Weertman, J. (1961). Mechanism for the formation of inner moraines found near the edge of cold ice caps and ice sheets. *Journal of Glaciology*, **3**, 965–978.

Weertman, J. (1972). General theory of water flow at the base of a glacier or ice sheet. *Reviews of Geophysics and Space Physics*, **10**, 287–333.

Weertman, J. (1986). Basal water and high-pressure basal ice. *Journal of Glaciology*, **32**, 455–463.

Wehmiller, J. F. (1984). Relative and absolute dating of Quaternary mollusks with amino acid racemization: evaluation, applications, and questions. In *Quaternary Dating Methods, Developments in Palaeontology and Stratigraphy, 7* (W. C. Mahaney, ed.) pp. 171–193, Amsterdam: Elsevier.

Wehmiller, J. F. (1986). Amino acid racemization geochronology. In *Dating Young Sediments* (A. J. Hurford, E. Jager and J. E. M. Ten Cate, eds) pp. 139–158, CCOP Technical Publication No. 16.2

Wehmiller, J. F. (1990). Amino acid racemization: applications in chemical taxonomy and chronolostratigraphy of Quaternary fossils. In *Skeletal Biomineralization: Patterns, Processes and Evolutionary Trends, 1* (J. Carter, ed.) pp. 583–608, New York: Van Nostrand-Reinhold.

Wehmiller, J. F., Belknap, D. F., Boutin B. S., Mirecki, J. E., Rahaim, S. D. and York, L. L. (1988). A review of the aminostratigraphy of Quaternary mollusks from United States Atlantic Coastal Plain sites. In *Dating Quaternary Sediments* (D. J. Easterbrook, ed) pp. 69–110, Geological Society of America Special Paper, 227.

Wehmiller, J. F., York, L. L., Belknap, D. F. and Snyder, S. W. (1992). Theoretical correlations and lateral discontinuities in the Quaternary aminostratigraphic record of the U.S. Atlantic Coastal Plain. *Quaternary Research*, **38**, 275–291.

Weir, A. H., Catt, J. A. and Madgett, P. A. (1971). Postglacial soil formation in the loess of Pegwell Bay, Kent (England). *Geoderma*, **5**, 131–149.

Werner, A. (1990). Lichen growth rates for the northwest coast of Spitsbergen, Svalbard. *Arctic and Alpine Research*, **22**, 129–140.

West, R. G. (1967). The Quaternary of the British Isles, In *The Quaternary, vol 2* (K. Rankama, ed) pp. 1–87, New York: Wiley-Interscience.

West, R. G. (1977). *Pleistocene Geology and Biology*, 2nd edn. London: Longman, 440 pp.

West, R. G. (1989). The use of type localities and type sections in the Quaternary, with especial reference to East Anglia. In *Quaternary Type Sections: Imagination or Reality?* (J. Rose and Ch. Schlüchter eds) pp. 3–10, Rotterdam: A.A. Balkema Publishers.

West, R. G. and Donner, J. J. (1956). The glaciations of East Anglia and the East Midlands: a differentiation based on stone orientation measurements of the tills. *Quarterly Journal of the Geological Society of London*, **112**, 69–91.

West, R. G., Rose, J., Coxon, P., Osmaston, H. and Lamb, H. H. (1988). The record of the Cold Stages. In *The Past Three Million Years: Evolution of Climatic Variability in the North Atlantic Region* (N. J. Shackleton, R. G. West and D. Q. Bowen, eds) pp. 95–112, London: Cambridge University Press.

Westgate, J. A. and Gorton, M. P. (1981). Correlation techniques in tephra studies. In *Tephra Studies* (S. Self and R. S. J. Sparks, eds) pp. 73–94, Dordrecht: D. Reidel.

Westgate, J. A., Easterbrook, D. J., Naeser, N. D. and Carson, R. J. (1987). Lake Tapps tephra: an early Pleistocene stratigraphic marker in the Puget Lowland, Washington. *Quaternary Research*, **28**, 340–355.

Westgate, J. A., Stemper, B. A. and Péwé, T. L. (1990). A 3 m.y. record of Pliocene-Pleistocene loess in interior Alaska. *Geology*, **18**, 858–861.

Westgate, J. A., Walter, R. C and Naeser, N (eds) (1992). Tephrachronology: stratigraphic applications of tephra. *Quaternary International*, **13/14**, 203 pp.

Whalley, W. B. (1974). Observations on the Kaiserstuhl loess. *Geological Magazine*, **111**, 84–86.

Whalley, W. B. (ed.) (1978a). *Scanning Electron Microscopy in the Study of Sediments*. Norwich: Geo Abstracts, 414 pp.

Whalley, W. B. (1978b). An SEM examination of quartz grains from sub-glacial and associated environments and some methods for their characterization. *Scanning Electron Microscopy*, 1978, **I**, 355–358.

Whalley, W. B. (1979). Quartz silt production and sand grain surface textures from fluvial and glacial environments. *Scanning Electron Microscopy*, **1979**, 547–551.

Whalley, W. B. (1982). A preliminary scanning electron microscope study of quartz grains from a dirt band in the Tuto ice tunnel, northwest Greenland. *Arctic and Alpine Research*, **14**, 355–360.

Whalley, W. B. (1983). Desert Varnish. In *Chemical Sediments and geomorphology: Precipitates and Residua in the Near-surface Environment* (A. S. Goudie and K. Pye, eds) pp. 197–126, London: Academic Press.

Whalley, W. B. (1985). Scanning electron microscopy and the sedimentological characterisation of soils. In *Geomorphology and Soils* (K. S. Richards, R. R. Arnett and S. Ellis, eds) pp. 183–201, London: George Allen & Unwin.

Whalley, W. B. and Krinsley, D. H. (1974). A scanning electron microscope study of surface textures of quartz grains from glacial environments. *Sedimentology*, **21**, 87–105.

Whalley, W. B. and Langway, C. C. Jr. (1980). A scanning electron microscope examination of subglacial quartz grains from Camp Century core, Greenland – a preliminary study. *Journal of Glaciology*, **25**, 125–131.

Whalley, W. B. and Orford, J. D. (1982). Analysis of SEM images of sedimentary particle form by fractal dimension and Fourier analysis methods. *Scanning Electron Microscopy*, 1982, **II**, 639–647.

Whalley, W. B. and Orford, J. D. (1986). Practical methods for analysing and quantifying two-dimensional images. In *The Scientific Study of Flint and Chert* (G. de G. Sieveking and M. B. Hart, eds) pp. 235–242. Cambridge: Cambridge University Press.

Whalley, W. B. and Orford, J. D. (1989). The use of fractals and pseudofractals in the analysis of two-dimensional outlines: review and further exploration. *Computers and Geosciences*, **15**, 185–197.

Whalley, W. B., Marshall, J. R. and Smith, B. J. (1982). Origin of desert loess from some experimental observations, *Nature*, **300**, 433–435.

Whalley, W. B., Smith, B. J., McAlister, J. J. and Edwards, A. J. (1987). Aeolian abrasion of quartz particles and the production of silt-size fragments: preliminary results. In *Desert Sediments: Ancient and Modern* (L. Frostick and I. Reid, eds) pp. 129–138, Geological Society of London, Special Publication 35.

Whalley, W. B., Gellatly, A. F., Gordon, J. E. and Hansom, J. D. (1990). Ferromanganese rock varnish in North Norway: a subglacial origin. *Earth Surface Processes and Landforms*, **15**, 265–275.

Wheeler, D. A. (1984). Using parabols to describe cross-sections of glaciated valleys. *Earth Surface Processes and Landforms*, **9**, 391–394.

White, G. W. (1962). Multiple tills of end moraines, United States Geological Survey Prof. Paper 450–C, C96–C98.

White, G. W. (1974). Buried glacial geomorphology. In *Glacial Geomorphology* (D. R. Coates, ed.) pp. 331–349, SUNY at Binghamton.

Whiteman, C. A. and Rose, J. (1992). Thames river sediments of the British Early and Middle Pleistocene. *Quaternary Science Reviews*, **11**, 363–375.

Whitehouse, I. E. (1987). Geomorphology of a compressional plate boundary, Southern Alps, New Zealand. In *International Geomorphology 1986, Part 1* (V. Gardiner ed.) pp. 897–924, Chichester: John Wiley & Sons.

Whittecar, G. R. and Mickelson, D. M. (1979). Composition, internal structures, and an hypothesis of formation for drumlins, Waukesha County, Wisconsin, U.S.A. *Journal of Glaciology*, **22**, 357–370.

Whittington, R. J. (1977). A late-glacial drainage pattern in the Kish Bank area and post-glacial sediments in the Central Irish Sea. In *The Quaternary History of the Irish Sea* (C. Kidson and M. J. Tooley eds) pp. 55–68, Liverpool: Seel House Press.

Wickham, S. S., Johnson, W. H. and Glass, H. D. (1988). Regional geology of the Tiskilwa Till Member, Wedron Formation, Northeastern Illinois. Illinois State Geological Survey Circular 543, 35 pp.

Wigley, T. M. L., Ingram, M. J. and Farmer, G. (1981). *Climate and History: Studies in Past Climates and their Impact on Man*. Cambridge: Cambridge University Press, 530 pp.

Wilke, H. and Ehlers, J. (1983). The thrust moraine of Hamburg-Blankenese. In *Glacial Deposits in North-West Europe* (J. Ehlers, ed.) pp. 331–333, Rotterdam: A.A. Balkema Publishers.

Williams, G. E. (1975). Late Precambrian glacial climate and the Earth's obliquity. *Geological Magazine*, **112**, 441–544.

Williams, G. E. (1979). Sedimentology, stable-isotope geochemistry and palaeoenvironment of dolostones capping Late Precambrian glacial sequences in Australia. *Journal of the Geological Society of Australia*, **26**, 377–386.

Williams, G. E. (1985). Solar affinity of sedimentary cycles in the Late Precambrian Elatina Formation. *Australian Journal of Physics*, **38**, 1027–1043.

Williams, G. E. (1986). Precambrian permafrost horizons as indicators of palaeoclimate. *Precambrian Research*, **32**, 233–242.

Williams, G. E. (1989). Late Precambrian tidal rhythmites in South Australia and the history of the Earth's rotation. *Journal of the Geological Society of London*, **146**, 97–111.

Williams, G. J. (1974). *Economic Geology of New Zealand*. The Australasian Institute of Mining and Metallurgy, Australia, Monograph Series 4, 490 pp.

Williams, H. and King, A. F. (1979). Trepassey map area, Newfoundland. *Geological Survey of Canada Memoir, 389*, 24 pp.

Williams, R. B. G. (1975). The British climate during the Last Glaciation: an interpretation based on periglacial phenomenon. In *Ice Ages: Ancient and Modern* (A. E. Wright and F. Moseley, eds) pp. 95–120, Geological Journal, Special Issue 6, Liverpool: Seel House Press.

Williams, R. B. G. (1987). Frost weathered mantles on the Chalk. In *Periglacial Processes and Landforms in Britain and Ireland* (J. Boardman, ed.) pp. 127–133, Cambridge:

Cambridge University Press.

Williams, R. E. and Farvolden, R. N. (1967). The influence of joints on the movement of ground water through glacial till. *Journal of Hydrology*, **5**, 163–170.

Willman, H. B. and Frye, J. C. (1970). Pleistocene stratigraphy of Illinois. *Illinois State Geological Survey Bulletin 94*, 204 pp.

Willman, H. B., and Payne, J. N. (1942). Geology and mineral resources of the Marseilles, Ottawa, and Streator Quadrangles. *Illinois State Geological Survey Bulletin 66*, 388 pp.

Wilson, A. T. (1964). Origin of ice ages: an ice shelf theory for Pleistocene glaciation. *Nature*, **201**, 147–149.

Wilson, P. (1978). A scanning electron microscope examination of quartz grain surface textures from the weathered Millstone Grit (Carboniferous) of the southern Pennines, England: a preliminary report. In *Scanning Electron Microscopy in the Study of Sediments* (W. B. Whalley, ed.) pp. 307–318, Norwich: Geo Abstracts.

Wilson, P. (1979a). Surface features of quartz and sand grains from the Brassington formation. *Mercian Geologist*, **7**, 19–30.

Wilson, P. (1979b). Experimental investigation of etch pit formation on quartz sand grains. *Geological Magazine*, **116**, 477–482.

Wilson, P. (1980). Surface textures of regolith quartz from the Southern Pennines. *Geological Journal*, **15**, 113–129.

Wilson, P., Bateman, R. M. and Catt, J. A. (1981). Petrography, origin and environment of deposition of the Shirdley Hill Sand of Southwest Lancashire, England. *Proceedings of the Geologists' Association*, **92**, 211–229.

Wilson, T. R. S., Thomson, J., Hydes, D. J., Colley, S., Culkin,, F and Sørensen, J. (1986). Oxidation fronts in pelagic sediments: diagenetic formation of metal-rich layers. *Science*, **232**, 972–975.

Wintle, A. G. (1973). Anomalous fading of thermoluminescence in mineral samples. *Nature*, **245**, 143–144.

Wintle, A. G. 1982. Thermoluminescence properties of fine-grain minerals in loess. *Soil Science*, **134**, 164–170.

Wintle, A. G. (1990). A review of current research on TL dating of loess. *Quaternary Science Reviews*, **9**, 385–397.

Wintle, A. G., Shackleton, N. J. and Lautridou, J. P. (1984). Thermoluminescence dating of periods of loess deposition and soil formation in Normandy. *Nature*, **310**, 491–493.

Wise, S. M. (1980). Cesium–137 and Lead–210: A review of the techniques and some applications in geomorphology. In *Timescales in Geomorphology* (R. A. Cullingford,

D. A. Davidson, and J. Lewin, eds) pp. 109–127. New York: John Wiley & Sons.

Wisniewski, E. (1965). On a method of examining the microstructure of ground–moraine clay (in Polish). *Czasopismo Geograficzne*, **36**, 291–294.

Wold, B. and Østrem, G. (1979). Subglacial constructions and investigations at Bondhusbreen, Norway. *Journal of Glaciology*, **23**, 363–379.

Woldstedt, P. (1952). Die Entstehung der Seen in den ehemals vergletscherten Gebieten. *Eiszeitalter und Gegenwart*, **2**, 146–153.

Wolf, T. C. W. and Thiede, J. (1991). History of terrigenous sedimentation during the past 10 My in the North Atlantic (ODP Legs 104, 105 and DSDP Leg 81). *Marine Geology*, **101**, 83–102.

Wood, J. (1973). Stratigraphy and depositional environments of upper Huronian rocks of the Rawhide Lake – Flack Lake area, Ontario. *Geological Association of Canada Special Paper, 12*, pp. 73–95.

Woodland, A. W. (1970). The buried tunnel valleys of East Anglia. *Proceedings of the Yorkshire Geological Society*, **37**, 521–578.

Woodworth-Lynas, C. M. T. (1988). Ice scours in the geological record: why they are not seen. Program with abstracts, Joint Annual Meeting, Geological Association of Canada, Mineralogical Association of Canada, Canadian Society of Petroleum Geologists, Canadian Geophysical Union, Memorial University of Newfoundland, May 23–25, **13**, A137.

Woodworth-Lynas, C. M. T. (1990). Observed deformation structures beneath relict iceberg scours. In *Ice Scouring and the Design of Offshore Pipelines* (J. I. Clark, I. Konuk, F. Poorooshasb, J. Whittick and C. M. T. Woodworth-Lynas, eds), Proceedings of an invited workshop, April 18–19th, Calgary, Alberta. Canada Oil and Gas Lands Administration and Centre for Cold Ocean Resources Engineering, pp. 103–125.

Woodworth-Lynas, C. M. T. and Guigné, J. Y. (1990). Iceberg scours in the geological record: examples from glacial Lake Aggasiz. In *Glaciomarine Environments: Processes and Sediments* (J. A. Dowdeswell and J. D. Scourse, eds), Geological Society of London, Special Publication No. 53, pp. 217–223.

Woodworth-Lynas, C. M. T. and Landva, J. (1988). Sediment deformation by ice scour. Contract report for Supply and Services Canada.

Woodworth-Lynas, C. M. T., Christian, D., Seidel, M. and Day, T. (1986). Relict ice scours on King William Island, N.W.T. In *Ice Scour and Seabed Engineering* (C. F. M. Lewis, D. R. Parrott, P. G. Simpkin and J. T. Buckley, eds), Environmental Studies Revolving Funds Report No.

049, pp. 64–71.

Woodworth-Lynas, C. M. T., Josenhans, H. W., Barrie, J. V., Lewis, C. F. M. and Parrott, D. R. (1991). The physical processes of seabed disturbance during iceberg grounding and scouring. *Continental Shelf Research*, **11**, 939–961.

Worsley, P. (1967). Problems of naming the Pleistocene deposits of the north-east Cheshire Plain. *Mercian Geologist*, **2**, 51–55.

Worsley, T. R., Nance, R. D. and Moody, J. B. (1986). Tectonic cycles and the history of the Earth's biochemical and paleoceanographic record. *Paleoceanography*, **1**, 233–263.

Wright, A. E. and Moseley, F. (eds) (1975). *Ice Ages: Ancient and Modern*. Geological Journal, Special Issue 6, Liverpool: Seel House, 320 pp.

Wright, H. E. Jr. (1957). Stone orientation in the Wadena drumlin field, Minnesota. *Geografiska Annaler*, **39**, 19–31.

Wright, H. E. Jr. (1962). Role of the Wadena Lobe in the Wisconsin glaciation of Minnesota. *Geological Society of America Bulletin*, **73**, 73–100.

Wright, H. E. Jr. (1973). Tunnel valleys, glacial surges, and subglacial hydrology of the Superior Lobe, Minnesota. In *The Wisconsinan Stage* (R. F. Black, R. P. Goldthwait and H. B. Willman, eds), Geological Society of America Memoir No. 136. pp. 251–276.

Wright, R. and Anderson, J. B. (1982). The importance of sediment gravity flow to sediment transport and sorting in a glacial marine environment: Eastern Weddell Sea, Antarctica. *Geological Society of America Bulletin*, **93**, 957–963.

Wright, W. B. (1937). *The Quaternary Ice Age*. 2nd edn, London: MacMillan, 478 pp.

Wu, P. and Peltier, W. A. (1983). Glacial isostatic adjustment and the free air gravity anomaly as a constraint on deep mantle viscosity. *Geophysical Journal of the Royal Astronomical Society*, **74**, 377–449.

Yaalon, D. H. (1969). Origin of desert loess. Abstracts. 8th INQUA Congress, Paris, 2, p. 755.

Yaalon, D. H. and Ganor, E. (1973). The influence of dust on soils during the Quaternary. *Soil Science*, **116**, 146–155.

Yeend, W. and Shawe, D. R. (1989). Gold in placer deposits. *United States Geological Survey Bulletin*. **1857(G)**, 1–13.

Yeo, G. M. (1981). The Late Proterozoic Rapitan glaciation in the northern Cordillera. In *Proterozoic Basins of Canada* (F. H. A. Campbell, ed.), Geological Survey of Canada. Paper 81–10, pp. 25–46.

Yokoyama, Y., Reyss, J. -L. and Guichard, F. (1977).

Production of radionuclides by cosmic rays at mountain altitudes. *Earth and Planetary Science Letters*, **36**, 44–50.

Yoshida, M. and Watanabe, O. (1983). The genesis of englacial till along the shear planes of Urumqui No. 3. Glacier Tien Shan. *Journal of Glaciology and Cryopedology*, **5**, 201–207.

Young, G. M. (1970). An extensive early Proterozoic glaciation in North America? *Palaeogeography, Palaeoclimatology, Palaeoecology*, **7**, 85–101.

Young, G. M. (1973). Tillites and aluminous quartzites as possible time markers for middle Precambrian (Aphebian) rocks of North America. In *Huronian Stratigraphy and Sedimentation* (G. M. Young, ed.), Geological Association of Canada Special Paper, 12, pp. 97–127.

Young G. M. (1981). The Early Proterozoic Gowganda Formation, Ontario, Canada. In *Earth's pre-Pleistocene glacial record* (M. J. Hambrey and W. B Harland, eds) pp. 807–812, Cambridge: Cambridge University Press.

Young, G. M. (1982). The Late Proterozoic Tindir Group, east-central Alaska: evolution of a continental margin. *Geological Society of America Bulletin*, **93**, 759–783.

Young, G. M. (1983). Tectono-sedimentary history of Early Proterozoic rocks of the northern Great Lakes region. In Early Proterozoic geology of the GreatLakes region (L. G. Medaris Jr., ed.) pp. 15–32, *Geological Society of America, Memoir 160*.

Young, G. M. (1988). Proterozoic plate tectonics, glaciation and iron-formations. *Sedimentary Geology*, **58**, 127–144.

Young, G. M. (1991). The geologic record of glaciation: relevance to the climatic history of Earth. *Geoscience Canada*, **18**, 100–108.

Young, G. M. and Gostin, V. A. (1988a). Stratigraphy and sedimentology of Sturtian glacigenic deposits in the western part of the North Flinders Basin, South Australia. *Precambrian Research*, **39**, 151–170.

Young, G. M. and Gostin, V. A. (1988b). Temperate glaciation in a rift setting: Late Proterozoic (Sturtian) deposits of the North Flinders Basin, South Australia. *Geological Society of America: Abstracts with Programs* **20**, A135.

Young, G. M. and Gostin, V. A. (1989a). An excetionally thick upper Proterozoic (Sturtian) glacial succession in the Mount Painter area, South Australia. *Geological Society of America Bulletin*, **101**, 834–845.

Young, G. M. and Gostin, V. A. (1989b). Depositional environment and regional stratigraphic significance of the Serle Conglomerate: a Late Proterozoic submarine fan complex, South Australia. *Palaeogeography, Palaeoclimatology, Palaeoecology*, **71**, 237–252.

Young, G. M. and Gostin, V. A. (1991). Late Proterozoic (Sturtian) succession of the North Flinders Basin, South Australia; an example of temperate glaciation in an active rift setting. In *Glacial Marine Sedimentation; Paleoclimatic Significance* (J. B. Anderson and G. M. Ashley, eds), Geological Society of America Special Paper, 261, pp. 207–222.

Young, G. M. and McLennan, S. M. (1981). Early Proterozoic Padlei Formation, Northwest Territories, Canada. In *Earth's pre-Pleistocene glacial record* (M. J. Hambrey and W. B. Harland, eds), pp. 790–794, Cambridge: Cambridge University Press.

Young, G. M. and Nesbitt, H. W. (1985). The Gowganda Formation in the southern part of the Huronian outcrop belt, Ontario, Canada: stratigraphy, depositional environments and regional tectonic significance. *Precambrian Research*, **29**, 265–301.

Young, G. M., Jefferson, C. W., Delaney, G. D. and Yeo, G. M. (1979). Middle and Late Proterozoic evolution of the northern Canadian Cordillera and Shield. *Geology*, **7**, 125–128.

Young, J. A. T. (1974). Ice wastage in Glenmore, upper Spey Valley, Inverness-shire. *Scottish Journal of Geology*, **10**, 147–158.

Young, J. A. T. (1975). Ice wastage in Glen Feshie, Inverness-shire. *Scottish Geographical Magazine*, **91**, 91–101,

Young, J. A. T. (1980). The fluvioglacial landforms of mid-Strathdearn, Inverness-shire. *Scottish Journal of Geology*, **16**, 209–220.

Zeuner, F. E. (1959). *The Pleistocene Period*. London: Hutchinson, 447 pp.

Zhang, J. and Crowley, T. J. (1989). Historical climate records in China and reconstruction of past climates, *Journal of Climate*, **2**, 833–849.

Zheng, B. X. (1989). The influence of Himalayan uplift on the development of Quaternary glaciers. *Zeitschrift für Geomorphologie Supplementband 76*, 89115.

Zilliacus, H. (1976). De Geer-moräner och isrecessionen i södra Finlands östra delar. *Terra*, **88**, 176–184.

Zilliacus, H. (1981). De Geer-moränera på replot och Björkön i Vasa skärgård. *Terra*, **93**, 12–24.

Zilliacus, H. (1987). De Geer moraines in Finland and the annual moraine problem. *Fennia*, **165**, 145–239.

Zilliacus, H. (1989). Genesis of De Geer Moraines in Finland. *Sedimentary Geology*, **62**, 309–317.

Zotikov, I. A. (1986). *The Thermophysics of Glaciers*. Dordrecht: D. Reidel, 275 pp.

Zumberge, J. H. (1955). Glacial erosion in tilted rock layers. *Journal of Geology*, **63**, 149–158.

INDEX

Aavatsmarkbreen, Spitsbergen 21
Aber, Scotland 260, 281
Aberdeen, Scotland 163
ablation moraine 63, 133
ablation 130, 138
ablation zone 137
abrasion 25, 90, 214, 368, 414
    wind 215, 217, 226
absolute age 380
accelerator-mass spectrometer 377
accretionary tectonics 251
active ice flow, landforms 29–46
acoustic sediment tracing 197, 199
adhesion structures from aeolian processes 227
    ripples 227
    warts 227
Adelaide Geosyncline 240, 247, 248
aeolian dust 221
aeolian effects (SEM) 226, 370–371
aeolian polishing of grains 368
aeolian sediments 214–215, 256, 280, 312, 370–371, 392, 393,
    399, 413
    placers 463
aeolian silt 213
aeolian transport processes 214
age estimate 380
airphoto interpretation in geomorphic mapping 278, 279
Africa 1, 242, 248, 370
Alaska 19, 56, 58, 78, 91, 131, 147, 150, 157, 196, 222, 229,
    233, 242, 246, 248, 288, 312, 395, 402, 410, 441, 448, 450,
    451, 454, 463, 464
    placers 464–468
        map 465
Alberta 41, 78, 91, 147, 296–298, 311, 373, 384, 459
    placers 459
Aleutian Mountains, Alaska 464

algae (red) 186
Algeria 172, 173
Alice Creek, B.C. 456, 457
Aldun River, Russia 463
Allen Park, Ontario 111
allokinetic discontinuities 29, 36, 40, 41, 46, 61
    classification 43
Allstar Mine, B.C. 456, 458
alluvial fans 215
alluvial fan/fan-delta placers 442, 445–447, 453, 472
    debris flows 445–446, 474
alluvial placer deposits 442–447, 455, 462, 470, 472, 477
    alluvial fan/fan-delta 442, 445–447, 453, 462
    creek and gulch 442–444, 455, 457, 462, 472
    river 442, 444–445, 455, 459, 462, 469
Altonian 311
Amazon River 183, 184
Amersfoort Interstadial 273
amino acid geochronology 393–396
    amino acid ratios 396
    epimerization process 394
    leucine 394
    isoleucine 394
    protein breakdown 394
    racemization process 394
    rate of racemization/epimerization 395
amino acid stratigraphy 181, 377
amorphous silica see silica
amplitude of sea level change (past/future) 318
Amur River, Russia 463
Ancocala-Ananea Basin, Peru 469
Andes 1, 469
Anglian Glaciation 256, 265, 279, 304
anisotropic defects 48
anisotropy of magnetic susceptibility 409
anorthosites 251

Antarctica 21, 74, 85, 92, 175, 176, 183, 184, 185, 186, 196,
    216, 217, 218, 221, 222, 242, 243, 245, 246, 247, 248, 270,
    298, 302, 311, 317, 319, 372, 391, 411
    continental shelf 246
    East 196
    West 19, 31, 61
Antarctic diatom ooze belt 189
Antarctic soils 373
apatite 391
Appalachian Mountains 113, 118, 309, 415, 416, 422, 425, 435
    placers in Québec 462
aquiclude 53
aquifers 51, 93
Ar/Ar dating see K/Ar dating
ARC models 333
Archean Era 192, 193, 241, 248, 249, 252, 416, 428, 431
arctic brown soil 304
Arctic Ocean 183, 185, 187, 196, 208, 209
Arctic diatom ooze belt 189
arêtes 64
Arena valley, Antarctica 391
Argentina 233, 337, 469
argillans 48, 61, 351, 354
argillasepic fabric 347
argillic horizon 311
argillic soil 304
argillites 191, 192
argon/argon dating see K/Ar dating
arid polar glaciers 298, 299
    model 299
aridity 213
Arizona 464
arroyos 215
arsenic 434, 437
artesian effects 110
Arthur, Ontario 258, 259
aschromite 422
Asia 19, 20, 141, 214, 215, 217, 219, 221, 229, 242, 248, 369,
    371, 373, 396, 410, 441, 451
    placer deposits 462–464
    map 463
Ash at Ashton 408
ash layers (volcanic) 209, 310
asthenosphere 315, 318, 324, 329
Athabasca Glacier, Alberta 78
Atlantic Ocean:
    North Atlantic 185, 187, 188, 191, 196, 198, 202, 205, 208,
        260, 379, 403
    crust 251
'Atlantic'-type ocean 250, 251
Atlin area, B.C. 454, 458, 459
atmospheric pressure systems 225, 227
attrition from aeolian processes 214
aufeis 125
auriferous gravel 452, 456, 457, 458, 462
auriferous till 447, 448, 450, 462

Austfonna Ice Cap, Svalbard 198, 199
Australia 1, 164, 175, 176, 195, 240, 241, 243, 244, 246, 247,
    248
autokinetic discontinuities 29, 36, 37, 41, 46, 48–49, 61, 71
    classification 43

backwasting see ice marginal backwasting/downwasting
Baffin Bay 196
Baffin Island 81, 89, 217, 312, 373, 395
Baker Lake, N.W.T. 117, 123, 432
Balanides 200
ball and pillow structures 44, 46, 51
Ballarat Mine, B.C. 447, 471
Balloch, Scotland 260
Baltic Ice Lake 152
Baltic Shield see Fennoscandian Shield
Baltic shorelines 326
banding (structures) 41, 44, 337
Barbados 317
barchan dunes 222
Barents Sea 179, 180, 183, 185, 195, 198–206
Barents Sea Ice Sheet 196, 199
Barents Sea/Northern Norwegian Shelf 198–206
    model of lithofacies associations 210
Barents Sea Trough 203
Barham Soil 304, 308, 313
barnacles 186, 200
basal crevasses 81, 82, 123, 131
basal crevasse infillings 80
basal ice flow 21, 426
basal ice velocities 19, 426
basal till 59, 146, 183, 201, 202, 416, 418
basalt 388
Bavaria 7, 111, 237
beach grains 370
bearing capacity failure in ice scoured sediments 164
Bear Island Fan (model) 203, 205
    debris lobes 203
    ice berg furrows 203
    moraine ridge 203
    trough fill 203
Bear Island Trough 197, 199, 200
Beauceville, Québec 461
Beaufort Sea, Arctic Canada 162, 165, 176
Bebdouro Formation, Brazil 170
bedding planes in sediment 36, 46, 48, 111, 343
bedform development see subglacial bedform development
bedform elongation ratio 67
bed limits 36
bedrock forms see plucked bedrock forms and streamlined
    bedrock
Beggs Gulch, B.C. 442
benthic calcareous tests 185
benthic foraminifers 185, 190, 195, 197, 207, 208, 271, 403
    Cd/Ca ratios 190, 191

benthic organisms 186, 190, 197, 202, 207, 208, 211, 395
Bering Sea 448
Berthelsen's 'empty' locality 270
Biferten Gletscher, Switzerland 73, 74
Billingen moraine, Sweden 152
bimasepic fabric 347
biogenic enrichment of magnetite 410
biogenic opal production 189
    undersaturation of sea water 189
biogenic silica 190
biogenic sediments 186, 383
biomass 261
biomineralization 394
biostratigraphy 181, 255, 261
biotite 387
bioturbation 60, 181, 189, 207, 208
biozones 262
    pollen assemblages 262
Bingham visco-plastic material 23
Birch Creek, B.C. 459
bird footprints 172
birefringence 343, 347
black beach sands 450
'blade-shaped' fragments in loess 214
Blairgowrie, Scotland 107
Blattnick moraines 64, 67, 160
block emplacement (bedrock) 50
'block-in-matrix' mélange 45
Bloody Canyon, California 154, 390, 400
Bloomington moraine, Illinois 150
Boothia Peninsula, N.W.T. 435
Blomstrandbreen, Spitsbergen 78, 79
blow-out dunes 222
Blue Glacier 19
Bodaibo River, Russia 464
Boden See see Lake Constance
Bolivia 469
bones 394
Bornhóved tunnel valley 93
Boston Harbour, Mass. 315
bottom waters 189, 197, 206, 207
    carbon dioxide content 189, 206, 207
    corrosive effects 210
    currents 200
    fauna 186, 190, 197, 202
    oxygen content 189, 197, 206, 207, 208
    temperatures 185
    ventilation 197
boudins 48, 49
boudinage 31, 46
boulder beds 38
boulder belts/aprons 64
Boulder Creek, B.C. 459
boulder pavements 38, 290
boulder train tracing 433
Boulton's glacio-isostatic spatial/temporal model 324–325, 333

Boulton's model of facies sequences in nearshore
    glacio-isostatically dominated sea-level cycle 332
Bradtville, Ontario 40
brackish water 61
Bråsvellbreen, Svalbard 167
Brazil 164 170, 175, 242, 248, 469
brecciation 31, 36, 41, 46, 49, 344
brine formation 189
brines (metal-charged) 248
Britain 7, 9, 19, 36, 38, 39, 83, 84, 85, 90, 92, 107, 133, 134,
    146, 148, 150, 162, 228, 233, 255, 256, 258, 259, 260, 262,
    266, 275, 276, 277, 279, 304, 305, 307, 309, 313, 315, 326,
    329, 330
    shoreline diagram 328
British Columbia 58, 59, 87, 89, 90, 91, 441, 442, 445–452,
    454–461, 470–476
    placer deposits 454–459
British Isles see Britain
British Ice Sheet 196, 208, 316, 324
brittle fracture 368, 369, 372
brittle-ductile shear 31, 54
Breiðamerkurjökull, Iceland 19, 78, 79
brickearths (Kent) 228
Broggi Glacier, Peru 371
Brooks Range, Alaska 464
Broomfield, England 304
Brorup Interstadial 273
Bruarjökull, Iceland 78
Brückner, E. 7
Brunhes Chron 406
    excursions 406
Brunhes/Matuyama reversal 270, 272, 404
Bs horizon 304
Buckeye Formation, Antarctica 175
Bull Lake Glaciation 306, 309, 390, 397
bulletstones 290
Bullion Mine, B.C. 471, 472
buried ice 58, 125, 138
buried placer deposits 470–476
    exploration methods 474–475
    gulch deposits 474
    mining methods 475476
    modern valleys 472
    preservation 475
    trunk-valleys 472
buried soils see paleosol
buried valley see tunnel valleys
Burns Creek, B.C. 449
Burra Group (Sturtian) 248
burrows 203
Bylot Island, N.W.T. 115, 117, 427, 428

^{13}C isotopes 186, 402
^{14}C dating see radiocarbon dating
'C' meltwater channels 22

cadmium 424
cadmium/calcium rations *see* Cd/Ca ratios
Cairngorm Mountains, Scotland 90
calcareous tests 185
calcareous tills 337
calcitan 351
calcium carbonates 187, 208
calcrete 214
calibrated age 380
California 154, 306, 390, 400, 404, 464
    placers 464
        map 465
calving 208
calving bays 81
calving margins 165
calving rates 167
Cambrian Period 242
Canada 1, 9, 11, 19, 25, 26, 27, 29, 36, 37, 38, 40, 41, 42, 44,
    50, 58, 59, 66, 68, 74, 75, 78, 81, 84, 85, 86, 87, 89, 90, 91,
    92, 93, 94, 95, 98, 99, 101, 102, 103, 112, 113, 114, 115,
    117, 119, 120, 122, 123, 124, 125, 128, 131, 144, 147, 150,
    152, 153, 158, 162, 163, 165, 170, 171, 172, 175, 176, 191,
    192, 196, 203, 216, 217, 222, 229, 240, 247, 249, 253, 255,
    257, 258, 259, 266, 291-292, 296–298, 312, 315, 328, 331,
    367, 372, 373, 383, 384, 389, 395, 411, 413, 414, 416, 417,
    418, 420–431, 433–436, 438, 441, 442, 443, 445, 446, 447,
    448, 449, 450, 451, 452, 453, 454, 455, 456, 457, 458, 459,
    460, 461, 462, 470–476
Canadian eastern continental shelf 93, 94, 162, 176, 177, 203,
    451
Canadian Interior Plains 150
    Holocene stream gravel 459
    placer 459
Canadian Placer deposits 452–462
    Cordillera 452–459
    Interior Plains 459
    Central Canada 461–462
    Maritime Provinces 462
Canadian Shield 26, 86, 113, 292, 411, 412, 414, 425, 435
Cangalli Formation, Bolivia 470
Canoe Brook, Vermont 384
Canon Fiord, Ellesmere Island 216
Cape Cod, Mass. 395
carbon dioxide *see* CO_2
Carbon isotopes 185–186
carbonate compensation depth (CCD) 186, 187
carbonate cycle 186
carbonate-diamicton association 186
carbonate flux 189
carbonate fossils 393
carbonate production and dissolution 181, 189–190, 197, 206,
    207, 236, 248
    SEM-based dissolution indices 190
    skeletal 186, 187
carbonate sedimentation 186, 187, 188, 200, 202, 248
carbonate shell production 196, 198, 208, 388

carbonate as cement in loess 227, 236
carbonate in soils 305, 412, 424
Carboniferous Period 172, 173, 174
    placers 462
'card-house' structures 40, 61
Cariboo Mining area, B.C. 445, 446, 447, 448, 454–458, 470, 474
case hardening (of boulder surfaces) 217
Cascade Mountains 312, 404, 405, 464
Caspian Sea, C.I.S. 176
cassiterite 422
cataclasis 51
catastrophic discharge 72, 77, 99, 100
catena (soil) 306
cation-ratio dating 373, 398–399
    leaching of Na, Mg, K, and Ca 399
Catfish Creek Till 257
cathodoluminescence (CL) 359
cavernous weathering 217
cavetto 102
cavitation 102
CCD *see* carbonate compensation depth
Cd/Ca ratios 190, 191
Cenozoic 179, 185, 197, 198
    placers 464
    plate tectonics 196
Cenozoic Glaciations 195
'centipede' type solution mark (grains) 371
Cerler lateral moraine complex, Pyrenees, Spain 154
Chagvan Bay, Alaska 467
chalk fragments 208
    ice rafted chalk 208
Champaign-Urbana, Illinois 151
channel fill 142
chattermarks 98, 239, 361, 367, 368, 369
Chaudière, Québec 461, 462
Chelmsford, England 265
Chemical Index of Alteration (CIA) 249
chemical and biological dating techniques 381, 393–399
    amino acid 393–396
chemical stratigraphy 262, 264
chert 248
Chibougamau Formation, Québec 241
Chile 68, 469
Chimney Bluffs 30
China 19, 20, 214, 215, 217, 228, 233, 237, 242, 248, 369, 396,
    410
China Creek, Vancouver Island, B.C. 459
Chinatown Silt 396
chlorite 412, 414
chlorite-epidote schists 422
Christochina district, Skagway, Alaska 467
chromium 412, 420, 422, 423, 428, 431
    marine placers 462
chronostratigraphy 261, 263, 280–283
    British Isles and N.W. European stratigraphy 282
    terminology 280

CIA *see* Chemical Index of Alteration
Cincinnati, Ohio 387
circular disintegration ridges 158, 159
  'doughnuts' 159
  Dekalb mounds 64, 113, 159
  model of formation 159
circulation patterns and system 185, 189, 196
cirque glaciers 138
cirques 64, 217
C.I.S. 74, 176, 233, 448, 451, 462–464
Clay Accumulation Index 311
clay balls 47, 60
clay clots 47
clay eluviation 46, 61
clay illuviation 312
clay minerals 214, 305, 312, 338, 344, 347, 348, 351, 370, 412, 414
clay pebbles 343
clay shards 44
clay skins *see* argillans
clay translocation 301, 306, 307, 311, 351, 354
  rates 311, 312
clasts
  abraded 60
  striated 59, 60, 177, 298
clast fabric 37, 38, 41, 46, 57, 59, 60, 61, 62, 71, 75, 9, 132, 142, 143, 144, 145, 146, 147, 181, 239, 243, 288, 296, 298
  dip 38
  eigenvalues 142, 144, 147
  'herring bone' patterns 75
  inherited 38
  micro-fabric 38, 146
  replication 42, 46
clastic volcano 53
Clear Creek, Yukon Terr. 452, 453, 475
climate 139
Climatic Optimum 226
climatostratigraphy 262
Clyde Beds, Scotland 260
Clyde, New York State 112
Clyde Tunnel Valley, Scotland 92
CO_2 (atmospheric) 250, 251, 252
CO_2 content of groundwater 388
CO_2 content of sea water 186
coal fragments 208
coastal and marine placers 448–451, 459, 468, 469
  beach and strandline deposits 450, 459
  buried marine scarps 450
  continental shelf deposits 462
  drowned fluvial deposits 450
  hydrodynamic processes 451
  lag deposits 450, 451
  offshore paleochannel placers 451
  Russian marine placers 450
coastal flooding 317, 318
cobalt 423, 424, 431

Cobalt Lake, Ontario 461
Cobequid Bay, Nova Scotia 163, 172, 175
coccoliths 187, 189, 196
coccolithus pelagicus 187
Cochrane Till 119
codes (lithofacies) 289
Colchester Formation, England 268
cold-based subglacial conditions 18, 25
cold-climate conditions 213
collapse structures (from melt out) 141, 142, 144, 147, 149, 159
colluvial placers 451, 455, 462, 470
colluvial reworking involved in loess production 237
colluvium 38, 141
Colorado 147, 222, 223, 224, 225, 312, 464
Comfortless Glacier, Spitsbergen 393
comma forms 102
comminution *see* abrasion
Commonwealth of Independant States *see* CIS
compound discontinuities 54
compressional folds 147
compressive ice flow 138
conchoidal fractures 98, 101, 215, 372
Connecticut valley 384
continental crust (emergent) 252
continental shelf 181, 195, 202, 316
  gold deposits 462
  topography 200
continental slope/channels 192, 203
  gullies 200, 203
  sediment gravity flows 192, 203, 247 244, 245, 246, 296
  slide scars 203, 204
contraction cracking (salt induced) 215
Cook Inlet, Alaska 467
copper 414, 423, 424, 432
Copperneedle Esker, N.W.T. 433
coquina 201
coral 317, 393, 403
  terraces 403
Cordillera of North American 248, 425, 436
Cordilleran Ice Sheet 392, 454
corrasion 102
correlated age 380
correlation dating methods 381, 402–410
  paleomagnetism 405–410
  stable isotopes 402–403
  tephrochronology 403–405
corrugated moraines 64
cosmogenic-isotope dating 377, 381, 390–391
  surface exposure 390
Cougar Creek rhyolite flow 397
County Line Silt 396
County Harbour, Nova Scotia 451, 462
couplet *see* varve
cover moraines 64, 213
coverloam 228
coversands 224, 225, 226, 227, 234, 256, 370–371

Cowan's mélange concept 45
  'block-in-matrix' type 45
Crary Trough, Antarctica 183
creek and gulch placers 442–444, 457, 462, 472
  paleogulch placers 442, 472
crescentic furrows 102, 104
crescentic gouges 98
crescentic troughs 63, 83, 84
Cretaceous foraminifers 208
Cretaceous Period 242, 427, 428, 459
Cretaceous pyrite coals 459
crag and tail 64, 83–84, 85, 99, 300
  crescentic hollows 83, 84
crevasse fillings 64, 117
crevasse patterns 158
critical hydraulic gradient 22
critical lodgement index 33–34
Croftamie, Scotland 260, 281
Cromer Till, England 39, 50
cross-bedded sands and gravels 243, 245, 444
cross-cutting lineations 65, 66
cross-lamination 296
cross-valley moraine 51, 64, 81, 153
crushing and grinding mechanisms (grains) 213, 214, 215, 229,
    236, 368, 371, 414
cryogenic disruption (turbation) 41, 49, 351
cryostatic pressures 44
crystal growth (weathering) 214
crystal growth facets 367
Cumbria, England 146, 148
current regimes 185
curvilinear scour marks 162, 170
  berms (raised embankments) 162, 177
  central trough 162
cutans see argillans
cyclopels 244, 247

Dalradian Subgroup 194
dam see ice dams
Damara Orogen, South Africa 248
Darcy's Equation 22
dating methods and techniques:
  absolute dating 310
  amino acid stratigraphy 181, 377
  cation-ratio 373, 398–399
  cosmogenic-isotope 377, 381, 390–391
  dendrochronology 382–383
  electron spin resonance (ESR) 393
  historical records 381–382
  lead-210 (^{210}Pb) dating 389
  lichenometry 261, 278, 310, 373, 397–398
  obsidian-hydration 390, 396–397
  optically stimulated luminescence (OSL) 393
  oxygen isotope stratigraphic record 179, 181
  potassium-argon (K/Ar) 310

radiocarbon dating (^{14}C) 179, 181, 240, 310, 316,.377, 379,
    381, 385–387
radiometric 259, 261, 310
relative dating 310
  techniques 312
rock varnish 373, 398–399
rubidium/strontium 240
thermoluminescence (TL) 259, 310, 377, 392–393
thorium decay 181, 388
thorium/uranium (^{230}Th/^{234}U) dating 389, 403
uranium-lead 241
uranium-series dating 377, 381, 388–390
varve chronology 383–384
weathering rind 259, 261, 312, 373, 401
dead ice 293
décollement
  in ice 21
  in sediment
debris flow 56, 57, 58, 92, 138, 141, 142, 144, 146, 147, 192,
    203, 204, 205, 243, 244, 245, 256, 292, 309, 368, 454, 474
  tubulent flows 146
  viscosity 143
debris lobes (marine) 203
debris plumes 60, 210
debris release (flux) 32, 125
debris rheology 23, 24, 58
debris velocity 24, 45
debris thickness 24
decalcification 301, 307
decay curves (isotopes) 385
Deep Rose Lake, N.W.T. 120
  case study 120–130
Deep Sea Environment 206–211
deep-sea marine record (dating) 377
Deep Water Formation 183185
  brine formation 189
  circulation patterns 189
  oxygen-rich, dense water 189, 197
deep water lacustrine environments 192
deflation processes (aeolian) 215, 217
deformable beds (deforming sediment)10, 19, 45, 74, 91, 436
  deforming layer thickness 19, 24
  deforming layer surface velocity 24, 45
  deposition 45
  ice overburden loading 29, 50
  homogenisation 23
  syndepositional 29
deformation till 31, 143, 291
deformation structures 23, 29, 31, 57, 72, 76, 141, 146, 147, 164,
    165, 177, 269
  flow 143
  non-pervasive 23
  pervasive 23, 48, 61, 150, 157, 165, 291
De Geer moraine 51, 63, 64, 81–83, 153
  clast fabric 82
  hypotheses of formation 82–83

deglaciation (nature and style) 154
deglaciation sequences (glaciomarine) (models) 205–208
De Kalb mounds 64, 113, 159
delataic foreset beds 152, 153
deltaic sediments 60, 125
dendrochronology 382–383
    use of tree rings 382, 383, 386, 387
Denekamp Interstadial 273
Denmark 92, 142, 143, 268, 370
depositional remanent magnetism (DRM) 405
deserts 215, 226
desert plains and peidmonts 215
desert-related processes 213
desiccation (wetting and drying cycles) 215
Des Moines Lobe, Iowa 141
Developments in Glacial Studies 4–6
Devensian Glaciation 7, 19, 162, 256, 262, 273, 279, 281, 307,
    309, 324, 326, 330
Devils Lake Creek, B.C. 473
Devonian Period 242
dewatering 49, 54, 61
    pipes 54
diachronic unit 262, 264
diagenesis 10, 136, 190, 239, 257, 420–423
diagenetic (marine) environments 190
diamicton 9, 29, 30–42, 48,, 93, 113, 119, 124, 139, 141, 142,
    143, 149, 152, 158, 183, 195, 196, 200, 201, 202, 203, 206,
    207, 208, 209, 244, 246, 256, 287–288, 291, 292, 296, 297,
    338, 425, 454
    banding 208, 337
    beds 143
    carbonate content 208, 424
    clots 173
    diamicton-carbonate association see carbonate diamicton
        association
    discontinuities 46–54
    inclusions 46
    layered 288
    laminae 144
    massive 139
    mélange 30, 37, 44, 45–46, 47, 60, 61, 65, 143
    montane 296
    provenance 30, 75
    shell fragments 200
    stone concentrations 143
    stone layers 143
    stratified 143, 151, 298
    subaquatic 59, 183, 195, 196, 200, 201, 202
    supraglacial 134, 141, 143, 297
        stacked sequences 141, 142
diamictites 170, 171, 175, 176, 183, 186, 191–195, 240, 241,
    242–245, 247, 248
    basal 192
    bioturbated 175, 176
    geochemistry 245
    massive 193, 195

provenance 245
    stratified 193, 244
diamonds 120, 431, 455
diapirism 41, 44, 51, 54
    folding 53, 61
diatoms 189, 190
diatom ooze belt 189
dike see dyke structures
Dimlington, England 19, 258
Dimlington Stadial (chronozone) 263, 279, 281
dish and pillar structures 44, 51
disintegration ridges 64, 130, 131–134, 157–159
dilatancy 74, 76
dispersal patterns 423–430
    boulder train tracing 433
    'conveyor-belt' effect 436
    curve 424, 426, 436
    'head' 423, 425, 426
    ice flow 'shifts' 427
    influencing factors 424–430
    magnitude 426
    scale 431–434
    'tail' 424, 425, 426
    3-D plot 425
dispersal train 412, 413, 424, 425
    amoeboid pattern 434, 435
    fan-shaped 434
    'negative' 431
    ribbon-shaped 434
dissolution of quartz in tropics 214
dolomicrite 248
dolostones 247–248
    remobilization of carbonate 248
Donegal, Ireland 11
'doughnuts' 159
downwasting see ice marginal backwasting/downwasting
drag folds 239
'draw-down' effects 29
Drente Plateau, The Netherlands 351
drift exploration (prospecting) 11, 120, 411–439
    boulder transport 411
    chemical partitioning 414–420
    dispersal patterns 423–430
    drift geochemical maps 418, 419
    geochemical patterns 412–423
    geochemical signatures 412–413
    glacial dispersal 412
    low-temperature processes 419
    mineral enrichment 412
    till partitioning 415–418, 420
    till weathering 420–423
drift plain 149, 150
DRM see depositional remanent magnetism
dropstones 60, 61, 147161, 170, 173, 176, 181, 192, 196, 201,
    202, 206, 207, 208, 209, 239, 240
drowned shorelines 321

drumlins 11, 29, 35, 36, 63, 64, 65, 68, 71, 72–77, 79, 80, 93, 94, 113, 114, 119, 128, 132, 243, 275, 291, 434, 436, 438
   boulder cores 75
   boulder dykes 75
   catastrophic discharge 77
   clast fabric 75
   dilatancy 74, 76
   dimensions 72
   helicoidal flow 77
   'herring bone' fabric pattern 75
   hypothesis of formation 76
   internal sedimentology 291
   localized freezing 77
   megadrumlins 67, 75
   model of formation 74
   model of internal sedimentology 75
   morphology 76
   porewater dissipation 77
   radiating pattern 74
   rock-cored 76
   streamlined hills 67
   subglacial deformation 77
drumlinoids 64, 75, 76
Dryas Stadial
   Early 226
   Older 227
   Younger 152, 224, 258, 260, 261, 275, 282, 317, 379
dry-based subglacial conditions 18
Dry Valleys, Antarctica 31, 217, 221, 248
Dubawnt Lake, N.W.T. 438
Dubawnt group 417, 428, 429, 430, 431
ductile banding 44
ductile extrusion 60
ductile shear 31, 54
Dublin Gulch, Yukon Terr. 452
Durango, Colorado 147
Duncan Creek, Yukon Terr. 452
dunes *see* sand dunes
dune sand 372
Dwyka Formation 172, 173, 174, 175
dyke structures 145, 146

Ea horizon 304
earthquakes *see* neotectonics
Earth's Glacial record 2
East Africa 370
East Anglia, England 39, 50, 265, 266, 313
Eastern Townships *see* Québec
echinoderms 207
Ecuador 469
edge abrasion (grains) 368–369
edge rounding (grains) 371
Edinburgh, Scotland 84
Edinburgh Castle, Scotland 83
Edmonton, Alberta 41

Eemian Interstadial 273, 328, 354
effective stress 22, 77, 109, 135
eigenvalues 37, 38
Eightmile moraine, Montana 154
electron spin resonance (ESR) dating 393
Elk Creel Tills 408
Ellesmere Island 19, 216, 395
Elliot Lake, Ontario 192
Elster Glaciation 7, 93
emergent continental crust 252
Emsland, Germany 234
end moraine 55, 63, 119, 147, 148–155, 287–288, 400
   palimpsest 148
   rock-cored 148
   stratified 153
Endrick Valley, Scotland 260
energy dispersive X-ray spectrometers (EDS or EDX) 359
engelman spruce 383
englacial debris 55, 59, 80, 82, 138, 217, 221, 296, 298
   foliation 221
   thrust zones 82, 147
englacial meltwater channels 106, 137, 159, 298
englacial piezometric surface 110
englacial water table 110
England 19, 38, 39, 50, 146, 148, 150, 228, 253, 256, 262, 265, 266, 268, 279, 304, 309, 326
English Midlands 309
Enkelsbukta, Spitsbergen 393
Eocene Period 196, 204, 372
epigenetic deformation 269
epimerization process 394
equatorial divergences 189
equigeopotential surface 315
erosion marks 64, 92, 101
erosional notch 332
Esera valley, Pyrenees, Spain 154
eskers 26, 29, 44, 64, 82, 109, 110, 113–130, 152, 153, 239, 420, 421, 422, 434, 436
   beaded 64
   bifurcating ridge 122
   crevasse fillings 64, 123
   delta 116, 125, 127, 152
   dimensions 116
   esker-delta complex 128, 152, 153
   englacial formation 123
   engorged 64
   fans 64, 116, 152, 153
   fossil eskers 239
   ice-cored 115
   internal structures 119, 129
   kames 64, 113, 143, 155–157, 448
   kettles 64, 121, 149, 152, 153
   kettle lakes 125, 126
   maps 117, 118
   marginal development 122
   model of esker-outwash sequence 126, 152, 153

morphology 115, 117, 129
provenance 123
squeeze-up 64, 131, 133, 167
subaqueous eskers 64, 153
subaqueous fans 64, 116, 152, 153, 245, 246, 247
superimposed 114
transport pathways 124
Espanola, Ontario 95, 101
ESR *see* electron spin resonance
Estonia 74
European Alps 7, 148, 229, 312
european loess sheets 213, 214, 215, 225
eustatic changes 2, 11, 251, 315–333, 321, 448, 451
   amplitude of sea level change (past/future) 318
   curves 316, 317, 324, 329, 330
   regressions 329, 332, 448
   transgressions 329, 332, 448
evaporites 248
'event stratigraphy' 262
extraglacial terrain 221

fabric *see* clast fabric
facies *see* lithofacies
facies models 285–300
facies types and characteristics 294–295
'faint' young sun 252
Farmdale Soil 304, 305
fan deltas 110, 125, 151, 152, 153
fan and interfan (facies) associations 193, 194
faults:
   bedrock 54, 87, 100
   sediment 46, 48, 51, 141, 142, 200
fault wedges 41
Fennoscandian Ice Sheet 11, 117, 196, 208, 213, 316, 319, 329,
   330, 331
Fennoscandian Nordkaloot Project 431
Fennoscandian Shield 113, 214, 411, 425
ferri-argillans 354
ferricretes 214
Filchner Ice Shelf, Antarctica 183
finely-graded stratification *see* laminations 161
finger lakes 64
Finland 67, 70, 73, 74, 76, 96, 99, 100, 105, 131, 132, 152, 160,
   241, 275, 326, 327, 330, 431, 433
   placer deposits 464
   shoreline diagram 326
Firth of Clyde, Scotland 260
fissile structures 36, 46, 48, 200, 296, 343
fission track dating 391–392, 403, 407
fjords 64, 81, 84, 85–92
   cross-profile 88–90
   cross-profile evolution 89, 91
   formation 90–92
   geologic control 87
   hydraulic erosion 90, 91, 94

   pre-glacial control 88
   relationship between ice-contributing area and fjord
      morphology 90
   riegels 85, 92
   rock mass strength 88
   sea-entrance 92
flame structures 46, 51, 146
Flandrian Interglacial Stage 281, 305, 328, 329, 330
Flinders Range 246
floating ice margins 29, 51, 59
flocculation 49, 61
Florencia Bay, Vancouver Island, B.C. 459
floodplain deposition 236, 237
flow lenses 60
flow till 32–35, 38, 58–59, 65, 110–112, 126, 139, 296, 297,
   347, 416
   flow nose 58, 59, 244
   flow plug 58, 144
   stacked 58
   subaquatic 245
   supraglacial 80
   wispy lamination 244
fluid stressing 102
fluidisation of sediment 44, 46, 50, 51, 108, 146
fluidised debris 100, 108
fluted moraine 35, 64, 65, 67, 68, 72, 74, 75, 77–81, 94, 133,
   167, 201, 275
   clast fabric 79, 80
   dimensions 77, 78
   grain size 80
   hypotheses of formation 80
   inter-flute troughs 80
   internal sedimentology 79
   megafluting 63, 67, 79
   squeezed debris 80, 131, 133, 159, 167
fluvial sediments 256, 413
   drowned placers 450
fluvial terraces 125, 307
fluvial transport 214, 368
fluvioglacial sediment *see* glaciofluvial sediment
folds in sediment 39, 41, 44, 48, 61, 200
foliation in sediment 36, 41
Foinaven, Scotland 275
foraminifers 185, 187, 189, 190, 195, 196, 197, 200, 202, 203,
   208, 270, 271, 403
foraminiferal muds 206, 207, 208, 209
foraminanro oozes 207
forebulge uplift 321, 323
Forth (river) 326
fossil content of sediment 181, 393
fourier analysis 362
Fox Island River, Newfoundland 462
fractal analysis 362
fracture patterns in sediment 25, 36, 38, 46
Fram Strait 187, 189, 196, 206, 208
France 143, 233, 263, 273

Franklin Furnace, New Jersey 417
fracture toughness 97
Fraser Glaciation, B.C. 454
Fraser Lake, Ontario 461
Fraser River, B.C. 91, 445, 457
freeze/thaw cycles 11, 214
freezing conditions 10, 45
French River, Ontario 102
friction cracks 98, 99, 100
frost creep 235, 302
frost-riven (shattered) clasts 59
frost wedge 351
frosted sand grains 225, 226
    distribution map in Europe 226
frozen bed subglacial conditions 18, 20, 25, 76, 94, 135
fuchsite 420
furrow 102

Gallery Gold mine, Lightning Creek, B.C. 474
Ganlose, Denmark 143
Garbh Eileach, Scotland 193
garnet 361, 367
Gartocharn Till, Scotland 260, 281
Garvellach Islands, Scotland 170, 171, 193
Gaspé, Québec 461
Gauss Chron 406
Gays River, Nova Scotia 462
Geikie, A. 7
gelifluctate 304
geochemical patterns 412–423
    chemical partitioning 414–420
    dispersal patterns 423–430
    drift geochemical maps 418, 419
    low-temperature processes 419
    till partitioning 415–418, 420
    till weathering 420–423
geochemistry 10, 48, 135, 136, 181, 245, 249
    trace element geochemistry 249
geochronology 263, 280–283, 377–410
    deep-sea marine record 377
    glacial sediments 377–410
    methods 381, 409
        chemical and biological 381, 393–399
        correlation methods 381, 402–410
        geomorphic 381, 399–402
        isotopic 381, 384–391
        radiogenic 381, 391–393
        sidereal 381–384
    relationship between datable materials and glacigenic
            sediments 378
    resolution 379
    terminology 280, 380
geochronometry 254
geologic control 87, 88
geological setting for placer mining 477
Geological Survey of Canada (GSC) 439

Geological Survey of Finland 439
geomorphic dating techniques 381, 399–402
    landform modification 401–402
    rock and mineral weathering 400–401
    soil-profile development 399
    weathering indices 401
    weathering parameters 400
Georgian Bay, Ontario 100, 103
geotechnical properties of sediment 40, 42, 136
Germany 6, 7, 74, 92, 93, 111, 150, 233, 234, 237, 337, 347, 372
Ghulkin Glacier, Pakistan 219
Gilbert Chron 406
Gilbert River, Québec 462
Gilbert-type deltas 152
Gipping Glaciation 7
glacial anticyclone 213
glacial chutes 64
glacial and glaciofluvial placers 447–448, 449, 455, 456, 457,
        461, 462, 469, 470, 477
    till 447
    glacigenic debris-flows 447, 454
    reconcentration by postglacial streams 448, 454
Glacial Lake Agassiz see Lake Agassiz
glacial outburst flood see jökulhlaups
glacial stages 262
    interstadials 262
    stadials 262
glacial stratigraphy 253–284
    aminostratigraphy 395
    biostratigraphy 255, 261
    chemical stratigraphy 262, 264
    chronostratigraphy 261, 263, 280–283
        British Isles and N.W. European stratigraphy 282
        terminology 280
    climatostratigraphy 262
    correlation 257
    disparate sequences 255
    'event stratigraphy' 262
    geochronology 263, 280–283, 409
        terminology 280
    geochronometry 280–283
    glacio-sedimentology 254
    hiatus 259, 261, 270, 343
    isotopic stratigraphy 262, 264, 270–274
    kineto-stratigraphy 256, 264, 268–270
    lithostratigraphic units of the southern Laurentide Ice Sheet
        258, 287–288
    lithostratigraphy 257, 258, 262, 264–268
    magnetostratigraphy 181, 264
    morphostratigraphy 254, 262, 274–279, 402
    pedostratigraphy 264, 399
    preservation potential 255
    proxy evidence 261, 262
    spatial fragmentation 255
    stratigraphic nomenclature 261–264
    ocean records 255, 270

glacial trimlines 397
glacial troughs 63, 64, 85–92
  bedrock sills 86
  cross profile 88–90
  longitudinal profile (thalweg) 86
  riegels 85, 86
Glacier National Park, U.S./Canada 383
Glacier Peak, Washington 404
glacier surface profile 45, 157
glacigenic sediment flows *see* flow till
glacioaeolian processes, sediments and landforms 213–237
  atmospheric pressure systems 225, 227
  distribution of wind-eroded surfaces 217, 225
  facies 229–237
    principal facies of periglacial sands in W. Europe 234, 235
    structures 234
  grain size 214, 220, 221, 228
  impact velocities 217, 220
  landform-sediment facies associations 231
  mineralogical composition of loess 233
  niveo-aeolian deposits 222
    sructures 223
  origin of loess 215, 236
    sequence of stages in formation of W. European loess 236
  seasonal aeolian deposition in periglacial environments 232
  sediment production and sources 214–215
  sediments and landforms 221–229
  silt production 215
  wetting and drying events 227
glacio-eustasy 315–333, 316–318, 321, 448, 451
  amplitude of sea level change (past/future) 318
  curves 316, 317, 324, 329, 330
glaciofluvial 27, 29, 38, 42–45, 71, 80, 83, 93, 94, 106, 109,
    110, 111, 142, 144, 155–157, 217, 288, 298, 304, 322, 442
  model of ice-contact and proglacial morphosequence 156
  outwash 93, 123, 124–125, 126, 128, 133, 148, 152, 173, 184,
    217, 222, 243, 245, 247, 256, 288, 298, 306, 309, 322,
    390, 448
glacio-isostasy 201, 315–333, 318–321
  Boulton's spatial/temporal model 324–325
  models 319, 320, 321, 322, 324–325
  proglacial depression 320, 323, 330
  shoreline formation 320
  stages 323
  Walcott's model 319
Glaciolacustrine Environments 29, 42–45, 129, 142, 144,
    161–178, 217, 227, 243, 290, 296, 307, 442
  distinction between glaciolacustrine and glaciomarine 161
  ice scour 161–178
  lake plain 150, 156
  soft sediment striated surfaces 165–168, 172
Glaciomarine Environments 59, 179–211
  biogenic opal production 189
  biogenic sediments 186
  bottom waters 189, 206, 207
  bottom water temperatures 185

brine formation 189
carbon isotopes 185
carbonate dissolution 181, 189–190, 197
carbonate flux 189
carbonate sedimentation 186, 202, 248, 251
CO_2 content of surface water 186
circulation system 185
current regimes 185
deltas 152, 322
dissolved iron 189
ice berg sedimentation 198, 206, 207
ice proximal deposits 199
lithofacies associations 206, 207, 208, 285–300
meltwater plumes 187, 199, 245
model of glacial/interglacial sedimentation on passive margin
    200
passive margin environments 195–211
planktonic/benthic ratios 190
Pre-Cenozoic glaciomarine sequences 191–195
oxygen isotopes 185–186
redox potential 211
regressive glaciomarine sediments 201
salinity 185, 198, 202, 203, 205, 210, 211
sea ice sedimentation 198
SEM-based dissolution indices 190
shelf sediments 186
stratified upper water 185
trough fill 199
turbid surface water plumes 198
glaciomarine mudstones 247
glaciomarine sediments 59, 243
  classification 181–191
  facies 332–333
  grain size 165, 183
  deep ocean sediments 179
  pelagic carbonate 190, 196
  pelagic sedimentation 181
  sedimentation rates 181, 183
  stratiform glacigenic sediments 198
glaciotectonics 41, 46, 49, 50, 80, 82, 111, 150, 152, 239, 243,
    256
Glen Muick, Scotland 90
Glen Turret, Scotland 134
gleying 301
global ice volume 185
gold 120, 419, 422, 436, 437, 442, 450, 451, 452, 454, 458
  coarse 451
  detrital 444
  lode 454
  nuggets 454, 460
  point-source concentrations 451
Gondwanaland 242
Goodnews Bay, Alaska 467
Gorcon Lake Formation 246
Gormack, Scotland 107
Göteborg moraine 52

Gowganda 1, 191–194, 240, 247
    'Upper Diamictite' 194
Graham Island, B.C. 450
grain crushing and grinding mechanisms 213, 214, 215, 229,
        236, 368, 369, 371, 414
grain edge abrasion 368–369
grain edge rounding 371
grain microfractures 214, 371
grain size see particle size
Grampian Mountains, Scotland 134
Grand Banks 165, 203
Grand Banks Slide (1929) 205
Grande Pile, France 273
Grassy River, Ontario 461
gravel lag 201, 444, 445
gravitational slumping 61, 245, 247
gravity anomalies 319
gravity flowtill see flow till
gravity flow deposits (marine) 183, 192, 247, 203, 204, 205, 244,
        245, 246, 292
'Great Breccia', Port Askaig 194, 195
Great Britain see Britain
Great Lakes 19, 36, 37, 38, 44, 99, 119, 162, 191, 249, 257, 258,
        266, 288, 379
Great Slave Lake, N.W.T. 162, 176
Green Bay Lobe, Wisconsin 73, 148, 149, 287–288
    map 149
Green Creek Valley, California 400
Greenhouse Climatic State 2
greenhouse effect (enhanced) 2, 249, 317, 331
    super greenhouse effect 252
Greenland 74, 171, 172, 213, 217, 367
    continental shelf 176
Greenland Ice Sheet 85, 196, 255, 270, 274, 319, 369, 371
Greenland-Scotland ridge 196
Grenville, Ontario 417, 418
Grenville Channel, B.C. 89
Grinell Glacier, Montana 78
GRIP ice core, Greenland 274
grooves 64, 85, 170
grounding line 11, 50, 51, 59, 61, 62, 81, 82, 112, 167, 201, 245
grounding line ridges 64, 82, 243
ground moraines (non-linear unoriented forms) 130, 150
    Blattnick moraines 64, 67, 160
    corrugated moraines 64
    cover moraines 64
    hummocky moraine 26, 64, 90, 130, 133–134, 143, 153,
            157–159, 275, 301, 302
    ice-pressed ridges 64
    map of east-central Illinois 151
    Pulju moraines 64, 67, 130, 131–132
    till plains 64, 150
    till veneers 64
    Veiki moraines 64, 113, 130, 132–133, 160
ground ice lenses 352
grounded ice margins 153, 181

groundwater 422
    ^{14}C effects 386
    solute-rich 48
grus 367, 372
GSC Lake, N.W.T. 128
Guelph-Paris moraine, Ontario 148
Gulf of Alaska 243, 246
Gulf of Bothnia 257, 326, 331
Gulkana Glacier, Alaska 288
gullying (marine shelf) 200, 203
Günz Glaciation 7

'H' beds see Hard or Rigid ('H' beds)
Haggart Creek, Yukon Terr. 452
Haliburton, Ontario 145, 292–296
Halland, Sweden 52
Hanover, Ontario 42, 111
Hard or Rigid ('H') beds 20–22, 25, 77, 84, 85, 94, 99, 292–296
Hardangerfjord, Norway 88
'hard-water' effect (^{14}C) 387
Hastings, Ontario 292
Hatfield, England 38
headwalls 64
heavy minerals 427, 441, 448, 451
Hebrides, Scotland 330
Hecate Strait, Queen Charlotte Islands, B.C. 451
Hector Lake, Alberta 384
helicoidal flow 77
Helsinki, Finland 327
hematite 312, 412, 414
Hengelo Interstadial 273
Henninga Lake, N.W.T. 432
Hertford, England 38
Hesle Till, England 304
hiatus (depositional) 259, 261, 270, 343
High Lodge, England 256
Himalaya Mountains 312
Hindu Kush Mountains 373
historical records (dating) 381–382
Hoekstra drumlin site, Ontario 291–292
Holmströmbreen, Spitsbergen 226
Holocene 165, 186, 190, 195, 198, 200, 206, 229, 262, 273, 282,
        326, 328, 354
holostratotype 264
Holsteinian Interglacial 93
homogenisation due to deformation 23, 164
hornblende etching 312
Horsefly River, B.C. 456, 457
Horton fluvial system 122
Huronian Age glaciation 95, 241
hydration (weathering) 214
hydraulic conductivity 22
hydraulic erosion 90, 94, 105, 130
hydraulic gradient 22, 130
hydraulic pumping 61, 62

hydro-isostatic loading 316, 317, 318
hydrological cycle 316
hydrostatic pressure 22, 84, 90, 94, 99, 110, 113, 116, 130
hydrothermal activity 248
hyper-srid environments 221
hypostratotype 264
Hudson Bay 26, 27, 119, 257, 331, 395, 413, 428, 429, 430, 431, 435
Hudson Bay Lowlands 222, 435
humic sands 304
hummocky moraine 26, 64, 90, 130, 133–134, 143, 153, 157–159, 275, 301, 302
  formation 134
  models of formation 134, 159
  morphology 133
hummocky stratification 176
Hunza Valley, Pakistan 219
Huron-Erie ice lobe, Indiana 149
Hurwitz Group 241
hypotheses of drumlin formation 76

ICE-1 model 333
ice (debris-rich) 38, 116, 138, 355
  stagnant 144, 157
Ice Age Concept 7
ice basal traction 48
ice/bed decoupling 22
ice bergs 60, 161, 200, 208, 244
  furrows 203
  grounded 162
  influx 182
  ploughing 202
  rafting 182, 183, 206, 207, 245
  sedimentation 198
  turbation 200
ice contact debris fan 145, 152
ice contact delta 111, 116, 125, 128, 151, 152, 153
  model 152
ice contact and proglacial glaciofluvial morphosequence model 156
ice contact slopes 143, 144, 160
ice contact rim 160
ice-cored moraine 55
ice cores 274, 390, 403
ice dammed lakes 146, 150, 154
ice disintegration forms 63, 64, 130, 131–134
  Pulju moraines 63, 64, 67, 131–132
    clast fabric 132
    formation 131
    morphology 132
ice floes 168
ice-free enclaves 217
Icehouse Climatic State 2, 250
ice keel 162
  debris-impregnated 177
  erosion 162

homogenization of structures 165
keel loading 162, 165
keel turbates 162, 164, 165, 177, 178
map 166
Iceland 6, 19, 74, 78, 79, 217, 222, 229, 328, 372, 403
Iceland-Faroe Ridge 180
ice marginal backwasting / downwasting 137, 144, 149
ice marginal landforms 137–160
  boulder belts/aprons 64
  end moraines 55, 63, 64, 147, 148–155, 306, 310, 390, 400
    stratified 153
  kame moraines 64
  meltwater systems 155
  recessional moraines 64, 148
  push moraines 51, 63, 64, 150, 167, 347, 351
  Stauchmoräne 64
  terminal moraines 64, 93, 130, 148, 436
ice margins floating 29, 59
ice mass 'draw-down' effects 29
ice overburden loading 29, 50
ice-pressed ridges 64
ice pressure ridges 161
ice proximal deposits 199
ice-pushed ridges 51, 63, 64, 150, 167, 347, 351
Ice rafted detritus (IRD) 173, 176, 177, 182, 185, 191, 193, 194, 195, 196, 197, 198, 200, 206, 207, 208, 247, 255
ice retreat (nature of) 137
ice scour 161–178
  criteria for recognition 164, 169
  fossil scour marks 165, 167
  furrows 168, 175
  grooves 170
  homogenization of structures 165
  ice floes 168
  keel erosion 162, 176
  keel loading 162, 165
  lithified sediments 168
  map 166
  ridge-and-groove microtopography 167, 170, 174, 175, 177
  seasonal ice 162
  scour marks 162, 175, 176
    cross-sections 163
    incision depth 164
    orientations 168
  scour 'pockets' 162
  sediment liquefaction 164
  structures 162, 164, 175
  sub-scour structures 168, 175
ice sheet centres 71
ice sheet models 16, 25, 27, 319–321
ice sheet stability 10
Ice Shelf Environments 59, 181
ice shelf 182, 243, 244
  striations 167–168
ice shelf water (ISW) 184
ice streams 24, 62, 119, 436

Ice Stream B 19, 61
ice surface profiles 45, 157
ice thermomechanical behaviour 24
ice-walled lakes 133
    plains 159
ice wedges 46, 194, 239, 304
Idaho 312, 404
    placer map 465
IFO Lake, N.W.T. 128
igneous intrusions 83
Illinoian Glaciation 7, 309, 390, 397
Illinois 93, 147, 148, 150, 151, 274, 288, 289, 305, 306, 309,
    314
    Quaternary stratigraphy 309
imbrication (in sediments) 150
immobile sediments (within deforming bed state) 45
impact breakage 214
impact velocites (aeolian) 217, 220
    drag velocities 220
    fluid threshold 217, 220
    impact threshold 217, 220
impact Vs 370
inclusions within sediment 36, 38, 39, 46
India 241
Indiana 93, 147, 148, 149, 396
indicator (drift exploration) 423
injection structures 38, 49, 50
Inoceramus prisms 208
insect content of sediments 262
International Stratigraphic Guide 261
inverse grading 144, 146, 245
Ireland 11, 74, 291, 337
interglacial deep water formation see Deep Water Formation
interlobate moraines 64, 148, 157
interstadial soils 308
inter-stratal disruption 41
intraclasts 39, 40, 47, 49, 60, 267
    rafting 44, 49, 82, 92, 119, 135, 245, 267
Iowa 140, 141, 142, 159, 407
Ipswichian Interglacial 273, 330
IRD see Ice rafted detritus
iron 423, 424
iron (dissolved in sea water) 189
iron-formations 248–249
iron hydroxide 347, 351, 412, 416
iron oxide 347, 351, 370, 412, 416
iron precipitates 343
iron-rich brines 248
iron-staining 101, 206
Isaacs Harbour, Nova Scotia 462
isoleucine 394
Isortoq, Baffin Island 81
Isfallsglaciären, Sweden 78
Isle of Skye, Scotland 276, 277
isostatic changes see glacio-isostasy 11, 116, 144, 246, 315–333,
    318–321

Boulton's spatial/temporal model 324–325
    models 319, 320, 321, 322, 324–325
    proglacial depression 320, 323, 330
    shoreline formation 320
    stages 323
isotope stages 209
isotopic dating methods 384–391
    argon-argon 381
    cosmogenic 377, 381, 390–391
    decay curves 385
    isotope half-lives table 384
    potassium-argon (K/Ar) 310, 381
    radiocarbon 179, 181, 240, 310, 316, 377, 379, 381, 385–387
    uranium series 181, 241, 377, 381, 388–390
isotopic profiles in basal ice 21
isotopic stratigraphy 262, 264, 270–274
Italy 233
Itararé Subgroup, Brazil 175

James Ross Island, Antarctica 74
Jan Mayen Ridge 180, 209
Jaramillo subchron 408
Jasper, Alberta 147, 296–298
Jemez Mountains, New Mexico 404
Jenny's Concept of Soil Formation 310
Jervis Inlet, B.C. 89
Johnstown End Moraine, Wisconsin 148, 149, 287
joints in sediment 36, 38, 46, 48, 54, 347
joints (bedrock) 99, 100, 424
jökulhlaups 99, 162
Jutland, Denmark 370

K/Ar dating (also Ar/Ar dating) 310, 381, 387–388
    decay table 388
K-feldspar 387
Kalix Till 49
Kalixpinmo Hills 64
'kame and kettle' topography 133, 134, 149, 152, 155–157, 448
kames 64, 113, 143, 155–157, 448
    kame fields 157
    moulin kames 64
kame deltas 64
kame moraines 64
kame terraces 64, 129, 144, 155–157, 296
    gradients 129, 156
    pitter surfaces 156
Kansan Glaciation 7, 407, 408
Kansas loess, Kansas 228, 233
kaolinite 414, 431, 432
Karakoram-Himalyan Mountains 214, 217, 219, 221, 222, 237
Karoo Basin, South Africa 175
katabatic winds 217
Kebnekajse, Sweden 78
keel see ice scour

Keewatin (District of) 68, 84, 113, 114, 120, 122, 123, 125, 323, 416, 417, 418, 419, 421, 426, 430, 431, 432, 433
Keewatin Ice Divide 26, 27, 117, 118, 122, 438
Keitele, Finland 76
Kenhardt, South Africa 175
Kennebec delta, Maine 330
Kent, England 228
Kent brickearths 228
Kenya 1
Kerguelen Plateau 196
Kesgrave Group, England 268, 304, 313
kettle holes 44, 64, 121, 149, 152, 153, 156
Kettle Interlobate moraine, Wisconsin 157
kettle lakes 125, 126, 293
Kianta moraine 63
Kiliminjaro, Kenya 1
kineto-stratigraphy 256, 264, 268–270
  'empty' locality 270
  'missing' unit 270
King William Island, N.W.T. 176
kink band arrays 48, 49, 59
Kirkham moraine, Cumbria 148
Kirkhill, Scotland 304
Kikrland Lake mining area, Ontario 461
Kiruna, Sweden 131
Klaza glacial period, Yukon Terr. 453
Klondike 452
  auriferous gravel 452
  'White Channel gravel' (Pliocene) 452
Knapatorpet, Sweden 143
'knock and lochan' 86
Kolyma River, Siberia 463
komatiite 416
Krattelstön, Sweden 70
Kuibis Series, Namibia 168, 170, 173
Kumara, Westland, New Zealand 469
Kuskokwim Bay, Alaska 467
Kuusamo, Finland 73

Labrador 68, 328, 367, 372, 434
Labrador Sea 165, 170, 171, 182
Lac Troie, Québec 86
lacustrine sediments 38, 119, 138, 256, 262, 386, 408
lacustrine terraces 296, 307
lake chains 64
Lake Agassiz (glacial) 162, 163, 164, 165, 176, 257
Lake Blane (glacial) 260, 275
Lake Bonneville (glacial) 391
Lake Constance 227
Lake Erie 36, 37, 38, 44, 162, 176
Lake Hitchcock (glacial) 384
Lake Huron 99, 249, 258
Lake Iroquois (glacial) 162
Lake Manitou, Ontario 461
Lake Michigan 19

Lake Michigan Lobe 149, 288
Lake Missoula (glacial) 307
Lake Nipigon, Ontario 436
Lake Ojibway (glacial) 162
Lake Ontario 19, 266
Lake Råstojaure, Sweden 131
Lake Superior 119, 162, 191
Lake Tapps, Washington 392
laminar structures 48
laminations (laminae) 36, 39, 40, 41, 112, 161, 181, 191, 244
laminated clays 36, 39, 40, 41, 112, 161, 181, 191, 202, 298, 452
laminated silts 161, 256, 298, 452
laminites 161
landforms
  classification 64
  'freshness' 279, 401–402
  subglacial 29–46
landform zones 25, 27, 85
  model of landform zonation 27
Landsat 66, 85, 91, 199
Lanzhou, China 228
Lanzhou loess 228
Lapland (Finland) placers 464
lateral moraines 146, 148, 153, 416
  cross-section of sediments 154
  looped moraines 64
  interlobate moraines 64, 148, 157
  perched moraines 64
  trimline moraines 64
  valley-side moraines 64
Laurentian Fan 203
Laurentide Ice Sheet 11, 19, 25, 85, 117, 118, 120, 141, 147, 148, 153, 203, 214, 258, 287-288, 292–296, 316, 319, 324, 329, 330, 395, 436, 438
Lava Creek B Ash bed (also Pearlette O tephra) 408
Law of Superposition 269
Lawson-type sediment flow classification 56–57, 142
  Type I 144, 146
  Type II 142, 145, 146
lead 423, 424
Lead-210 (^{210}Pb) dating 389
  pathways 389
Leaf Lake, Saskatchewan 459
Lena River, Russia 463
lectostratotype 264
Leech Creek, Vancouver Island, B.C. 459
lee-ridges (cones/moraines) 63, 64
lee-side deposition 44
lenticular cross-bedded sands 243, 245
leucine 394
Levene moraine, Sweden 152
'Libby half-life' (^{14}C) 385
lichen growth curves 398
lichenometry 261, 278, 310, 373, 397–398
  crustose lichens 397
  growth curves 398

lichenometry – *continued*
  rhizocarpon geographicum 398
  thalli 397, 398
Lightning Creek mining area, B.C. 455, 474
linear disintegration troughs 158
Lingle and Brown Equation 24
liquefaction of sediment 22, 44, 46, 50, 51, 54
lithofacies 136, 181, 203, 229–237, 242–249
  codes 289
  fan and interfan associations 193, 194
  glacioaeolian associations 229–237
  glaciomarine associations 206–208, 332–333
  homofacial associations 244
  model sequences 28
lithofacies associations (terrestrial) 285–300
  advance-retreat sequences 288–291, 297, 298
  facies models 285–300
  facies types and characteristics 294–295
  montane facies sequence 297–298
  polar ice cap facies 298–300
  variables influencing glacigenic depositional settings 286
lithosphere 319
lithostratigraphic units of the southern Laurentide Ice Sheet 258
lithostratigraphic terminology 261–264
lithostratigraphy 257, 258, 262, 264–268
Little Ice Age 1, 261 *see also* Neoglacial
Littorina Shoreline 326
Livingstone Creek, Yukon Terr. 452, 475
Livingstone Lake, Saskatchewan 75
load structures 44
loading 36, 50
local erosional forms (subglacial) 94–110
Lochem, The Netherlands 224
Loch Lomond Basin, Scotland 260, 281
  stratigraphy 281
Loch Lomond Stadial 90, 134, 258, 258, 260, 275, 276, 277, 280
lodgement till 31–40, 44, 62, 65, 130, 150, 192, 200, 256, 288, 291, 296, 300, 302, 355, 410, 425
  critical lodgement index 33–34
  geotechnical properties 40
  overconsolidation 40, 201
lodgement tillite 191, 192
loess 213, 225, 227, 230, 262, 302, 306, 370–371, 374, 379, 390, 392, 410
  carbonates 227
  colour 227
  glacial loess 214
  distribution in Europe 225
  mineralogical composition of loess 233
  mountain loess 214
  particle size 214, 220, 221, 228, 230, 233
  sequence of stages in formation of W. European loess 236
  sorting coefficient 230
Loess Plateau of China 215
loessite 246
  grain fabric 246

Long Valley – Mono Glass Mountain area 404
looped moraines 64
Loveland Silt 396
Lower Wright Glacier, Antarctica 21
Lowestoft Formation, Britain 265, 266
Lowestoft Glaciation, Britain 7, 304
Lowestoft Till 313
luminescence dating 259
lunate fractures 98
Lunteren, The Netherlands 338
Lynx Creek, Yukon Terr. 452
lysocline 198

'M' beds *see* Soft or Mobile ('M')
Maclaren Glacier, Alaska 288
Magadan area, Siberia 451, 463
magnetite 410, 420
magnetopolarity time scale 406
  excursions 406–407
magnetostratigraphy 181, 264
Main Lateglacial Shoreline, Scotland 260, 281
Main Perth Shoreline, Scotland 326
Main Postglacial Shoreline, Scotland 326, 330
Maine 153, 329, 330, 434
  sea-level curve 330
Makkovik Bank, Labrador Sea 170
Malaspina Glacier, Alaska 157
manganese 421, 424
manganese-brines 248
manganese hydroxide 351, 412, 416
manganese oxide 351, 398, 412, 416
manganese precipitates 343
Manitoba 29, 92, 114, 122, 162, 163, 165
'marbled' appearance of sediment in thin section 47
Marengo moraine, Illinois 150
marine channels 27
marine fan sedimentation 153
marine ice sheets 245
marine sediments 119, 262
marine shell beds (Scotland) 330
Maritime Provinces of Canada (placers) 462
Marlborough, New Zealand 468
masepic fabric 347
Massachusetts 315, 395
mass balance 148, 154
Masset Inlet, B.C. 451
mass wasting (mass movement) 139, 147, 149
Matanuska Glacier, Alaska 56, 58, 150, 288, 454
Matuyama Chron 406
Matuyama/Gauss reversal 270
Mauritania 1
Mayo, Yukon Terr. 452
McKee Creek, B.C. 459, 460
McLeod River, Alberta 459

McMurdo Sound, Antarctica 196
megadrumlins 67
megafluting 63, 67, 79
Meguma group, Nova Scotia 462
mélange *see* diamicton mélange
melt out till 31, 32–35, 38, 40–42, 65, 80, 110, 126, 130, 149,
    150, 159, 292, 296, 297, 298, 299, 300, 355, 454
    englacial 40, 41, 42
    geotechnical properties 42
    sublimation till 31, 41, 298, 299
    supraglacial 40
melted-bed subglacial conditions 18, 25
meltwater:
    discharge 24, 60, 90, 247
    erosion marks 64, 92, 98, 101
    flow separation 100
    flux 22, 24
    flux rate 24, 60
    fountains 110, 117
    glacial chutes 64
    p-forms 64, 85, 92, 98, 99, 101, 155
    pipe flow 41, 109, 117
    pondings 41
    potholes 64, 99, 102, 105
    production 22, 27
    s-forms 64, 99, 100, 102, 103, 105, 130, 155
    sheet flow 22
    stream competency 42
    thaw consolidation 41
meltwater channels 22, 41, 42, 60, 94, 99, 105–110, 122, 130,
    152, 153, 247, 298
    'C' channels 22
    chutes 64, 110
    closure rates 108
    englacial 106, 159
    full pipe flow 109, 117
    ice marginal 155, 156
    interlobate streams 113
    longitudinal profile (thalweg) 99
    maps 107, 108
    'N' channels 22
    pradoliny 64
    'R' channels 22, 108, 109
    subglacial 60, 64, 84, 99, 105–110, 122, 130, 152, 153, 155,
    175, 246, 288
    submarginal 64, 107, 155
    urstromtaler 64
    till channels 108, 109
    tunnels 42, 64, 90, 92–94, 113, 118, 122, 123, 195, 245, 238,
    453, 461
meltwater fountains 110, 117
meltwater plumes 187, 199, 210, 245
Mendenhall Glacier, Alaska 78
Merrimack delta, Maine 330
Mesa Falls Ash bed (also Pearlette S tephra) 408
metabasalts 422

Metazoan fossils 241
Meteor Lake, Ontario 461
Meule Creek, Québec 462
Middleton Island, Alaska 246
Midland Valley, Scotland 258
mica:
    comminution by wind 226
    weathering 312
Michigan 148, 150, 153, 241
'microblock' texture 369
microfauna 317
microfluting 48
microfractures (grains) 214, 371
micromorphology 40, 302, 335–355
    birefringence 343, 347
    'card-house' structures 40
    erratics 343
    faults 343
    features 337–354
    fine-grained casings 337
    folds 343
    foliation 40
    diagenetic features 337
    impregnation methods 336
    intraclasts 343
    mineralogy 337
    oriented domains 347, 351
    nomenclature 336–337
    rotational features 339, 343, 344, 347
    sampling procedures 335–336
    structure 343–347
    texture 337–343
micromorphology nomenclature:
    argillasepic 347
    bimasepic 347
    domains 337, 347
    masepic 347
    plasma 337
    plasmic 337, 343, 347–351, 355
    silasepic 347
    skeleton grains 337
    skelsepic 343, 347, 351
    strial 347
    unistrial 347
Milankovitch solar forcing 251, 270, 386, 403
Milford Sound, New Zealand 88
Milk River, Alberta 459
Mindel Glaciation 7
mineral prospecting *see* drift exploration
mineralogy 181
Minnesota 153
miocline 251
mirabilite 217
Mississippi 233
Mississippi River 396
Mississippi Valley 214

mixtites 194
  stratified 194
Mobile subglacial beds *see* Soft or Mobile 'M'
Moershoofd Interstadial 273
Mohawk Bay, Ontario 36, 37, 38, 44
Mohawk Valley, New York 408
molluscs 186, 200, 201, 393, 394, 395
Mongolia 371
Montana 78, 91, 154, 311, 390, 397, 401, 464
  placer map 465
montane facies sequence 297–298
moraine:
  ablation 63, 133
  Blattnick moraines 64, 67, 160
  basin 60
  cross-profile 402
  cross-valley 51, 81, 153
  De Geer 51, 63, 64, 81–83, 153
  end 55, 63, 64, 147, 148–155, 306, 310, 390, 400
    stratified 153
  fluted 35, 64, 67, 68, 72, 74, 75, 77–81, 94, 133, 167, 201, 275
  hummocky 26, 64, 90, 130, 133–134, 143, 153, 157–159, 275, 301, 302
  ice-cored 55
  interlobate 64, 148, 157
  kame 64
  Kianta 63
  lateral 64, 146, 148, 153
  looped 64
  medial 146, 153
  perched 64
  Pulju 63, 64, 67, 130, 131–132
  push 51, 63, 64, 150, 167, 347, 351
  radial 63, 64
  ribbed 64, 153
  Rogen 35, 63, 64, 65, 68–72, 74, 75, 79, 80, 120
  Sevetti 63, 113
  shear 64
  shelf 60
  squeeze 64, 167
  subaqueous 64, 203
  submarine 60
  surge 167
  terminal 51, 64, 93, 130, 148, 390, 436
  thrust 64
  trimline 64
  valley-sided 64
  Veiki 63, 64, 113, 130, 132–133, 160
  washboard 51, 81, 153
moraine development model 52–53, 55
moraine-lake plateaus 159
moraine plateaus 159
moraine ridge cross-profile 402
morphological development of soil with weathering 303, 399
morphostratigraphy 254, 262, 274–279, 402
Morro Chapéau Formation, Brazil 170

moulins 159
Mount Mazama, Washington 404
Mountrail County, North Dakota 158
Mount St. Helens, Washington 404
mud flows 235
mudstones 193, 240, 246–247
  glaciomarine 247
  varved 247
Munich area, Bavaria 237
muschelbrüch 102, 105
Mya truncata 200
mylonised bedrock 87
mylonised (glacial) zones 48
Myrdalsjökull, Iceland 74

'N' meltwater channels 22
Nama Syatem, Namibia 168, 170
Namibia 168, 170
Nansen glacial period, Yukon Terr. 453
Nattavaara, Sweden 133
natural remanent magnetism (NRM) 406
nearest neighbour analysis 135
Nebraska 225, 233, 407
Nebraskan Glaciation 7, 407, 408
Nelson, New Zealand 468, 469
Neoglacial 1, 261, 373, 383 *see also* Little Ice Age
neogloboquadrina pachyderma 187, 190
neostratotype 264
neotectonics 54, 61, 170, 214
  marine slide triggers 205, 245
Netherlands 51, 150, 224, 225, 227, 337, 338, 344, 347, 351, 372
Nevada 464
New Brunswick 255, 417, 462
New England, U.S.A. 147, 157
Newfoundland 165, 462
New Jersey 417
New Mexico 404
New York State 30, 112, 148, 331, 408
New Zealand 1, 85, 88, 214, 221, 233, 312, 448
  placer 468–469
Niagara-on-the-Lake, Ontario 112
nickel 423, 424, 425, 426, 427, 431, 433
Nigardsbreen, Norway 99, 373
Nithsdale, Scotland 108
nivation basins 64
niveo-aeolian deposits 222
  sedimentary structures 223
Nome, Alaska 448, 450, 451, 464
non-pervasive deformation 23, 48
Nordaustlandet, Svalbard 198, 199
Norfolk, England 228
North America 7, 63, 113, 117, 132, 141, 159, 161, 213, 229, 248, 257, 261, 274, 307, 309, 328, 329, 331, 370, 379, 387, 403
North American Stratigraphic Code (NASC) 261, 379

North Dakota 68, 75, 147, 158, 387
North Flinders Basin 245
North Park, Colorado 223, 224
North Saskatchewan River, Alberta/Sask. 459
North Sea Basin 92, 208, 329
North Sea Drift Formation 266
North West Territories 29, 68, 84, 113, 114, 115, 117, 120, 121, 122, 123, 125, 128, 162, 176, 196, 216, 241, 291, 328, 415, 417, 418, 419, 421, 426, 427, 428, 429, 430, 433, 438
Northern Ireland 315
Norway 68, 78, 88, 99, 105, 162, 163, 164, 189, 152, 200, 202, 203, 246, 275, 278, 312, 329, 373
    sea-level curves 329
Norwegian slope and shelf 179, 208, 209
Norwegian Basin 180
Norwegian Current 205, 209, 210
Norwegian-Greenland Sea 179, 180, 183, 185, 187, 191, 194, 196–198, 203, 204, 206, 255, 403
    model of deglaciation sequence 206–208
    radiographs of sediment 206, 207
Nouveau-Québec 27, 86, 434
Nova Scotia 93, 94, 163, 172, 175, 416, 434, 451, 462
NRM see natural remanent magnetism
numerical age 380
nunataks 59, 137
N.W.T. see North West Territories

^{18}O isotope 402
obsidian-hydration 390, 396–397
    hydration layer/rind 397
    rate of rind growth 397
oceanic circulation 181
Odderade 273
ODP 196, 271
    Leg 104 180, 196
    Leg 105 182
    Leg 120 196
offshore paleochannel placers 451
offshore pipelines and ice berg scour 165
Ohio 147, 148, 150, 309, 387
Olduvai subchron 408
Oligocene Epoch 196, 204
olivine 390
Omineca tectonic belt, B.C. 452
Ontario 1, 9, 11, 36, 37, 38, 40, 42, 44, 50, 68, 75, 92, 95, 99, 100, 101, 102, 103, 111, 112, 144, 145, 148, 150, 153, 162, 163, 191, 192, 240, 247, 253, 257, 258, 259, 266, 291–292, 417, 418, 431, 436
    placers 461
Ontario Geological Survey (OGS) 439
opal (biogenic) production 189
opaline silica 189
ophiolite 416
optically stimulated luminescence (OSL) dating 393
Ordovician Period 1, 173, 239, 242

Oregon 404, 464
organic carbon dating 312
organic content of sediment 181, 208, 210
organic-rich horizon 301
orogenic events 195
orthoconglomerates 244, 245–246
OSL see optically stimulated luminescence
Oslo, Norway 329
Otago, New Zealand 468
Otter Creek, B.C. 459, 476
outlet ice lobes (glaciers) 24, 217
outwash (sandur) 93, 123, 124–124, 126, 128, 133, 148, 152, 173, 194, 217, 222, 243, 245, 256, 288, 298, 306, 322, 390, 443
    braided 124
    braided channels 124
    fans (sandur) 152, 153
    graded surfaces 277, 278
    heads of outwash 148, 153
    sandur plain 156
    subaqueous 151, 194, 199, 247
    terraces 124–125, 129, 144, 309
overconsolidation 40, 200, 201
overprinting 54, 76, 135
overriding by ice 141
overshearing by ice 29, 141
oxidation 301
oxygen isotope stratigraphic record 179, 181, 182, 262, 330
oxygen isotopes 182, 185–186, 195, 214, 317, 330
    curves 317

p-forms 64, 85, 92, 98, 99, 101, 155
Pacific Ocean 186, 372
pack ice 165, 190, 208, 210
Pagoda Formation, Antarctica 175
Pakistan 214, 217, 219, 221, 237
paleogulch placers 442, 472
paleolatitudes, low 244
paleomagnetic dating 405–410
    anisotropy of magnetic susceptibility 409
    Curie temperature 405
    depositional remanent magnetism (DRM) 405
    magnetopolarity time scale 406
    natural remanent magnetism (NRM) 406
    post-depositional remanent magnetism (p-DRM) 406
    reversals 406, 407
    thermal remanent magnetism (TRM) 405
    time-stratigraphic correlation for central New York State 408
paleorotation of the earth 247
paleosol 259, 301–315, 368, 373, 396, 399
    age indicators 310–312, 373
    definition 302
    Illinois quaternary soils 309
    interstadial soils 308
    morphological development of soil with weathering 303, 399

paleosol – *continued*
  paleosols in British Quaternary succession 308
  parent materials 306–307
  relative dating of soils 312
  stadial soils 308
  stratigraphic markers 309–310
  temperate soils 308
Paleozoic 119, 179, 416, 431, 435
parabolic dunes 222, 224
paradigm shifts in glacial geo(morphol)ogy 480
paraglacial sediments 225
Parana Basin, Brazil 175
parastratotype 264
parent materials 306–307
particles:
  flocculation 49, 61
  particle-to-particle buoyancy 62
  size 31, 57, 60, 80, 139, 165, 181, 183, 214, 220, 221, 228,
    230, 233, 296, 305, 337, 414–418
    sorting coefficient 230
  surface morphology 11, 181, 214
partings 48
passive ice flow:
  landforms 110–135, 130, 133
Patagonia 68, 85, 274, 469
patterned ground 194, 239
p-DRM *see* post-depositional remanent magnetism
Peace River, Alberta 459
Pearlette O tephra (also Lava Creek B Ash bed) 408
Pearlette S tephra (also Mesa Falls Ash bed) 408
'pebble nests' 247
Pebbley Beach Formation, Australia 175, 176
pedological enrichment of magnetite 410
pedological processes 54, 302
pedostratigraphy 264
pelagic calcareous tests 185
pelagic carbonates 190, 196
pelagic sedimentation 181
  fluxes 181, 208
Penck, A. 7
penetration hardness 97
Pennines, England 146
Pennsylvania 148, 311, 312
Peoria Loess 396
perched aquifer 51, 93
perched lake plains 159–160
perched moraines 64
percussion erosional forms (subglacial) 95–98
percussion gouges 97
periglacial environments and seasonal aeolian deposition 232
  principal facies of periglacial sands in W. Europe 234, 235
periglacial processes 194, 213, 232, 275, 302, 304, 307, 309,
  313, 351, 372, 424
periglacial river sediments 215
permafrost contraction cracks 172, 235
permafrost ice wedges 194, 235

permafrost sand wedges 235
Permian Period 164, 186, 175, 176
Permo-Carboniferous Glaciation 173–175, 242, 251
pervasive deformation 23, 48, 61, 150, 157, 165, 291
Peru 371, 469
Peterborough, Ontario 75, 291–292
Petrof Glacier, Alaska 78
Petteril Valley, Cumbria,England 146
phosphate retention 312
phyllosilicates *see* clay minerals
phytoplankton 189
  blooms 189
'piano-key' tectonics 329
picea engelmannii 383
Pieksämäki, Finland 76
Pine Creel, B.C. 458
Pinedale Glaciation 306, 397
'pinning points' 243
pipe channels 46, 109, 117
Pitstone Soil 308
Pitz Lake, N.W.T. 431, 432
placer deposits 441–478
  alluvial placer deposits 442–447, 455, 462, 470, 472, 477
  buried placer deposits 470–477
  colluvial placers 451, 455, 462, 470, 477
  coastal and marine placers 448–451, 459, 462, 468, 469, 477
  geological setting 477
  glacial and glaciofluvial placers 447–448, 455, 456, 457, 461,
    462, 469, 470, 477
  linear pay streaks 444
  map of main placer mining areas in W. Canada 443
  paleochannel placers 447, 449, 454, 459, 462
  regional distribution of placers in glaciated terrains 451–470
  river placers 442, 444–445, 458, 459, 462, 469, 477
  sedimentary environments 442–451
placer mining areas in W. Canada 443
plains (till) 35
planktonic foraminifers 185, 187, 190, 196, 208, 271, 403
planktonic organisms 186
'plano-convex lenses' 296
plant succession 301
plasmic fabric 337, 343, 347–351, 355
plastic deformation of rock 97
plate tectonics 242
  accretionary tectonics 251
  emergent continental crust 252
  model relating glaciation to plate
  rift basins 249, 251
  tectonics 250
platinum 419, 450, 459, 467
platy structures 48
playa basins 215
Pleniglacial 226
Pliocene Epoch 196, 452
ploughing of clasts 38, 80
ploughing by ice bergs 202

plucked bedrock forms:
  arêtes 64
  cirques 64
  finger lakes 64
  fjords 64, 81, 84, 85–92
  headwalls 64
  lake chains 64
  nivation basins 64
  roches moutonées 63, 64, 85, 94, 98–99, 100, 105, 239, 243
  rock basins 64
  tarns 64
  troughs 64, 85–92
  u-shaped valleys 63, 64, 88–90, 239
    cross-profile 88–90
plucking 90
plumes (debris) 60, 210
plumes (meltwater) 187, 199, 245
podzolisation 301
Poland 74, 92, 233
polar deserts 215
polar ice cap facies 298–300
polar subglacial bed types 20, 25
polarity excursions 406, 407, 408
pollen assemblages 262, 273
pollen in soil 312
polygonal network of sandstone wedges 170, 171
polymictic clasts 242, 246
polymict sediments 168
polyphase beds 24, 25, 29
polythermal bed conditions 18–24, 25, 27, 63, 72
Poplar River, Saskatchewan 459
Porcupine Creek, Alaska 467
porewater 10, 23, 31, 41, 48, 50, 51, 61, 76, 110, 136
  expulsion 44
  pressure 31, 41, 146, 347, 368
  sulphate reduction 208
porewater dissipation 77
porosity of sediments 23
Port Askaig tillite 9, 170, 171, 172, 193–195, 266
  floating ice model 194–195
  'Great Breccia' 194, 195
  grounded ice model 194–195
  mixtites 194
  tunnel deposits 195
Port au Port Bay, Newfoundland 462
Port Burwell, Ontario 42
Portimo, Finland 70
Port Stanley Till 40
post-depositional remanent magnetism (p-DRM) 406
postglacial rebound 323
potassium-argon dating see K/Ar dating
potholes 64, 99, 102, 105
Potwar Plateau, Pakistan 237
pradoliny 64
Precambrian Period 172, 173, 186, 266, 427
  'soda ocean' 186

Precambrian diamictite 95, 186
Precambrian Glacial Record 3, 179
Pre-Cenozoic glaciomarine sequences 191–195
pre-deformation structures 23
pre-glacial control 88
Pre-Pleistocene Glacial Environments 239–252
  paleolandforms 243
  record of glaciations 242
pre-Reid Glacial advance, Yukon Terr. 453
primary tills 31
Princess Louise Inlet, B.C. 89
Princeton district, B.C. 459
Problems and Perspectives in Past Glacial Environments
    479–485
  englacial environments 483
  glaciomarine/glaciolacustrine environments 484
  paradigm shifts 480
  proglacial environments 483
  subglacial environments 481–483
  supraglacial environments 483
Proglacial Environments 55, 120, 222, 288, 296, 393, 413
  lakes 144, 146
  model of basin-fill sedimentation 293
  ponded drainage 144, 146
proglacial deltaic terraces 156
proglacial depression 320, 323, 330
proglacial lake 11, 116, 127, 144, 256, 260, 292, 379
proglacial and ice contact glaciofluvial morphosequence model
  156
proglacial sediments 137
progradation wedge of sediment 200
Prospecting in Areas of Glaciated Terrain (PAGT) series 438
protactinium-231 (^{231}Pa) 388
protein breakdown 394
Proterozoic Eon 179, 165, 168–173, 194, 195, 239, 241,
  244–249, 251
provenance 30, 75, 123, 239
proximal subaquatic diamicton mélanges 60, 61–62
proxy evidence 261, 262
Prytz Bay 196
psuedomorphs of ice veins 172, 235
Puget Ice Lobe 19
Puget Lowland, Washington 392
Pulju moraine 63, 64, 67, 130, 131–132
  clast fabric 132
  formation 131
  morphology 132
push ridges 51, 63, 64, 150, 167, 347, 351
Pyrenees Mountains, Spain 154
pyrite 422, 423, 454

'Q' beds see Quasi-Rigid/Quasi-Soft beds
Qilian Shan, Tibet 221
Qinghai-Xizang Plateau see Tibetan Plateau
quartz-albite-sericite schists 422

quartz grain micrographs (SEM) 364–367
quartz particles 214
Quasi-Rigid/Quasi-Soft ('Q') beds 20, 24, 25, 29, 76, 84
Quaternary stratigraphy 253
Québec 27, 29, 68, 71, 86, 241, 255, 415, 416, 420, 422–423, 424, 425, 426, 427, 428, 434, 435, 448
    placers 461–462
Queen Charlotte Islands, B.C. 451, 459
Queen Charlotte Sound, B.C. 451
Queen Maud Gulf, N.W.T. 120
Quesnel River, B.C. 456, 457, 458, 472

'R' meltwater channels 22, 108, 109
Ra moraine, Norway 152
racemization process 394
radial moraine 63, 64
radiation-balance modelling 215
radiocarbon dating 179, 181, 240, 310, 316, 377, 379, 381, 385–387
    atmospheric ^{14}C 387
    calibration 386
    errors 386–387
    ground water effects 386
    isotopic enrichment 385
    'hard-water' effect 387
    'Libby half-life' 385
    marine organism 'reservoir' effects 387
    production of ^{14}C 386
radiogenic dating 381, 391–393
    fission track 391–392, 403, 407
    methods 391
radiographs of glaciomarine sediment 206, 207
radiolarian assemblages 190
radiometric dating 259, 261
rafting 44, 82, 119, 135, 245, 247
raised beaches 315, 321, 372
raised plateaus 159–160
    moraine-lake plateaus 159
    moraine plateaus 159
    ice-walled lake plains 159
rainbeat 235
rain-out (of sediments) 60, 112, 175, 243, 244
rainwash 235
rapid ice flow see ice streams or surge
Rapitan Group, Alaska 248
Rankin Inlet, N.W.T. 114
Ranua, Finland 70
Rannoch Moor, Scotland 134
rapakivi-type granite 251
rat tail 104
rate of racemization/epimerization 395
Rb-Sr dating see rubidium-strontium dating
rebound (glacio-isostatic) 323
    rates 331
recessional moraines 64, 148

red algae 186
Red Deer River, Alberta 459
Red Head, Nova Scotia 462
redox potential 211
Redwater River, Alberta 459
reefs 83
reef-crest corals 317
Reefton, New Zealand 468, 469
refuges 262
regelation 32
regional areal erosion (subglacial) 85
regional distribution of placers in glaciated terrains 451–470
regional linear erosional forms (subglacial) 85–94
regressive glaciomarine sediments 201
    deep banks 201
    deep troughs 201
    shallow banks 201
regs 220
relative age 380
relict paleosol see paleosol
Rennie Lake esker system, N.W.T. 118
resedimentation 135, 200, 243, 244, 245, 259, 337
residual rebound 323
restrained rebound 323
Rhine River valley 236
rhizocarpon geographicum 398
rhomohedral pattern of moraine ridges from surge event 167
Rhu Sands and Gravel, Scotland 260
rhyolite 391, 396, 422
rhythmically bedded sediment 60, 111, 144, 146, 207, 211, 244, 247, 292, 339, 343
rhythmites 383
ribbed moraine 64, 153
Riedel shears 48, 59
riegels 85
rift basins 248, 251
    margins 245
Rigid subglacial beds see Hard or Rigid ('H') beds
Rimrock, Baffin Island 81
rinnen see tunnel valleys
rinnentaler see tunnel valleys
Rio do Sul Formation, Brazil 175
rip-up clasts 245, 248
ripple-drift cross-lamination 296
Riss Glaciation 7
river placers 442, 444–445, 458, 462, 469, 477
    alluvial fill 444
    bars 444, 445
    braided environment 445, 452
    channel dunes 444
    chute cut offs 444
    detrital gold 444
    lag deposits 444, 445
    linear pay streaks 444
    terrace gravel sequences 445, 457
Rivière des Plantes, Québec 462

roches moutonées 63, 64, 85, 94, 98–99, 100, 105, 239, 243
    formative processes 99
    friction cracks 98, 99, 100
    tensile stress 99
rock basins 64
rock drumlins (roc drumlins) 64, 75, 99, 102, 104
rockfalls 141, 397
rock flour 215
rock glaciers 397
rock mass strength 88
rock stacks 217
Rocky Brook, New Brunswick 417
Rocky Mountains 91, 147, 309, 383, 384, 389
Rocky Mountain Trench 91
Rogen moraine 35, 63, 64, 65, 68–72, 74, 75, 79, 80, 120
    models of formation 71, 72
'roller' stones in pothole 105
Romerike, Norway 163, 164
Rosenheim, Bavaria 111
Ross Ice Shelf, Antarctica 183
roundness of clasts/particles 147, 362, 371
Roxanna Silt 396
Royal Mile, Edinburgh, Scotland 83, 84
rubidium/strontium dating 240
Ruby Creek, B.C. 459, 461
Russia 74, 176, 448, 451, 462–464

s-forms 64, 99, 100, 102, 103, 105, 130, 155
    cavetto 102, 104
    comma forms 102, 104
    furrow 102, 104
    model of crescentic furrow formation 104
    muschelbrüch 102, 104, 105
    pothole 64, 99, 102, 105
    rat tail 104
    sichelwannen 102, 104105
    spindle flutes 102, 104, 105
    transverse trough 102
Saale Glaciation 7, 93, 150, 343
Sable Island, Nova Scotia 93, 94
Saglek Bank, Labrador Sea 171
Sahara 239
Saharan Africa 1, 172, 173
Saints Rest, New Brunswick 462
salinity 185, 198, 202, 205, 210, 211
Salmon River, California 464
Salmon Springs Drift, Washington 392
Salpausselkä moraines, Finland 152
    model 152
saltation 214, 220, 221
salt concentration (enrichment) 215, 248
salt-induced contraction cracking 215
saltwater intrusion 317
salts in soils 302

salt weathering 213, 215
sand dunes 213, 220, 222
    barchan 222
    blow-out 222
    parabolic 222, 224
    sedimentary structures 223
    transverse 222
    whaleback 222
sand dykes (or dikes) 41, 145
sand laminae 142
sand lenses 202, 296
Sandgigjukvisi, Iceland 78
sand shadows 222
sand sheets 220, 225, 227
    sub-horizontal planar bedding 225
sandstones 244, 245–246
sand wedges 227
sandur see outwash
sandur plain see outwash
Sandy Glacier, Antarctica 217, 221
Sangamon interglacial 258, 304, 305, 306, 309, 310
sanidine 387
San Juan Mountains, Colorado 147
saprolites 255
Saskatchewan 68, 75, 92, 291, 459
    placers 459, 460
Savante Lake, Ontario 461
Sawmill Canyon, California 154, 400
Scandinavia 6, 85, 99, 152, 160, 257, 275, 315, 326, 329
scanning electron microscopy see SEM
Scarborough Bluffs, Ontario 9, 162, 163, 266
Schmidt equal-area projections 37, 79
Schmidt hammer 261, 278
Scotian Shelf, Nova Scotia 93, 94, 203, 451
Scotland 7, 9, 36, 83, 84, 85, 89, 92, 107, 134, 157, 162, 163, 170, 193, 255, 258, 259, 260, 266, 275, 276, 277, 281, 315, 326, 329, 330, 337
    sea level curve 330
    shoreline diagram 327
Scottish Readvance Ice 148
scoured terrain 86
sea ice cover 185, 200
sea ice rafted clasts 182, 206, 207, 208
    iceberg rafted 182
    melt out sediment 200, 256, 298
sea ice sedimentation 198
sea level change see also eustatic changes
    amplitude past/future 318
    causes 315
    curves 316, 317, 324, 329, 330
sea water composition 185, 186, 189, 198, 202, 203, 205, 210, 211, 317
secondary tills 31
sediment porosity 23
sedimentary environments of placer deposits 442–451
    alluvial placer deposits 442–447

seismic events *see* Neotectonics
seismic sediment tracing 197, 199
SEM (scanning electron microscopy) 215, 239, 357–376
  aeolian textures 370–371
    'upturned plates' 371
  beach textures 372
  chattermarks 361, 367, 368, 369
    trails 369, 371
  chemical effects 370
  crushing and grinding mechanisms 368, 371, 414
  crystal growth facets 367
  diagnostic features 363
  edge abrasion 368–369
  edge rounding 371
  etch pits 363, 370, 373
  etching process 369
etched Vs 372
  fabric analysis 374
  facets 177
  fissure/pore detection 374
  fluvial textures 371
    'centipede' type solution mark 371
  glacial textures 364
  glaciomarine textures 372
  grain selection 362
  image capture and processing 360
  image analysis 360
  impact Vs 370
  microfractures 214, 371
  operation 357–360
  particle characterization 360–361
  particle outlines 362
  periglacial textures 372
  quantitative estimation 362
    form, shape, angularity, roundness 362
  quartz grain micrographs 364–367
  sample preparation 359–360
  sampling problems 361
  stepped fracture surfaces 215
  surface textures 215, 362
    'microblock' 369
    replication 363, 369
  till thickness 369
  transport pathways 363
  volcanic textures 372–373
  Vs 363, 370
  weathering effects 373
SEM-based dissolution indices 190
Serle conglomerate (Sturtian) 246
sesquioxide dating 312
Sevetti moraine 63, 113
Seward Peninsula, Alaska 402
shattered bedrock zones 87
shear fractures in sediment 41, 44
shear joints in sediment 36, 41, 347
shear moraines 64

shear planes in sediment 23, 29, 36, 48, 290, 338, 343, 374
  micro- 48, 54, 374
  macro- 48
sheared inclusions in sediment 36, 41, 337, 355
Shelbyville moraine, Illinois 150
shell beds (marine) (Scotland) 330
shell fragments in glacigenic sediments 200
Sherbrooke district, Québec 461
shield terranes 144–146
shoreline formation 320
  drowned 321
  raised 321
  updomed 321
Shreve's theory 20
Shoemaker Type II 22
shoreline diagrams 326
Siberia 448, 451, 462–464
sichelwannen 102
Sicily Island Loess 396
sidereal dating methods 381
  historical records 381–382
  dendrochronology 382–383
  varve chronology 383–384
sidescan sonographs 162
Sidujökull, Iceland 78
Sierra Nevada 312, 390, 399, 400, 464
silasepic fabric 347
silica (amorphous) content of sediment 181
silica (opaline) 189
silica blebs 248
siliceous tests 189
silt balls 60
silt droplets 351, 352, 354
silt pebbles 343
silt production 215
Silurian Period 172, 173
silver 461
Similkameen, B.C. 459
'sinkholes' 159
skeletal carbonate 186
skelsepic fabric 343, 347, 351
Skovde moraine, Sweden 152
slickensides 46, 48
slope failure 214
slumps (mass wasting) 141
slump structures 41, 143
slumping (gravitational) 61, 245, 247
smudges 290
snails 395, 396
snow 215
snow banks 397
snow meltwater flow 235
'soda ocean' 186
Soft or Mobile ('M') 20, 22–24, 76, 94
soils *see* paleosols
soil carbonates 305, 412, 424

soil catena 306
soil chemistry 302
soil colour 302, 312
soil horizon development 302, 312
soil profile development 312
solar forcing 2, 270
    Milankovitch-type 251, 270, 386, 403
solar precession:
    elliptic 248
    obliquity 248
solifluction debris 134, 235
Sørbreen, Spitsbergen 21
South Africa 172, 173, 174, 175, 241, 248
South America 1, 242, 371
    placers 469–470
South Australia 1, 195, 240, 241, 243, 244, 245, 246, 247, 248
South Dakota 147, 407, 464
South Island, New Zealand 85, 469
Southern Alps, New Zealand 1, 221, 468
Southern Lake, N.WT. 433
Spain 154
Spanish Mountain, B.C. 455, 471, 475
spatial variations in subglacial bed conditions 24–29
species colonisation 262
speleothems 393
Sperry Glacier, Montana 78
Spi Lake, N.W.T. 418, 419
spindle flutes 102
spinel 420
Spitsbergen see Svalbard
Spitsbergenbanken 200, 201, 202
'splash up' structure 240
Spokane Floods 307
sponges 207
Spruce Creek, B.C. 458, 460
'sputtering' (SEM) 360
squeeze moraines 64, 167
St. Croix moraine, Wisconsin 149
St. Germain I 273
St. Germain II 273
St. Lawrence valley 258, 389
St. Magliore, Québec 416
St. Georges, Québec 461
stacking (of sediment) 50, 71
stadial soils 308
stagnant ice blocks 156, 296
stagnation of an ice mass 141, 144, 157–159
Stauchmoräne 64
step fractures 239
stillstands 62, 116, 148, 202
Stillwater succession 241
stone layers in diamicton 143
stone pavments in diamicton 217
Storfjördrenna 197
Storbreen, Norway 78, 278

Storegga Slide 204, 205
    map 204
    sediment slabs 206
Story County, Iowa 140
Stouffville, Ontario 50
Strangford Lough, Northern Ireland 315
stratified mixtites 194
stratified sediment 71, 75, 77, 81, 130
    ice-contact 113, 128
stratified till 49, 60, 130, 298
stratiform glacigenic sediments (marine) 198
stratigraphic nomenclature 261–264
    diachronic unit 262, 264
    holostratotype 264
    hypostratotype 264
    lectostratotype 264
    lithostratigraphic terminology 265
    neostratotype 264
    parastratotype 264
    stratotype 263, 264, 279–280
stratigraphic relationship between datable materials and glacigenic sediments 378
stratigraphic resolution 11, 137, 379–380
stratotype 263, 264, 279–280
    holostratotype 264
    hypostratotype 264
    lectostratotype 264
    neostratotype 264
    parastratotype 264
streamlined bedrock forms:
    crag and tail 64, 83–84, 85, 99, 300
    erosion marks 64, 92, 98, 101
    flutes 84, 101, 167
    grooves 64, 85, 170
    p-forms 64, 85, 92, 95, 98, 99, 101, 155
    roches moutonées 63, 64, 85, 94, 98–99, 100, 105, 239, 243
    rock drumlins 64, 99, 102, 105
    s-forms 64, 99, 100, 102, 103, 105, 130, 155
    scallops 101
    scoured terrain 86
streamlined hills 67
streamlined landforms/bedforms
    bedform elongation ratio 67
    bedform length 67
    continuum 65, 67, 68
    drumlins 11, 29, 35, 36, 63, 64, 65, 68, 71, 72–77, 79, 80, 93, 94, 113, 119, 125, 132, 243, 275, 291, 434, 436, 438
    drumlinoids 64
    fluted moraine 35, 64, 65, 67, 68, 72, 74, 75, 77–81, 94, 133, 167, 201, 275
    lee-side moraines 64
    megadrumlins 67
    radial flow patterns 66, 74
    radial moraine 64
    streamlined hills 67

Streicek Mine, B.C. 471, 472, 473
Stevens Gulch, B.C. 471, 474
striae 59, 94–98, 130, 170, 239, 240, 241, 242, 426
    cross-cutting relationships 97
    curved 173
    leveés 97
    map 96
    percussion gouges 97
    SEM micrograph 97
strial fabric 347
striated surfaces (soft sediment) 165–168, 172
stromatolitic sediments 248
structural discontinuities in sediment 29, 35, 41, 46–55, 57
    classification 43
    fissility 36, 296
Sturtian 1, 195, 240, 241, 243, 244, 246, 248
subaerial debris flows 298
subaerial denudation 280
subaquatic diamictons 59, 245, 297
Subaquatic Environments 59–62, 191
    distal zone 60
    proximal zone 60, 61–62, 81
subaquatic diamicton mélanges 60, 61
subaqueous fans 60, 152, 153, 194 , 245, 246, 247
subaqueous moraines 64
subaqueous outwash 151, 194, 199, 243, 247
    model 152
subglacial abrasion 368
subglacial bed (interface) conditions 10, 18–24, 25
subglacial bed types 18–24, 25
    polar 20, 25, 94
temperate 20–24, 94
subglacial bedform development 27, 29–46, 62–84
    bedform classification 64
    interface conditions 69
subglacial cavities 42, 44, 58, 110, 112, 292
subglacial crevasses 159
subglacial debris rheology 23, 24
subglacial depression 323
subglacial erosional forms 84–110, 130
    Alpine-type landscape 85
    fjords 84, 85–92
    landscape zones 85
    local erosional forms 94–110
    regional areal erosion 85
    regional linear erosional forms 85–94
Subglacial Environments 15–136, 296
    models of subglacial processes 17, 25, 27, 55
subglacial glaciofluvial sediments 112–113, 134
subglacial glaciolacustrine sediments 112–113
subglacial hydraulic conditions 17, 25, 60, 84, 93, 94
subglacial lakes 99
subglacial landforms 29–46, 64
subglacial meltwater channels 60, 64, 84, 99, 105–110, 122, 130,
    152, 153, 159, 175, 246, 288, 296
    channel types 106, 131

channel wall collapse 110, 116, 298
    closed 106
        subglacial 106, 122, 130, 159
        subglacial chutes 110
        submarginal 106, 107, 110
    closure rates 108
    englacial superimposition 110
    full pipe flow 109, 117
    maps 107, 108
    open 106
        marginal 106, 110
        subglacial/proglacial 106
    till channels 108, 109
subglacial meltwater erosional forms 99–110, 130, 175
    fluidized debris 100, 108
subglacial meltwater floods 72, 77, 99, 100 see also catastrophic
    discharge jökulhlaups
subglacial non-linear, unoriented morainic forms 64, 130 see also
    ground moraine
Subglacial – Proglacial Transition Environments 55
subglacial rheology 19
subglacial thermal conditions 18, 25, 27, 63, 72, 91, 93
    models 18, 25, 27
subglacial transport 24
subglacial transverse moraines:
    De Geer moraines 51, 63, 64, 81–83, 153
    Kalixpinmo Hills 64
    ribbed moraine 64, 153
    Rogen moraines 64, 65, 68–72, 74, 75, 79, 80, 120
    shear moraines 64
    squeeze moraines 64, 159, 167
    thrust moraines 64
    washboard moraines 64, 81, 153
sublimation till 31, 41, 298, 299
submarine hydrothermal vents 418
subsidence structures (from melt out) 142, 159
Sudbury Formation, England 268
Sudbury, Ontario 461
Suffolk, England 228, 313
sulphides 412, 414, 418, 420, 422
sulphur 424
supercontinents 249–251
    model relating glaciation to plate tectonics 250, 329
supraglacial debris 56, 124, 125, 138, 141, 221, 288, 296, 425
supraglacial diamicton 134, 141
Supraglacial and Ice-Marginal Deposits and Landforms 137–160
    disintegration features 157–159
    end moraines 148–155
        stratified 153
    glaciotectonism 150, 152
    grain size 139, 140, 147
    high relief mountain terranes 146–147
    ice-marginal landforms 155–157
    landforms 147–160
    low relief terranes 141–144
    meltwater systems and channels 155, 159

Supraglacial and Ice-Marginal Deposits and Landforms –
  continued
  moderate relief terranes 144–146
    facies types 144–146
    push moraine 51, 63, 64, 150, 167, 347, 351
  perched lake plains 159–160
  raised plateaus 159–160
  schematic diagram 138
  sediment types 139–147
  sorting (grading) 139, 147
supraglacial grains 369
supraglacial lakes and ponds 144, 146, 147
surface tool mark 259
surface crevasse infillings 80
surge 11, 119, 141, 208
surge moraine 167
Susitna Valley placers, Alaska 467
Svalbard 21, 38, 58, 78, 79, 167, 183, 196, 198, 199, 203, 217,
    226, 248, 333, 337, 393
Sveg Till 49, 71
Sweden 19, 52, 65, 68, 70, 71, 78, 131, 132, 133, 143, 150, 152,
    157, 331, 337
Switzerland 6, 73, 74, 227, 256, 337, 416
syndepositional deformation 29, 30, 164, 165, 177, 269
syneresis 49, 61

taffoni 217
Tajikistan 233
Talsand, Emsland, Germany 234
Tamadjert Formation, Algeria 172, 173
Tangier Lake, Nova Scotia 416
tarns 64
Tay River, Scotland 32
Taylor Peninsula, New Brunswick 462
Taylor Valley, Antarctica 216, 217
Tayside , Scotland 107, 134
tectonic activity 195
tectonic shattering of quartz 215
tectonic setting of glacial sediments 249–251
  supercontinents 249–251
'Tehuelche-type gravels', Brazil 469
temperate soils 308
temperate subglacial bed types 20–24, 25, 94
  Hard or Rigid ('H') beds 20–22, 25, 77, 84, 85, 94, 99, 202
  Soft or Mobile ('M') 20, 22–24, 25, 76, 94
  Quasi-Rigid/Quasi-Soft ('Q') beds 20, 24, 25, 76, 84
tephra 372, 387, 392, 403, 407, 408
  ice cores 372
tephrochronology 403–405
  chemical composition 403
  field characteristics 403
  index of refraction 403
  mineralogy 403
  phenocryst characteristics 403
  shard morphology 403
  trace element composition 403

ter Izard, The Netherlands 351
terminal moraine 51, 64, 93, 130, 148, 436
terrestrial lithofacies associations *see* lithofacies associations
Tertiary Period 197, 242, 310, 451
Tertiary placers, B.C. 451, 455, 456, 457, 468, 470
tests 185, 189
Thames River, England 304
thaw consolidation 41
thermal convection (sea) 185
thermal remanent magnetism (TRM) 405
thermal stratification in lakes 384
thermoluminescence (TL) dating 259, 310, 377, 392–393
  acquisition and reduction of natural TL signal 392
  dosage 392
thermokarst 63
Thetford Mines, Québec 422, 423, 424, 425, 426, 427, 428
thin section examination *see* micromorphology
thorium-230 (^{230}Th) 388
thorium decay 181, 388
thorium/uranium (^{230}Th/^{234}U) dating 389, 403
threshold velocities of wind 215, 220, 221
thrust block ridges 63
thrust faults 239
thrust folds 147
thrust moraines 64
thrust slices 50, 53
Tibetan Plateau (Qinghai-Xizang) 214, 215
tidewater glaciers 11, 59 181, 200
Tien Shan Mountains 19, 20, 369
till 27, 29, 30–42, 149, 199, 200, 203, 215, 217, 256, 287–288,
    296, 304, 307, 337, 347, 351, 374, 413, 414, 421, 447
  aqua 60
  auriferous sediment incorporation 447, 448, 450, 462
  balls 44
  banded 338
  basal 59, 183, 201, 202
  calcareous 337, 351, 424
  colour 258
  critical lodgement index 33–34
  deformation 31
  flow 32–35, 38, 58–59, 65, 92, 110–112, 126, 139, 296, 297,
    347, 416
  folding 71
  geotechnical properties 40, 42
  glacioaquatic 60
  interbedded 407
  lodgement 31–40, 44, 62, 65, 130, 150, 192, 200, 256, 288,
    291, 296, 297, 300, 302, 355, 410, 425
  matrix 24
  melt out 31, 32–35, 38, 40–42, 54, 65, 80, 110, 126, 130,
    149, 150, 159, 292, 296, 297, 298, 299, 300, 355
  overconsolidation 40, 201
  para 60
  pebbles 338, 339, 343, 344
  pellets 149, 239, 247
  plains 35, 63–65

till – *continued*
  primary 31
  provenance 30, 75
  resedimented 38
  secondary 31
  sheets 257
  slumped 38
  stacked 71
  stratified 49, 130, 298
  subaqueous 60, 297
  subglacial 59, 139, 141, 149, 150, 183, 201, 202, 416, 418
  sublimation 31, 41, 298, 299
  supraglacial 139, 297
  waterlain 59, 60, 110–112, 343
  weathering of till 420–423
till balls 44
till channels 108, 109
till delta 61, 62
till fabric *see* clast fabric
till floes 347
till partitioning 415–418, 420
till pellets 239, 247
till plain (drift plain) 64, 150
till 'prism' 31
till sheets 257, 396
till thickness 369
till tongue 62
till veneers 64
till weathering 420–423
till wedge 80
tillite 38, 161, 168, 172, 177, 243, 355, 369
  ice keel tillites 177
  lodgement 191, 192
Timiskiming District, Ontario 461
Timmins mining area, Ontario 461
tin 422
tin-bearing granite 417
Tipuani-Mapiri Basin, Bolivia 470
titanium-rich sands 451
TL *see* thermoluminescence
Toop Nugget mine, B.C. 456, 457
tooth enamel 393
topographic inversion 159
Toronto 9
Torpa Ridge, Sweden 52
trace element geochemistry 249
Transantarctic Mountains, Antarctica 175
translocation 301, 306, 307, 311, 351, 354
  rates 311, 312
transport pathways 11
transverse dunes 222
transverse furrow 102
transverse subglacial landforms *see* subglacial transverse
  moraines
travertine 393
tree-ring chronology *see* dendrochronology

Tregillus Lake, B.C. 449
trimlines 85
trimline moraines 64
TRM *see* thermal remanent magnetism
topical weathering of quartz 214
trough cross-stratified gravels 444, 454
trough fill (glaciomarine/marine) 199, 203
troughs 64, 85–92
Troutbeck Soil 308
Tulameen, B.C. 459
tunnel valleys 42, 64, 92–94, 113, 118, 122, 123, 245, 288, 453,
  461
  formation 94
  longitudinal profile 94
tunneldalen *see* tunnel valleys
tunneltaler *see* tunnel valleys
turbation by ice bergs 200
turbid surface water plumes 198
turbidite 204, 205, 246, 372
Turqueti Lake, N.W.T. 421
TUTO Tunnel, Greenland 369
Two Creeks, Wisconsin 139

U-Pb dating *see* Uranium-Lead dating
ultrabasic 426
ultramafic rocks 416, 419, 420, 431, 435
Ulveso Formation, Greenland 171
undersaturation of sea water 189
Ungava 29, 420
unistrial fabric 347
United Kingdom *see* Britain
United States of America 19, 29, 30, 56, 58, 68, 73, 74, 75, 78,
  91, 92, 93, 112, 131, 139, 140, 141, 142, 147, 148, 149,
  150, 151, 153, 154, 155, 157, 158, 159, 162, 222, 223, 224,
  225, 228, 229, 233, 237, 241, 242, 246, 248, 253, 274, 287,
  288, 289, 305, 306, 309, 310, 311, 312, 314, 315, 329, 330,
  331, 383, 383, 387, 389, 390, 391, 392, 395, 396, 397, 399,
  400, 401, 402, 404, 407, 408, 410, 411, 417, 434, 436, 441,
  448, 450, 451, 454, 463
  placers 464–467
'Upper Diamictite', Gowganda 194
Upton Warren Interstadial Complex, England 262, 308
'upturned plates' (grains) 371
uranium 412
uranium-lead dating 241
uranium-series dating 377, 388–390
  decay table 388
Urk, The Netherlands 347
urstromtaler 64
Urumqi Glacier No.1 19, 20, 369
U-shaped valleys 63, 64, 88–90, 239
  cross-profile 88–90
Uthusslön, Sweden 70
Uzbekistan 233

Valdez Creek, Alaska 464, 466
Valley Farm Soil 304, 308, 313
valley-fill sequences 296
valley glaciers 29, 59, 436
valley-sided moraines 64
valley train *see* outwash 156
Vancouver Island, B.C. 58, 59, 459
varnish dating *see* cation-ratio dating
varve chronology 383–384
varved clays 175, 227, 247, 408
    summer inflow events 384
    thermal stratification 384
varved argillites 175, 191, 239
varved mudstone 240, 247
varvites 161
vegetation colonization 135
Veiki moraine 63, 64, 113, 130, 132–133, 160
    clast fabric 132
    formation 133
    morphology 132
Velva, North Dakota 75
Vendian Tillite Group, Greenland 172
ventifacts 217, 225
Vermillion moraine, Minnesota 153
Vermillion River, Ontario 461
Vermont 155, 384
viscous magnetization *see* natural remanent magnetism
void ratio 62
volcanic ash 209, 310, 387, 393
volcaniclastic 429
volcanic cores (vents) 83, 84
volcanic glass 391, 404
volcanic textures (grains) 372–373
volcanigenic base metals 418
Vøring Plateau 180, 197, 206, 210

wadis 215
Wager Bay, N.W.T. 117
Walcott's model of glacio-isostatic displacement 319
Wales 326
Walker Lake, California 154
Walther's Law 254
Wanapietei River, Ontario 461
warm-based subglacial conditions 18, 25
warm-freezing 25
warm-melting 25
washboard moraine 51, 64, 81, 153
Washington State 19, 392, 404
Waterhen River, Saskatchewan 459
waterlain till 59, 60, 110–112
wavelength dispersive spectrometer (WDS) 359
weathering index 249
    Chemical Index of Alteration 249

weathering processes 214, 217, 301, 306, 310, 370, 419
    morphological development of soil 303, 399
    zones 140, 303
weathering effects (SEM) 373
weathering of till 420–423
weathering parameters in relative dating 400
weathering rind 259, 261, 312, 373
    thickness 401
Weddell Sea 183, 184, 189, 372
Wedron, Illinois 288, 289
Weischel Glaciation 7, 93, 186, 195, 200, 205, 234, 235, 302
Wells-Barkerville mining area, B.C. 455
Wempleton, Illinois 305, 314
Werenskjoldbreen, Spitsbergen 78
Westland, New Zealand 468, 469
Western Australia 241
Westerburg push moraine 51
West Runton, England 39, 50
West Spitsbergen Ice Sheet 324
wet-based subglacial conditions 18
whaleback dunes 222
Wheatcroft Lake arsenic dispersal train 434
Whitefish Falls, Ontario 95, 98
White River, Vermont 155
Wijnjewoude, The Netherlands 344
Wilderness Till, Scotland 260, 281
Wilton Creek, Ontario 101
wind patterns 217, 225
wind threshold velocites 215
Windermere Interstadial 279, 281
Winnebago County, Iowa 142
winnowing:
    aeolian 217, 221
    marine 200
Wisconsin 29, 73, 139, 148, 149, 157, 159, 297–288
Wisconsinan Glaciation 7, 93, 186, 195, 258, 309, 324, 389, 392, 395, 397, 436, 447
Witwatersrand 241
Wolstonian Glaciation 307, 309
wood (dating) 379
    growth rings 379
Woodfordian 311
Wright Creek, B.C. 459
Wright dry valley, Antarctica 217, 218, 221
Würm Glaciation 7
Wyoming 241, 306, 404

X-ray examination of thin sections 343
X-ray diffraction (XRD) 302

Yakataga Formation, Alaska 196, 246
Yakutat-Yakataga area, Alaska 467

INDEX

Yakutia, Russia 233
Yangtze River, China 215
yardangs 216
Yellowstone (west), Montana 154, 390, 397, 401, 404
Yentna-Cache Creek district placers, Alaska 467
Yoldia Shoreline 326
Yorkshire, England 304
Younger Coversand II 224

Younger Dryas 152, 224, 258, 260, 261, 275, 282, 317, 379
Yukon Territory 217, 229, 312, 441, 448, 451, 452, 453, 475
    Klondike 452

Zermatt 416
zinc 412, 414, 419, 423, 424, 430, 431, 432
zircon 239, 241, 391